KU-437-711

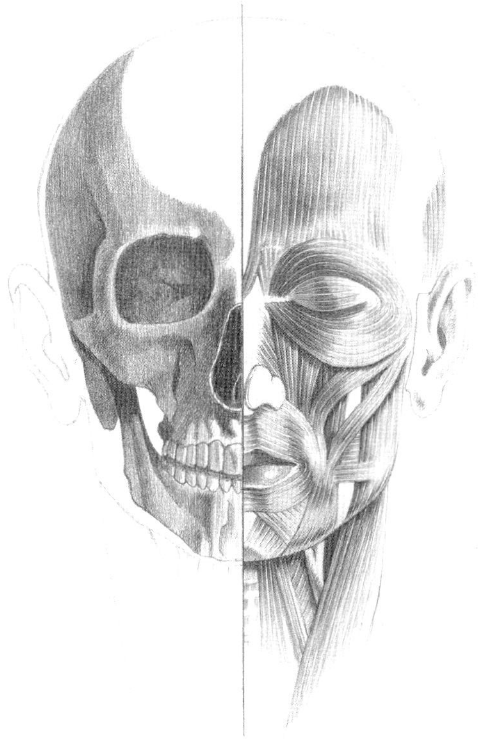

Human Anatomy for Artists

© 1999 Könemann Verlagsgesellschaft mbH
Bonner Strasse 126, D – 50968 Cologne

Publishing and Art Direction: Peter Feierabend
Project Management: Miriam Rodriguez Startz
Assistant: Dr. Nicolai Thesenvitz
Project Coordination: Könemann Music Budapest
Layout: Dezsö Varga
Typesetting: g.win/köln
Production: Mark Voges
Reproductions: typografik, Cologne

Original title: Menschliche Anatomie für Künstler

© 2000 for this English edition:
Könemann Verlagsgesellschaft mbH

Translation from German: Dr. Colin Grant in association with Goodfellow and Egan
Editing: Robin Campbell in association with Goodfellow and Egan
Typesetting: Goodfellow and Egan
Project Management: Jackie Dobbyne and Karen Baldwin for Goodfellow and Egan
Publishing Management, Cambridge
Project Coordination: Alex Morkramer
Production: Ursula Schümer
Printing and Binding: Reálszisztéma Dabas Printing House

Printed in Hungary
ISBN 3-8290-0573-3

10 9 8 7 6 5 4 3 2 1

HUMAN ANATOMY

FOR ARTISTS

DRAWINGS BY

ANDRÁS SZUNYOGHY

TEXT BY

DR. GYÖRGY FEHÉR

KÖNEMANN

CONTENTS

CONTENTS

INTRODUCTION

Artists such as Michelangelo, Leonardo da Vinci, Raphael, Titian or Albrecht Dürer studied anatomy because the proportions and movements of the human body are determined by the skeleton, the joints and the muscles. Awareness of the body's structure enhances the artist's perceptions and improve his eye for form and detail. Each individual has different posture and movement. It is often possible to recognize someone by his gait or posture. The sensory organs such as eyes, nose, ears and mouth, as well as skin character, stress the individual's external characteristics.

The slightest alteration in the pose of a model inevitably induces a chain reaction of muscular contractions, which serve to re-establish the body's balance. Even when a model is perfectly still, muscles are required to maintain balance.

In order to achieve life-like illustration and elaborate the individual characteristics of each individual body, it is therefore essential for artists to study human anatomy.

Human Anatomy for Artists should be a practical aid in acquiring the necessary knowledge. In order to make reading easy, the book starts with a brief summary of anatomical principles. These ease the artist into the subject matter and acquaint him with the movement apparatus of the human body. The second part of the book illustrates all the muscles of the human body, together with their joints and points of origin and insertion on the bone. It is followed by studies in movement, and general surveys of characteristic properties such as the body axis, constitution and proportions. The appendix contains a comprehensive table of muscles. This atlas of a border discipline – plastic anatomy – will be of equal value to artists and other interested readers.

FOUNDATIONS OF ANATOMY

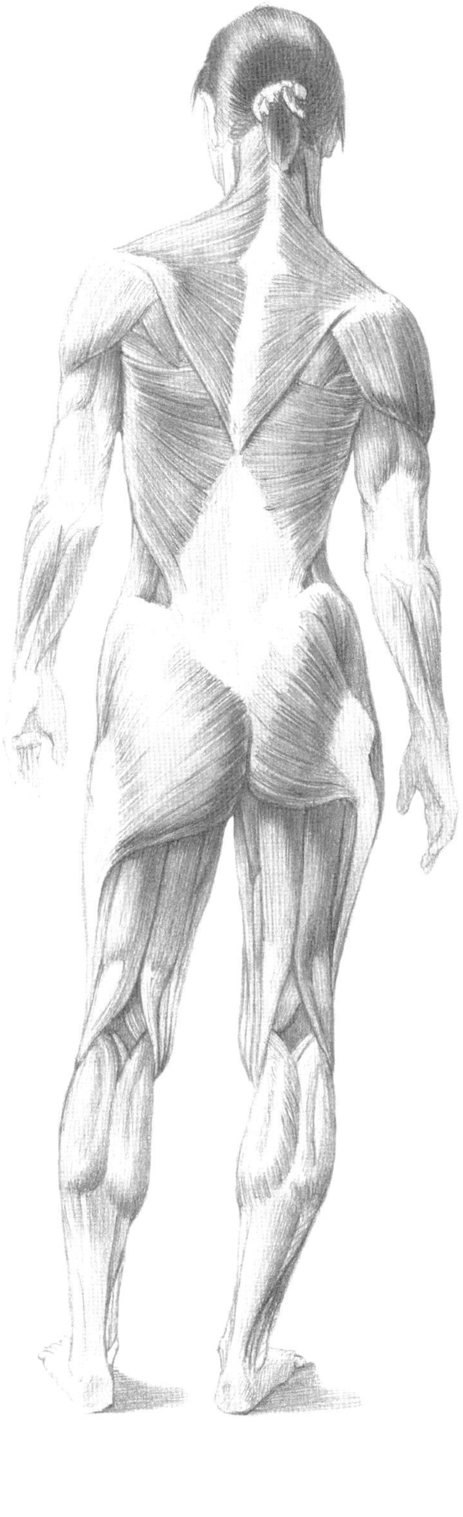

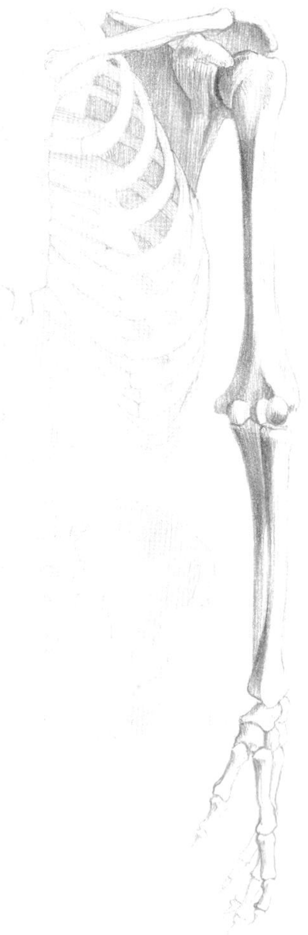

PARTS OF THE MOTOR EQUIPMENT

BONES

Together with the teeth, bones form the most solid structure of the body. Bone tissue is composed of bone cells and intercellular substance. Some 30% of this elastic connective tissue exists as collagens; the rest is a hard, anorganic substance (calcium phosphate, calcium carbonate and magnesium phosphate). Some bones possess a hard outer layer (cortex) and an internal structure made of a spongy substance which contains bone marrow.

BONE-RELATED TISSUE

CARTILAGE. Cartilage is a firm but very elastic tissue. Almost the entire skeleton of a new-born baby consists of cartilage, which gradually turns into bones as the child grows. The most important cartilage in adults is found at the end of the ribs (costal cartilage) and the joint surfaces of the arm and leg bones.

BONE MEMBRANE. The thin membrane that covers the bones is known as periosteum. It contains nerves and blood vessels through which it provides the bone with nutrients.

JOINTS. Joints are the flexible connections between two or more bones. The joint surfaces are covered in cartilage. The joint capsule is composed of taut, collagenous connective tissue and contains joint liquid, which lubricates the joint and makes it easily mobile.

MUSCLES

People need muscles in order to move. Muscles are organs that receive impulses and contract in response to stimuli. They form a major part of body mass. Some 200 to 250 muscles normally account for 36% to 45% of body mass. Muscles are frequently located in facing pairs; for example, when a muscle flexes the elbow, another extends it. Muscles are not only movement organs, however. They also determine the position and stability of the joints, carry the weight of an organism and protect its internal organs. They maintain the body's balance, and maintain its form, size and contours. Under a microscope we can see that conscious muscles (which can be deliberately controlled) are composed of horizontally striated muscle fibres. These are long, thread-like cells made up of a very high number of cell nuclei, and bundles of myofibrils, which contain actin and myosin. These are the substances that induce contractions.

MUSCLE-RELATED TISSUE AND STRUCTURES

Muscles are connected by tendon fibres (tendons, tendinous plates or septa).

MUSCLE FASCIAE. Muscles are surrounded by a coarse shell known as the muscle fasciae, which turn into one or more tendons that connect it to the skeleton. The entire surface of the body is covered by a dual layer of surface fasciae. Located between these layers are muscle plates, which are known as skin muscles. These muscles can lift the skin from the surface fasciae and allow the skin to fold.

TENDONS. A tendon is a shiny white end point of a muscle. It is composed of collagenous connective tissue and connects the muscle to the bone.

MUSCLE SEPTA. Septa, made of connective tissue, are located inside the muscle and divide it into sections. The muscle fibres are attached to the septa by the pointed end. This is known as pennatus (feather-like).

LIGAMENT. A ligament is a string-like or flat structure, composed of collagenous or, more rarely, elastic connective tissue. It connects mobile parts of the skeleton.

DEFINING BODY LOCATION

In order to describe the position of body organs, a system of axes is used that divides the body into several fictitious planes. The MEDIAM PLANE divides the body into right and left halves. SAGITTAL PLANES are located on both sides and run parallel to the median plane. By means of these planes it is possible to refer to structures as being either MEDIAL (in proximity to the median line) or LATERAL (further removed from the median line). In the corner to the right of the median plane along the longitudinal axis of the body the FRONTAL LEVEL is located. Accordingly, structures can be described as being DORSAL or POSTERIOR (closer to the back) or VENTRAL or ANTERIOR (closer to the abdominal wall). A HORIZONTAL PLANE forms a right angle with the vertical axis and divides the body into an upper (SUPERIOR) and a lower (INFERIOR) part.

In order to describe the position of the limbs we can use the terms PROXIMAL (near the center of the body) and DISTAL (further removed from the center). Structures located on the front side of extremities are said to be VENTRAL; those on the back are said to be DORSAL; or DORSAL or PLANTAR in pelvic limbs; the dorsum of the hand is known as DORSAL, the palm is known as PALMAR. Fingers and toes are described as being AXIAL when located near the body axis and ABAXIAL when further away.

Structures located on the head can be described as FRONTAL (anterior), OCCIPITAL (posterior), NASAL (medial) or TEMPORAL (lateral). Organs close to the body surface are said to be SUPERFICIAL and others PROFUNDUS (deep). The term EXTERNUS refers to structures outside a body region whereas INTERNUS denotes structures inside a given region.

Defining Body Location

Planes
1 Median Plane
2 Frontal Plane
3 Horizontal Plane

Relating to the head
4 Frontal: towards the forehead, on the side of the forehead
5 Occipital: towards the rear of the head, belonging to the rear of the head
6 Nasal: internal, relating to the nose
7 Temporal: external, relating to the temples

Relating to the body
8 Dorsal (posterior): located towards the back, belonging to the back
9 Ventral (anterior): located towards the abdomen, located towards the abdominal wall

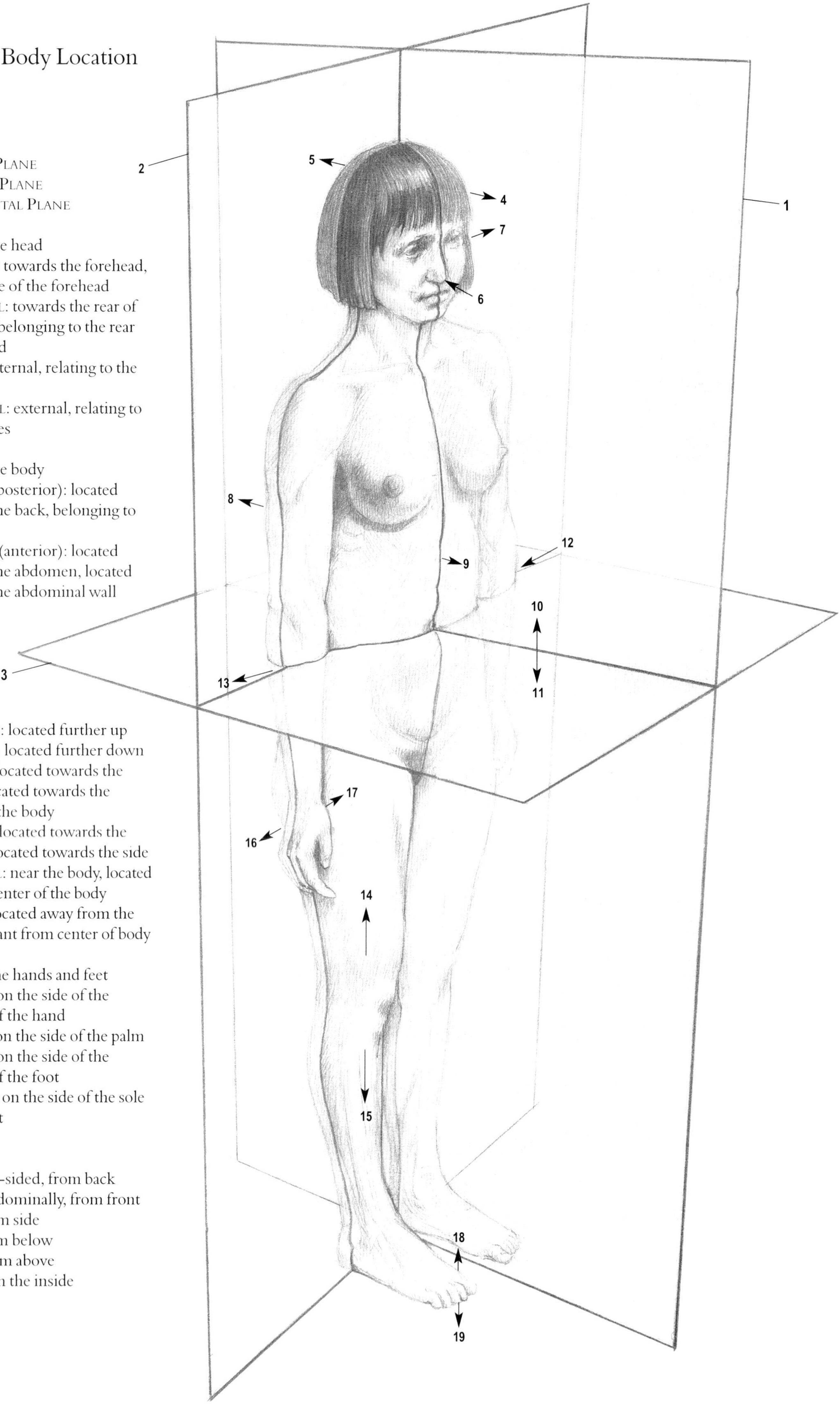

10 Superior: located further up
11 Inferior: located further down
12 Medial: located towards the inside, located towards the center of the body
13 Lateral: located towards the outside, located towards the side
14 Proximal: near the body, located towards center of the body
15 Distal: located away from the body, distant from center of body

Relating to the hands and feet
16 Dorsal: on the side of the dorsum of the hand
17 Palmar: on the side of the palm
18 Dorsal: on the side of the dorsum of the foot
19 Plantar: on the side of the sole of the foot

Views
Dorsal: rear-sided, from back
Ventral: abdominally, from front
Lateral: from side
Caudal: from below
Cranial: from above
Medial: from the inside

Movements of the Body

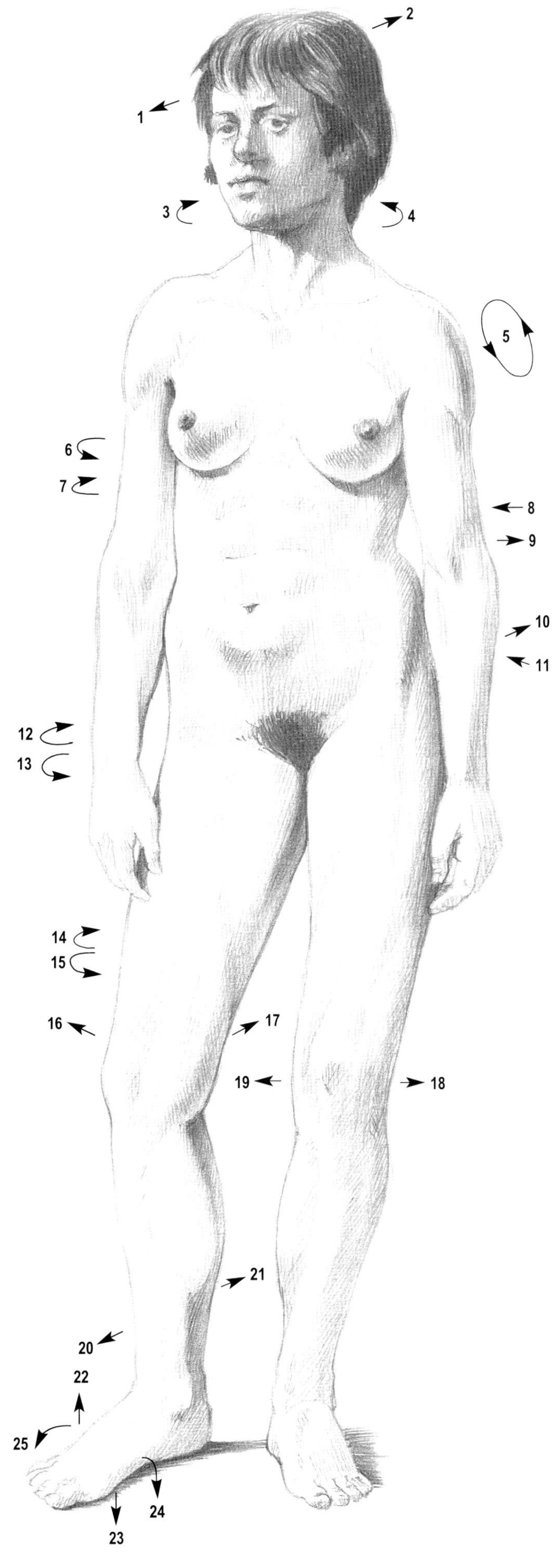

Possible Forms of Movement
Adduction: movement towards body
Abduction: movement away from body
Flexion: bending
Extension: stretching
Rotation: circular movement
Pronation: rotation towards body
Supination: rotation away from body
Inversion: bending towards body, titling
Eversion: bending away from body, tilting
Circumduction: conical or circular movement
Dorsal Flexion: bending of front of the foot on the side of dorsum
Plantar Flexion: bending of front of foot on the side of the sole

Movements of Individual Parts of the Body and Joints

Head and Neck
1 Flexion
2 Extension
3 Rotation (to the right)
4 Rotation (to the left)

Upper Extremity
5 Conical rotation
6 Rotation towards body
7 Rotation away from body
8 Movement towards body
9 Movement away from body
10 Extension
11 Flexion
12 Rotation away from the body
13 Rotation towards the body

Lower Extremity
14 Rotation away from body
15 Rotation towards body
16 Forward swing
17 Backward swing
18 Movement away from body
19 Movement towards body
20 Extension
21 Flexion
22 Flexion of the front of the foot dorsally
23 Flexion of the front of the foot plantarly
24 Flexion away from the body
25 Flexion towards from the body

Fig. 1
The Skeleton

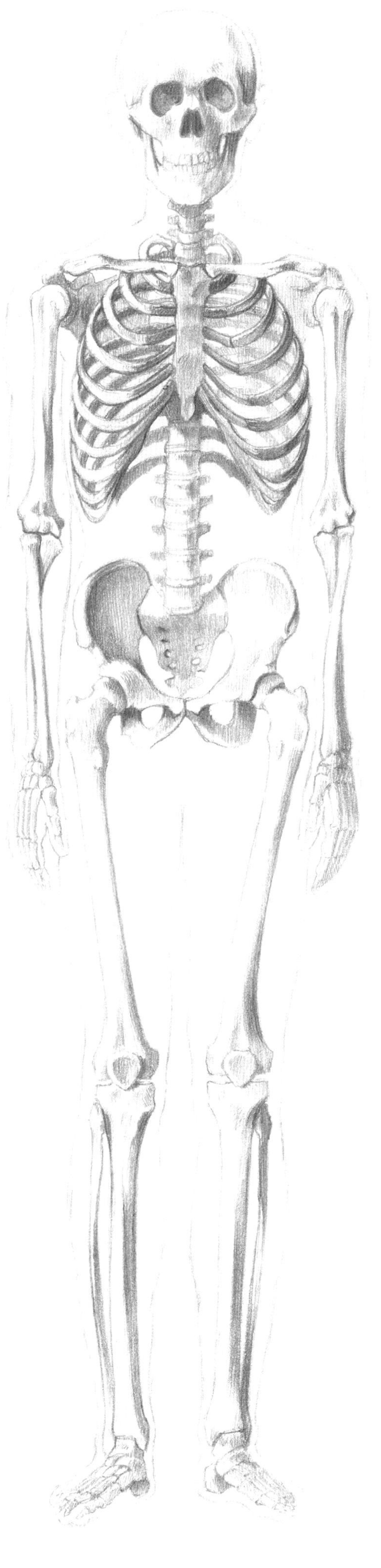

ventral aspect

The skeleton forms a solid internal frame for the body. It supports the internal organs and makes movements possible. Bones are either single or double levers moved by the muscles. In total there are 233 bones including near-identical pairs and singular bones in the median plane (vertebra). Since bones undergo constant disintegration and rebuilding their structure and form also change. They can either be linked by rigid connection by means of bony or cartilaginous joints, or flexibly linked by muscles or ligaments.

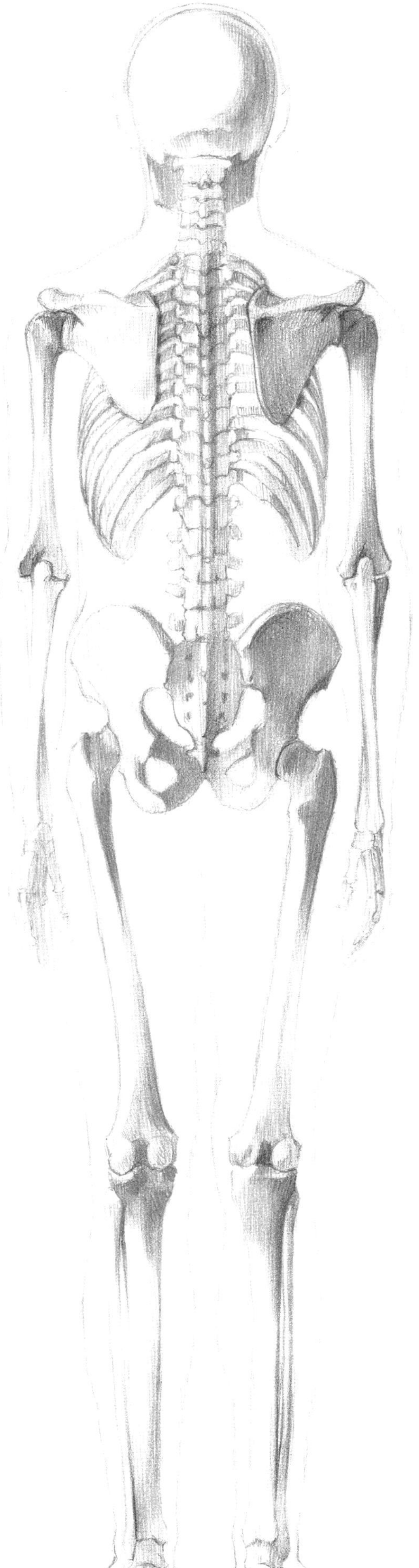

dorsal aspect

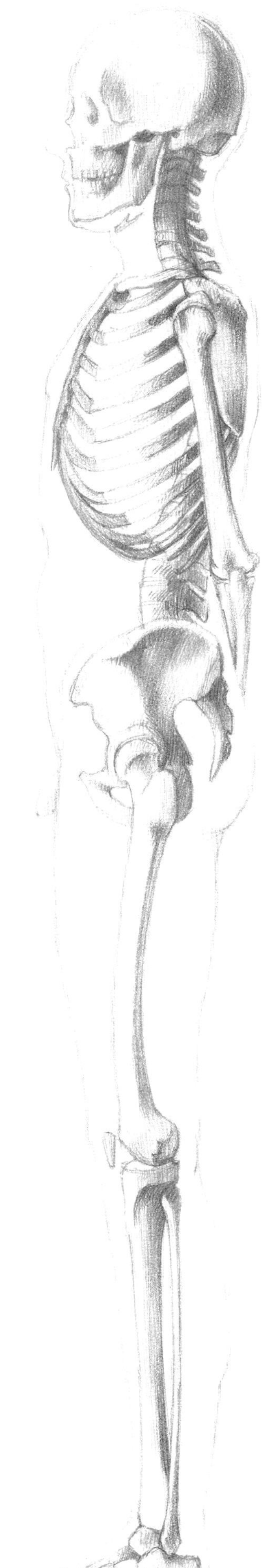

lateral aspect

Fig. 2
Bone Form

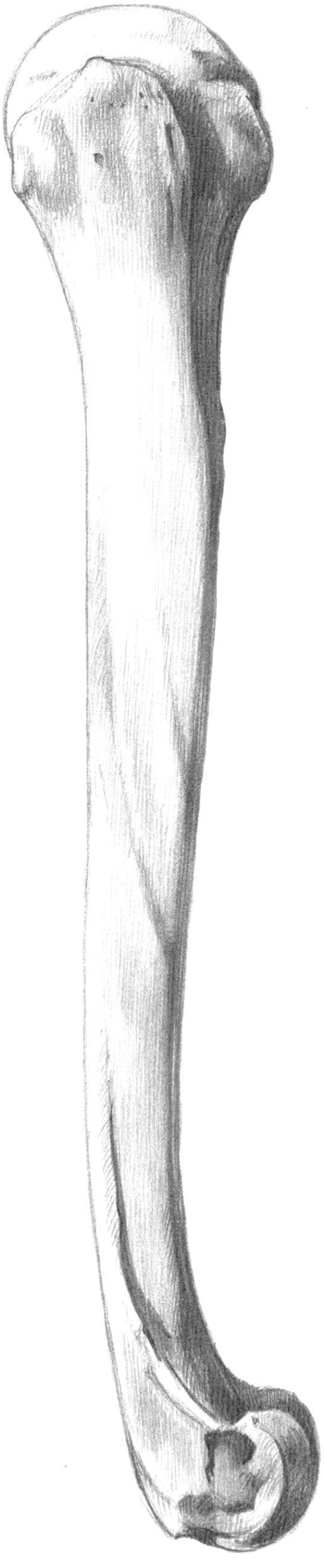

Upper arm bone
(long, tubular)

Rib
(long, flat)

Carpal bone

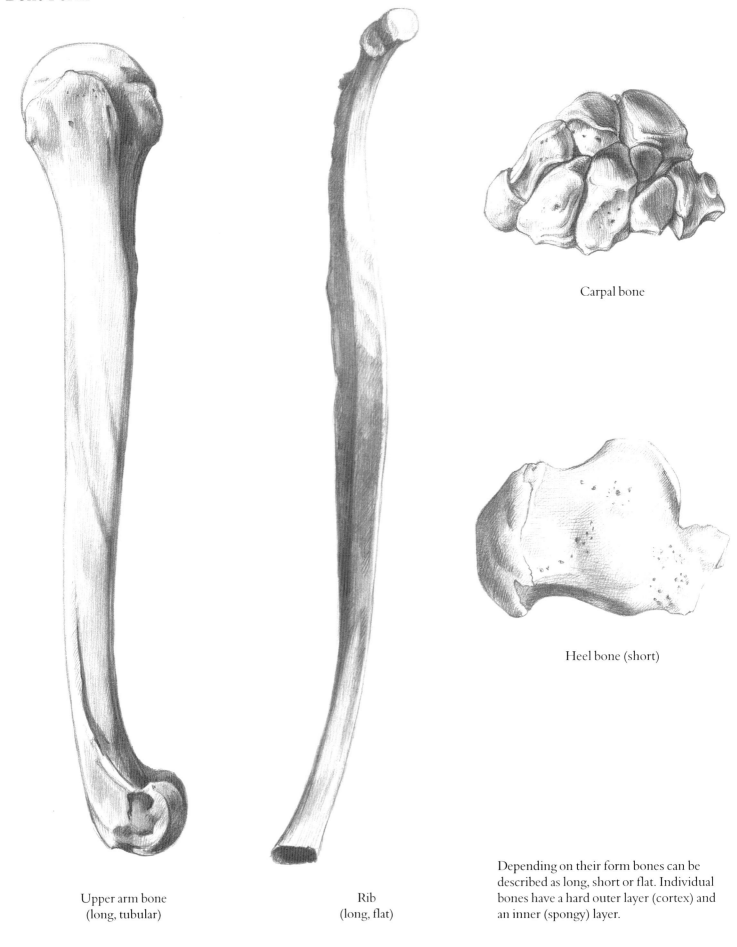

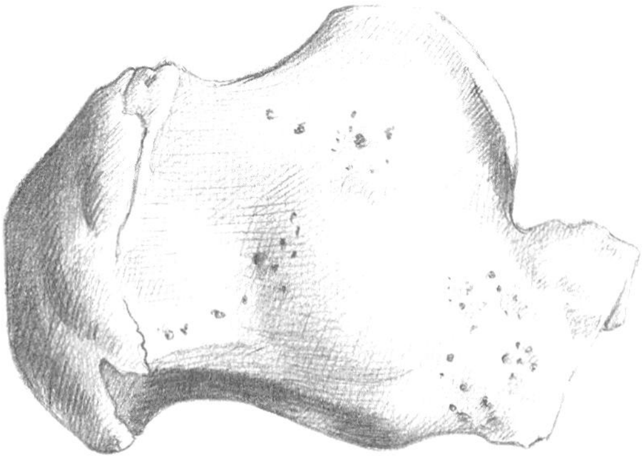

Heel bone (short)

Depending on their form bones can be described as long, short or flat. Individual bones have a hard outer layer (cortex) and an inner (spongy) layer.

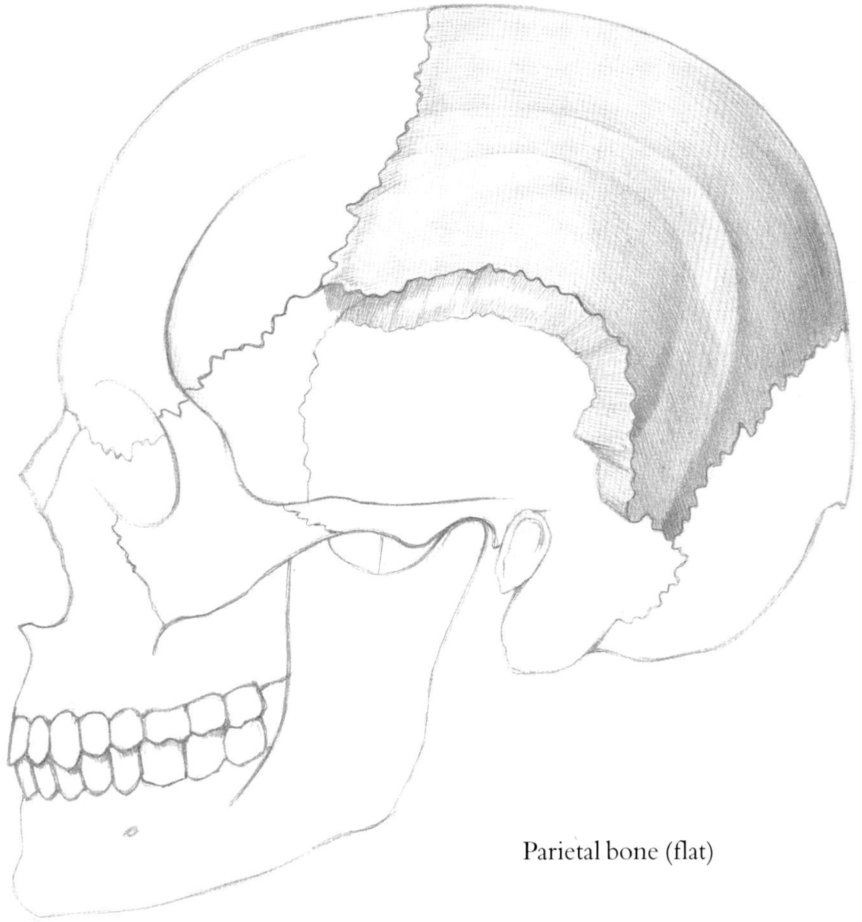

Parietal bone (flat)

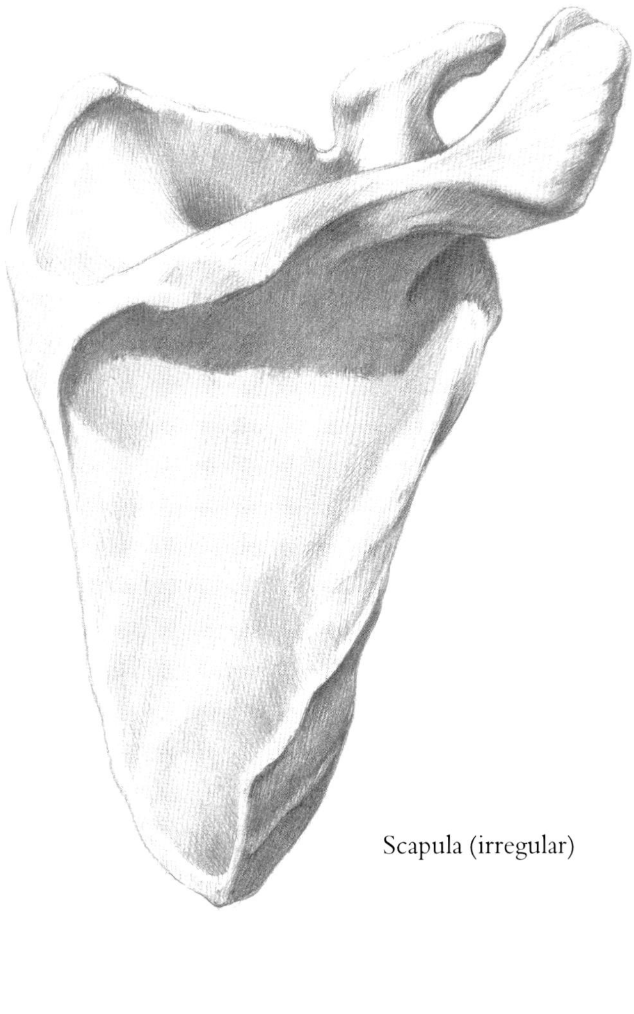

Scapula (irregular)

Fig. 3
The Functional Structure of Bones

Proximal part of the humerus

In spongy bones the bone supports are organised according to the direction and strength of the forces exerted upon them. They form what are known as trabeculae.

Fig. 4
The Form of Joint Surfaces

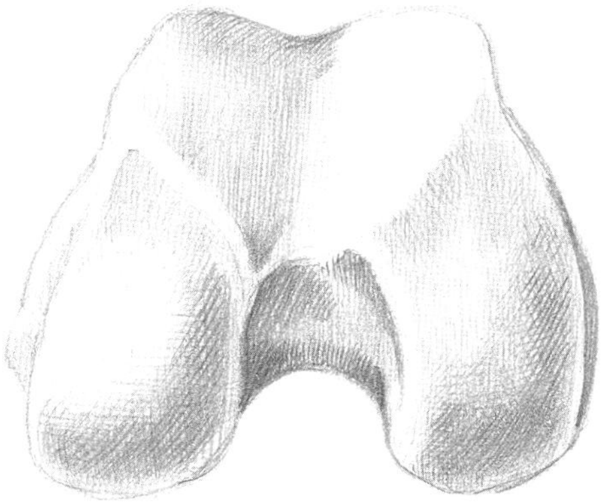

Articular knot

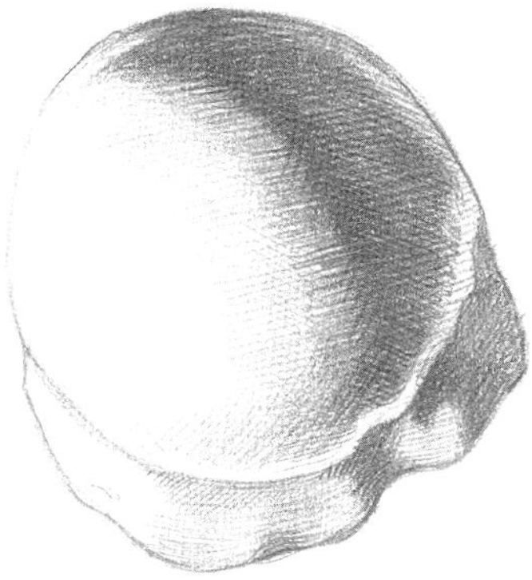

Articular head

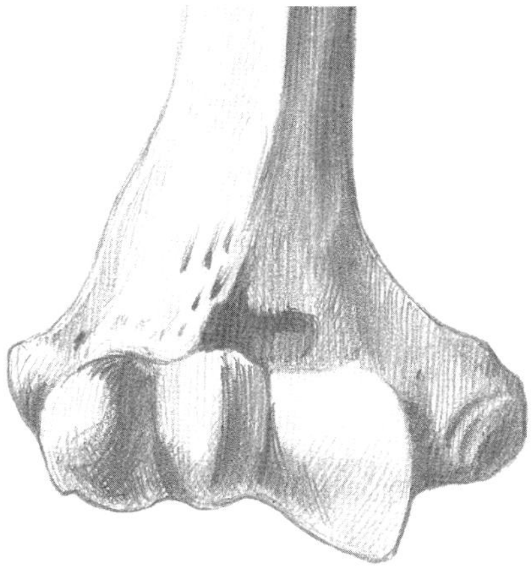

Articular condyle

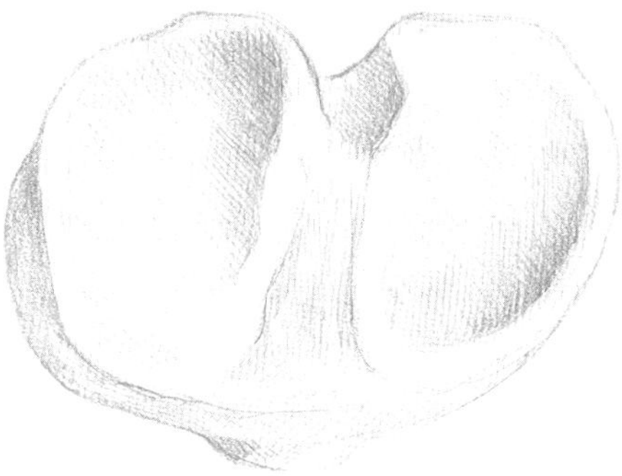

Glenoid cavity

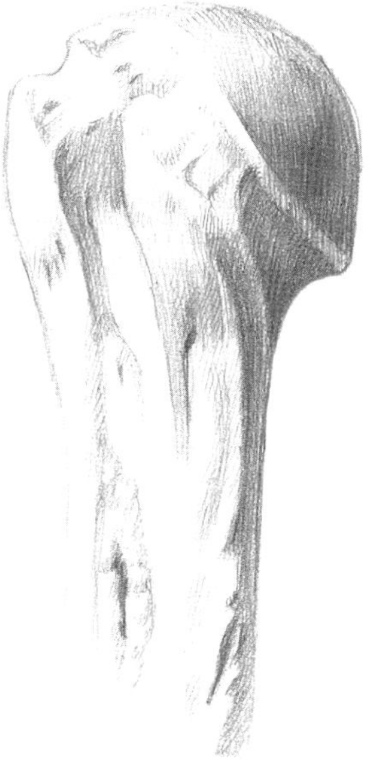

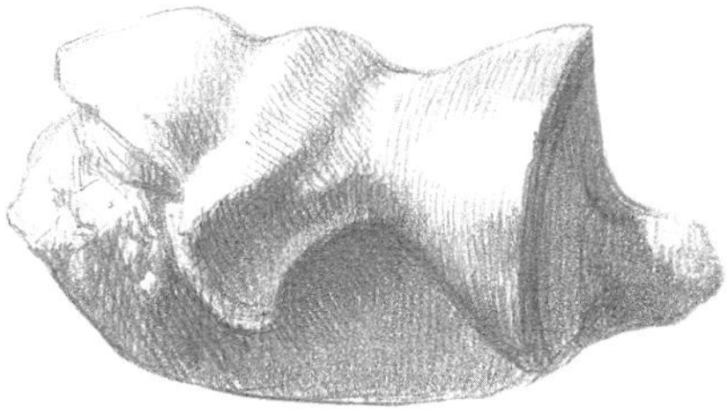

Articular fossa

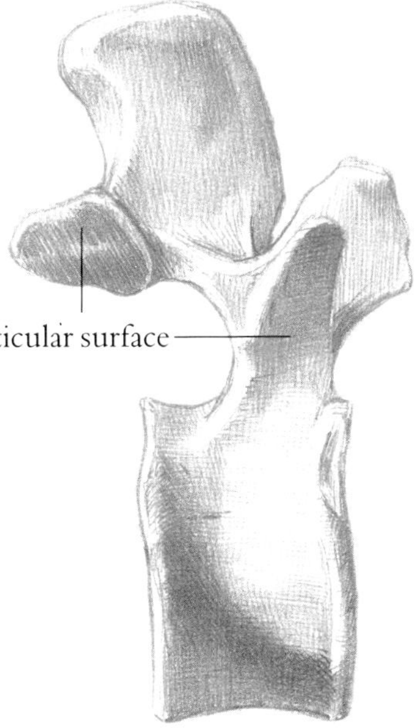

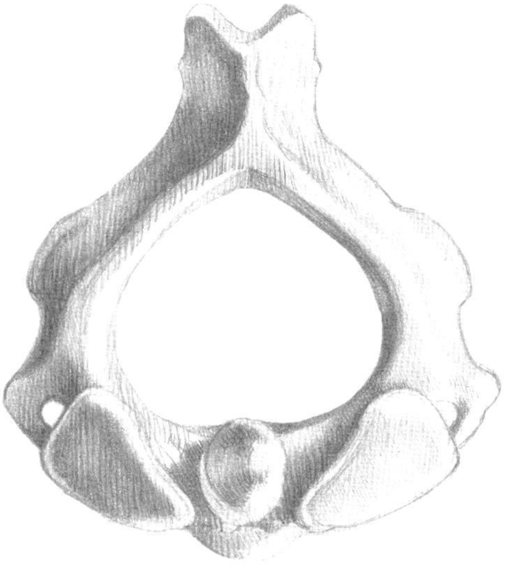

Articular process

Fig. 5
Origin and Insertion Surfaces
of Bone Muscles

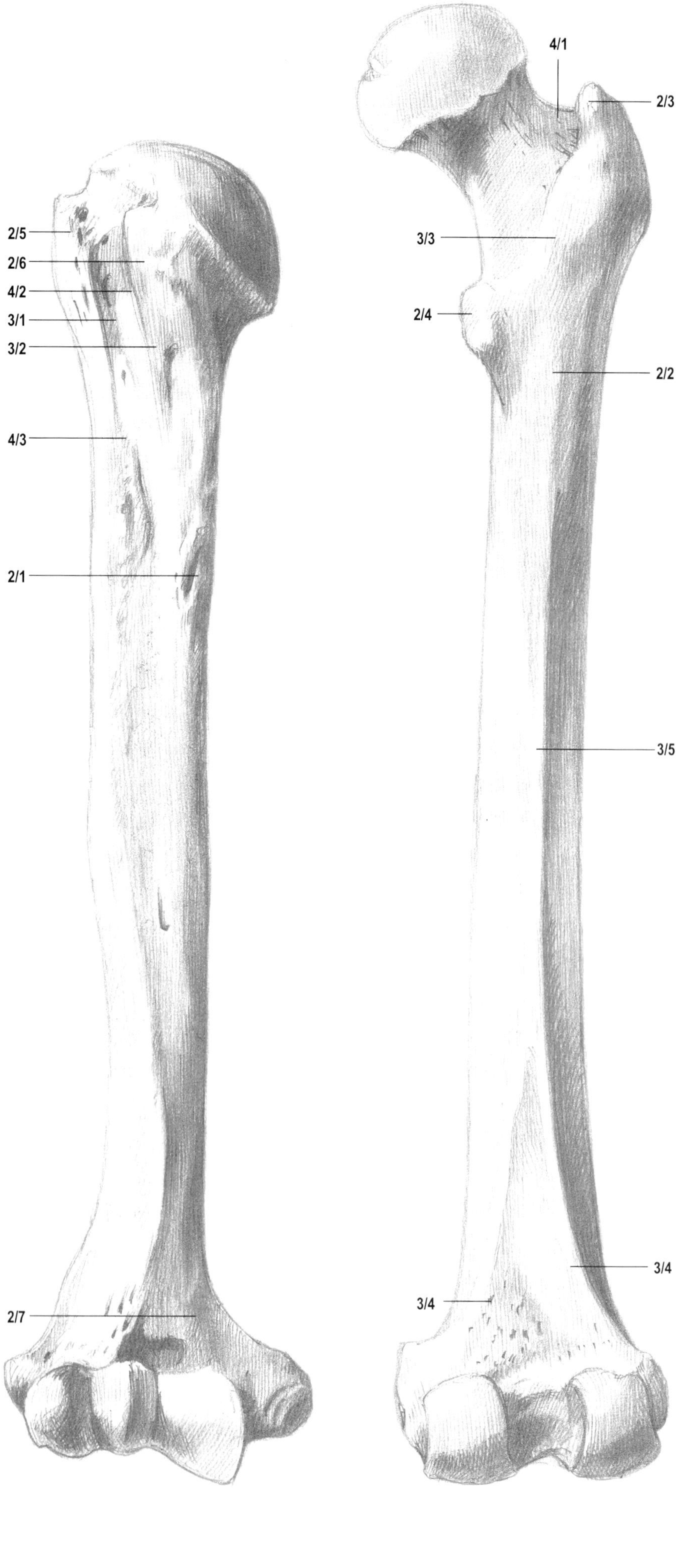

1 PROCESSUS MUSCULARIS
(Muscular process)
- **1/1** Coronoideus process (ulnar process)
- **1/2** Spinous process (spiny process)
- **1/3** Transversal process (transverse process)

2 CONDYLE, TUBEROSITAS, TUBER
- **2/1** Deltoid tuberositas (insertion point of the deltoid muscle)
- **2/2** Glutaeal tuberositas (humeral protuberance, insertion point of the glutaeal muscle)
- **2/3** Trochanter major (greater trochanter, insertion of most hip muscles)
- **2/4** Trochanter minor (lesser trochanter, insertion of the lumbar muscles)
- **2/5** Tuberculum majoris (greater humeral protuberance, insertion of the extensor muscles)
- **2/6** Tuberculum minus (lesser humeral protuberance, insertion of the extensor muscles)
- **2/7** Epicondylus extensorius et flexorius (extensor and flexor tuberosities for flexor muscles)

3 CRISTA (Muscular crest);
LINEA (Muscular line)
- **3/1** Crista tuberculi majoris humeri (Greater humeral crest)
- **3/2** Crista tuberculi minoris humeri (lesser humeral crest)
- **3/3** Crista intertrochanterica (between greater and lesser trochanter)
- **3/4** Linea supracondylaris medialis et lateralis (medial and lateral head line)
- **3/5** Linea aspera (rough longitudinal axis on the posterior surface of the humeral shaft)

4 FACIES MUSCULARIS
(Muscle origin and insertion surfaces)
- **4/1** Fossa trochanterica (depression near the greater trochanter)
- **4/2** Sulcus intertubercularis (sulcus between greater and lesser humeral crest for the tendon of the biceps)
- **4/3** Facies muscularis (muscle origin or insertion surface)

Fig. 6 Complex Joint (Hinge and Condyle Joint)

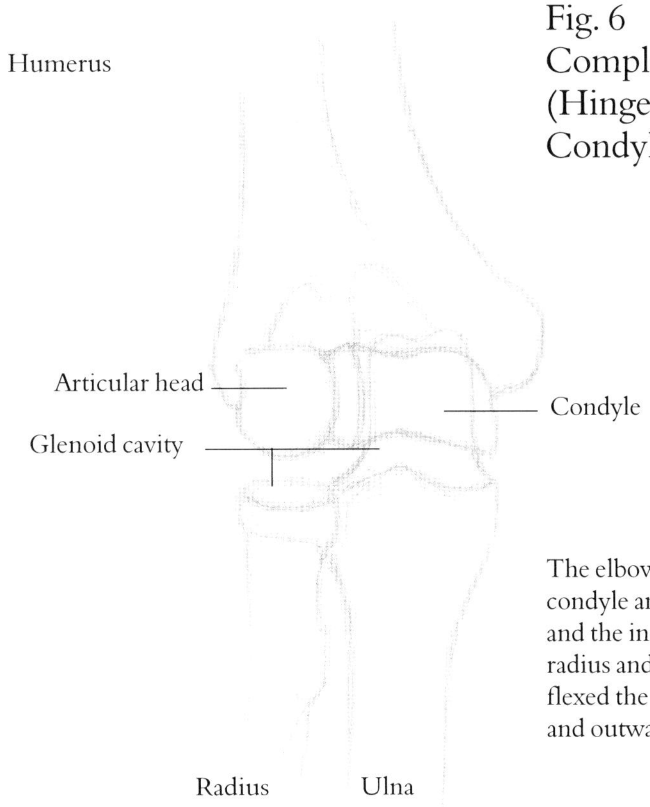

dorsal aspect

The elbow joint consists of the condyle and head of the humerus, and the incisura between the radius and ulna. When the elbow is flexed the hand rotates upwards and outwards.

Ligaments of the elbow joint

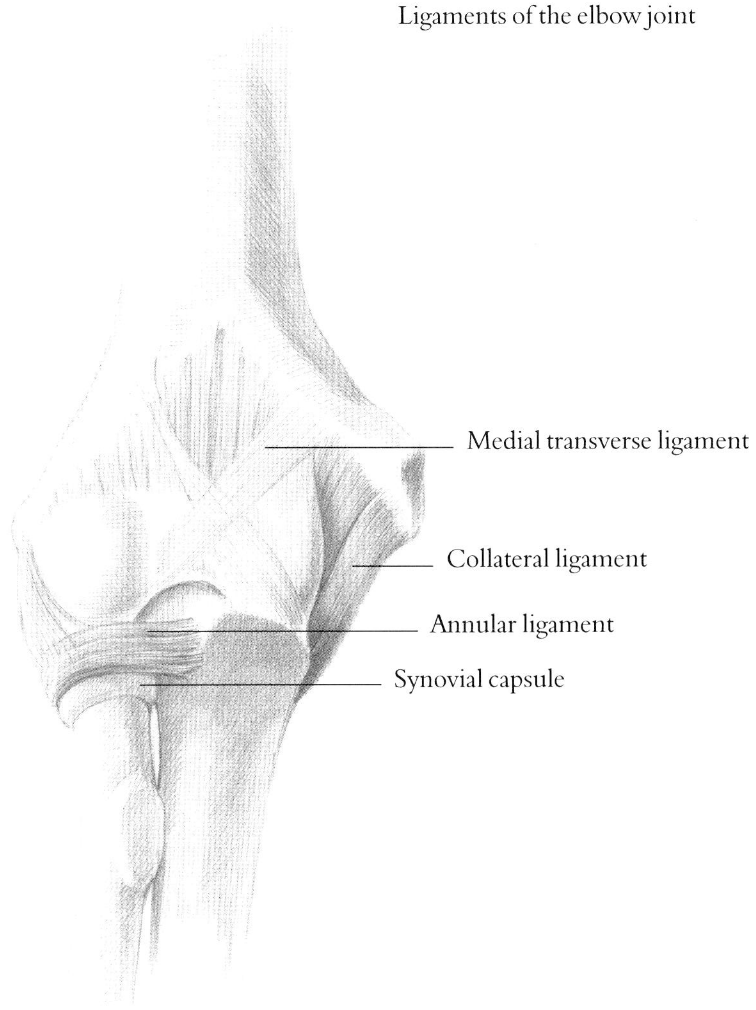

Fig. 7
The Finger Hinge Joint

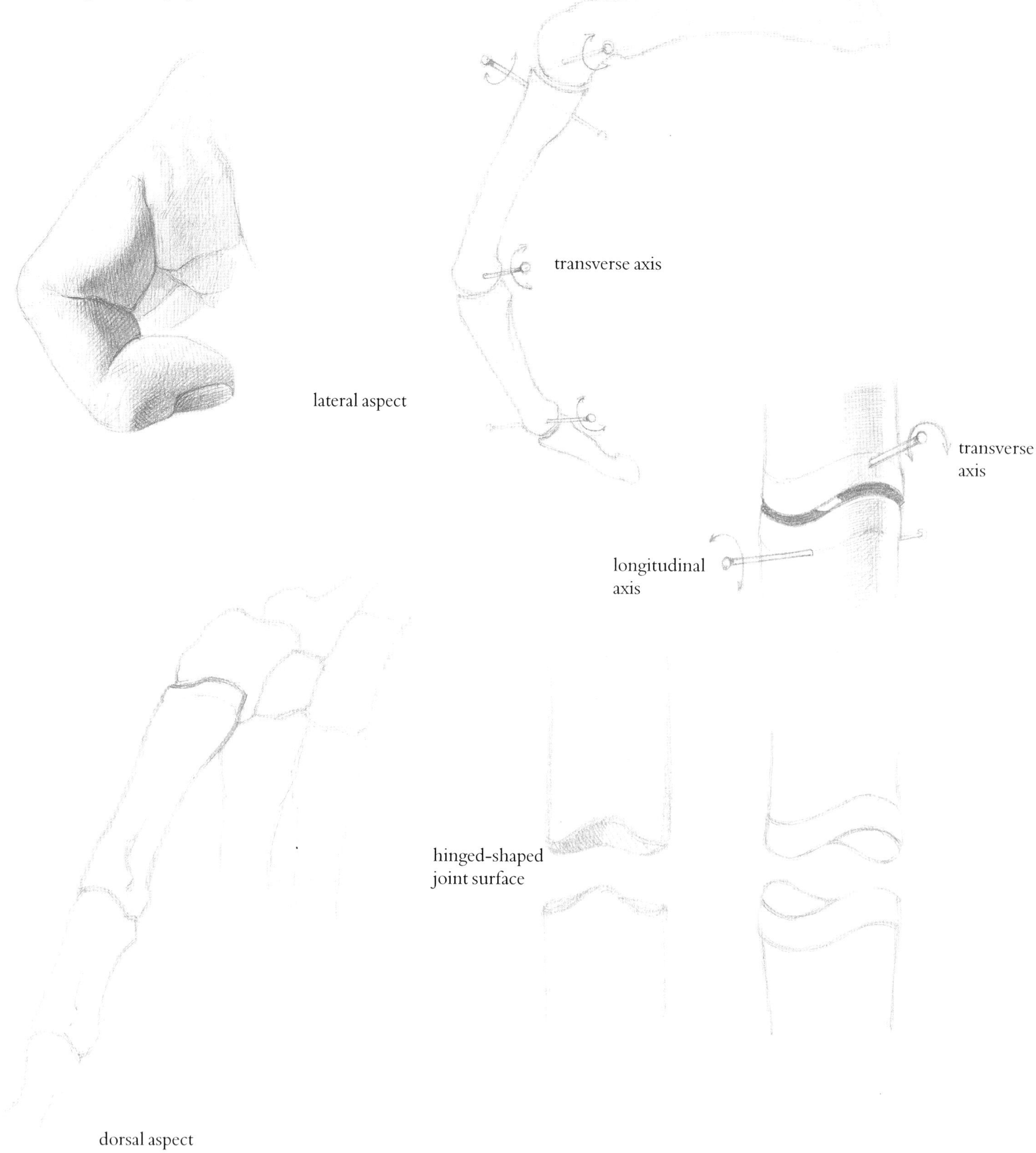

Here the phalanges of the finger are illustrated. They consist of the proximal and dorsal joint surfaces of the finger bones. These are biaxial joints, which not only bend and stretch, but also permit a slight sideways movement.

Fig. 8
The Ball-and-Socket Joint

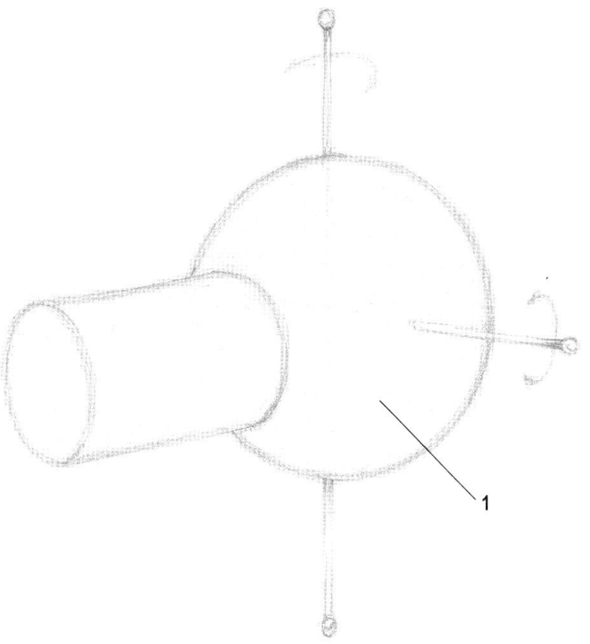

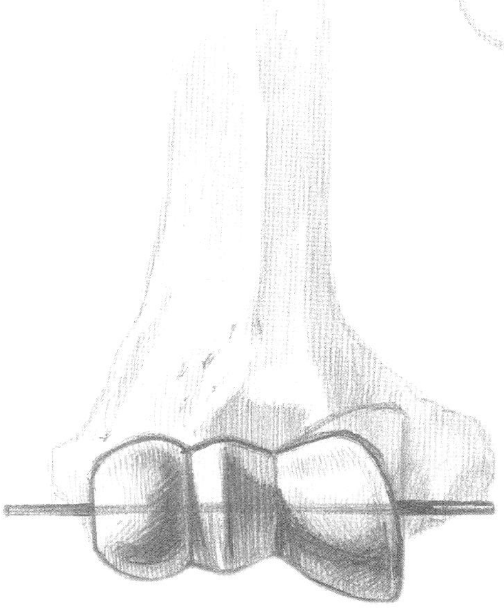

ventral aspect

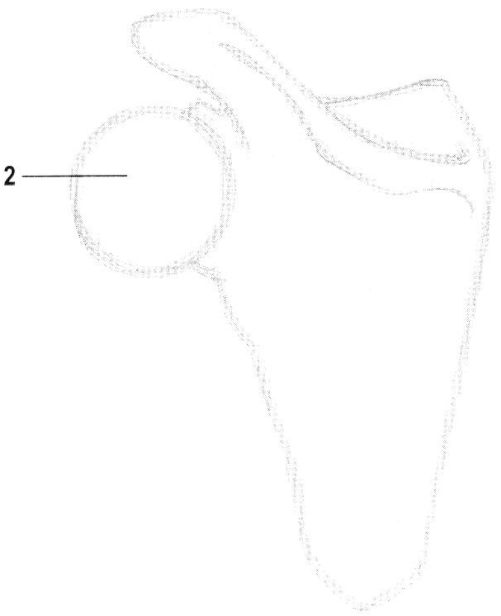

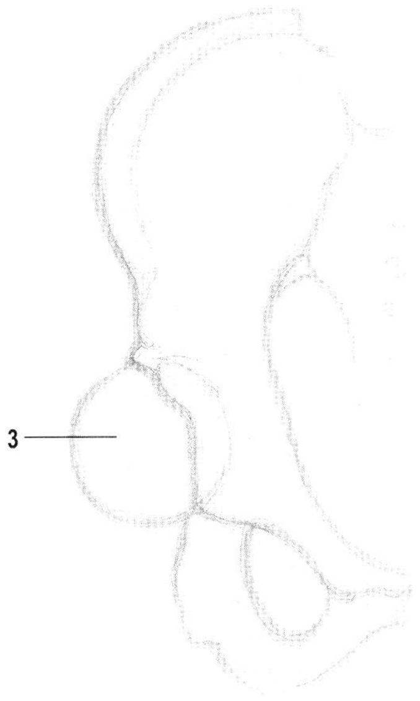

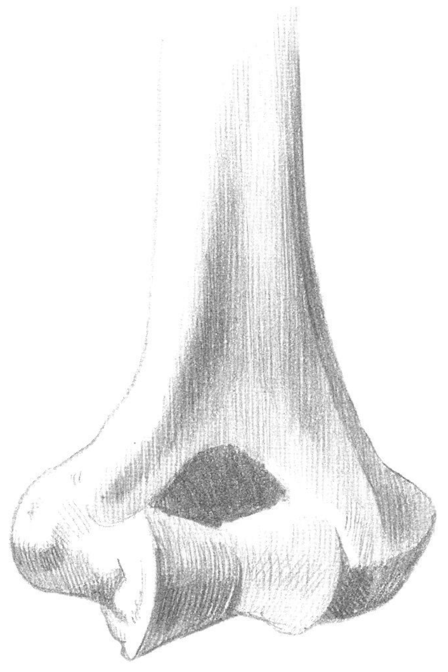

dorsal aspect

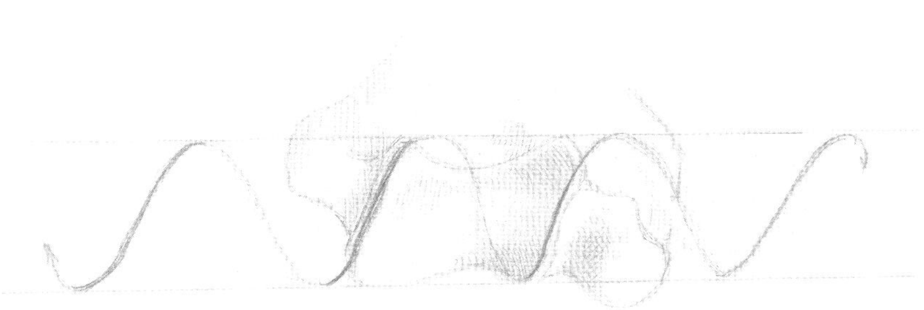

caudal aspect

The illustration shows the shoulder and hip joints. The shoulder joint consists of the round head of the humerus (**1**) and the glenoid cavity of the scapula (**2**). The hip joint consists of the semi-spherical head of the femur and the hip glenoid cavity (**3**).

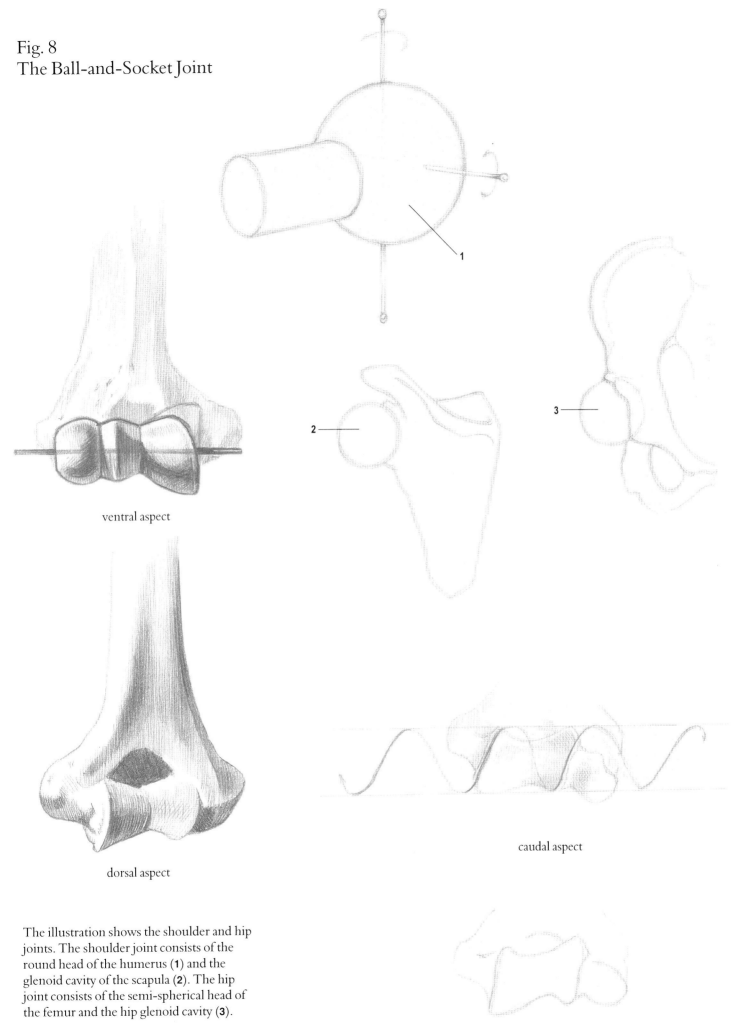

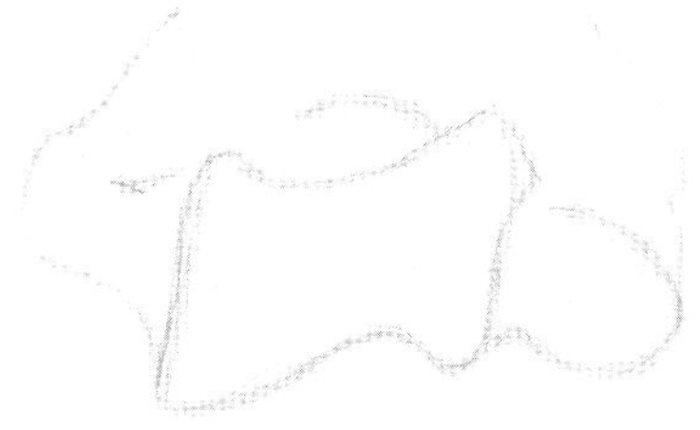

Fig. 9
Muscle Types

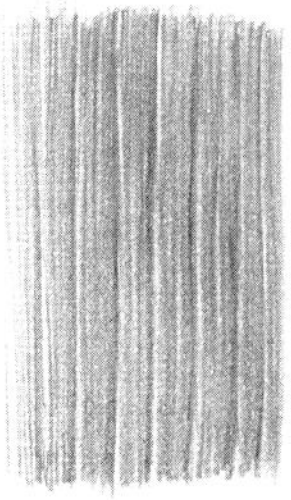

Quadrilateral

Strap

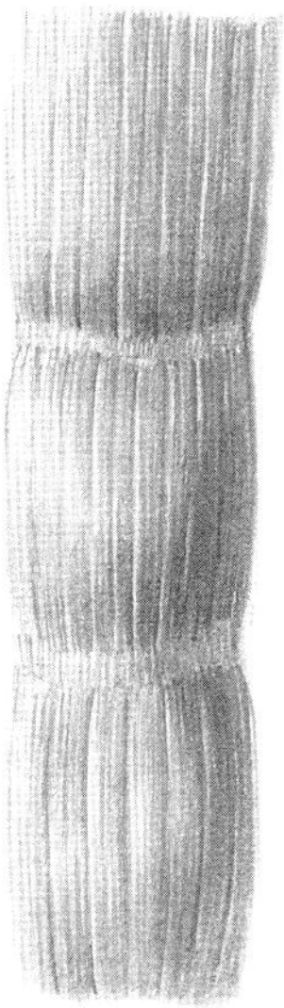

Strap with tendinous intersections

Unipennate

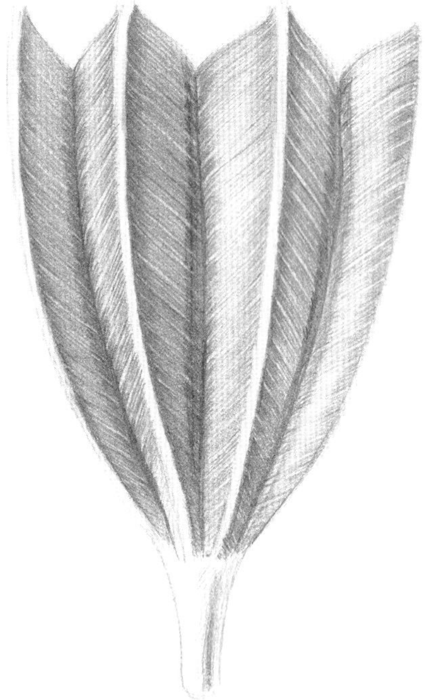

Multi-pennate

Triangular

Muscle plates

Fusiform

Biventer

Tricipital

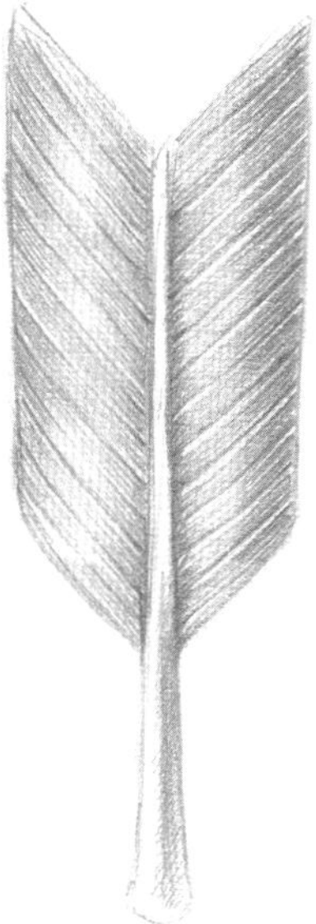

Bipennate

Fig. 10
Muscle Structure

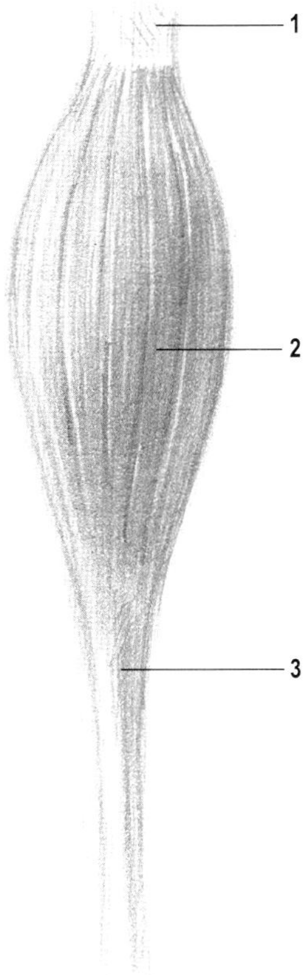

1 Muscle head
2 Muscle belly
3 End of the muscle with origin of tendon
4 Tendinous muscle fibres penetrate into the membrane and bone in the form of a fan

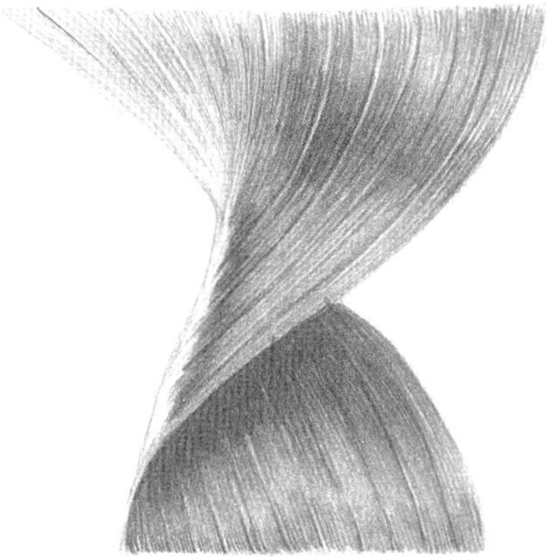

Spiral muscle plates

Spiral-strap

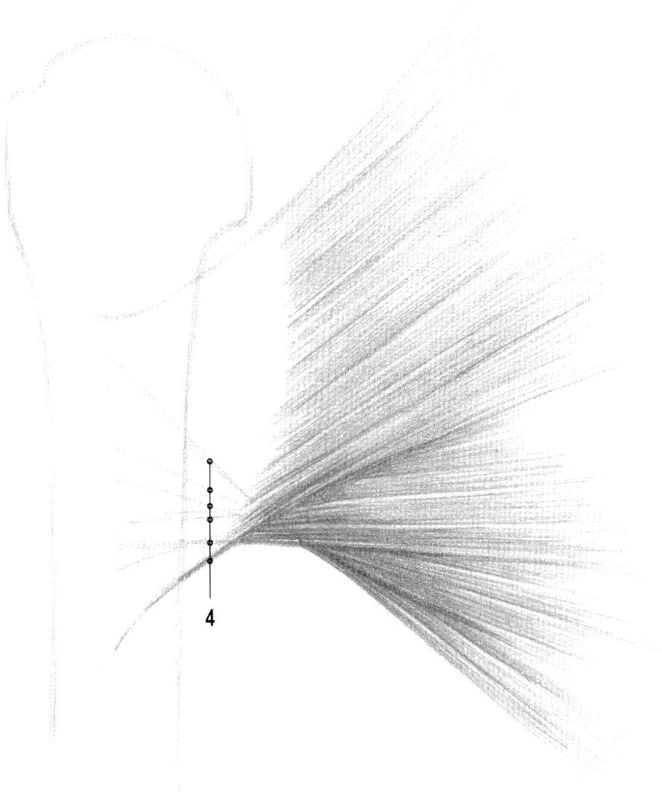

Fig. 11
Types of Origin and Insertion of a Muscle

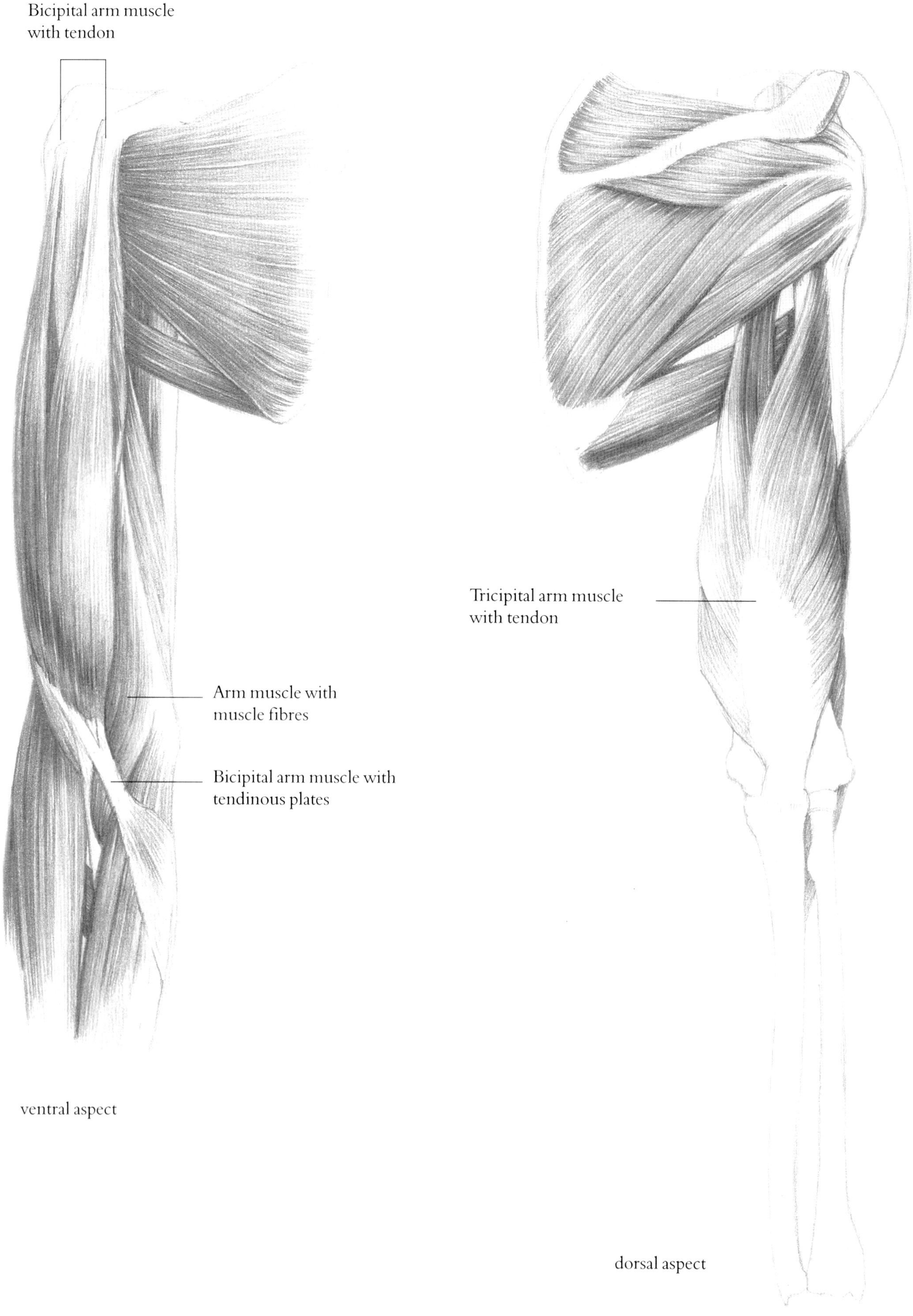

Fig. 12
How Muscles Act Together

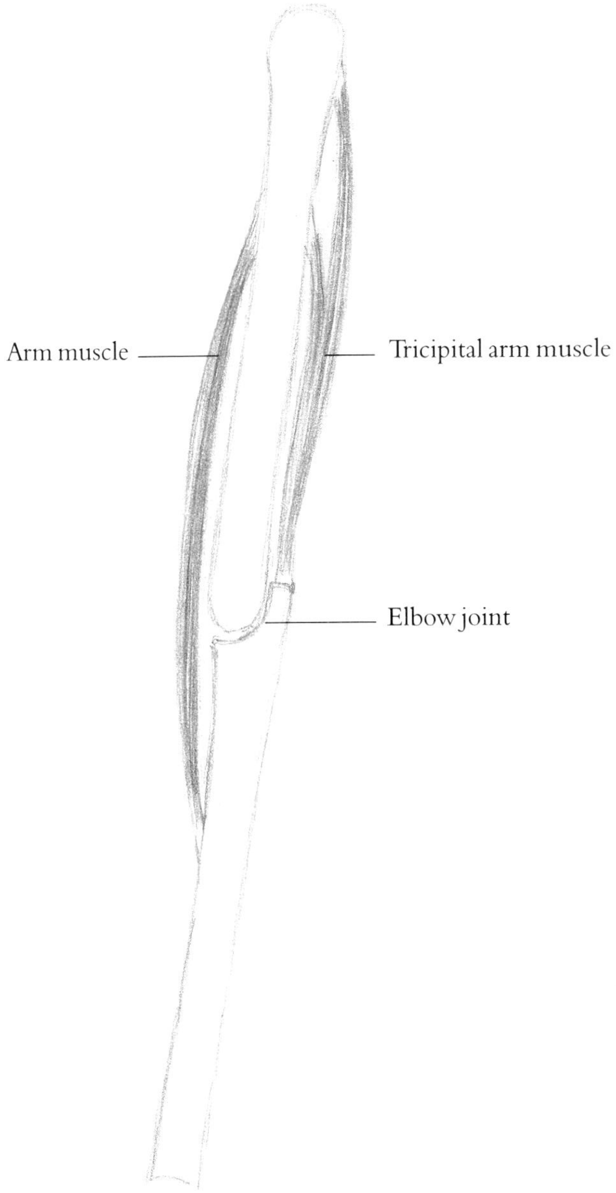

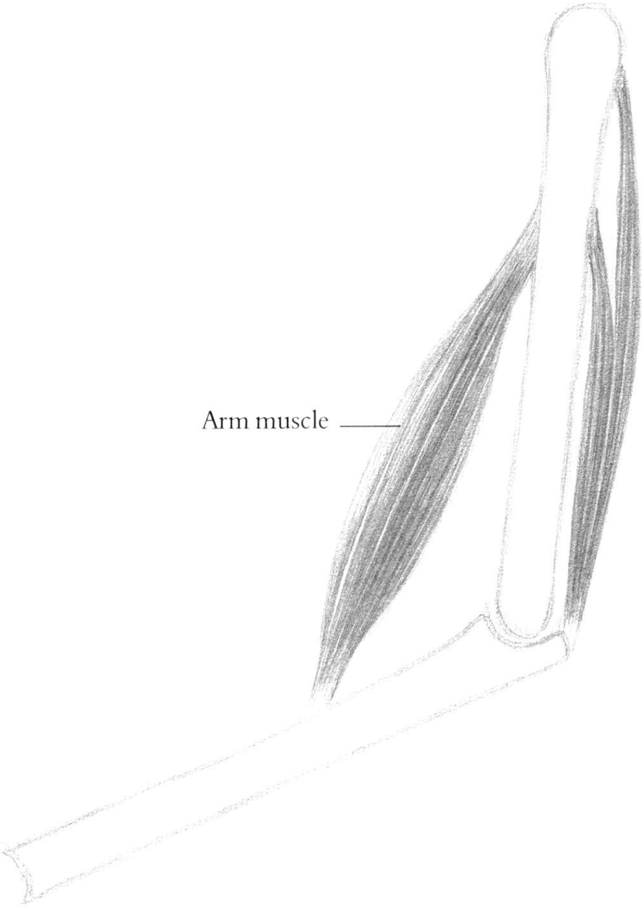

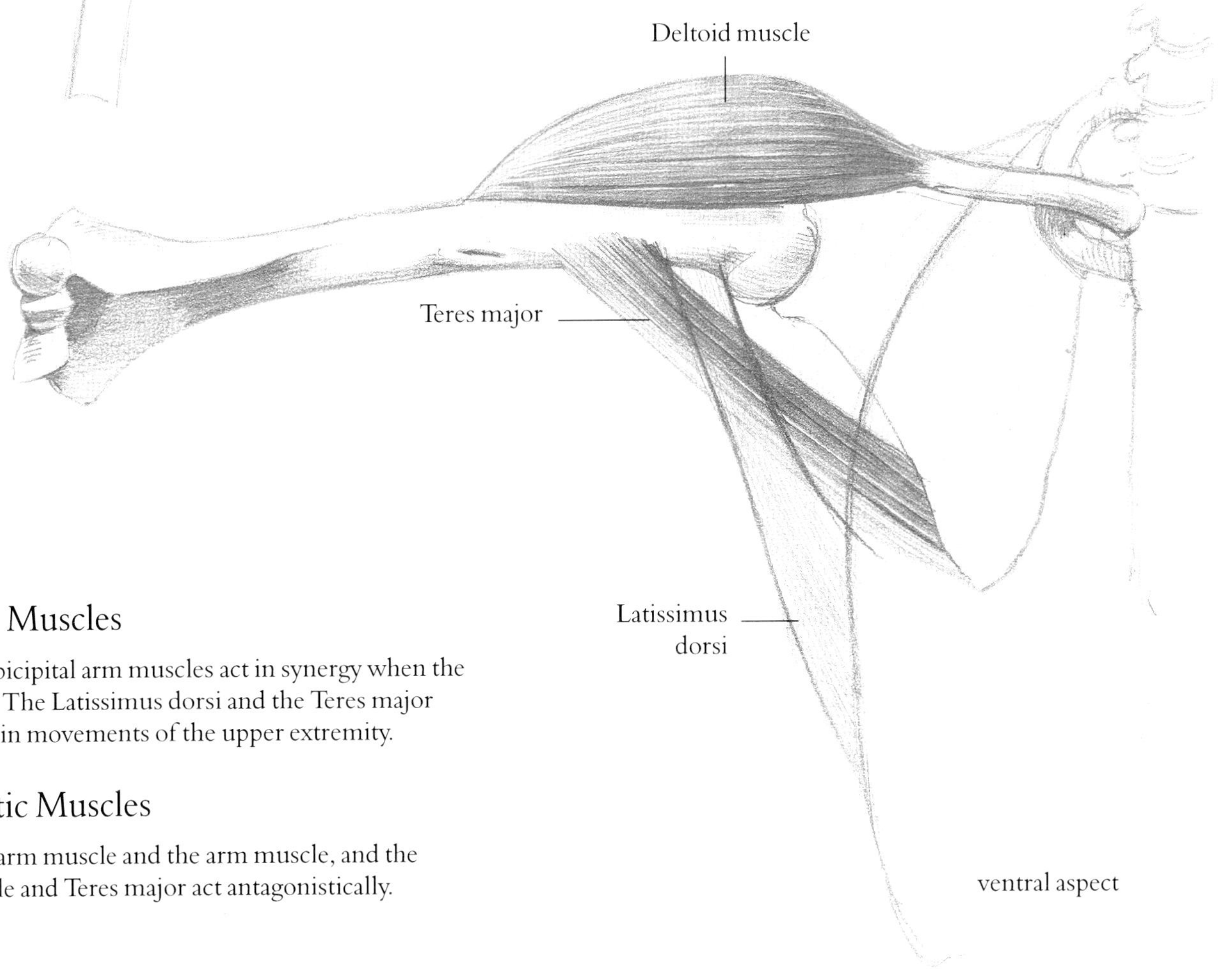

Synergetic Muscles

The arm and bicipital arm muscles act in synergy when the elbow is bent. The Latissimus dorsi and the Teres major act in synergy in movements of the upper extremity.

Antagonistic Muscles

The tricipital arm muscle and the arm muscle, and the Deltoid muscle and Teres major act antagonistically.

THE UPPER EXTREMITY

Fig. 13
The Acromioclavicular and
Arm Bones

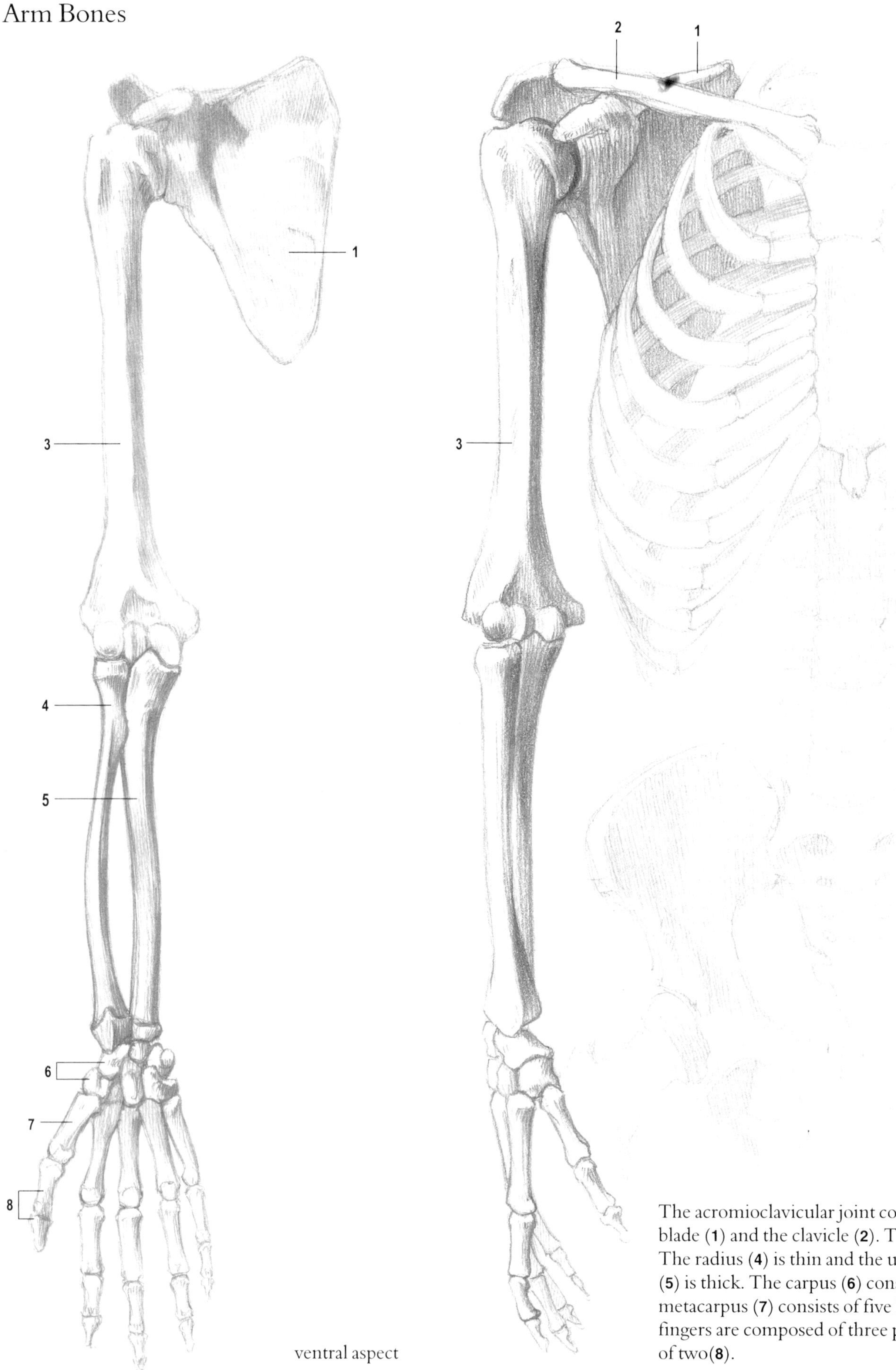

ventral aspect

The acromioclavicular joint consists of the shoulder blade (**1**) and the clavicle (**2**). The humerus (**3**) is long. The radius (**4**) is thin and the upper part of the ulna (**5**) is thick. The carpus (**6**) consists of small bones, the metacarpus (**7**) consists of five longer bones. The fingers are composed of three parts and the thumb of two(**8**).

Fig. 14
The Acromioclavicular Joints and Arm Joints

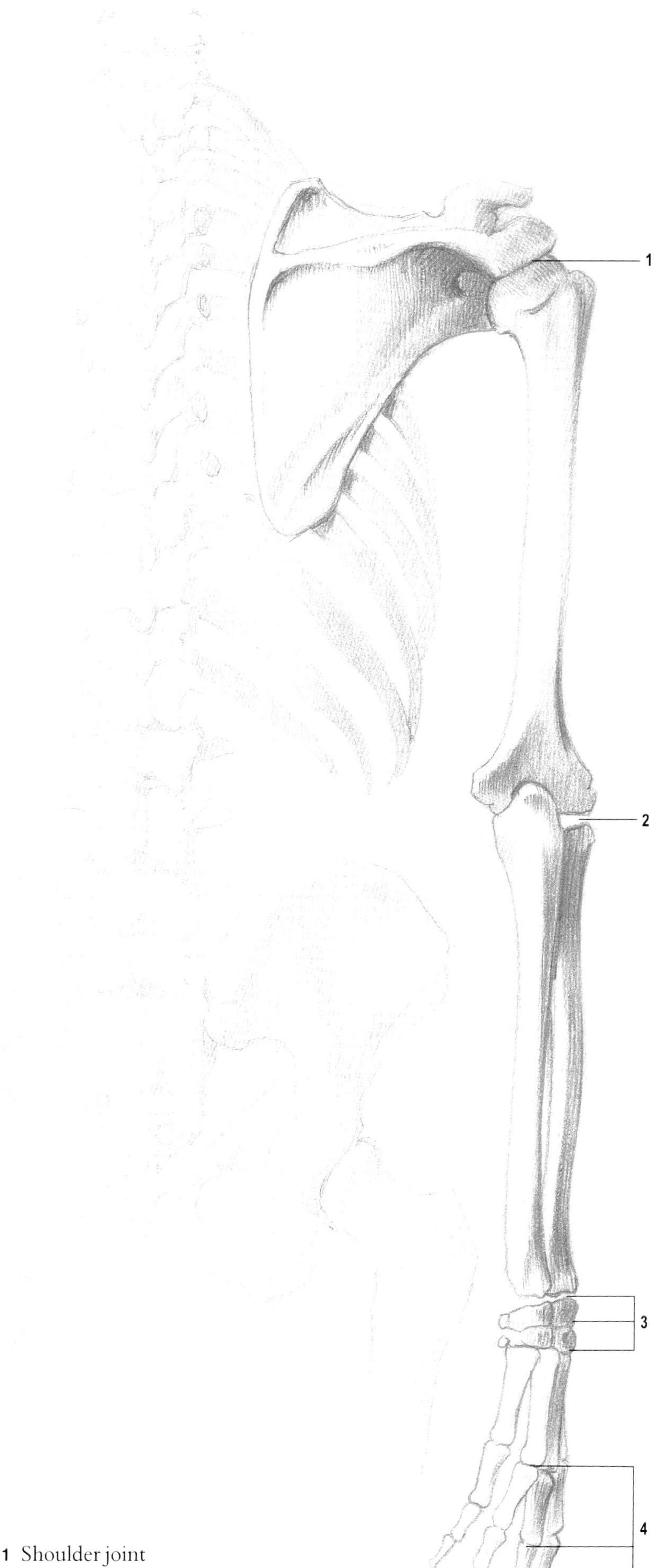

1 Shoulder joint
2 Elbow joint
3 Carpal joints
4 Finger joints

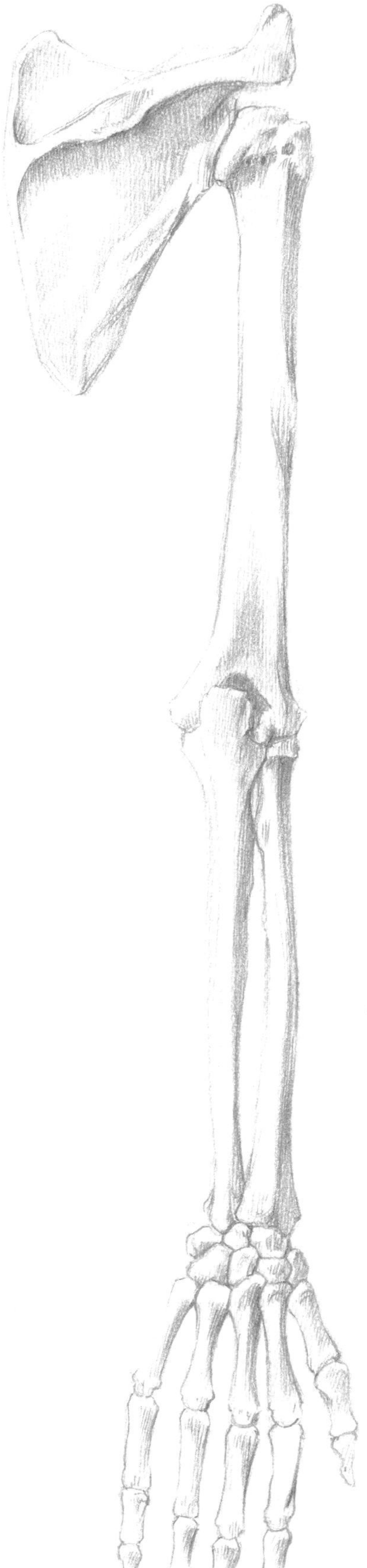

dorsal aspect

Fig. 15
The Shoulder Blade

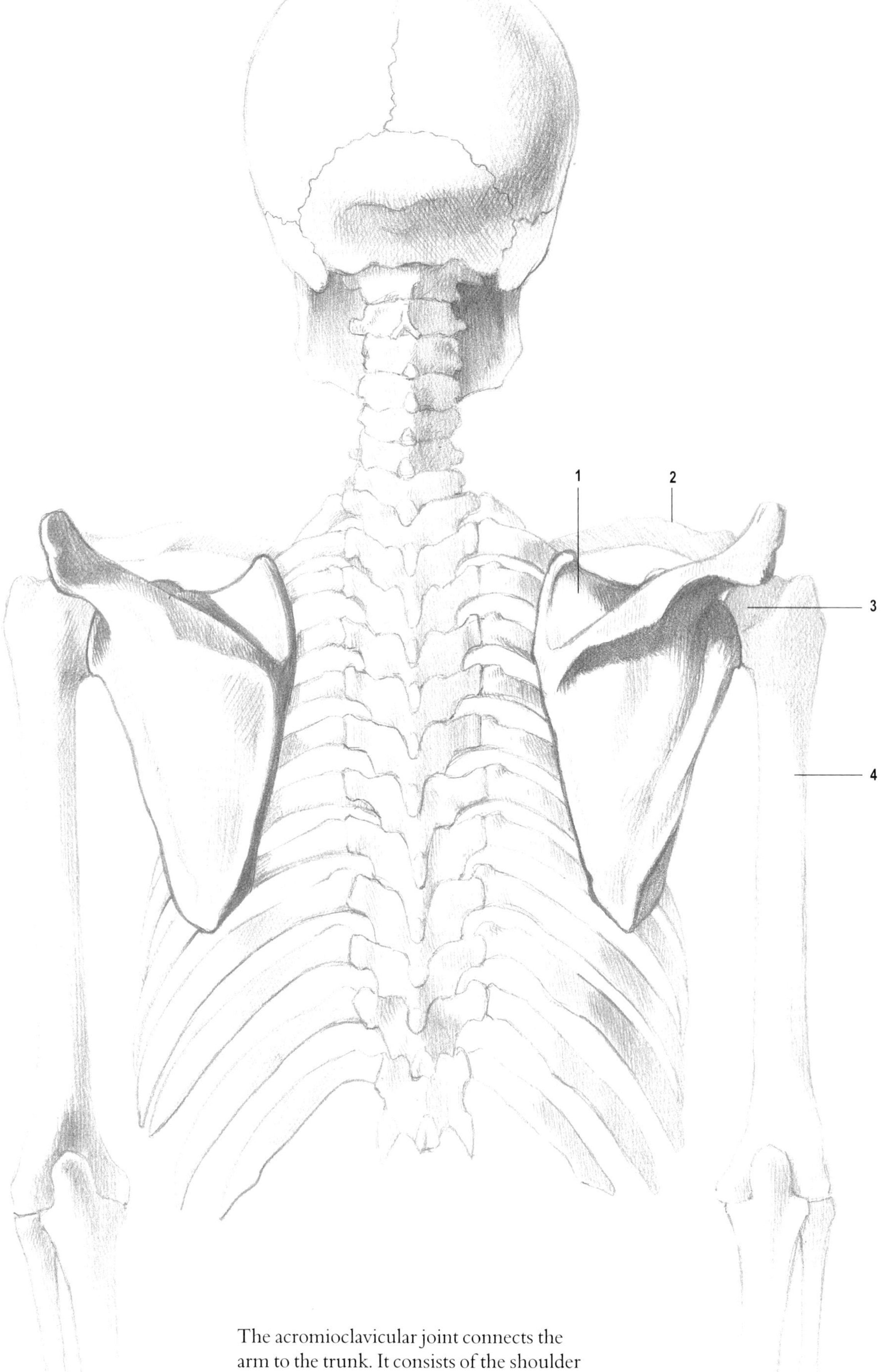

The acromioclavicular joint connects the arm to the trunk. It consists of the shoulder blade (**1**) and the clavicle (**2**). The shoulder blade and humerus (**4**) form the shoulder joint (**3**).

dorsal aspect

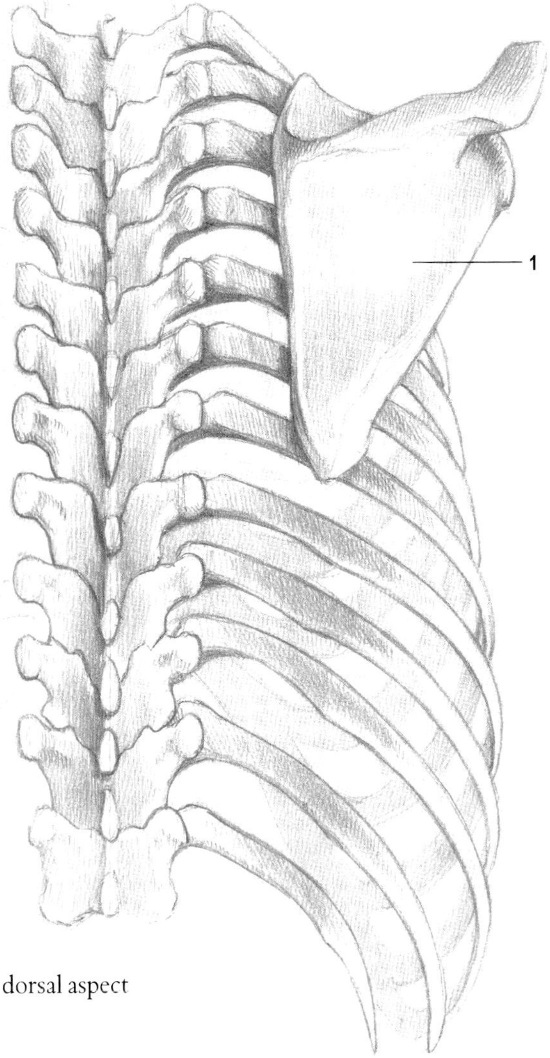

dorsal aspect

The shoulder blade borders the dorsal surface of the second to the seventh ribs on both sides of the spine.

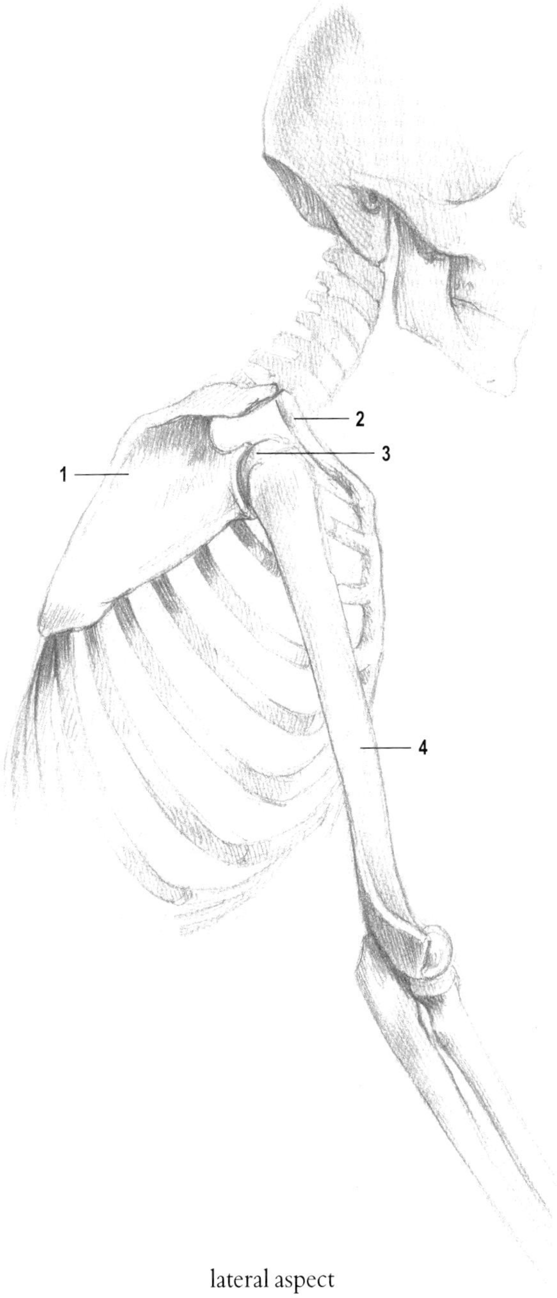

1 Shoulder blade
2 Clavicle
3 Shoulder joint
4 Humerus

lateral aspect

The Shoulder Blade
(cont.)

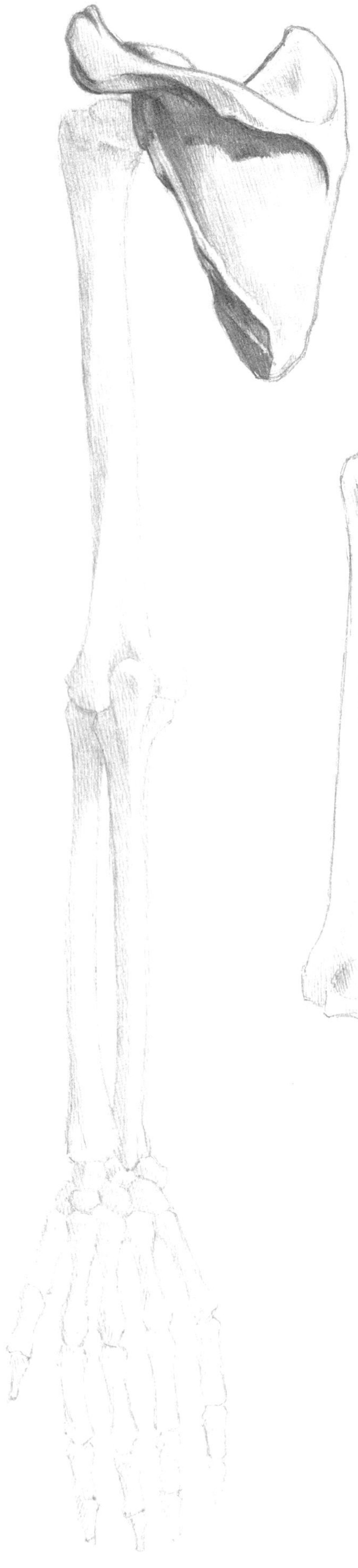

1 Shoulder blade
2 Clavicle
3 Shoulder joint
4 Humerus

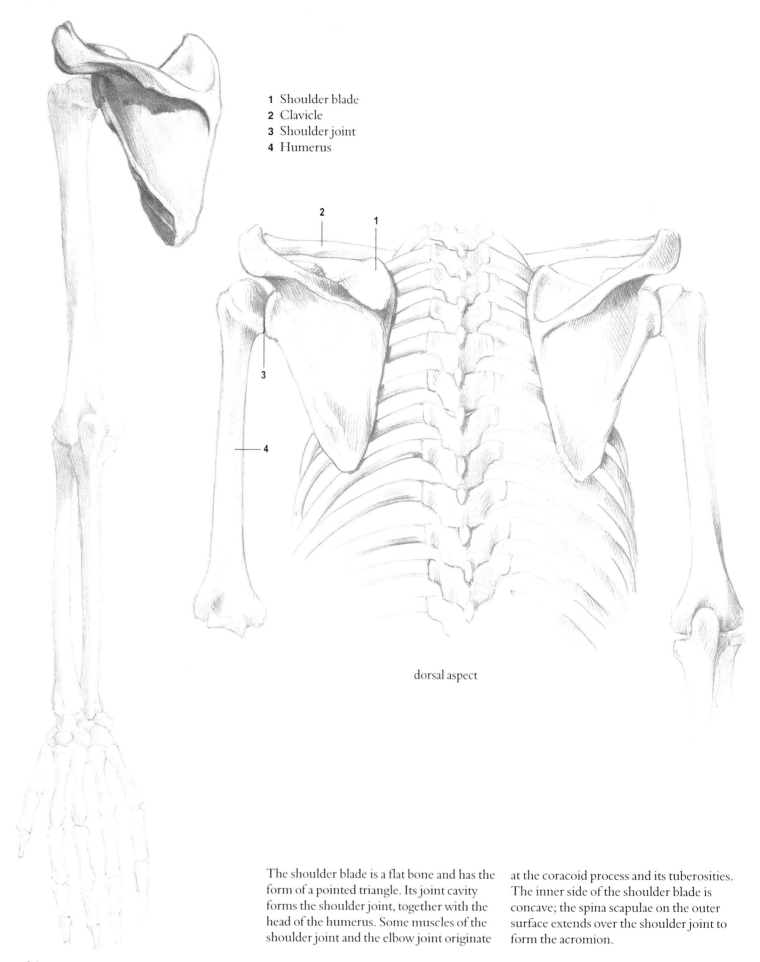

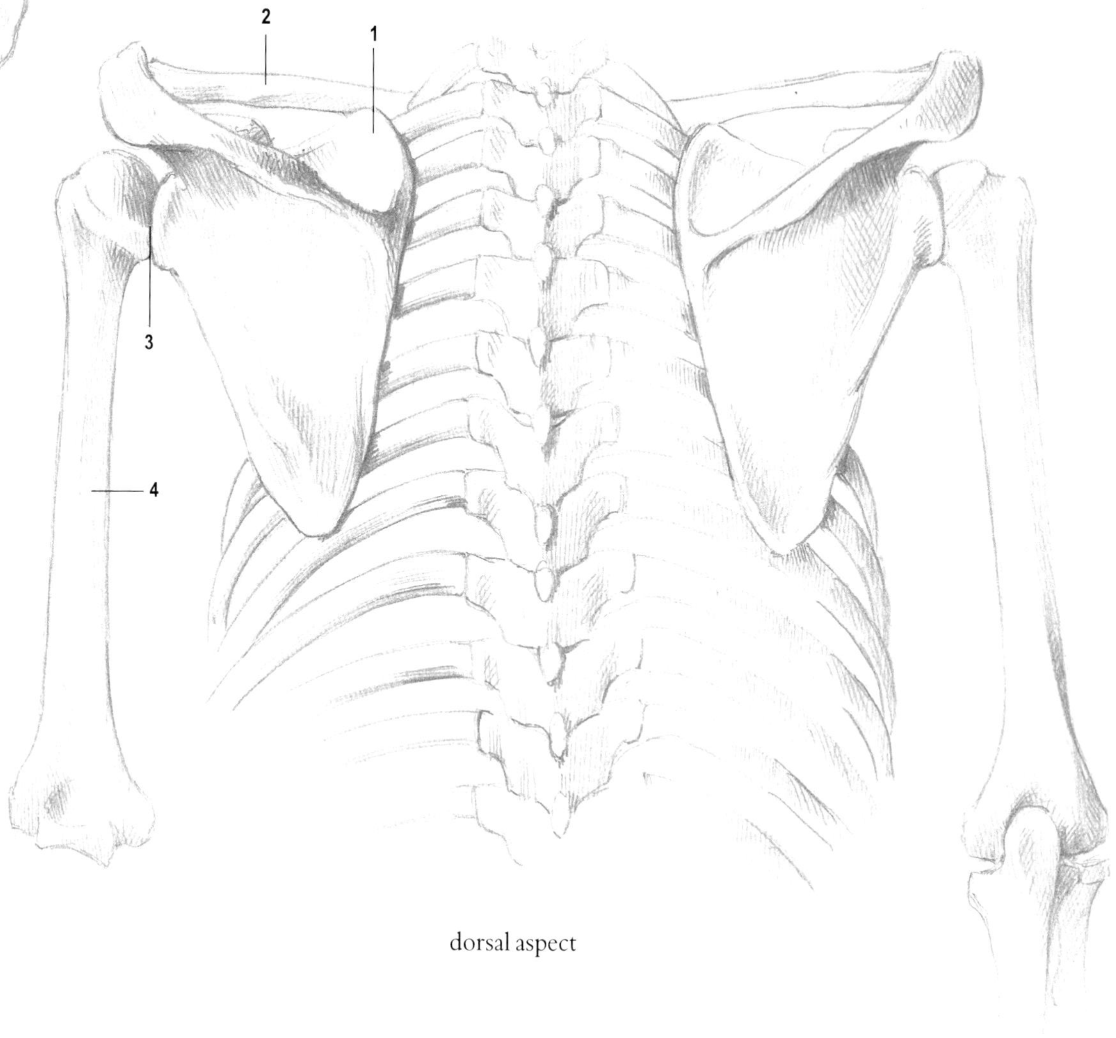

dorsal aspect

The shoulder blade is a flat bone and has the form of a pointed triangle. Its joint cavity forms the shoulder joint, together with the head of the humerus. Some muscles of the shoulder joint and the elbow joint originate at the coracoid process and its tuberosities. The inner side of the shoulder blade is concave; the spina scapulae on the outer surface extends over the shoulder joint to form the acromion.

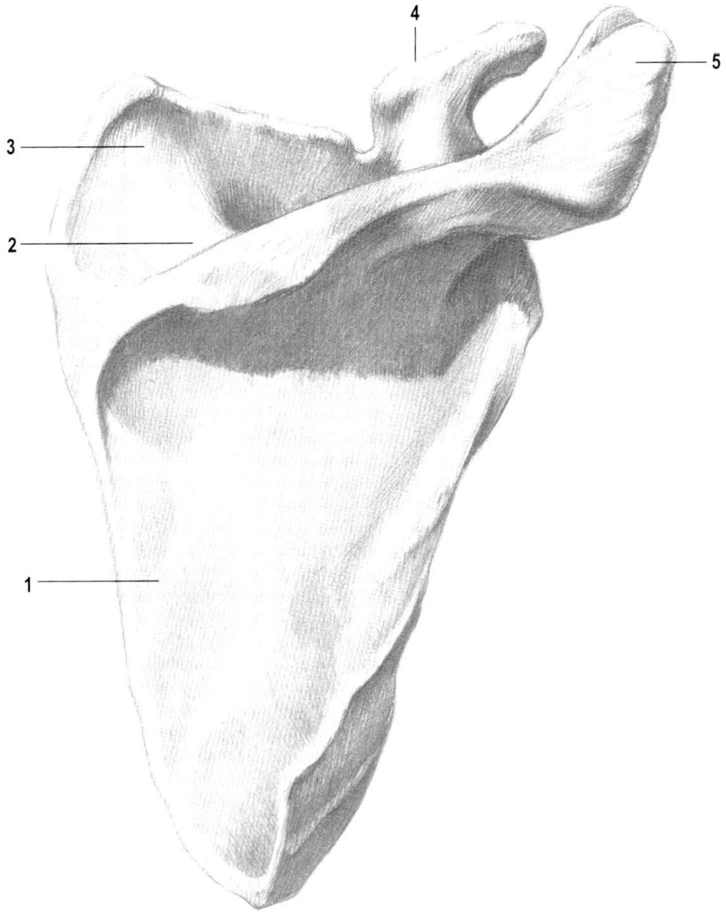

dorsal aspect

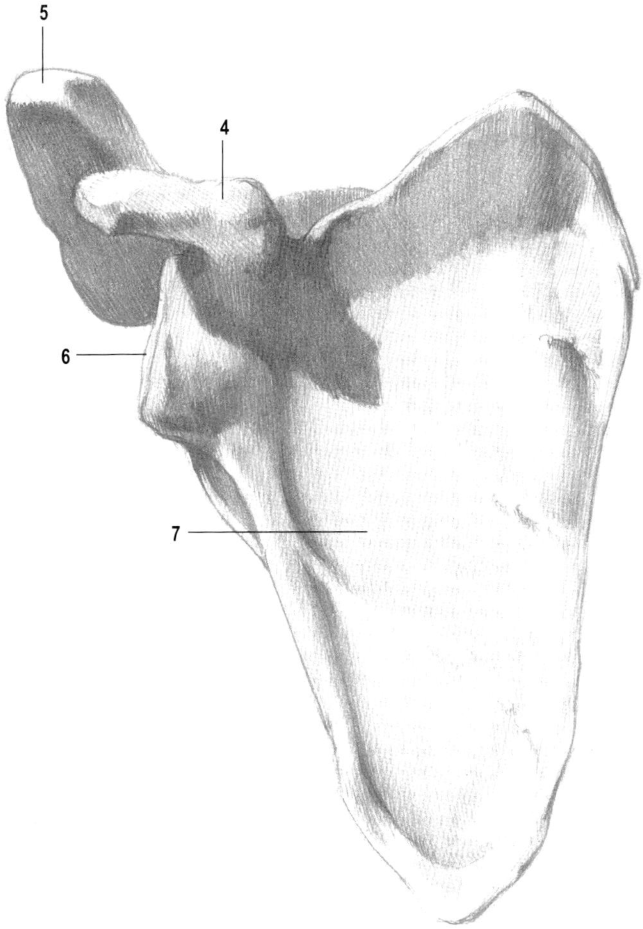

ventral aspect

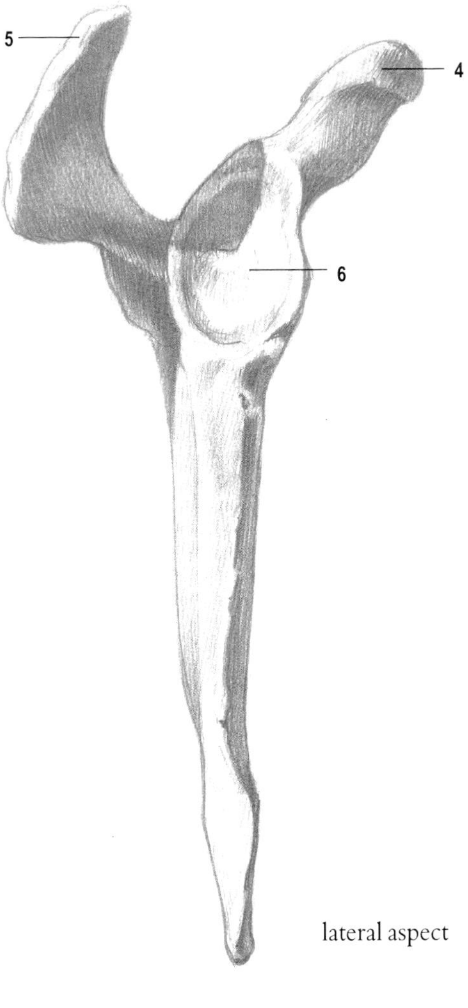

lateral aspect

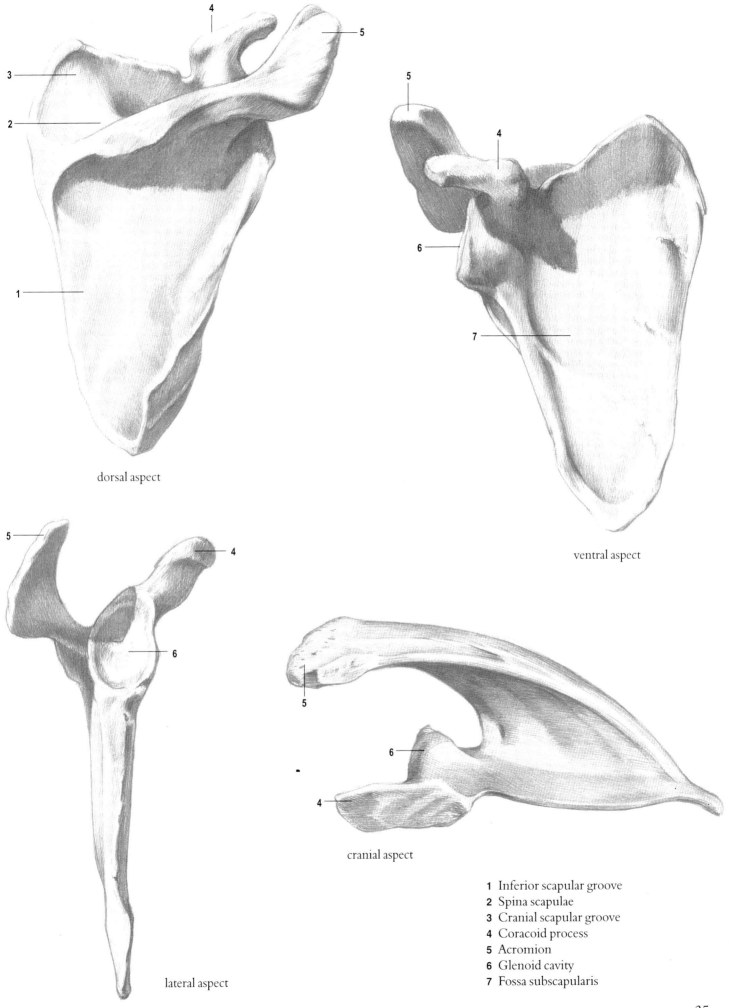

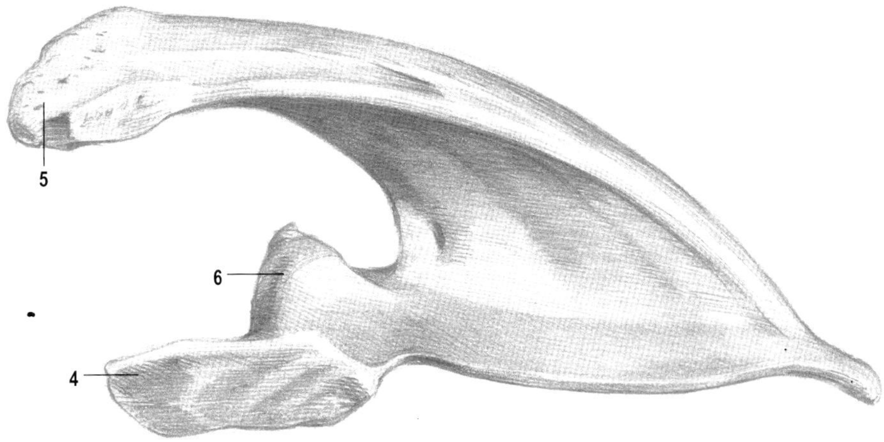

cranial aspect

1 Inferior scapular groove
2 Spina scapulae
3 Cranial scapular groove
4 Coracoid process
5 Acromion
6 Glenoid cavity
7 Fossa subscapularis

Fig. 16
The Bones and Joints of the Acromioclavicular Joint

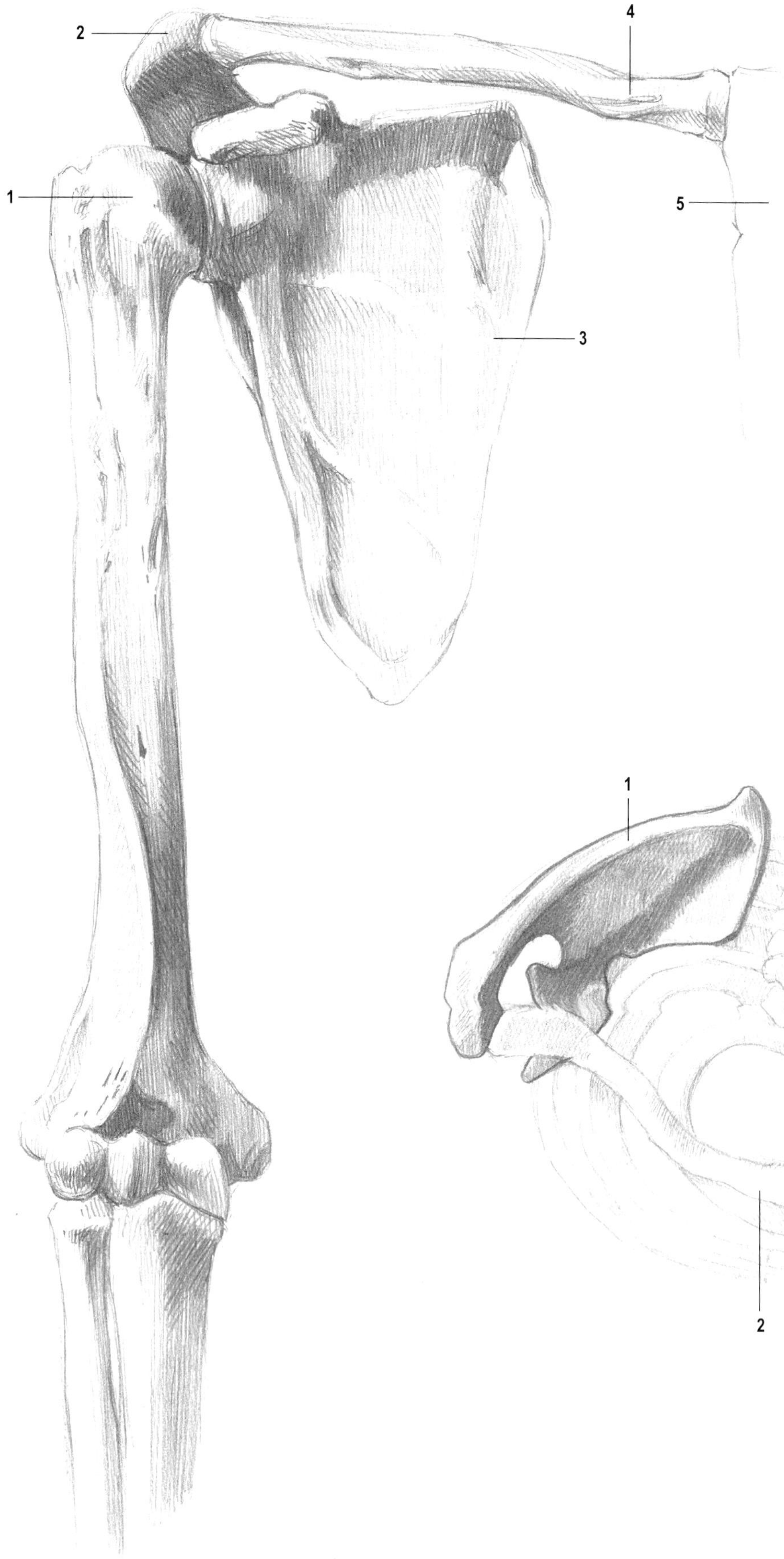

ventral aspect

BONES

1 Head of the humerus
2 Acromion
3 Shoulder blade
4 Clavicle
5 Sternum

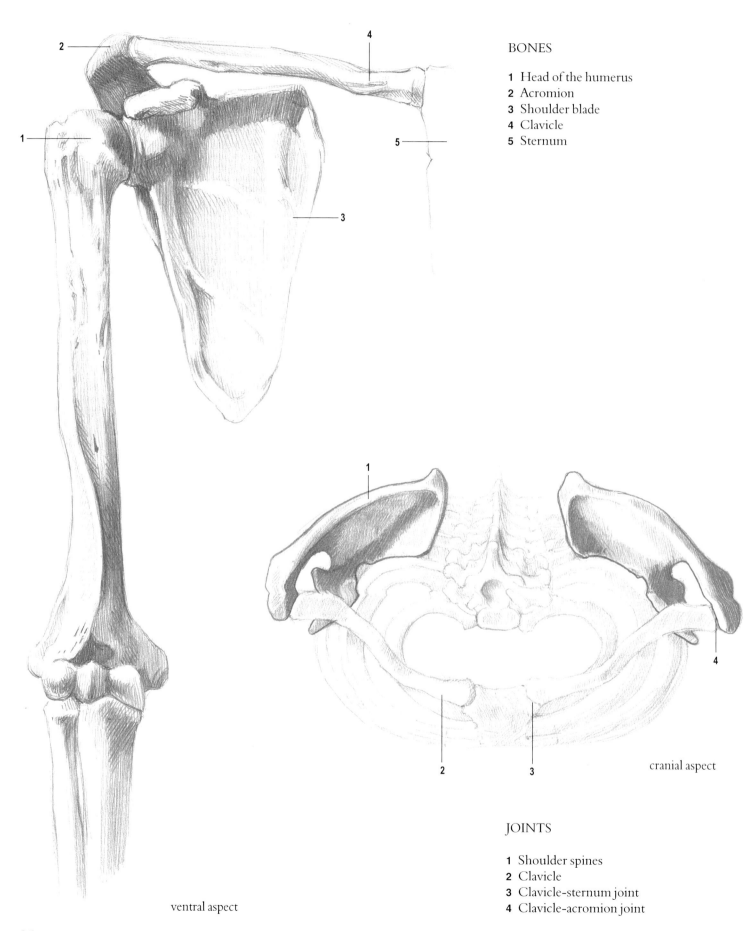

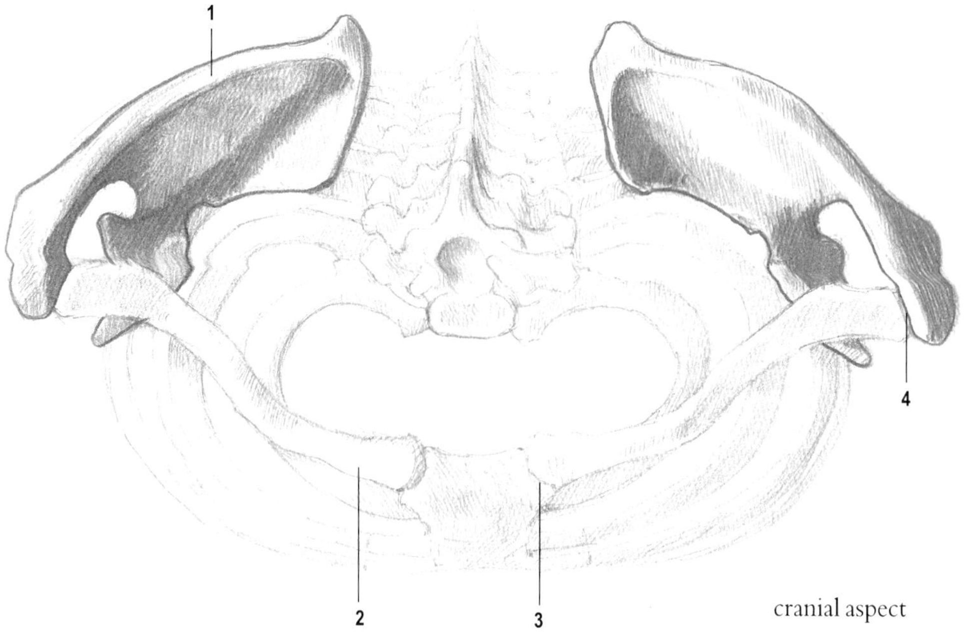

cranial aspect

JOINTS

1 Shoulder spines
2 Clavicle
3 Clavicle-sternum joint
4 Clavicle-acromion joint

Fig. 17
The Acromioclavicular Joints

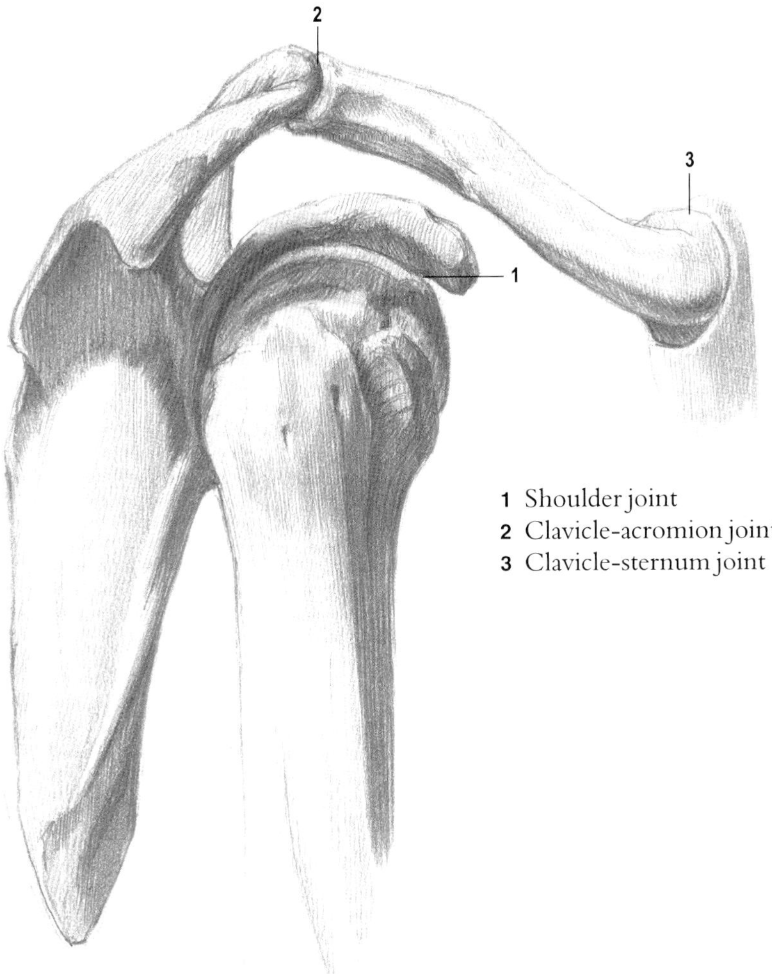

1 Shoulder joint
2 Clavicle-acromion joint
3 Clavicle-sternum joint

lateral aspect

ventral aspect

Fig. 18
The Clavicle

The clavicle is a long flat bone which forms an s-shape when viewed from above. Its medial extremity forms a joint with the body of the sternum, while its lateral extremity forms a joint with the acromion.

1 Sternal joint end of the clavicle
2 Body of the clavicle
3 Shoulder blade joint surface of the clavicle
4 Insertion of the subclavius muscle

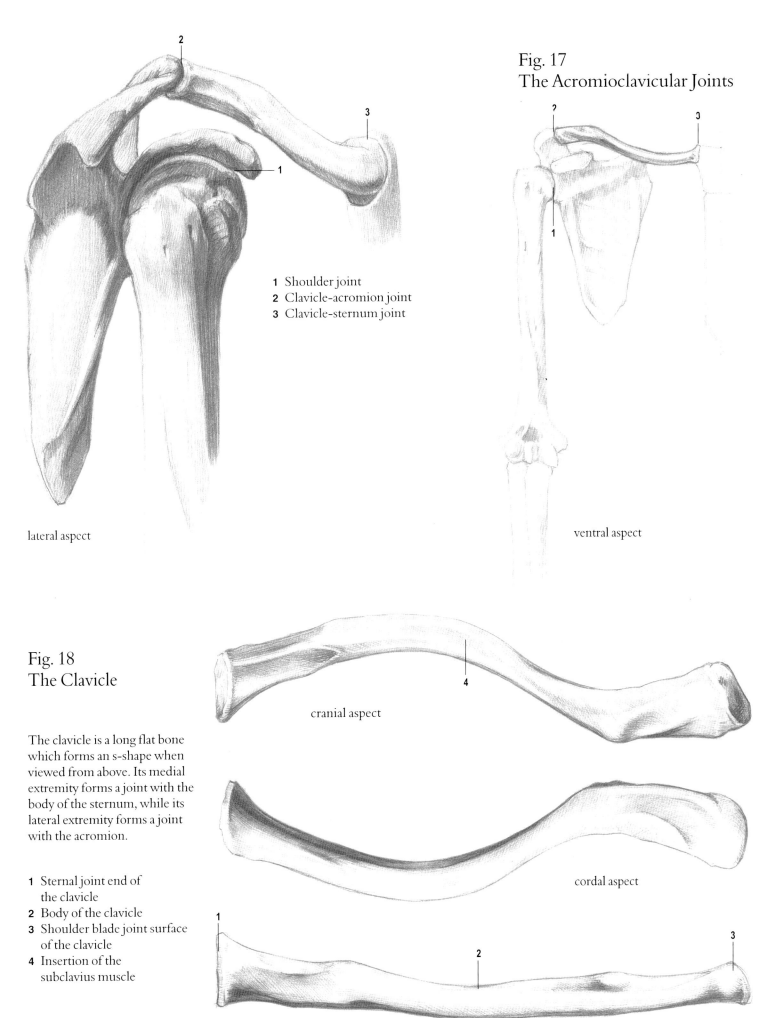

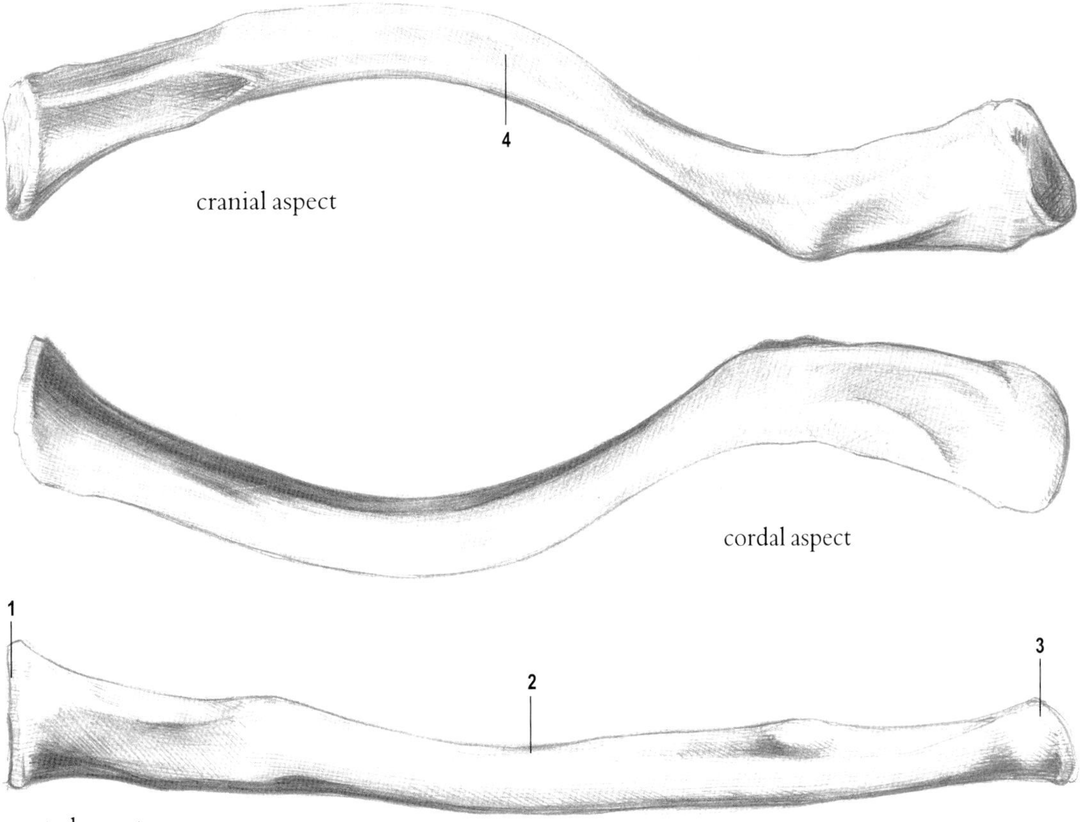

cranial aspect

cordal aspect

ventral aspect

Fig. 19
The Humerus

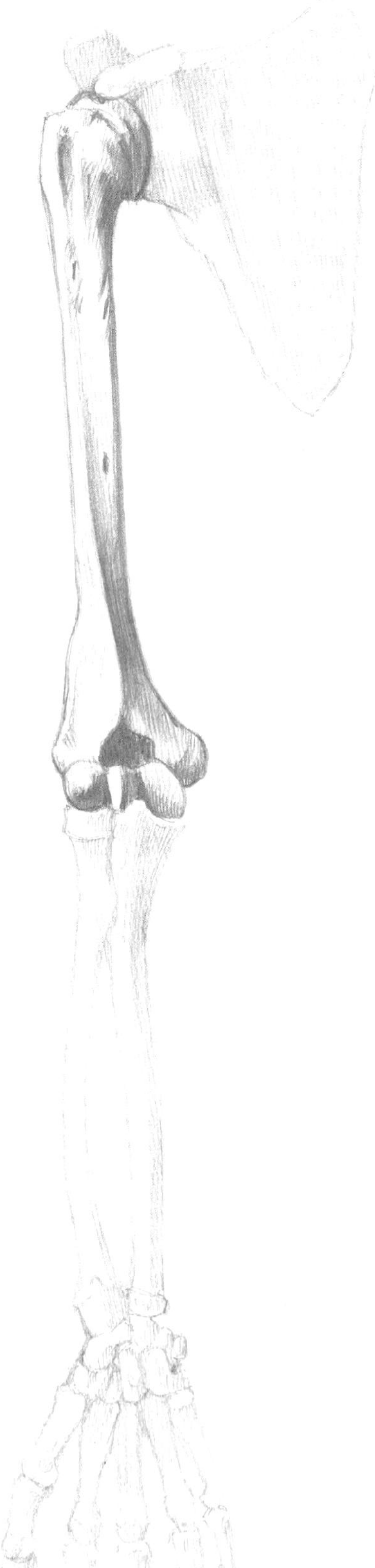

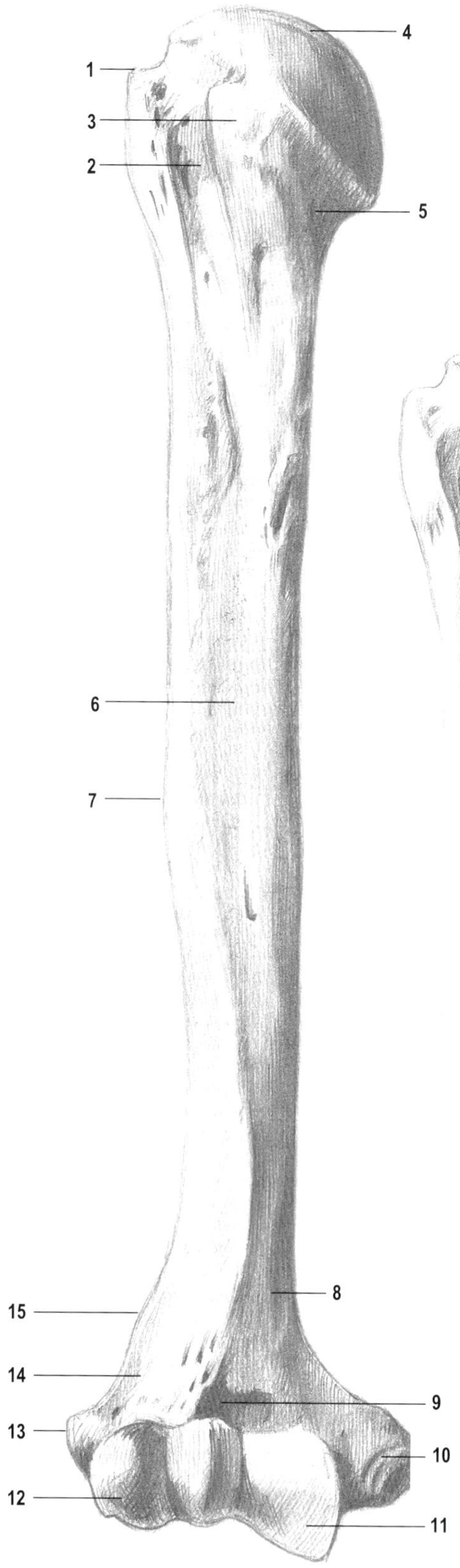

ventral aspect

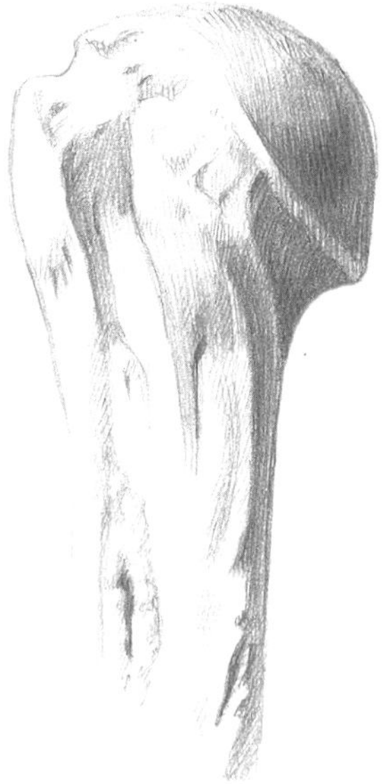

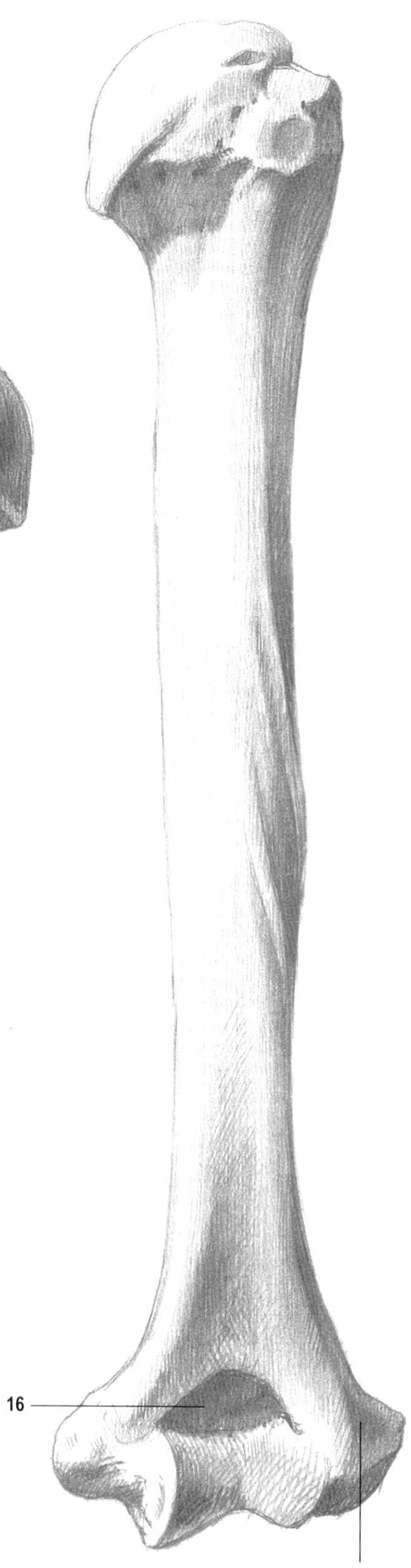

dorsal aspect

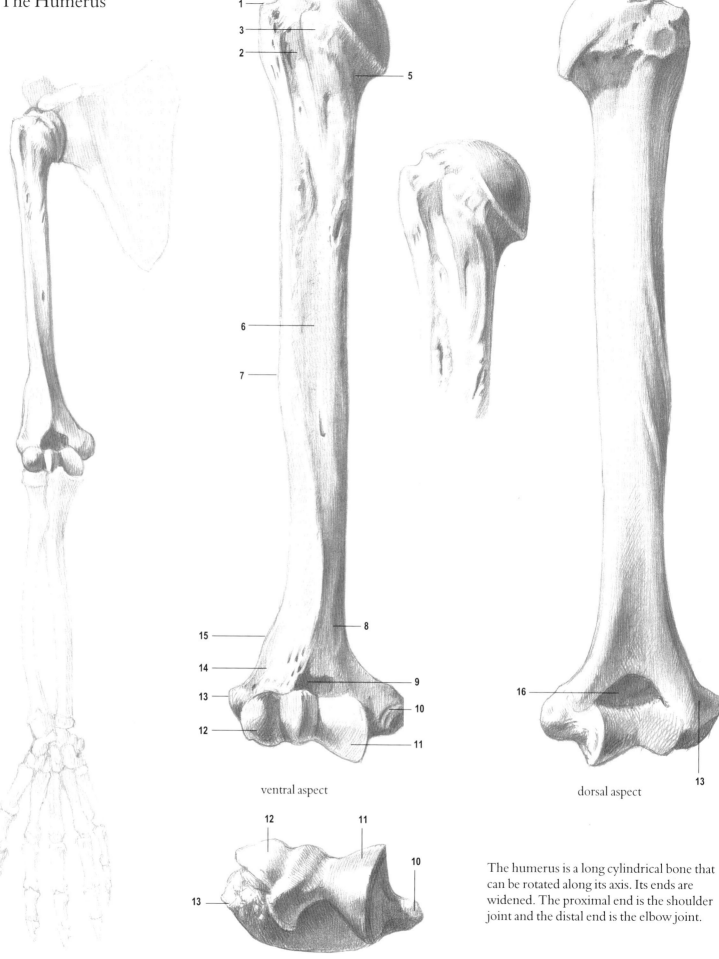

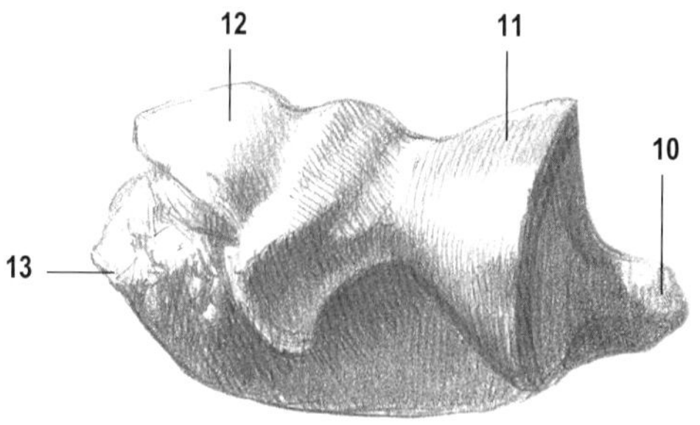

The humerus is a long cylindrical bone that can be rotated along its axis. Its ends are widened. The proximal end is the shoulder joint and the distal end is the elbow joint.

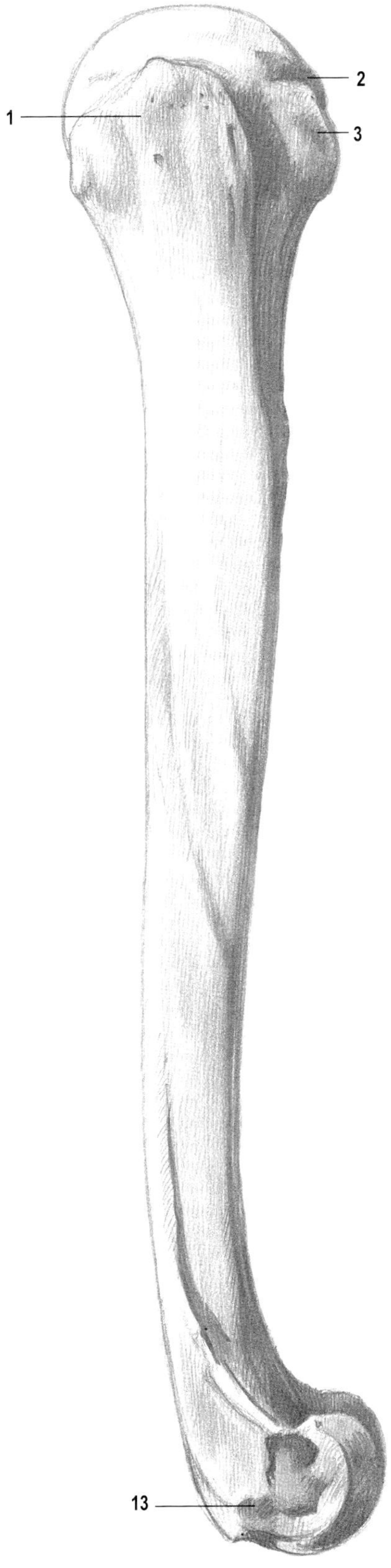

lateral aspect

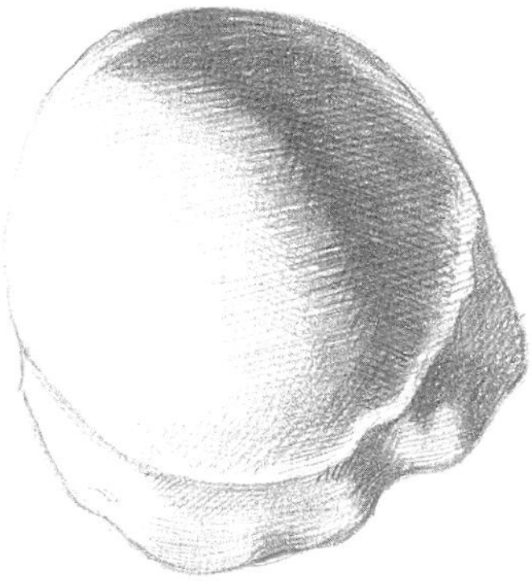

cranial aspect

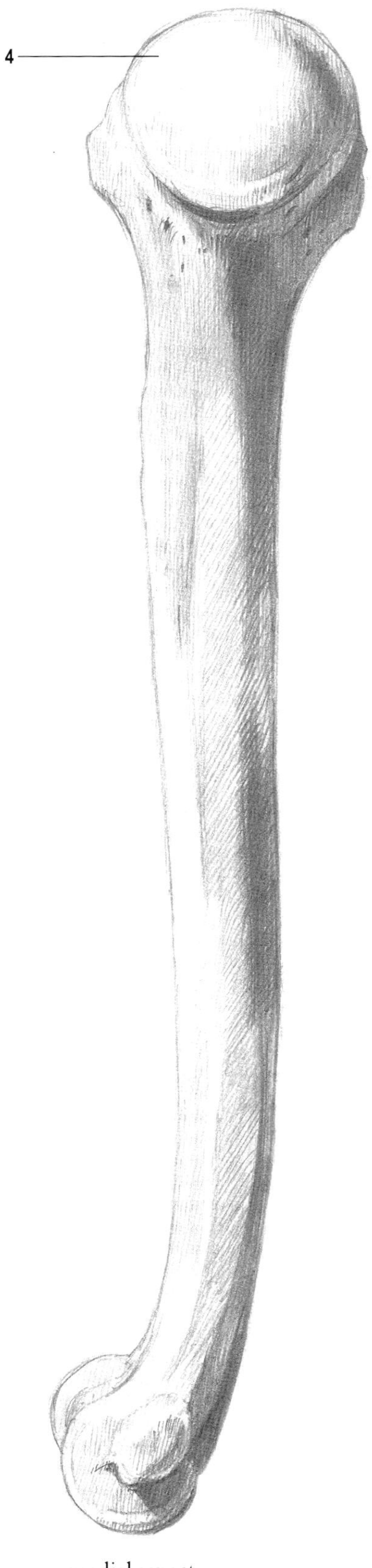

medial aspect

1 Greater protuberance
2 Biceps groove
3 Lesser protuberance
4 Head
5 Neck

6 Shaft
7 Deltoid tuberosity
8 Medial supracondylar ridge
9 Radial fossa
10 Medial epicondyle
11 Trochlea

12 Capitulum humeri
13 Lateral epicondyle
14 Fossa for the ulnar process
15 Lateral supracondylar process
16 Olecranon fossa

Fig. 20
The Radius

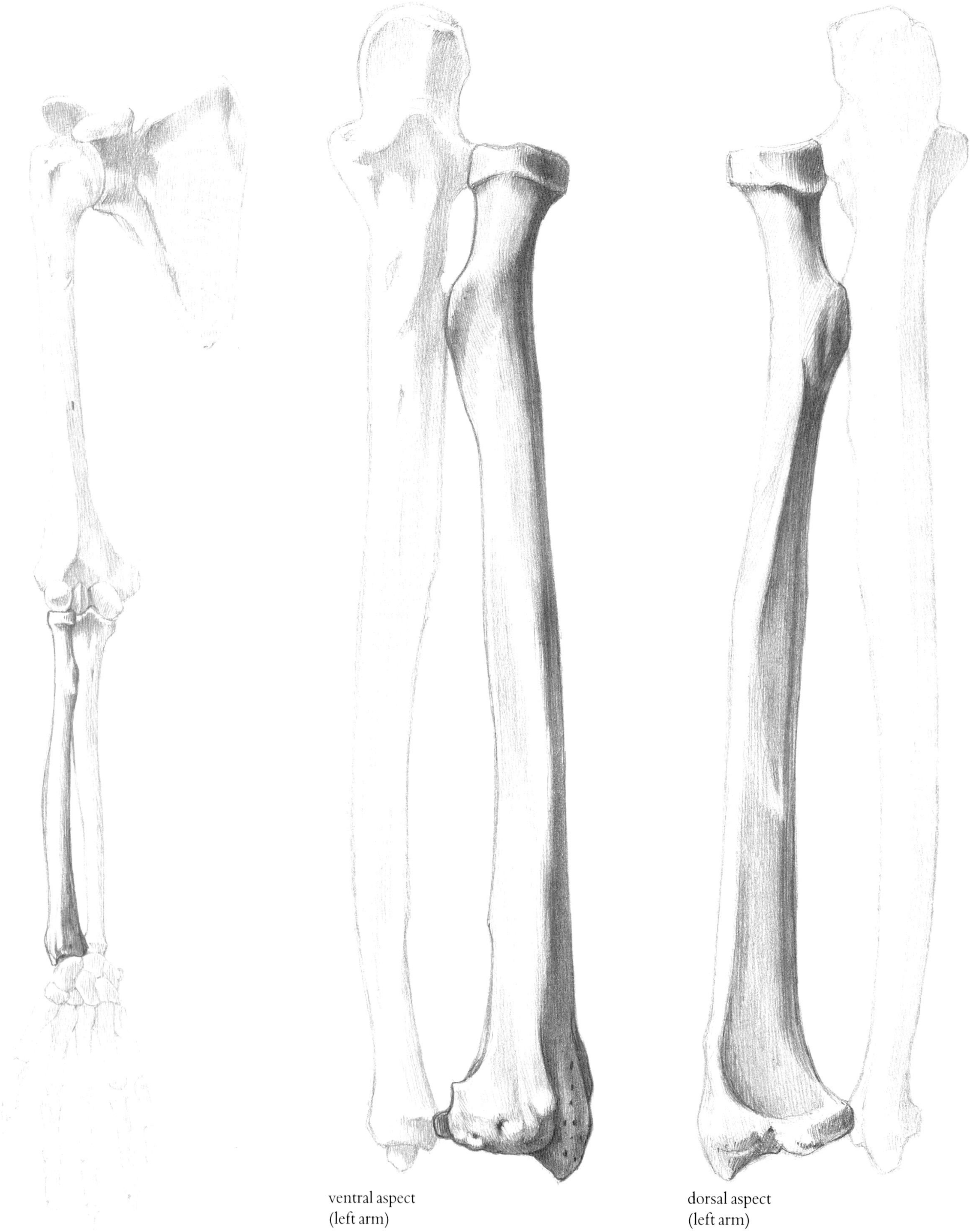

ventral aspect
(left arm)

dorsal aspect
(left arm)

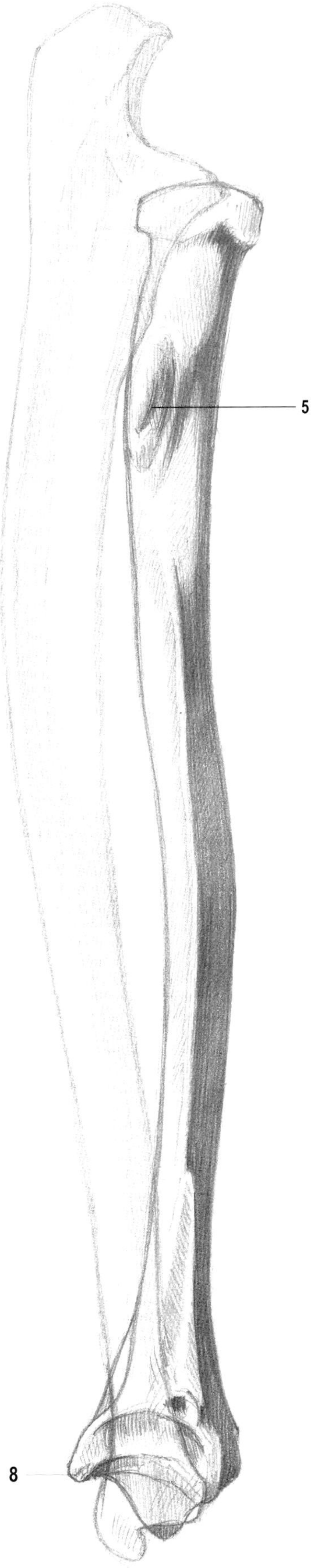

medial aspect
(left arm)

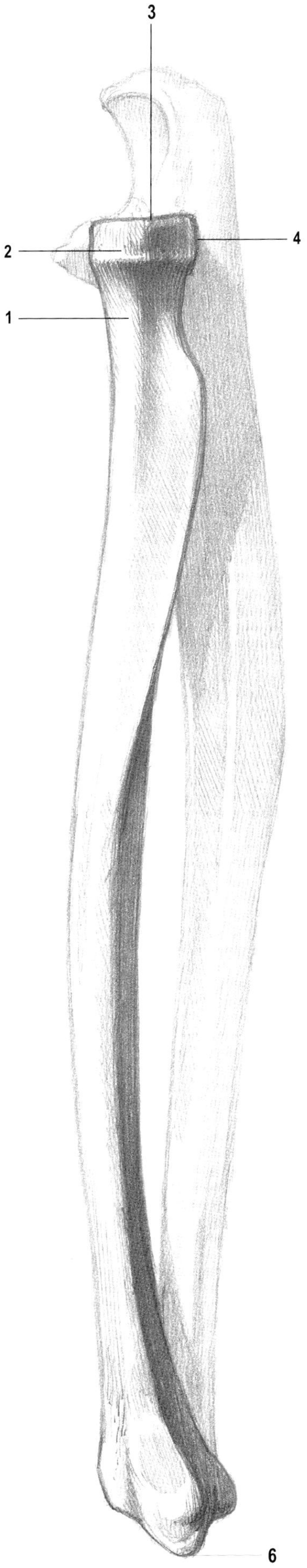

lateral aspect
(left arm)

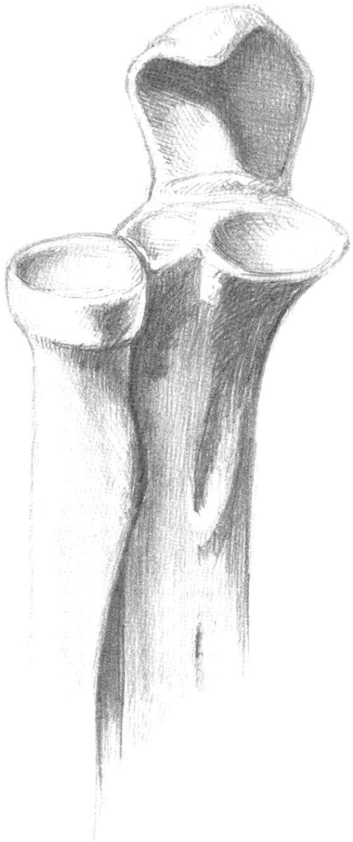

cranial aspect
(right arm)

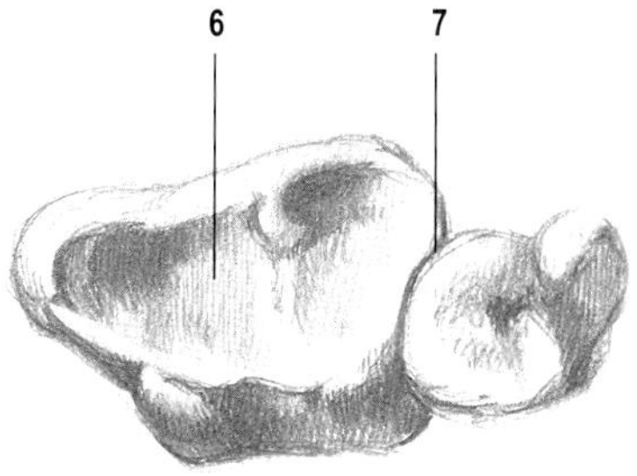

cordal aspect

The bones of the lower arm are the s-formed bones known as the radius and the ulna. Above the neck, (**1**) at the proximal end, the radius has a head (**2**) whose disc-shaped surface (**3**) forms a joint with the humerus, while its lateral joint surface (**4**) forms a joint with the ulna. The tubercle at the upper end of the shaft (**5**) is also the insertion of the biceps. The joint surface of the distal end (**6**) forms a mobile joint with the carpus. The ulnar head fits into the distal (**7**). The styloid process (**8**) is located on the medial side.

Fig. 21
The Ulna

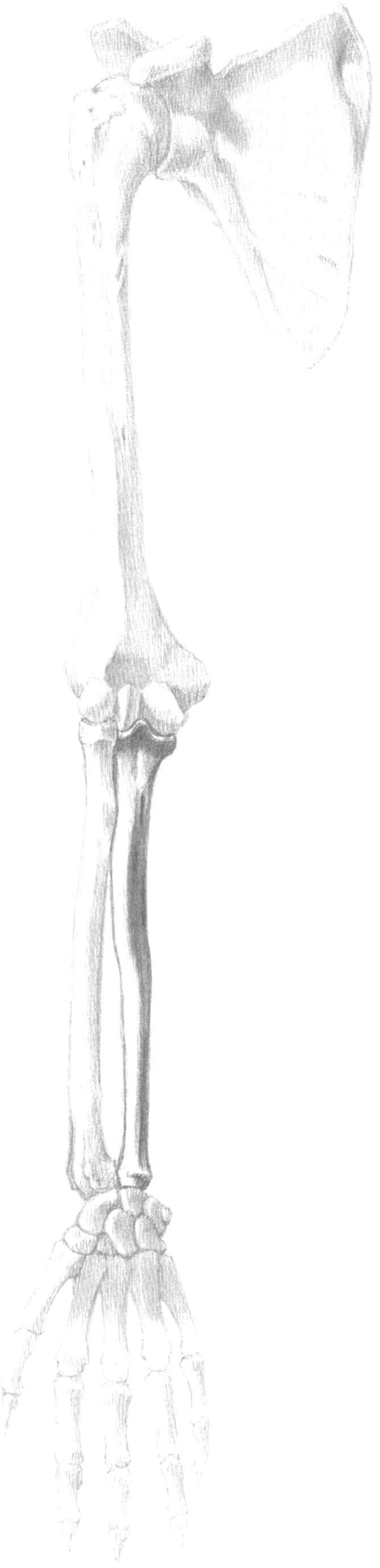

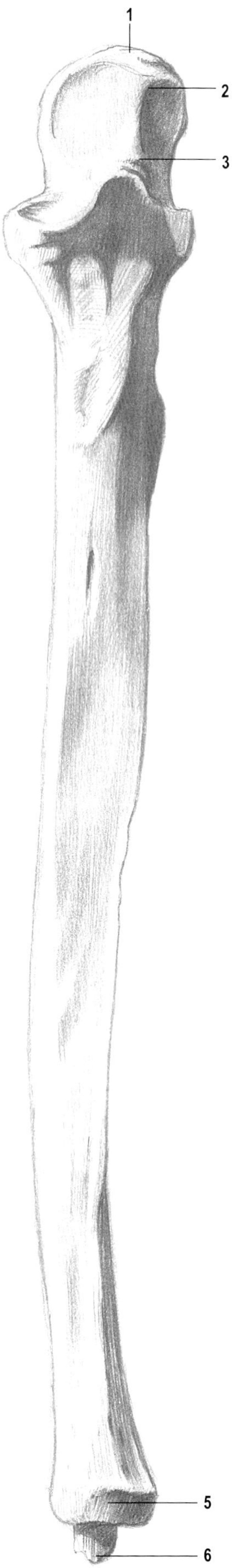

At its proximal end, the ulnar protuberance of the elbow (**1**) possesses an indentation (**2**) which forms a joint (**3**) with the condyle of the humerus. The supinator originates at the ulnar ridge (**4**). At the distal end of the ulna, the head (**5**) and styloid process (**6**) are to be found.

ventral aspect
(left elbow)

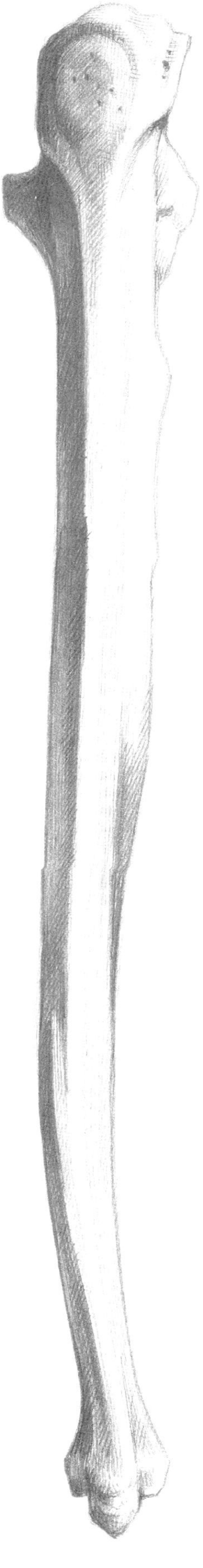

dorsal aspect
(right elbow)

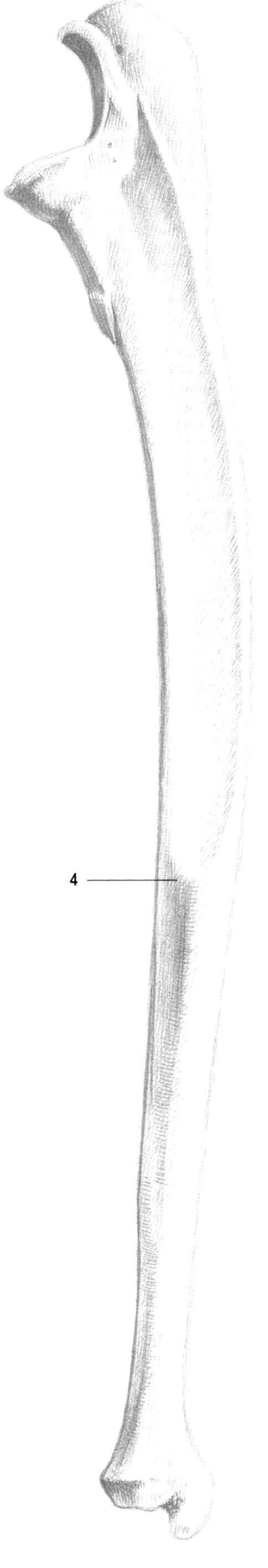

medial aspect
(right elbow)

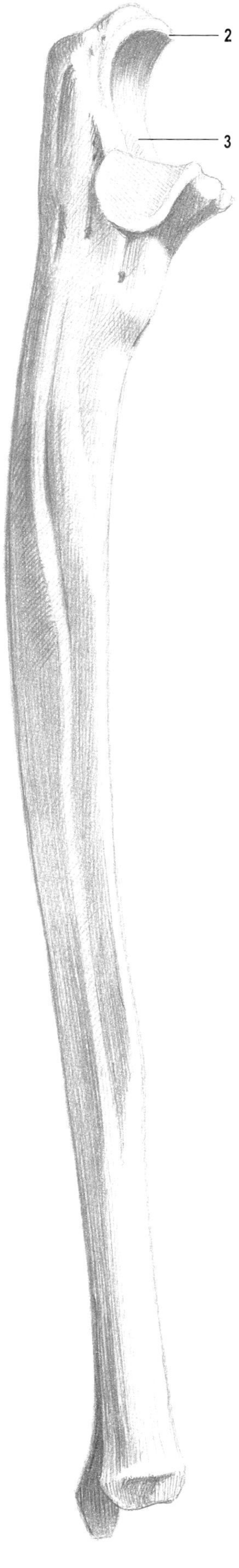

lateral aspect
(right elbow)

Fig. 22
The Bones of the Carpus

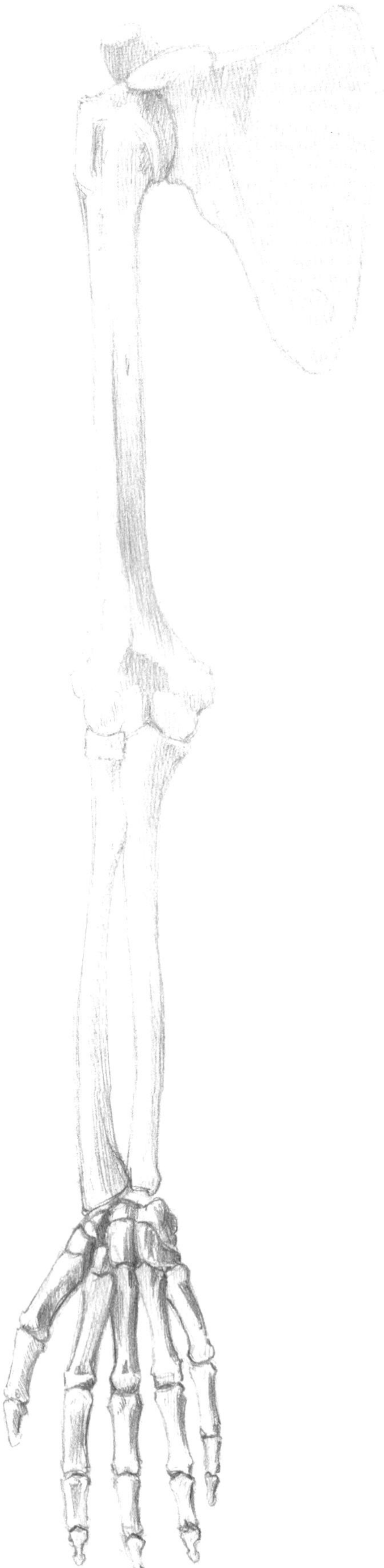

dorsal aspect

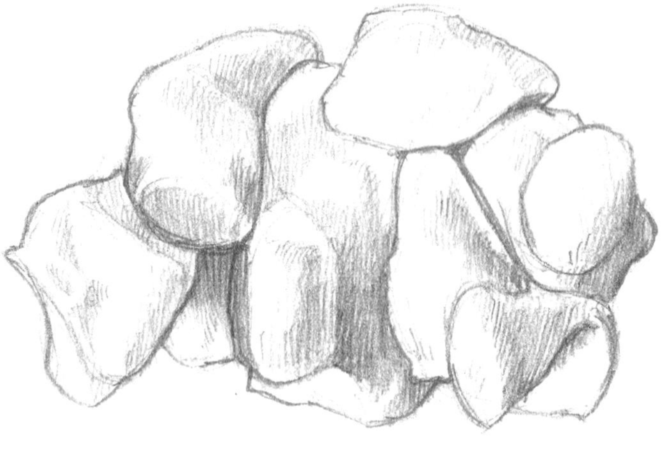

palmar aspect

Together the metacarpal bones, the phalanges and the bones of the carpus form the bones of the hand.

Fig. 23
Change in Position of Carpal
Bones in Hand Movements

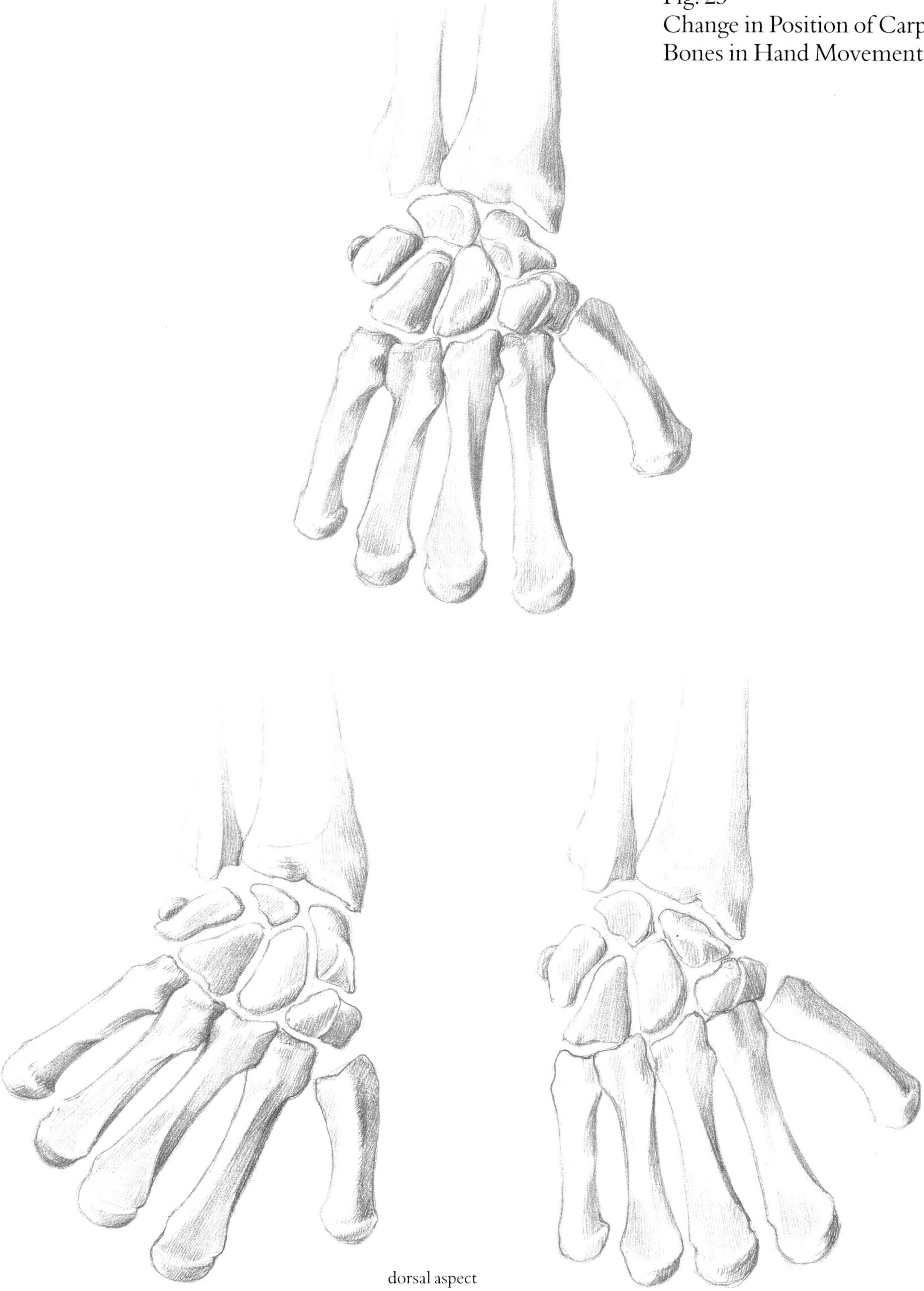

dorsal aspect

Fig. 24
The Bones of the Hand

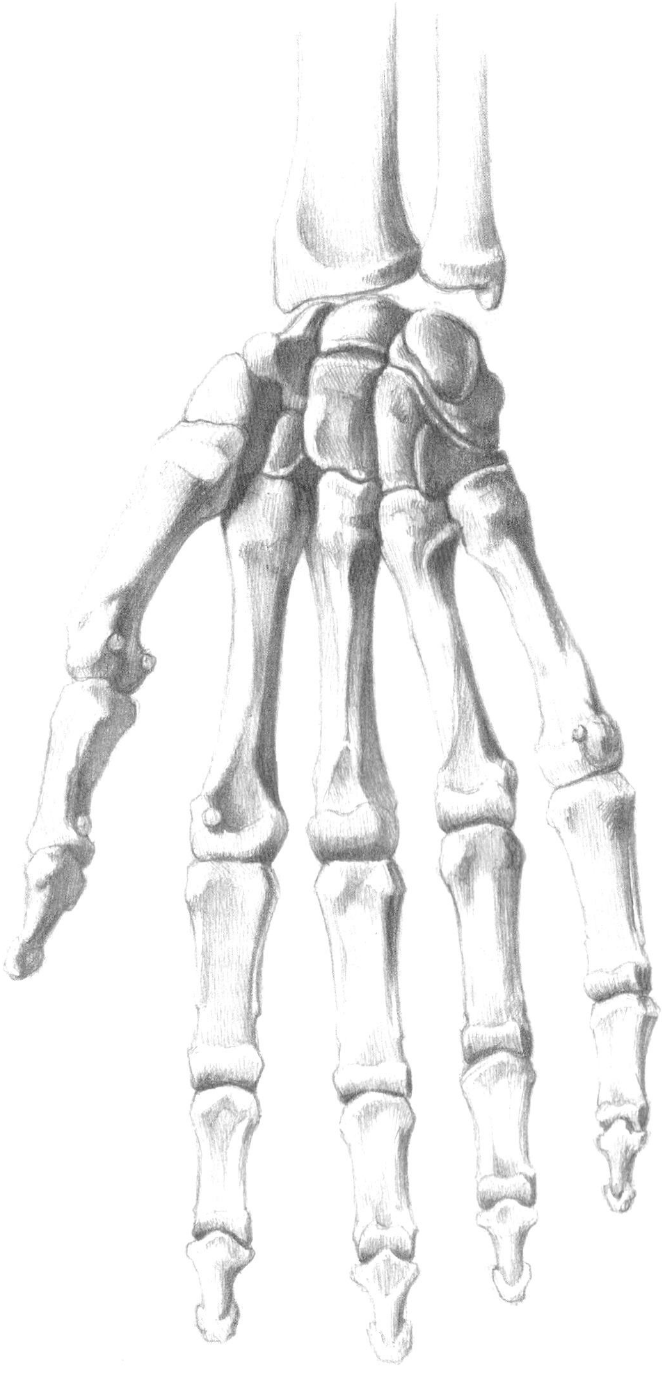

palmar aspect

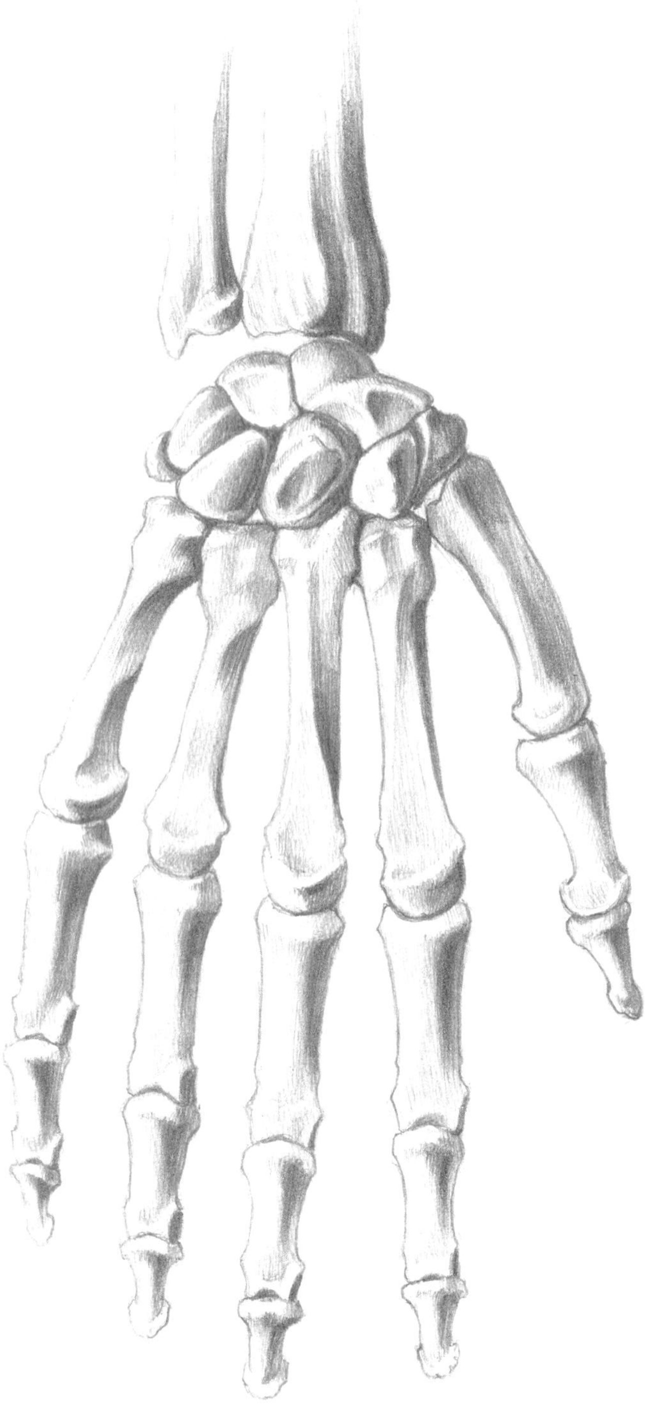

dorsal aspect

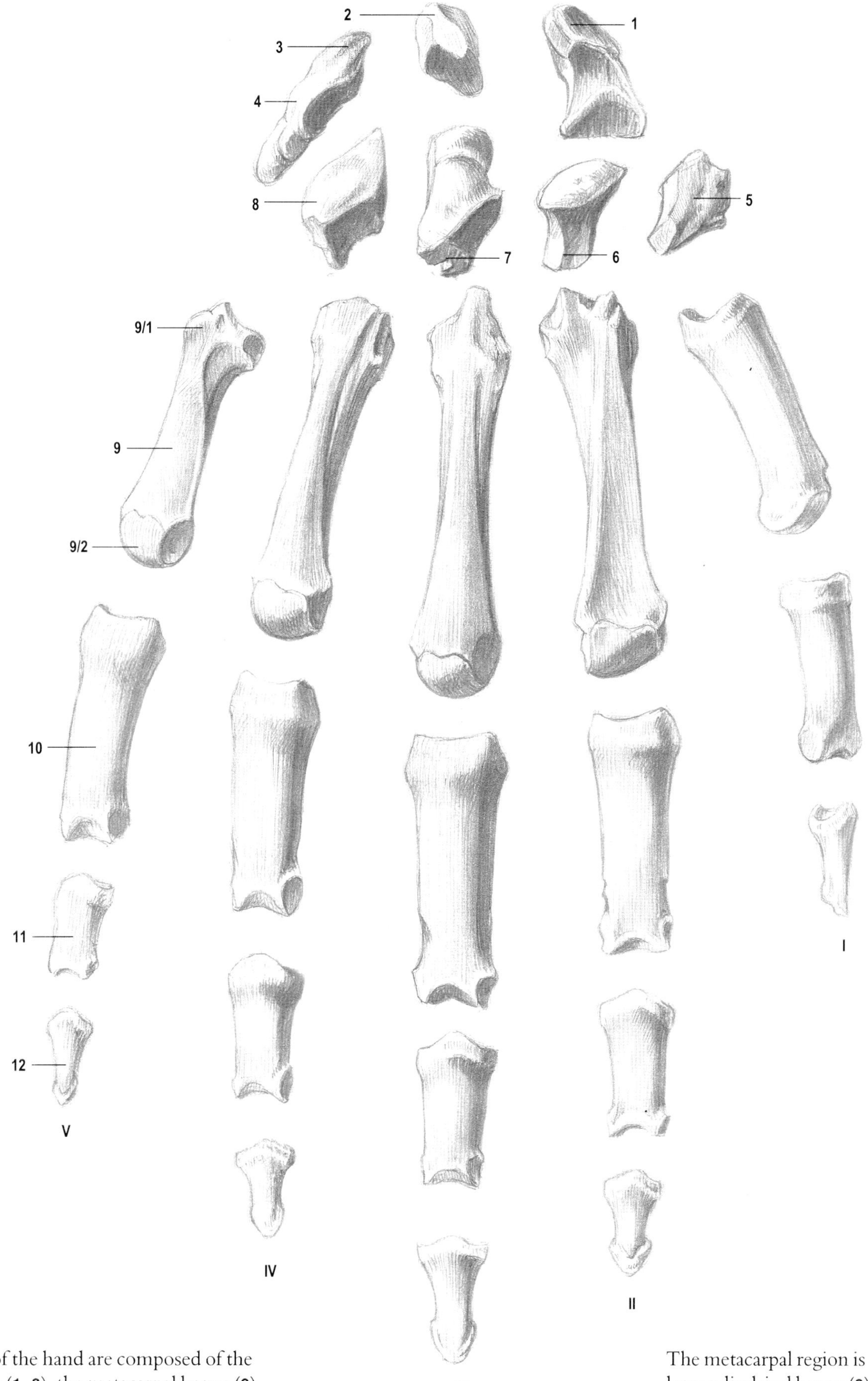

The bones of the hand are composed of the carpal bones (**1–8**), the metacarpal bones (**9**) and the phalanges (**10–12**). The carpal bones are organised in two rows.

Upper Row:
1 Tubercle of scaphoid, **2** Lunate
3 Triquetral, **4** Pisiform

Lower Row:
5 Trapezium, **6** Trapezoid
7 Capitate, **8** Hook of Hamate

The metacarpal region is composed of five long cylindrical bones (**9**) with thin shafts (**9/1**) and thicker joint ends (**9/2**). The finger consist of three phalanges: the proximal or first phalanx (**10**), the middle or second phalanx (**11**) and the distal or third phalanx (**12**). The thumb has only two phalanges.
I–V: Finger

Fig. 25
The Carpal Joints

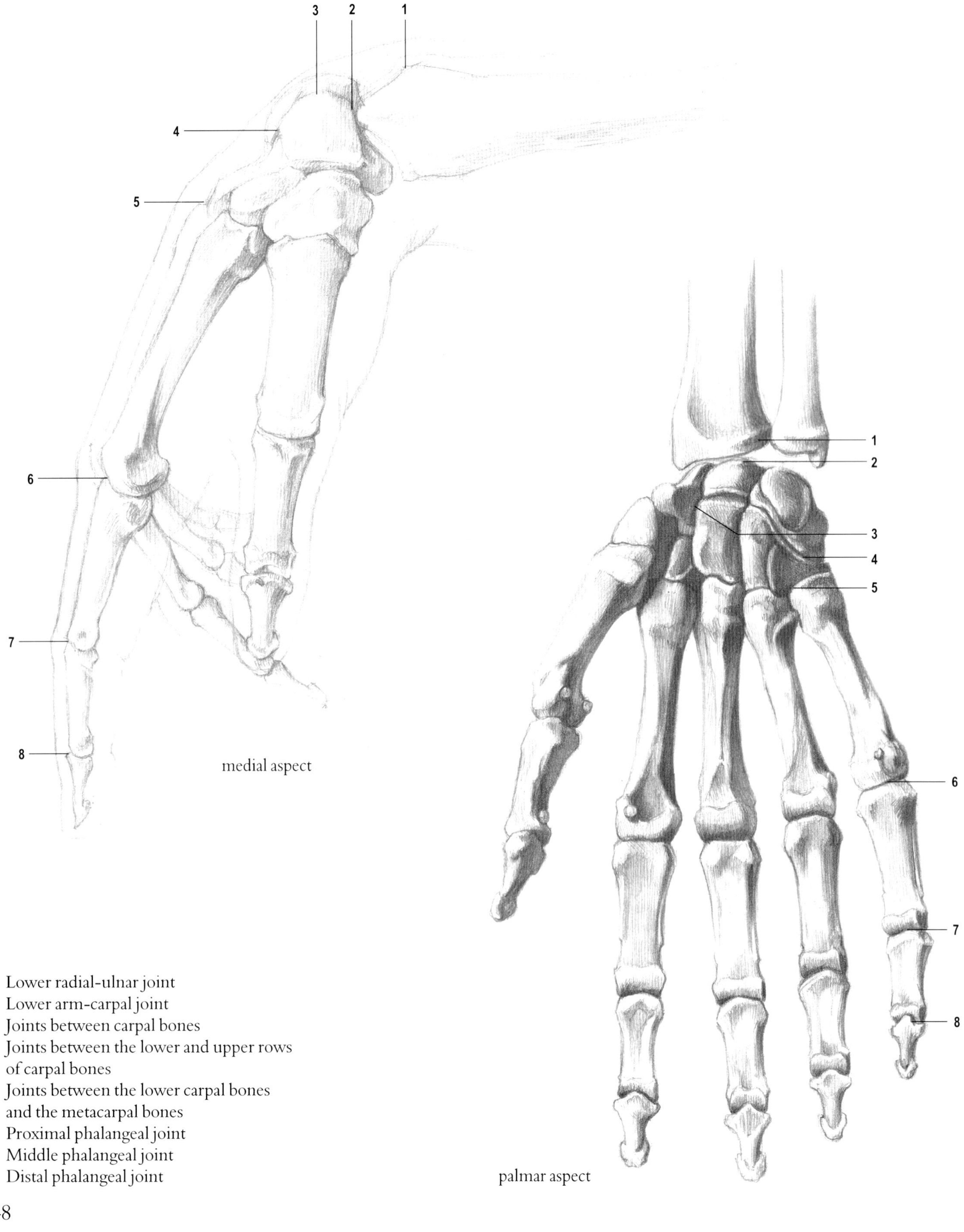

1 Lower radial-ulnar joint
2 Lower arm-carpal joint
3 Joints between carpal bones
4 Joints between the lower and upper rows of carpal bones
5 Joints between the lower carpal bones and the metacarpal bones
6 Proximal phalangeal joint
7 Middle phalangeal joint
8 Distal phalangeal joint

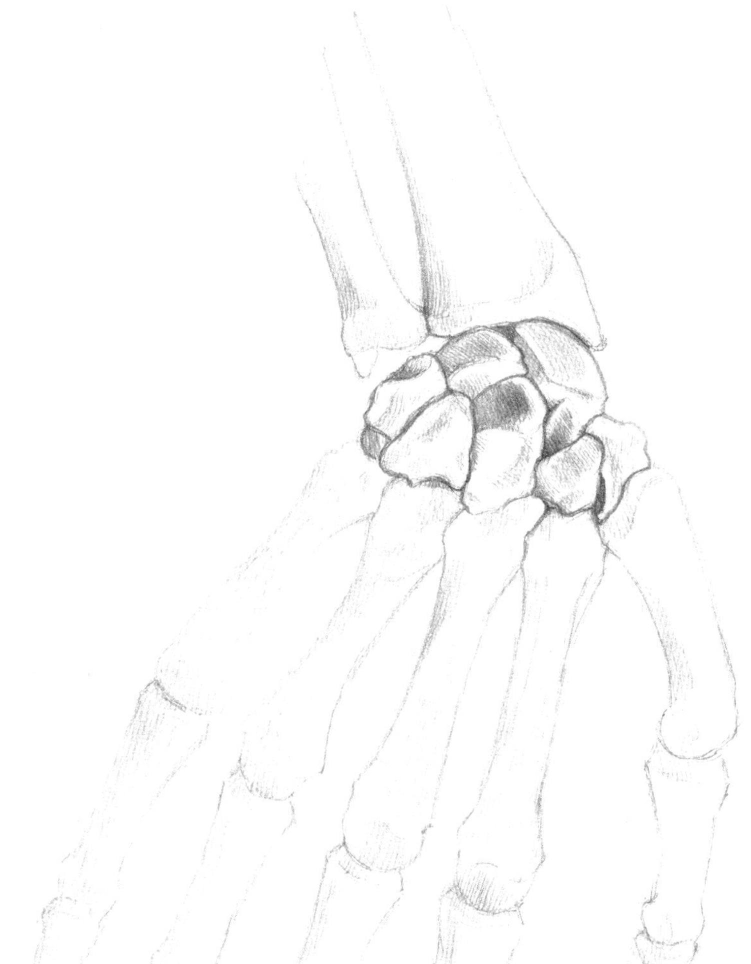

dorsal aspect

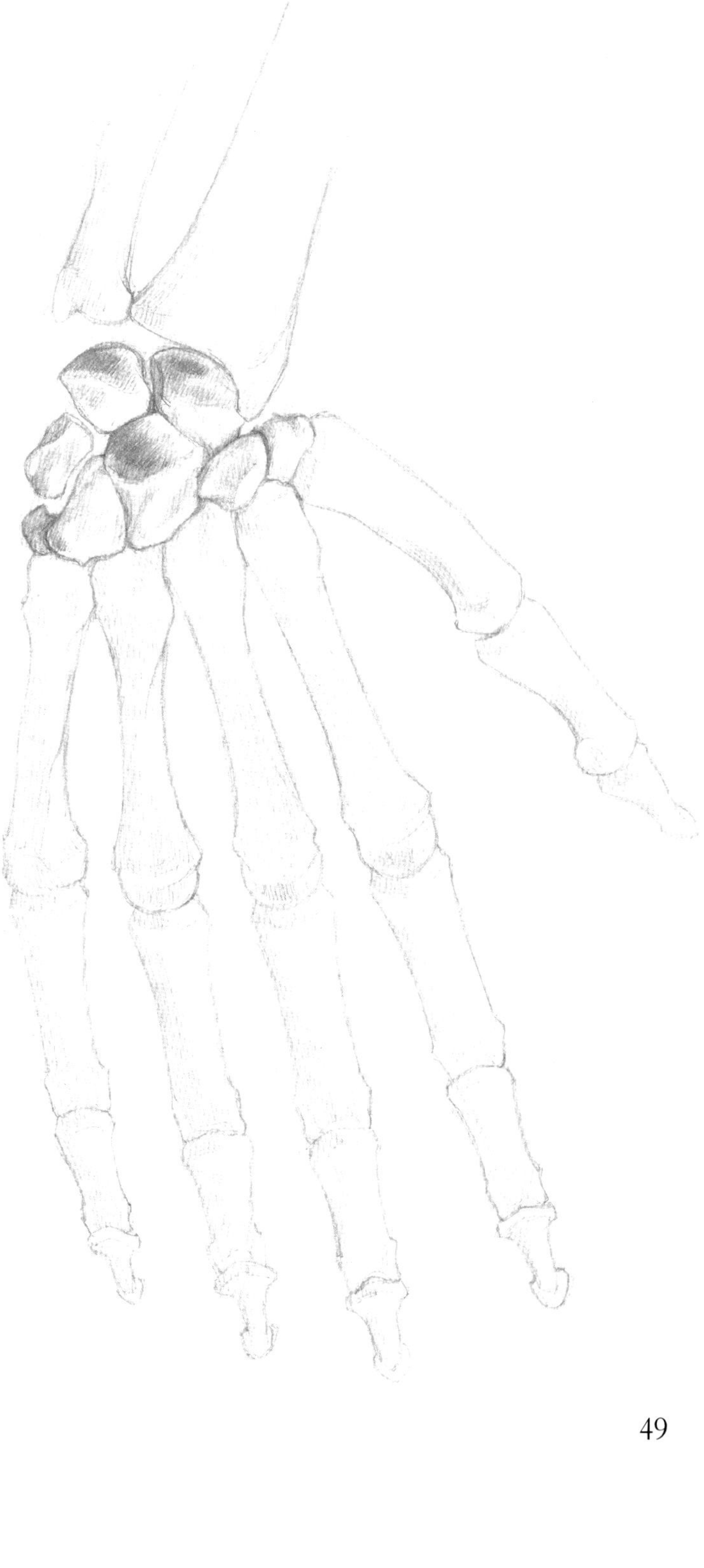

Fig. 26
The Ligaments of the Shoulder Joint and Lateral Clavicular Joint

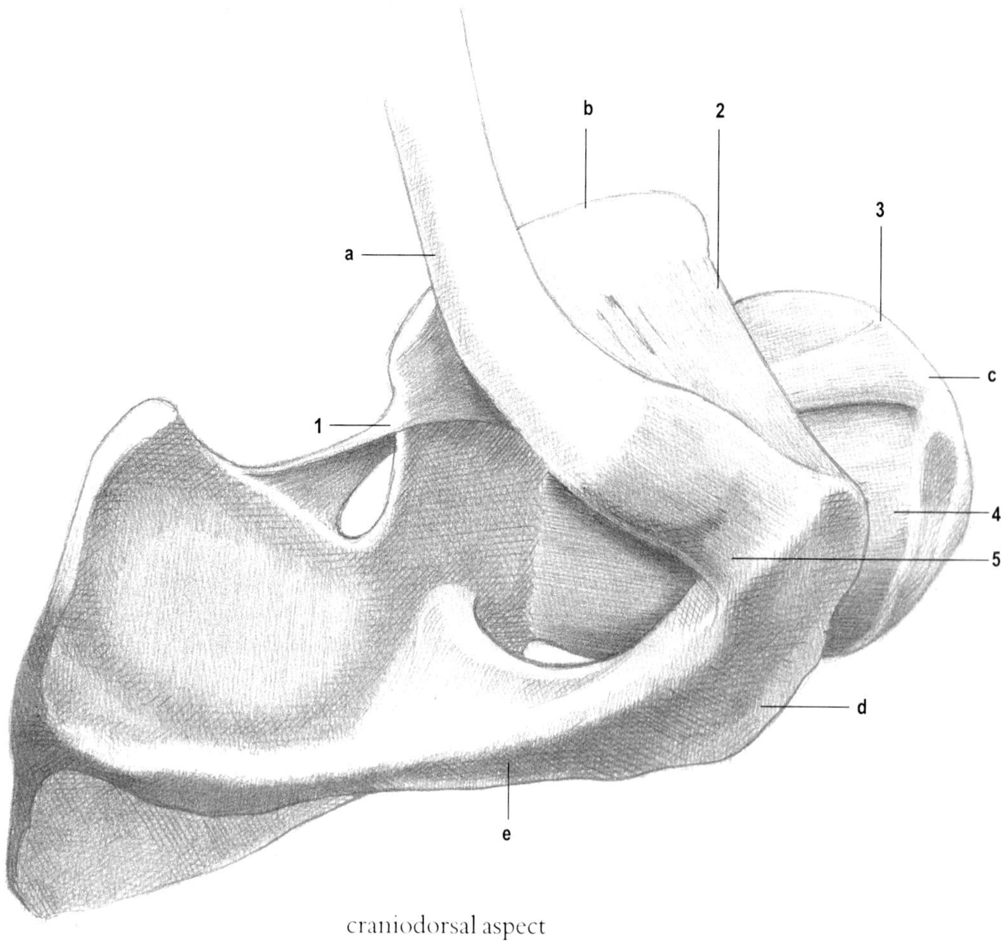

1 Transverse scapular ligament
2 Broad, flat covering between the coracoid process and acromion
3 Ligament between coracoid process and humerus
4 Shoulder joint capsule
5 Acromio-clavicular ligament

a Clavicle
b Coracoid process of shoulder blade
c Great tubercle of the bone
d Acromion
e Spinae scapulae

craniodorsal aspect

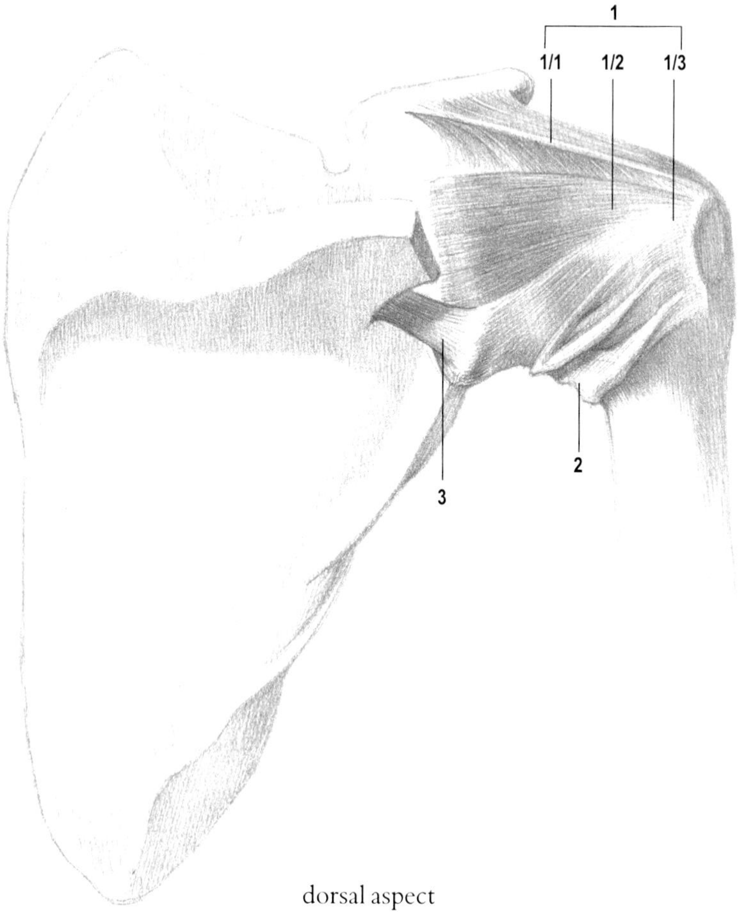

1 Humeroscapular tendon
- 1/1 upper ligament between humerus and shoulder blade
- 1/2 middle ligament between humerus and shoulder blade
- 1/3 lower ligament between humerus and shoulder blade

2 Joint capsule
3 Broad, flat covering between the coracoid process and acromion

dorsal aspect

Fig. 27
The Retraction and Protraction of the Scapula

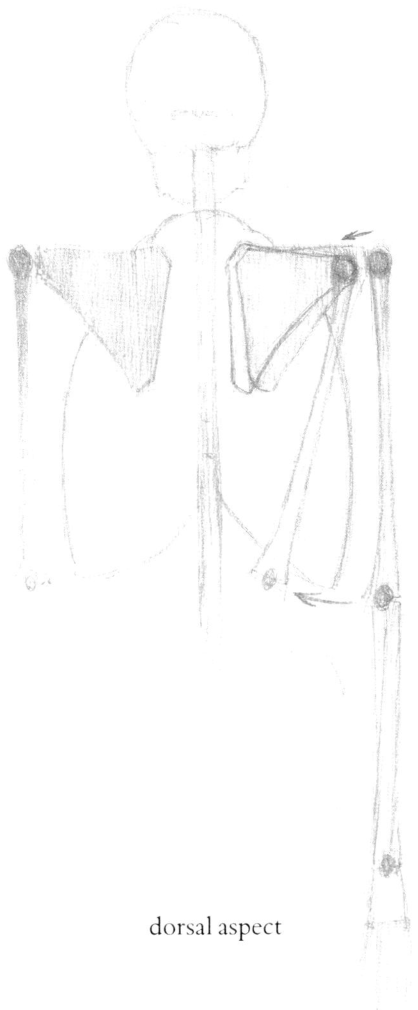

dorsal aspect

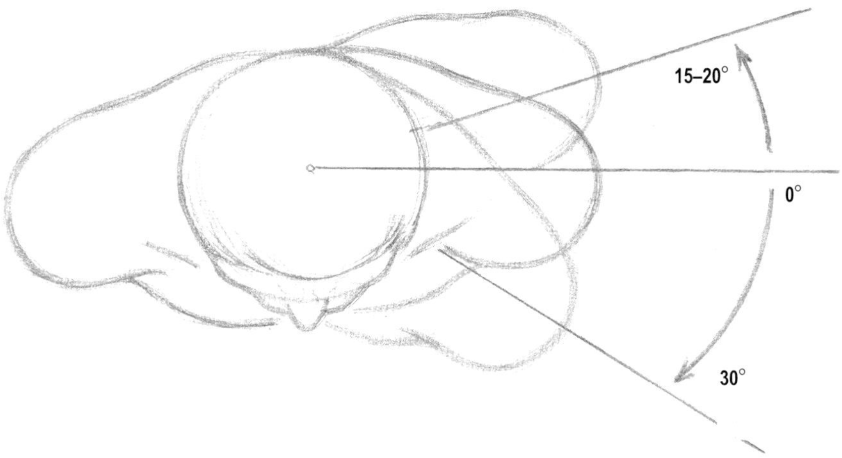

Through the action of the major (*28*) and minor (*29*) pectoral muscles, inferior clavicular muscles (*30*) and the anterior serratus muscle (*31*) the scapula is brought forward, away from the spine. The clavicle bends forwards (up to 30 degrees) in the sternoclavicular joint. The shoulder blade swings forwards.

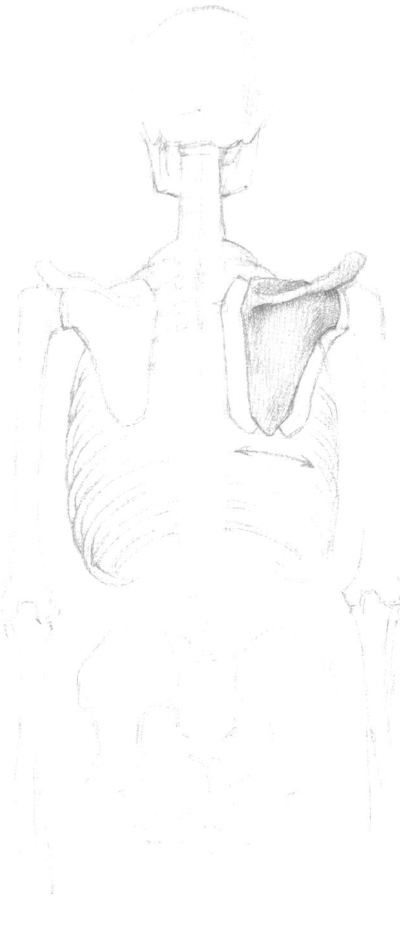

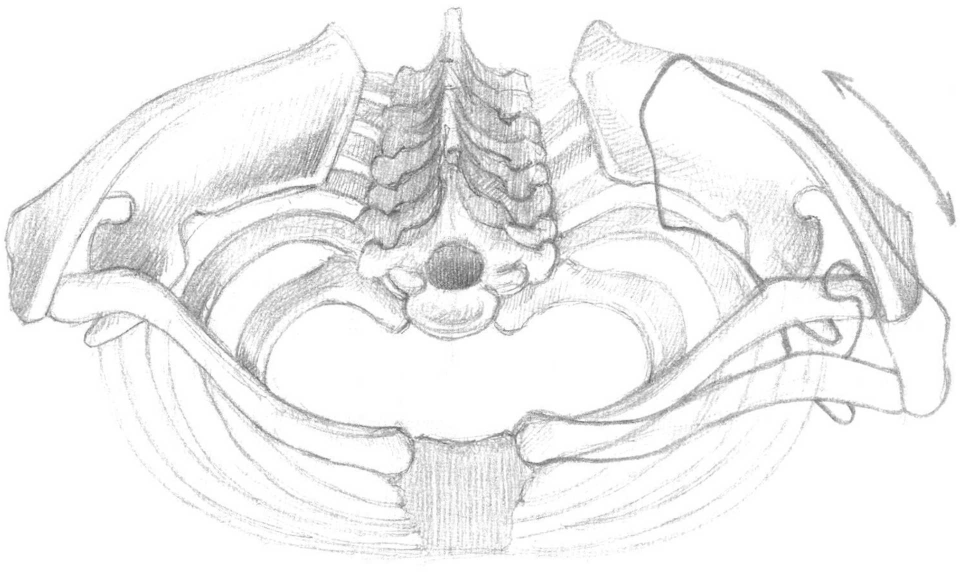

cranial aspect

In retraction the shoulder blade moves towards the spine. The sternocalvicular joint rotates backwards (between 15 and 20 degrees) and the shoulder joint swings backwards (up to 20 degrees). The retracting muscles are: the trapezius, rhomboids and the Latissimus dorsi (*21*).

Fig. 28
Elevation of the Arm

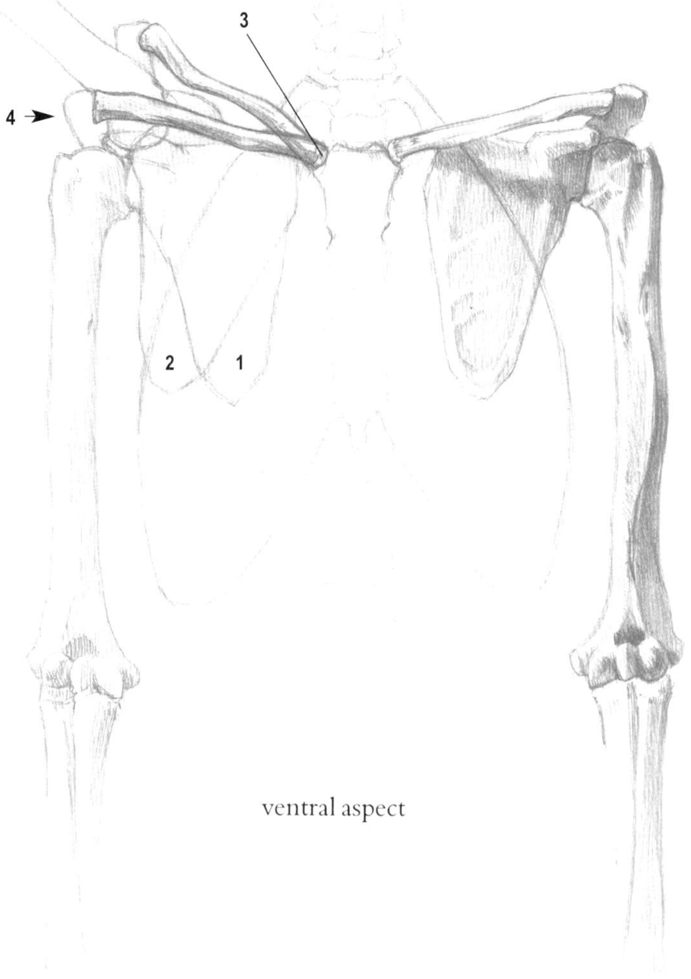

ventral aspect

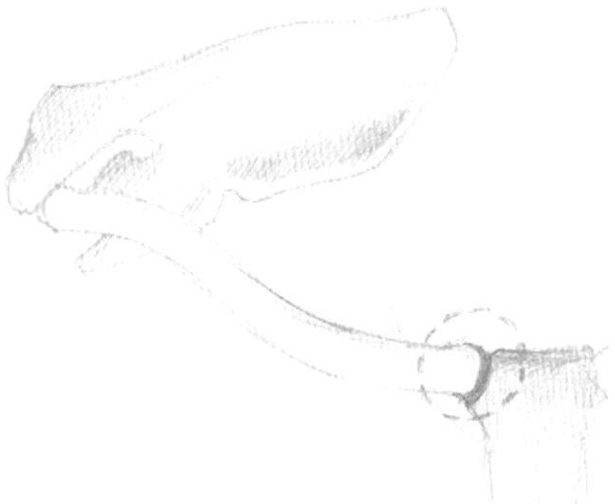

dorsal aspect

The shoulder blade (**1**) is loosely connected with the back region by means of flat muscles. When the arm is lifted, the lower corner of the shoulder blade (**2**) moves upwards laterally. The sternoclavicular joint (**3**) bends and the shoulder joint swings (**4**) upwards (between 30 and 40 degrees).

HIGH SCHOOL LIBRARY
American Community School
'Heywood' Portsmouth Road
Cobham, Surrey KT11 1BL

Fig. 29
The Levator Scapulae Muscles

When the arm is raised, the muscles of the armpit fossa bulge forwards. When the acromioclavicular joint is retracted, the lateral side of the pectoral muscle moves forwards.

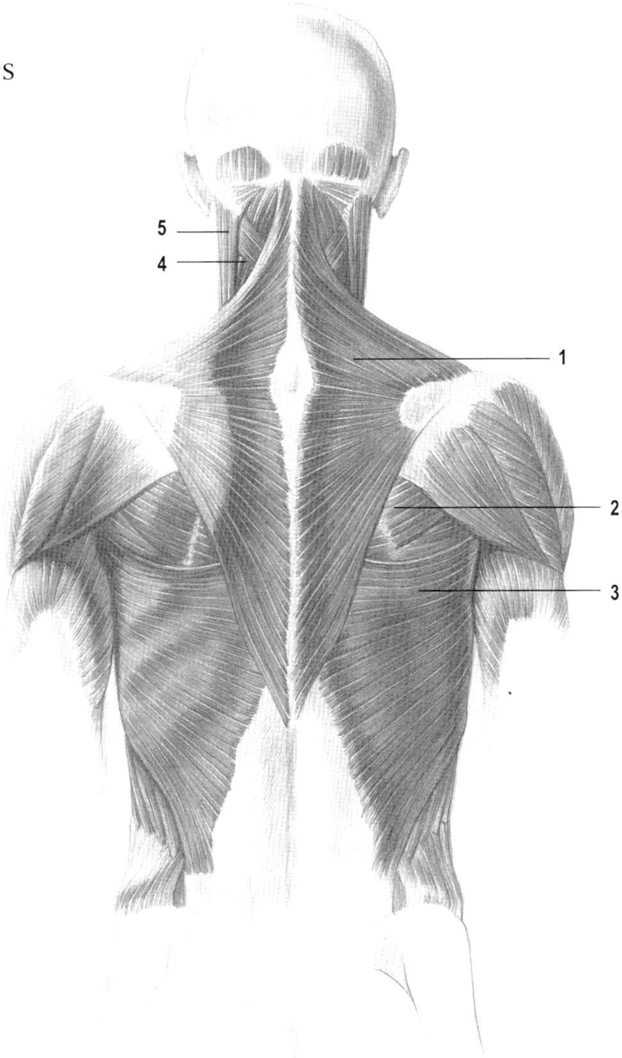

1 Trapezius (*20*)
2 Rhomboideus (*22*)
3 Latissimus dorsi (*21*)
4 Levator scapulae (*24*)
5 Sternocleidomastoideus (*11*) (clavicular part)

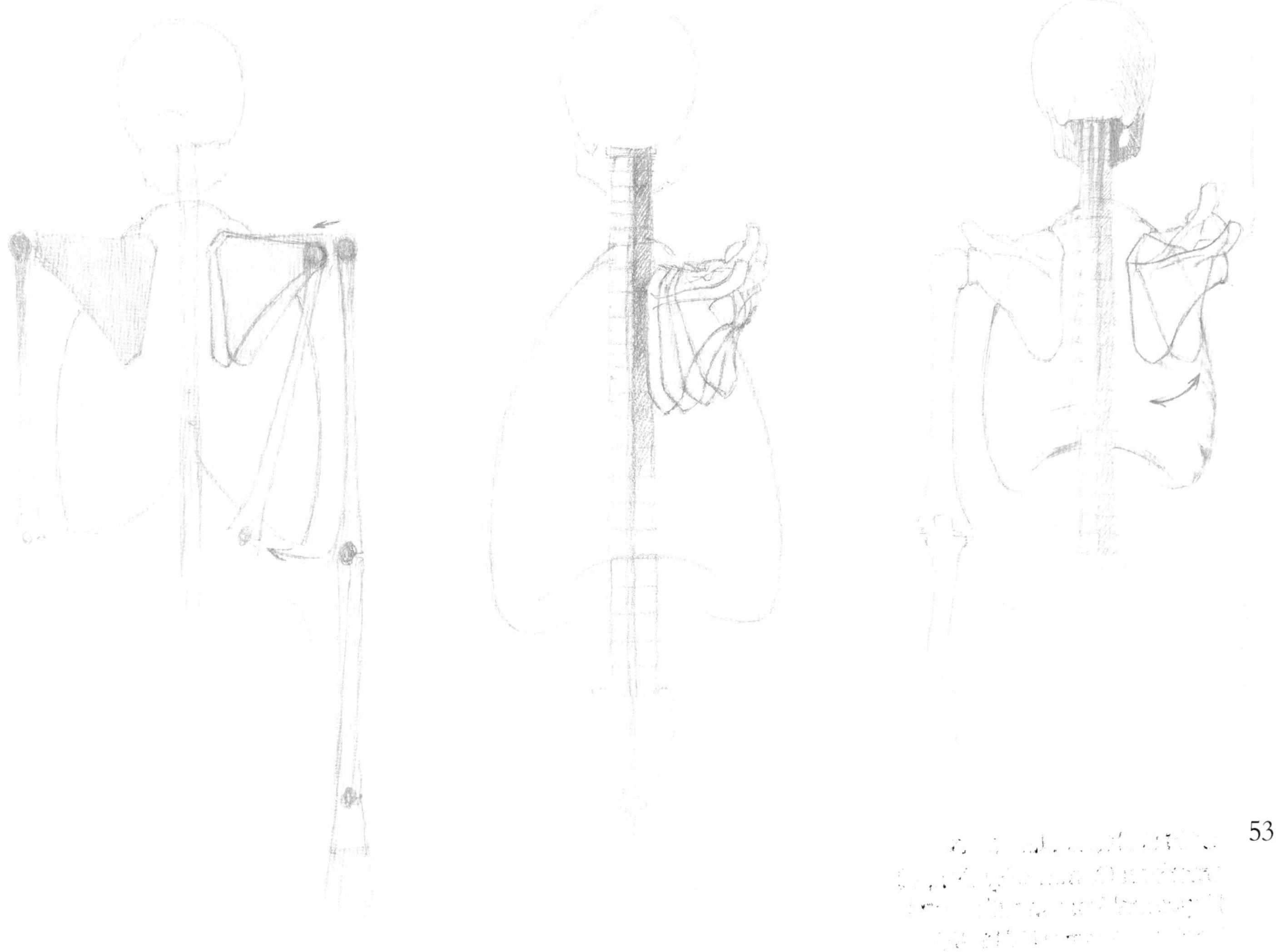

Fig. 30
Circular Rotation of the
Shoulder Joint

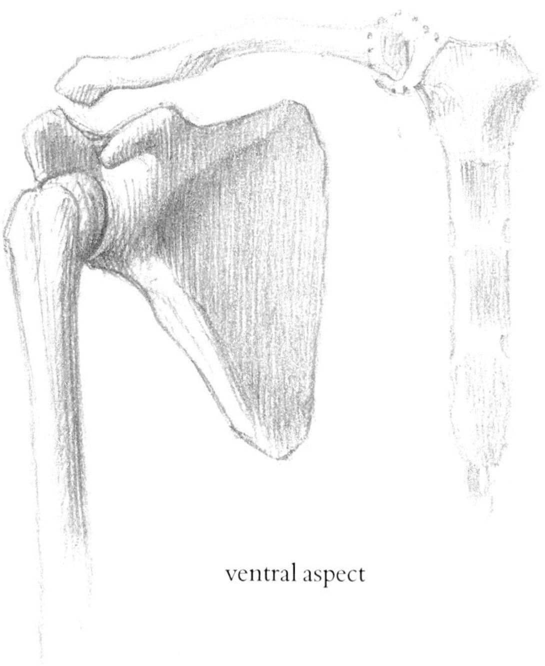

ventral aspect

The shoulder joints can move in many different directions (upwards by 30 to 40 degrees, backwards 20 degrees, forwards 20 degrees). When the arm or shoulder on its own is elevated, the clavicle rotates in the vertical axis through its joint with the sternum. When the arm is lowered, it rotates in the opposite direction.

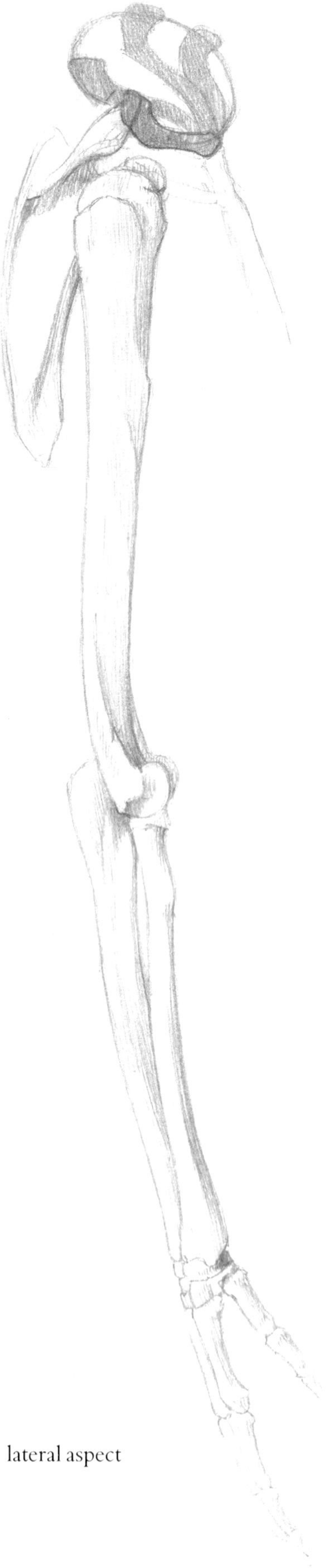

lateral aspect

A horizontal rotation of the arm against the abdomen (circumduction) is also possible.
Here, the clavicle performs a rotation in the sternoclavicular joint.

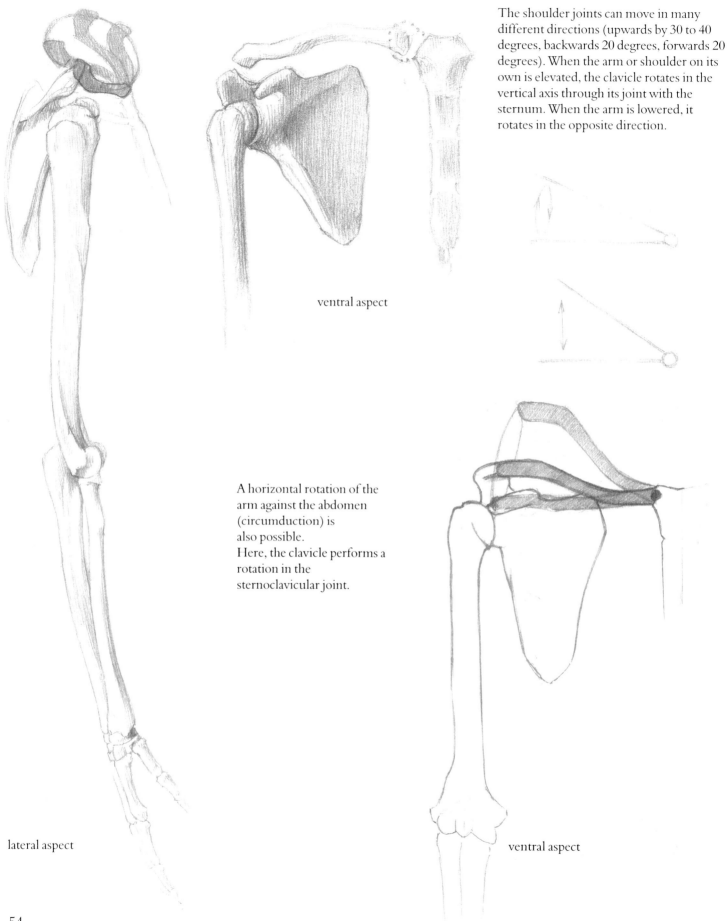

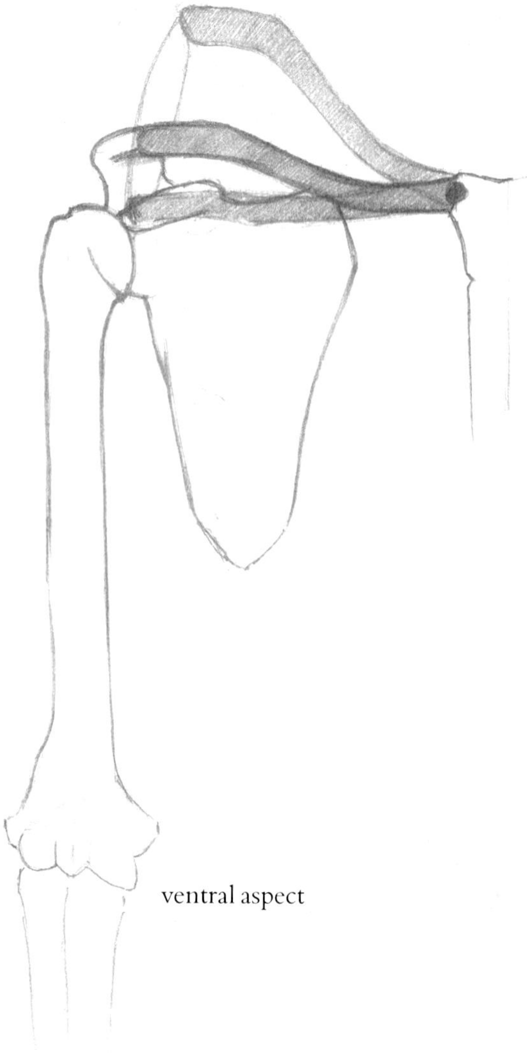

ventral aspect

Fig. 31
The Acromioclavicular Joints

1 Sternoclavicular joint
2 Acromioclavicular joint
3 Humeroscapular joint

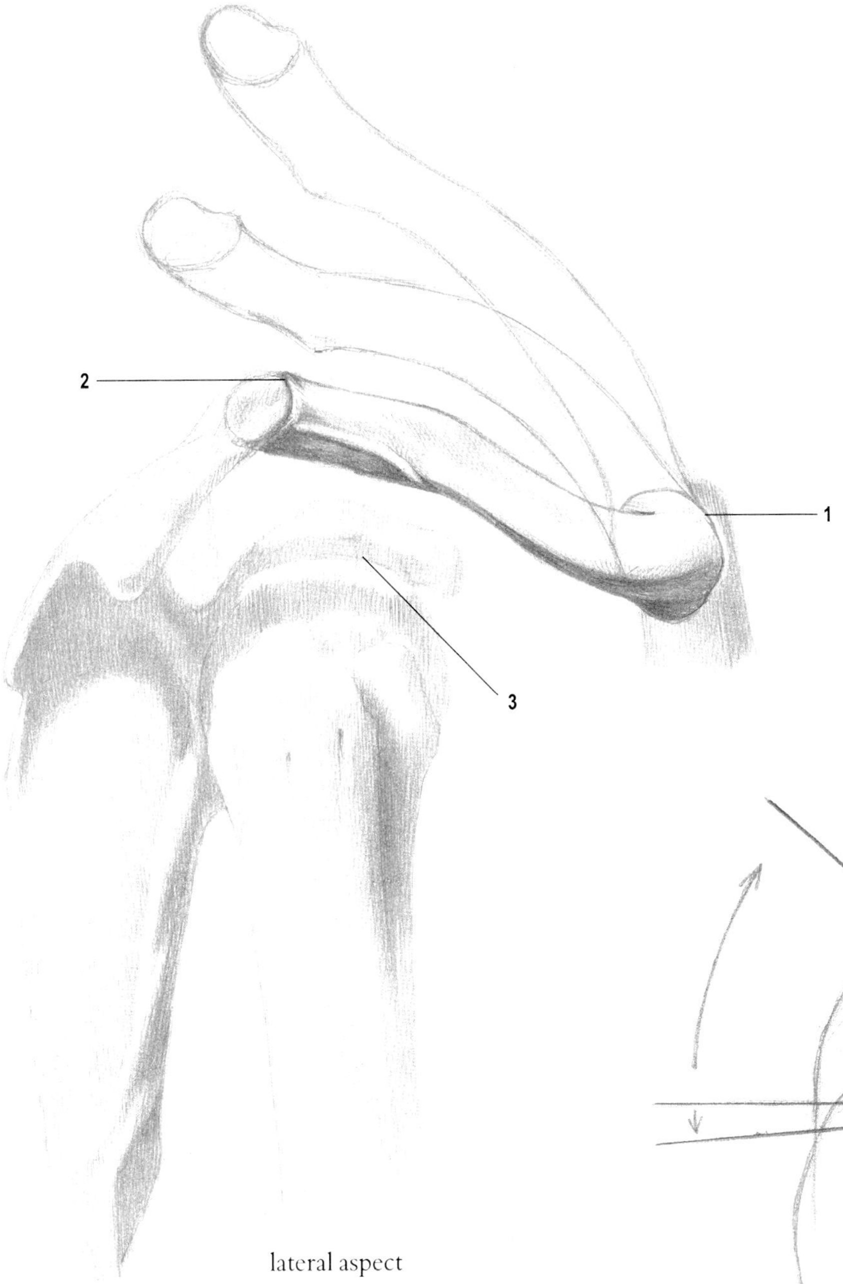

lateral aspect

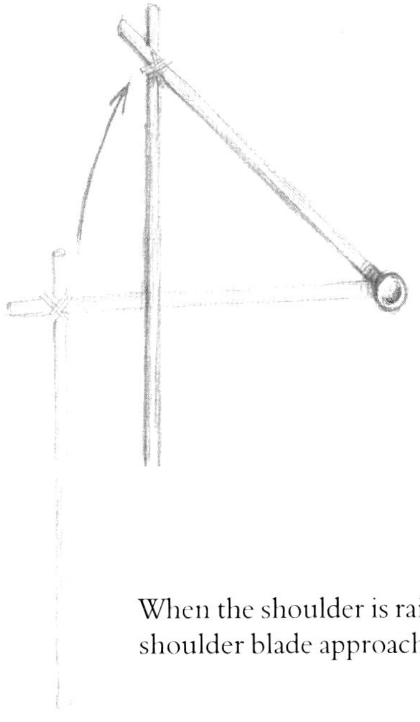

When the shoulder is raised, the shoulder blade approaches the sternum.

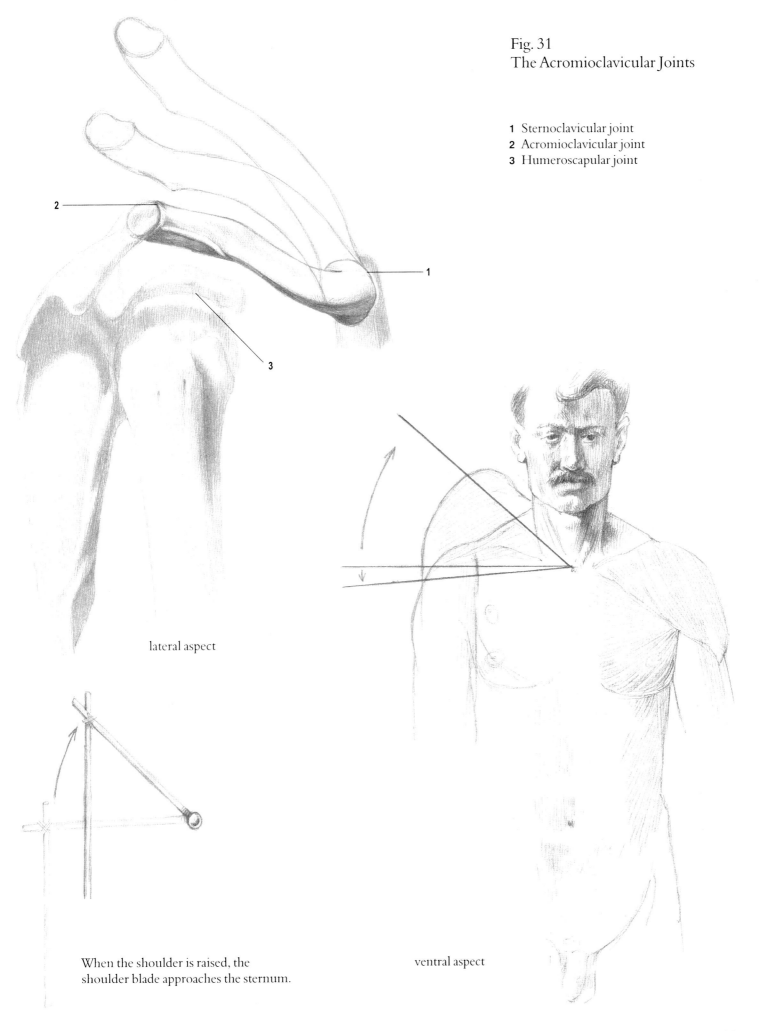

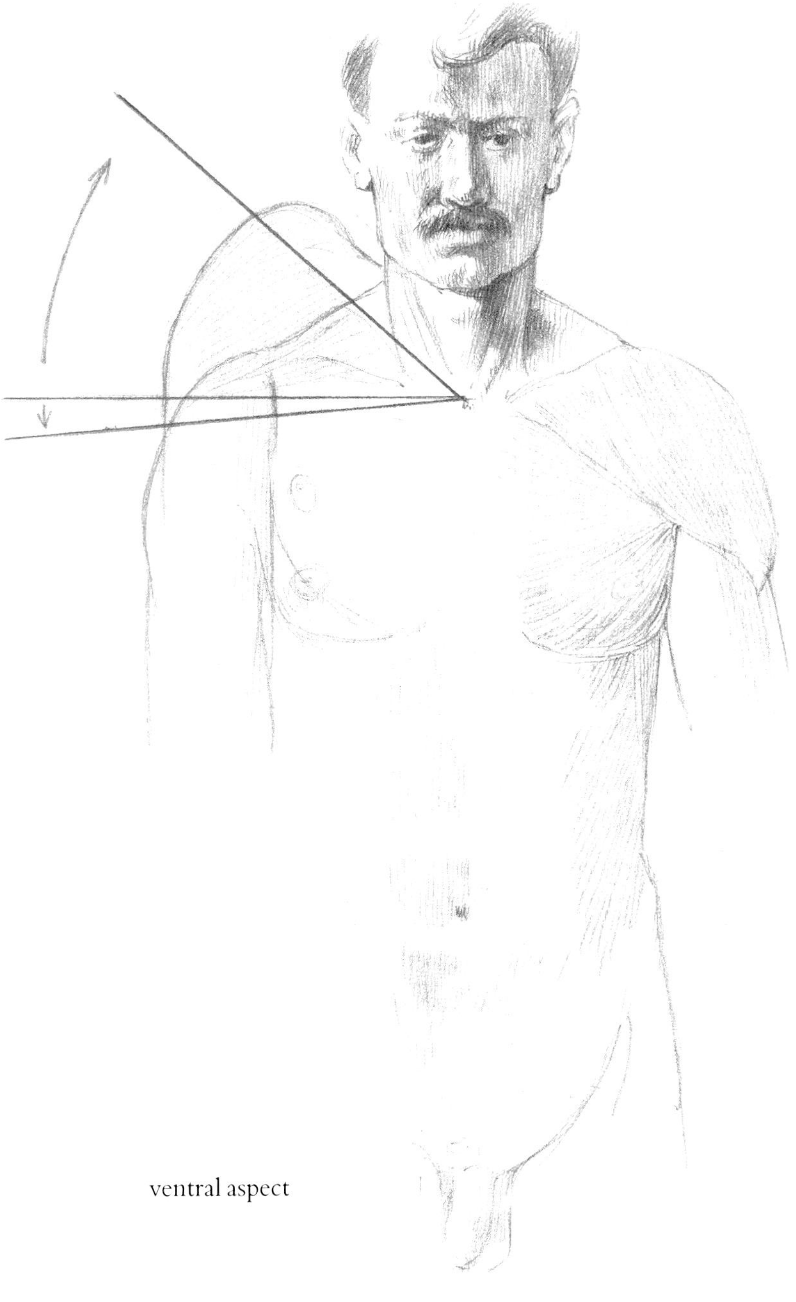

ventral aspect

Fig. 32
The Movements of the Shoulder Joint

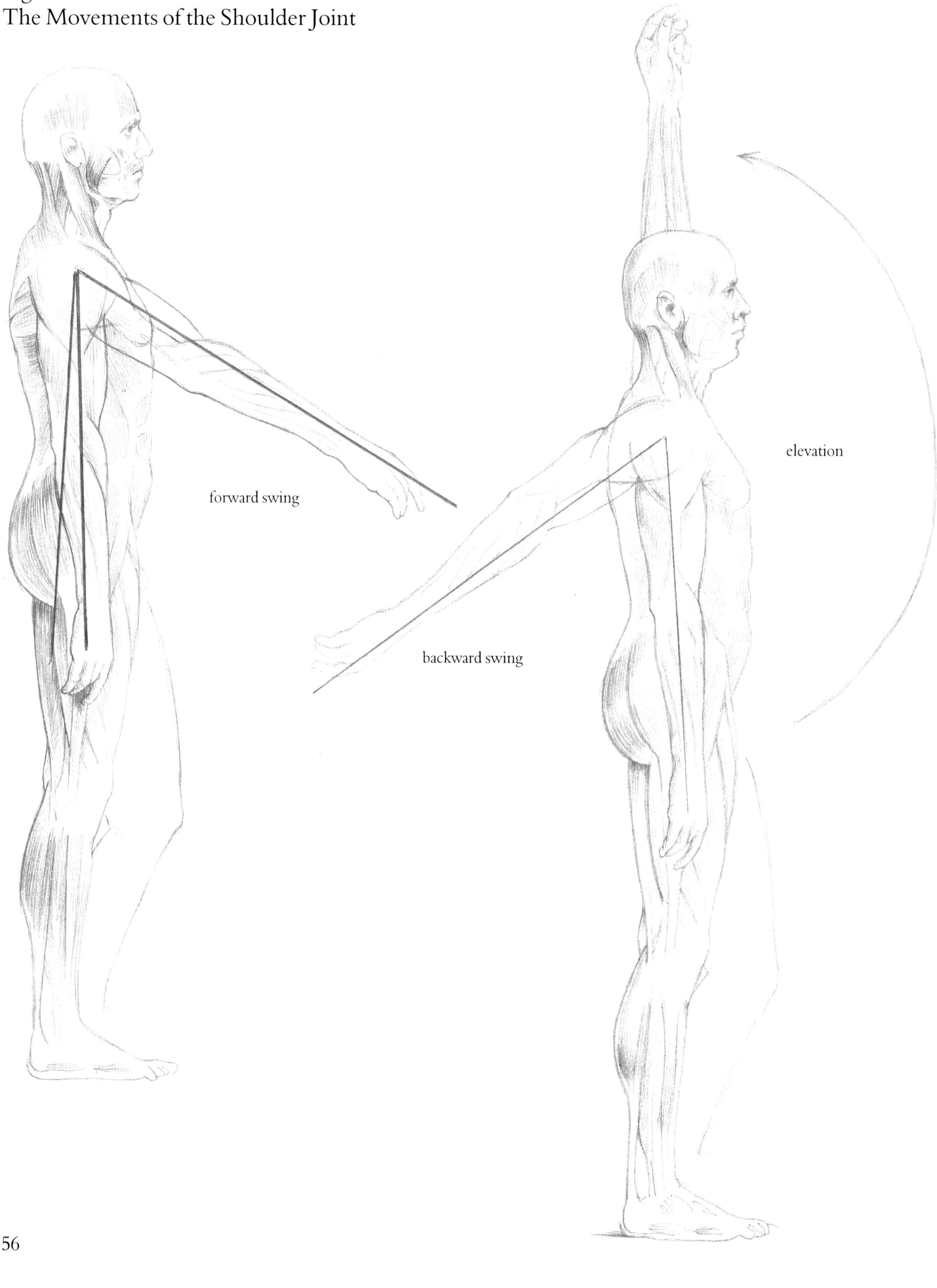

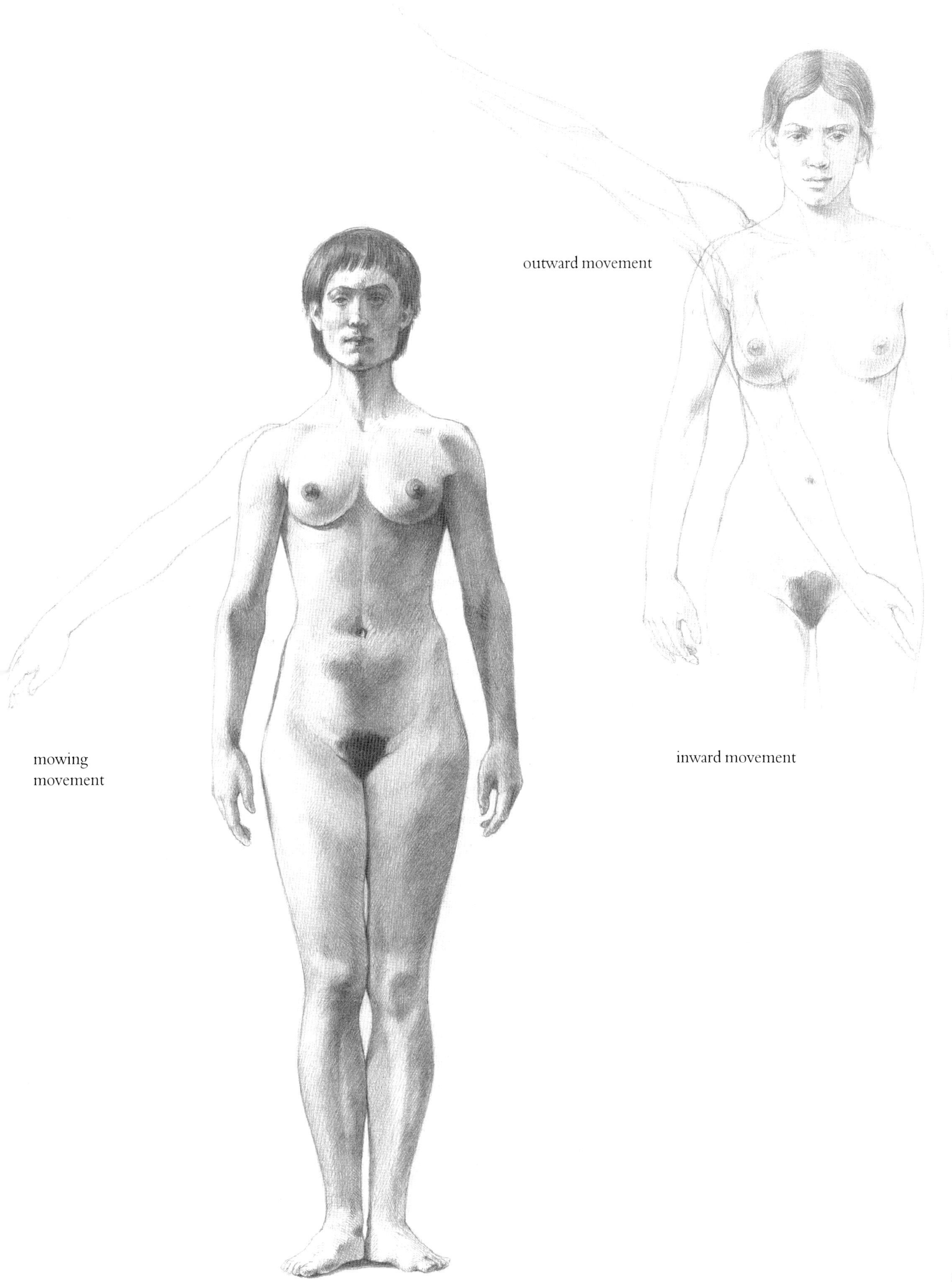
outward movement
mowing movement
inward movement

Fig. 33
The Axillary Fossa

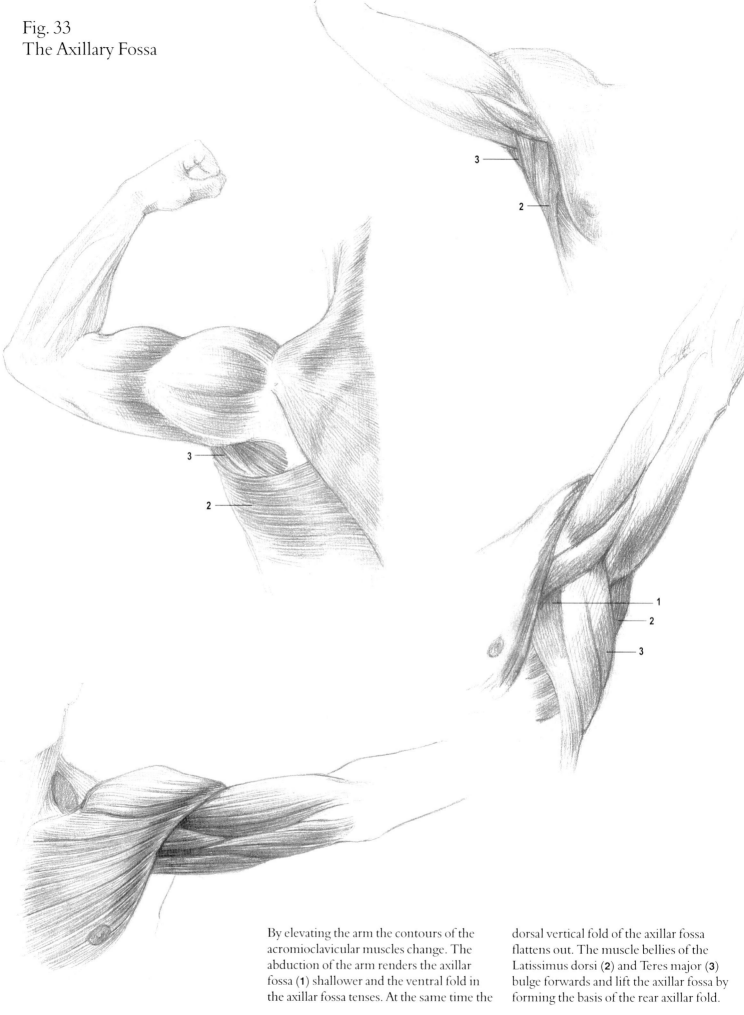

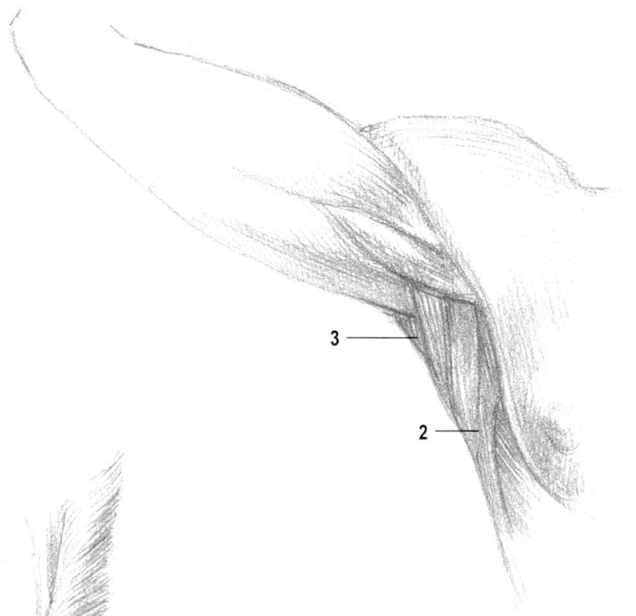

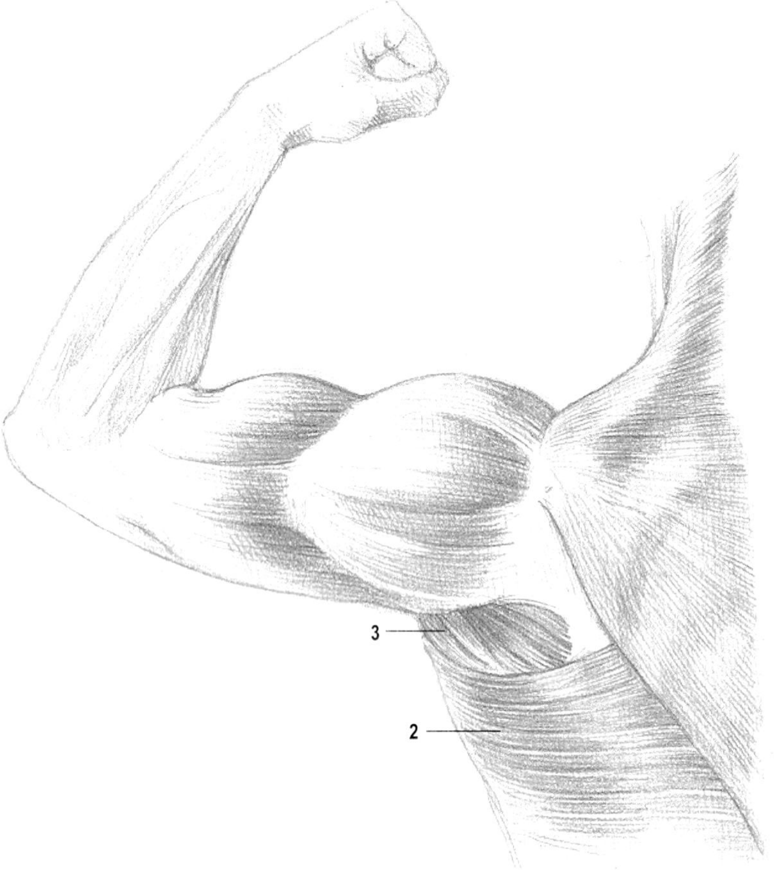

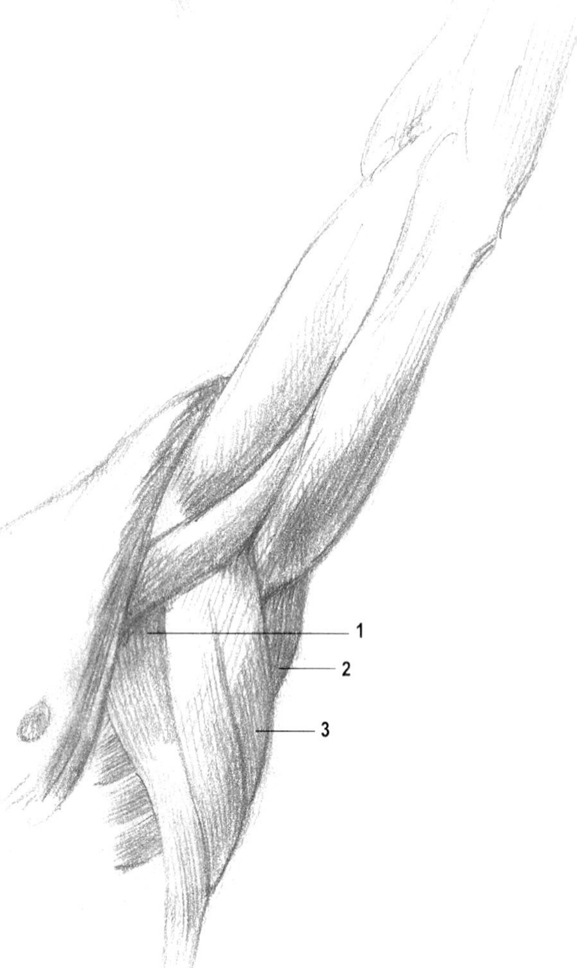

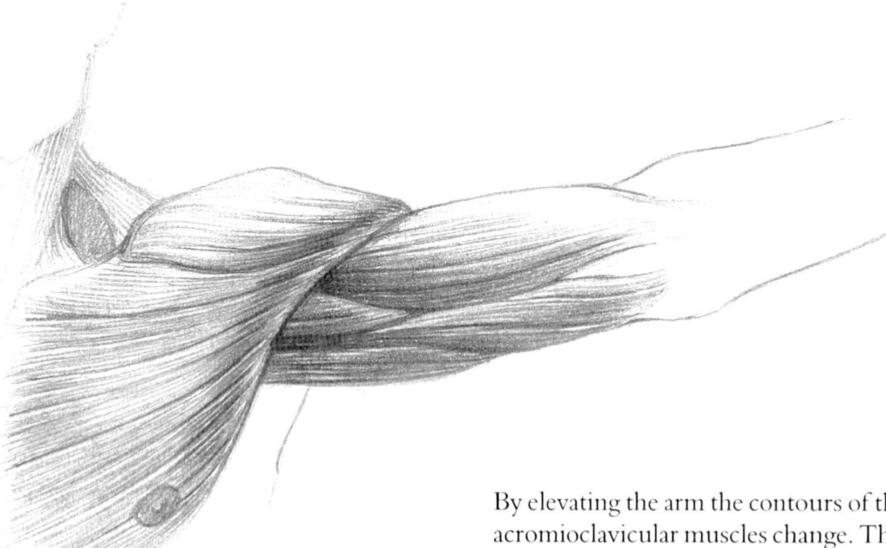

By elevating the arm the contours of the acromioclavicular muscles change. The abduction of the arm renders the axillar fossa (**1**) shallower and the ventral fold in the axillar fossa tenses. At the same time the dorsal vertical fold of the axillar fossa flattens out. The muscle bellies of the Latissimus dorsi (**2**) and Teres major (**3**) bulge forwards and lift the axillar fossa by forming the basis of the rear axillar fold.

Fig. 34
The Movement of the Humerus

Since the shoulder joint is a free joint, the humerus can move in various directions.

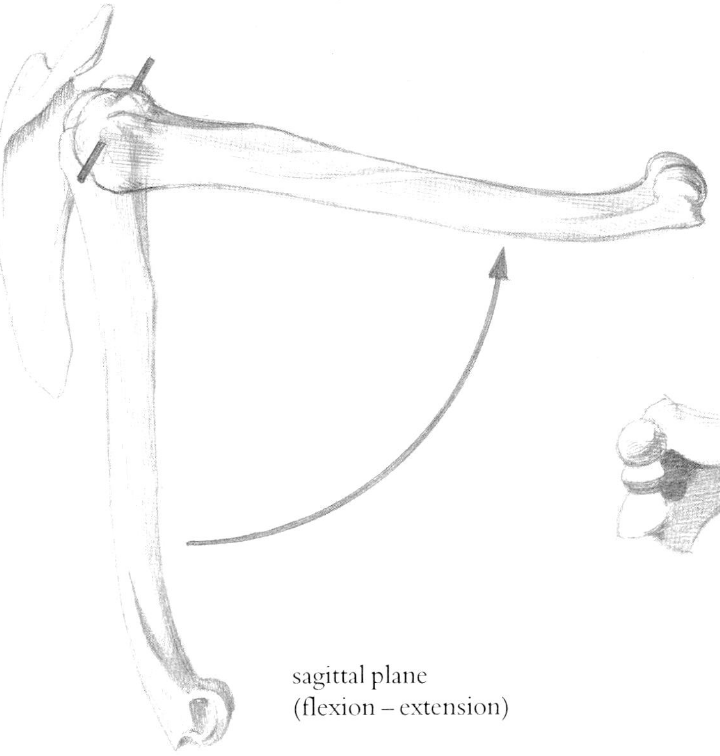

sagittal plane
(flexion – extension)

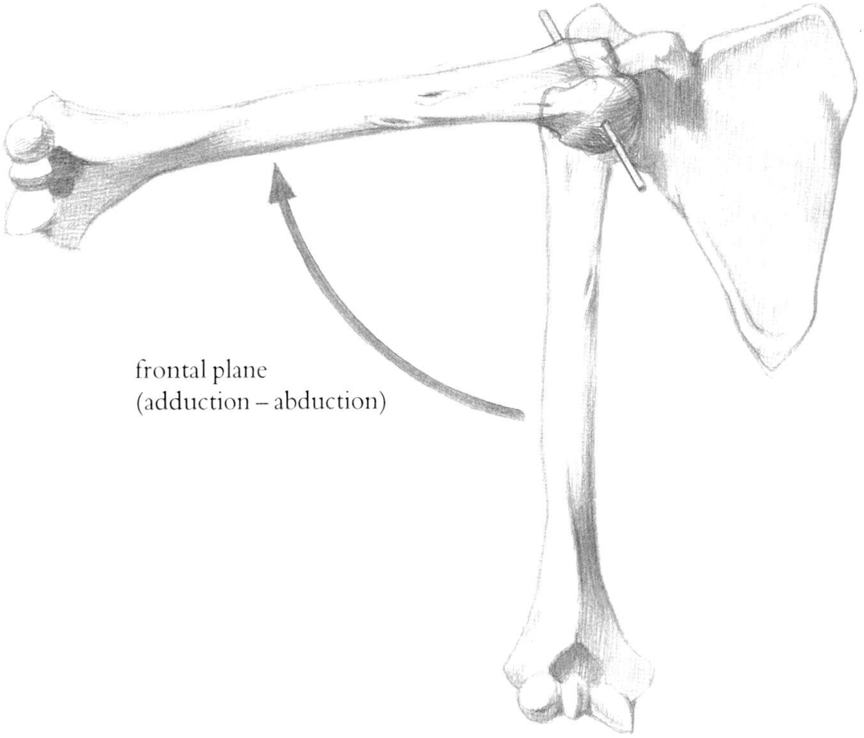

frontal plane
(adduction – abduction)

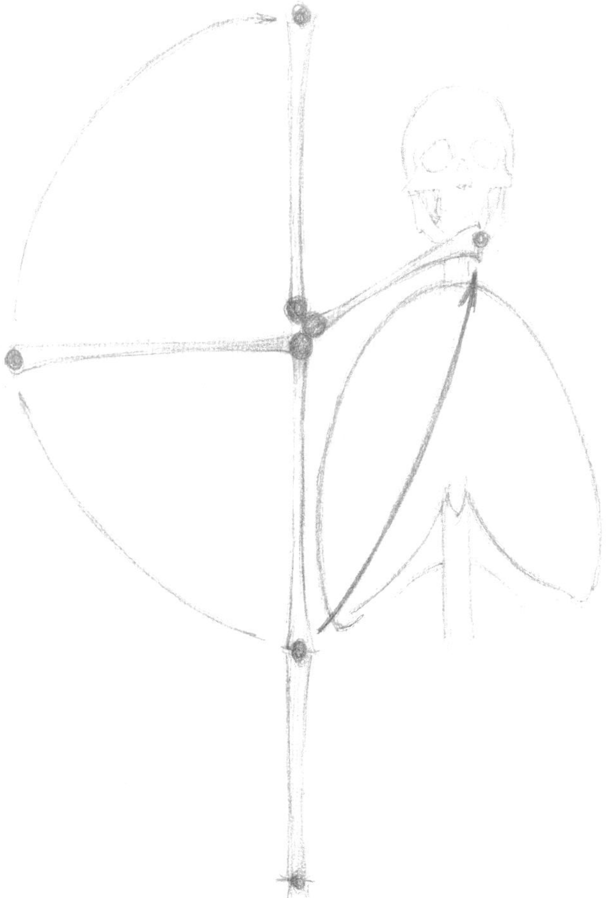

elevation

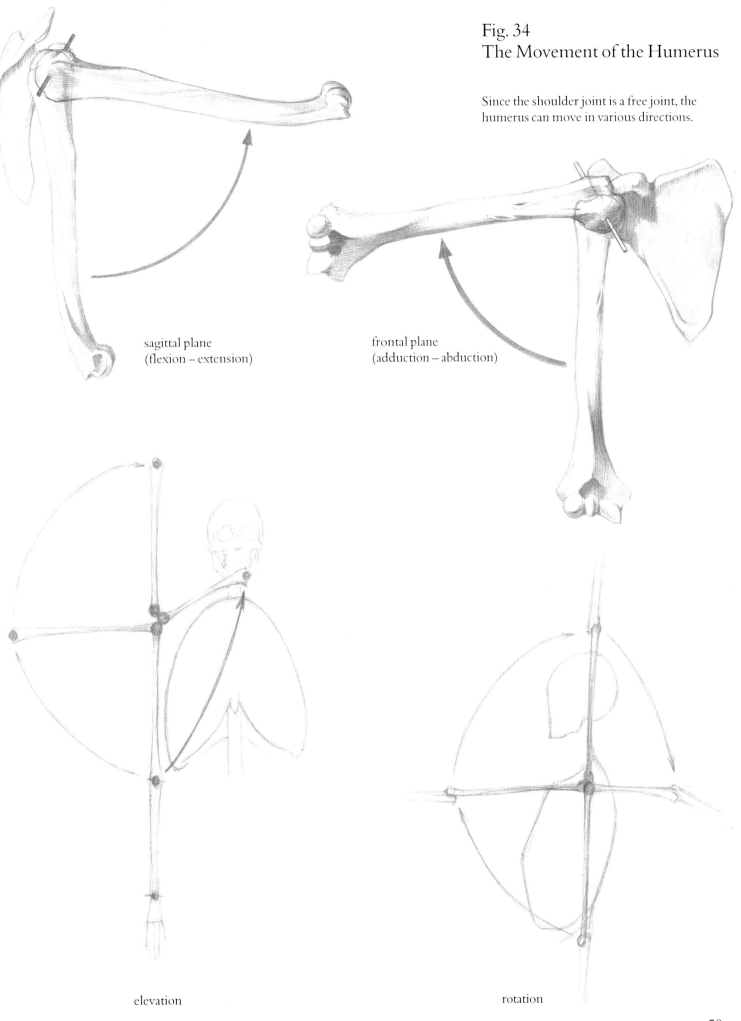

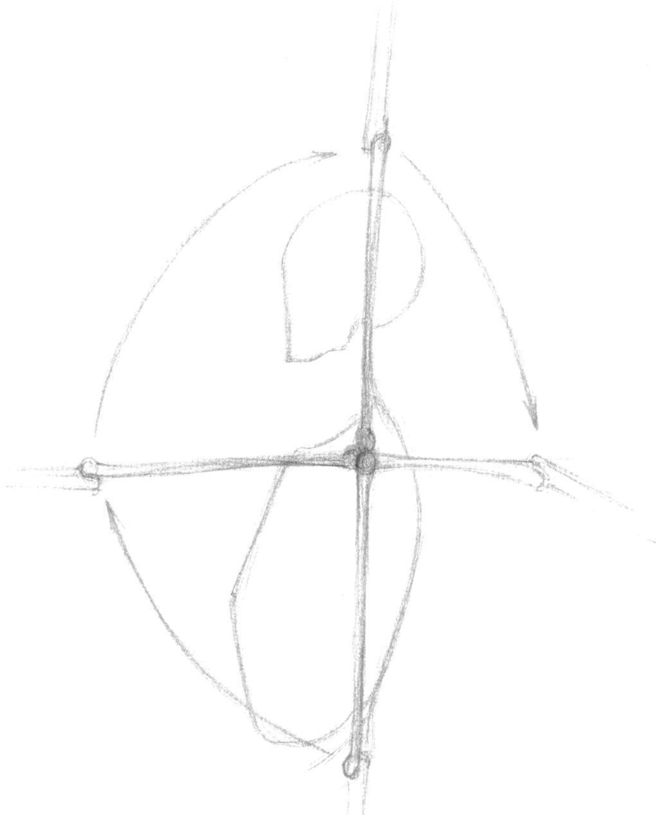

rotation

Fig. 35
The Ligaments of the Elbow Joint

The humero-ulnar joint and the humeroradial joint are a composite joint that enables the extension, flexion and rotation of the bones of the lower arm.

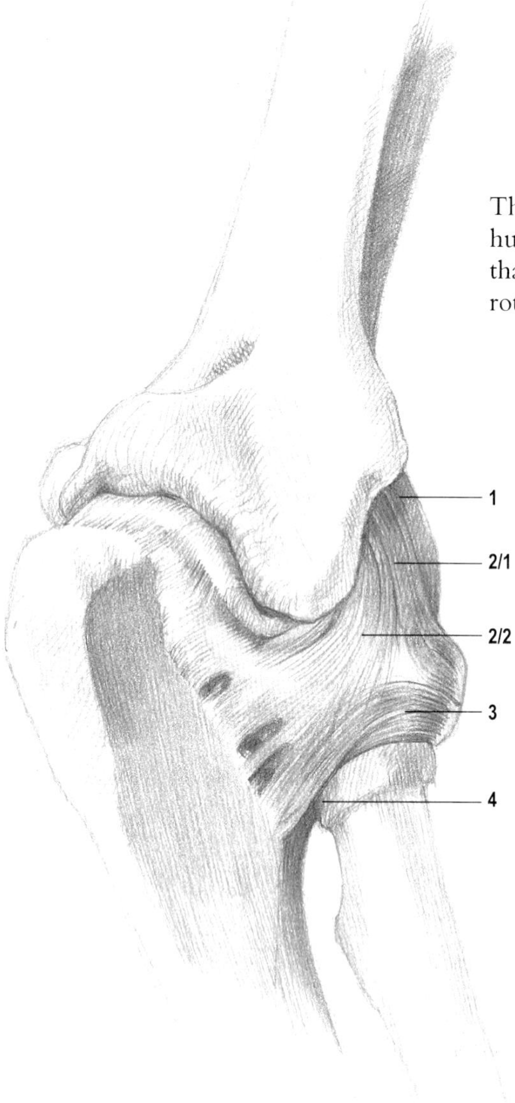

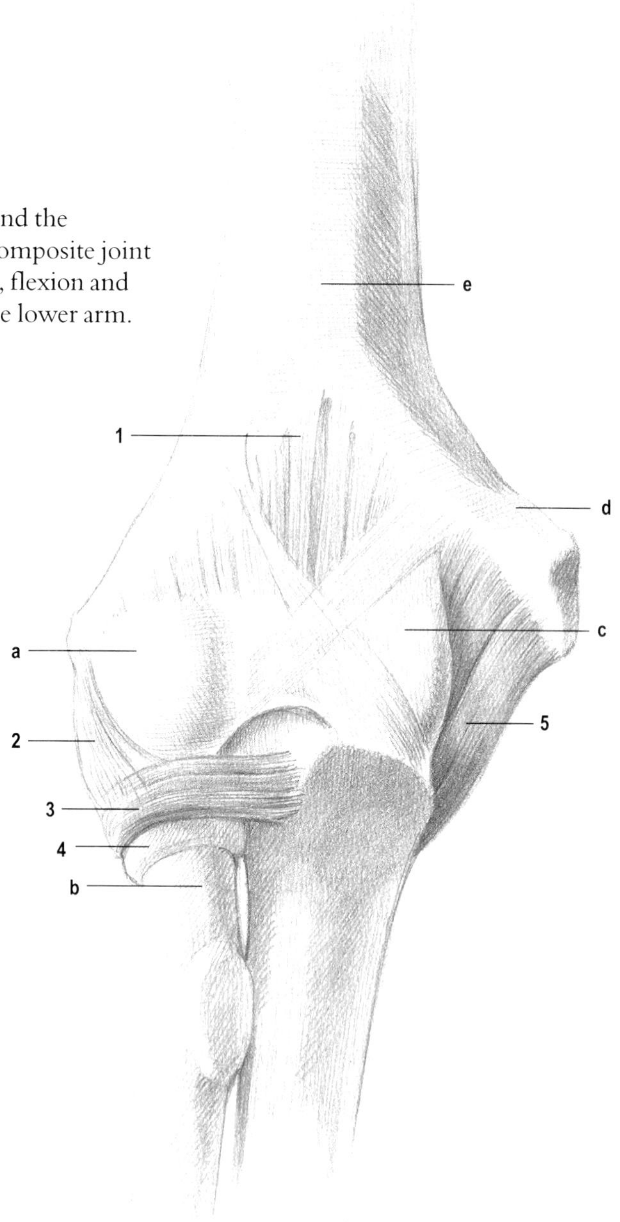

Lateral Dorsal Aspect

1 Joint capsule
2 **2/1** Ventral sheath of medial ligament
2/2 Dorsal sheath of medial ligament
3 Annular ligament
4 Upper ulnar-radial joint

Ventral Aspect

1 Joint capsule
2 Lateral ligament
3 Annular ligament
4 Head of joint capsule
5 Medial collateral ligament

a Head of humerus
b Neck of radius
c Humeral trochlea
d Humeral anconeus
e Humerus

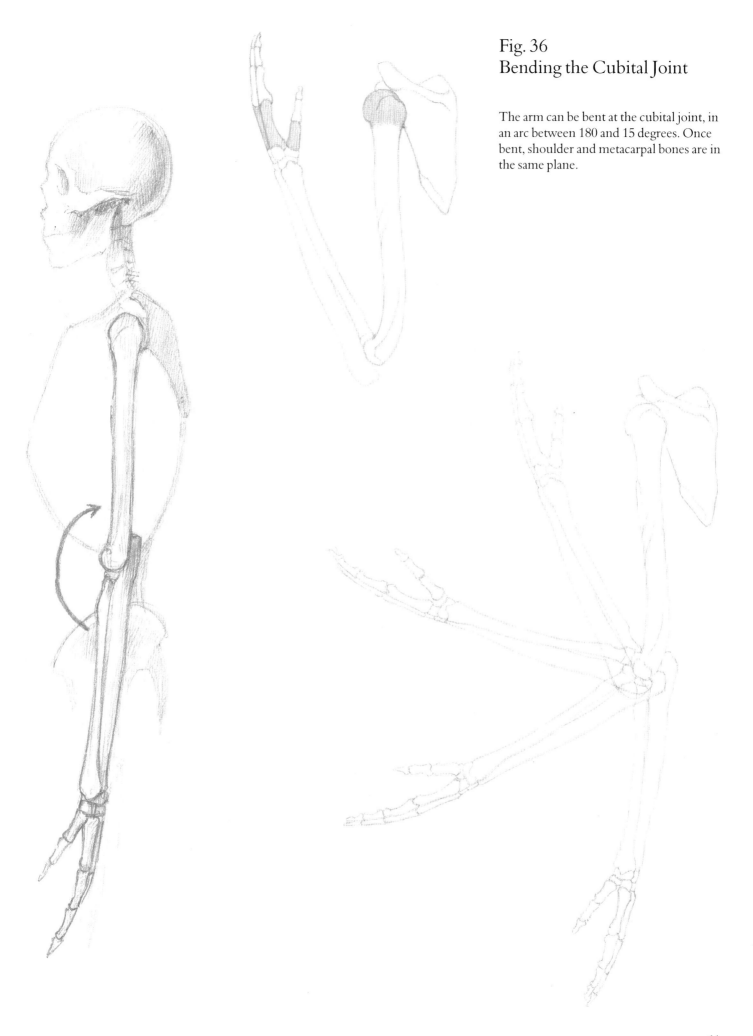

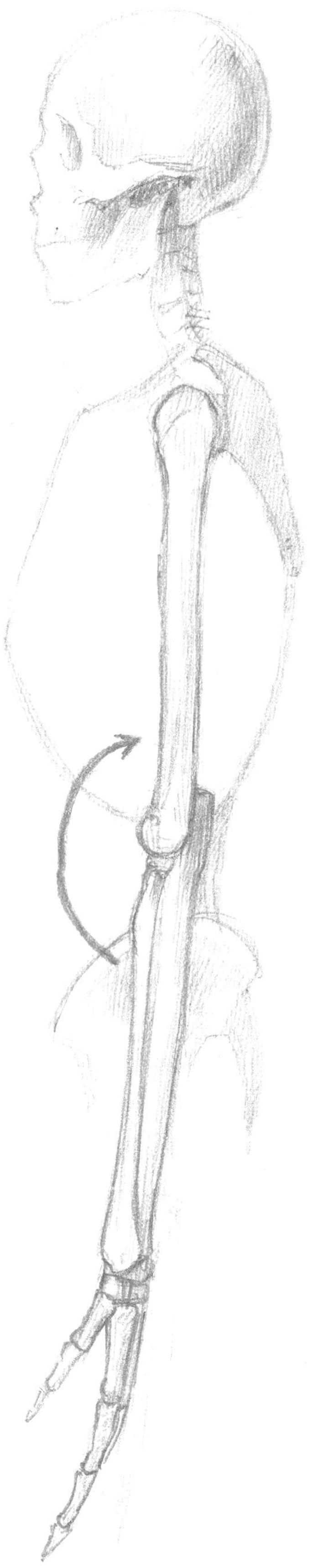

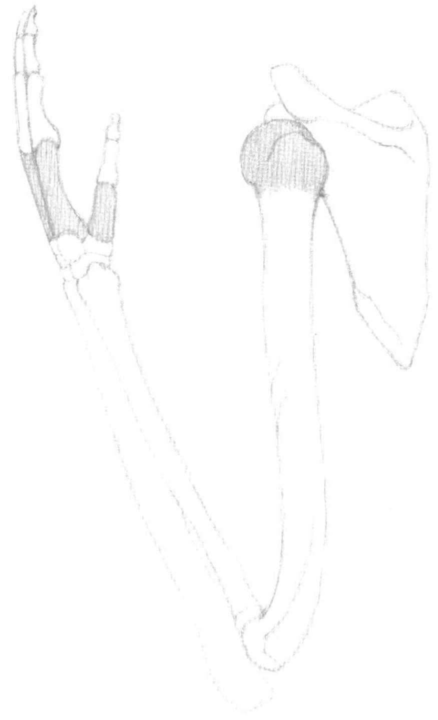

Fig. 36
Bending the Cubital Joint

The arm can be bent at the cubital joint, in an arc between 180 and 15 degrees. Once bent, shoulder and metacarpal bones are in the same plane.

Fig. 37
Movement of the Humerus

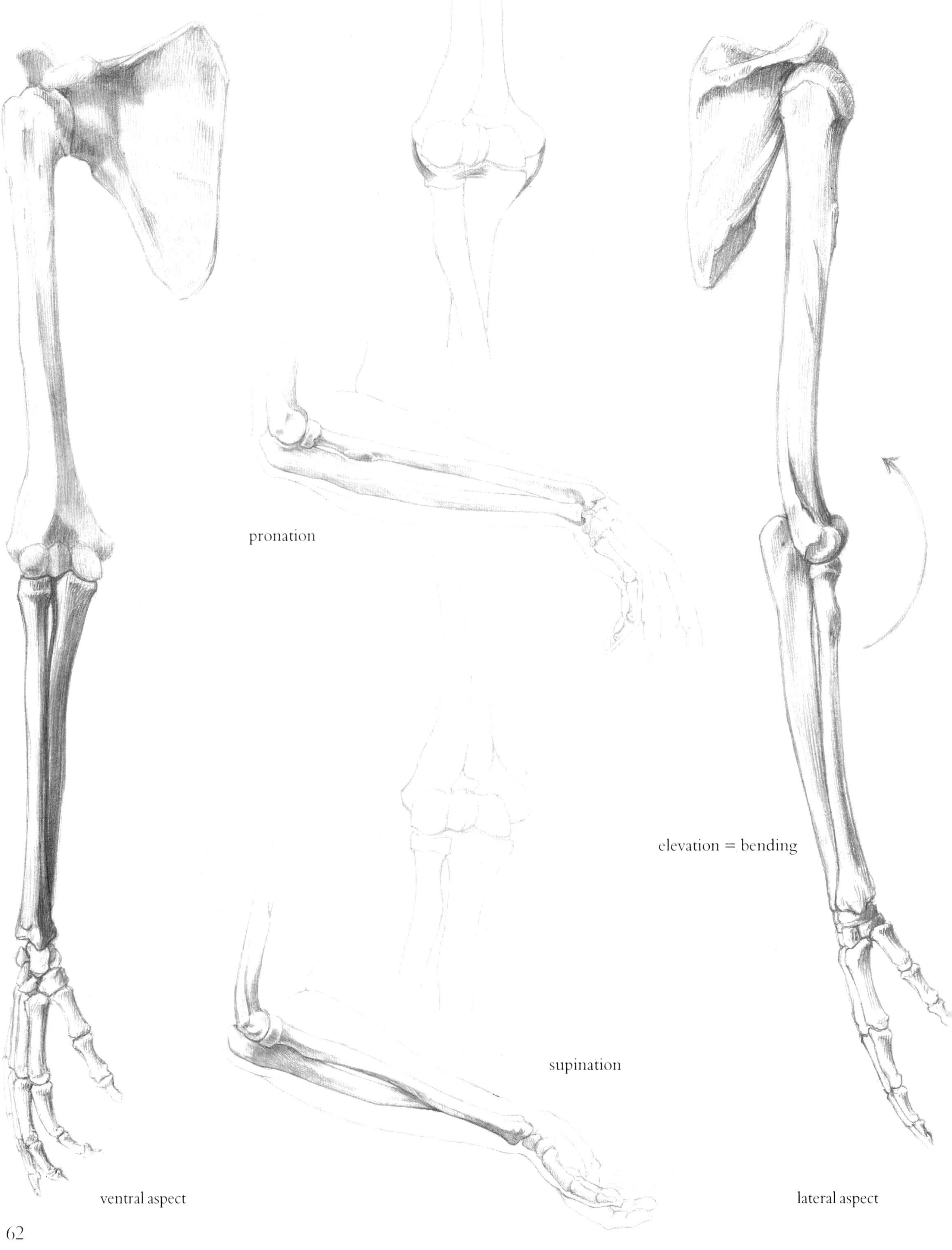

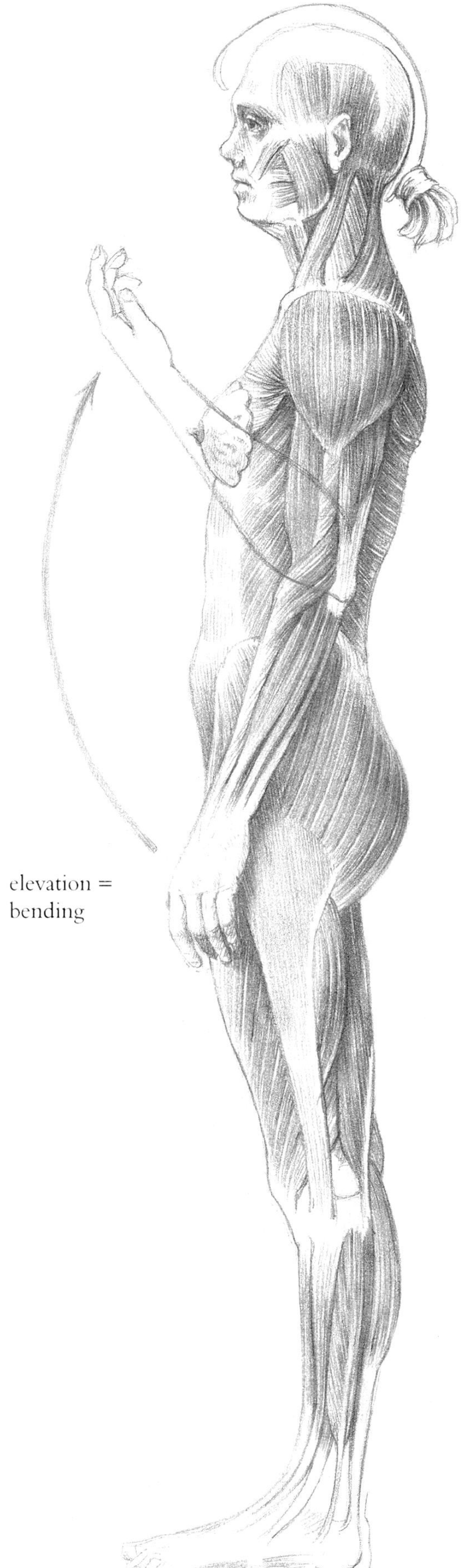

elevation =
bending

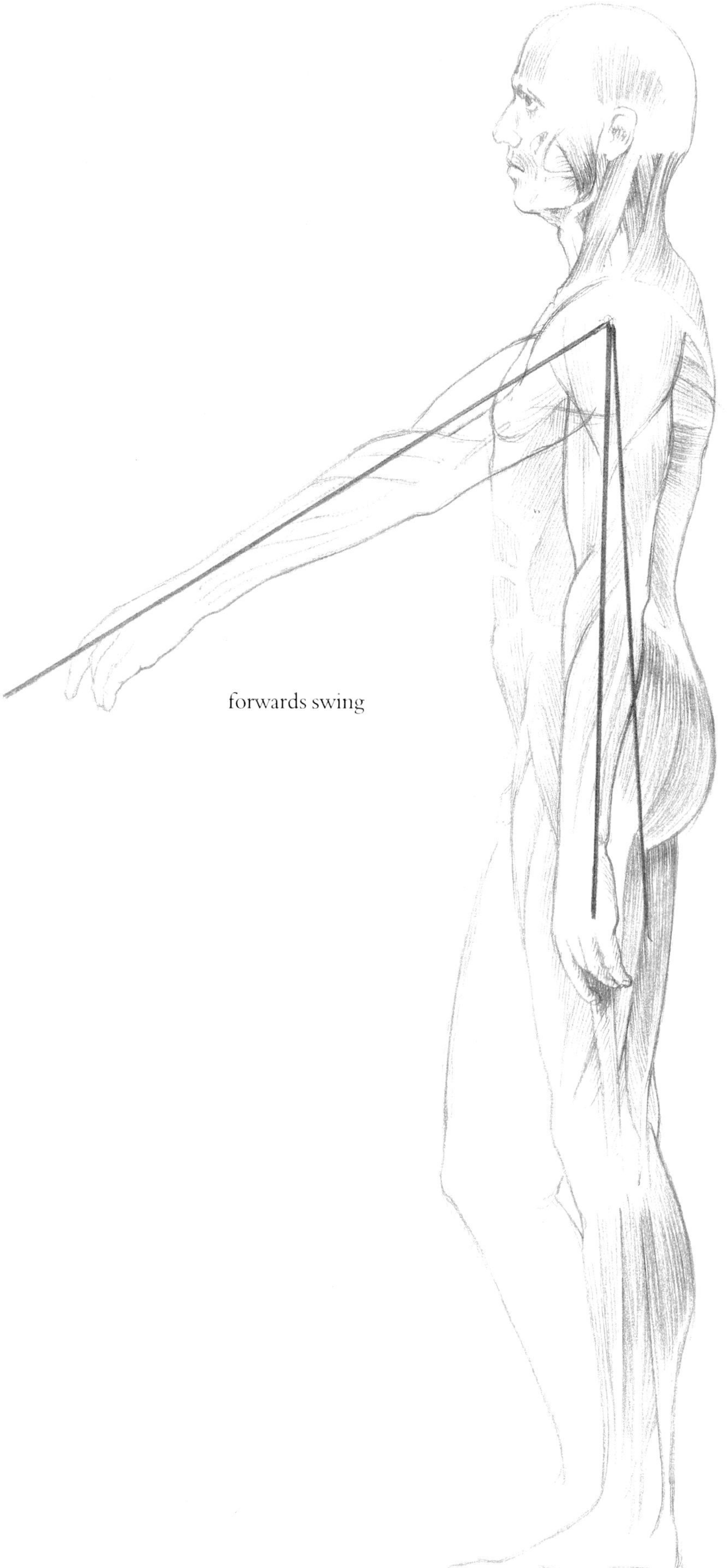

forwards swing

Fig. 38
Inwards Rotation and Outwards Rotation
(Pronation and Supination)

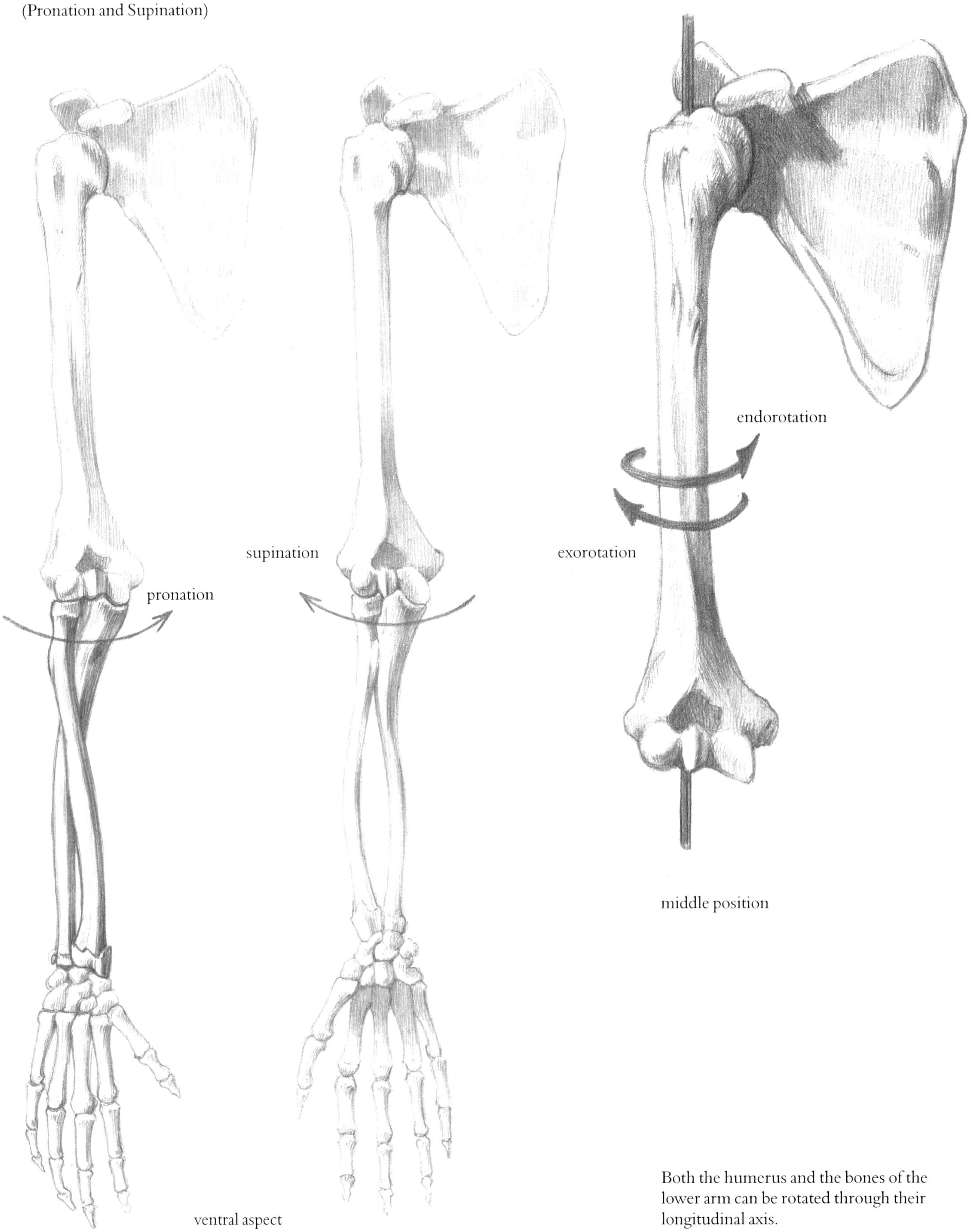

Both the humerus and the bones of the lower arm can be rotated through their longitudinal axis.

Fig. 39
Connections between the Bones of the Lower Arm

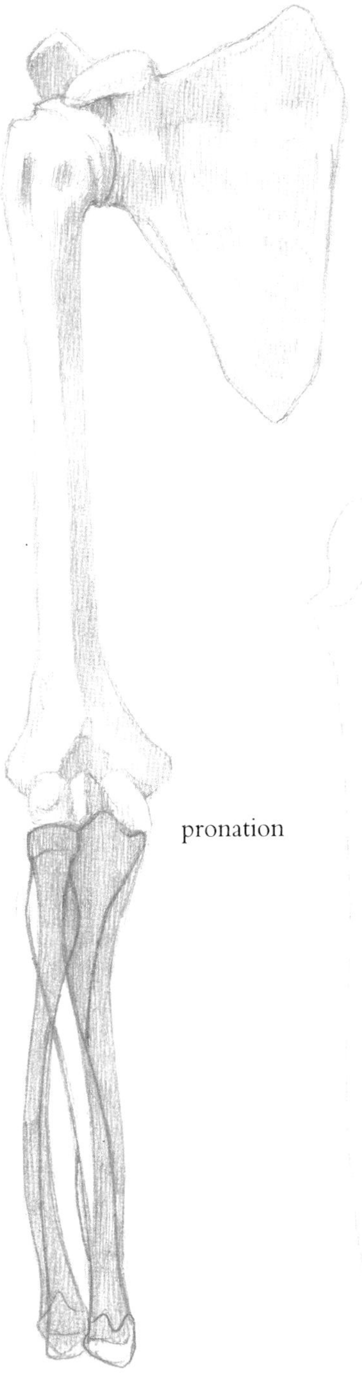

pronation

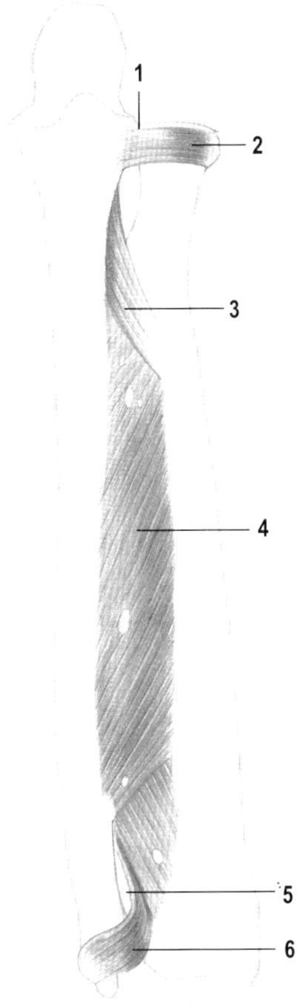

ventral aspect

The radius and ulna are connected by both joints and ligaments. Both bones can rotate in relation to each other.

1 Superior radioulnar joint
2 Annular ligament
3 Superior interosseous ligament
4 Interosseous membrane
5 Inferior radioulnar joint
6 Inferior annular ligament

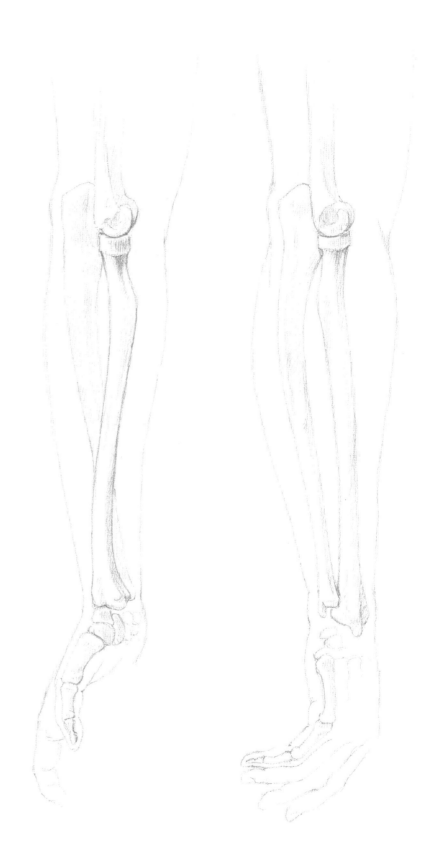

lateral aspect
(right arm in supination)

lateral aspect
(right arm in pronation)

Fig. 40
The Ligaments of the Carpal Joints

1 Anterior radioulnar ligament
2 Ulnar collateral ligament
3 Medial radiocarpal ligament
4 Transverse carpal ligament
5 Transverse metacarpal ligament
6 Transverse ligament of the proximal phalanx
7 Insertions of extensor tendons

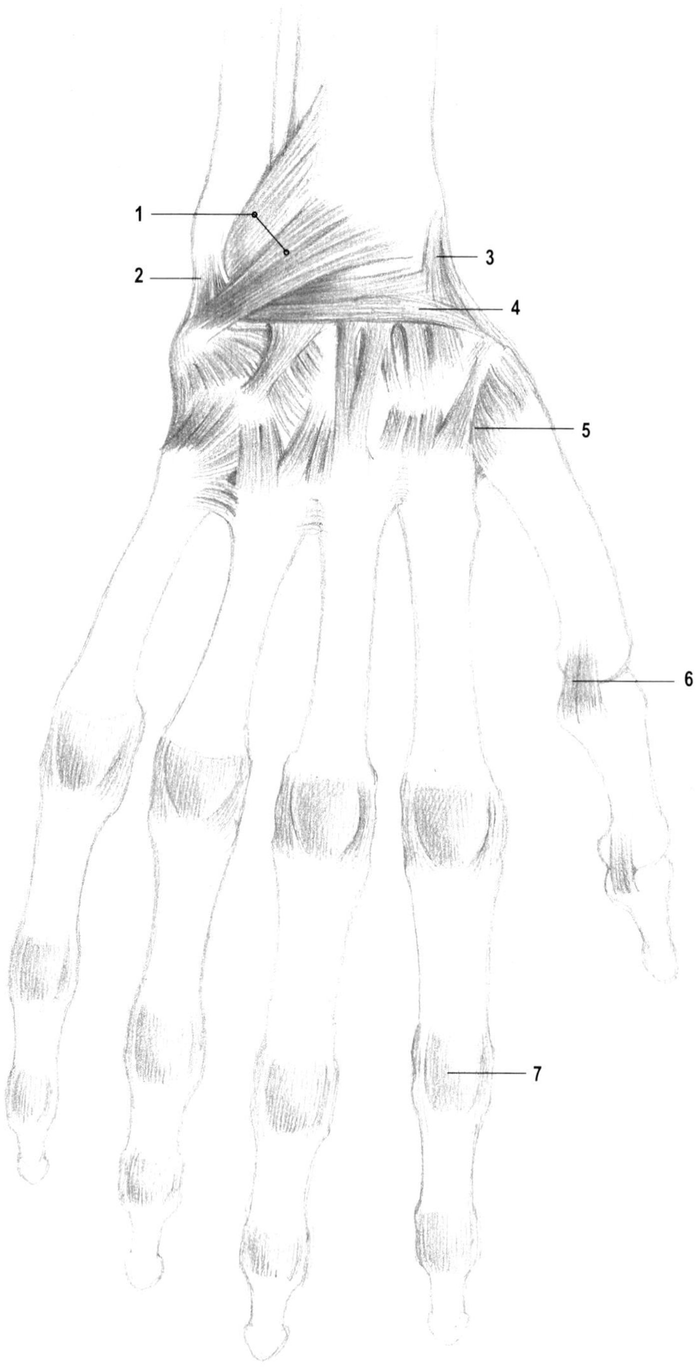

dorsal aspect

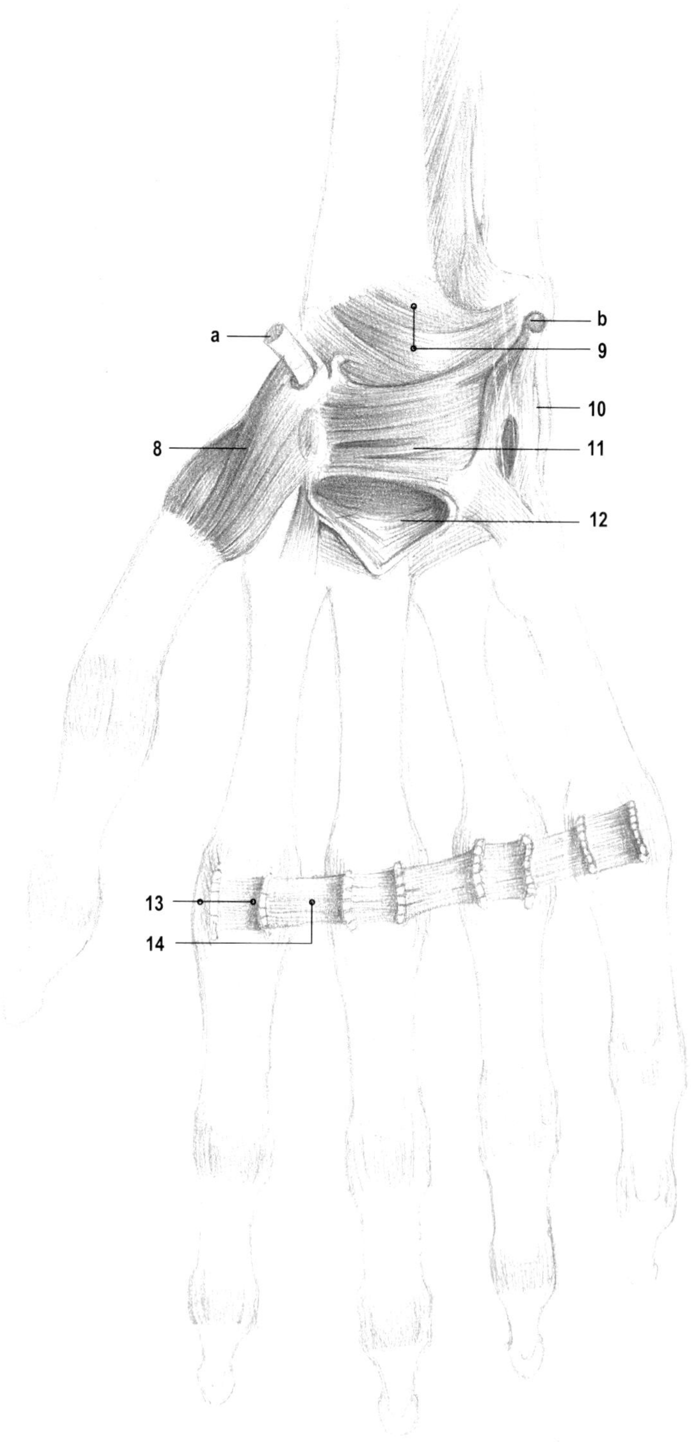

palmar aspect

8 Posterior metacarpal ligament
9 Posterior radioulnar ligament
10 Pisometacarpal ligament
11 Transverse sheath (fixes flexor tendons)
12 Annular channel
13 Tendon fossa of flexor muscles
14 Transverse metacarpal ligament

a Tendon of flexor carpi radialis (*53*)
b Tendon of flexor carpi ulnaris (*55*)

Fig. 41
Movements of the Hand

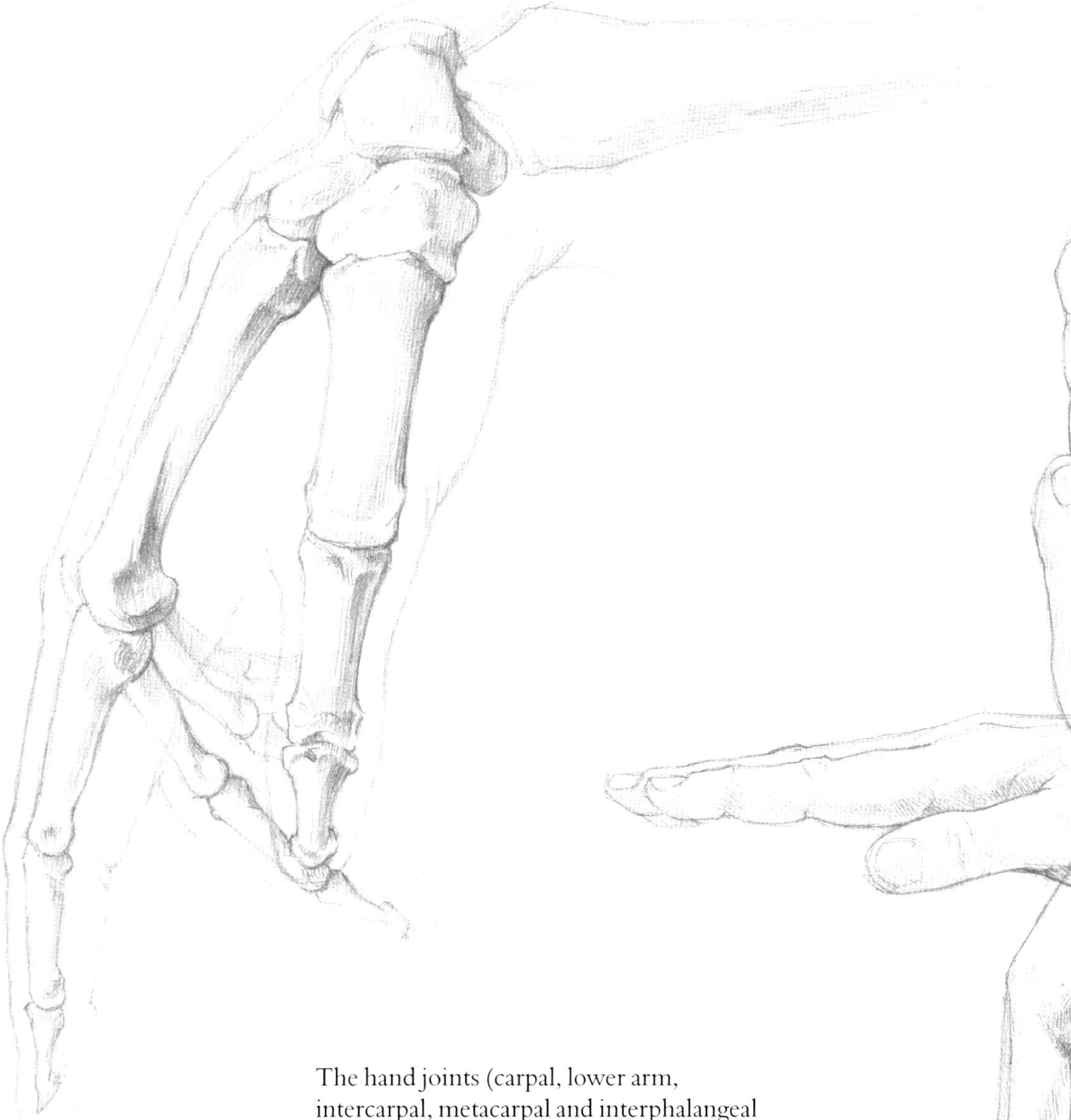

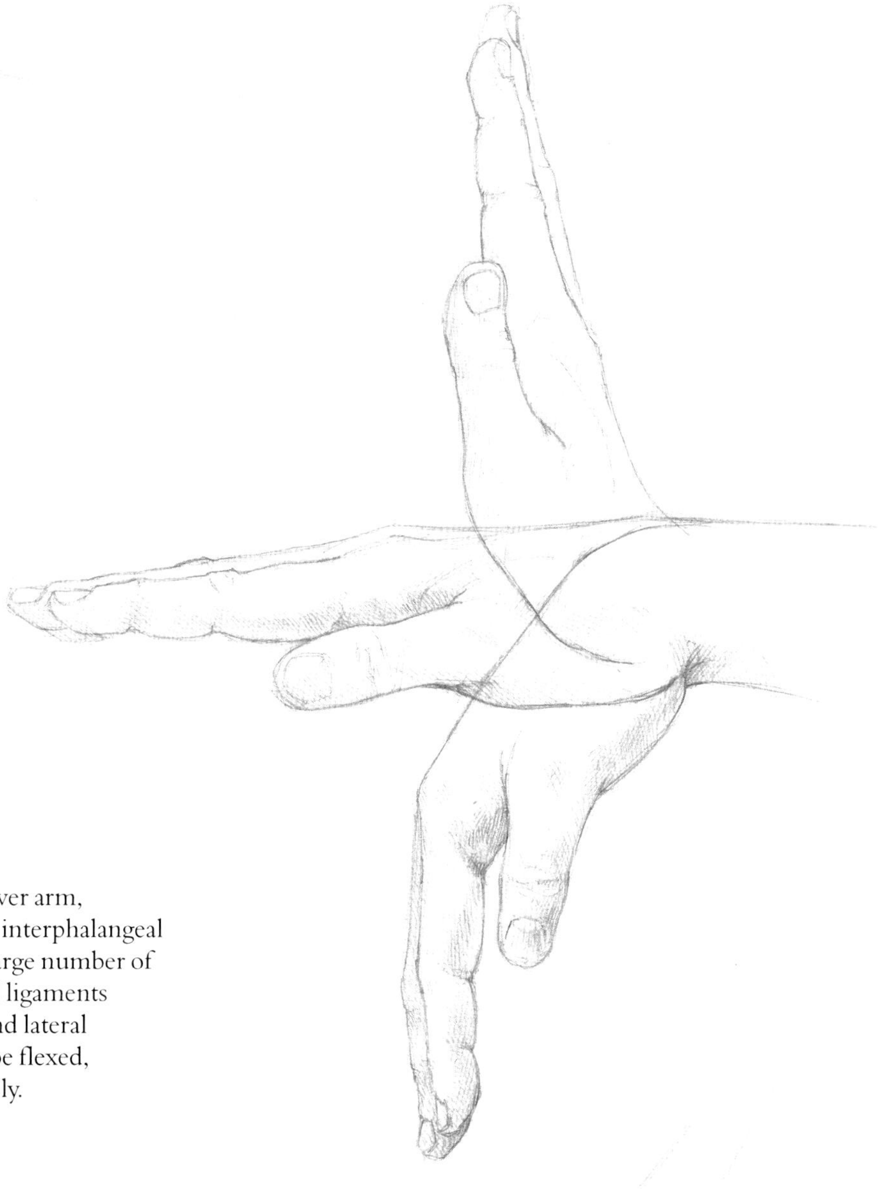

The hand joints (carpal, lower arm, intercarpal, metacarpal and interphalangeal joints) are connected by a large number of ligaments. These joints and ligaments enable flexion, extension and lateral movement. The hand can be flexed, extended and moved laterally.

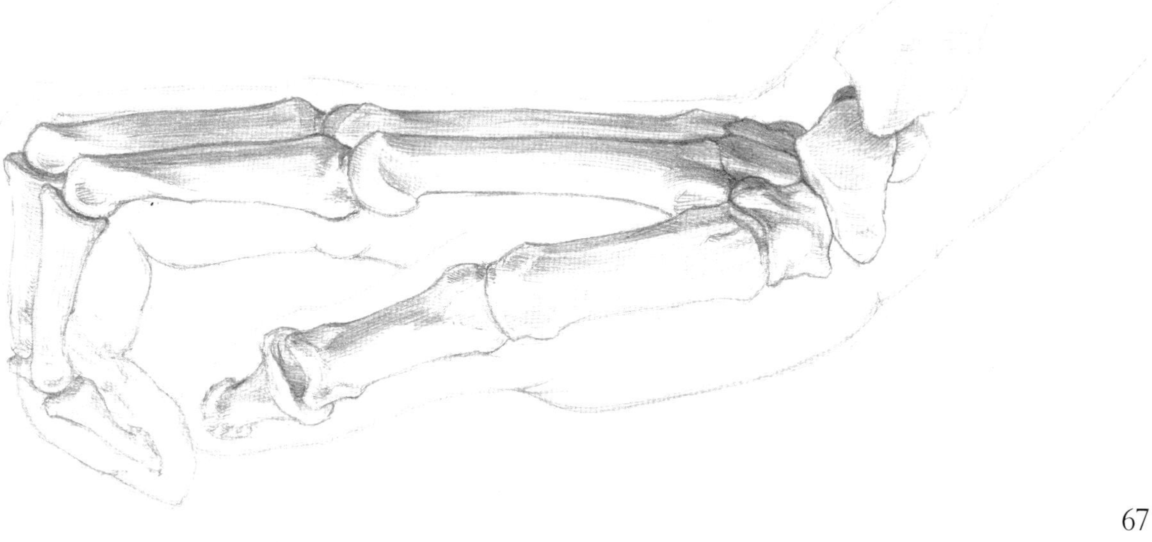

Fig. 42
The Finger Joints
(Saddle Joint)

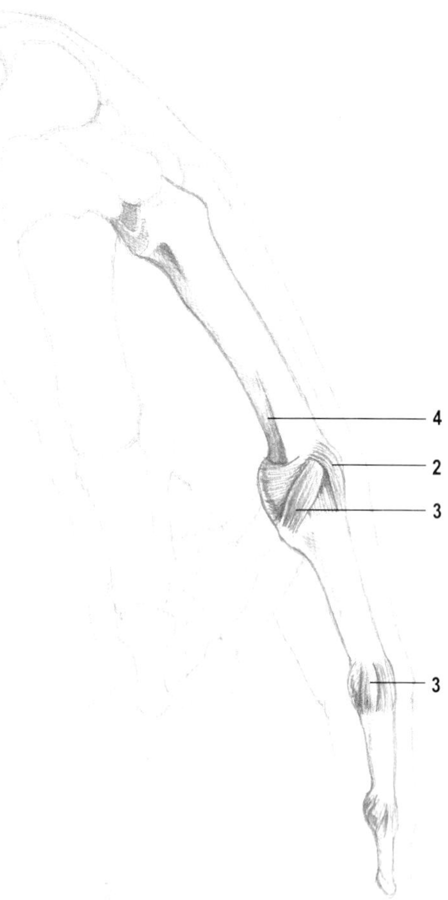

lateral aspect

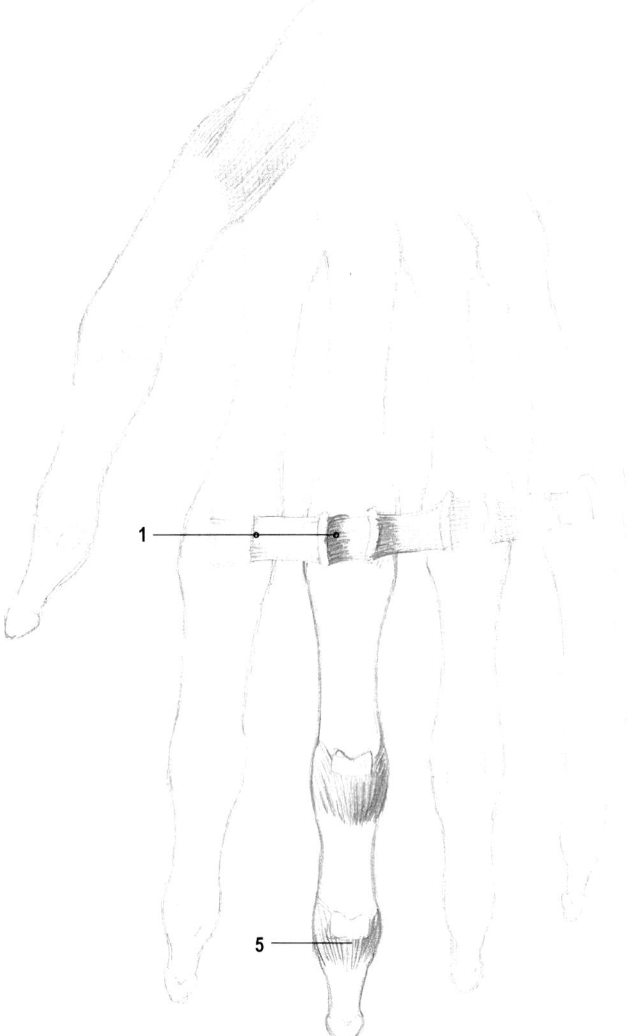

palmar aspect

1 Transverse metacarpal ligaments
2 Joint capsule
3 Collateral ligament
4 Palmar ligament
5 Insertion of flexor tendons

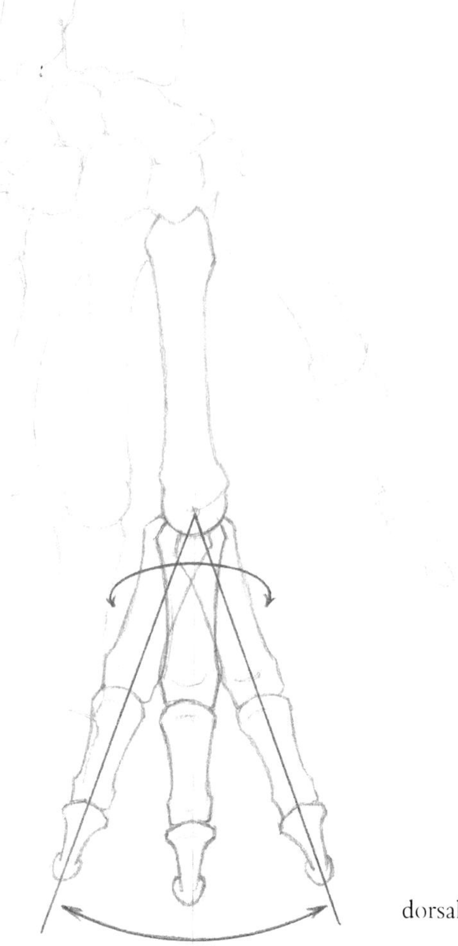

dorsal aspect

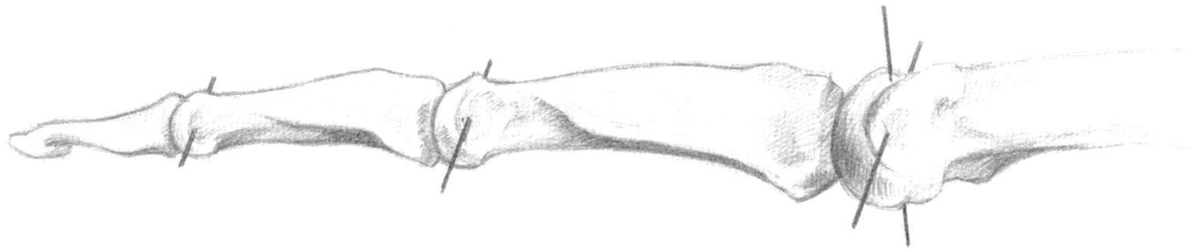

Fig. 43
Movements of the Fingers

The second to fifth fingers can move laterally or in a circle.
The fingers can be flexed to an angle of almost 45 degrees.

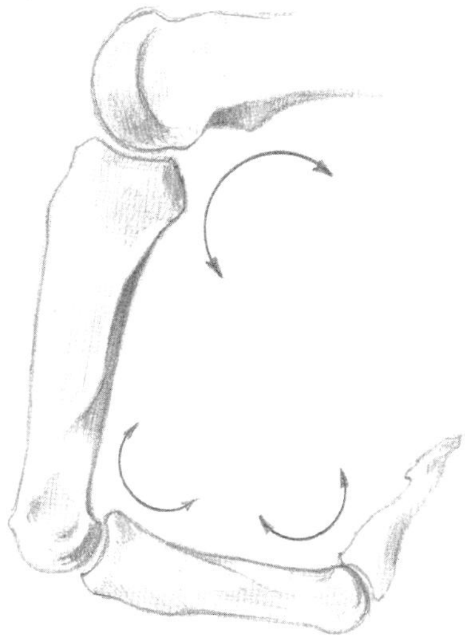

lateral aspect

The base of the little finger can be reached by the thumb.

The first phalanx is a third shorter than the metacarpal bones. The second phalanx is a third shorter than the first, and the third phalanx is a third shorter than the second.

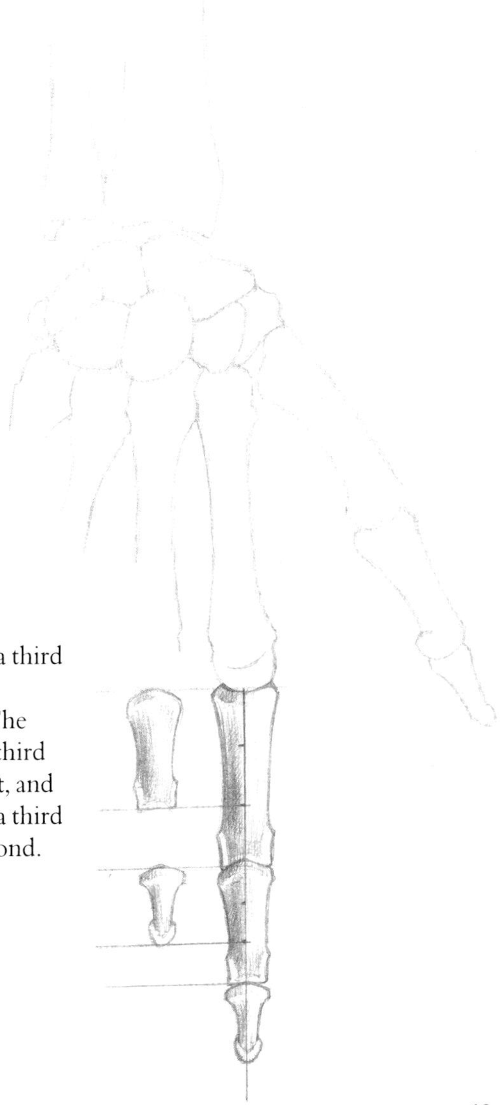

palmar aspect

Finger Movements
(cont.)

adduction

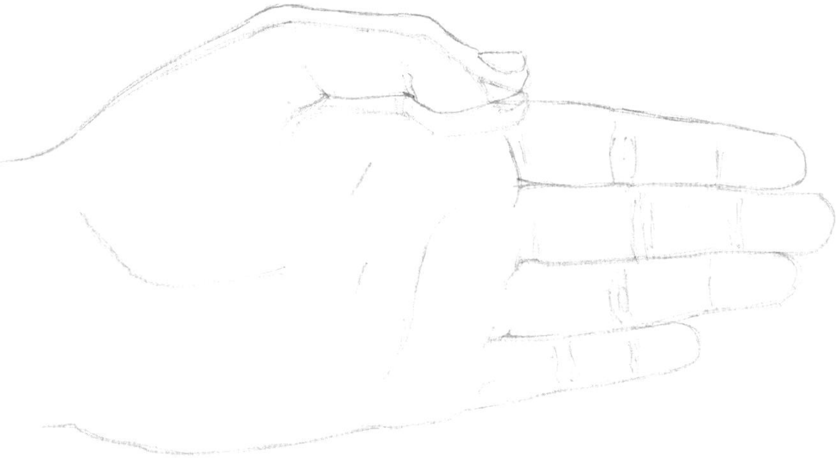

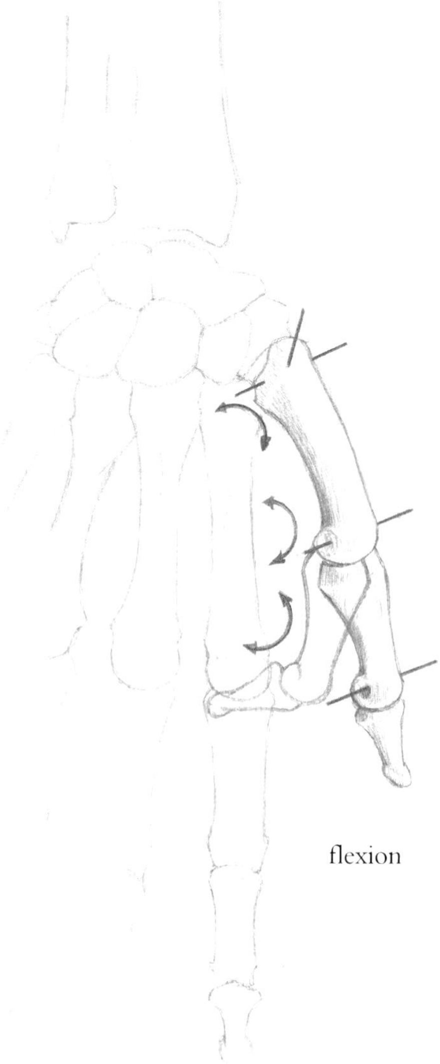

flexion

The thumb cannot be flexed to the same degree as the other fingers.

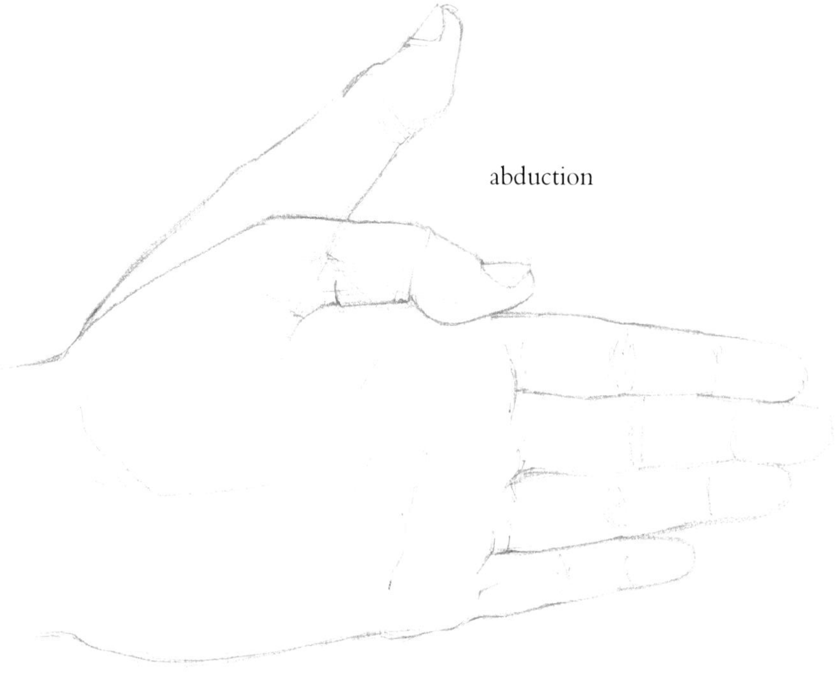

abduction

opposition

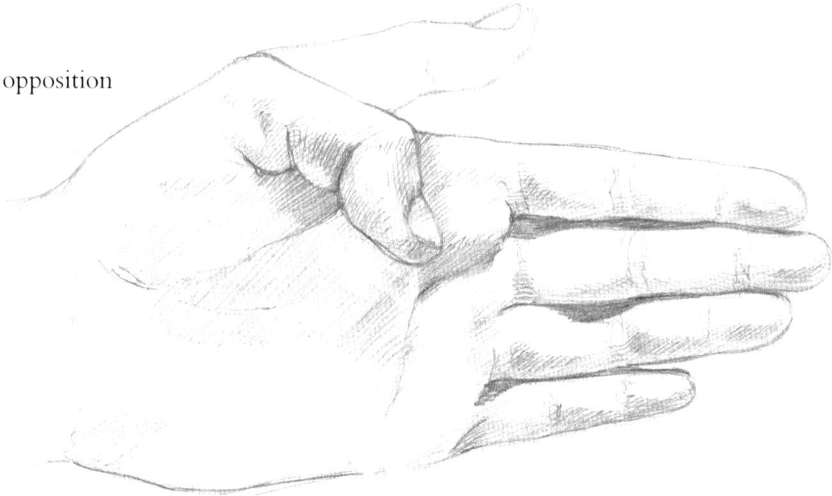

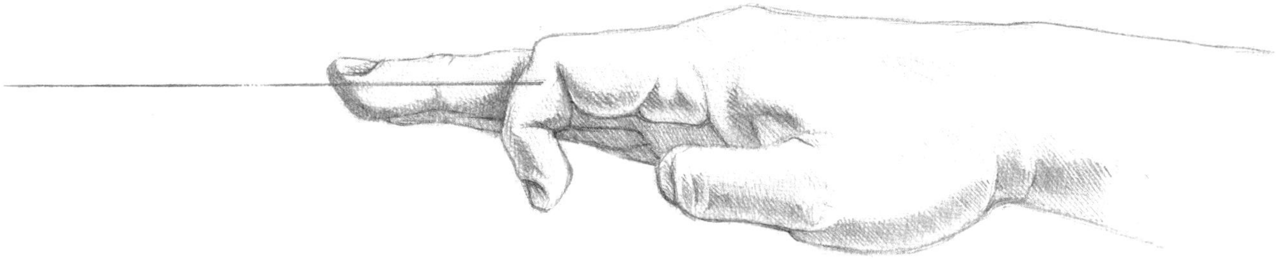

flexion

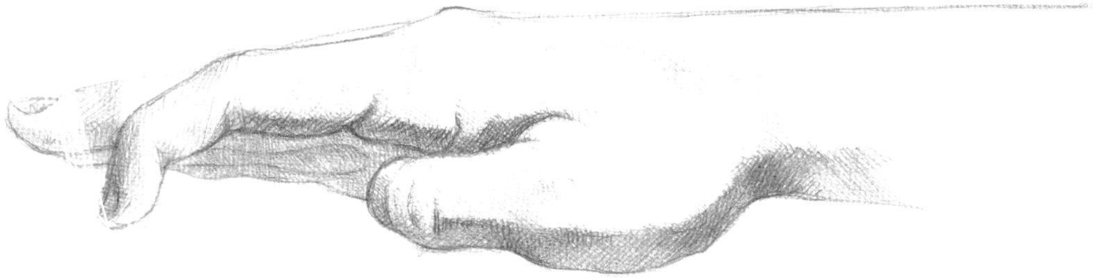

opposition

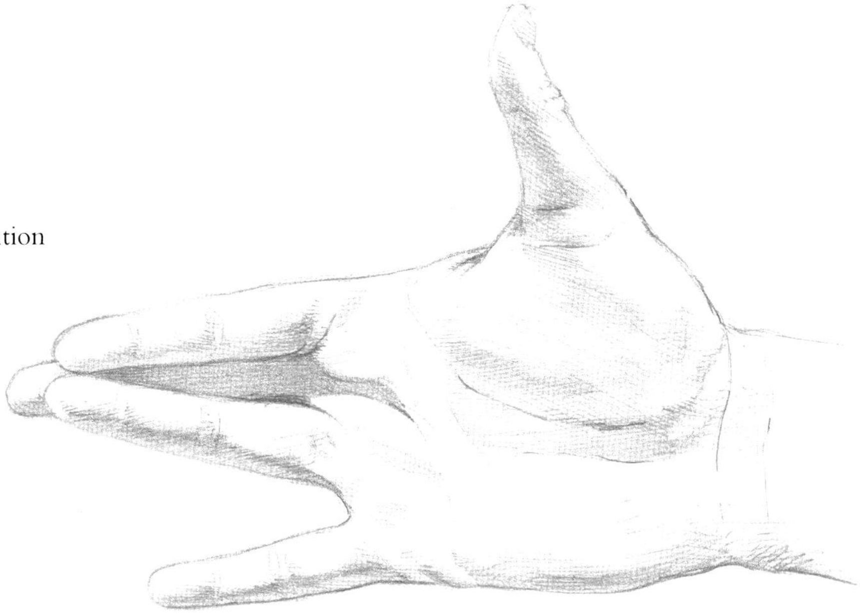

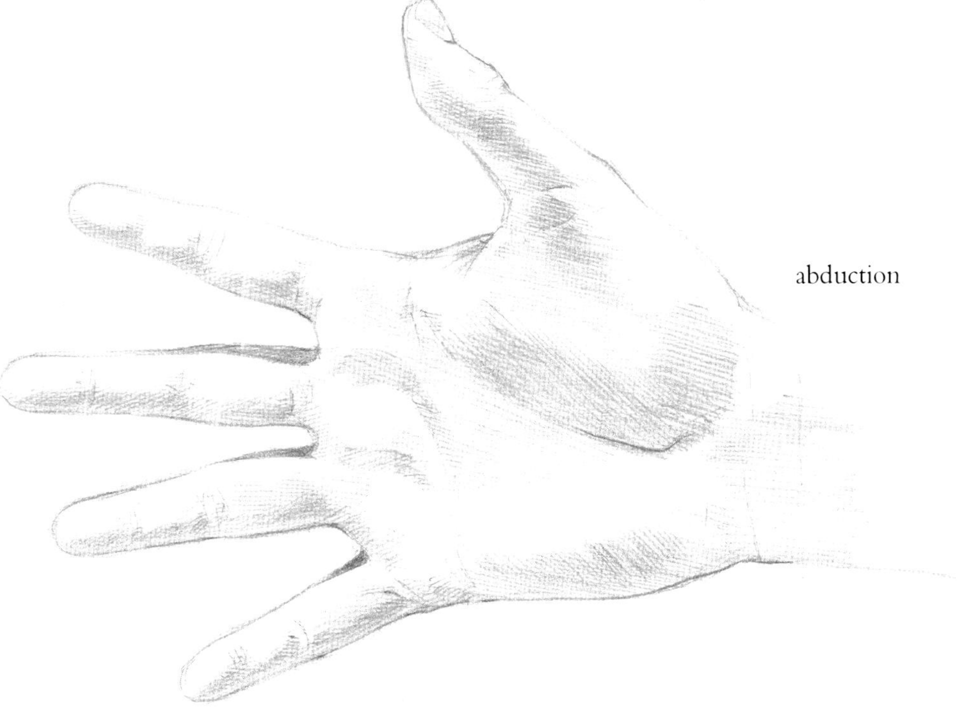

abduction

Fig. 44
The Muscles of the Neck and Shoulder

1 Digastricus (*16*)
2 Mylohyoid muscle (*18*)
3 Sternohoid (*12*)
4 anterior Scalenus (*8*)
5 Sternocleidomastoid (bicipital, *11*)
6 Omohyoideus (*13*)
7 Pectoralis major (*28*)
8 External oblique (*37*)
9 Anterior Serratus (*31*)
10 Head and neck parts of the striate muscle (*5*)
11 Trapezius (*20*)
12 Deltoid muscle (*41*)

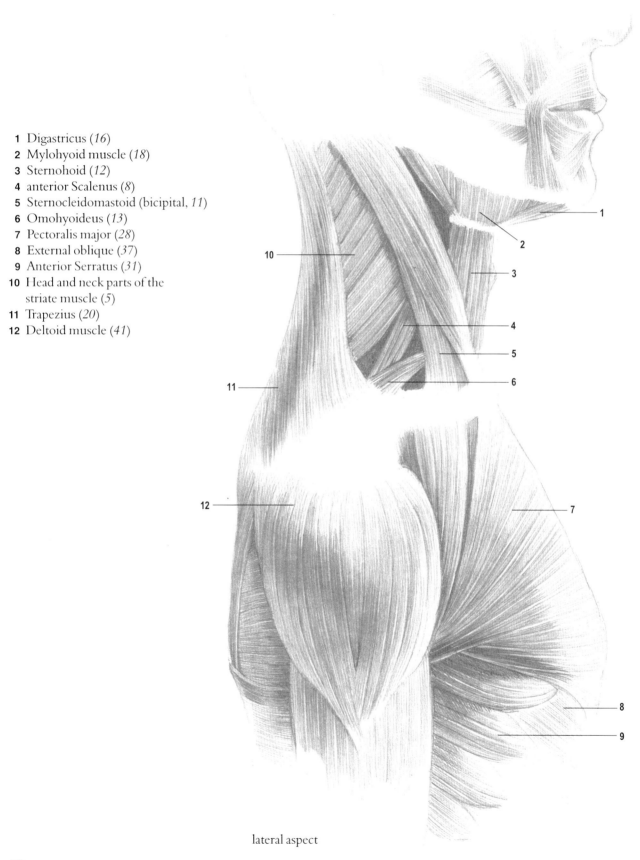

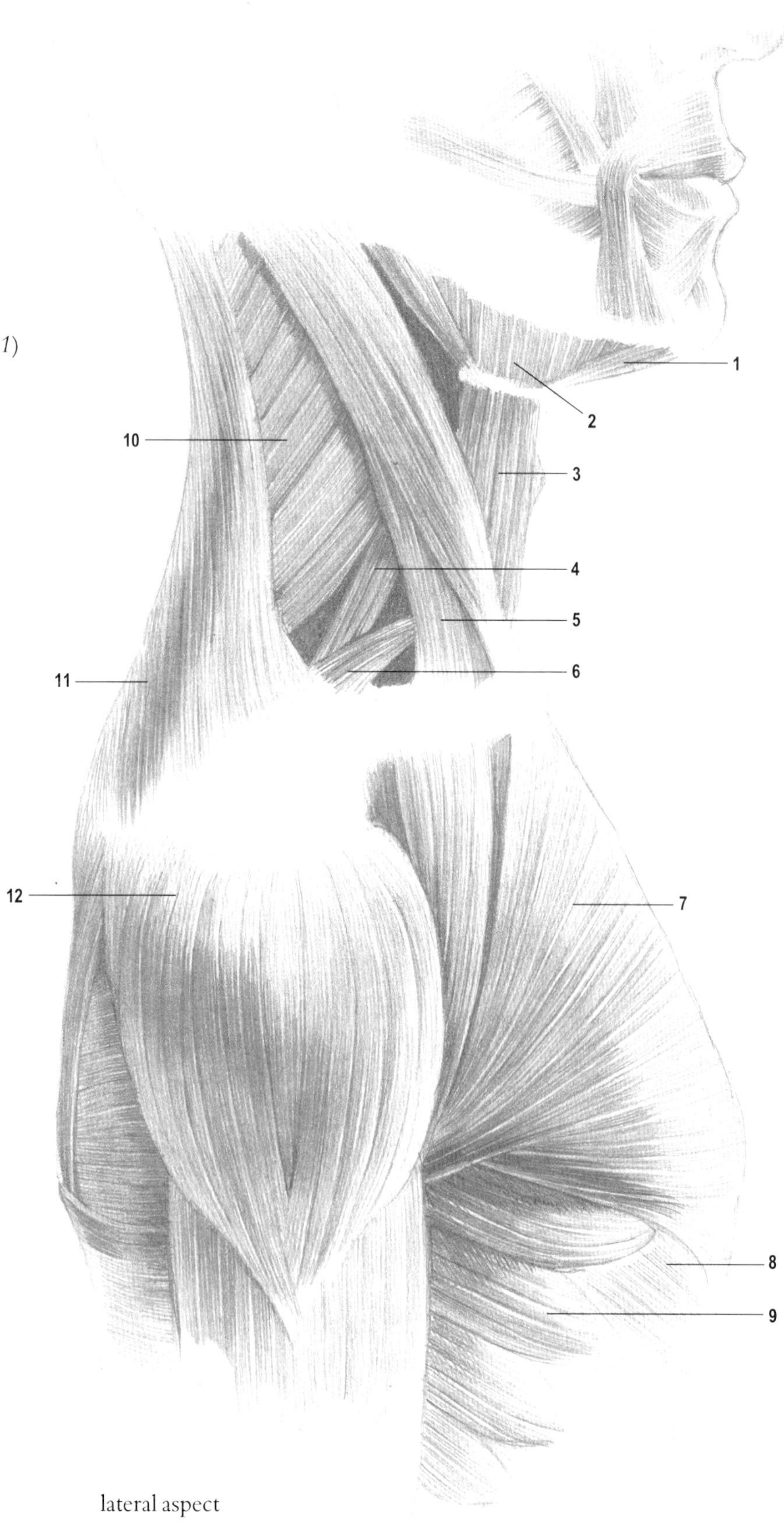

lateral aspect

Fig. 45
The Acromioclavicular and Trunk Muscles

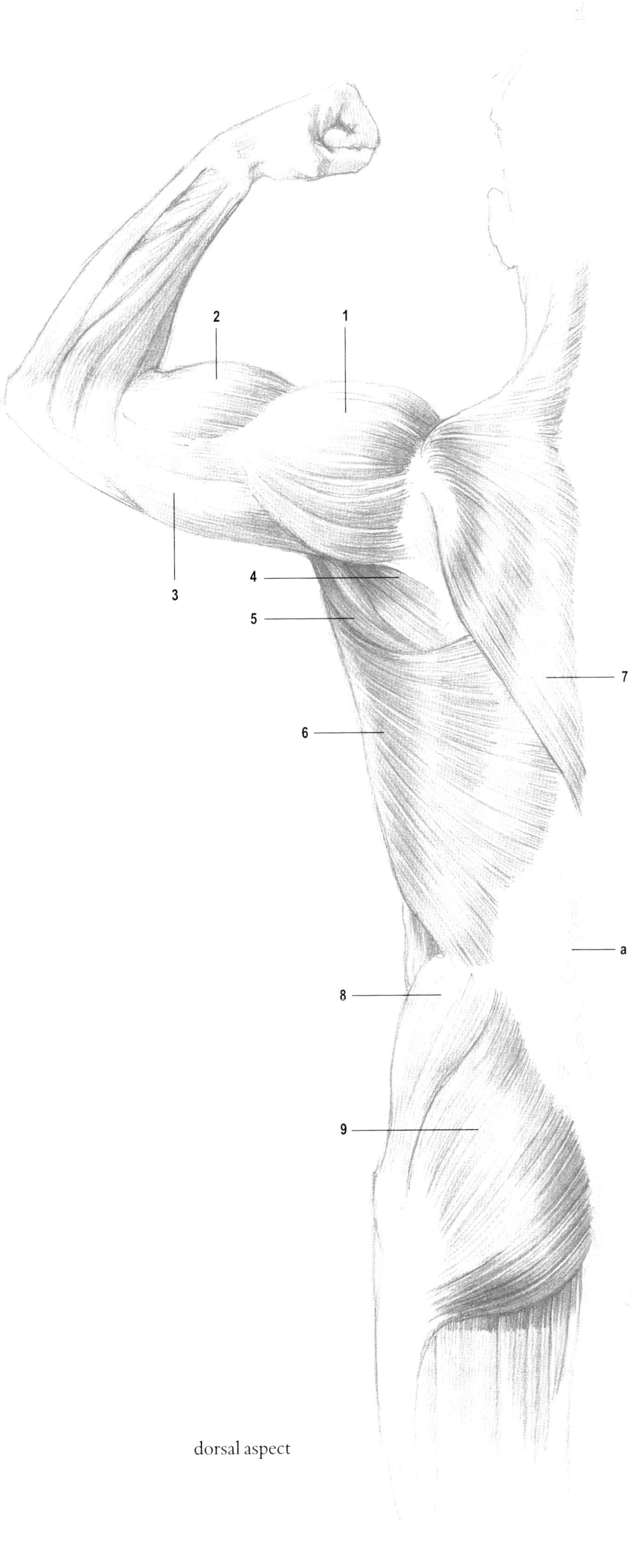

dorsal aspect

1 Deltoid muscle (*41*)
2 Biceps (*47*)
3 Triceps (*50*)
4 Teres minor (*44*)
5 Teres major (*45*)
6 Latissimus dorsi (*21*)
7 Trapezius (*20*)
8 Gluteus medius (*82*)
9 Gluteus maximus (*81*)

a Thoracolumbar fascia

Fig. 46
The Acromioclavicular Bones and Muscles

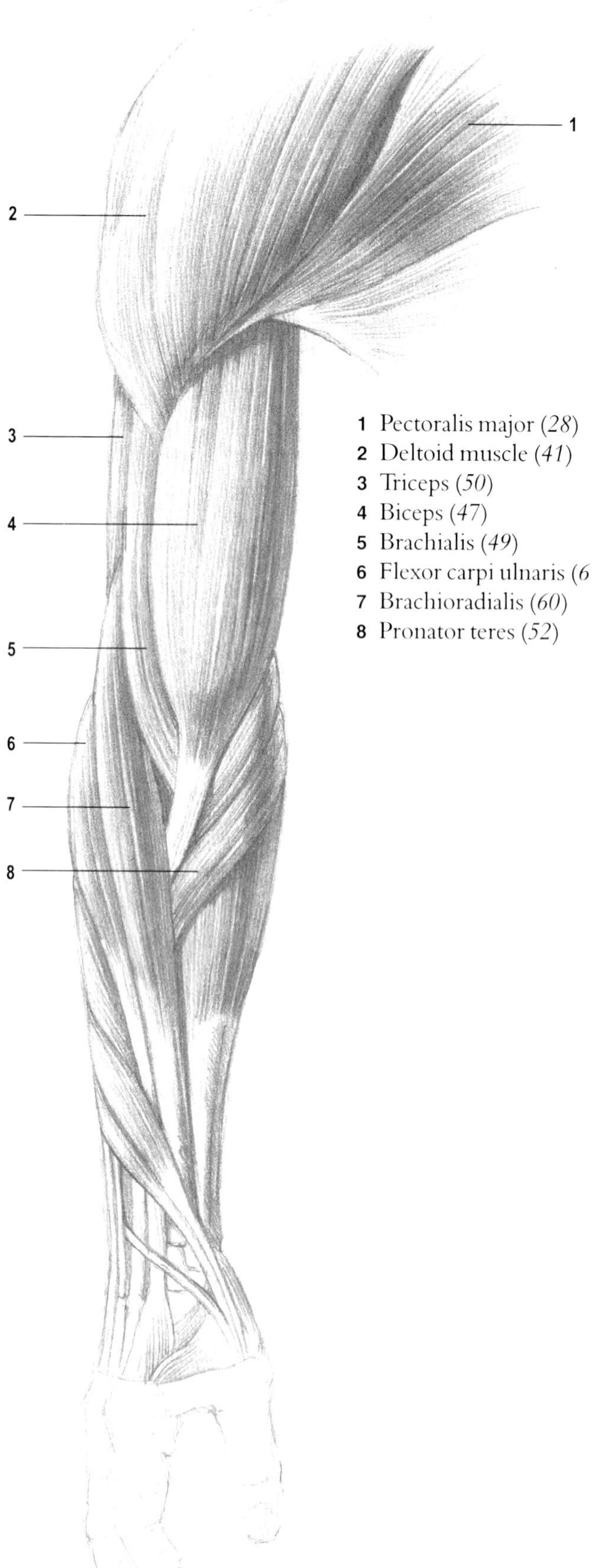

1 Pectoralis major (*28*)
2 Deltoid muscle (*41*)
3 Triceps (*50*)
4 Biceps (*47*)
5 Brachialis (*49*)
6 Flexor carpi ulnaris (*61*)
7 Brachioradialis (*60*)
8 Pronator teres (*52*)

ventral aspect

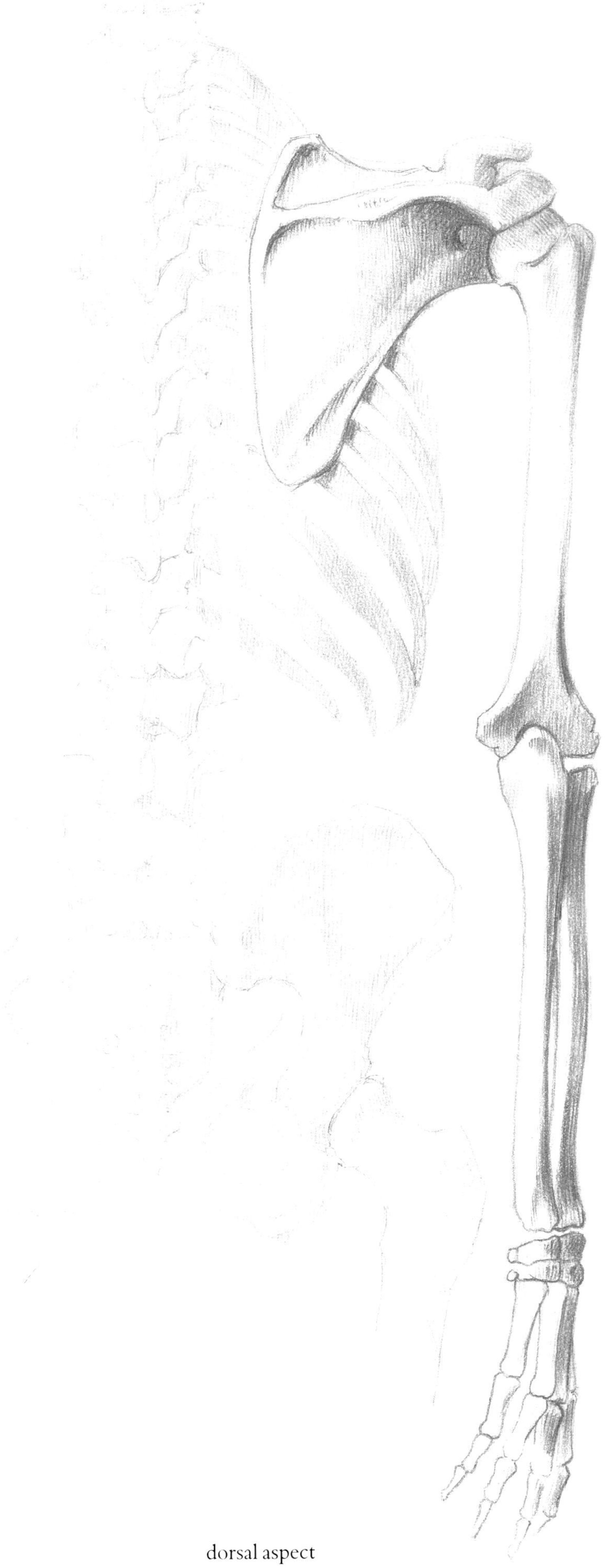

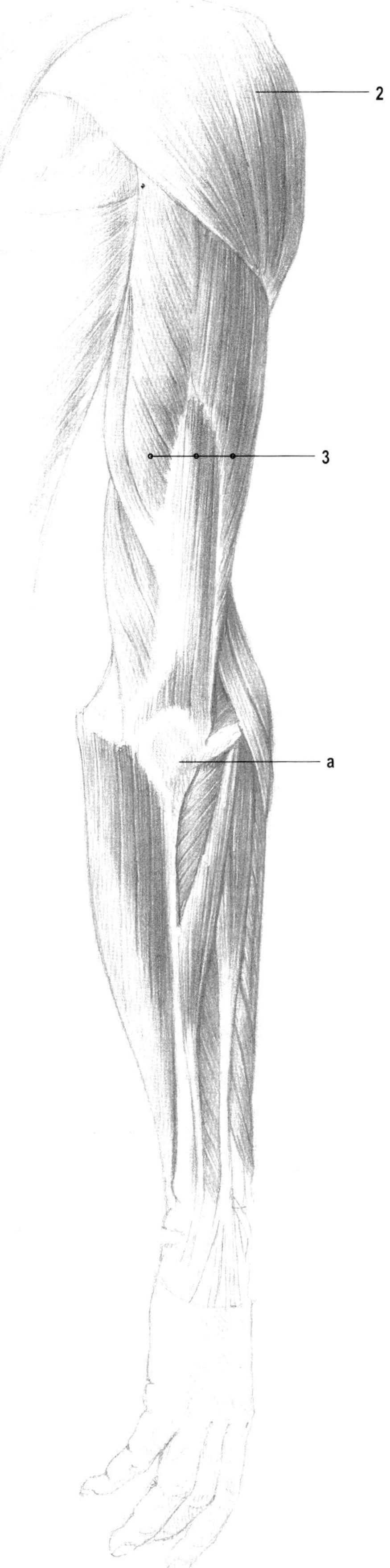

1 Trapezius (*20*)
2 Deltoid muscle (*41*)
3 Triceps (*50*)

a Olecranon

dorsal aspect

Fig. 47
The Scapular Muscles

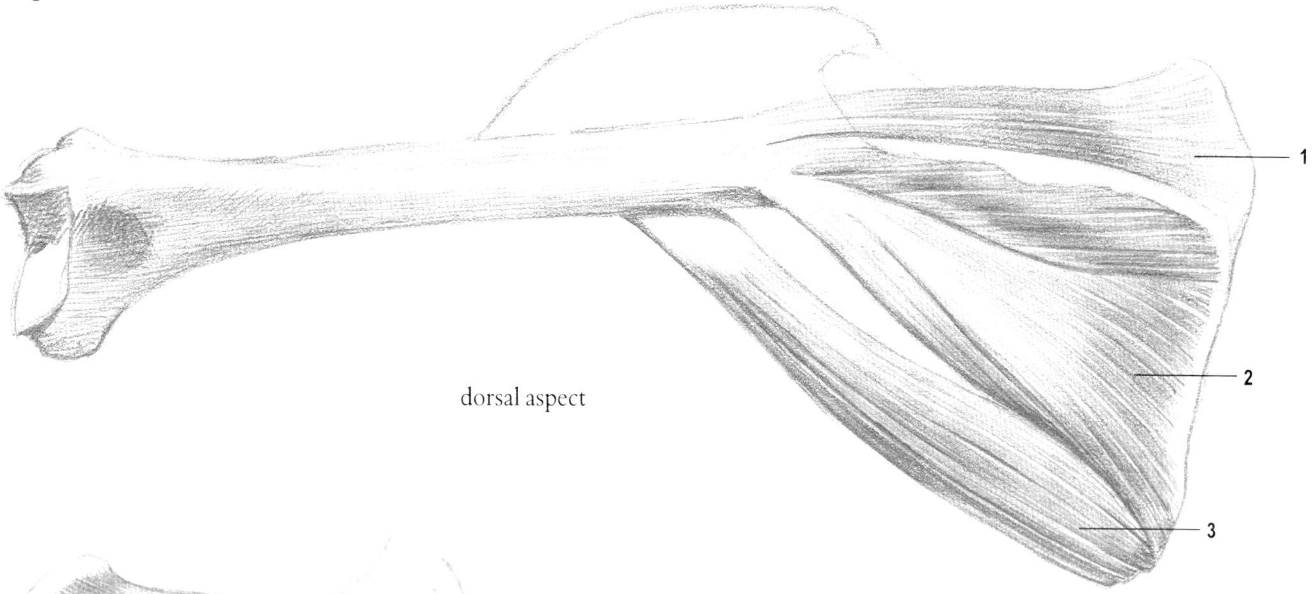

dorsal aspect

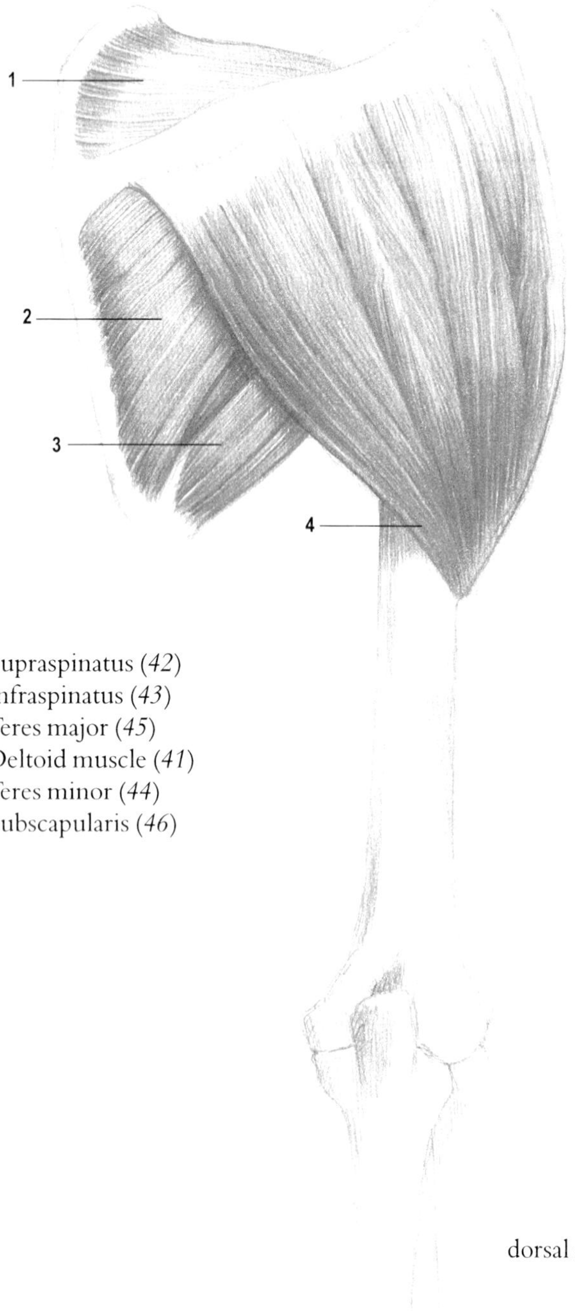

1 Supraspinatus (*42*)
2 Infraspinatus (*43*)
3 Teres major (*45*)
4 Deltoid muscle (*41*)
5 Teres minor (*44*)
6 Subscapularis (*46*)

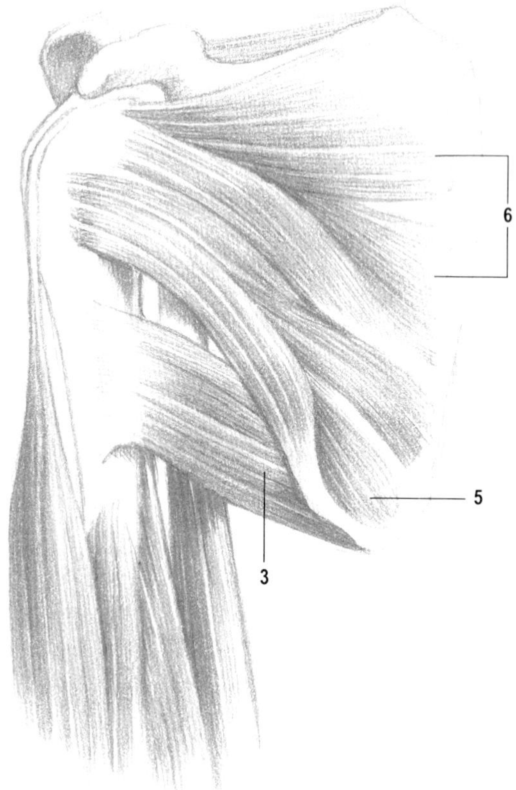

dorsal aspect

ventral aspect

Fig. 48
The Deltoid Muscle
(M. deltoideus, *41*)

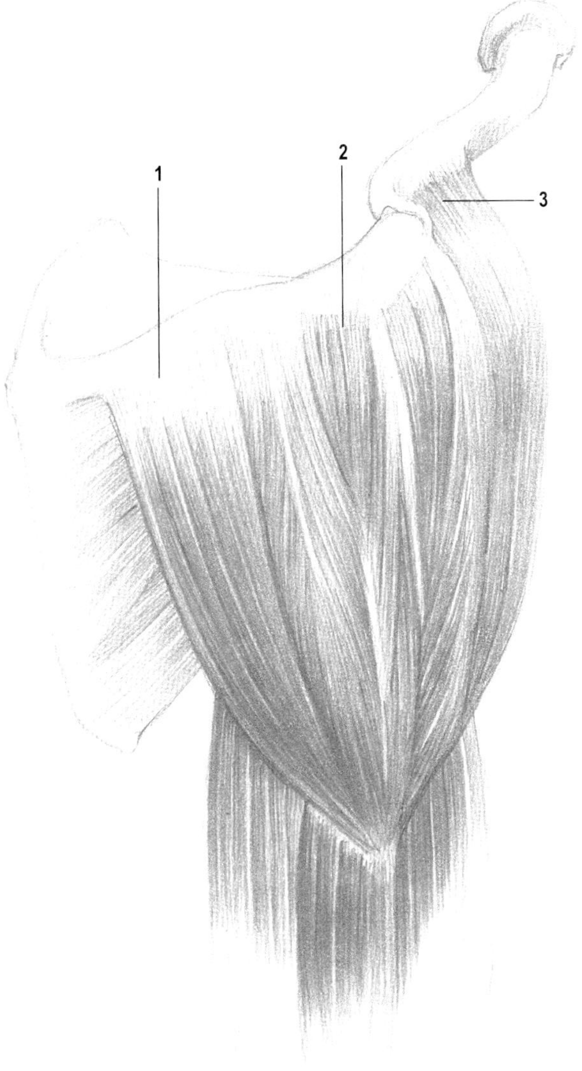

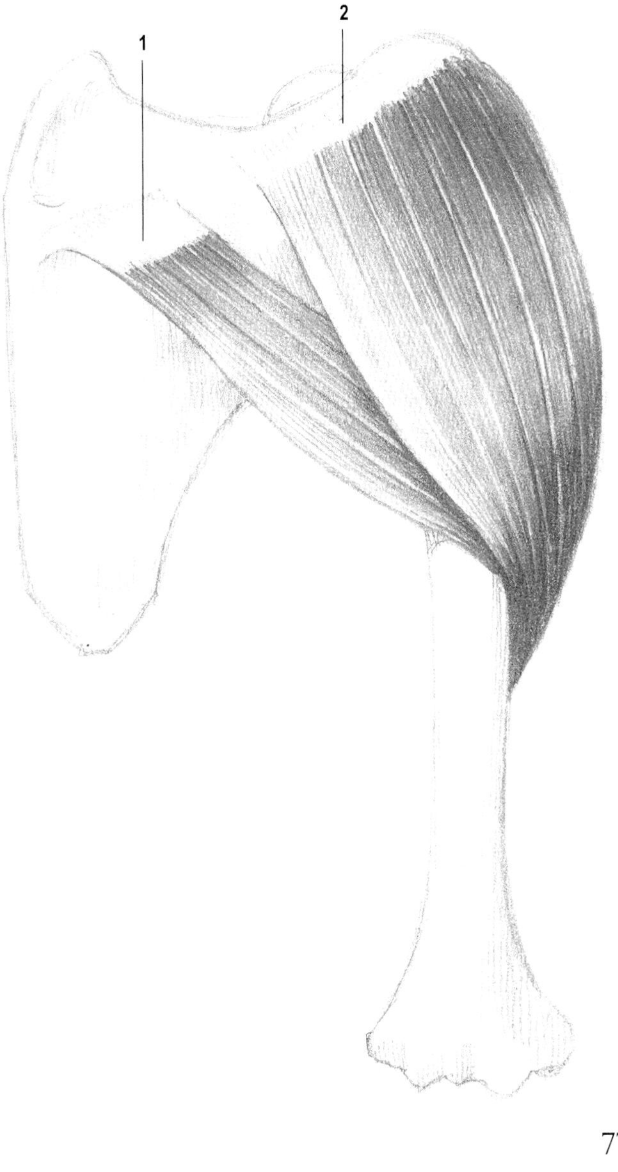

1 Pars scapularis
2 Pars acromialis
3 Pars clavicularis

dorsal aspect

The Deltoid Muscle
(cont.)

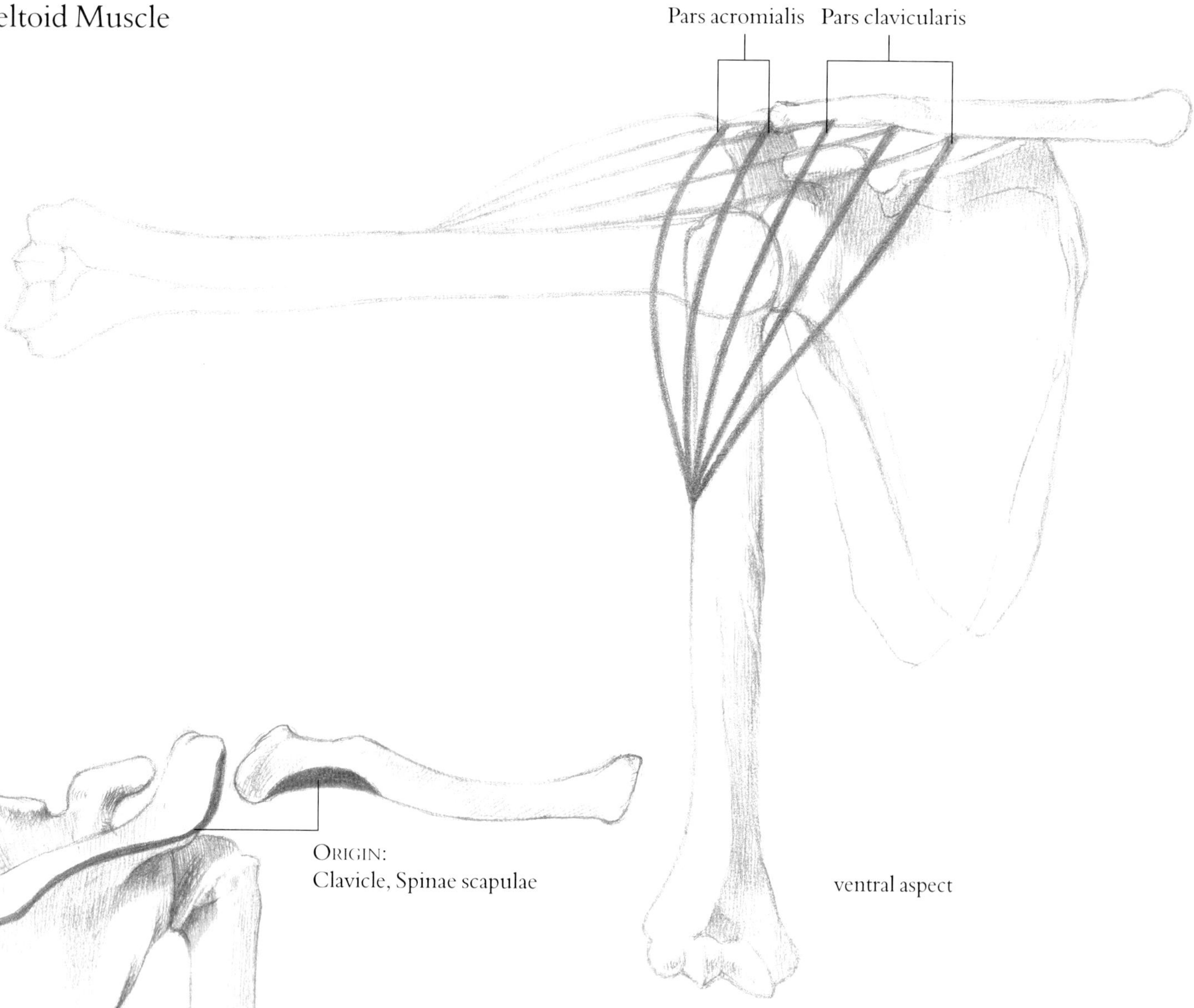

ventral aspect

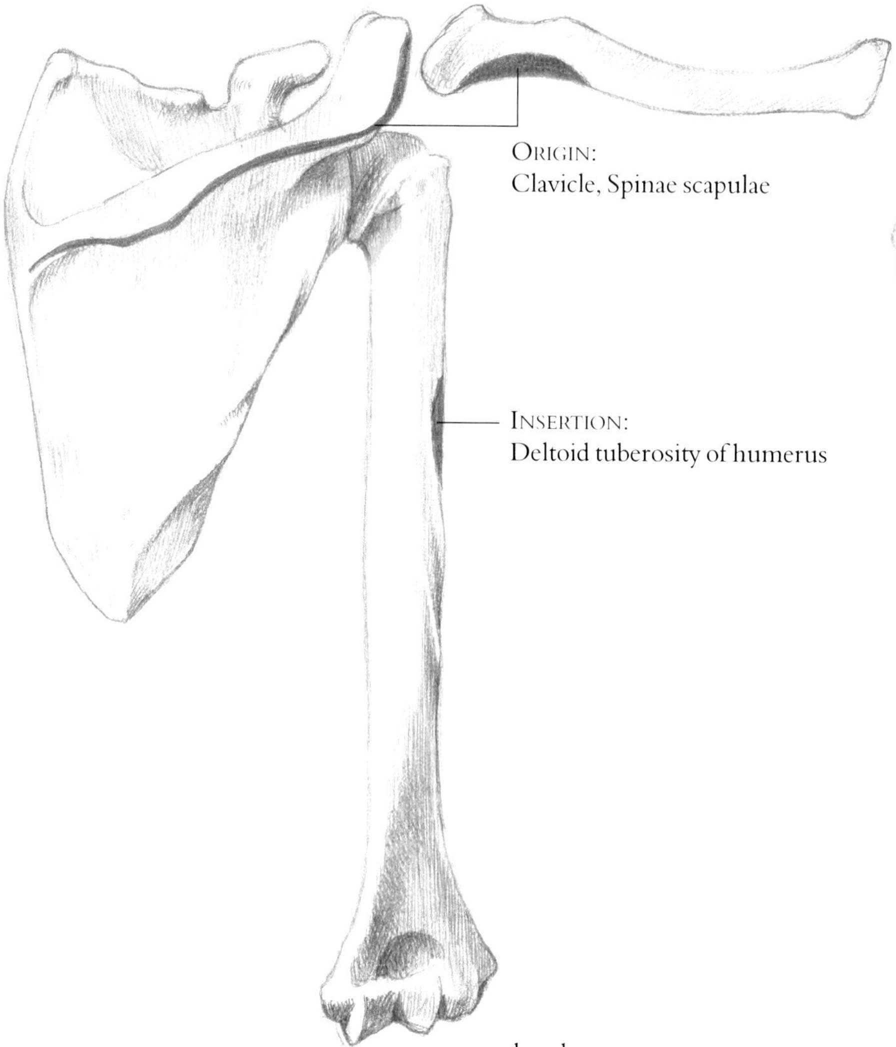

dorsal aspect

The muscle gives the shoulder its form.

1 Pars scapularis
2 Pars acromialis
3 Pars clavicularis

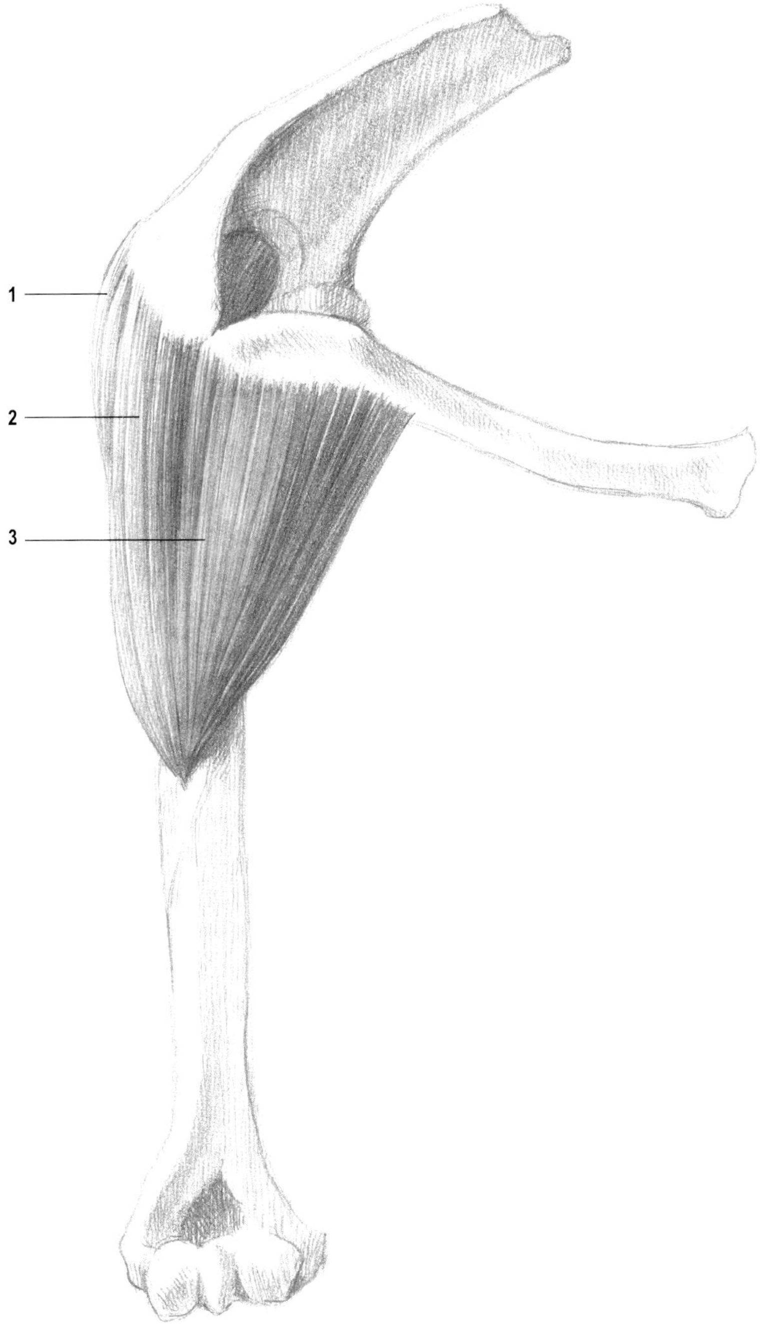

ventral aspect

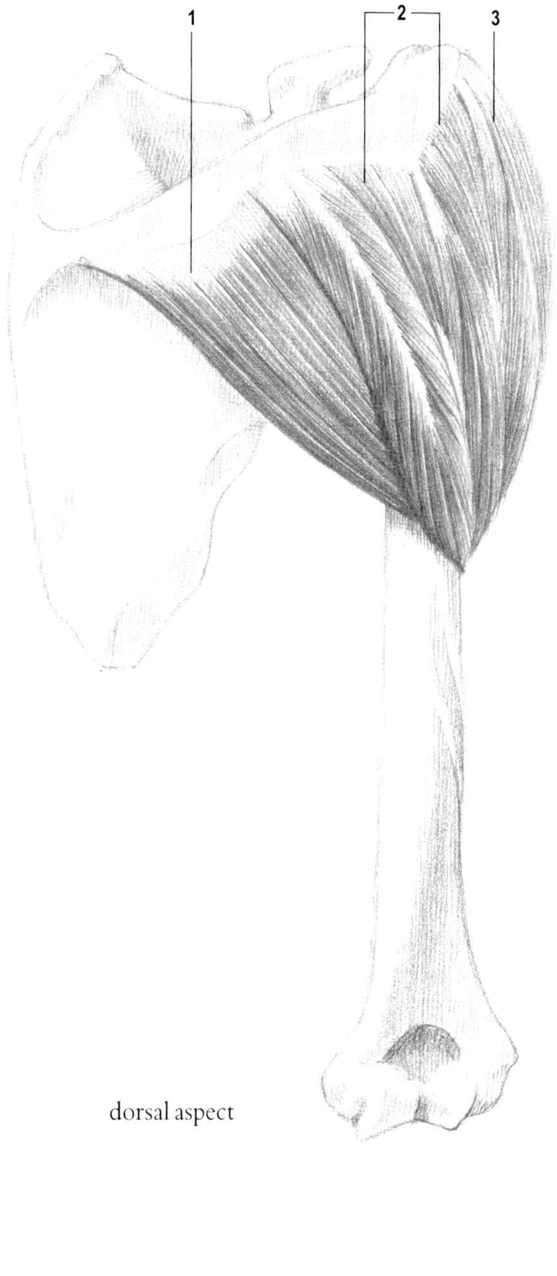

dorsal aspect

The Deltoid Muscle
(cont.)

FUNCTION:
Abducts the clavicular humerus (to 90 degrees); abducts and pulls backwards, rotates outwards and fixes the shoulder joint

ventral aspect

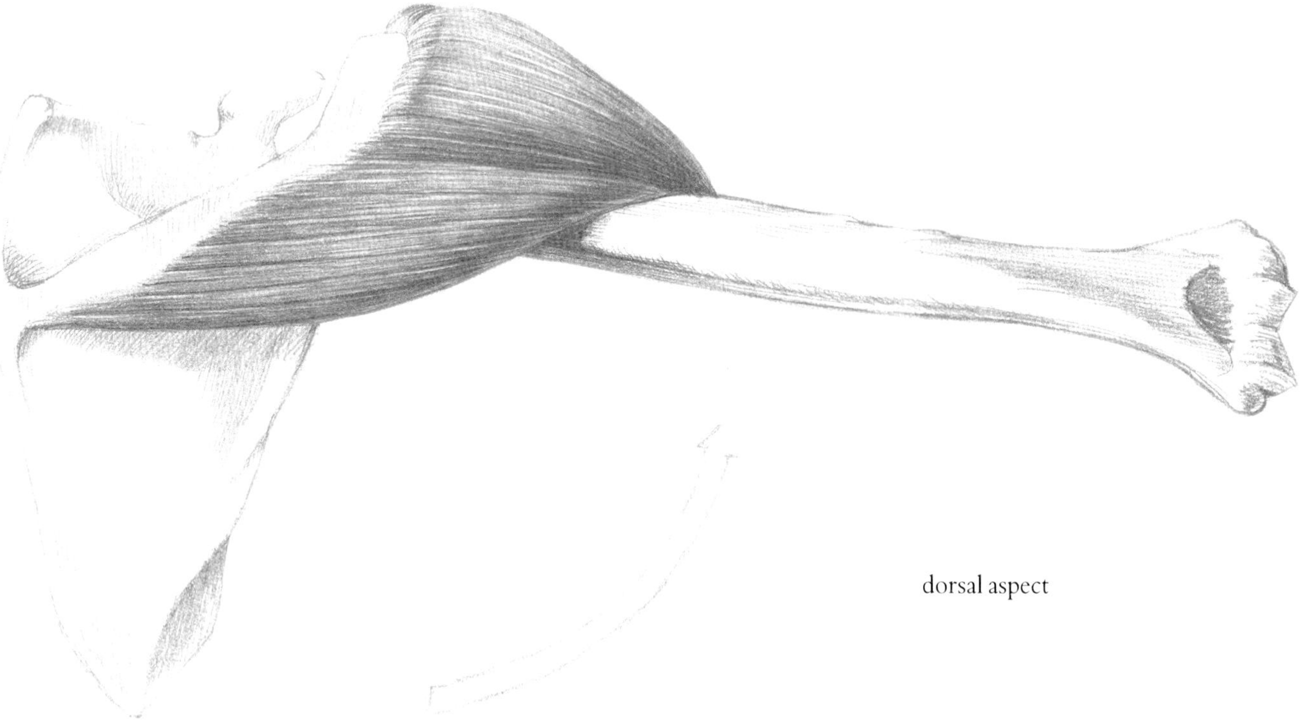

dorsal aspect

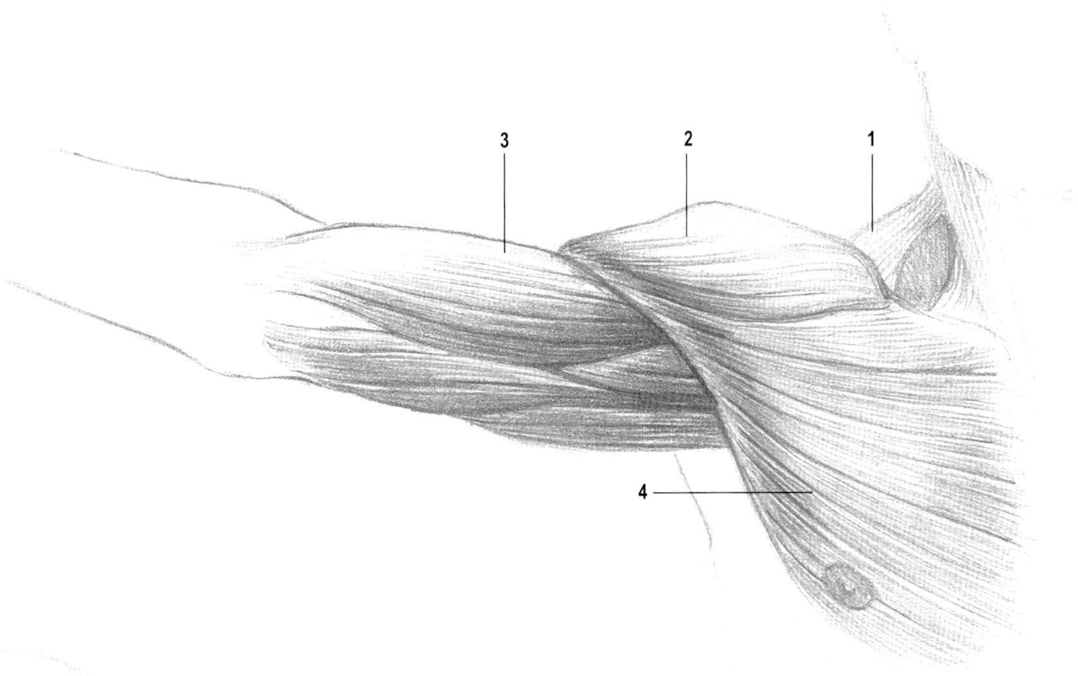

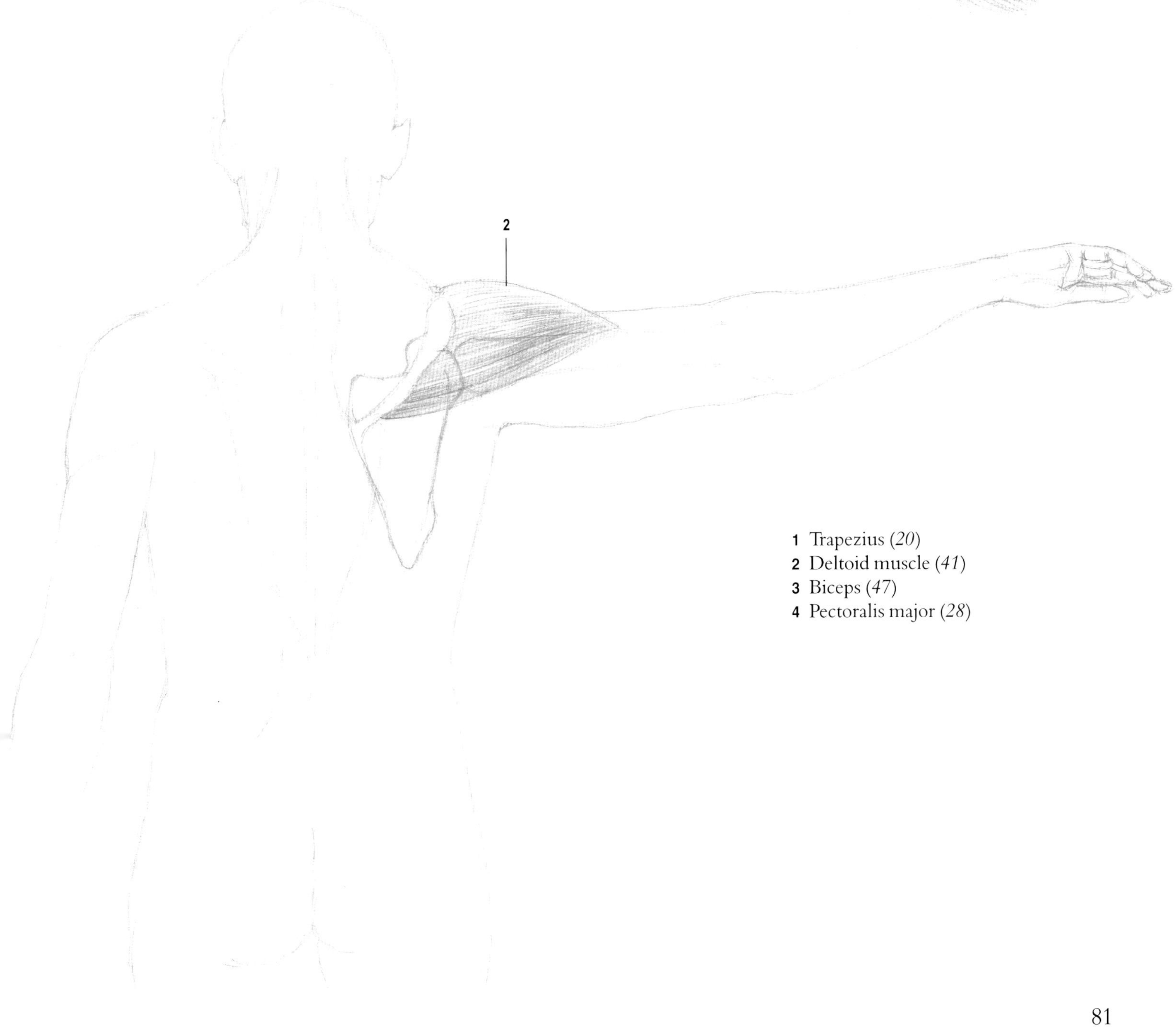

1 Trapezius (*20*)
2 Deltoid muscle (*41*)
3 Biceps (*47*)
4 Pectoralis major (*28*)

Fig. 49
The Supraspinous Muscle
(M. supraspinatus, *42*)

ORIGIN:
In the fossa of the spinae scapulae

INSERTION:
Below the acromion on the tuberculum majus through a tendon which adheres to the articular capsule

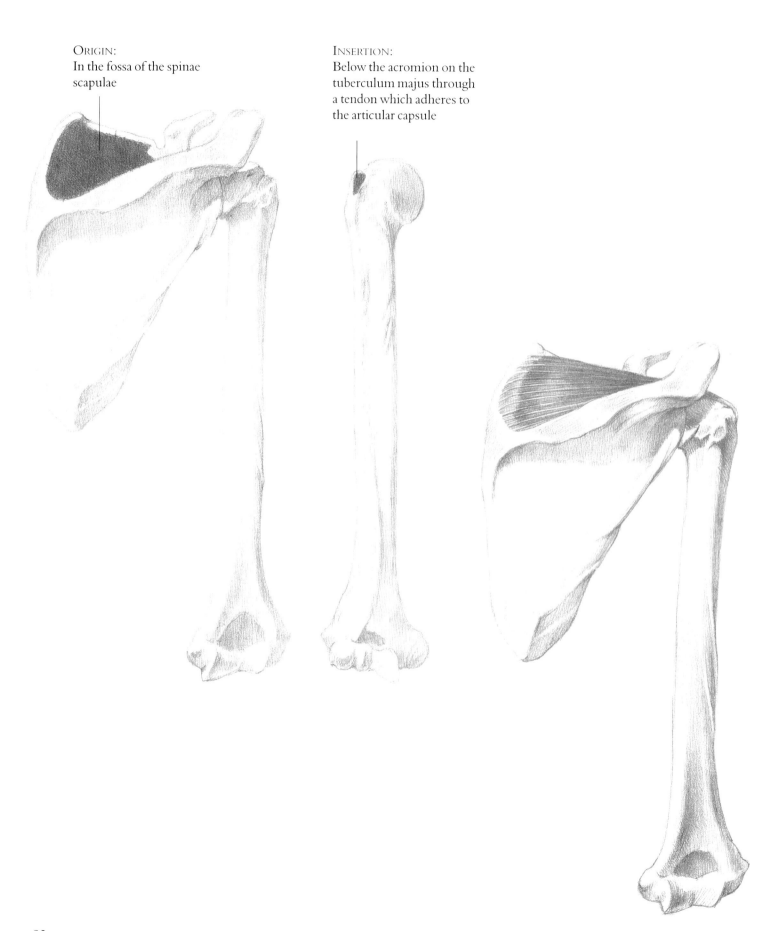

FUNCTION:
Rotates the upper arm

FUNCTION:
Abducts the humerus (to 90 degrees)

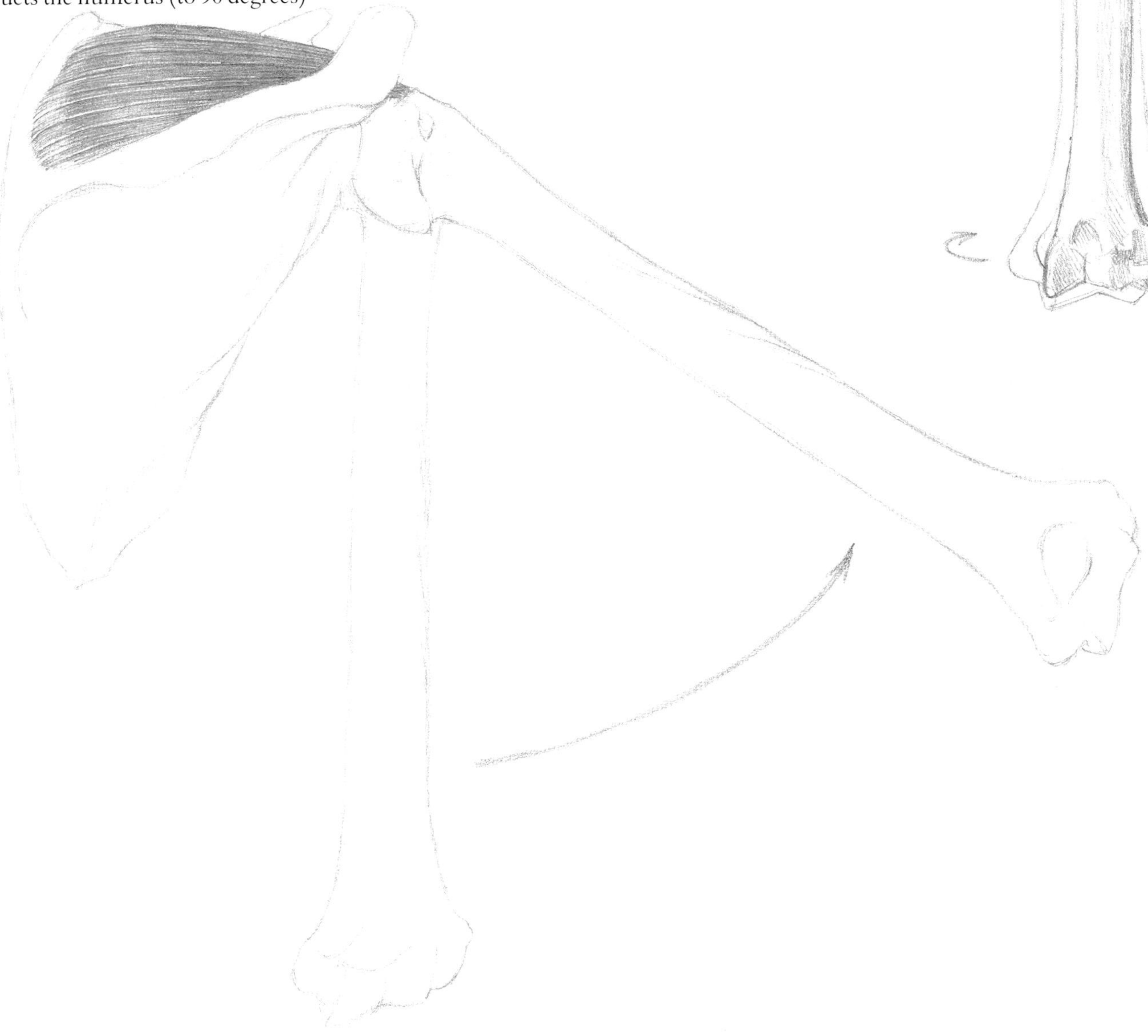

Fig. 50
The Infraspinous Muscle
(M. infraspinatus, *43*)

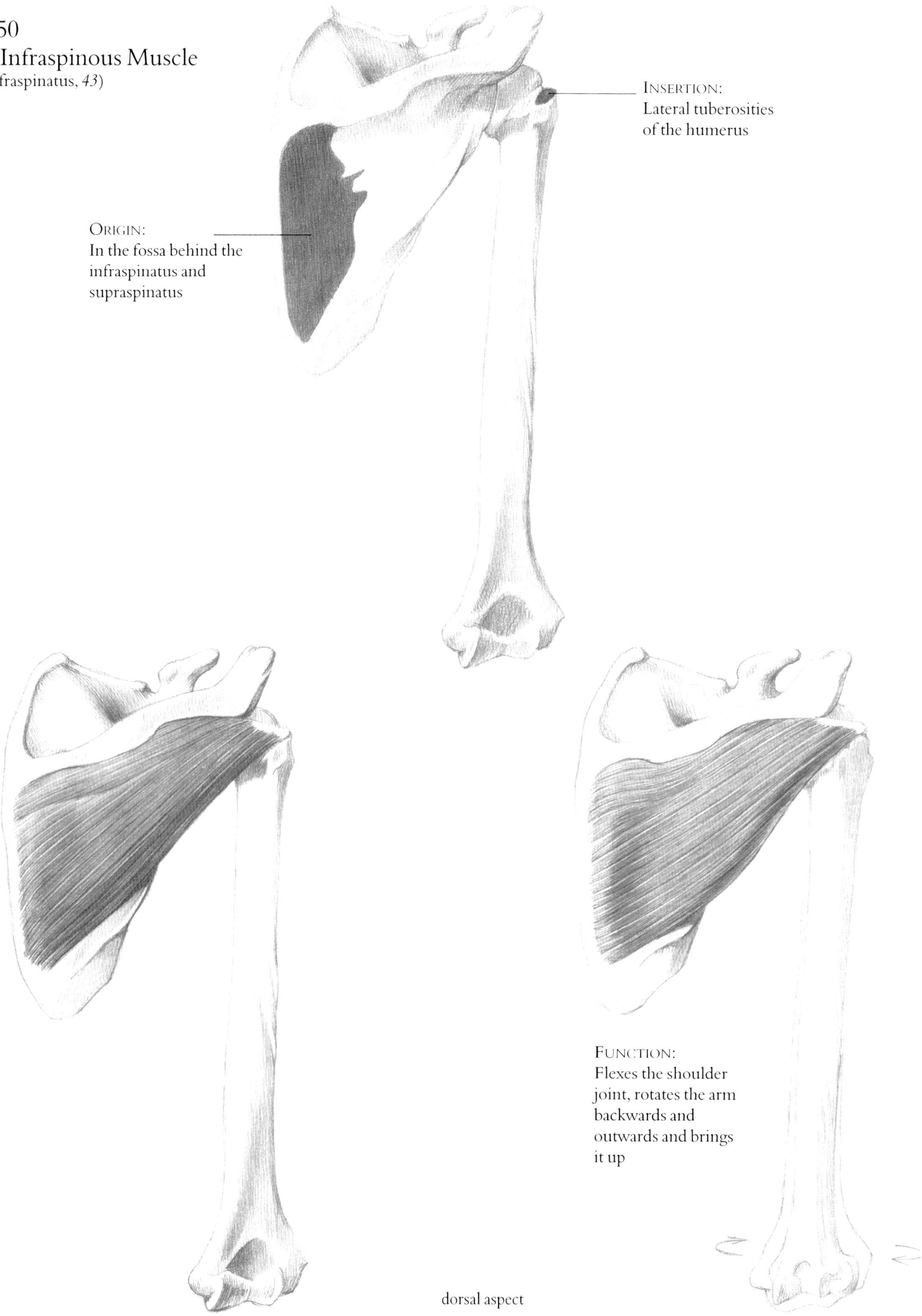
INSERTION:
Lateral tuberosities of the humerus
ORIGIN:
In the fossa behind the infraspinatus and supraspinatus
FUNCTION:
Flexes the shoulder joint, rotates the arm backwards and outwards and brings it up

dorsal aspect

Fig. 51
The Teres Major, *45*

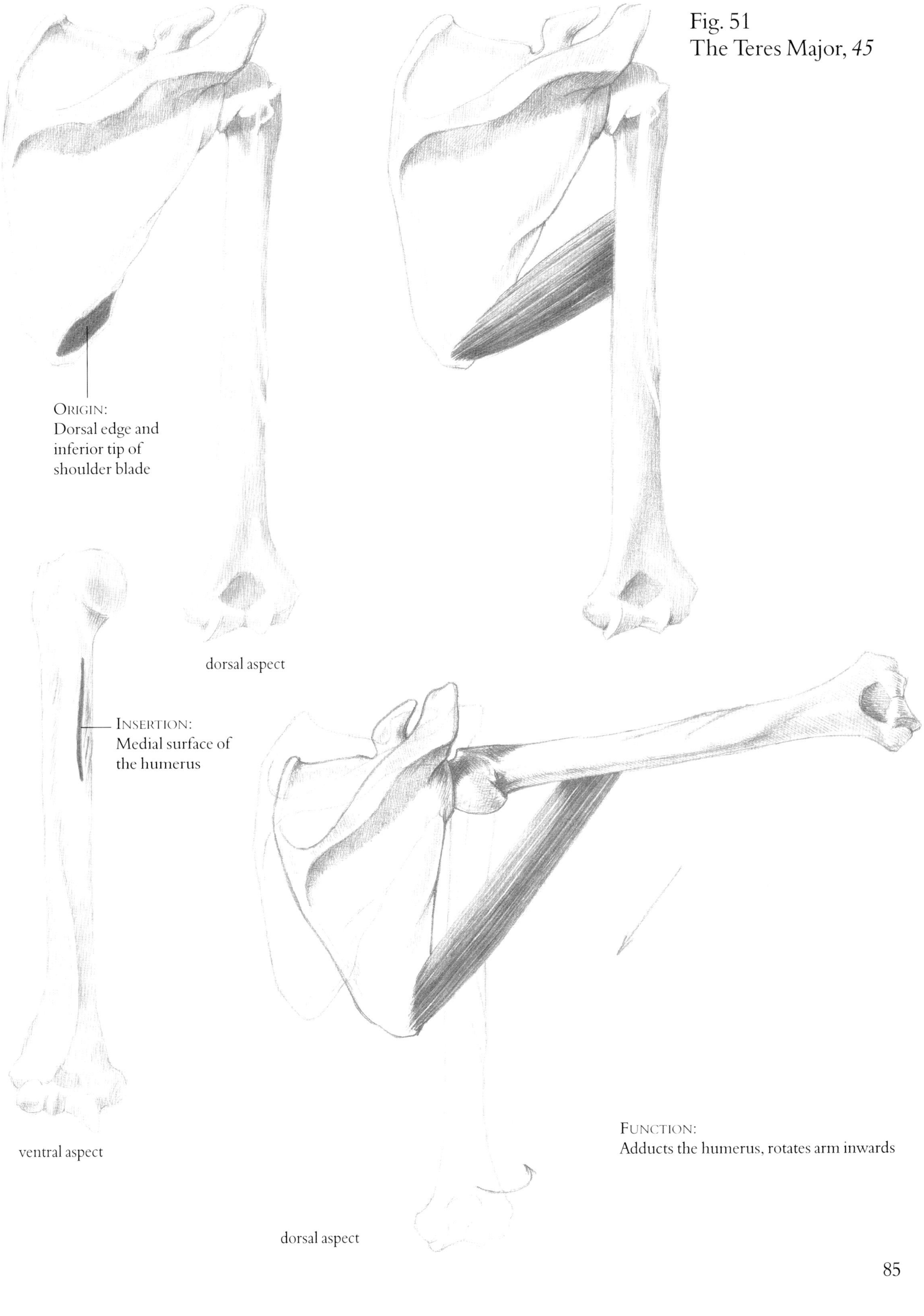

Fig. 52
The Teres Minor, *44*

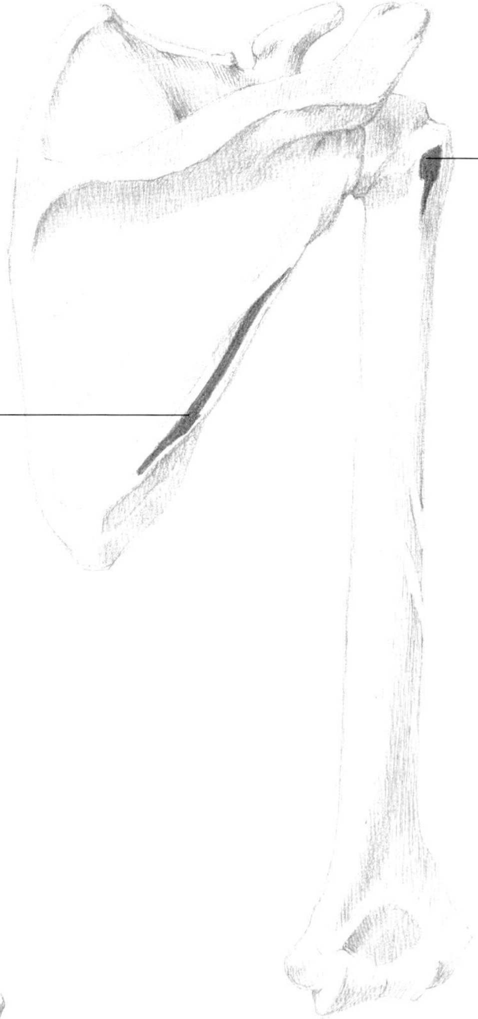

dorsal aspect

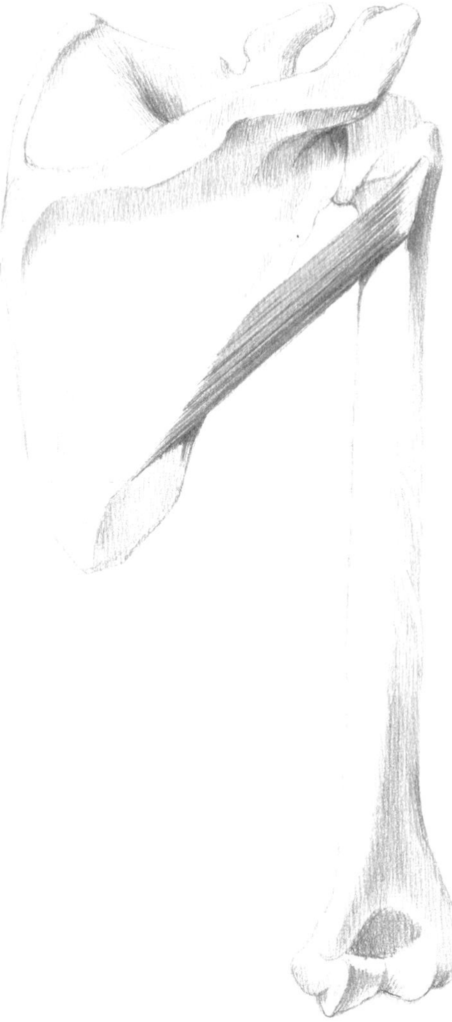

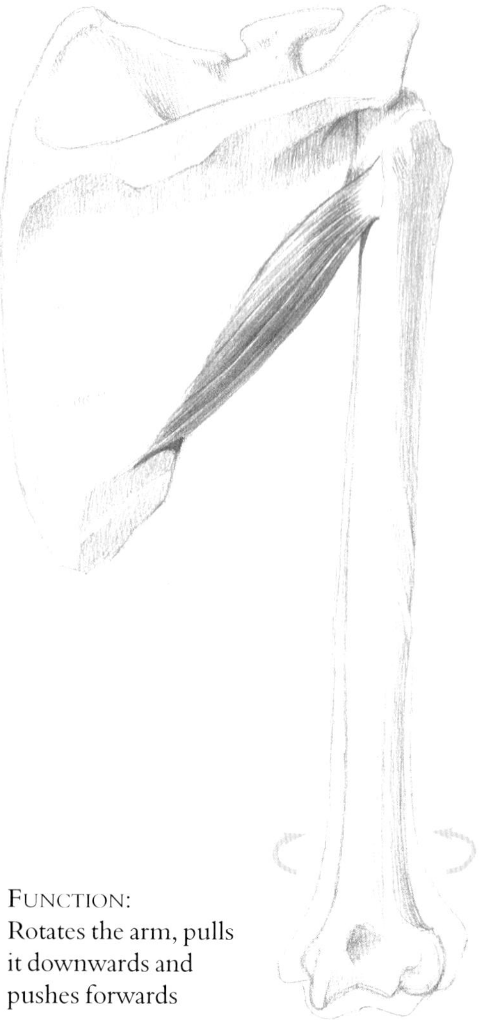

Fig. 53
The Subscapular Muscle
(M. subscapularis, *46*)

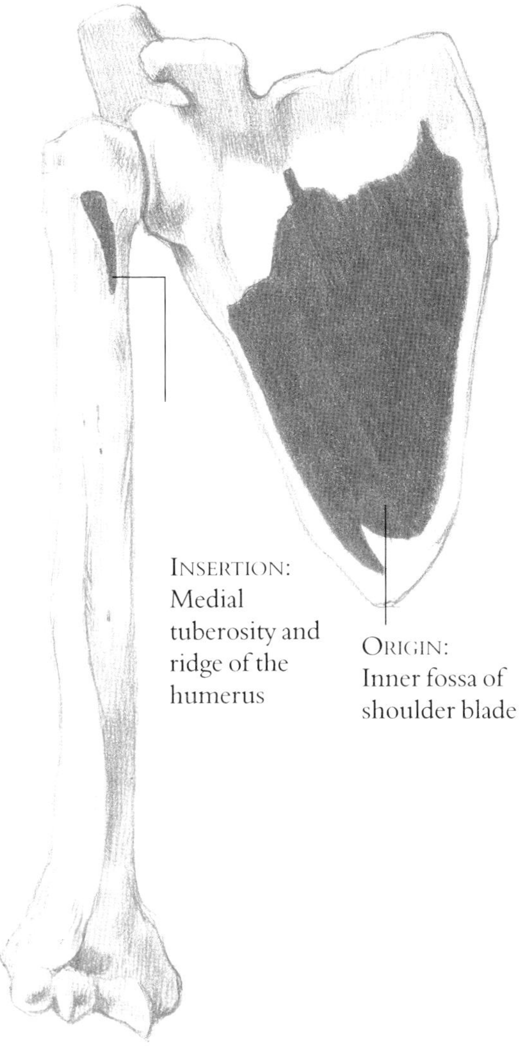

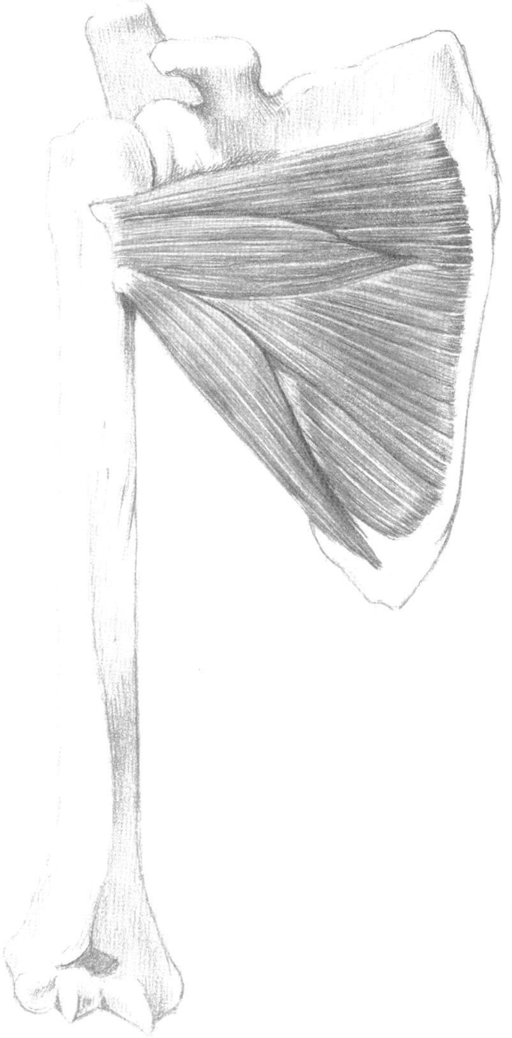

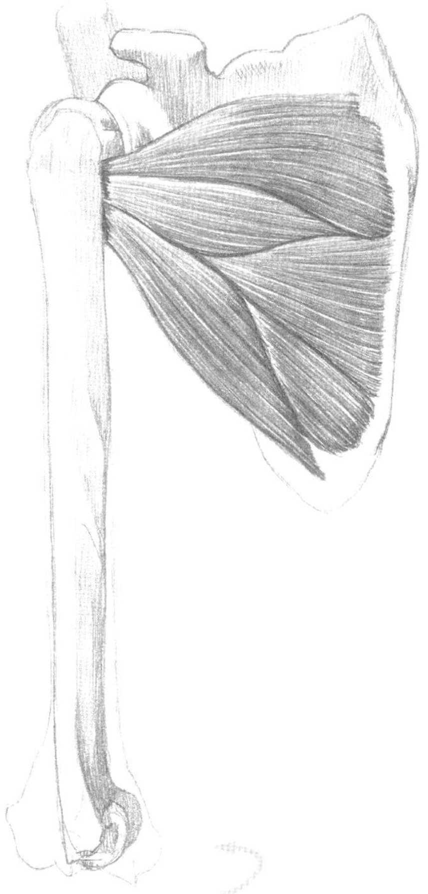

ventral aspect

FUNCTION:
Rotates the arm inwards, extends and pushes it forward

Fig. 54
The Coracobrachial Muscle
(M. coracobrachialis, *48*)

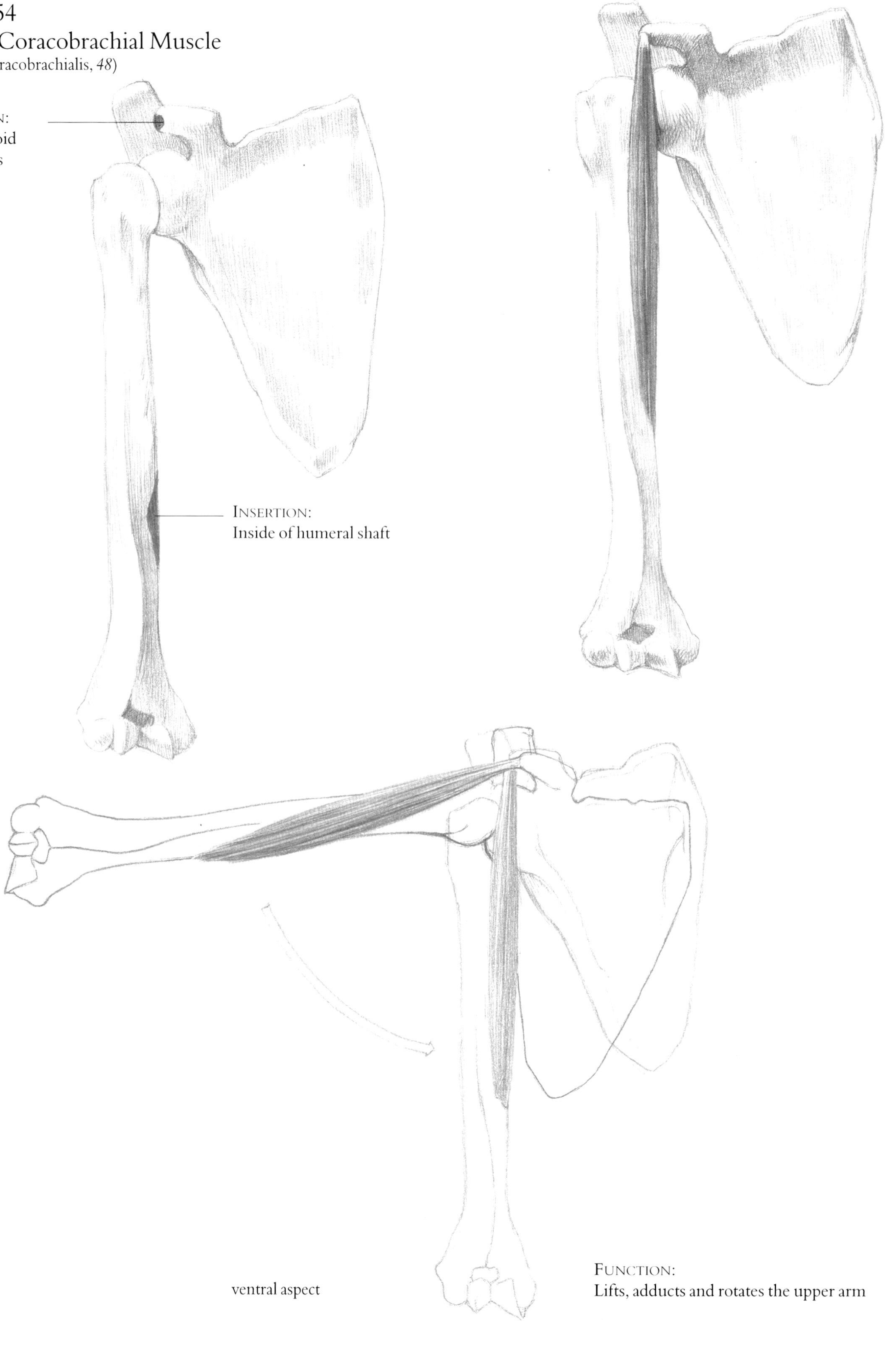

ventral aspect

FUNCTION:
Lifts, adducts and rotates the upper arm

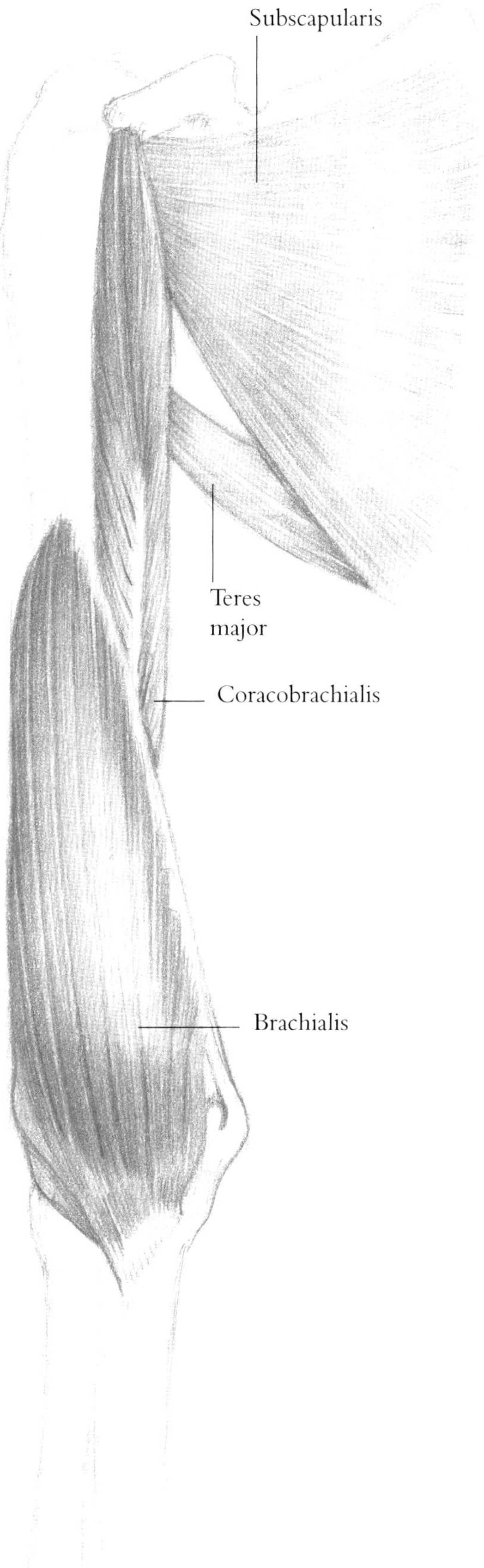

ventral aspect

THE MUSCLES OF THE ELBOW JOINT

Fig. 55
The Muscles of the Humerus

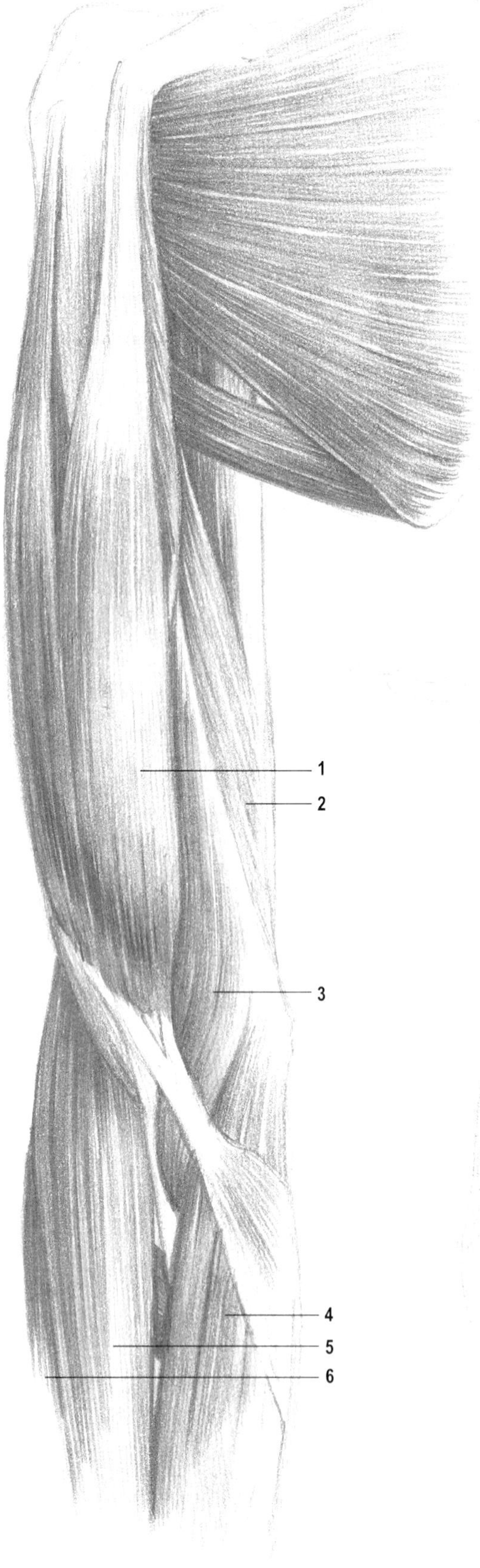

ventral aspect

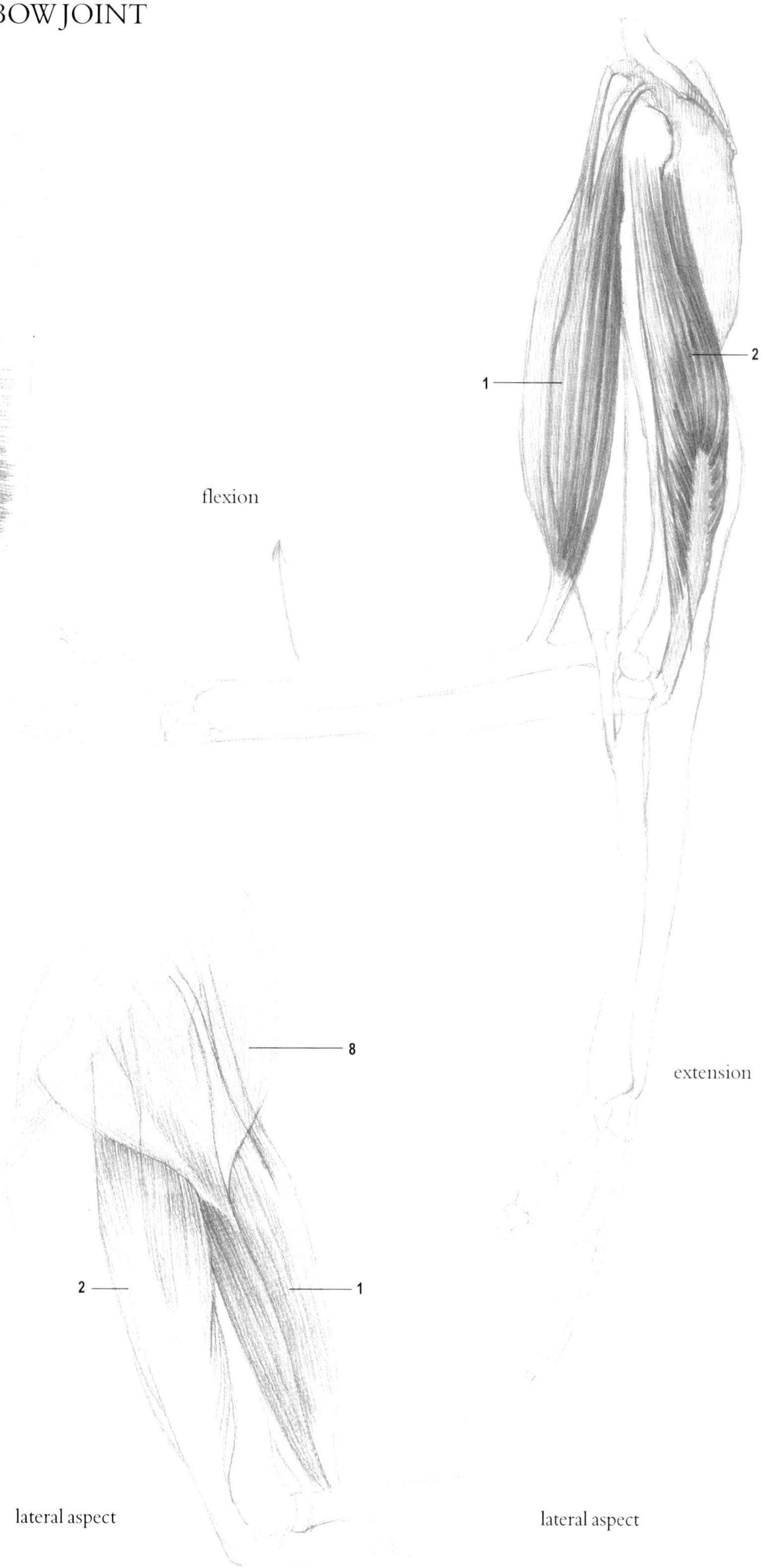

lateral aspect

lateral aspect

middle muscle layer

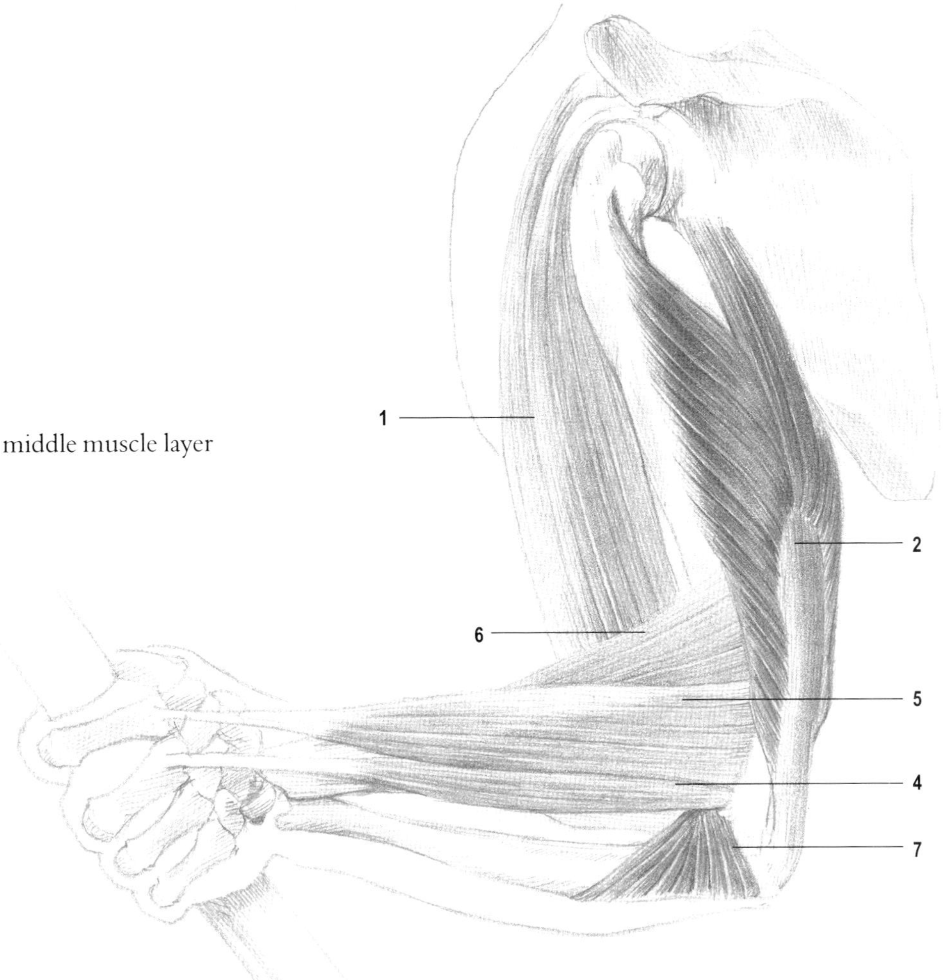

dorsal aspect

deep muscle layer

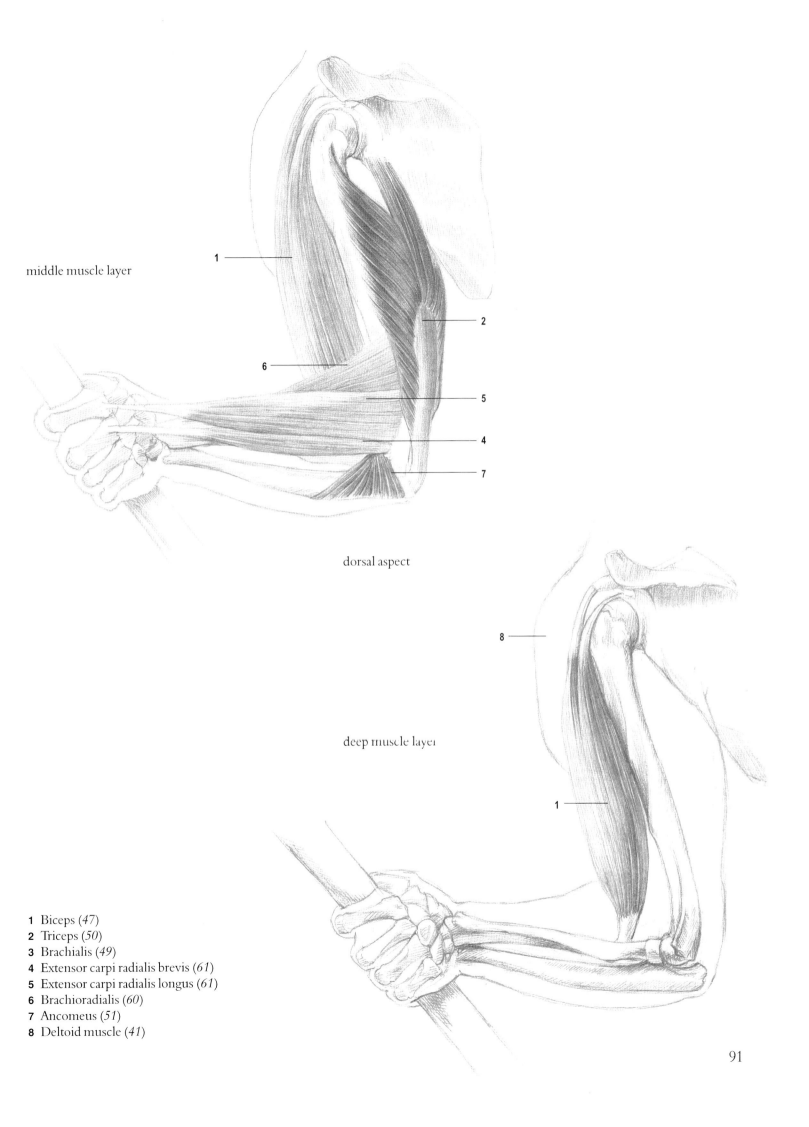

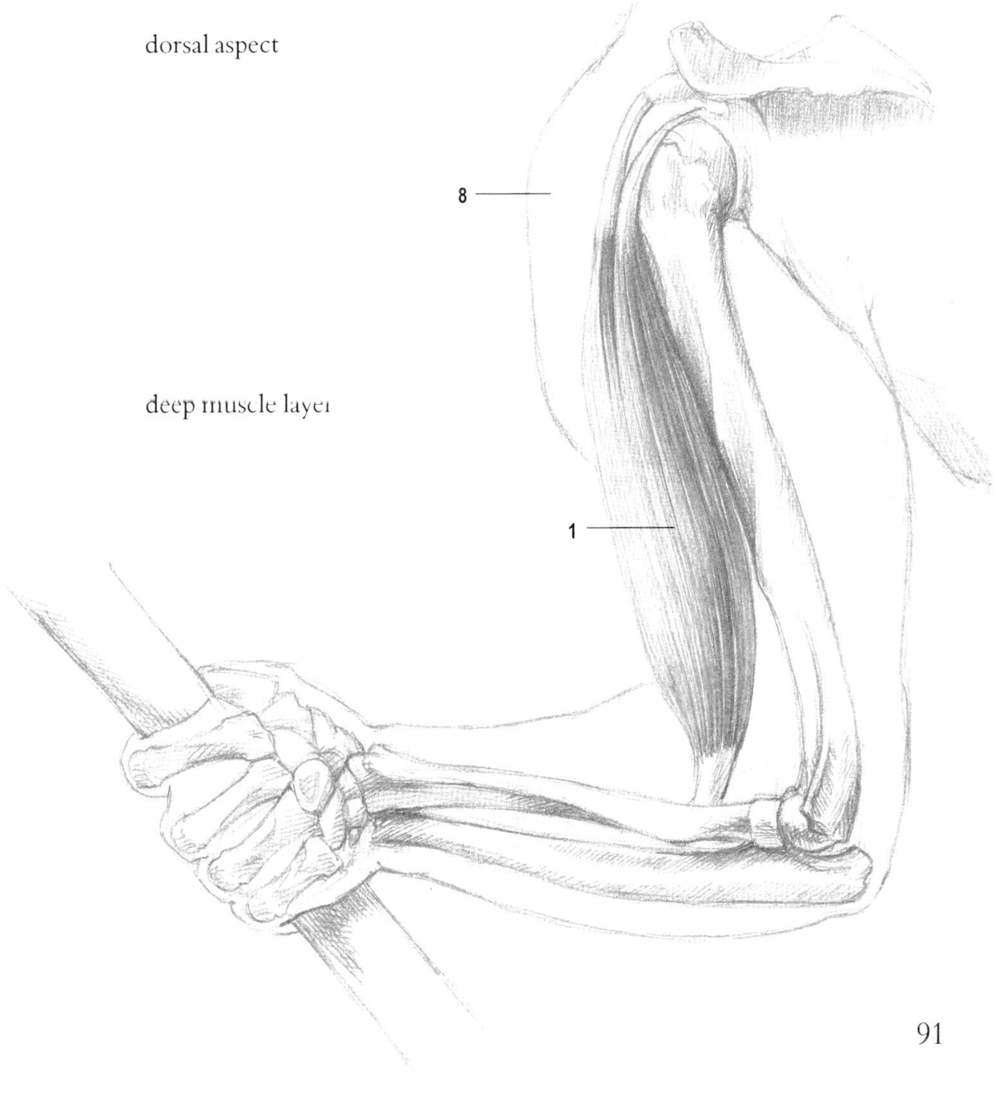

1 Biceps (*47*)
2 Triceps (*50*)
3 Brachialis (*49*)
4 Extensor carpi radialis brevis (*61*)
5 Extensor carpi radialis longus (*61*)
6 Brachioradialis (*60*)
7 Ancomeus (*51*)
8 Deltoid muscle (*41*)

Fig. 56
The Biceps
(M. biceps brachii, *47*)

ORIGIN:
Tuberosity of shoulder blade [long head (**1**)], and coracoid process [short head (**2**)]

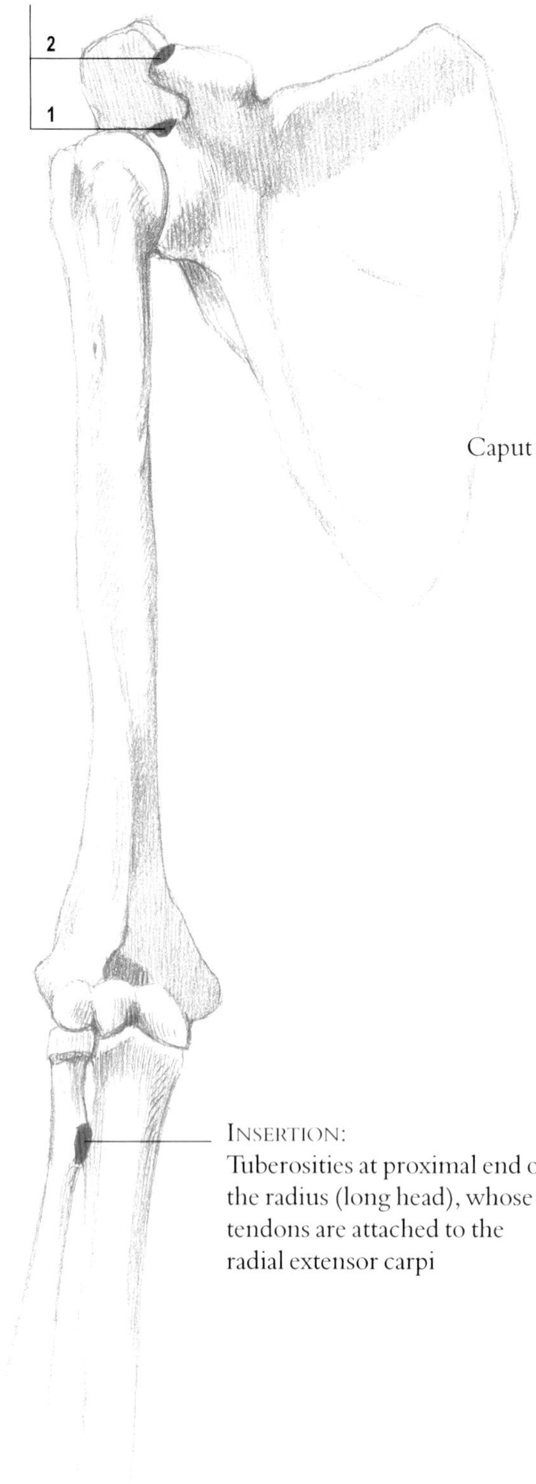

INSERTION:
Tuberosities at proximal end of the radius (long head), whose tendons are attached to the radial extensor carpi

ventral aspect

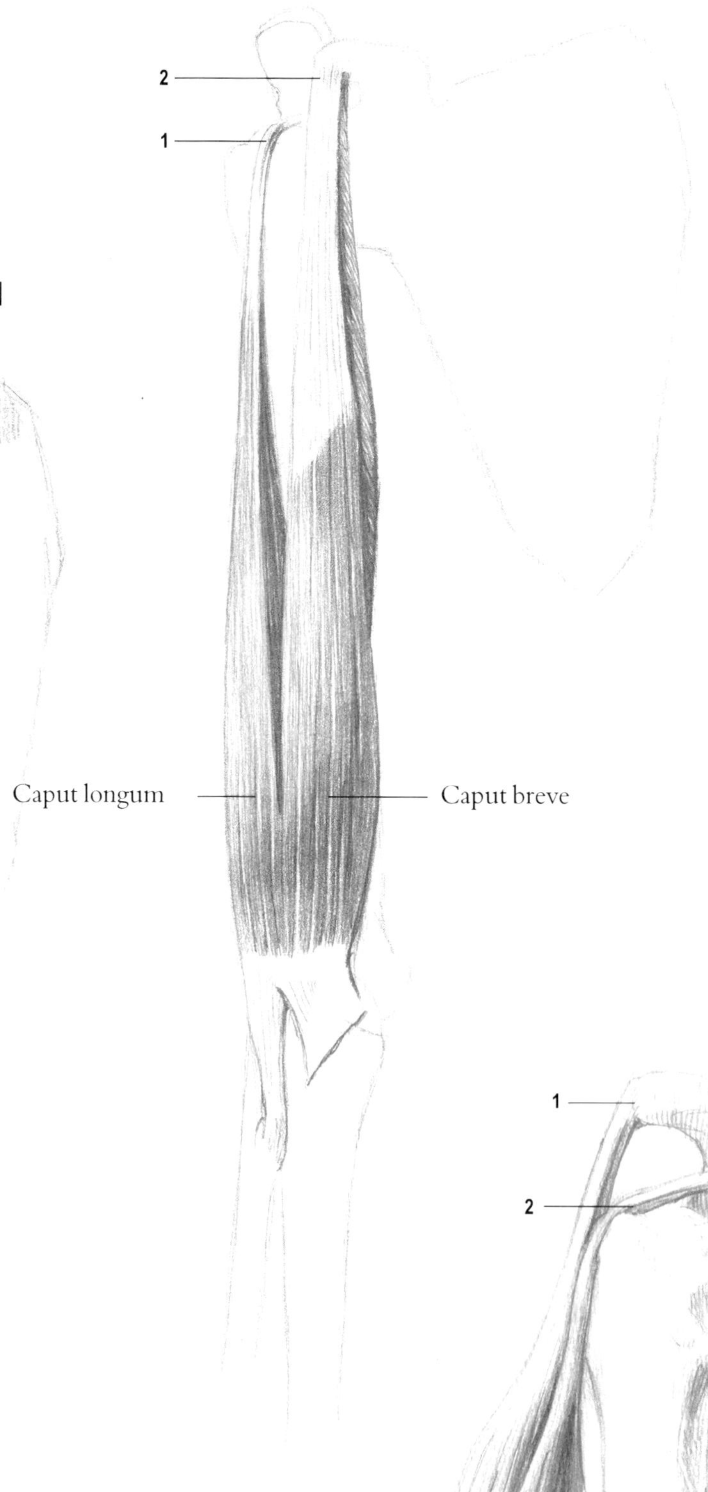

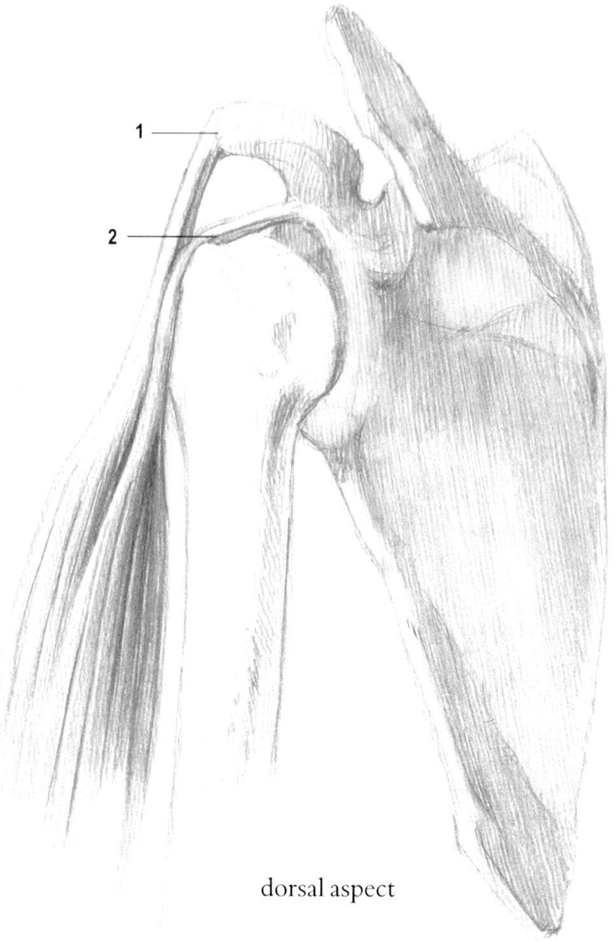

dorsal aspect

Function:
Flexes the arm in the elbow joint; extends the shoulder joint, fixes both and the carpus, supinates the lower arm

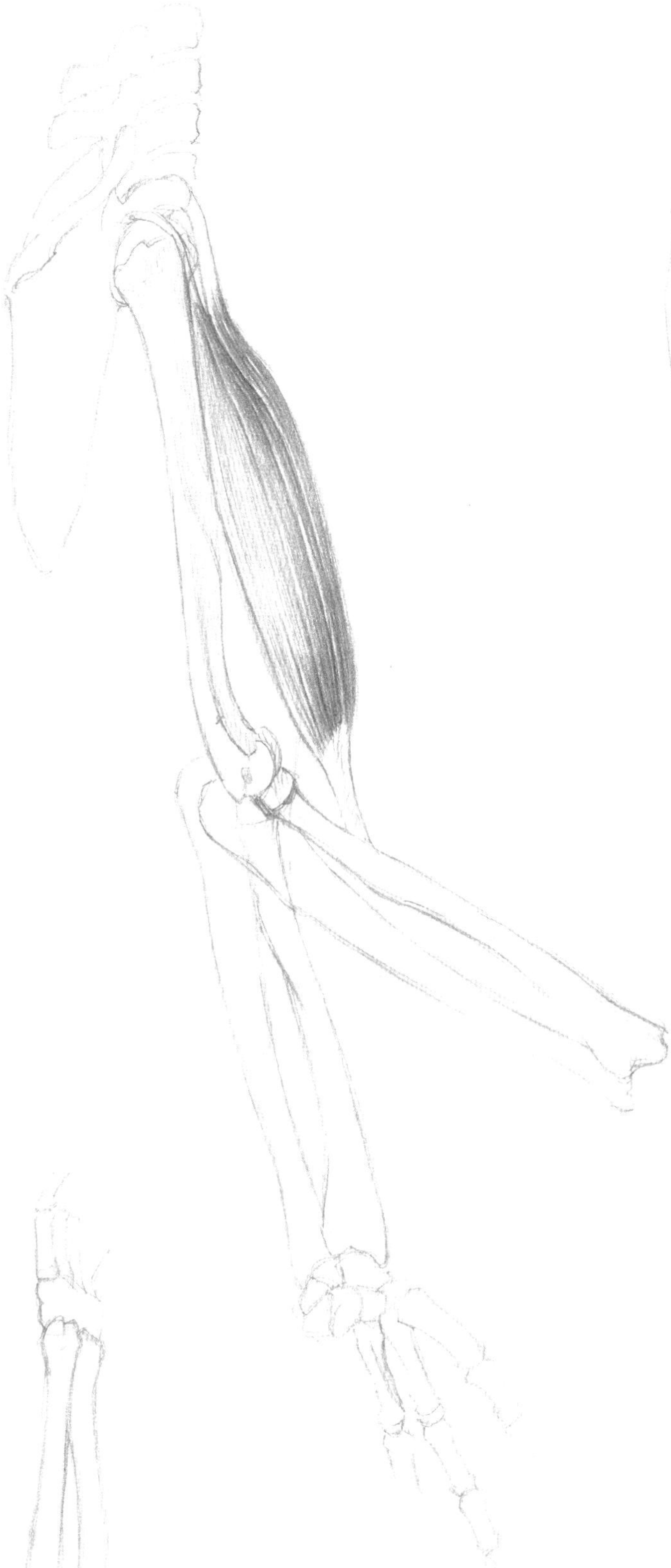

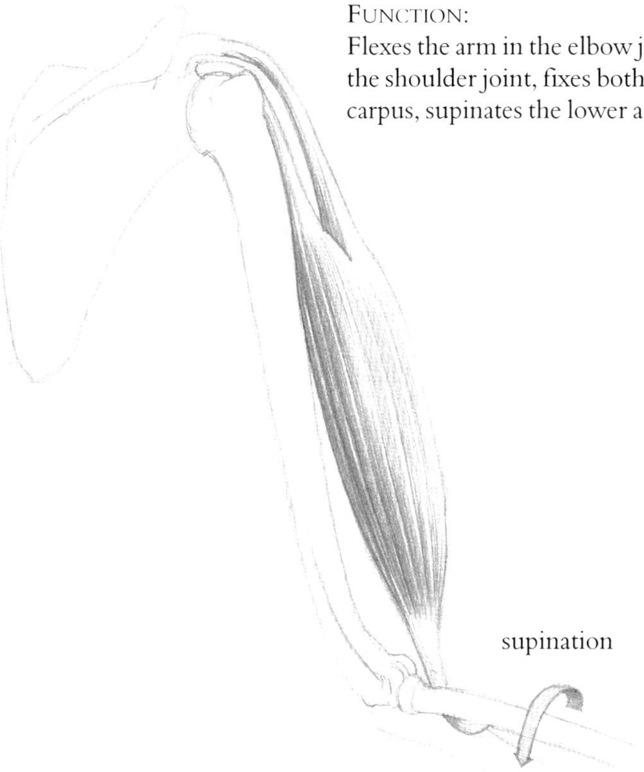

lateral aspect

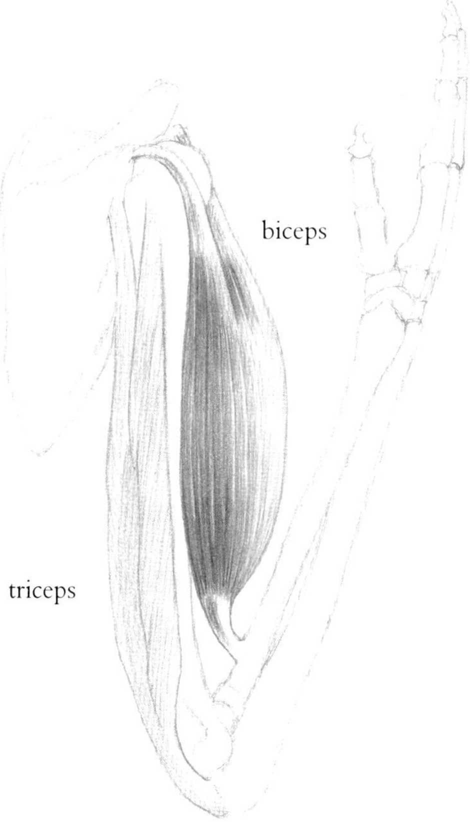

flexion

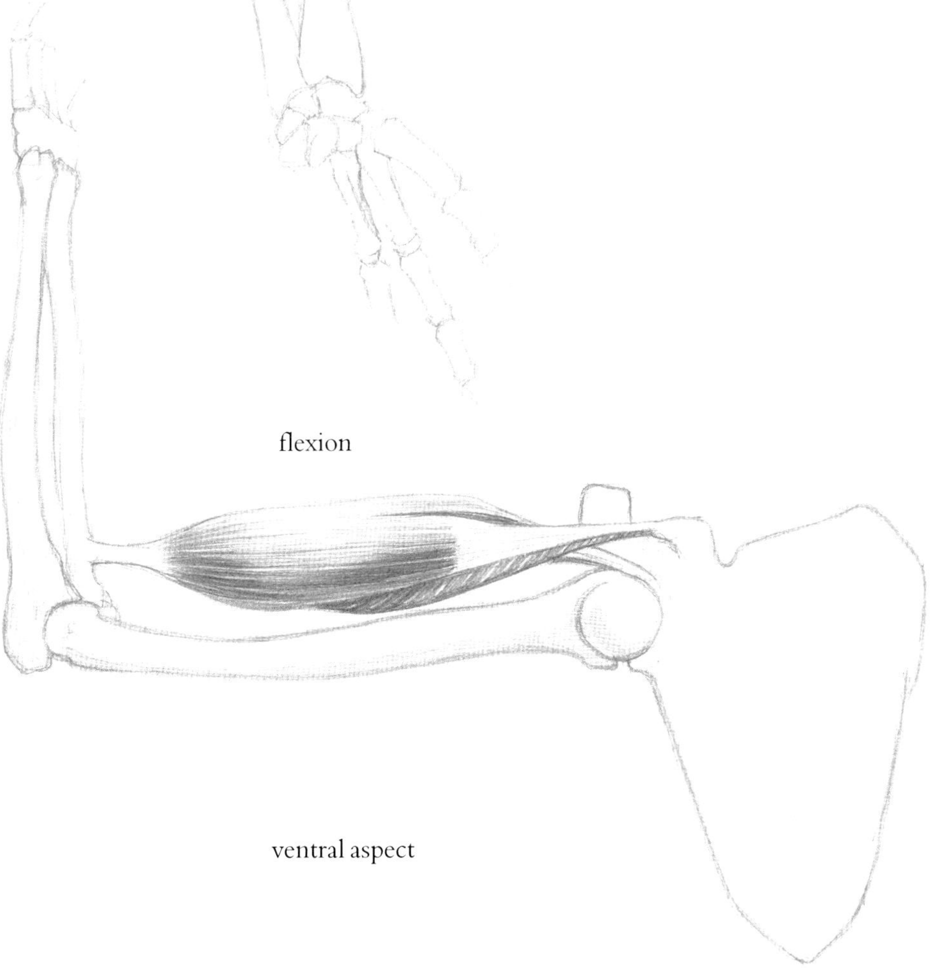

ventral aspect

The triceps extends the lower arm in the elbow joint.

Fig. 57
The Brachial Muscle
(M. brachialis, *49*)

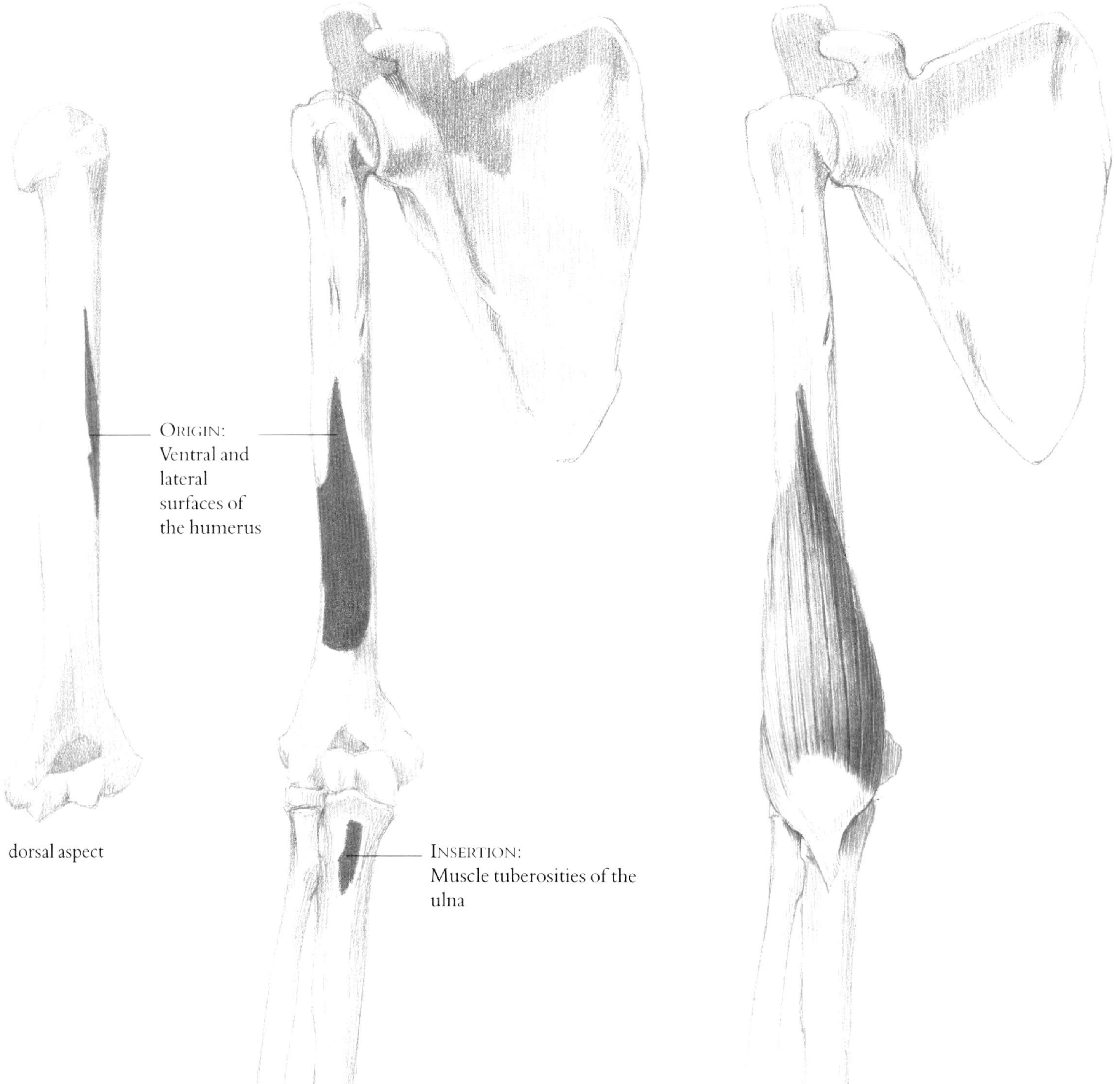
ORIGIN:
Ventral and lateral surfaces of the humerus
INSERTION:
Muscle tuberosities of the ulna

dorsal aspect

ventral aspect

The right hand drawing is the M. pronator teres (52)

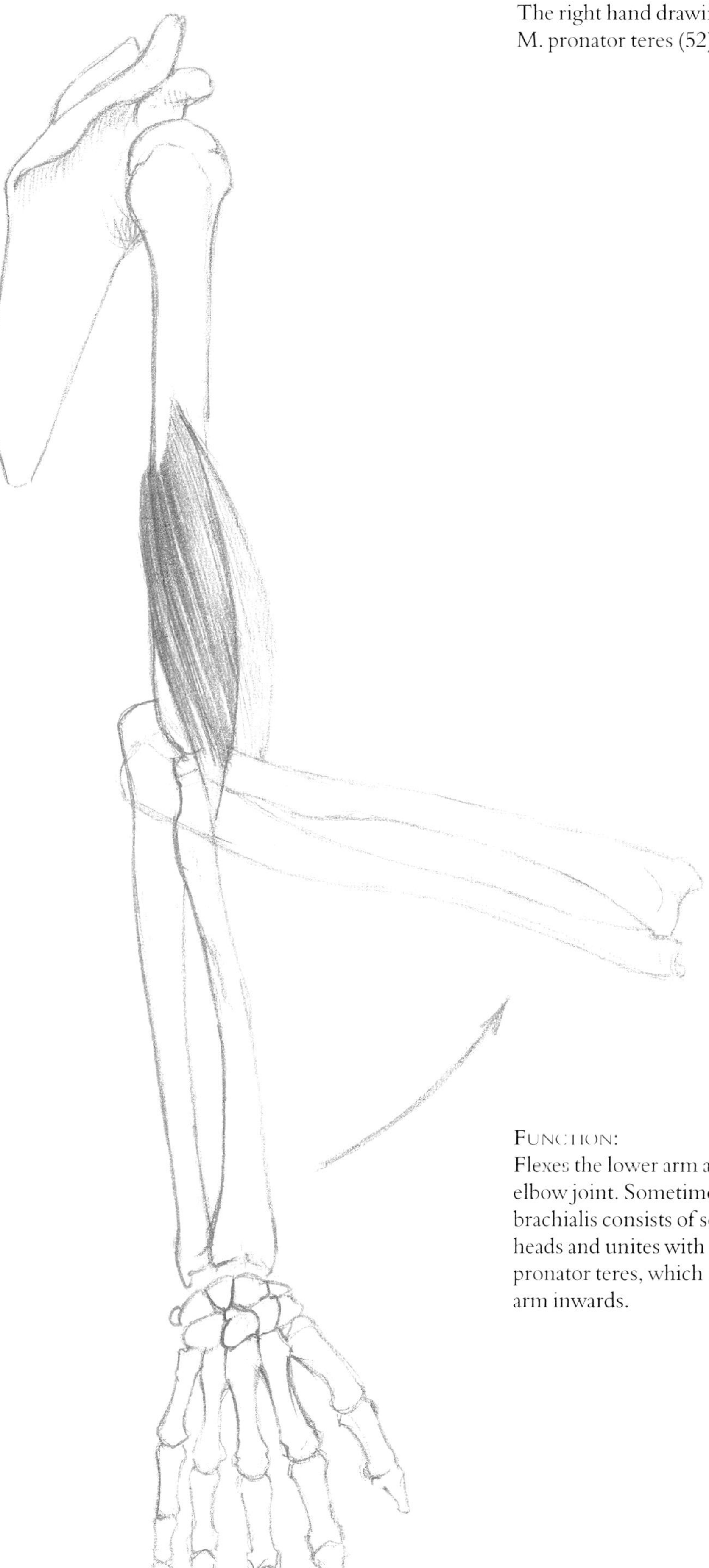

lateral aspect

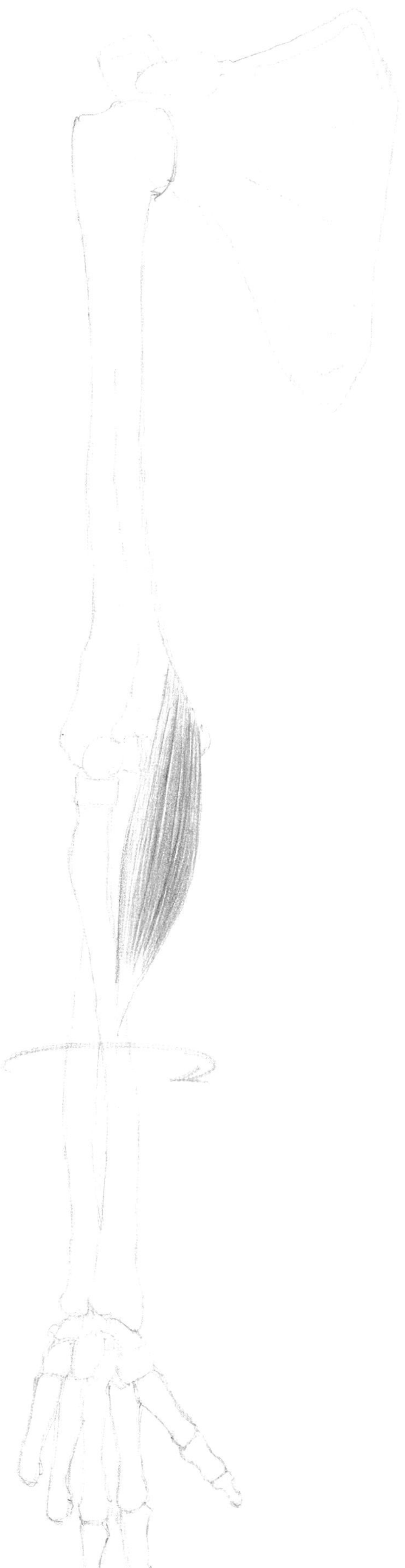

ventral aspect

FUNCTION:
Flexes the lower arm at the elbow joint. Sometimes the M. brachialis consists of several heads and unites with the M. pronator teres, which rotates the arm inwards.

Fig. 58
The Triceps
(M. triceps brachii, *50*)

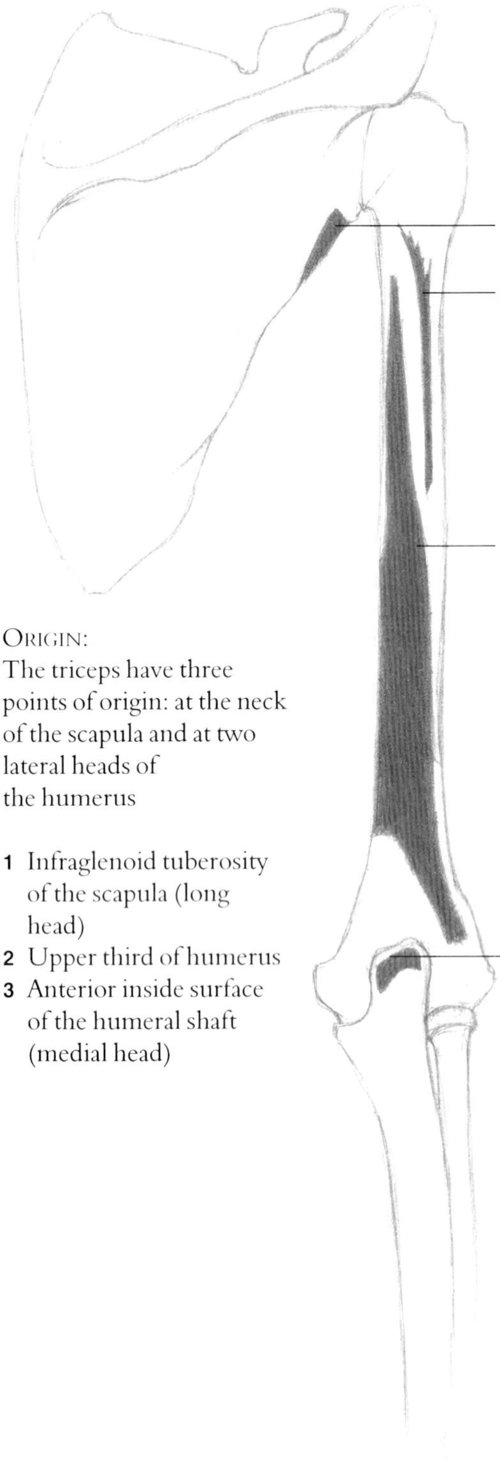

ORIGIN:
The triceps have three points of origin: at the neck of the scapula and at two lateral heads of the humerus

1 Infraglenoid tuberosity of the scapula (long head)
2 Upper third of humerus
3 Anterior inside surface of the humeral shaft (medial head)

INSERTION:
Coracoid process of the ulna

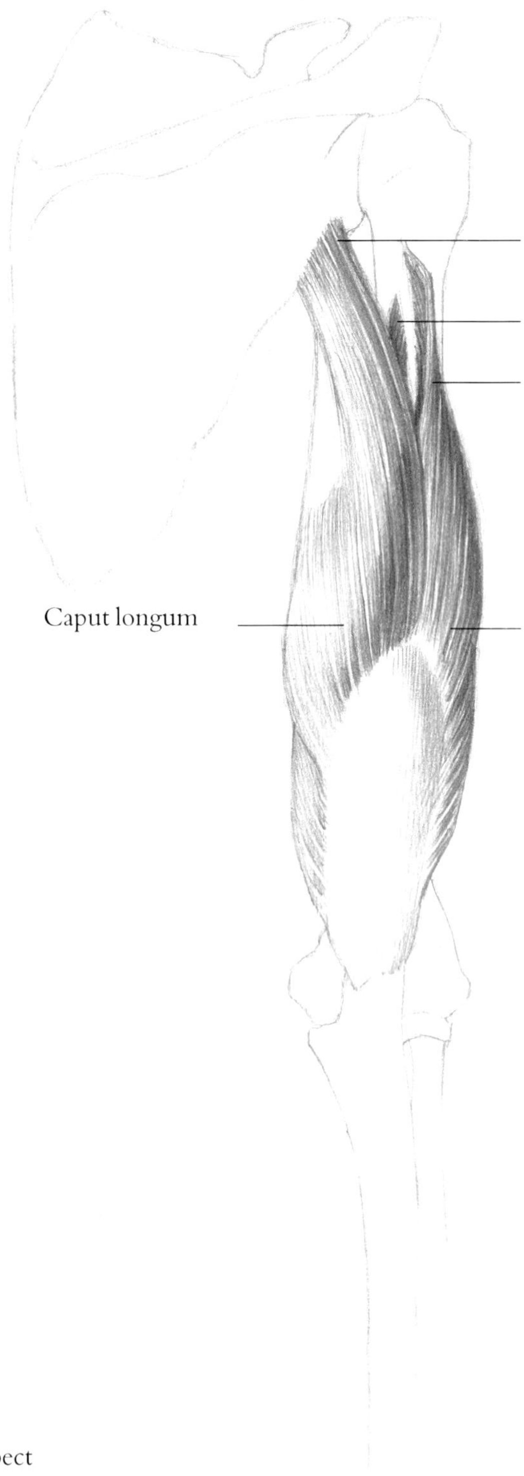

dorsal aspect

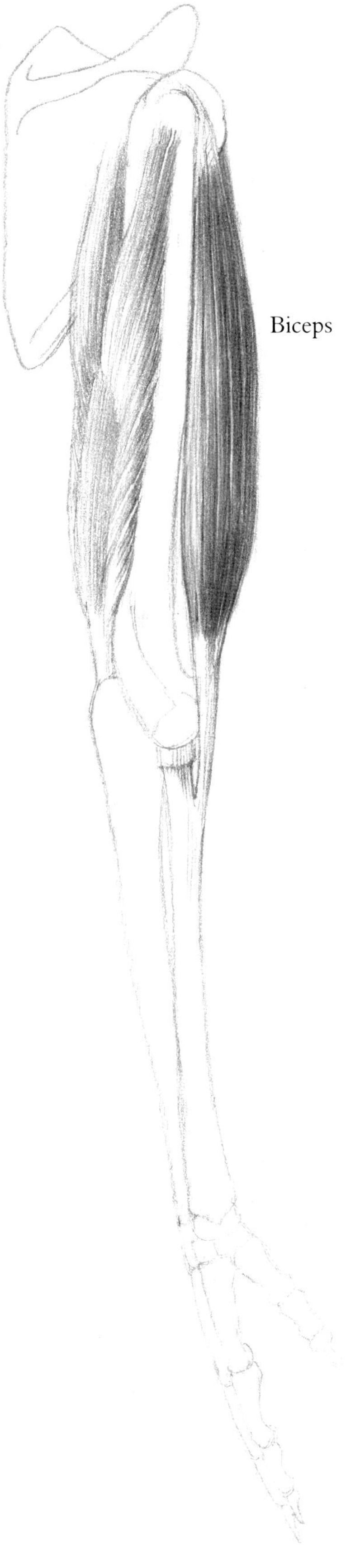

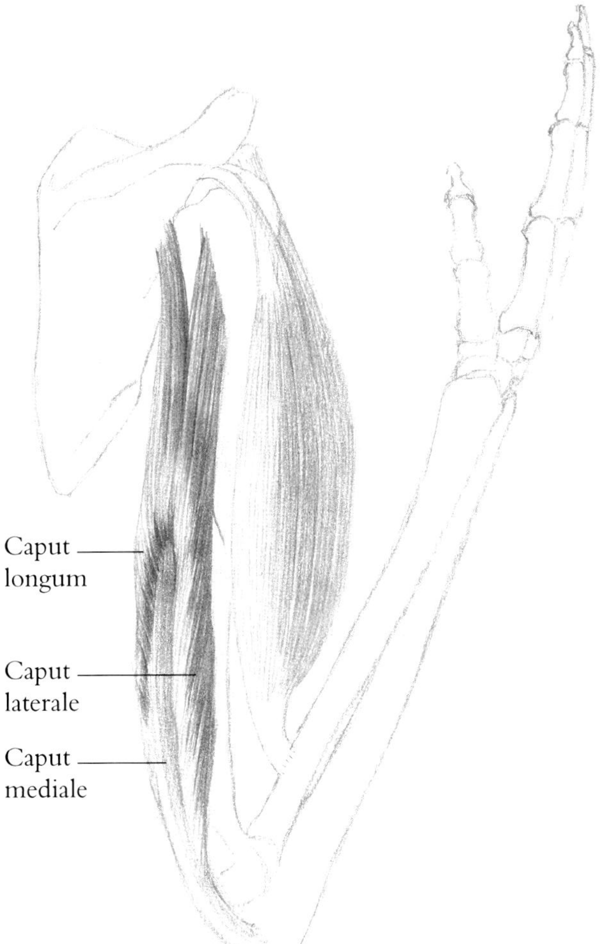

FUNCTION:
Flexes the lower arm at the elbow joint, pulls the medial part of the scapula upwards when the scapula is fixed, and pulls the humerus inwards

lateral aspect

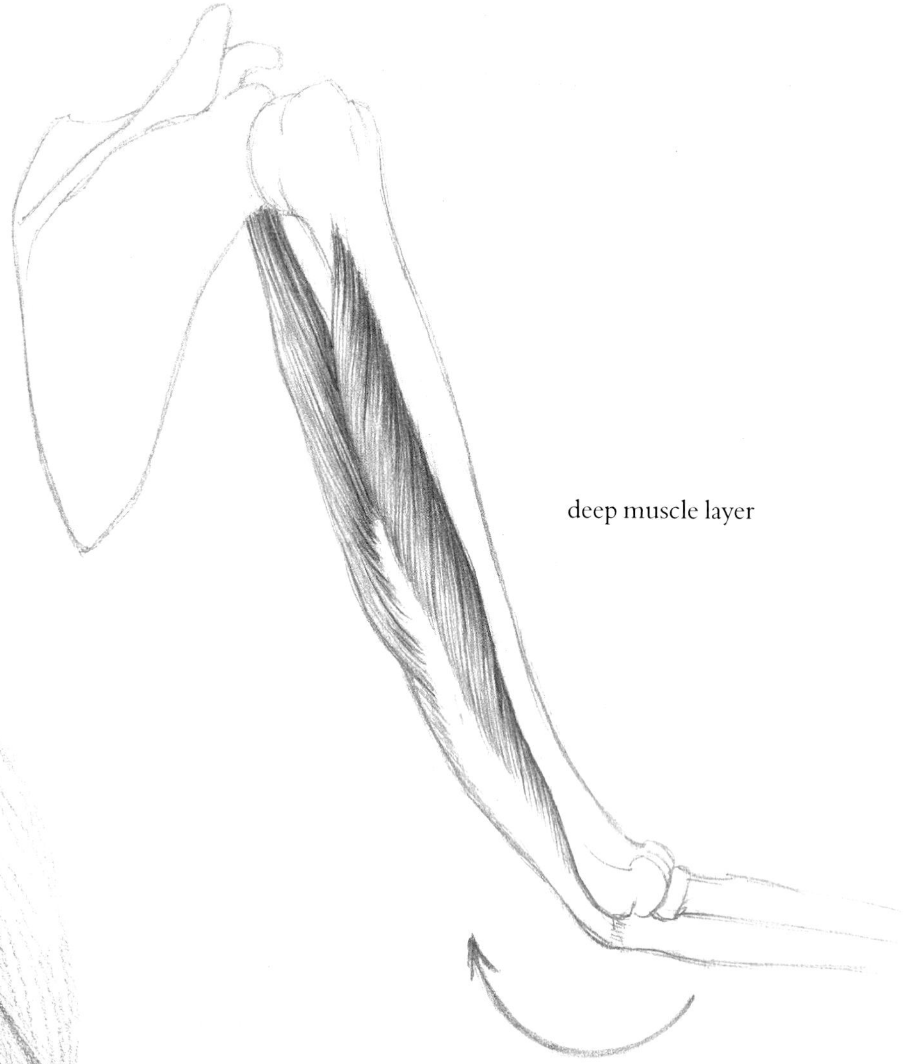
deep muscle layer

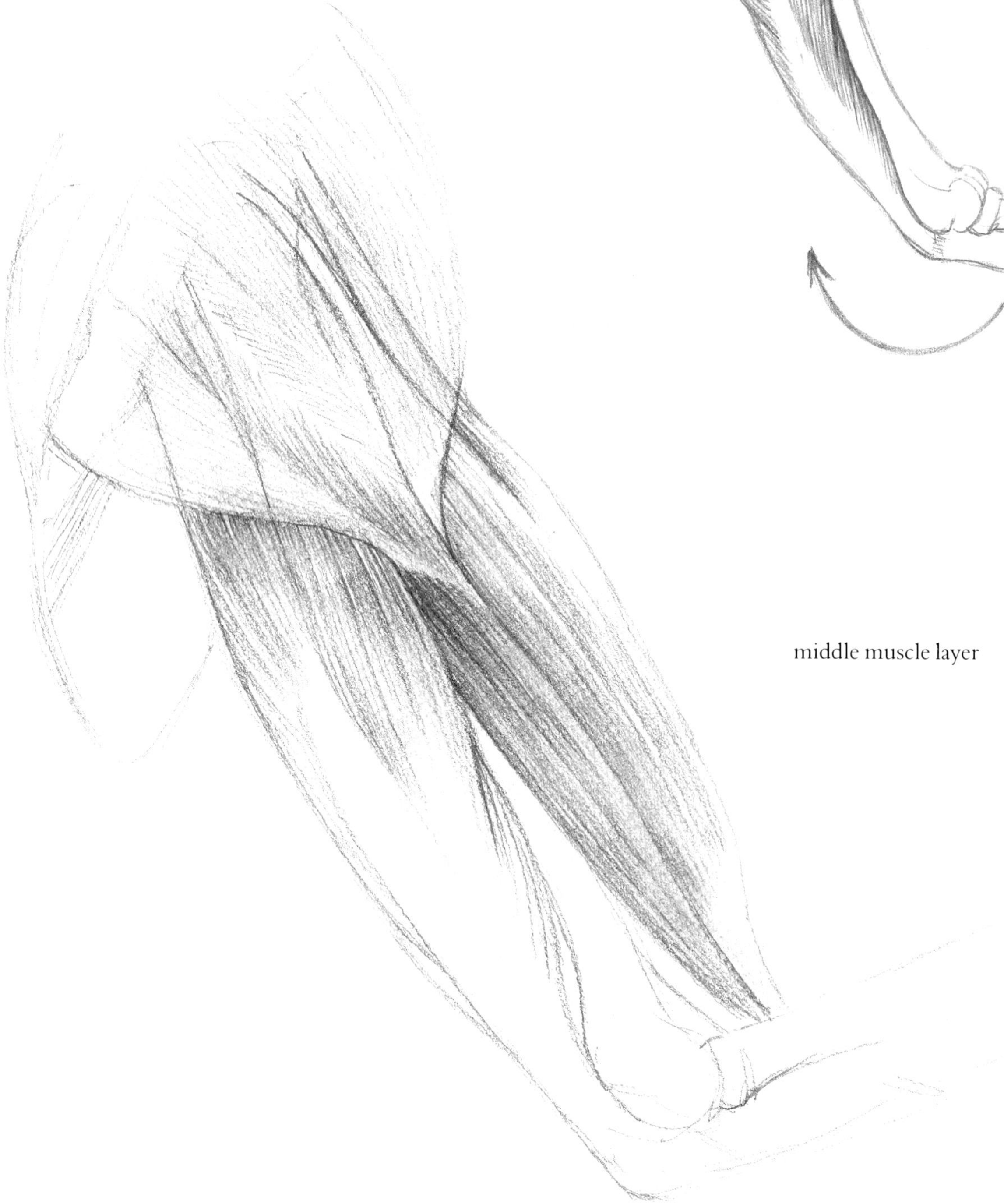
middle muscle layer

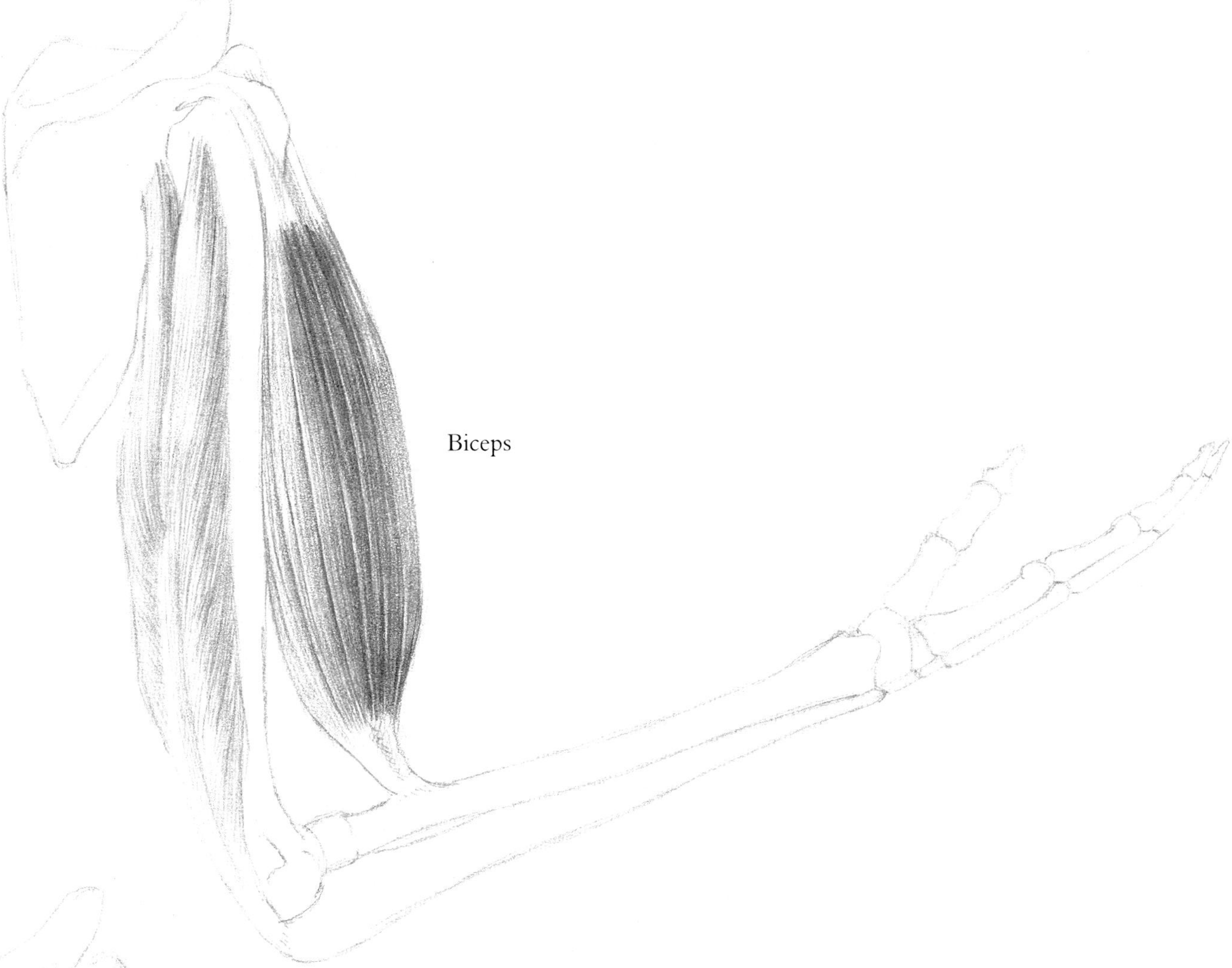

lateral aspect

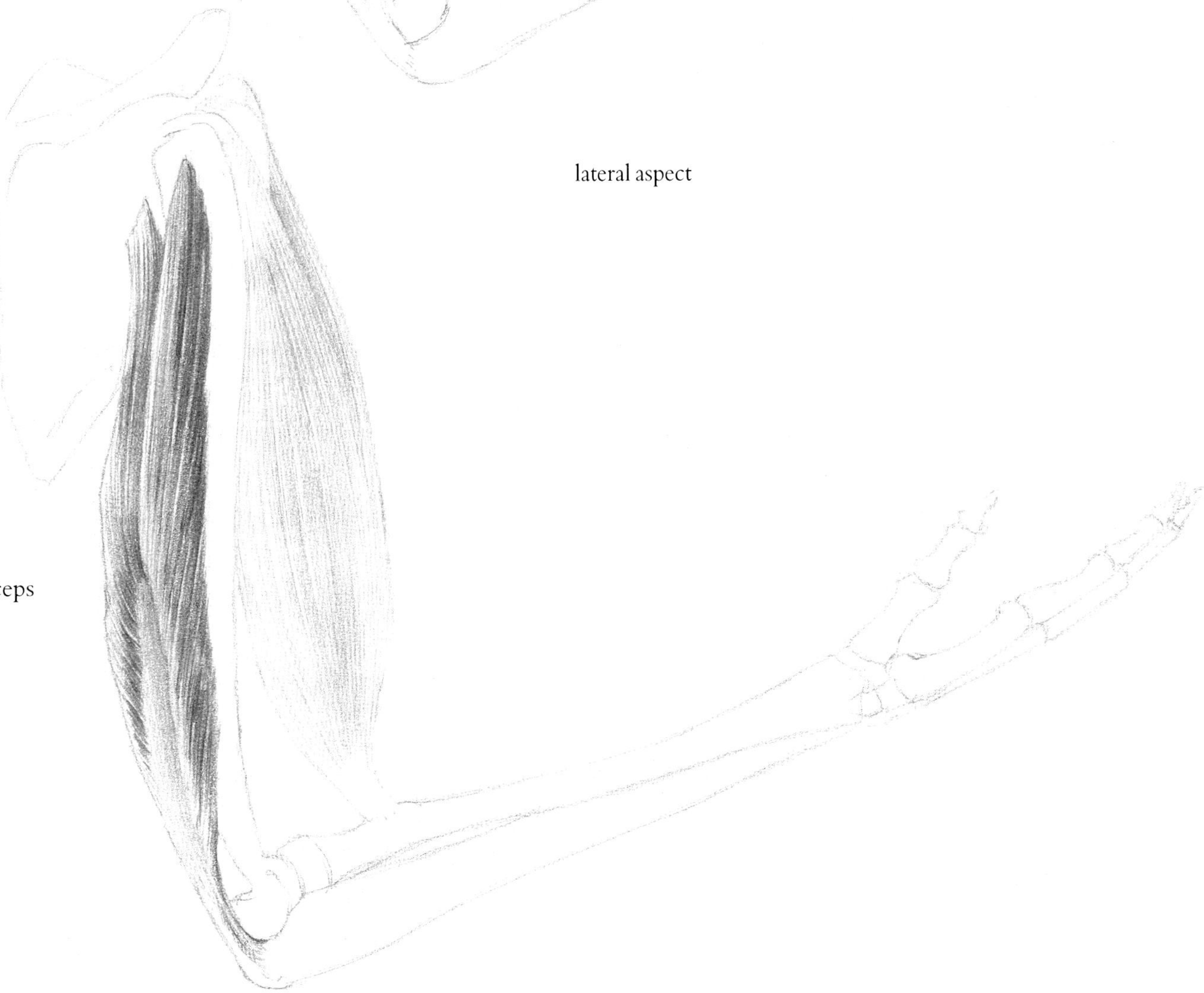

The Triceps
(cont.)

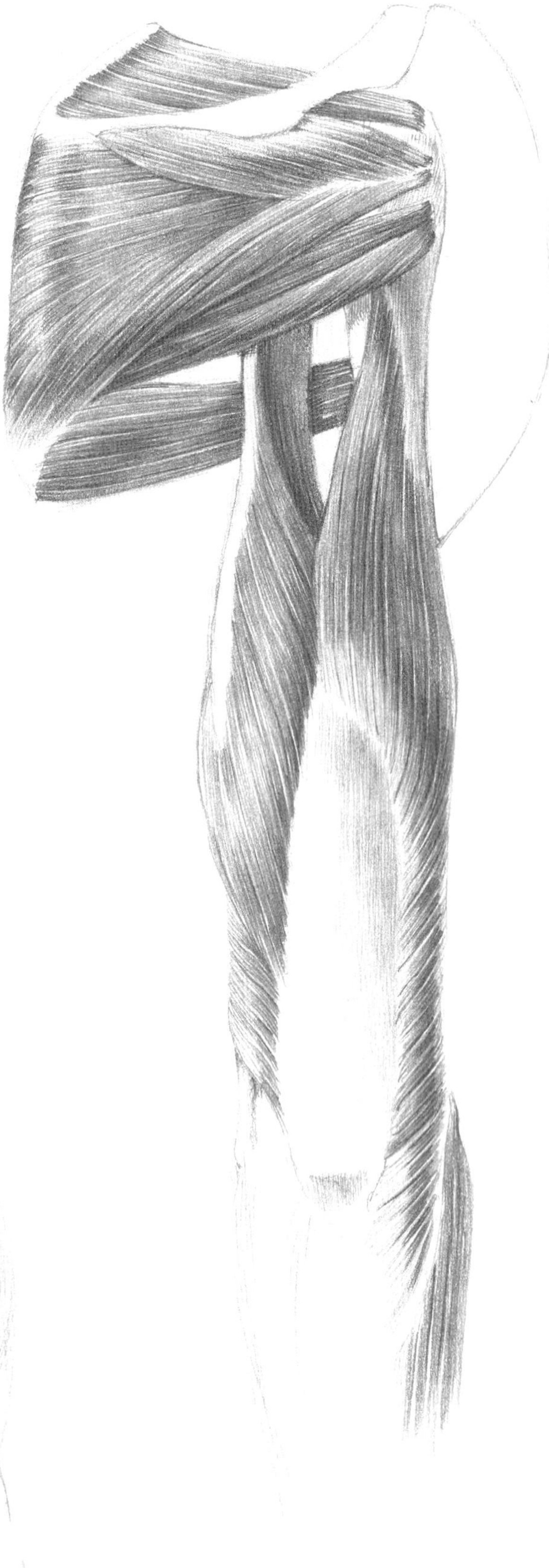

dorsal aspect

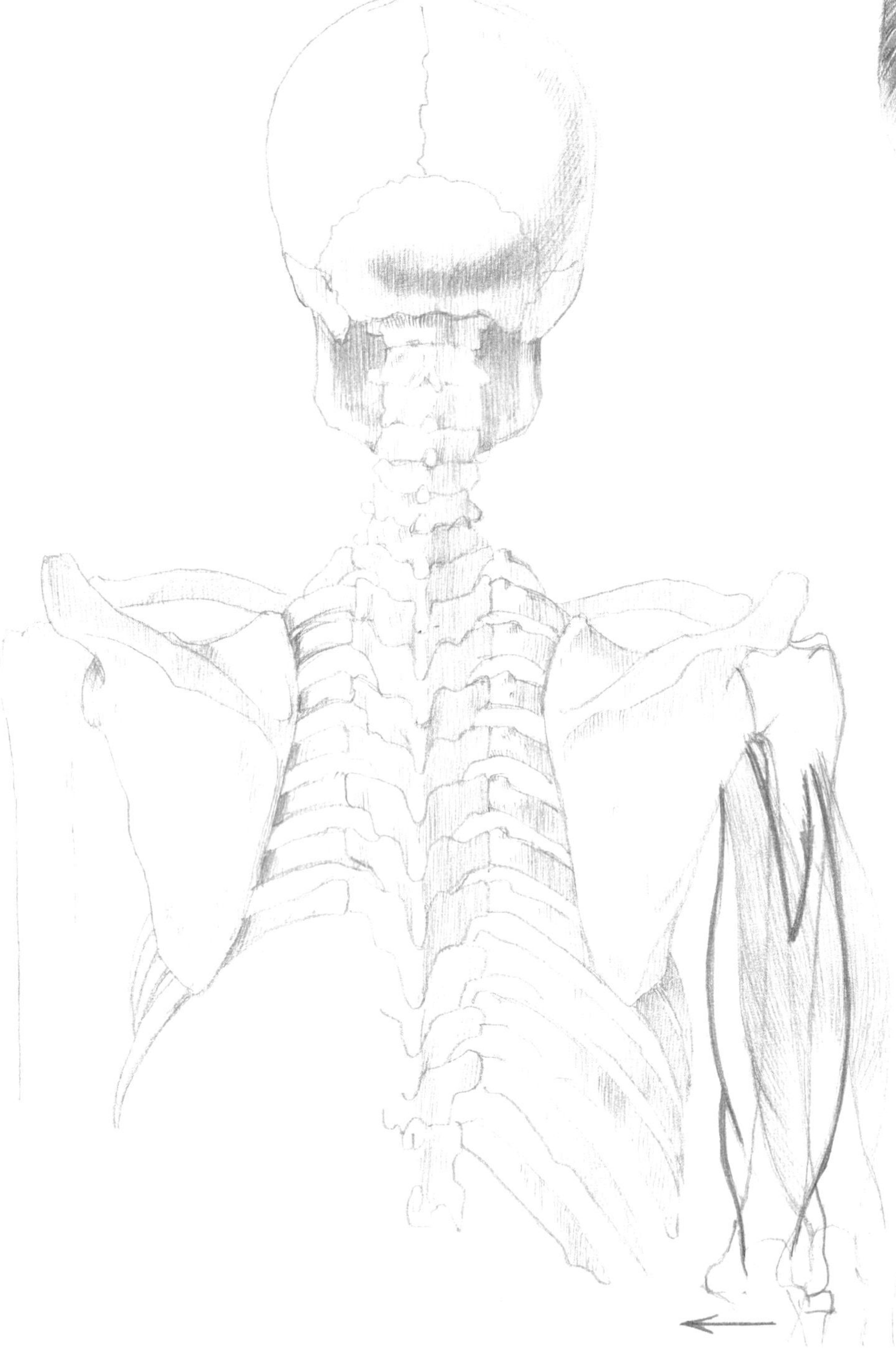

pulls the upper arm to the body

Fig. 59
The Acromioclavicular and Arm Muscles

The muscles of the arm are long and spindle-shaped. Their contours are visible under the skin when the arm moves.

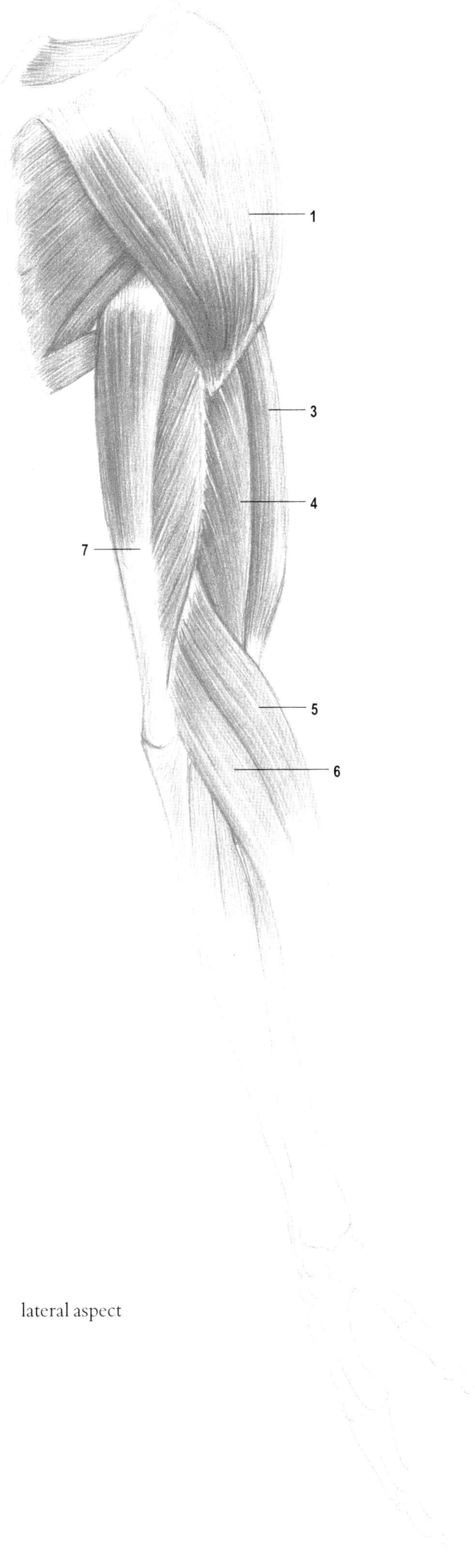

lateral aspect

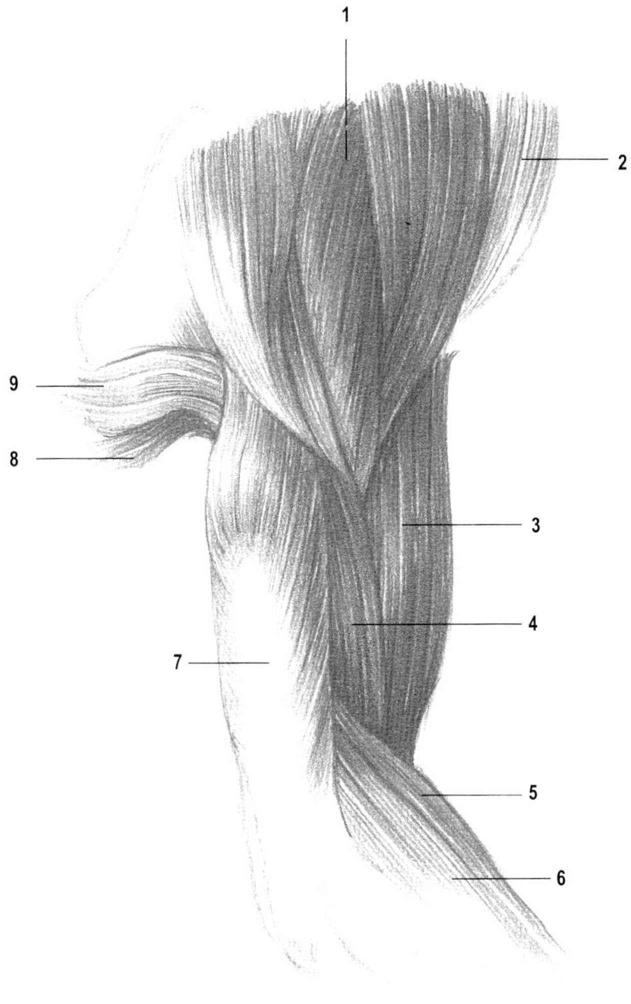

middle muscle layer

1 Deltoid muscle (*41*)
2 Pectoralis major (*28*)
3 Biceps (*47*)
4 Brachialis (*49*)
5 Brachioradialis (*60*)
6 Extensor carpi radialis longus (*61*)
7 Triceps (*50*)
8 Teres major (*45*)
9 Teres minor (*44*)

Fig. 60
Origins and Insertion Surfaces of Various Muscles in the Shoulder Area

O= Origin
I = Insertion

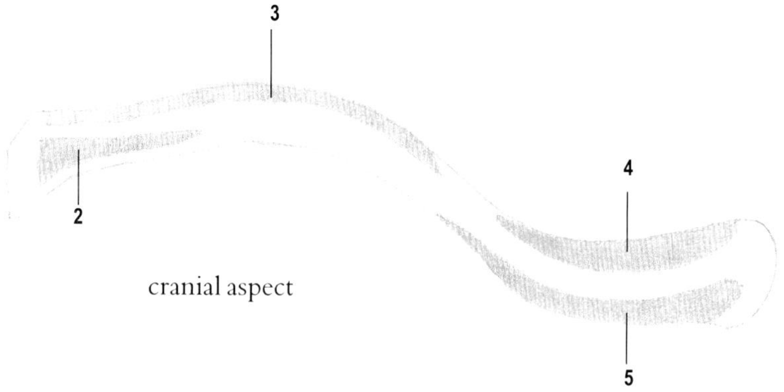

cranial aspect

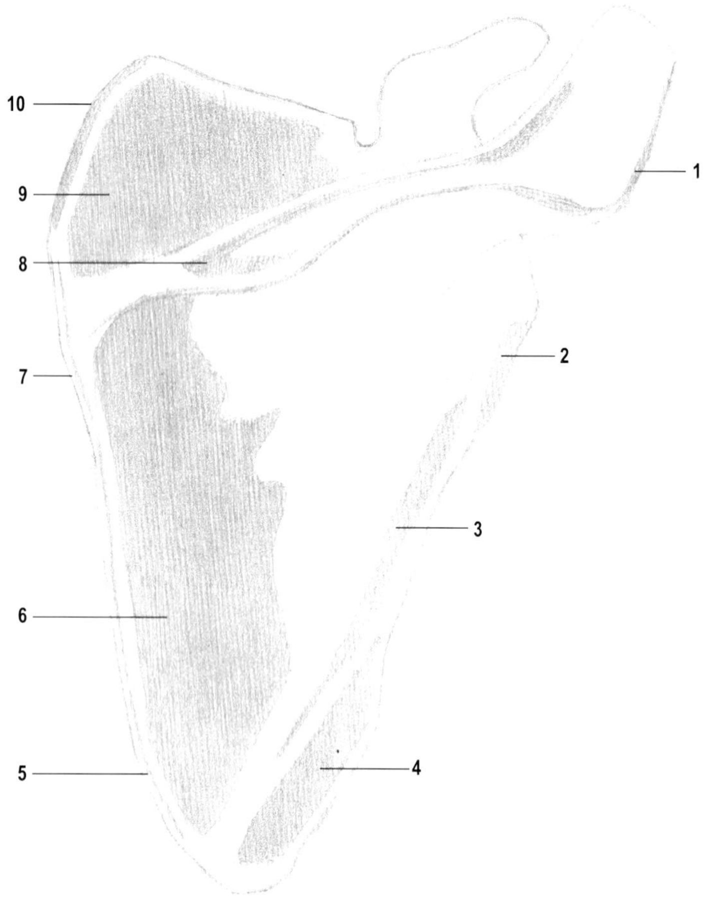

dorsal surface

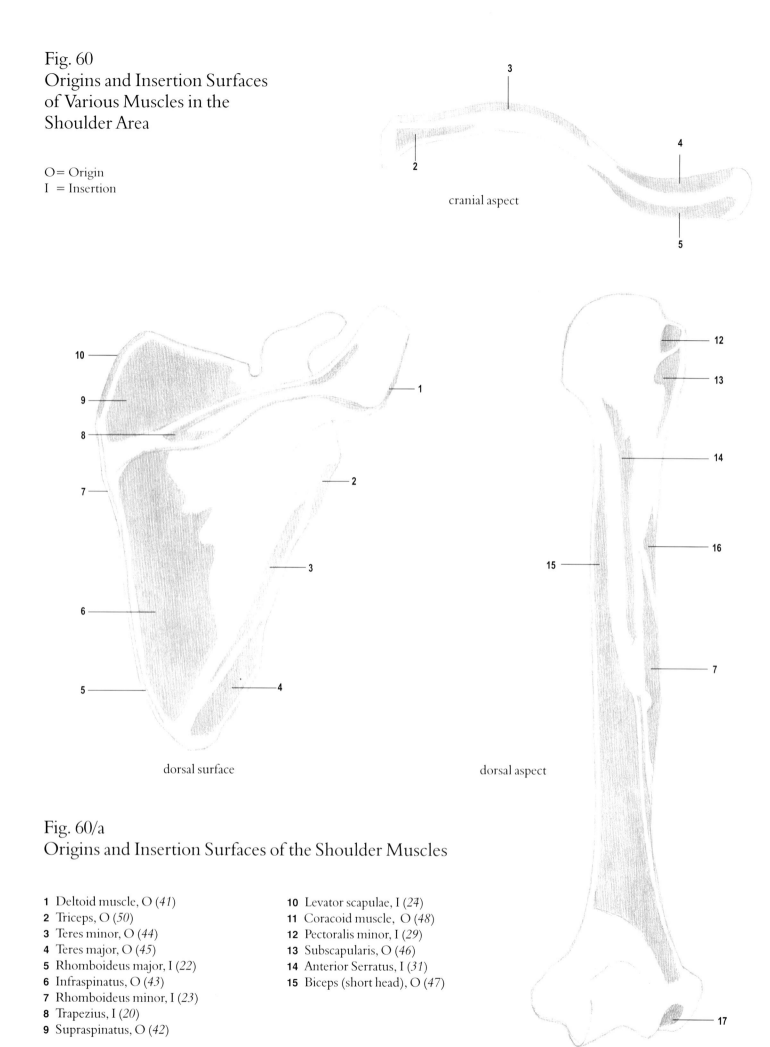

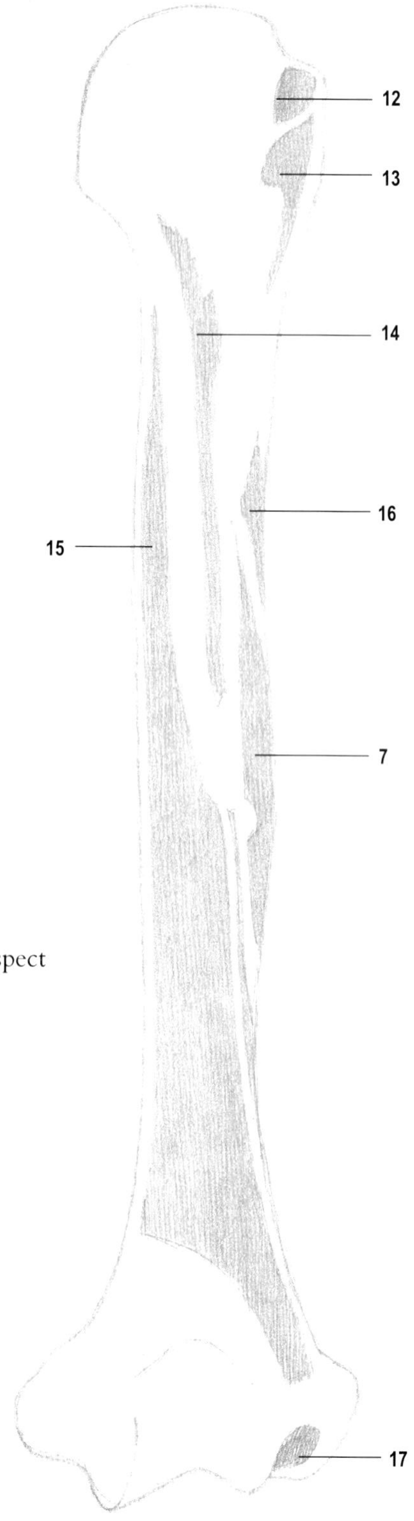

dorsal aspect

Fig. 60/a
Origins and Insertion Surfaces of the Shoulder Muscles

1 Deltoid muscle, O (*41*)
2 Triceps, O (*50*)
3 Teres minor, O (*44*)
4 Teres major, O (*45*)
5 Rhomboideus major, I (*22*)
6 Infraspinatus, O (*43*)
7 Rhomboideus minor, I (*23*)
8 Trapezius, I (*20*)
9 Supraspinatus, O (*42*)
10 Levator scapulae, I (*24*)
11 Coracoid muscle, O (*48*)
12 Pectoralis minor, I (*29*)
13 Subscapularis, O (*46*)
14 Anterior Serratus, I (*31*)
15 Biceps (short head), O (*47*)

Fig. 60/b
Origins and Insertion Surfaces of the Clavicular Muscles

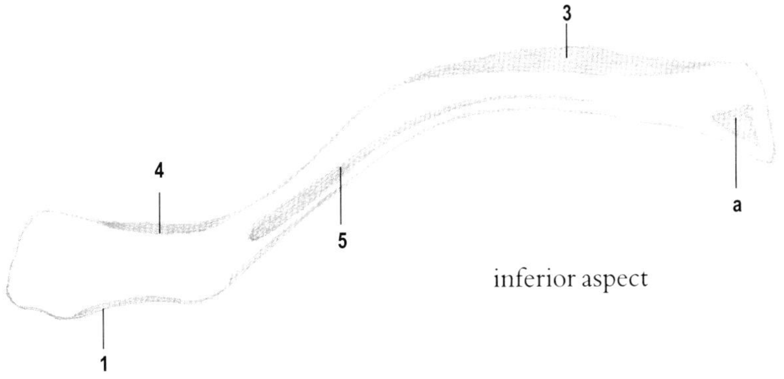

inferior aspect

1 Trapezius, I (*20*)
2 Sternocleidomastoideus, O (*11*)
3 Pectoralis major, O (*28*)
4 Deltoid Muscle, O (*41*)
5 Subclavius, I (*30*)

a Rib-clavicle tendon

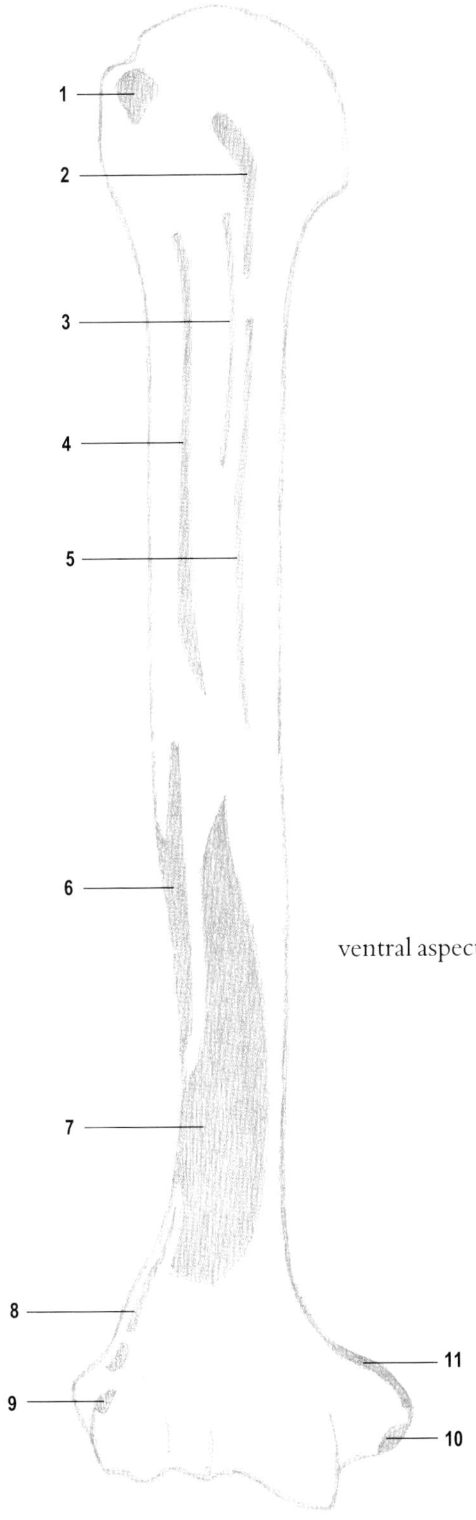

ventral aspect

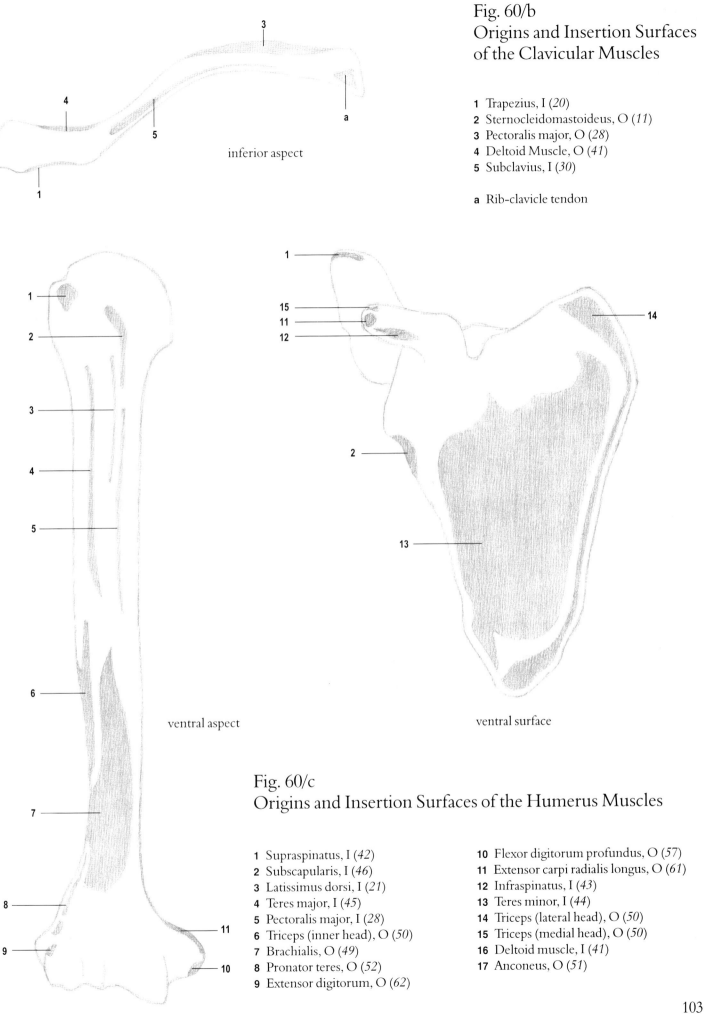

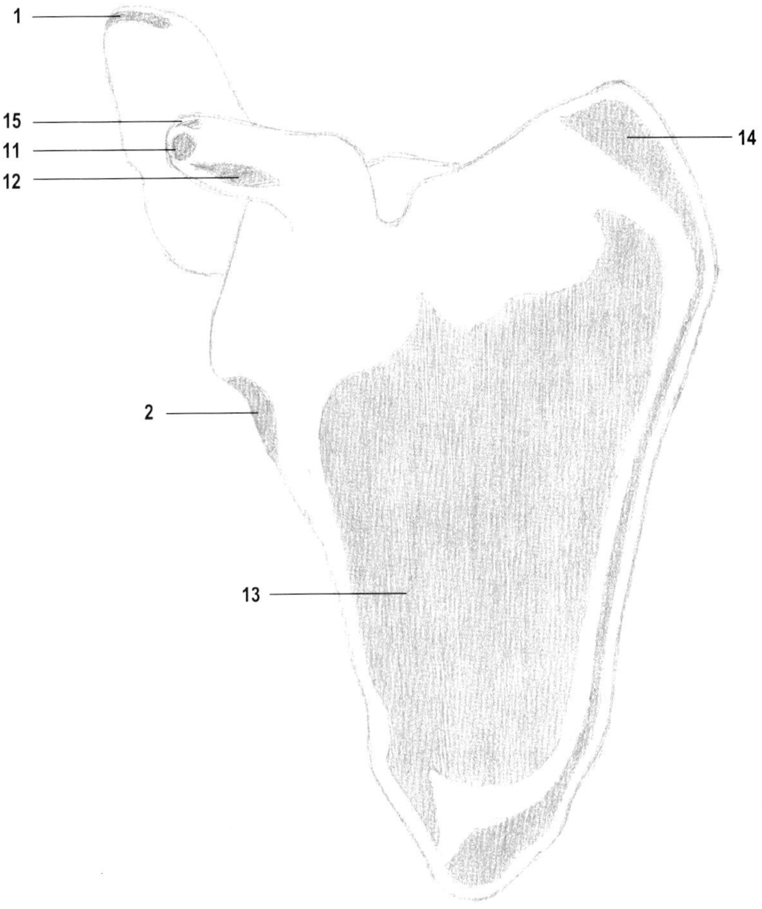

ventral surface

Fig. 60/c
Origins and Insertion Surfaces of the Humerus Muscles

1 Supraspinatus, I (*42*)
2 Subscapularis, I (*46*)
3 Latissimus dorsi, I (*21*)
4 Teres major, I (*45*)
5 Pectoralis major, I (*28*)
6 Triceps (inner head), O (*50*)
7 Brachialis, O (*49*)
8 Pronator teres, O (*52*)
9 Extensor digitorum, O (*62*)
10 Flexor digitorum profundus, O (*57*)
11 Extensor carpi radialis longus, O (*61*)
12 Infraspinatus, I (*43*)
13 Teres minor, I (*44*)
14 Triceps (lateral head), O (*50*)
15 Triceps (medial head), O (*50*)
16 Deltoid muscle, I (*41*)
17 Anconeus, O (*51*)

Fig. 61
The Anconeus, *51*

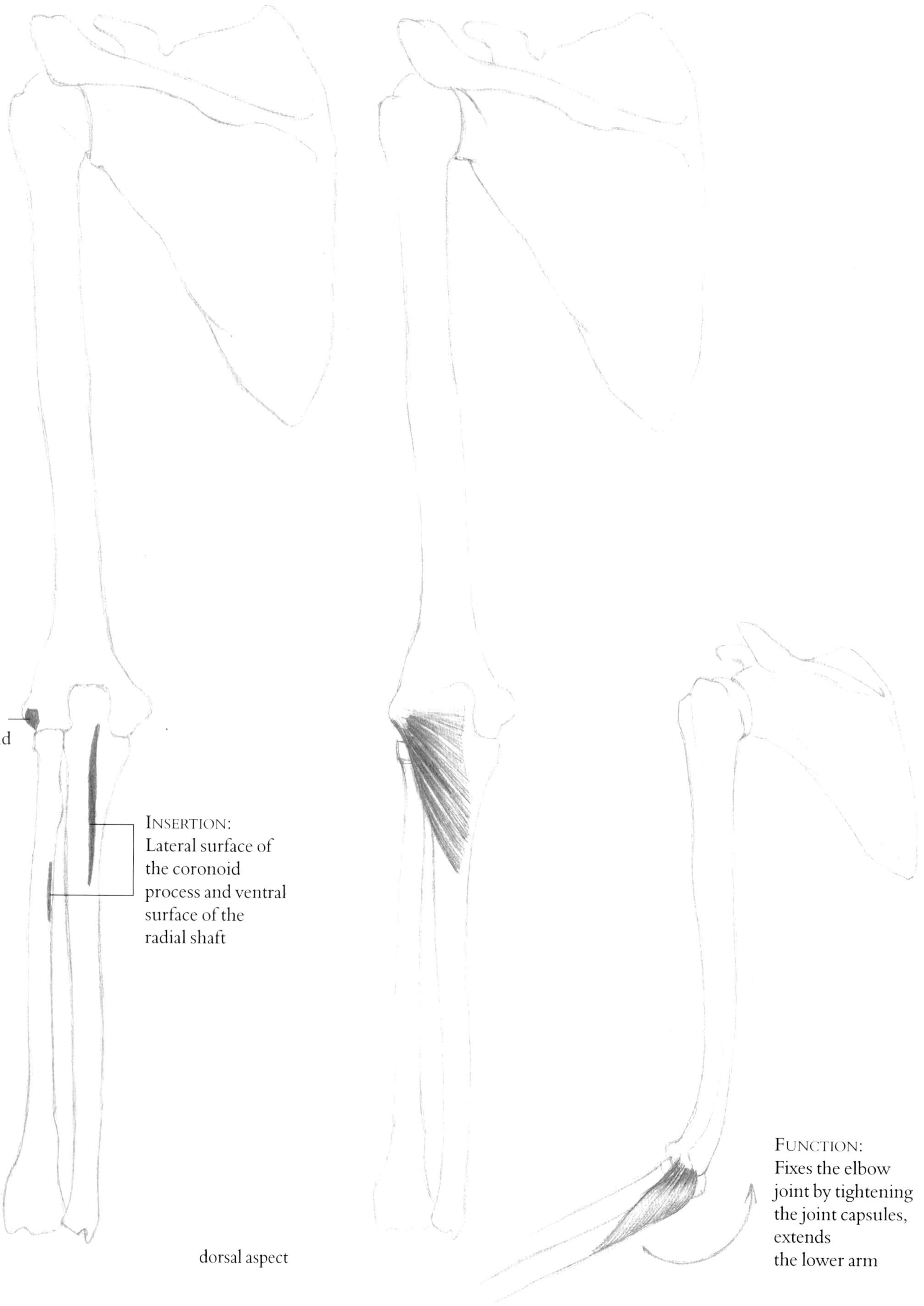
ORIGIN:
Dorsal surface and lateral anconeus extensor of the humerus
INSERTION:
Lateral surface of the coronoid process and ventral surface of the radial shaft
FUNCTION:
Fixes the elbow joint by tightening the joint capsules, extends the lower arm
dorsal aspect

Fig. 62
The Pronator Teres, *52*

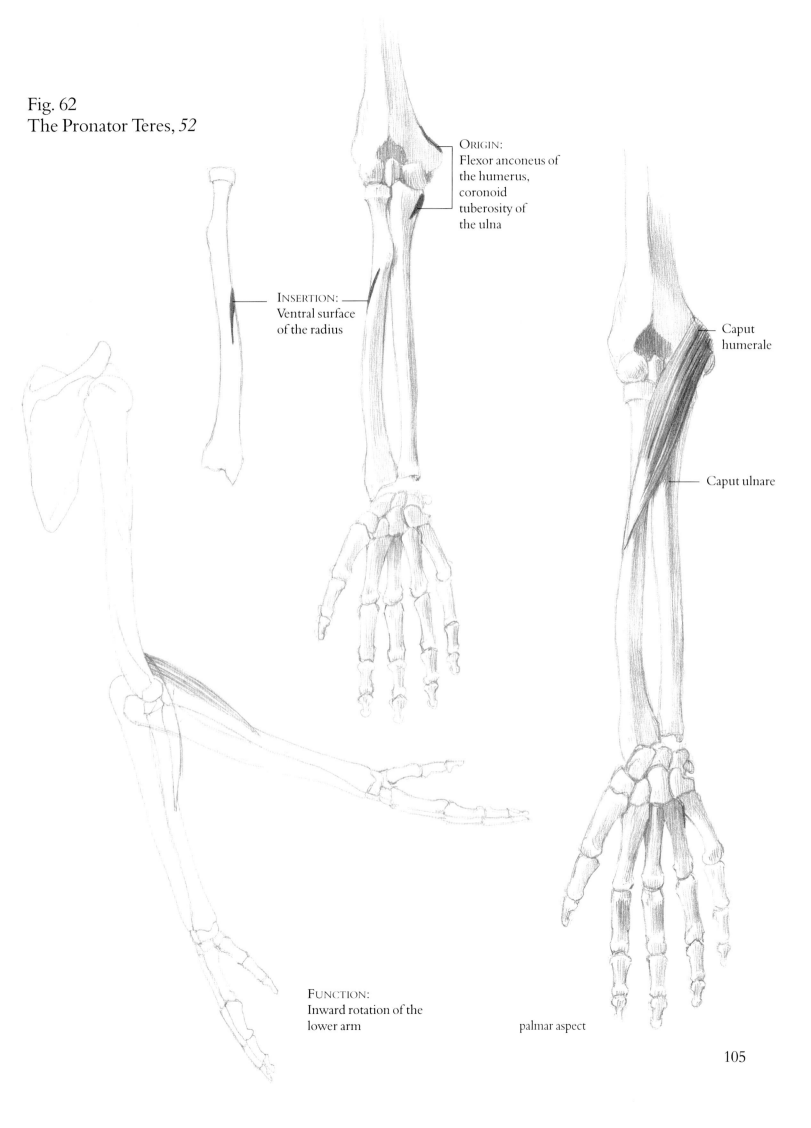

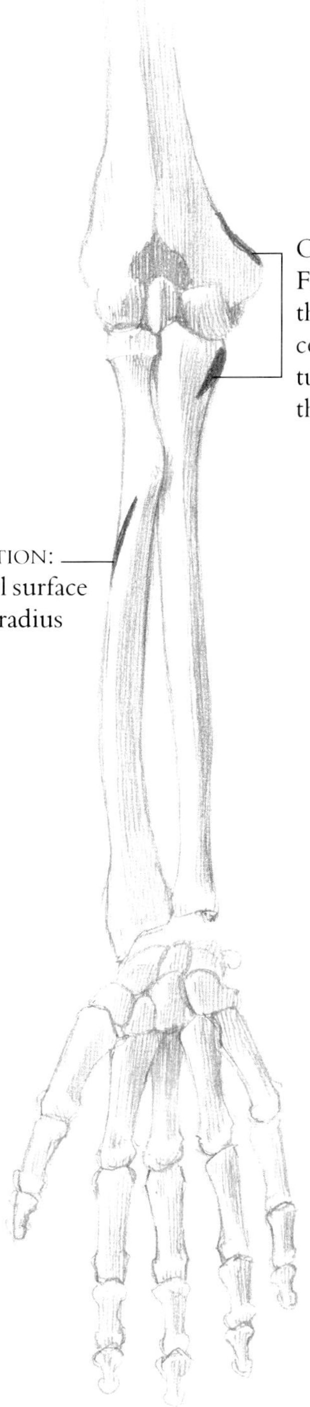
ORIGIN:
Flexor anconeus of the humerus, coronoid tuberosity of the ulna
INSERTION:
Ventral surface of the radius

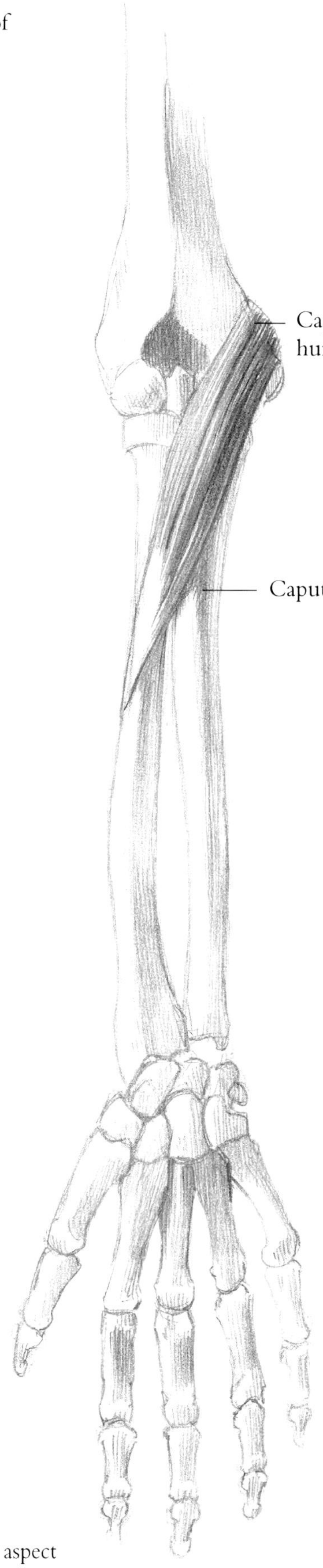
Caput humerale
Caput ulnare

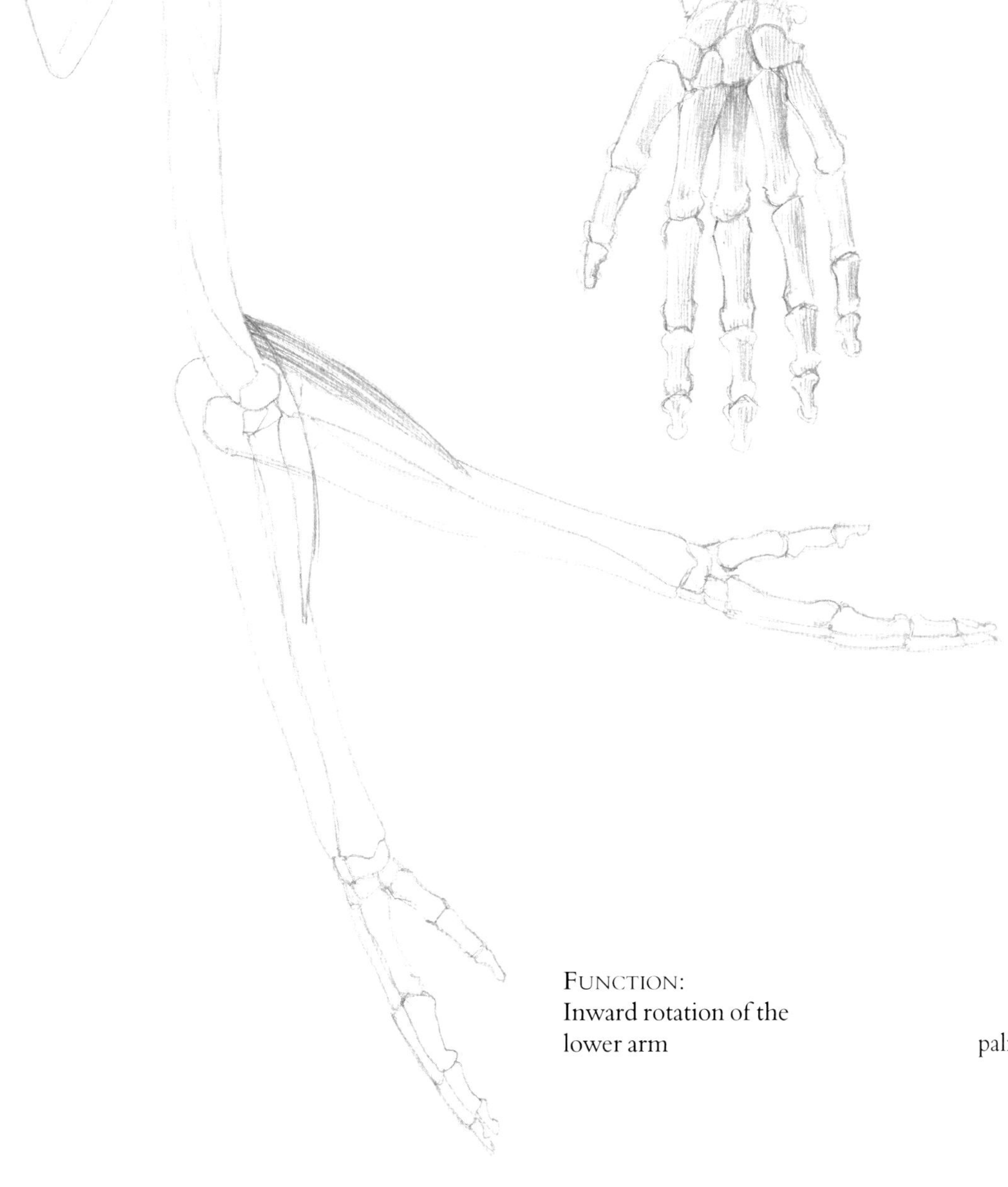

FUNCTION:
Inward rotation of the lower arm

palmar aspect

Fig. 63
The Pronator Quadratus, *59*

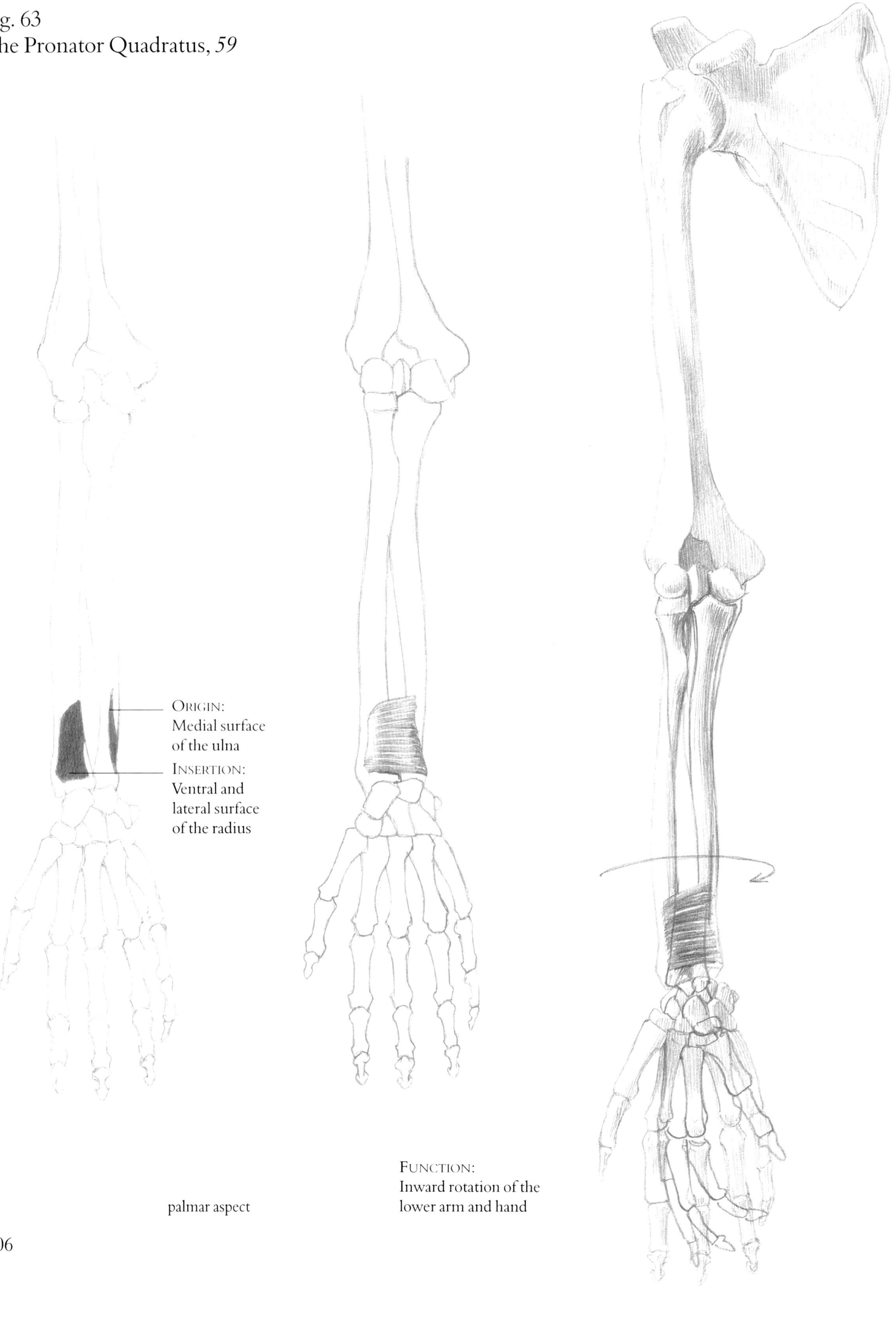
Origin:
Medial surface of the ulna
Insertion:
Ventral and lateral surface of the radius
Function:
Inward rotation of the lower arm and hand
palmar aspect

Fig. 64
The Long Flexor Muscle of the Thumb

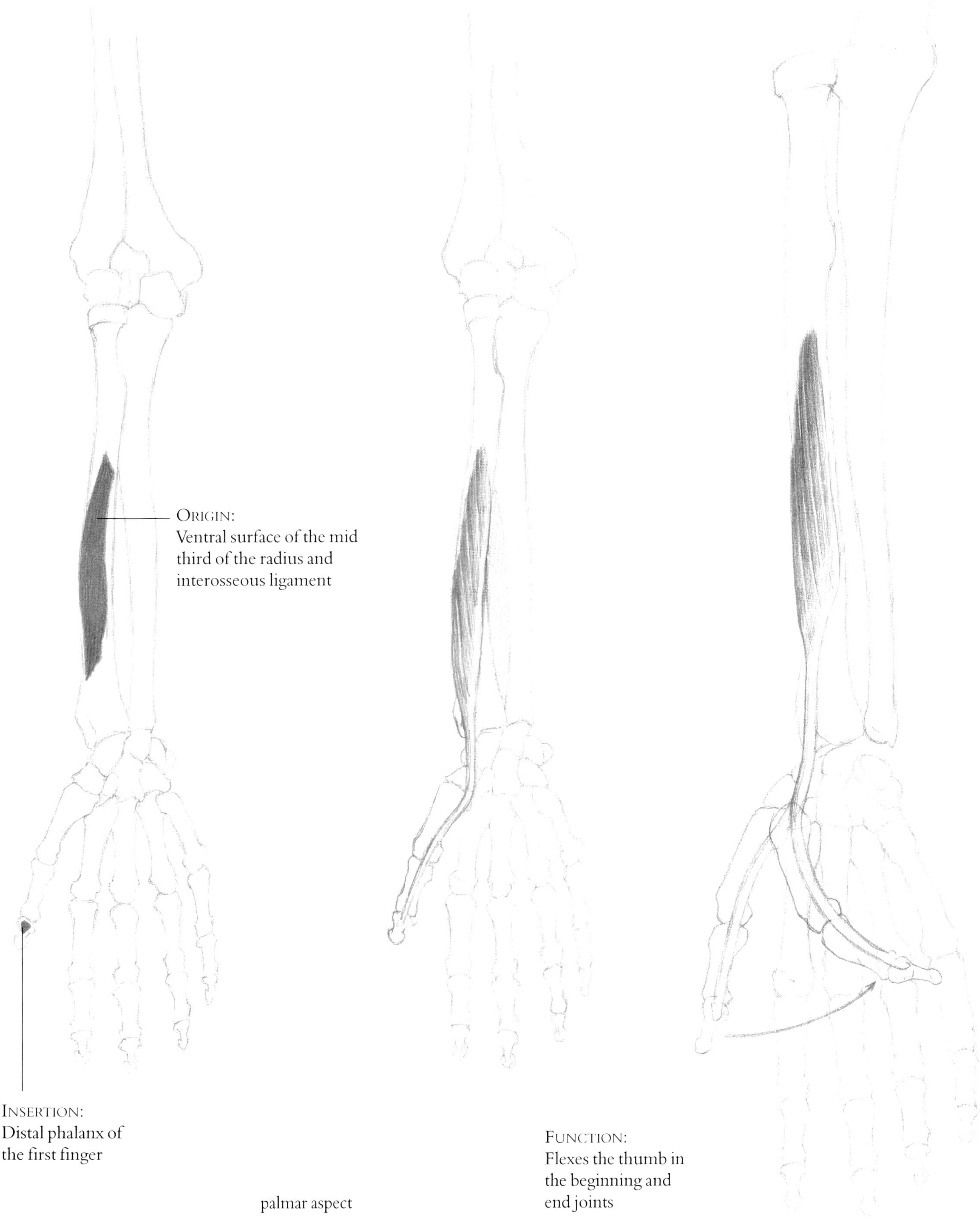
ORIGIN:
Ventral surface of the mid third of the radius and interosseous ligament
INSERTION:
Distal phalanx of the first finger
FUNCTION:
Flexes the thumb in the beginning and end joints
palmar aspect

Fig. 65
The Long Palmar Muscle

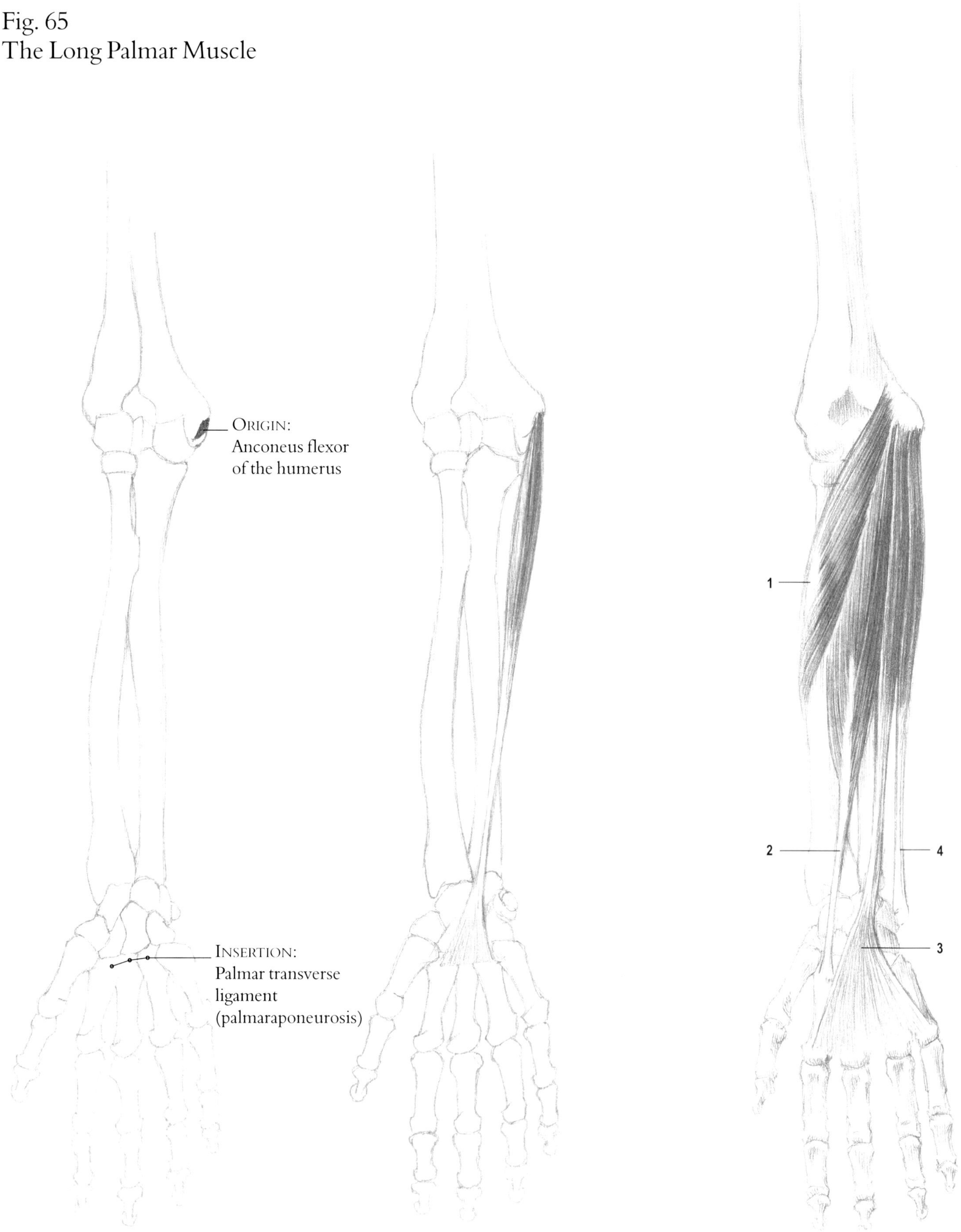
ORIGIN:
Anconeus flexor
of the humerus
INSERTION:
Palmar transverse
ligament
(palmaraponeurosis)
1
2
4
3

1 Supinator (*65*)
2 Flexor carpi radialis (*53*)
3 Palmaris longus (*54/1*)
4 Flexor carpi ulnaris (*55*)

palmar aspect

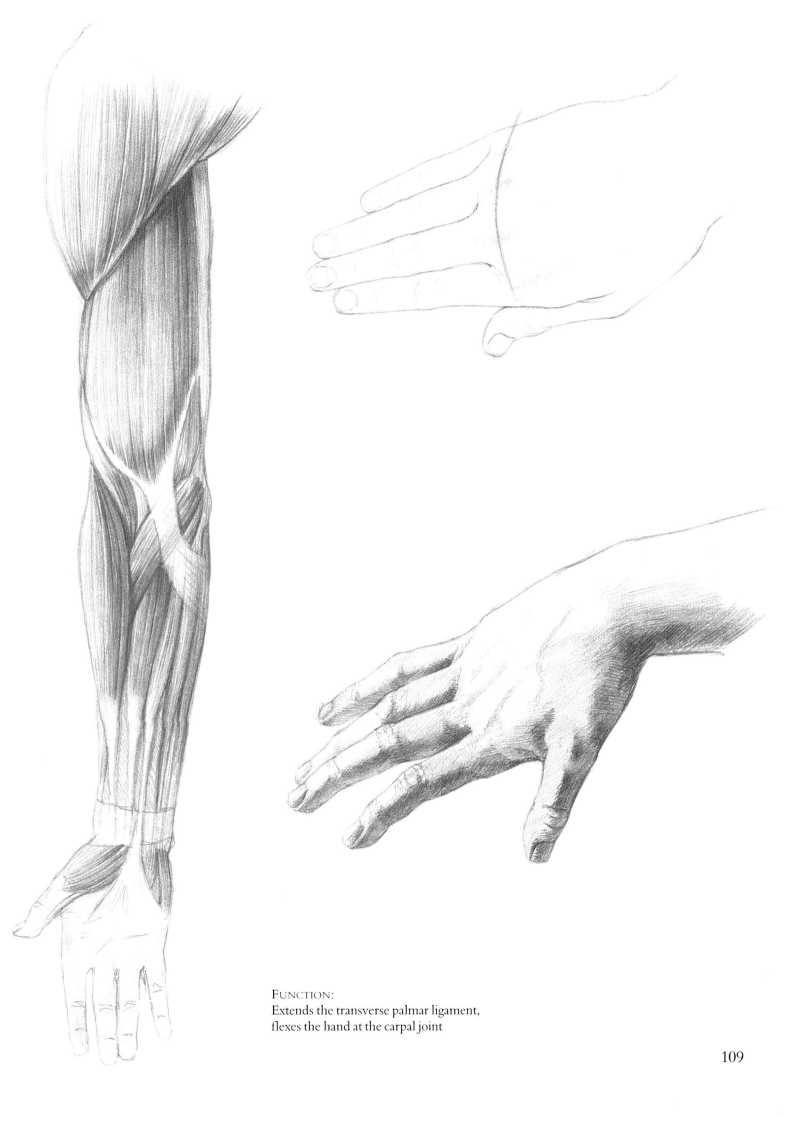

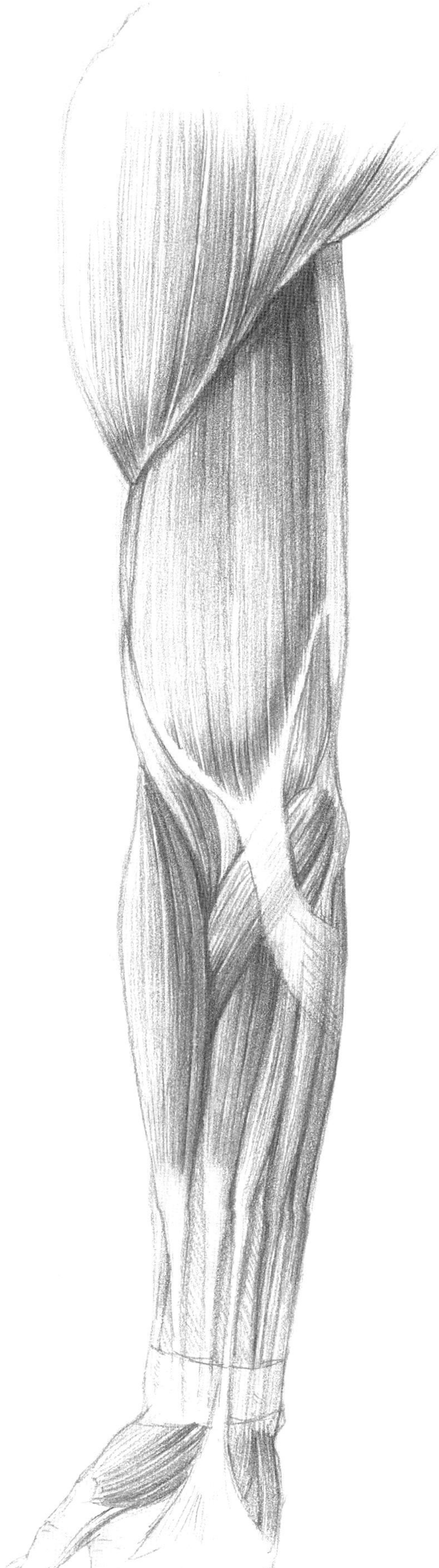

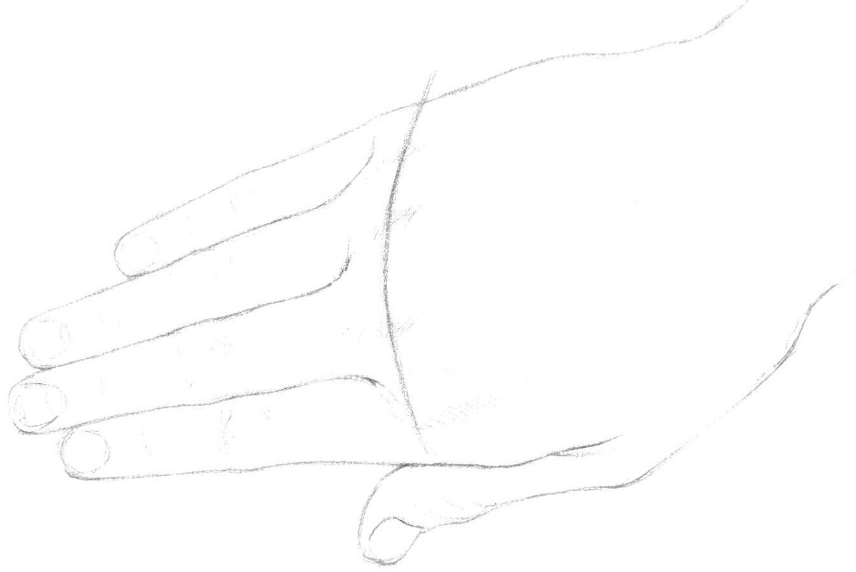

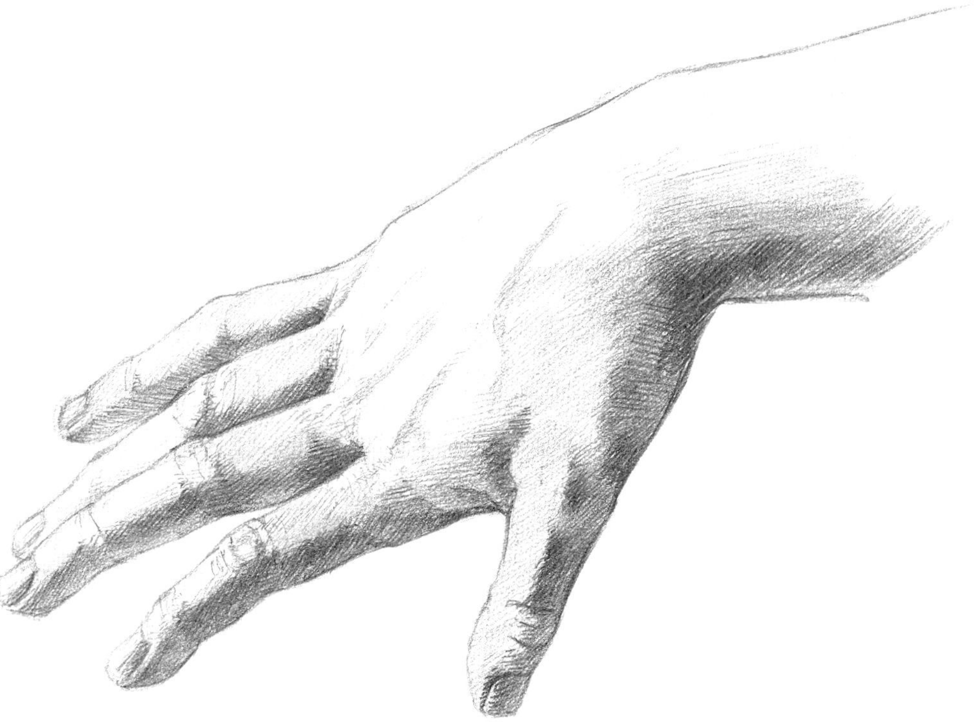

FUNCTION:
Extends the transverse palmar ligament,
flexes the hand at the carpal joint

Fig. 66
The Long Abductor Muscle of the Thumb

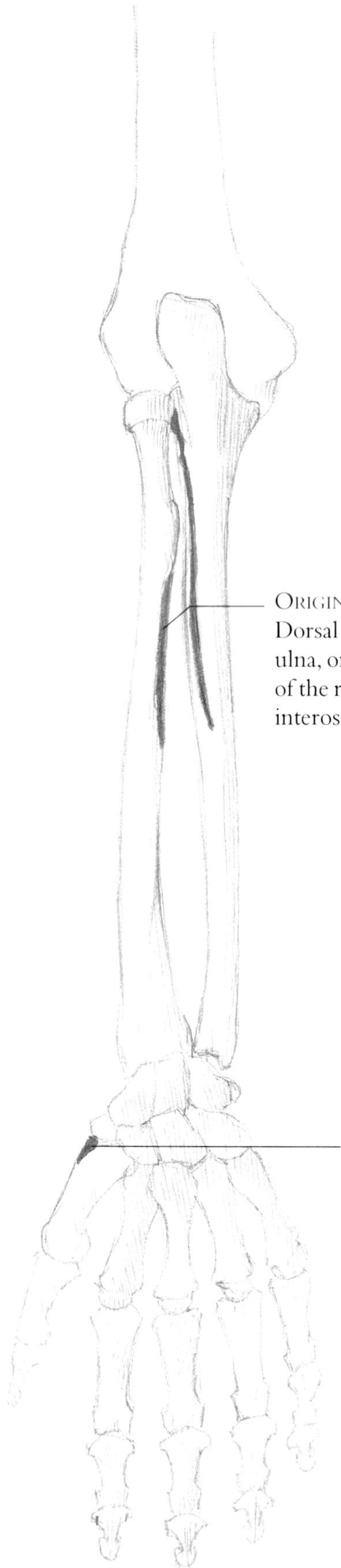

ORIGIN:
Dorsal surface of the ulna, on the mid third of the radius and interosseous ligament

INSERTION:
Dorsal surface of the first metacarpal bones

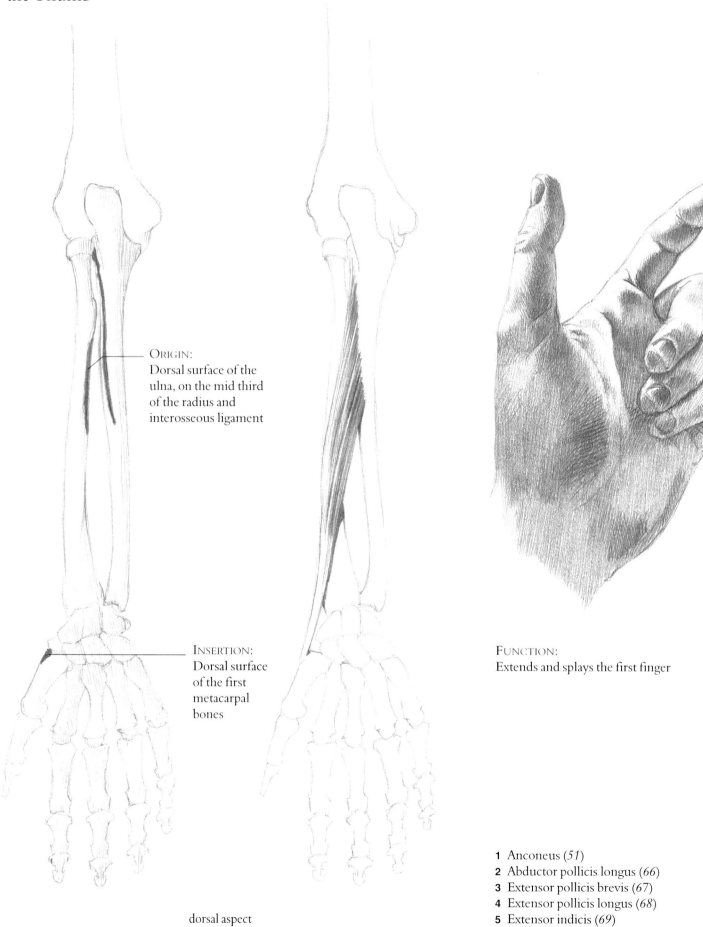

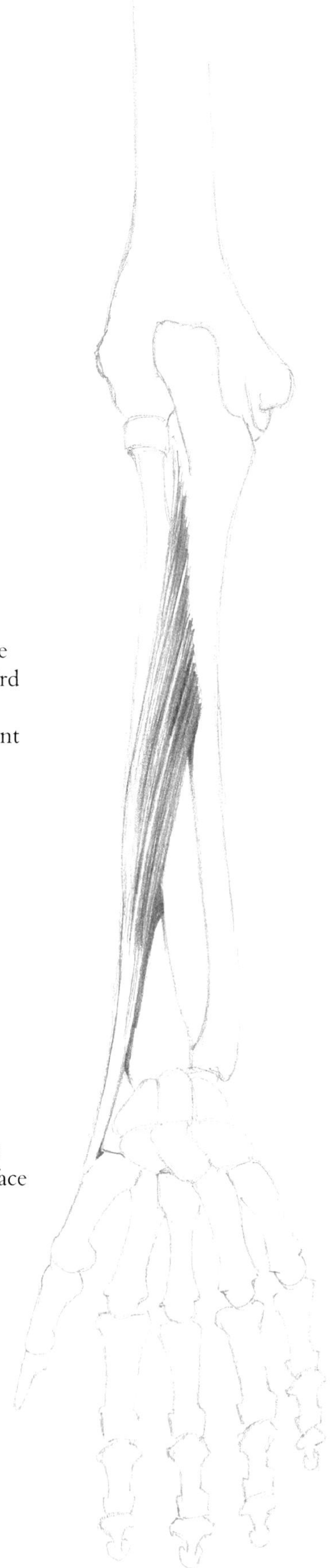

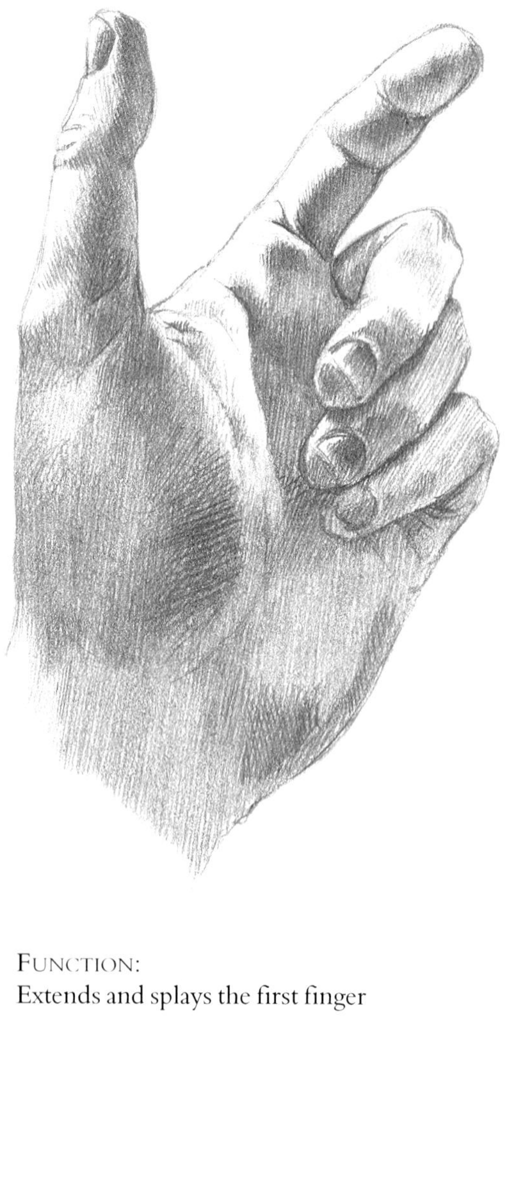

FUNCTION:
Extends and splays the first finger

dorsal aspect

1 Anconeus (*51*)
2 Abductor pollicis longus (*66*)
3 Extensor pollicis brevis (*67*)
4 Extensor pollicis longus (*68*)
5 Extensor indicis (*69*)

Fig. 67
The Radial Flexor Muscle of the Wrist

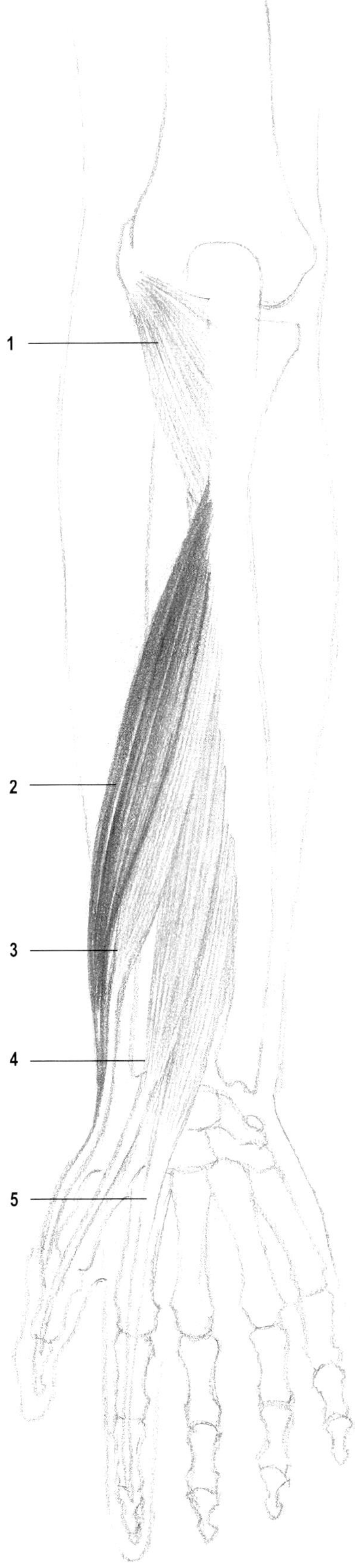

dorsal aspect

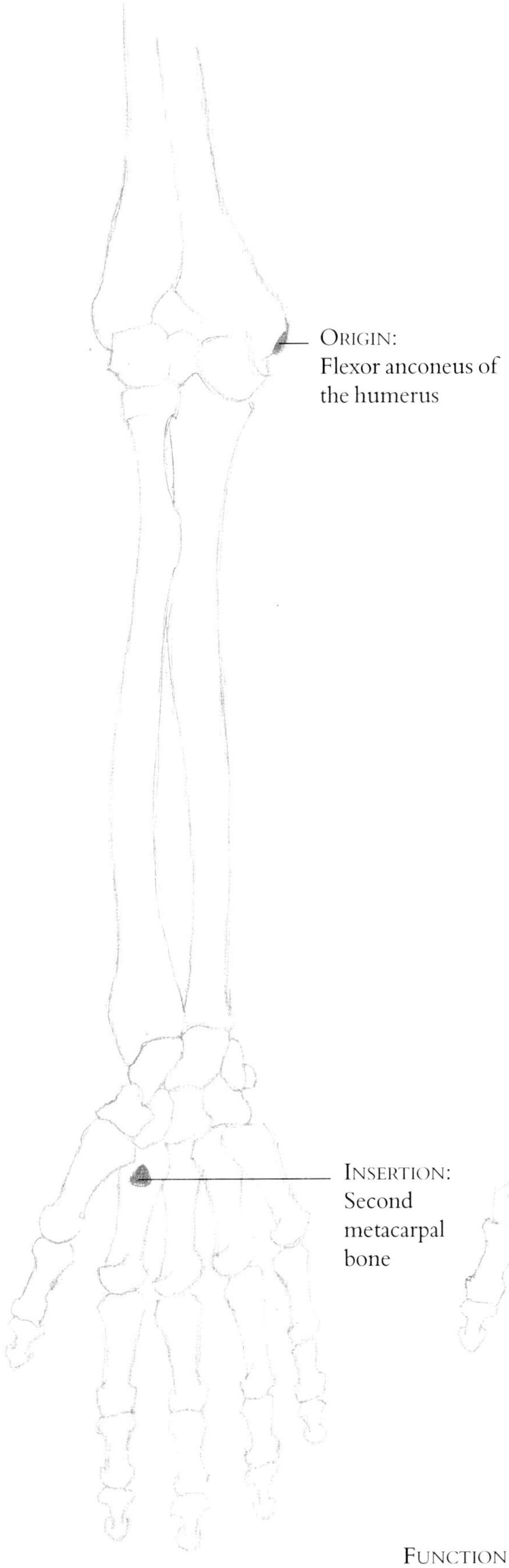

palmar aspect

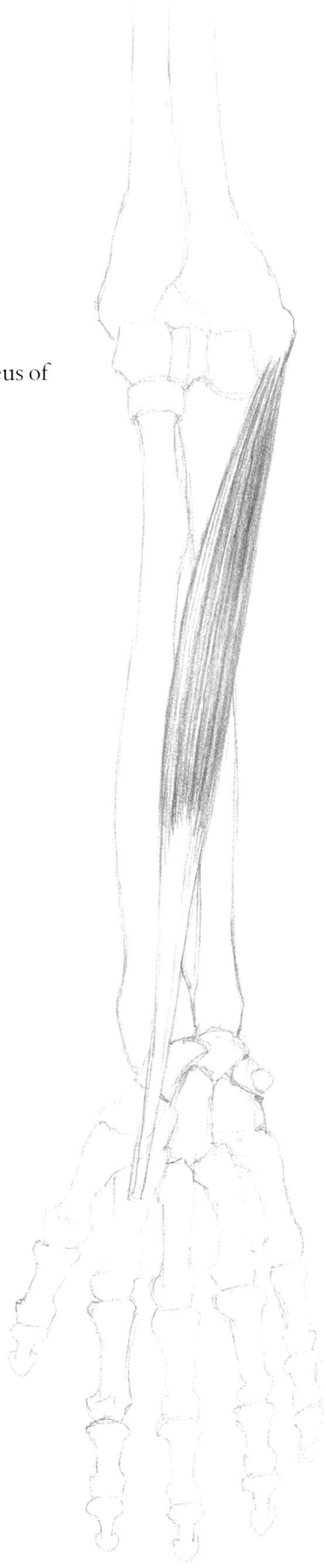

FUNCTION:
Flexes the hand toward the radial side, rotates towards the body

Fig. 68
The Ulnar Flexor Muscle of the Wrist

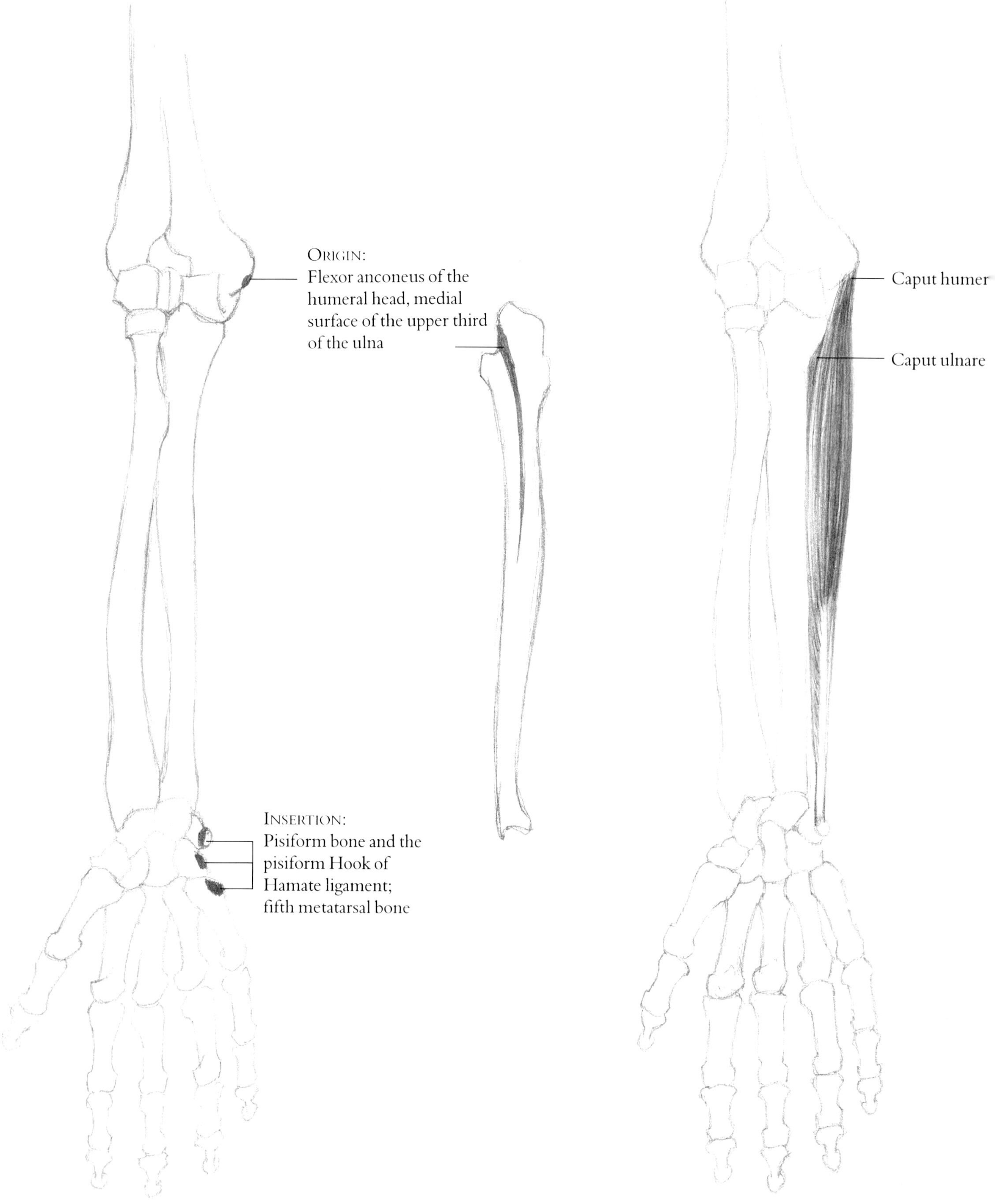

palmar aspect

Fig. 68/a
The Ulnar Extensor Muscle of the Wrist

(M. extensor carpi ulnaris, 64, see also p. 124)

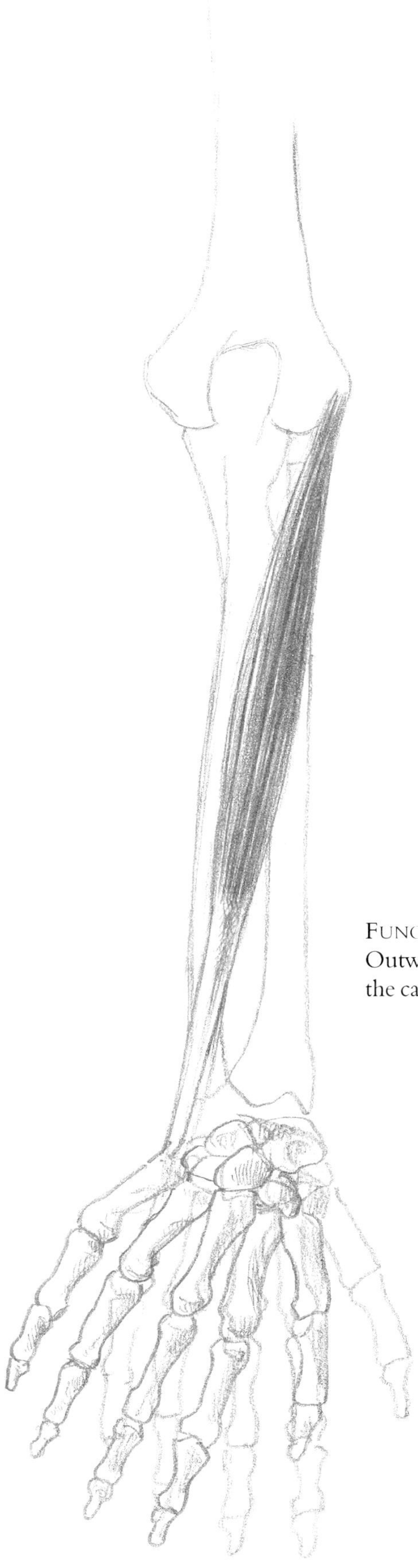

FUNCTION:
Outwards rotation of the carpal joint; flexes the carpus and pulls it towards the body

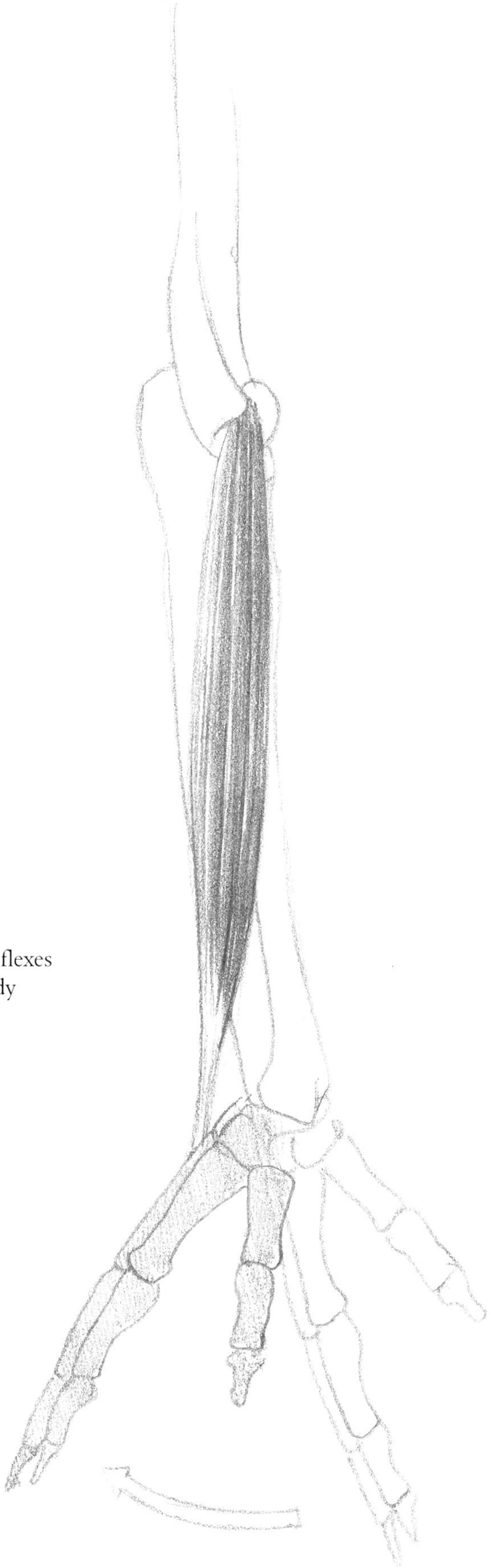

dorsal aspect

lateral aspect

Fig. 69
The Flexor Digitorum Superficialis, *56*

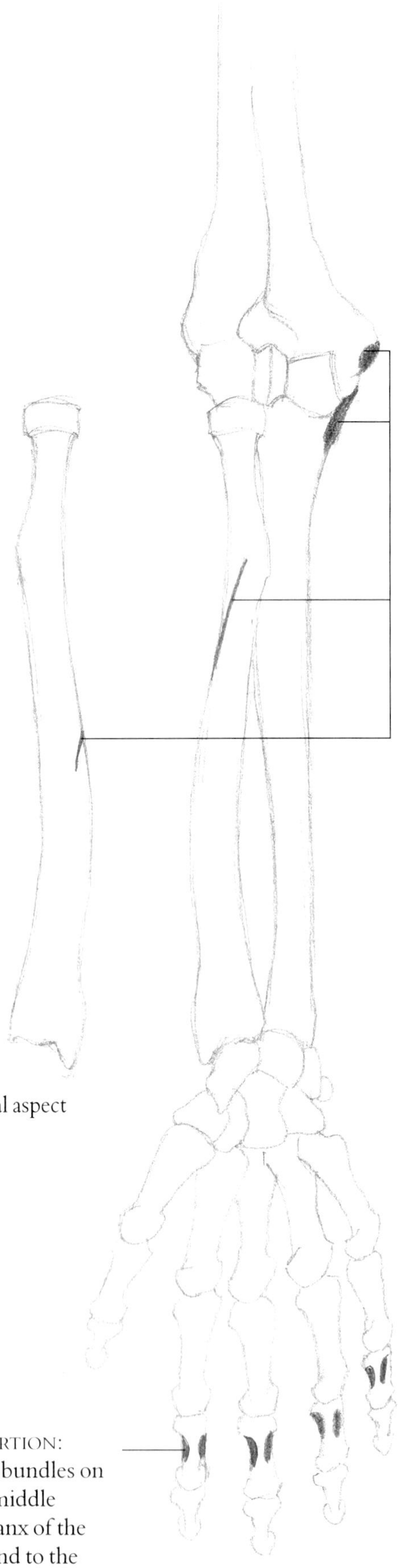

ORIGIN:
Flexor anconeus of the humerus, medial ligament of the elbow joint (humero-ulnar head), in the tendon below the coronoid process of the radius

dorsal aspect

INSERTION:
Two bundles on the middle phalanx of the second to the fifth fingers.

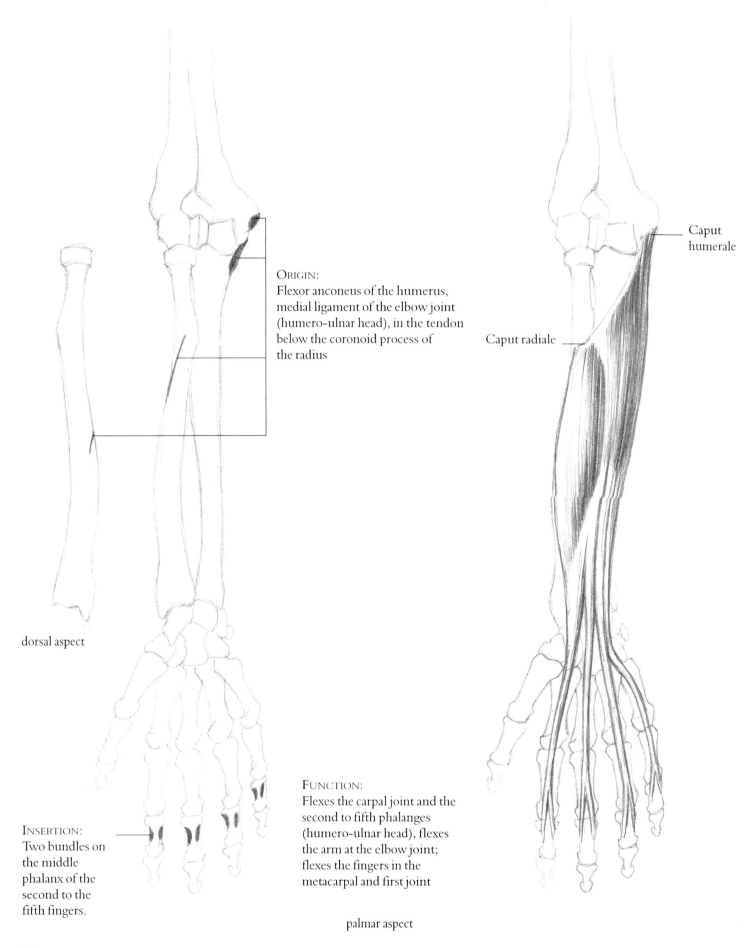

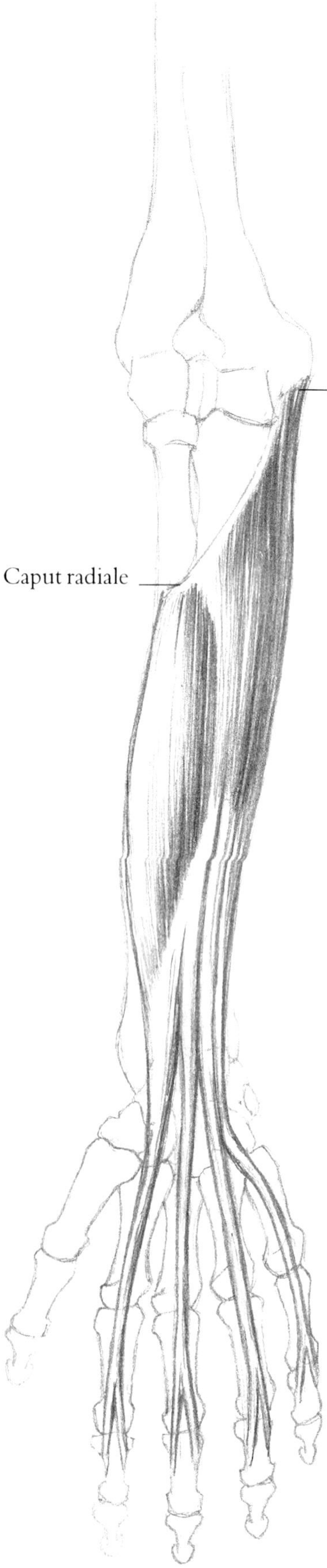

FUNCTION:
Flexes the carpal joint and the second to fifth phalanges (humero-ulnar head), flexes the arm at the elbow joint; flexes the fingers in the metacarpal and first joint

palmar aspect

Fig. 70
The Flexor Digitorum Profundus, *57*

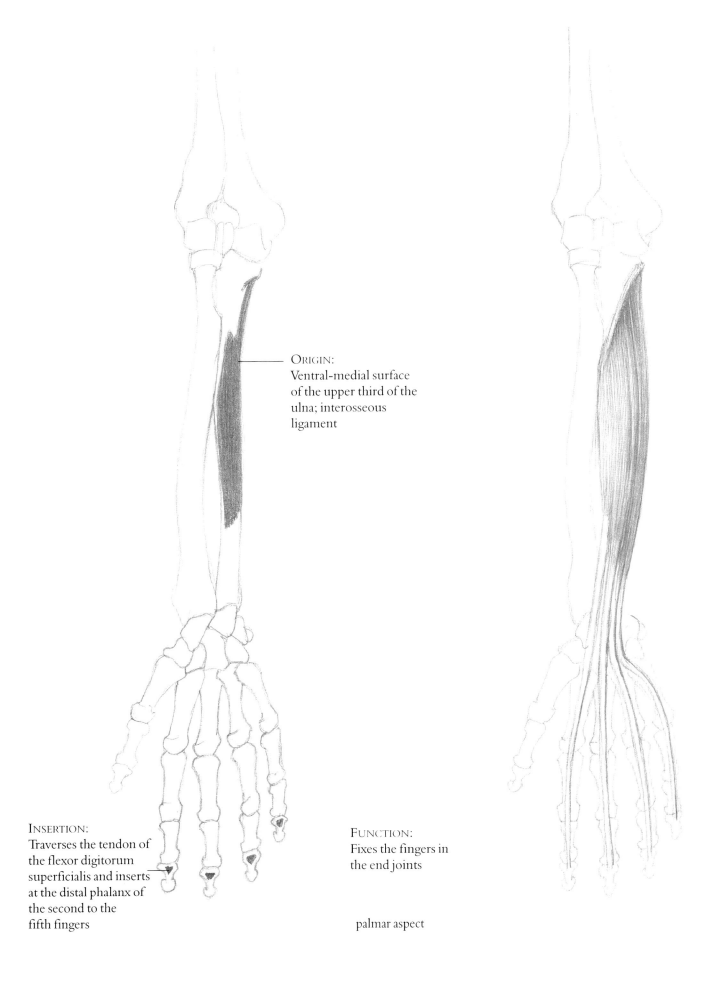

ORIGIN:
Ventral-medial surface of the upper third of the ulna; interosseous ligament

INSERTION:
Traverses the tendon of the flexor digitorum superficialis and inserts at the distal phalanx of the second to the fifth fingers

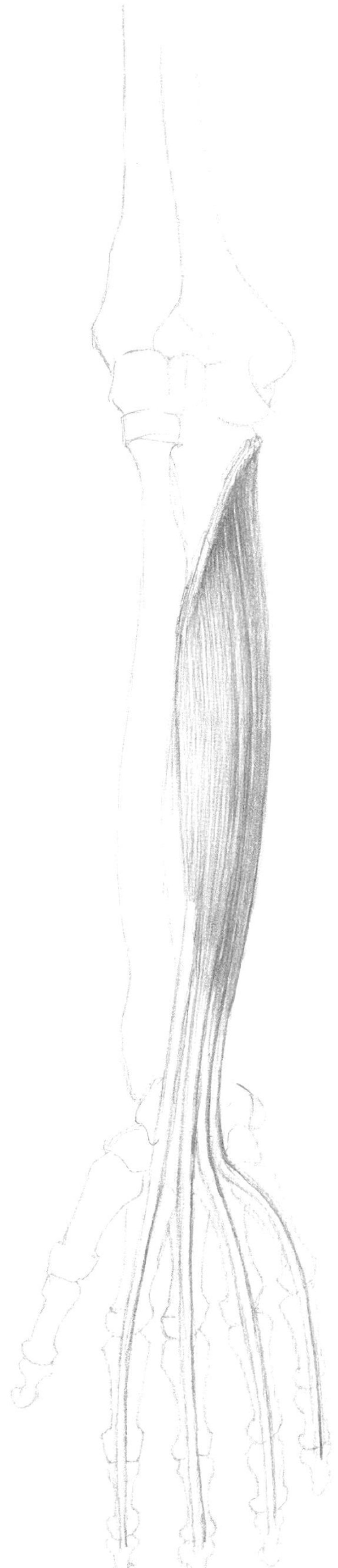

FUNCTION:
Fixes the fingers in the end joints

palmar aspect

Fig. 71
The Finger Flexors

1 Humeral head
2 Radial head
3 Tendon
4 Pronator quadratus
5 Abductor pollicis longus
6 Lumbricals

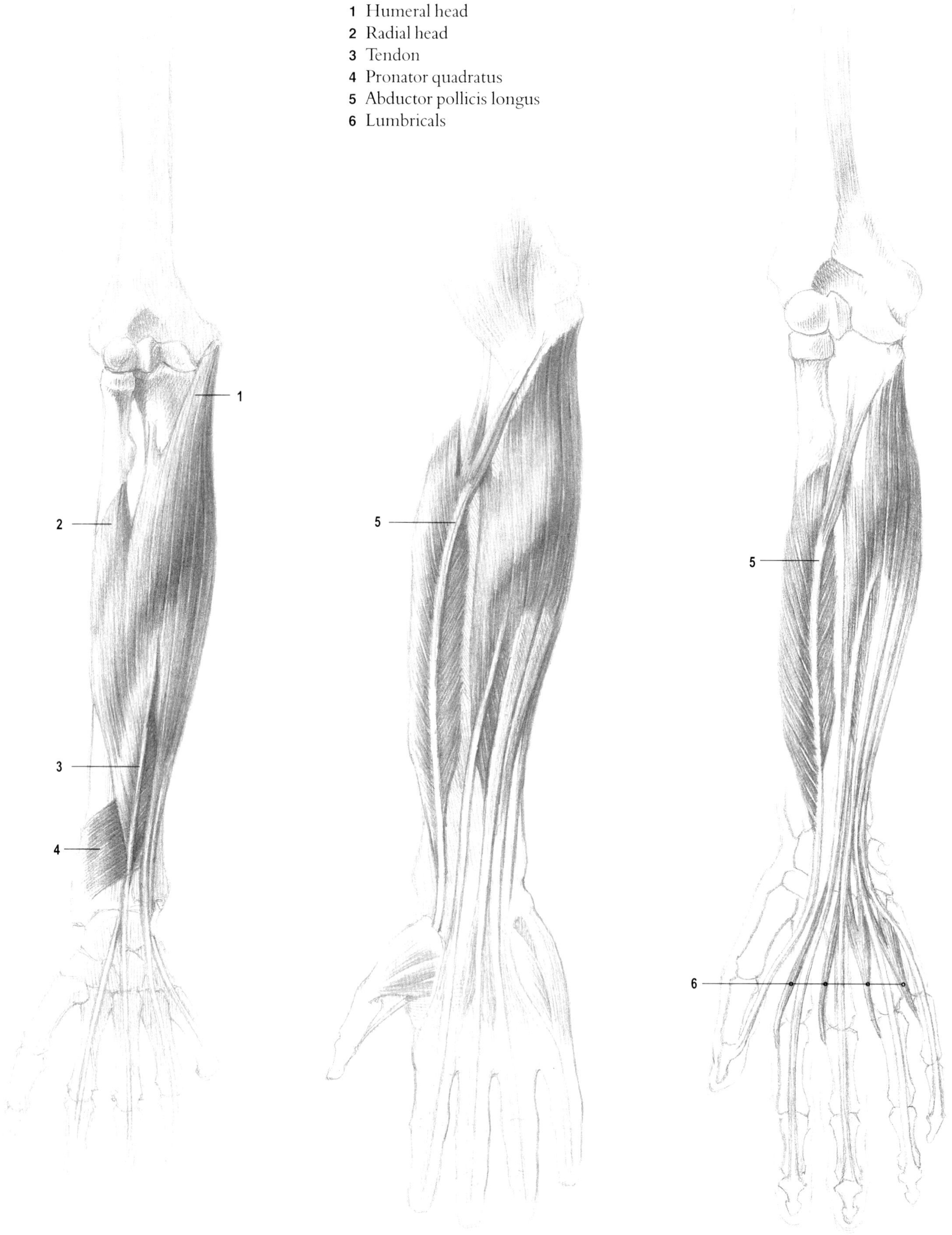

superficial muscle layer

middle muscle layer

deep muscle layer

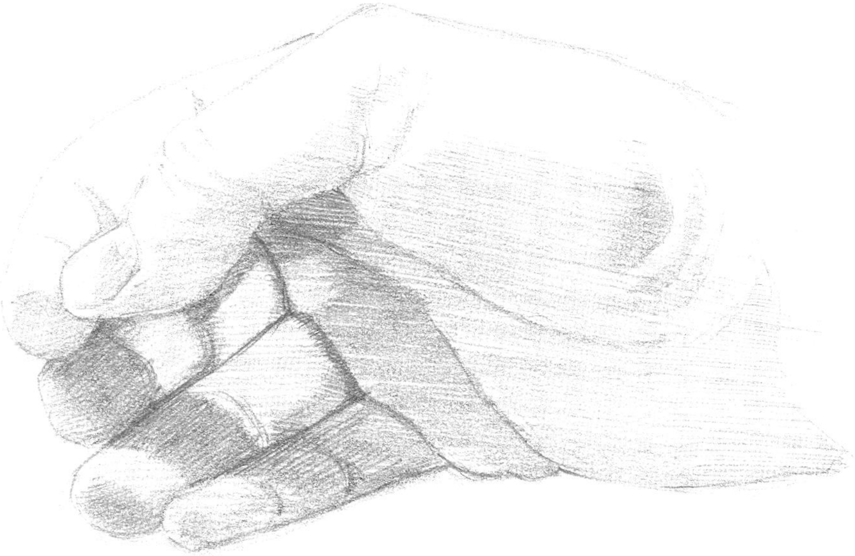

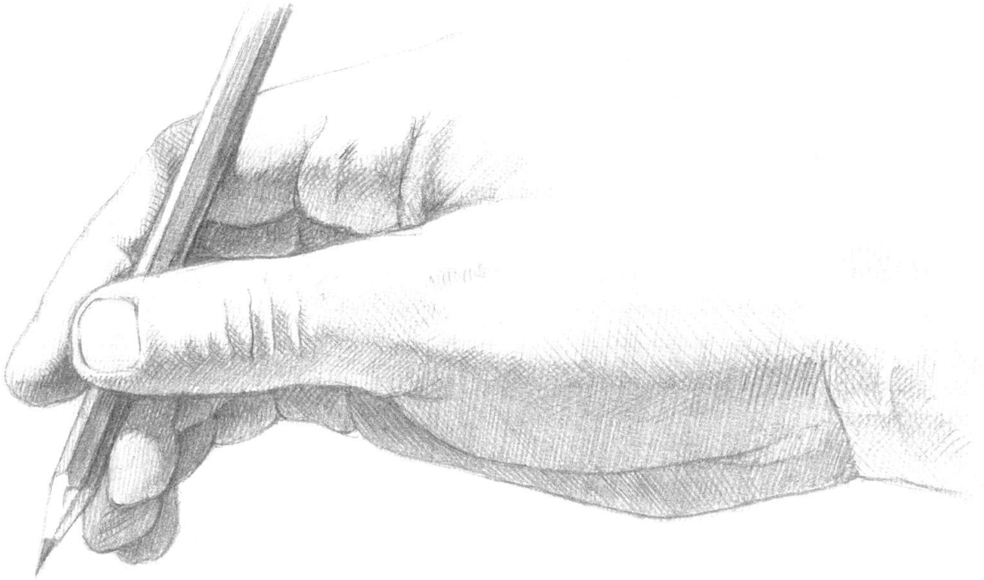

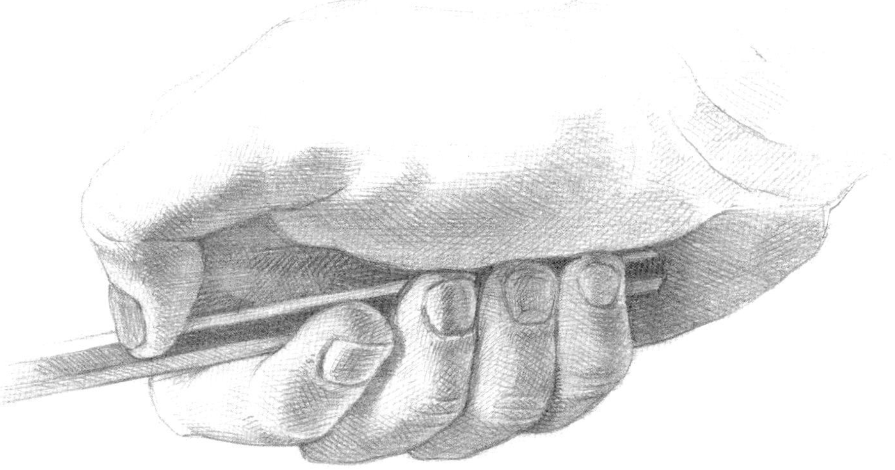

Fig. 72
The Extensor Carpi Radialis Longus, *61*

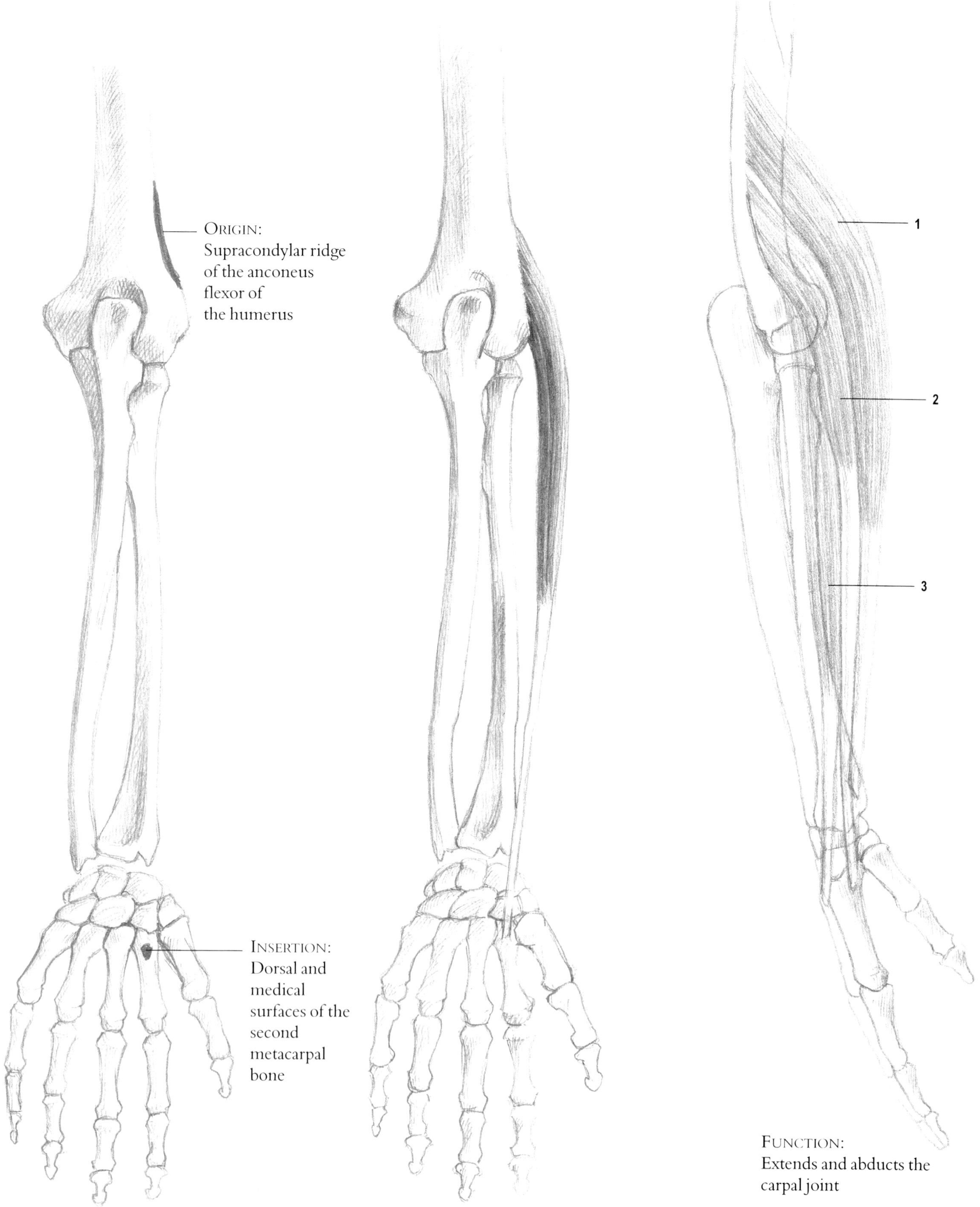
ORIGIN:
Supracondylar ridge of the anconeus flexor of the humerus
INSERTION:
Dorsal and medical surfaces of the second metacarpal bone
1
2
3
FUNCTION:
Extends and abducts the carpal joint

dorsal aspect

lateral aspect

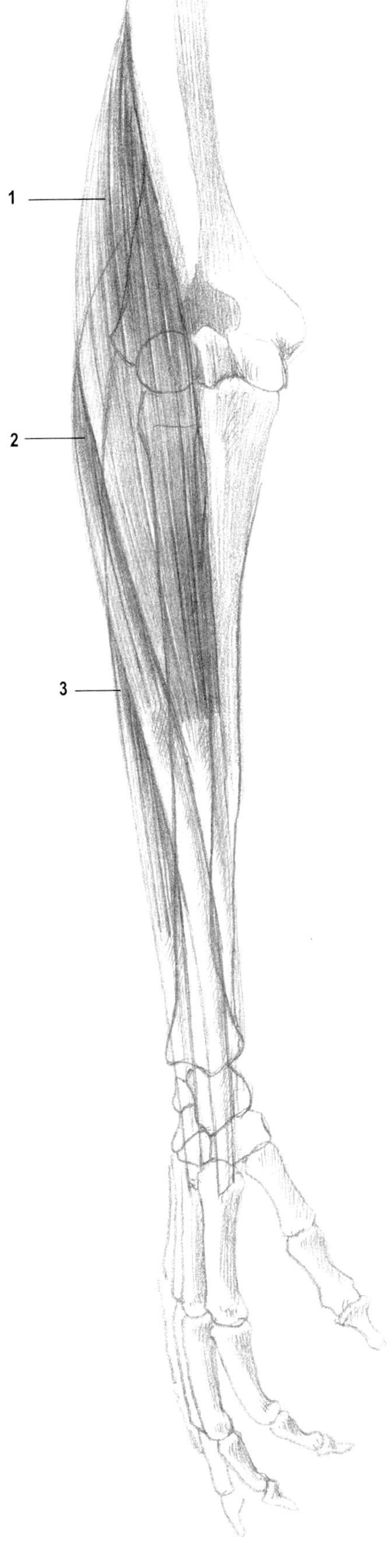

anteriolateral aspect

palmar aspect

lateral aspect

1 Brachioradialis (*60*)
2 Extensor carpi radialis longus (*61*)
3 Extensor carpi radialis brevis (*61*)

Fig. 73
The Brachioradial Muscle
(M. brachioradialis, *60*)

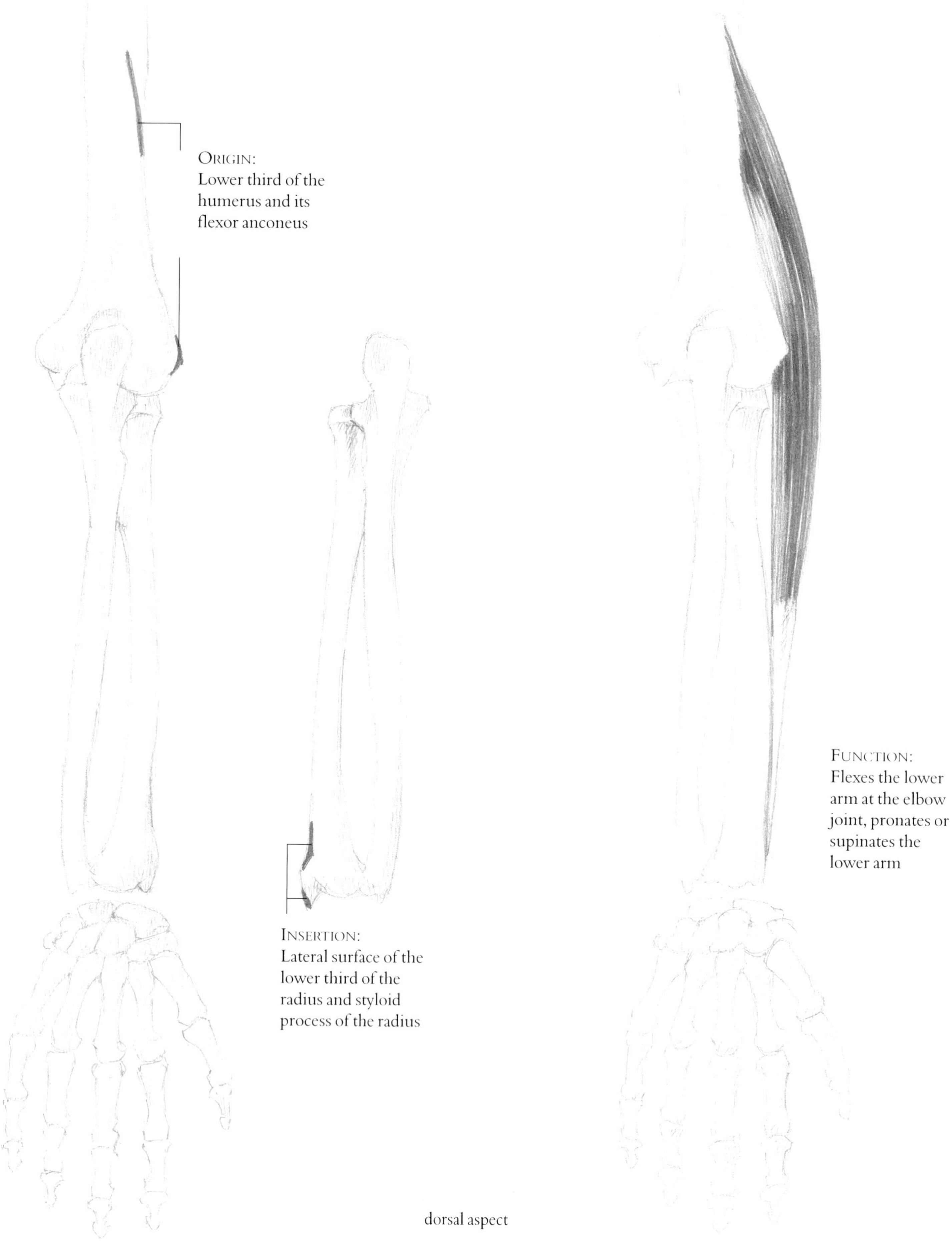
ORIGIN:
Lower third of the humerus and its flexor anconeus
INSERTION:
Lateral surface of the lower third of the radius and styloid process of the radius
FUNCTION:
Flexes the lower arm at the elbow joint, pronates or supinates the lower arm

dorsal aspect

Fig. 74
The Radial Extensor Muscle of the Wrist
(M. extensor carpi radialis, *61*)

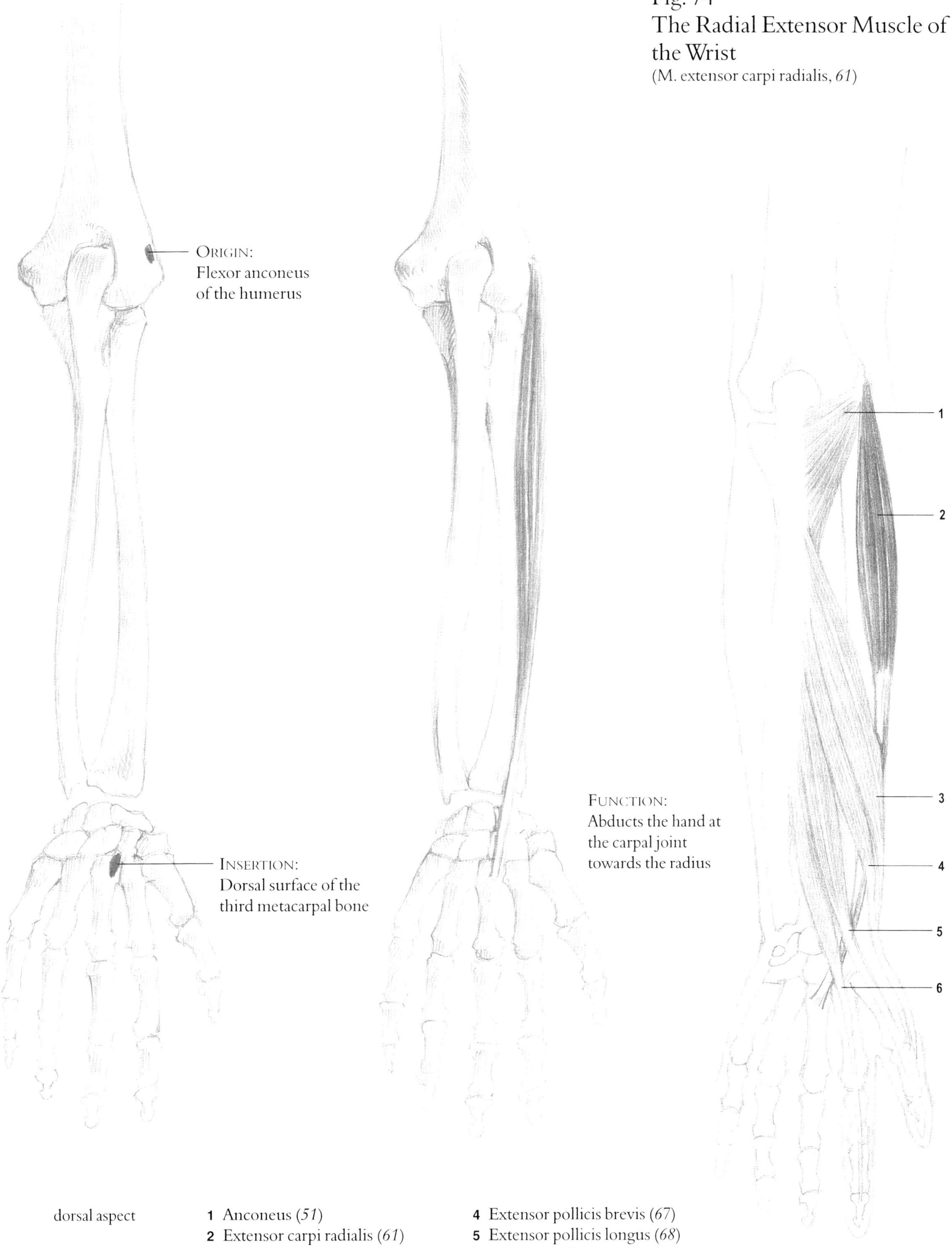

ORIGIN:
Flexor anconeus of the humerus
INSERTION:
Dorsal surface of the third metacarpal bone
FUNCTION:
Abducts the hand at the carpal joint towards the radius
1
2
3
4
5
6

dorsal aspect

1 Anconeus (*51*)
2 Extensor carpi radialis (*61*)
3 Flexor pollicis longus (*58*)
4 Extensor pollicis brevis (*67*)
5 Extensor pollicis longus (*68*)
6 Extensor indicis (*69*)

Fig. 75
The Extensor Muscle of the Fingers
(M. extensor digitorum, *62*)

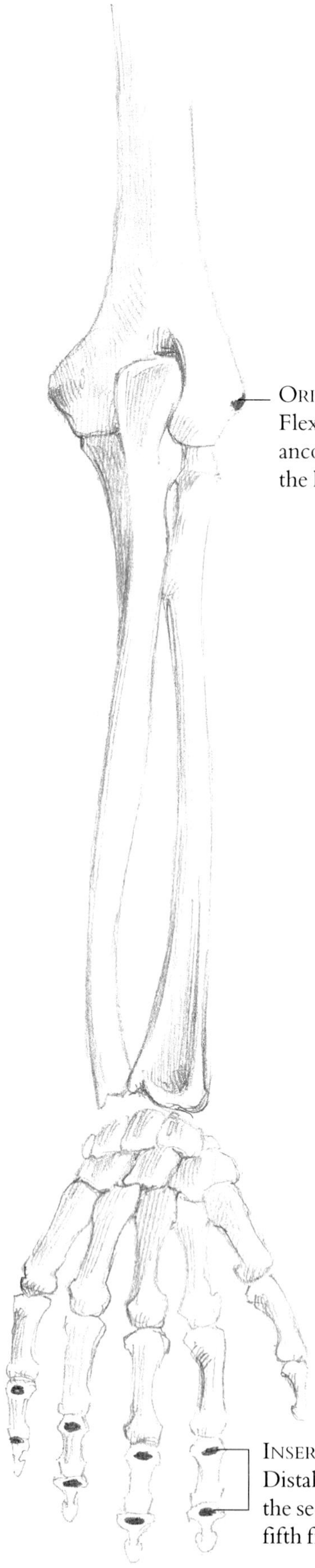

ORIGIN:
Flexor anconeus of the humerus

INSERTION:
Distal phalanx of the second to fifth fingers

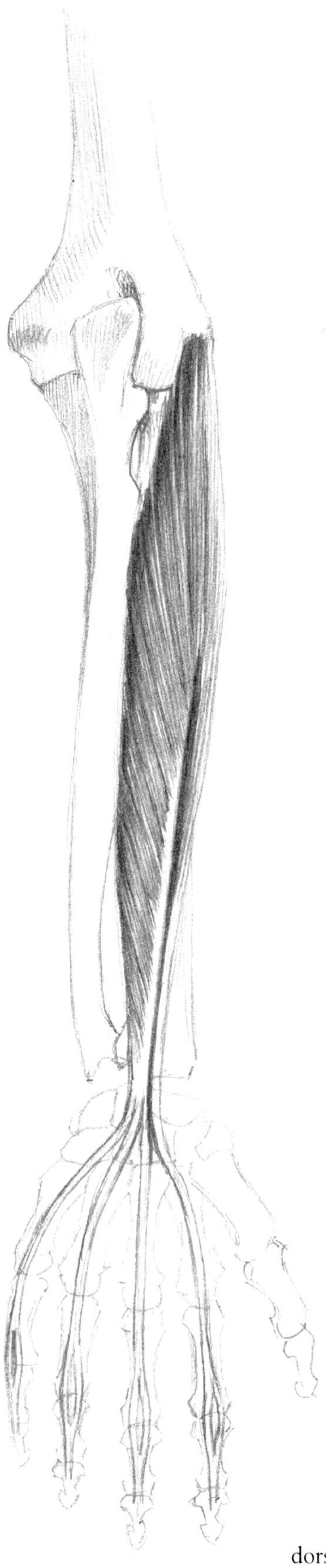

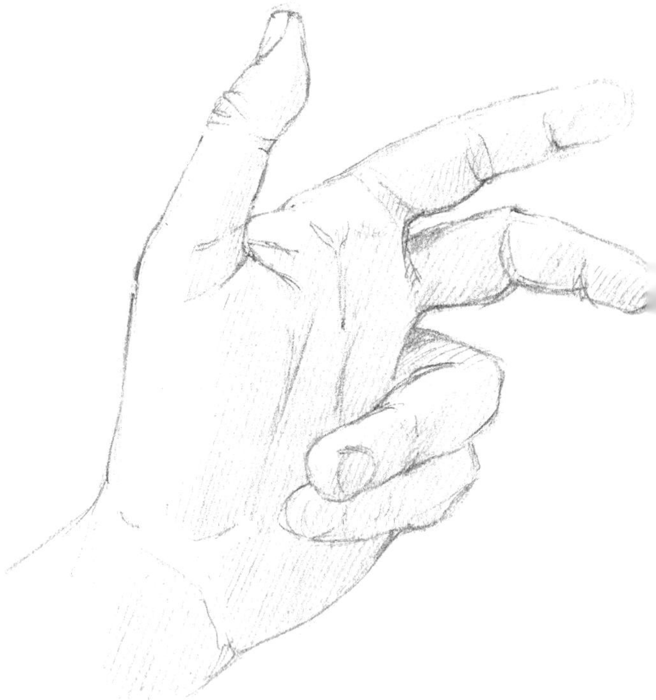

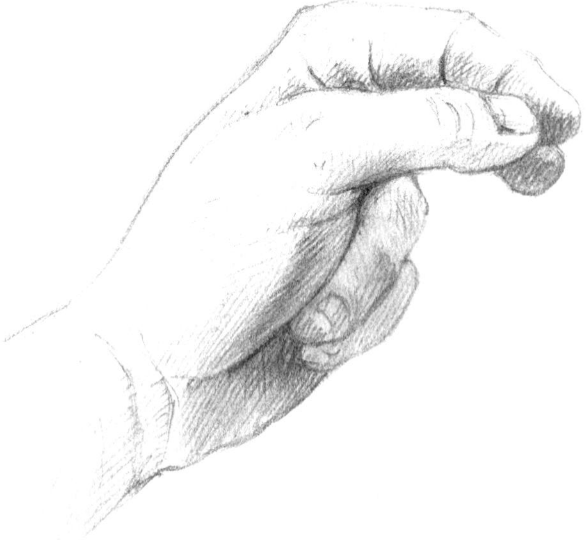

FUNCTION:
Extends the second to fifth fingers

dorsal aspect

Fig. 76
The Extensor Muscle of the Little Finger
(M. extensor digiti minimi, *63*)

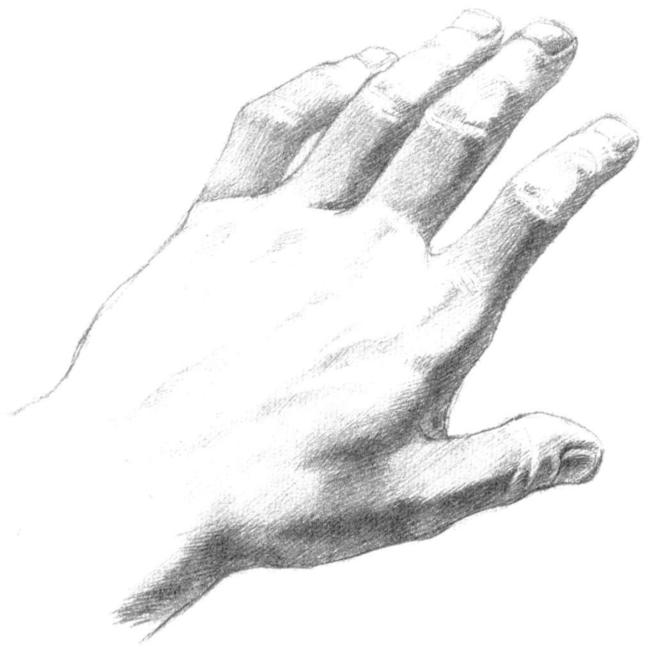

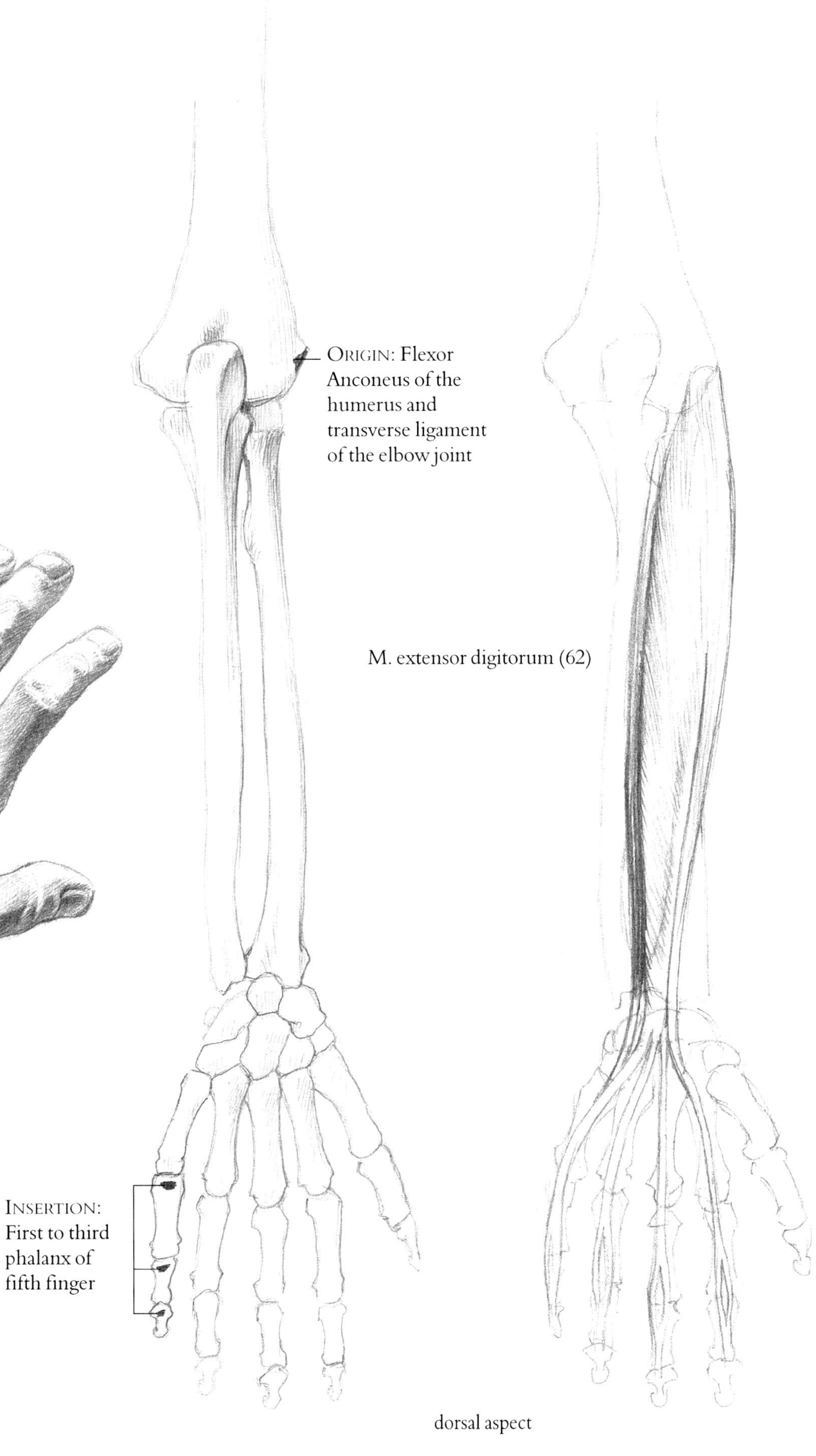

FUNCTION:
Extends and retracts the fifth finger

dorsal aspect

Fig. 77
The Ulnar Extensor Muscle of the Wrist
(M. extensor carpi ulnaris, *64*)

FUNCTION:
Ulnar abduction and extension in the carpal joint (with muscles *55* and *61*)

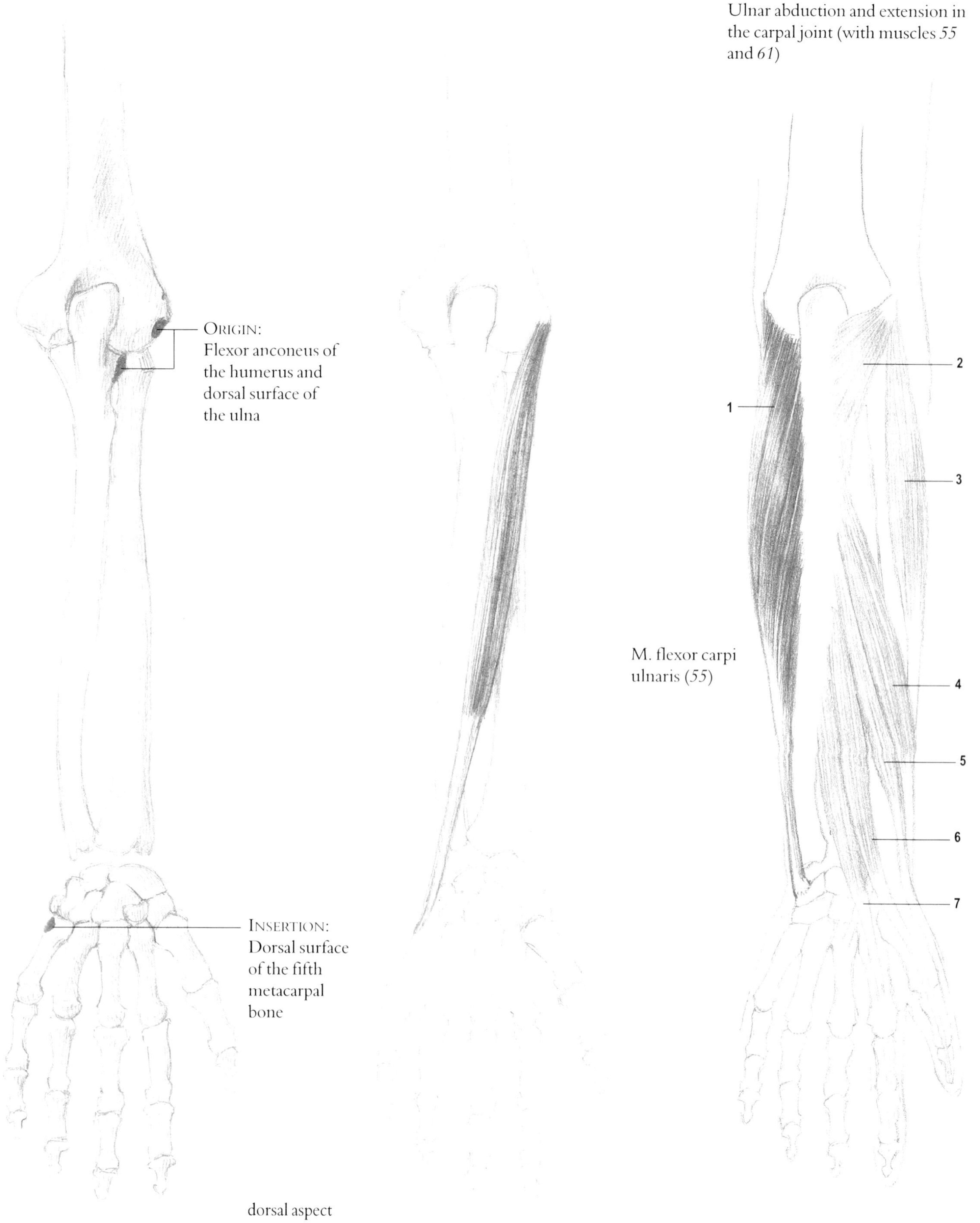

ORIGIN:
Flexor anconeus of the humerus and dorsal surface of the ulna
INSERTION:
Dorsal surface of the fifth metacarpal bone
M. flexor carpi ulnaris (*55*)
1
2
3
4
5
6
7

dorsal aspect

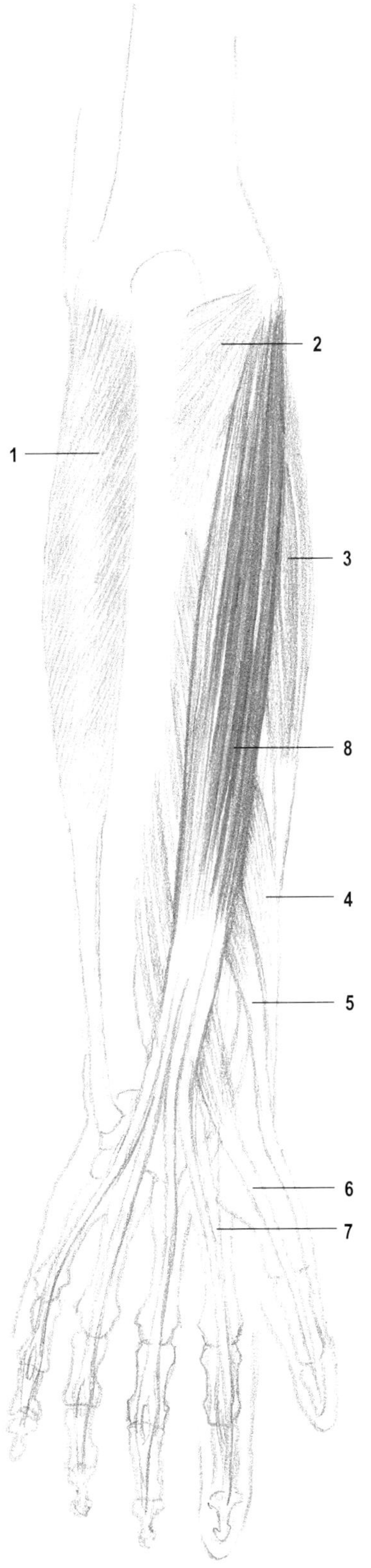

M. flexor carpi ulnaris (*55*)

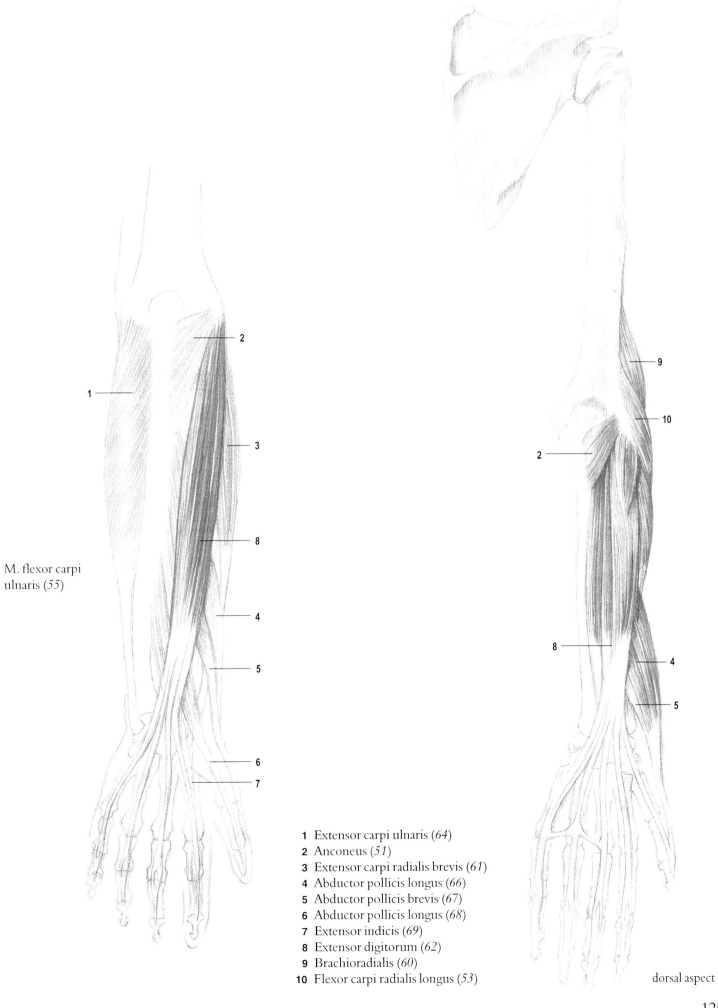

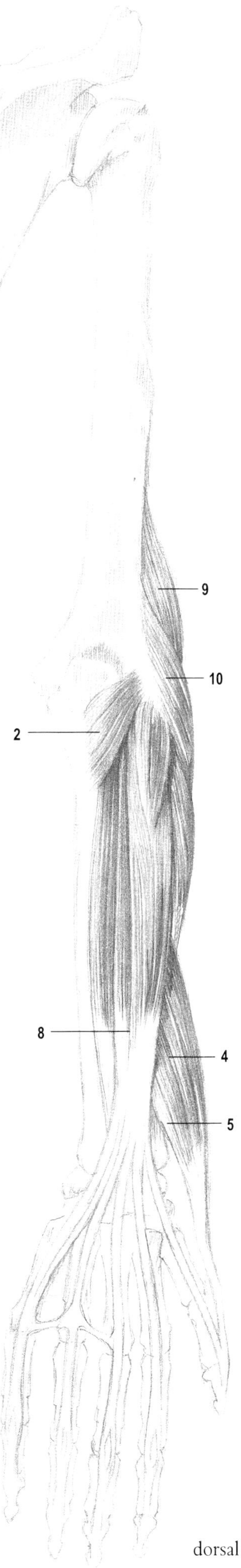

1 Extensor carpi ulnaris (*64*)
2 Anconeus (*51*)
3 Extensor carpi radialis brevis (*61*)
4 Abductor pollicis longus (*66*)
5 Abductor pollicis brevis (*67*)
6 Abductor pollicis longus (*68*)
7 Extensor indicis (*69*)
8 Extensor digitorum (*62*)
9 Brachioradialis (*60*)
10 Flexor carpi radialis longus (*53*)

dorsal aspect

Fig. 78
The Extensor Pollicis Brevis, *67*

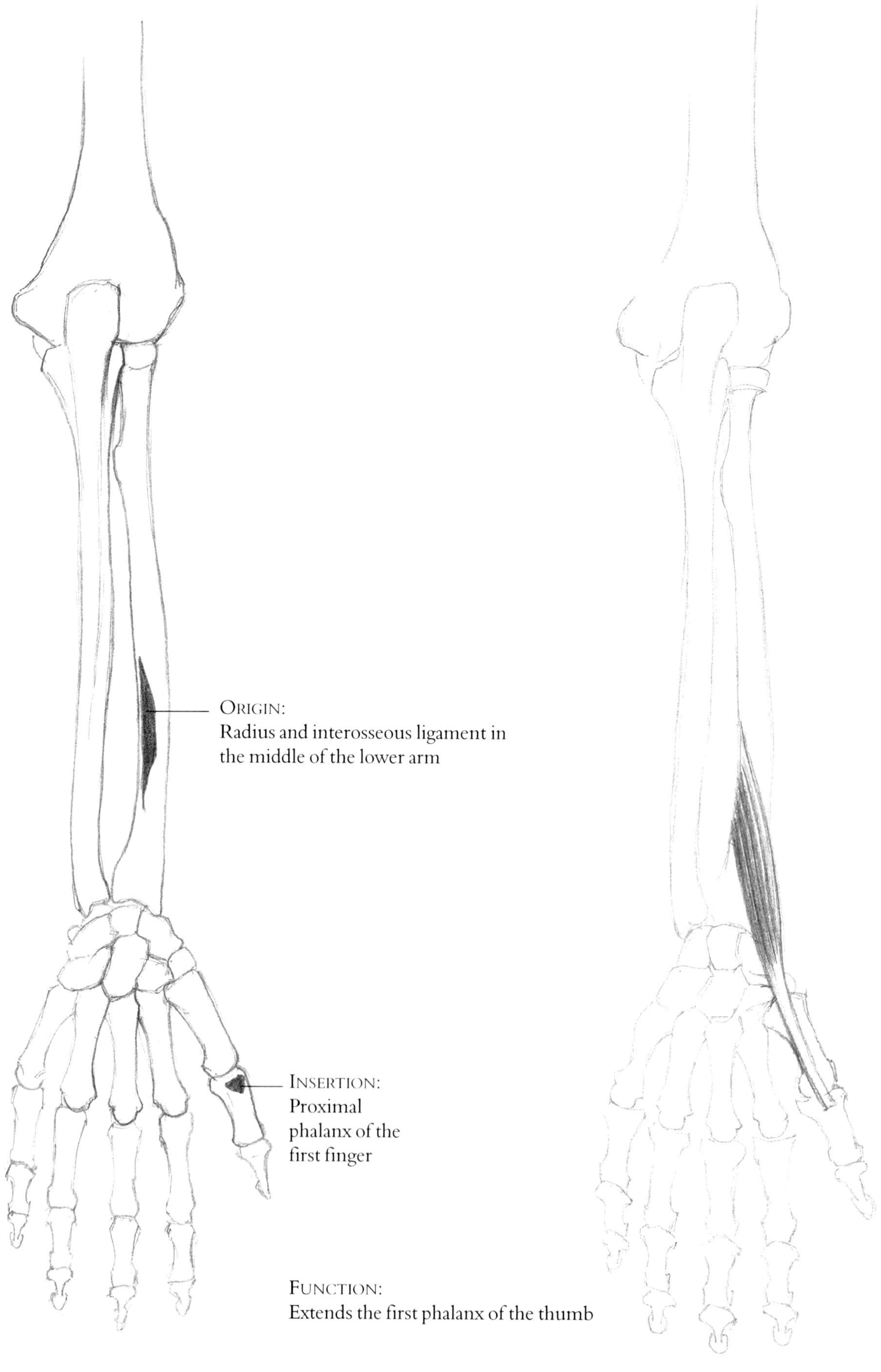
Origin:
Radius and interosseous ligament in the middle of the lower arm
Insertion:
Proximal phalanx of the first finger
Function:
Extends the first phalanx of the thumb

dorsal aspect

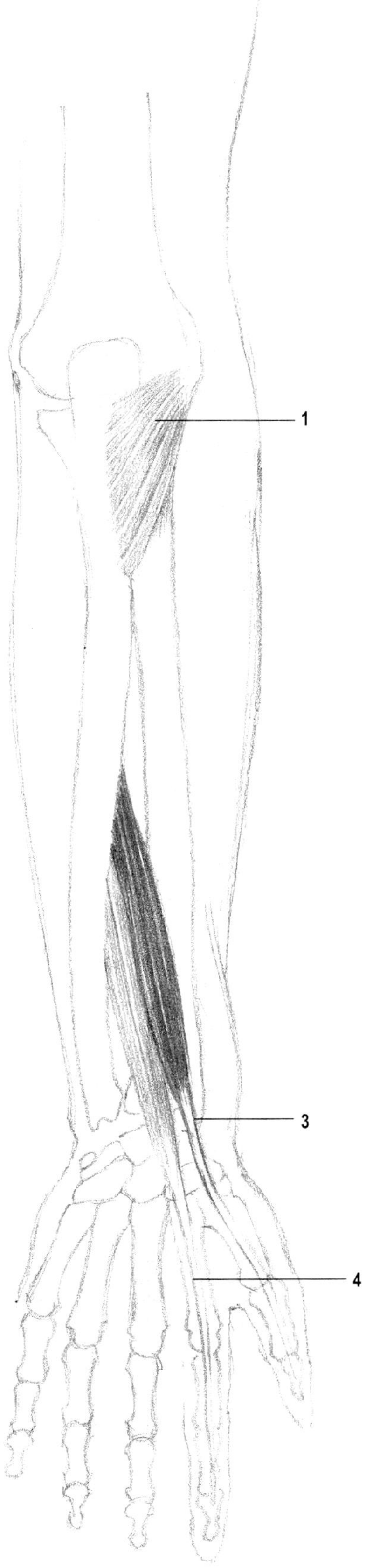

1 Anconeus (*51*)
2 Abductor pollicis longus (*66*)
3 Extensor pollicis brevis (*67*)
4 Extensor indicis (*69*)

dorsal aspect

Fig. 79
The Extensor Pollicis Longus, *68*

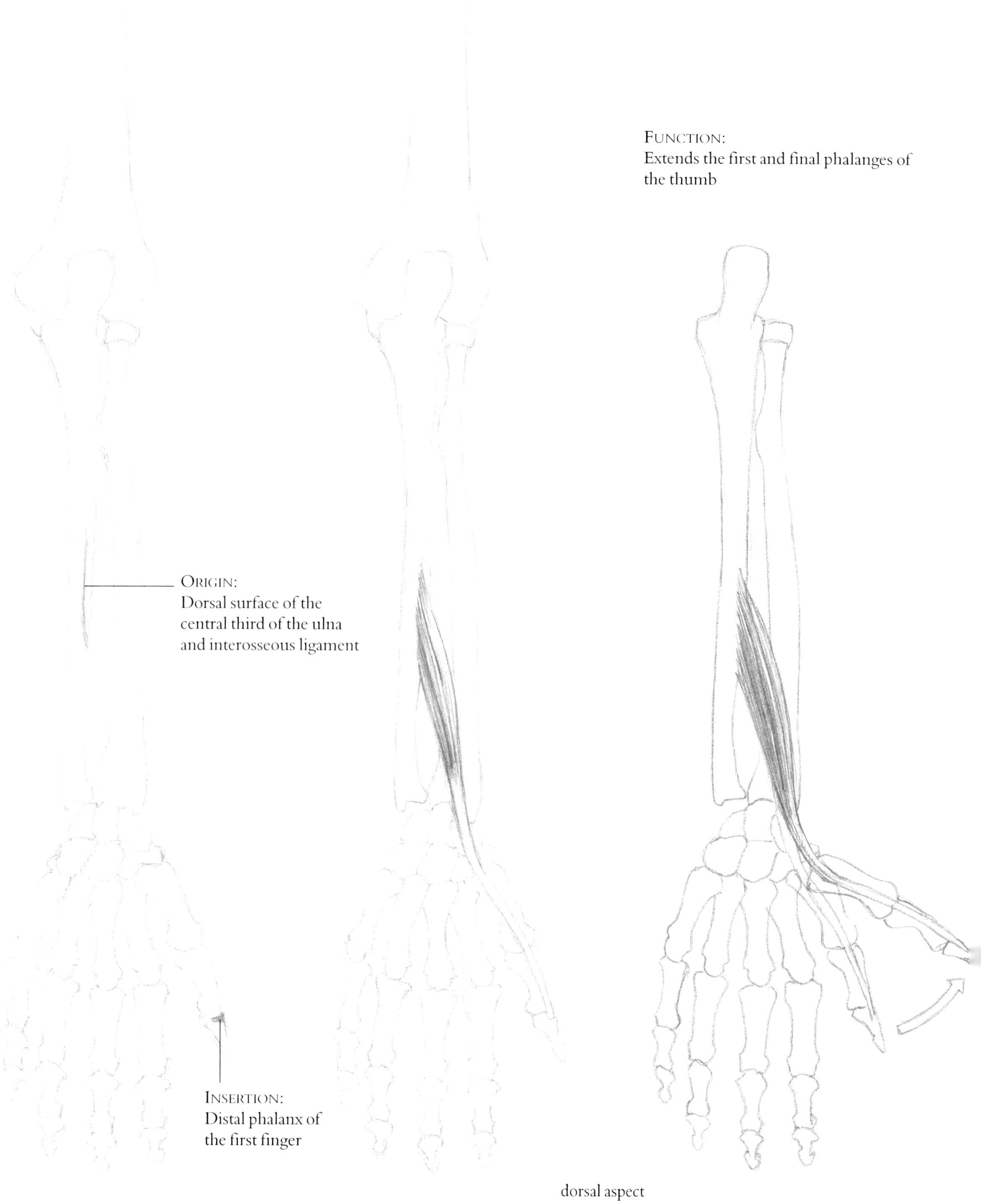
Function:
Extends the first and final phalanges of the thumb
Origin:
Dorsal surface of the central third of the ulna and interosseous ligament
Insertion:
Distal phalanx of the first finger
dorsal aspect

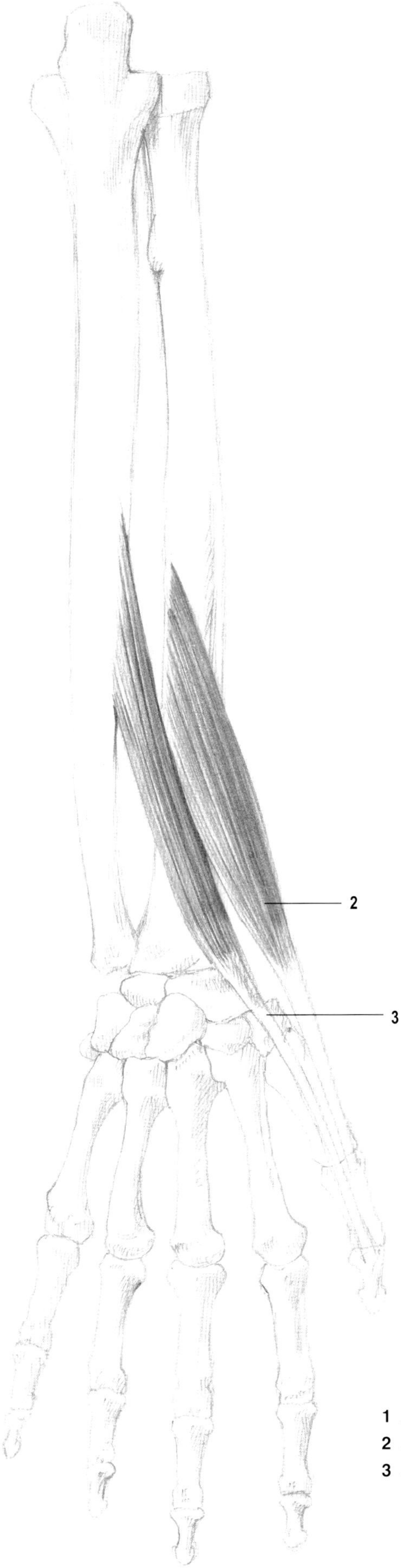

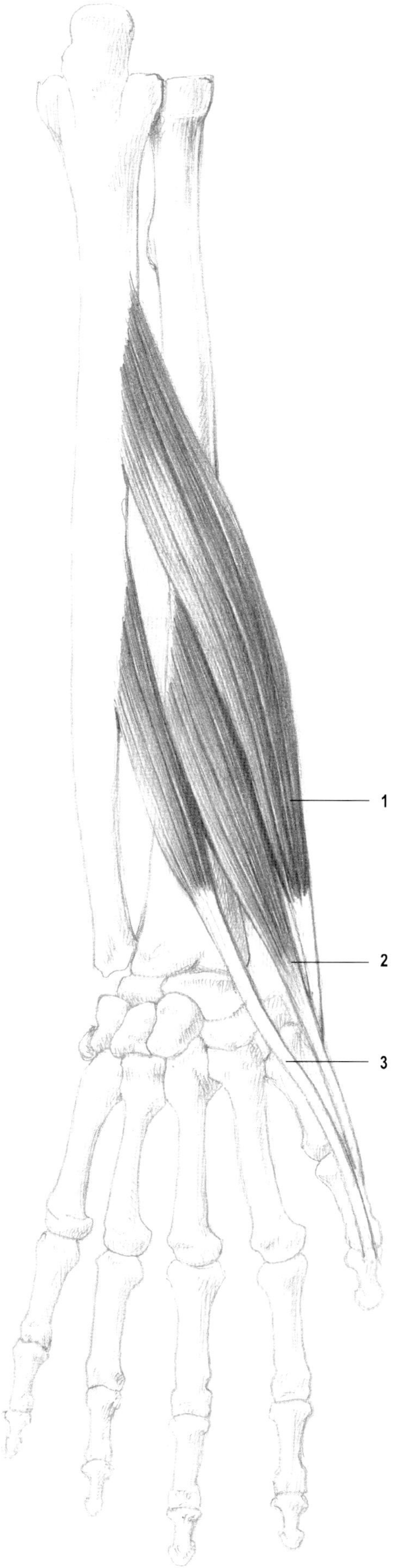

1 Abductor pollicis longus (*66*)
2 Extensor pollicis brevis (*67*)
3 Extensor pollicis longus (*68*)

dorsal aspect

Fig. 80
The Extensor Muscle of the Index

(M. extensor indicis, *69*)

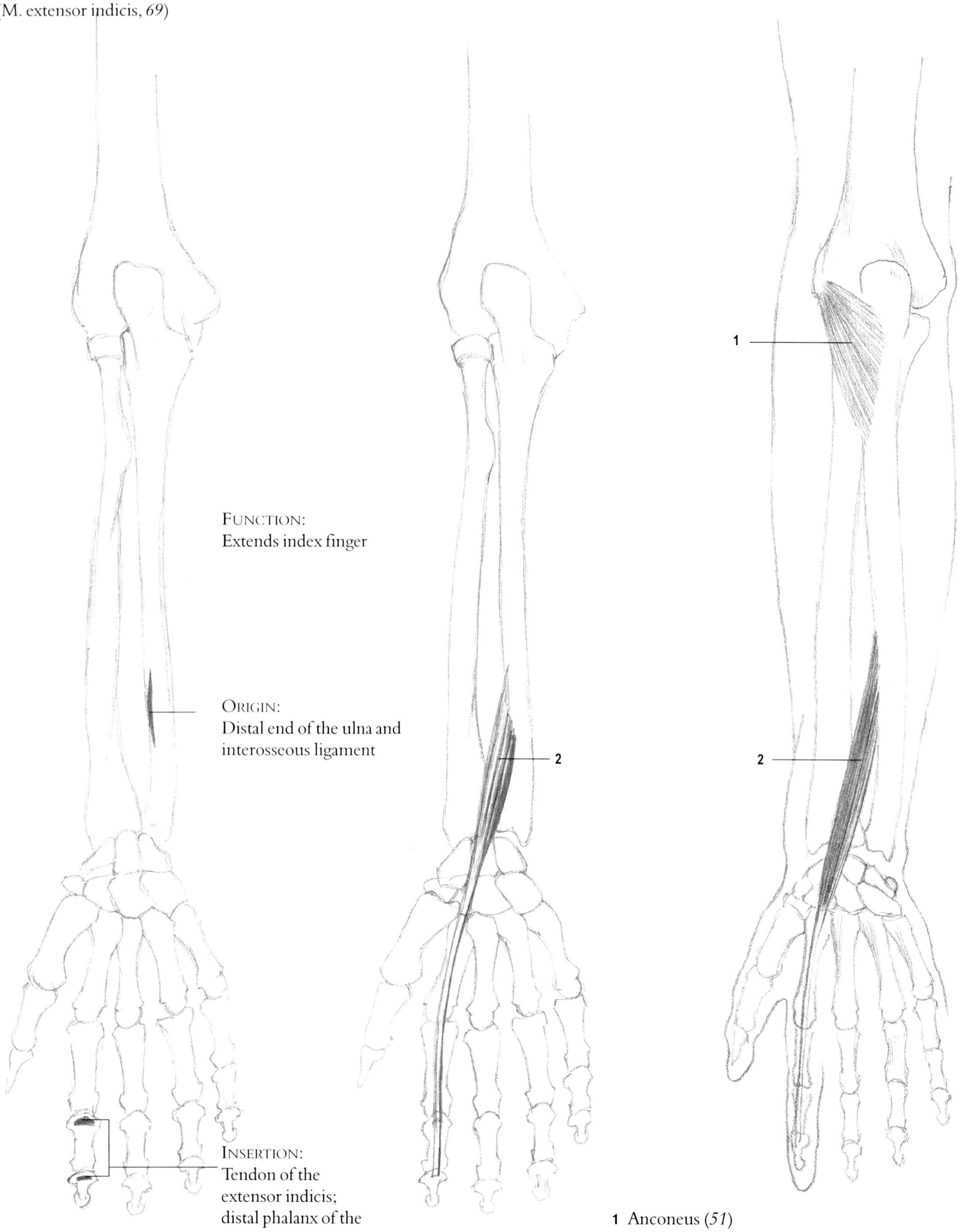

FUNCTION:
Extends index finger

ORIGIN:
Distal end of the ulna and interosseous ligament

INSERTION:
Tendon of the extensor indicis; distal phalanx of the second finger

dorsal aspect

1 Anconeus (*51*)
2 Extensor indicis (*69*)

Fig. 81
The Supinator, *65*

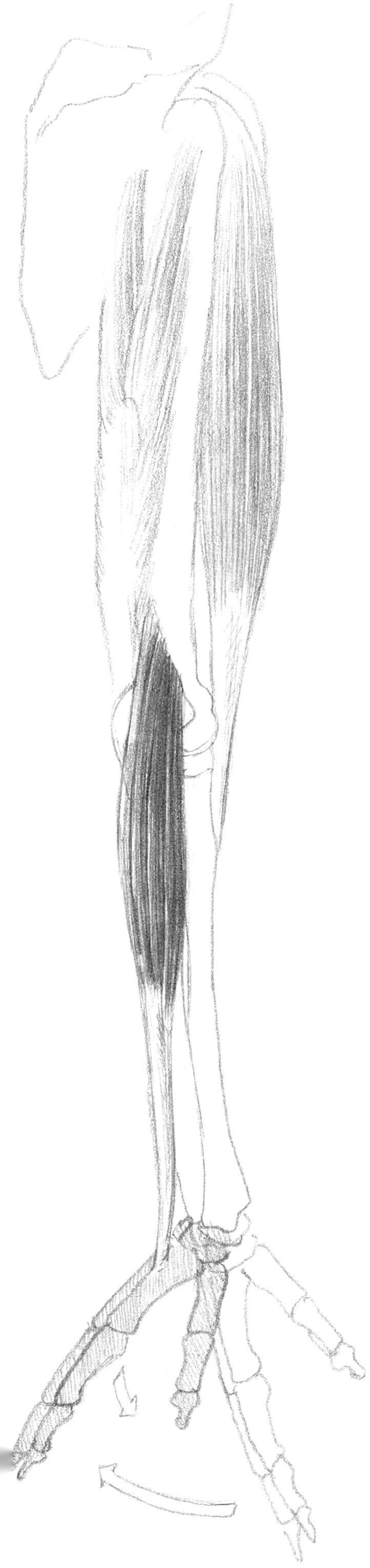

Origin:
Anconeus of the humerus and collateral ligament of the elbow joint

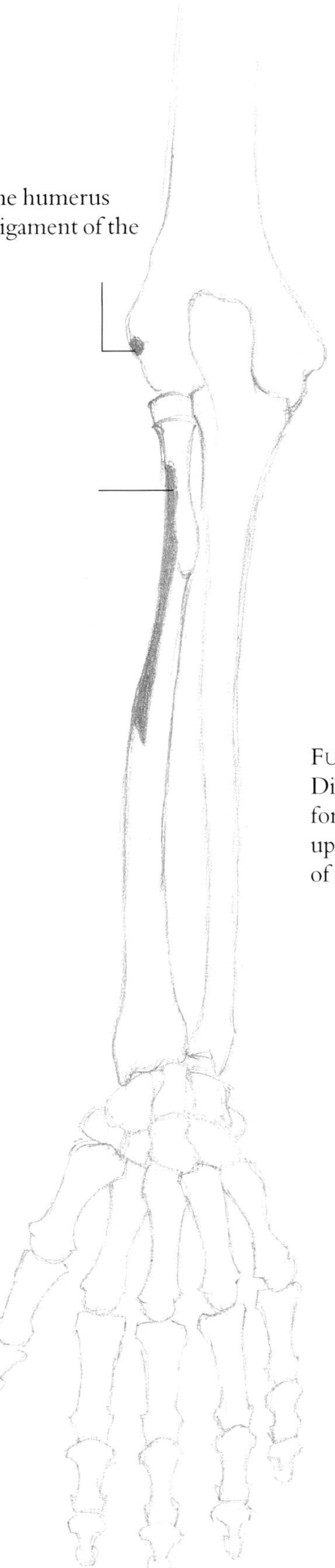

Insertion:
Lateral surface of the upper third of the radius, interosseous ligament, ulna

Function:
Directs palm forwards by upwards rotation of the lower arm

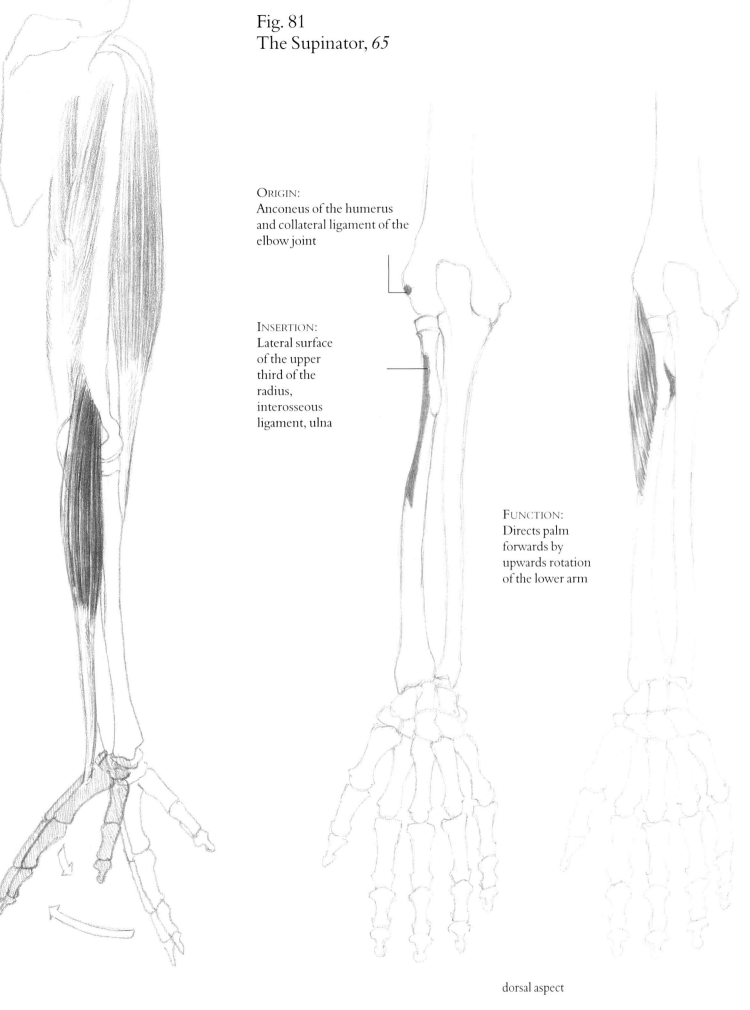

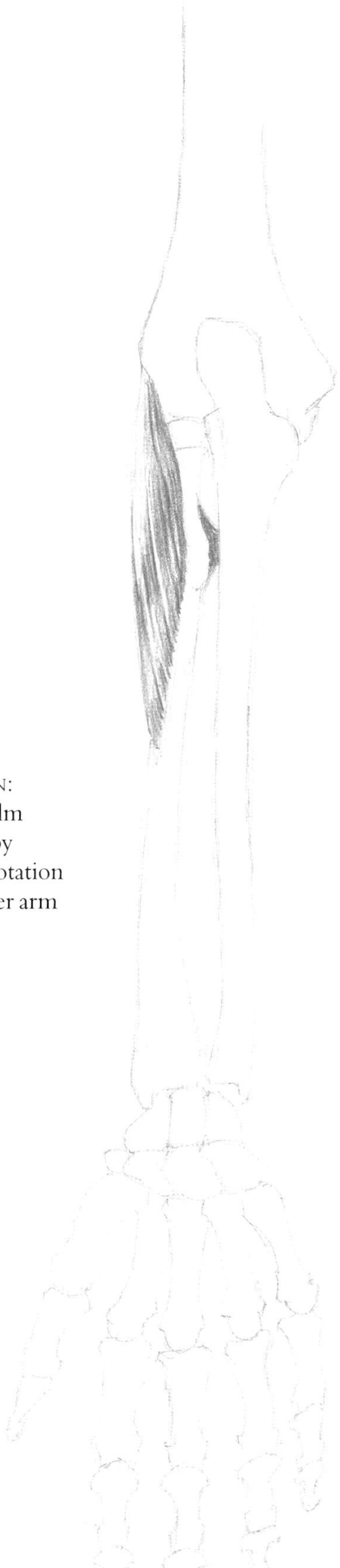

dorsal aspect

Fig. 82
The Muscles of the Lower Arm

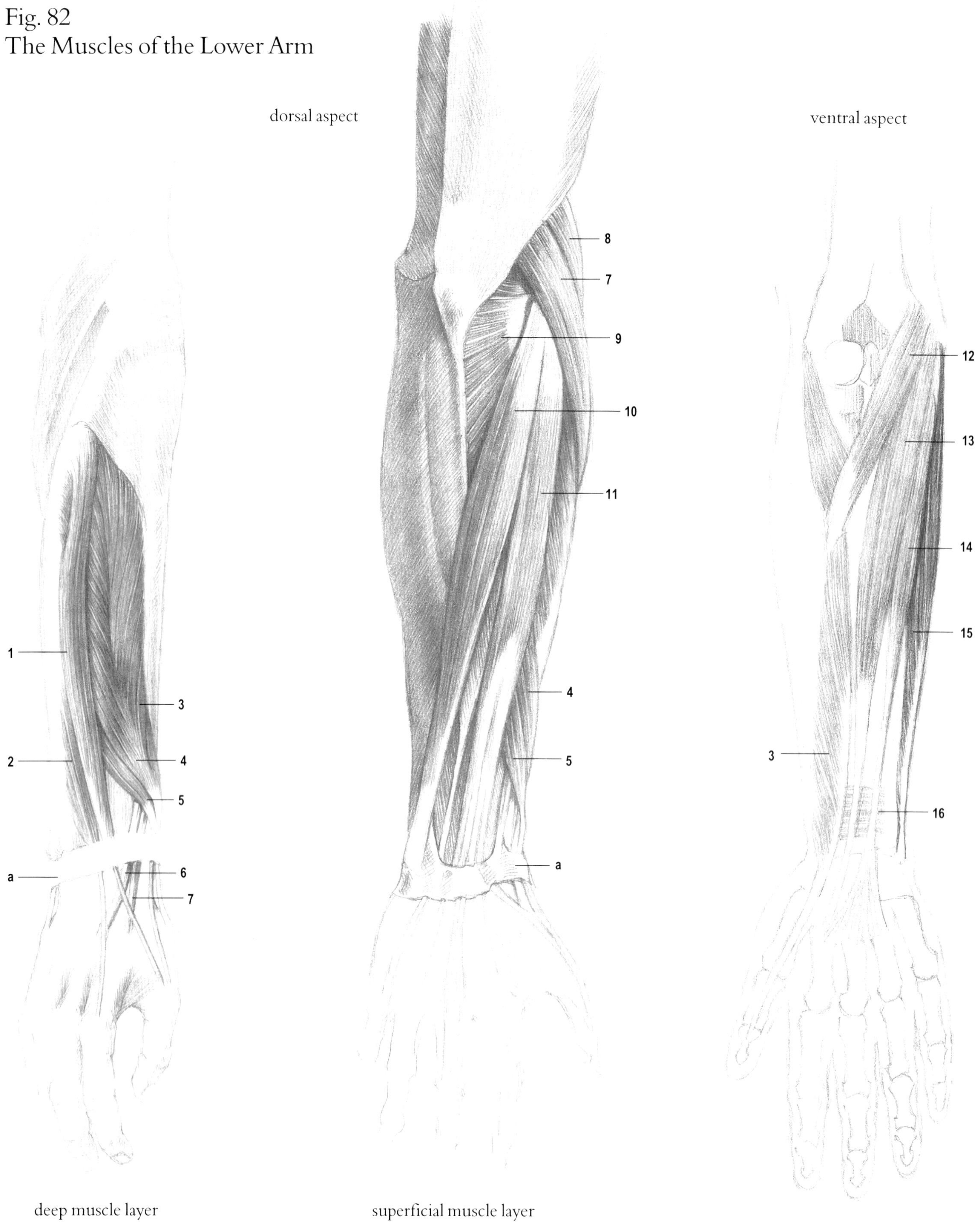

a Retinaculum extensorum
1 Abductor pollicis longus (*66*)
2 Extensor indicis (*69*)
3 Flexor pollicis longus (*58*)
4 Extensor pollicis brevis (*67*)
5 Extensor pollicis longus (*68*)
6 Extensor carpi radialis brevis (*61*)
7 Extensor carpi radialis longus (*61*)
8 Brachioradialis (*60*)

ventral aspect

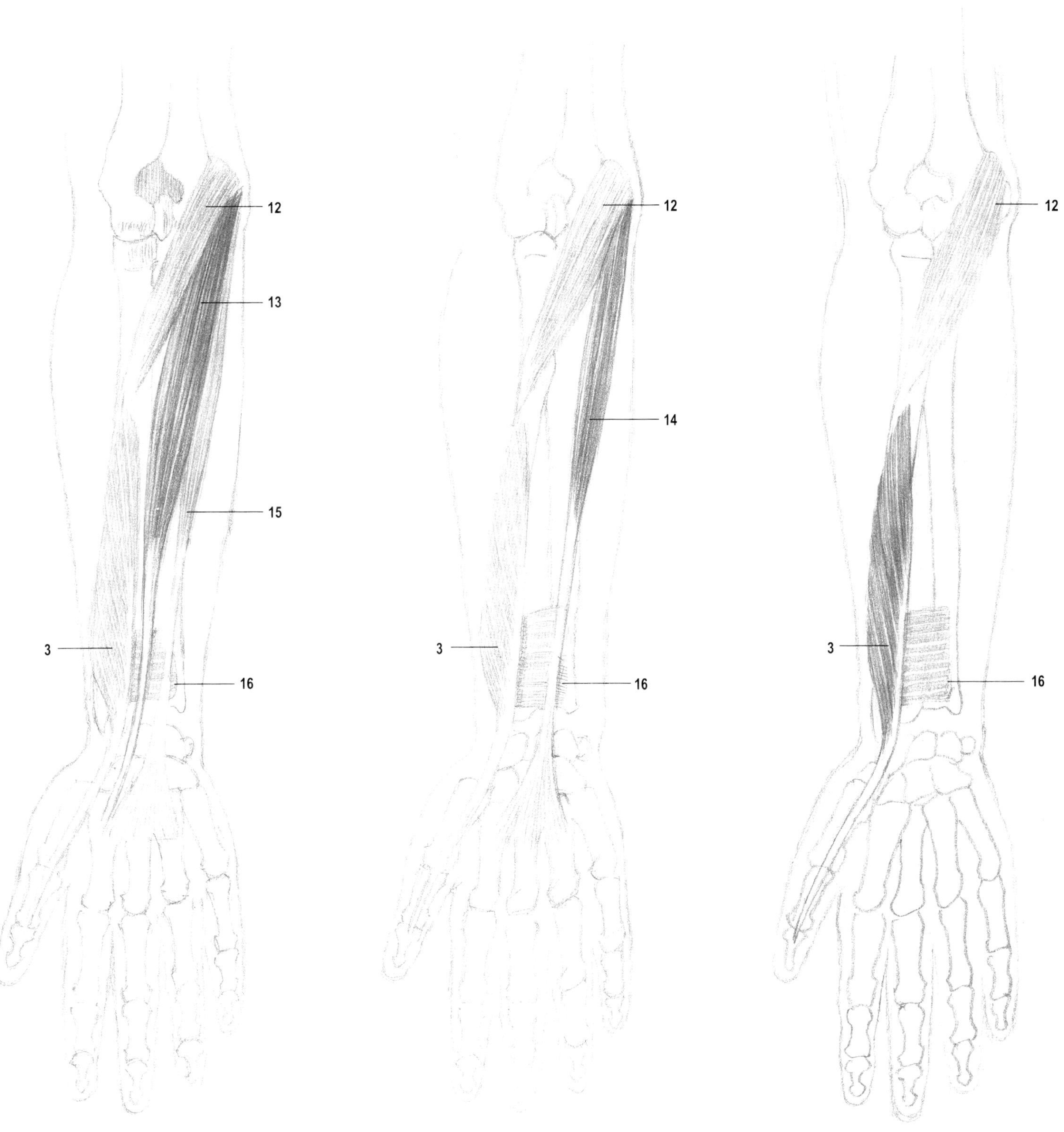

9 Anconeus (*51*)
10 Extensor carpi ulnar is (*64*)
11 Extensor digitorum (*62*)
12 Pronator teres (*52*)
13 Flexor carpi radialis (*53*)
14 Palmaris longus (*54/1*)
15 Flexor carpi ulnaris (*55*)
16 Pronator quadratus (*59*)

Fig. 83
The Adductor Muscle of the Thumb
(M. adductor pollicis, *73*)

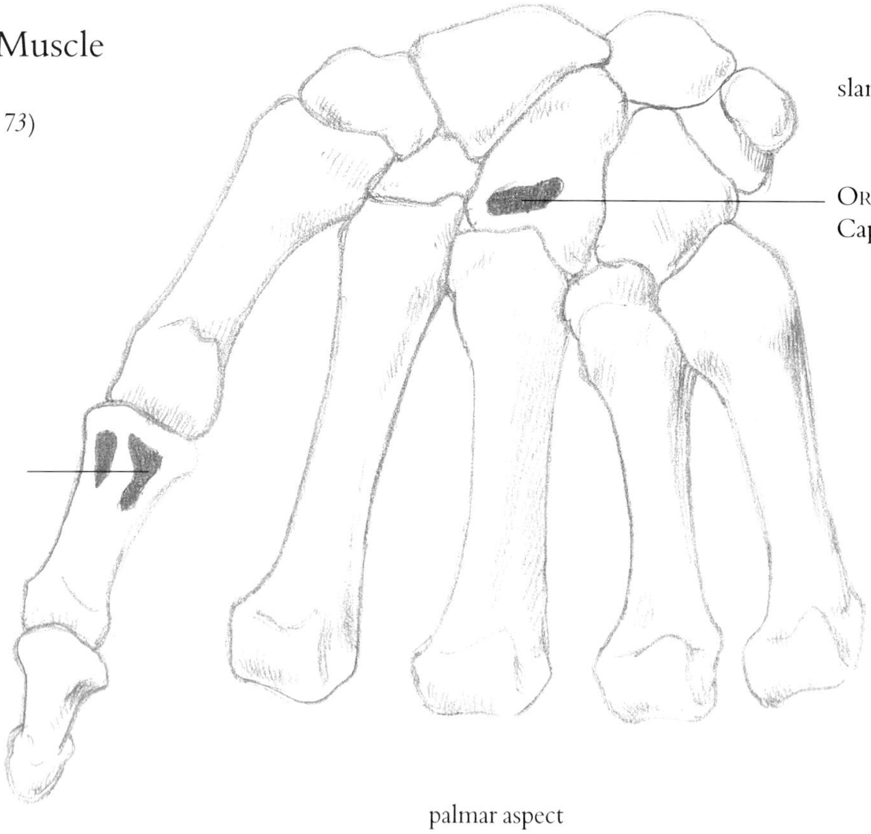

slanted head

ORIGIN:
Capitate

INSERTION:
Proximal phalanx of the first finger

palmar aspect

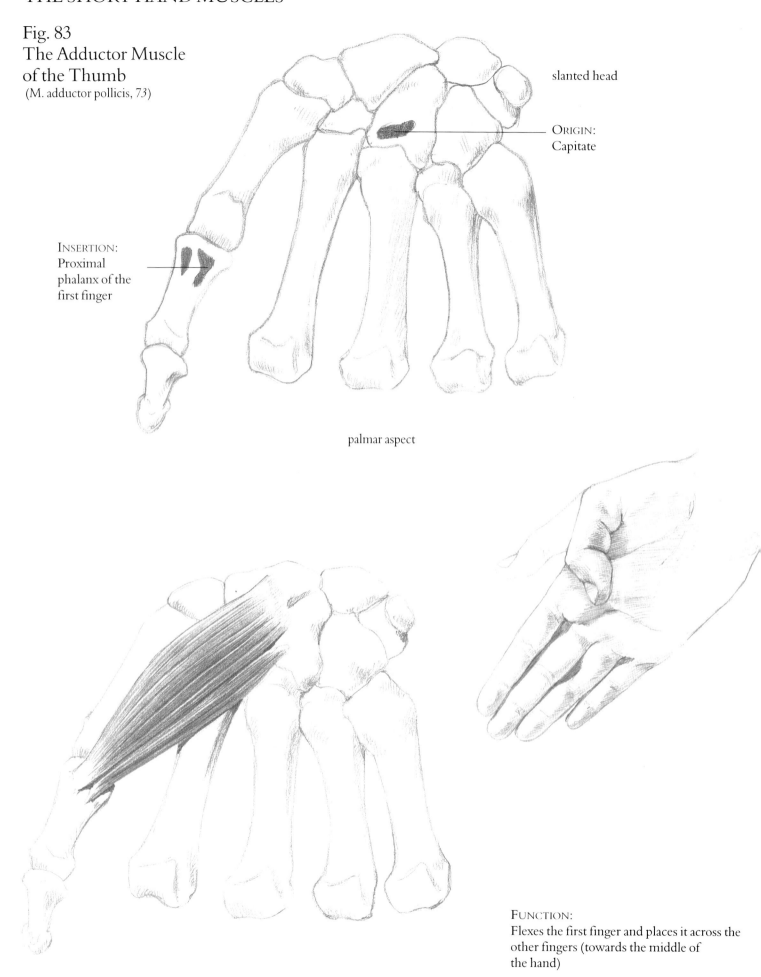

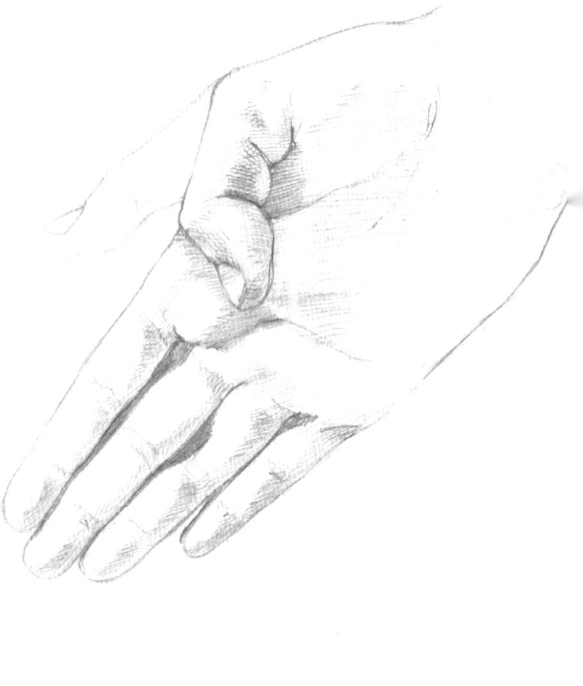

FUNCTION:
Flexes the first finger and places it across the other fingers (towards the middle of the hand)

Fig. 84
The Adductor Muscle of the Thumb
(cont.)

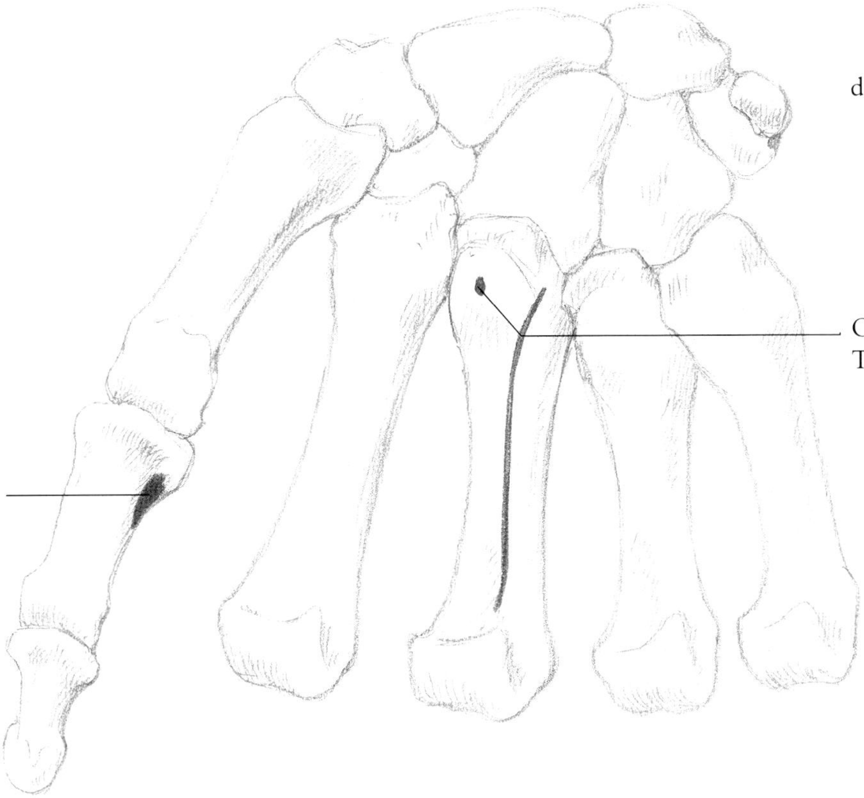

diagonal head

ORIGIN:
Third metacarpal bone

INSERTION:
Proximal phalanx of the first finger

palmar aspect

FUNCTION:
The hinge joint of the thumb is remarkable since it allows for an extraordinary range of movement, such as flexion, extension, abduction, adduction and even circumduction. The combination of these different movements enables the thumb to extend across the palm and to the fingers. The opposable thumb is unique to humans.

Fig. 85
The Short Flexor Muscle of the Thumb

(M. flexor pollicis brevis, *71*)

ORIGIN:
Carpal bones,
transverse ligament of
the carpal joint

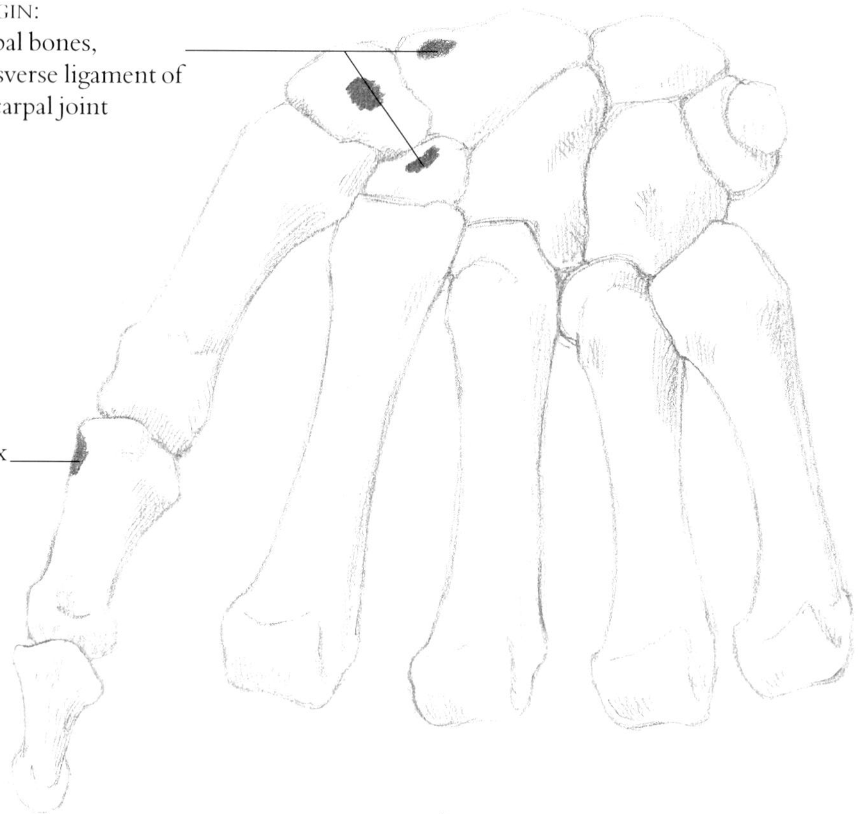

INSERTION:
Proximal phalanx
of the first finger

palmar aspect

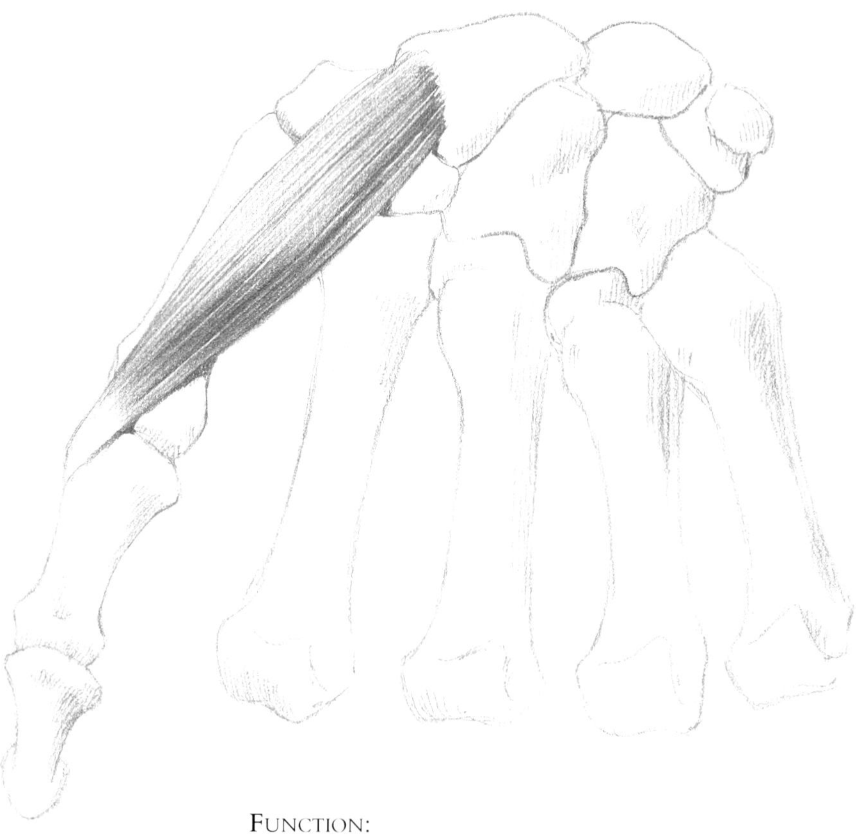

FUNCTION:
Flexes the thumb at its base joint

Fig. 86
The Opposing Muscle of the Thumb
(M. opponens pollicis, *72*)

ORIGIN:
Trapezium, transverse ligament of the carpus

INSERTION:
Lateral-palmar surface of the first metacarpal bone

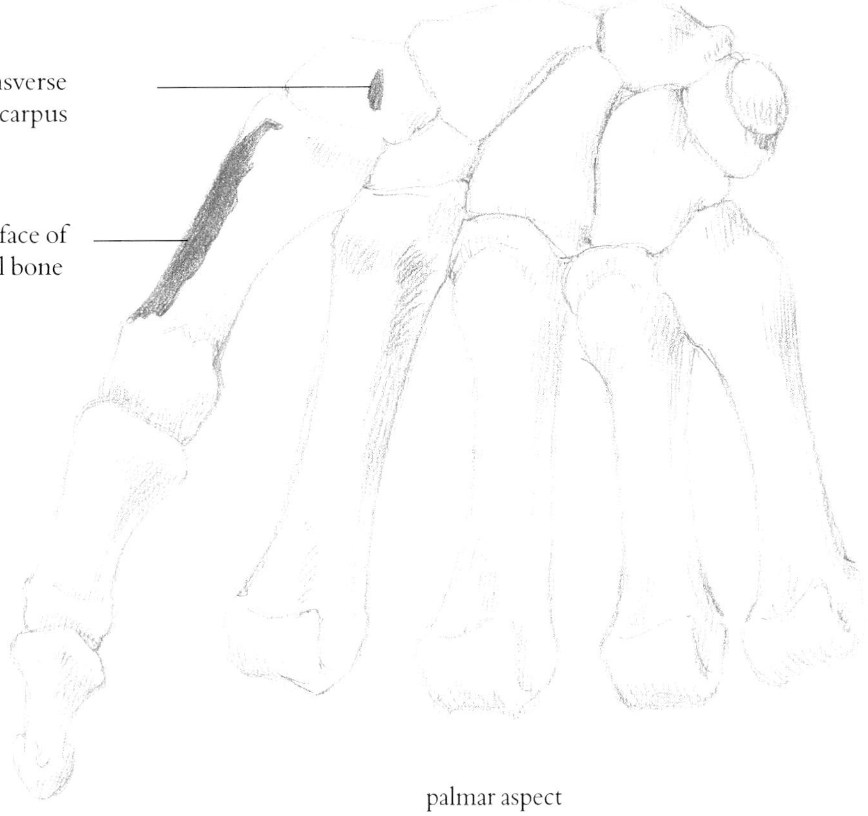

palmar aspect

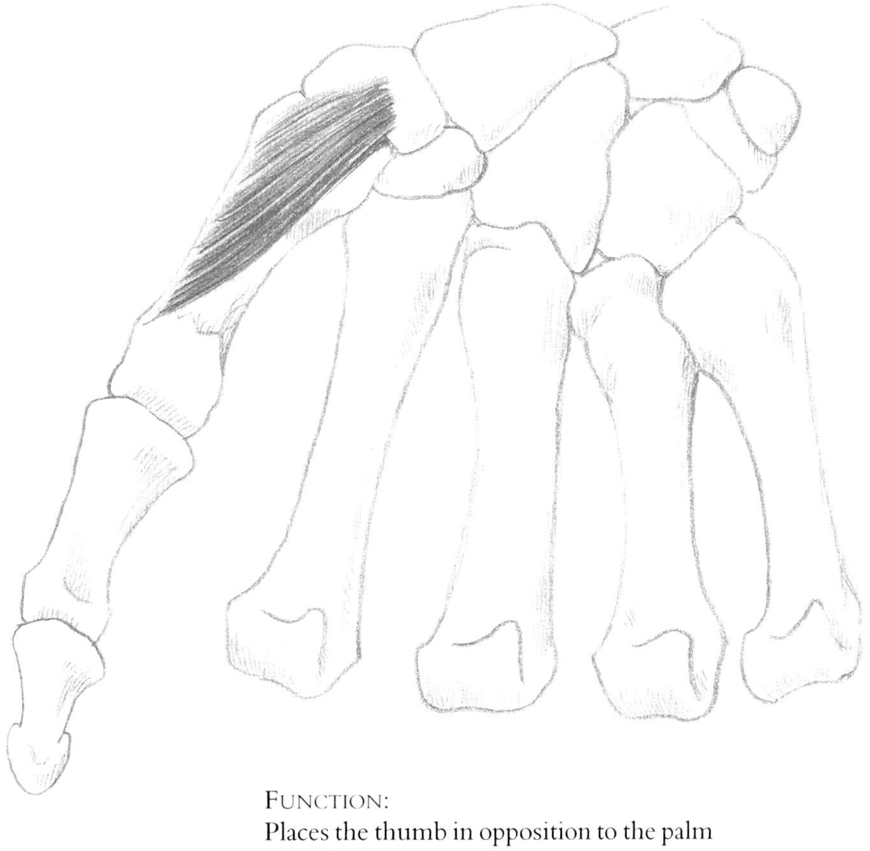

FUNCTION:
Places the thumb in opposition to the palm of the hand (claw position)

Fig. 87
The Movement of the Thumb

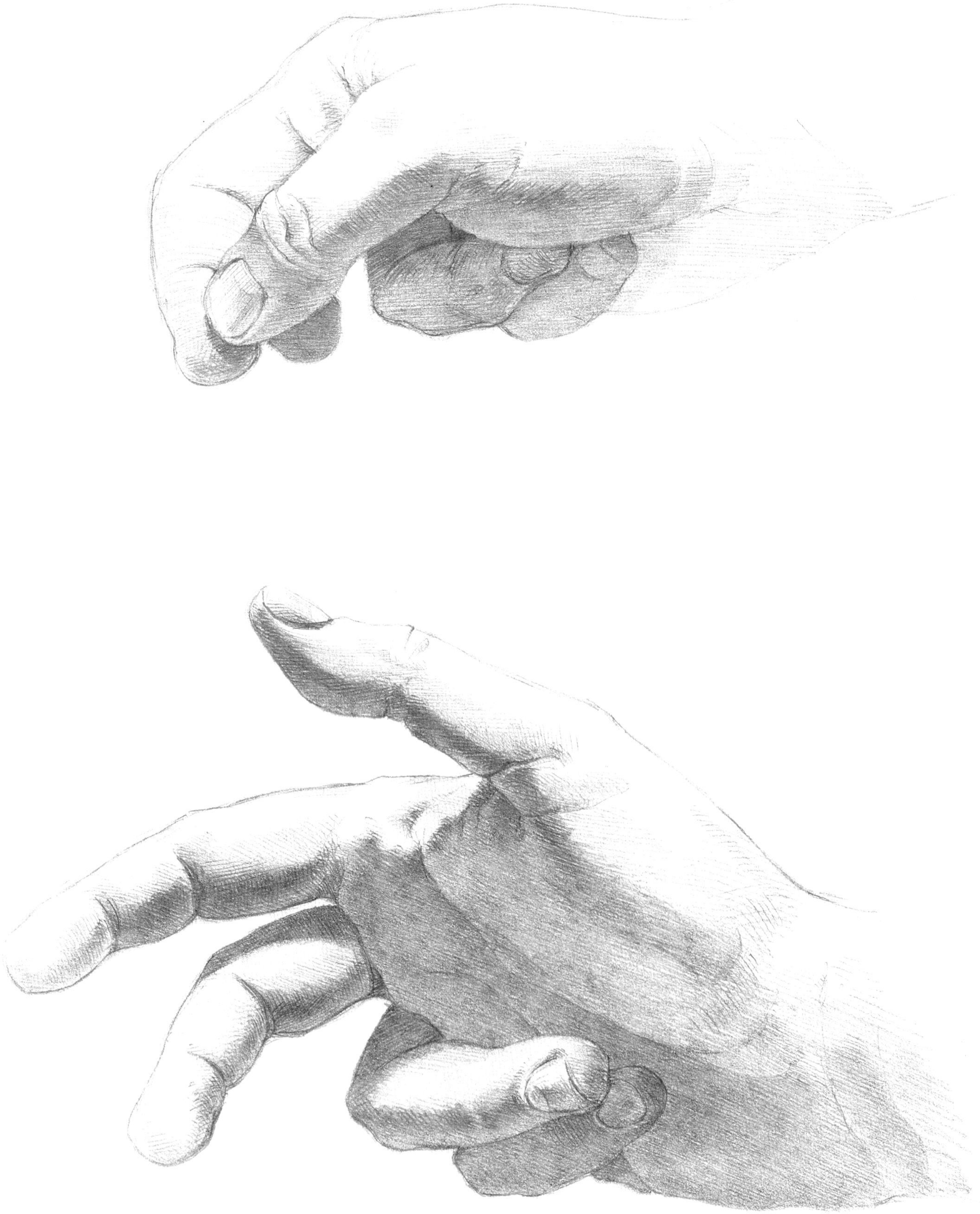

Fig. 88
The Opposing Muscle of the Little Finger

(M. opponens digiti minimi, *78*)

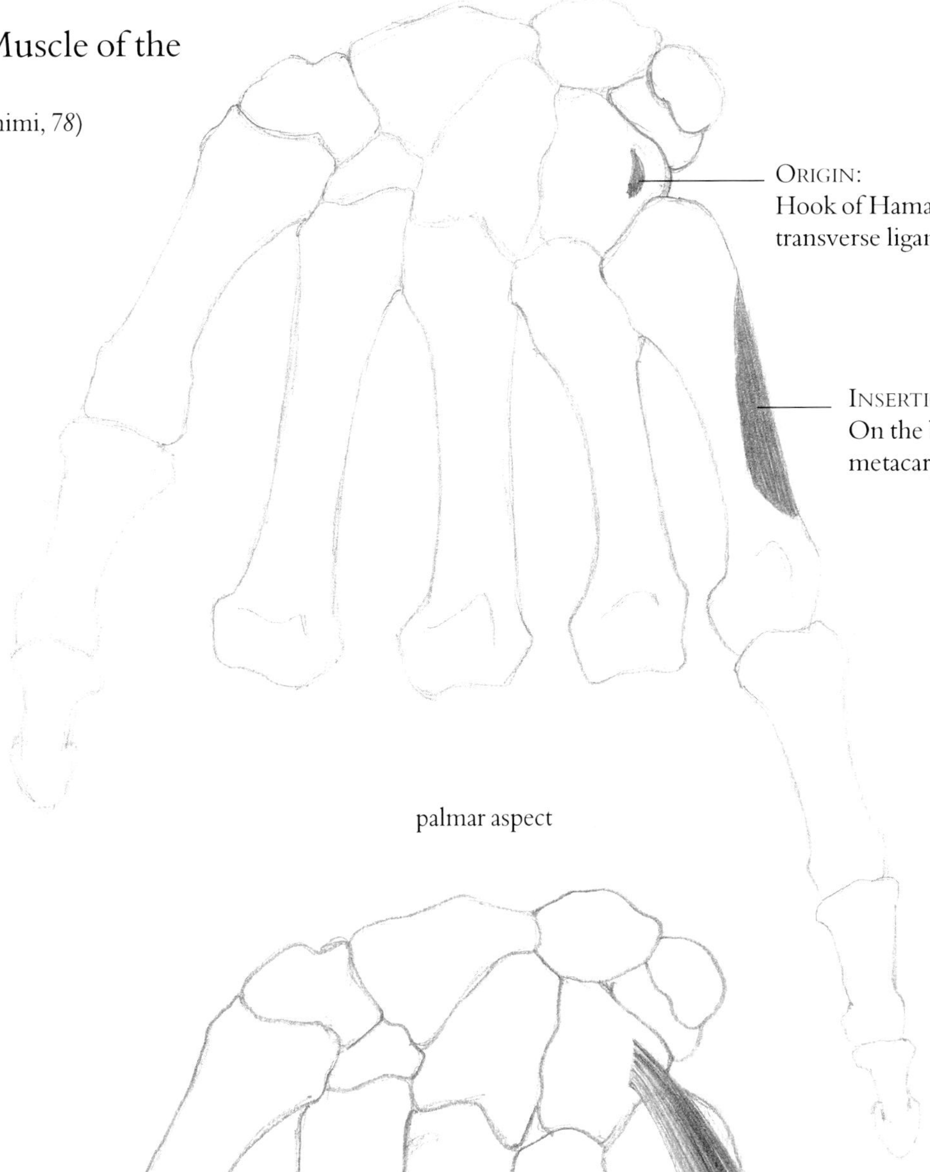

ORIGIN:
Hook of Hamate on the transverse ligament

INSERTION:
On the body of the fifth metacarpal bone

palmar aspect

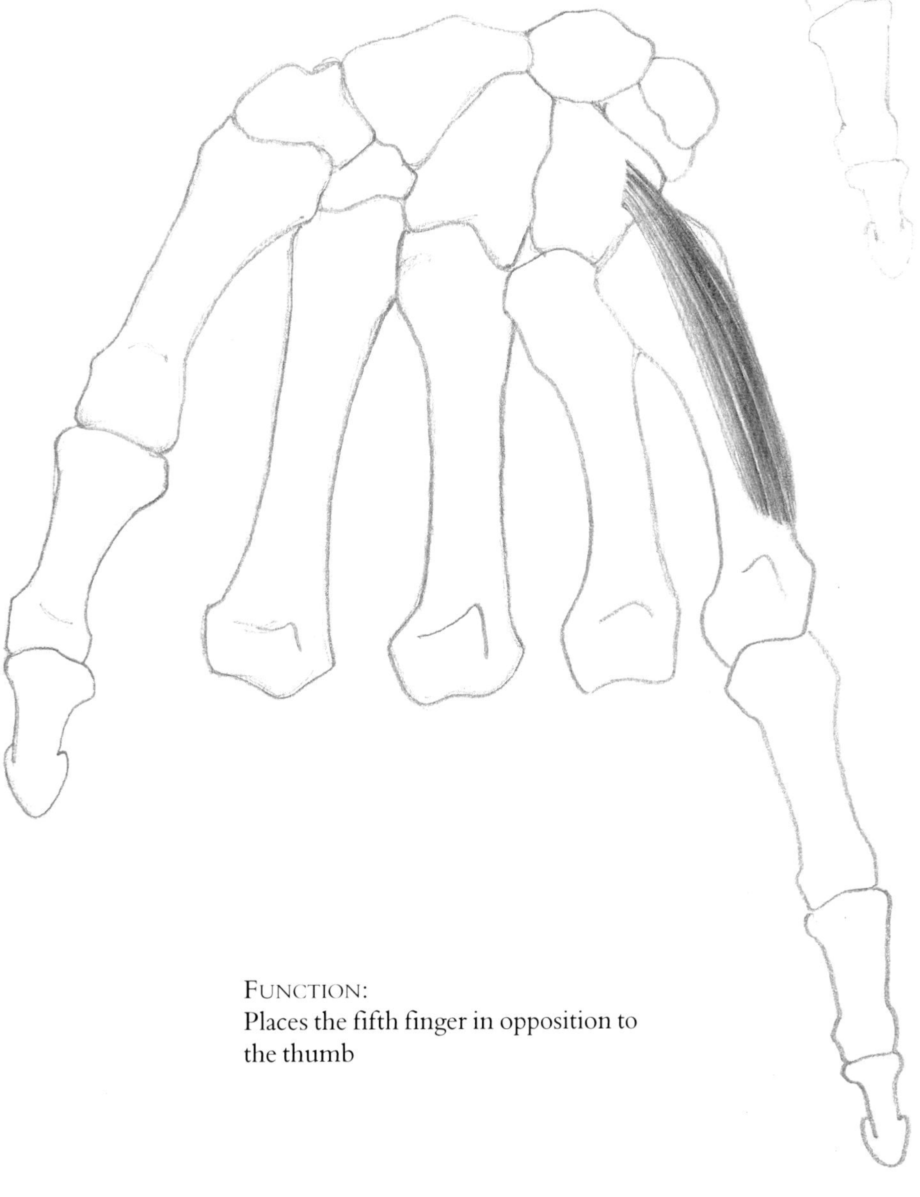

FUNCTION:
Places the fifth finger in opposition to the thumb

Fig. 89
The Adductor Muscle of the Little Finger
(M. adductor digiti minimi, 76)

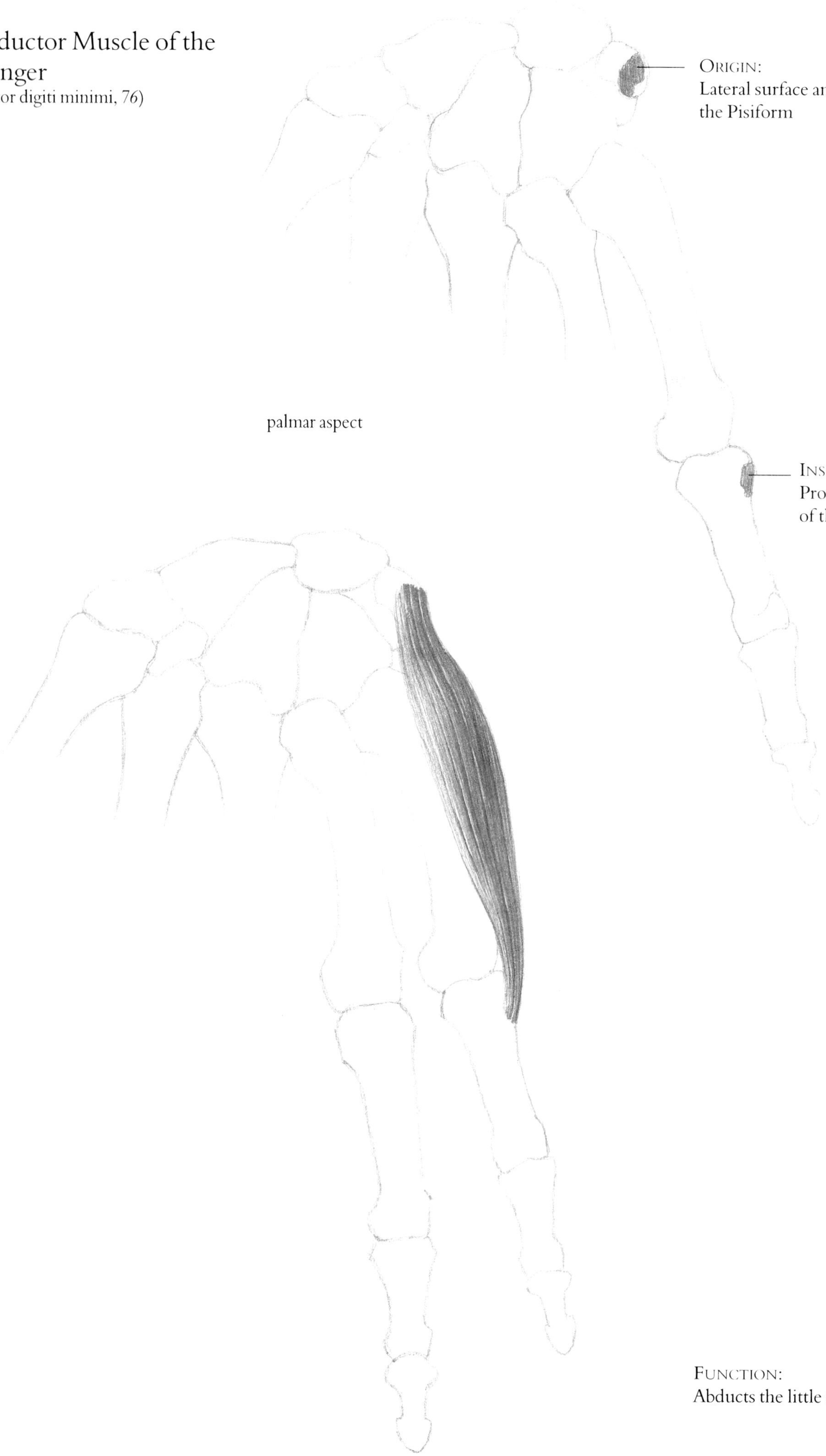

ORIGIN:
Lateral surface and ligaments of the Pisiform
palmar aspect
INSERTION:
Proximal phalanx of the fifth finger

FUNCTION:
Abducts the little finger

Fig. 90
The Short Flexor Muscle of the Little Finger
(M. flexor digiti minimi brevis manus, 77)

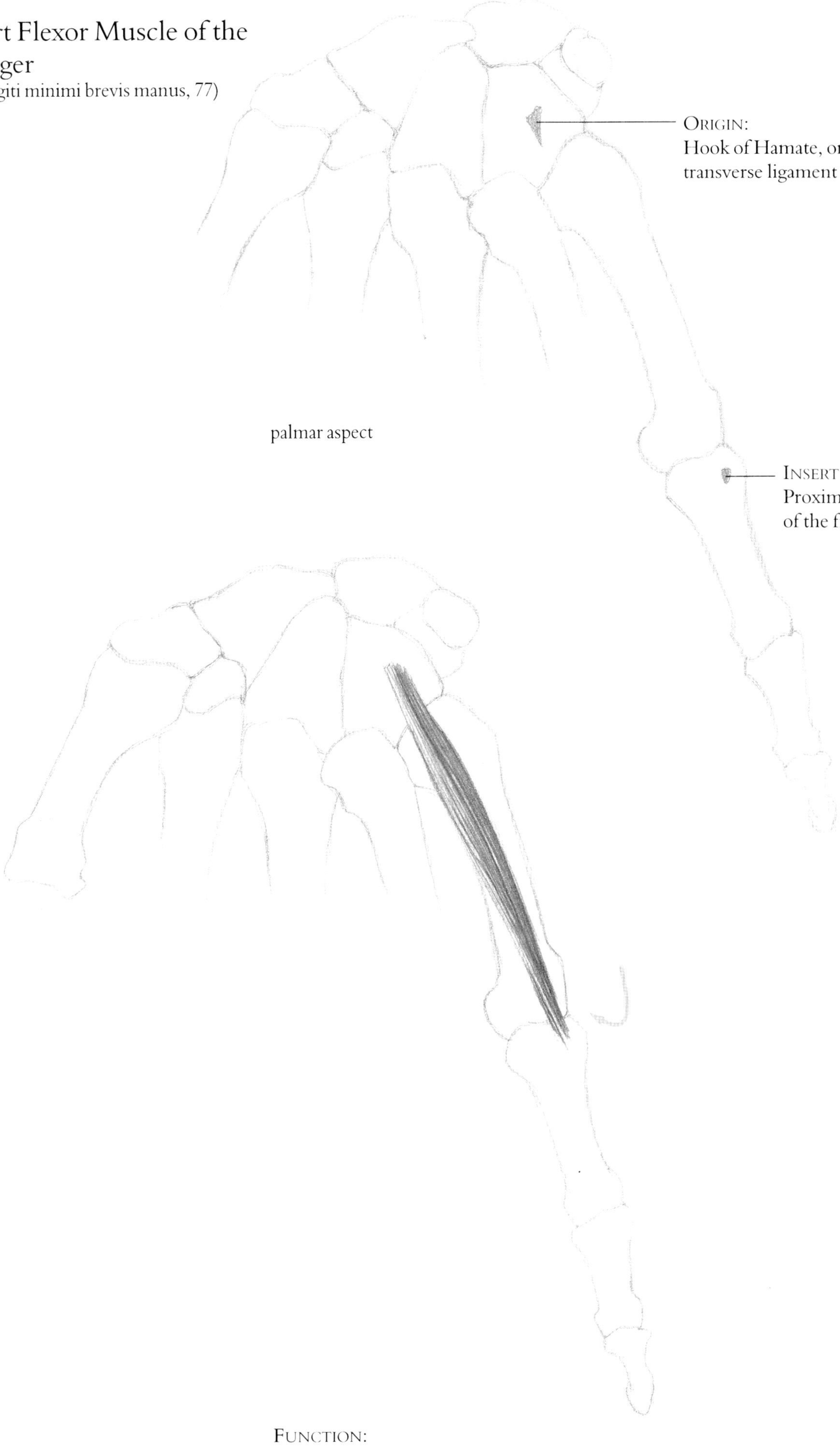
ORIGIN:
Hook of Hamate, on the transverse ligament
palmar aspect
INSERTION:
Proximal phalanx of the fifth finger

FUNCTION:
Flexes the little finger

Fig. 91
The Interosseous Muscles
(Mm. interossei, 75)

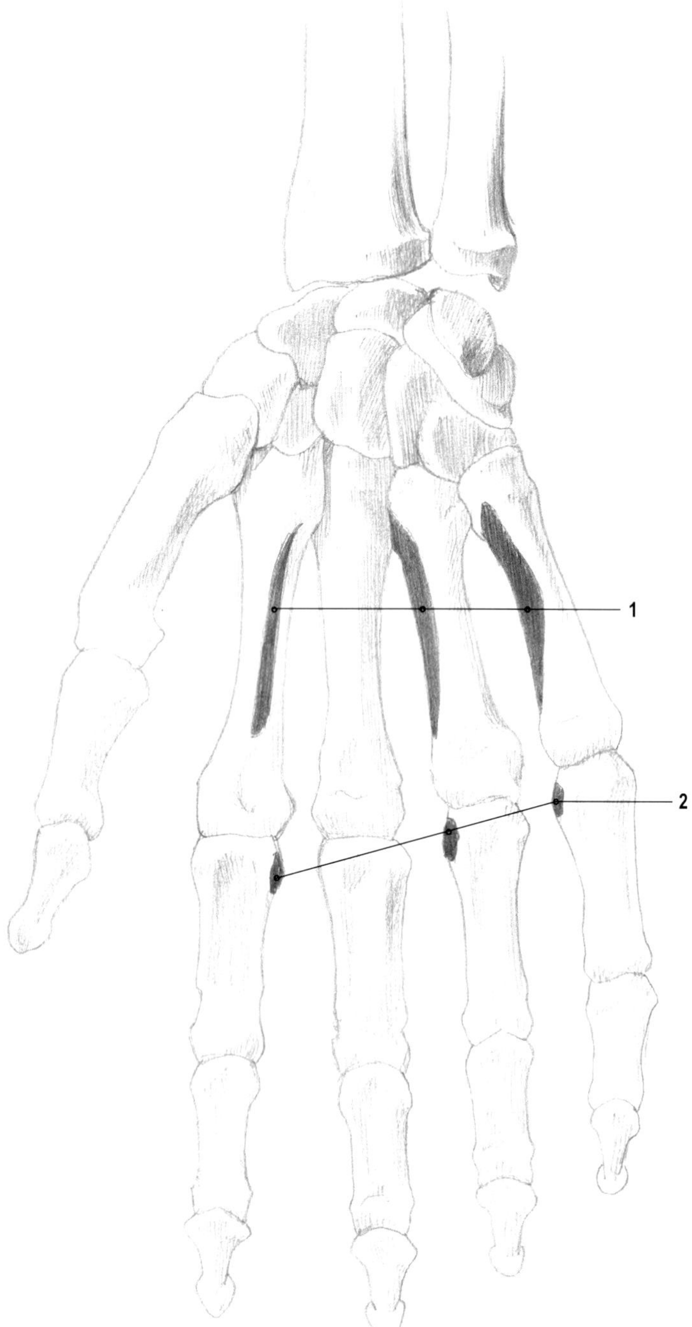

palmar interosseous muscles of the right hand (3)

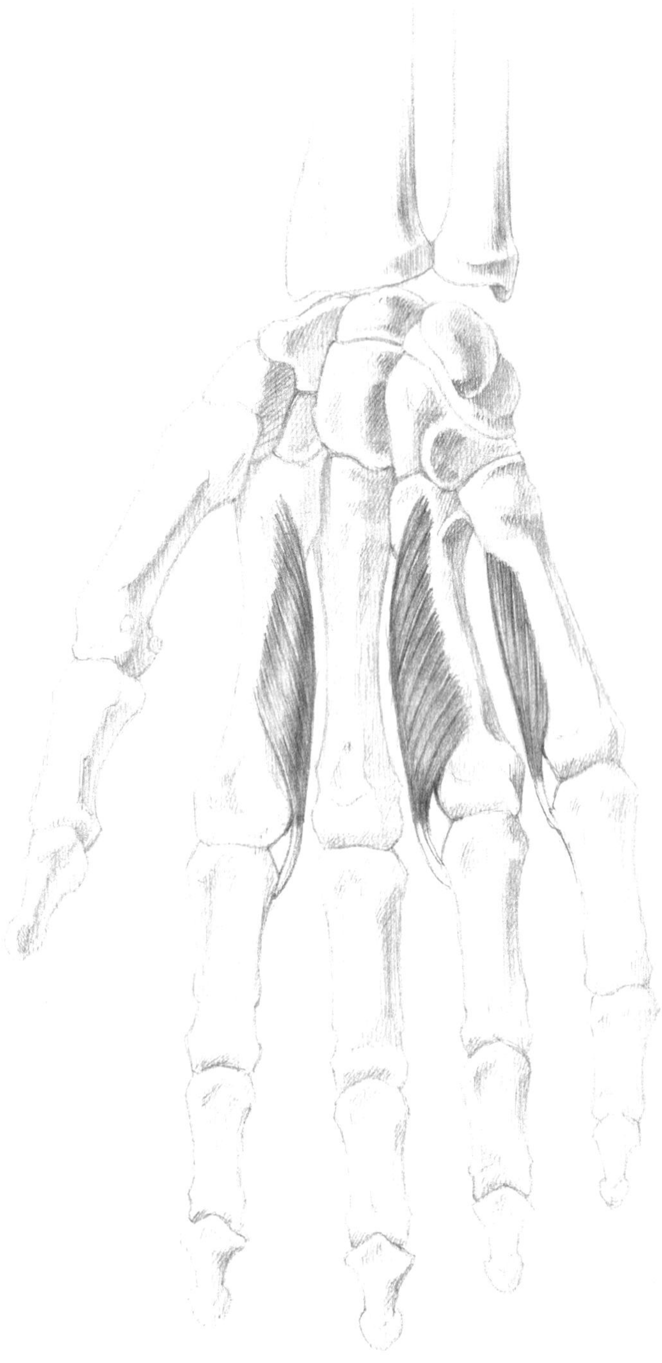

palmar aspect

dorsal interosseous muscles of the right hand (4)

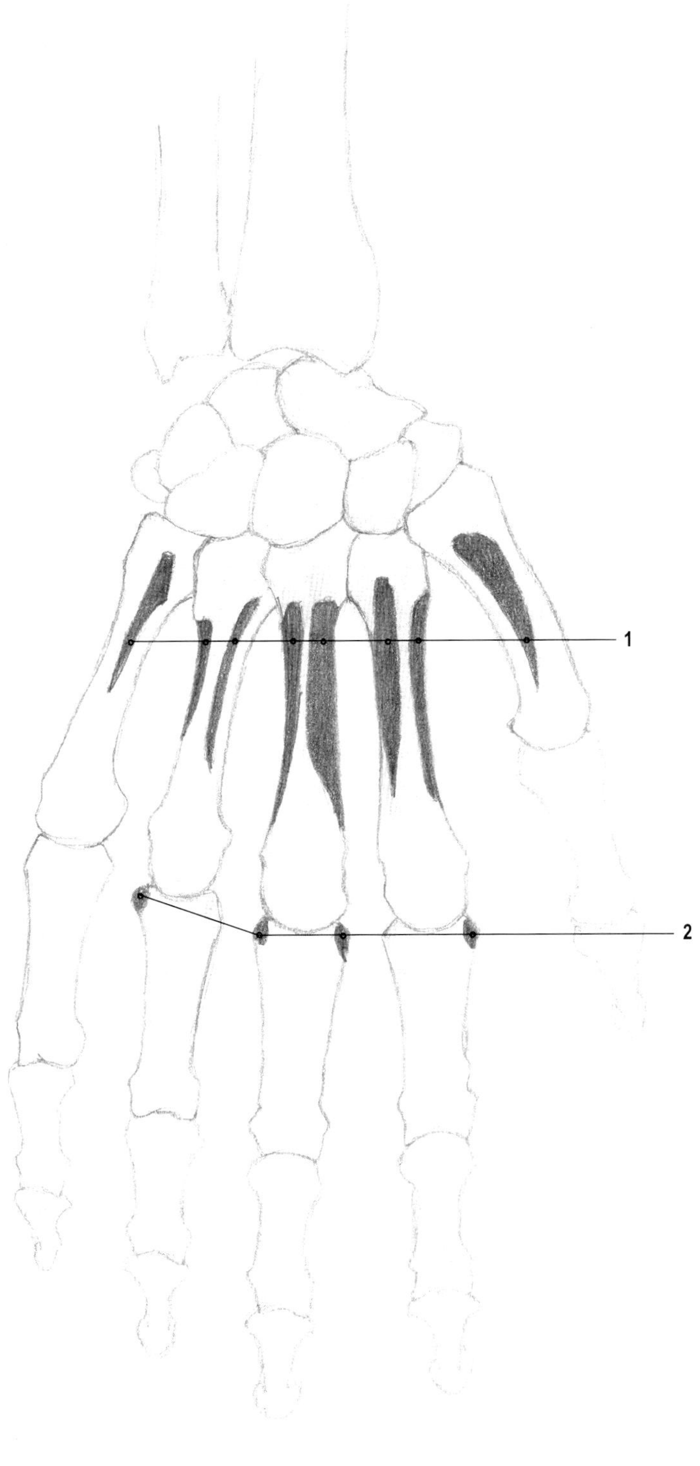

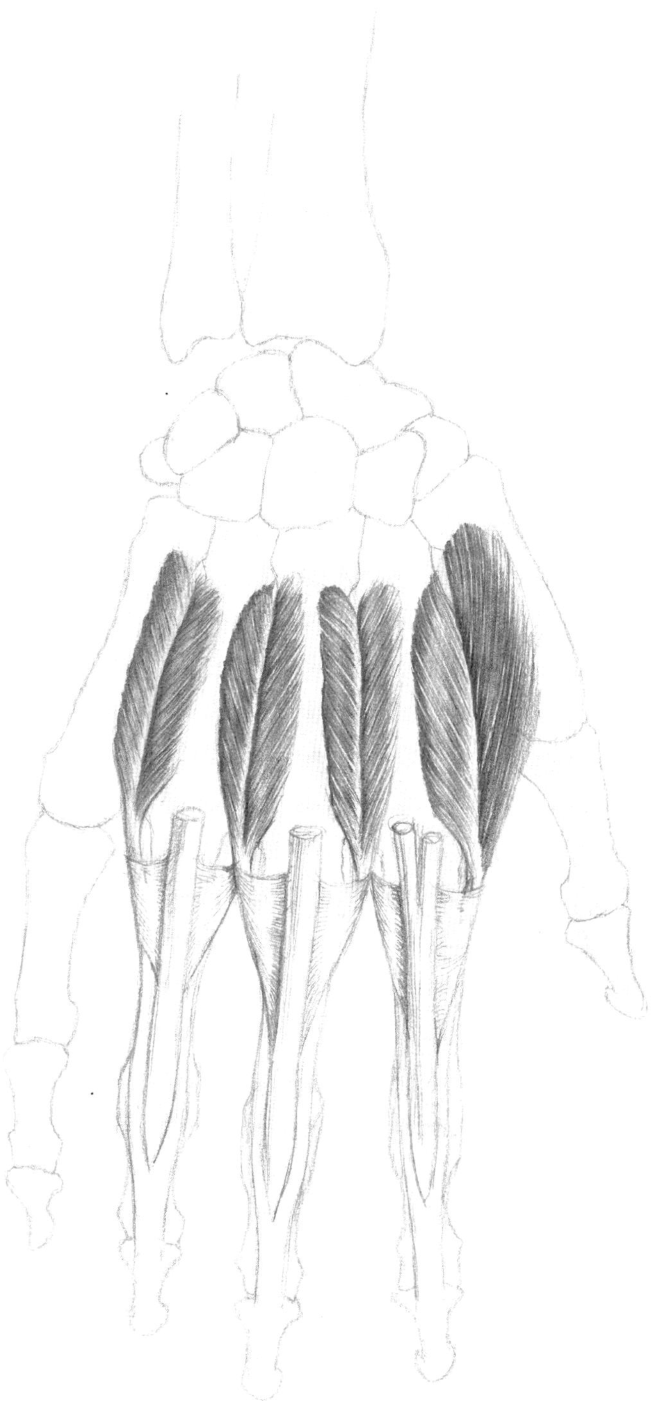

1 ORIGIN:
Interface of the metacarpal bones
2 INSERTIONS:
Proximal phalanges of the fingers

FUNCTION:
The dorsal interosseous muscles splay the fingers, the palmar interosseous muscles produce the closing of the fingers

dorsal aspect

Fig. 92
The Lumbricals
(Mm. lumbricales, *74*)

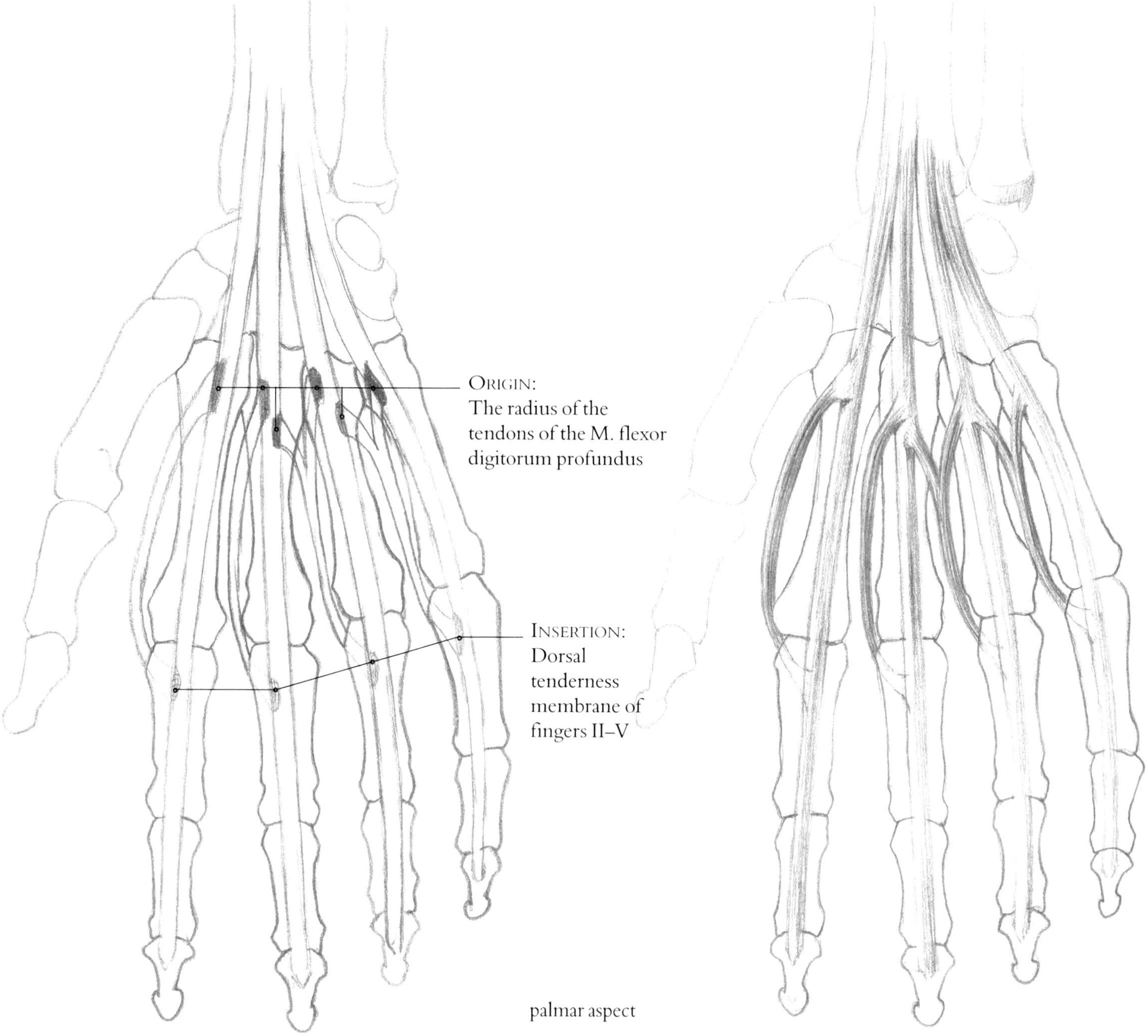

palmar aspect

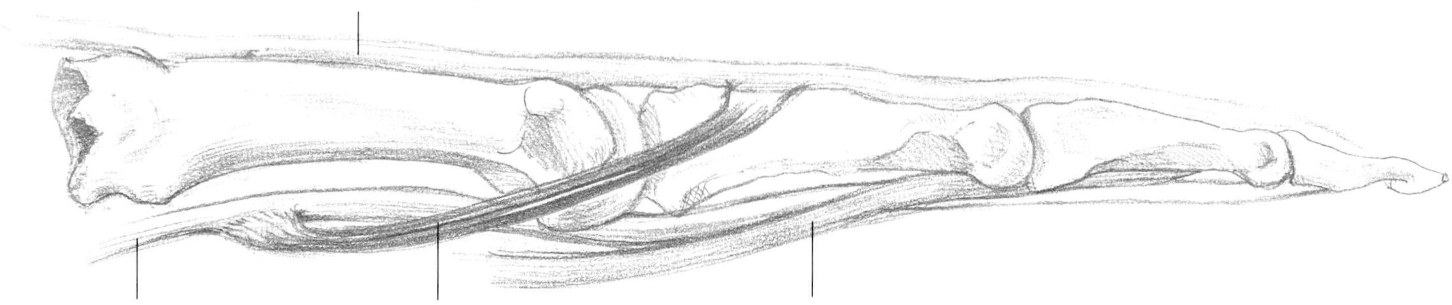

lateral aspect

dorsal aspect

FUNCTION:
Finger stretch

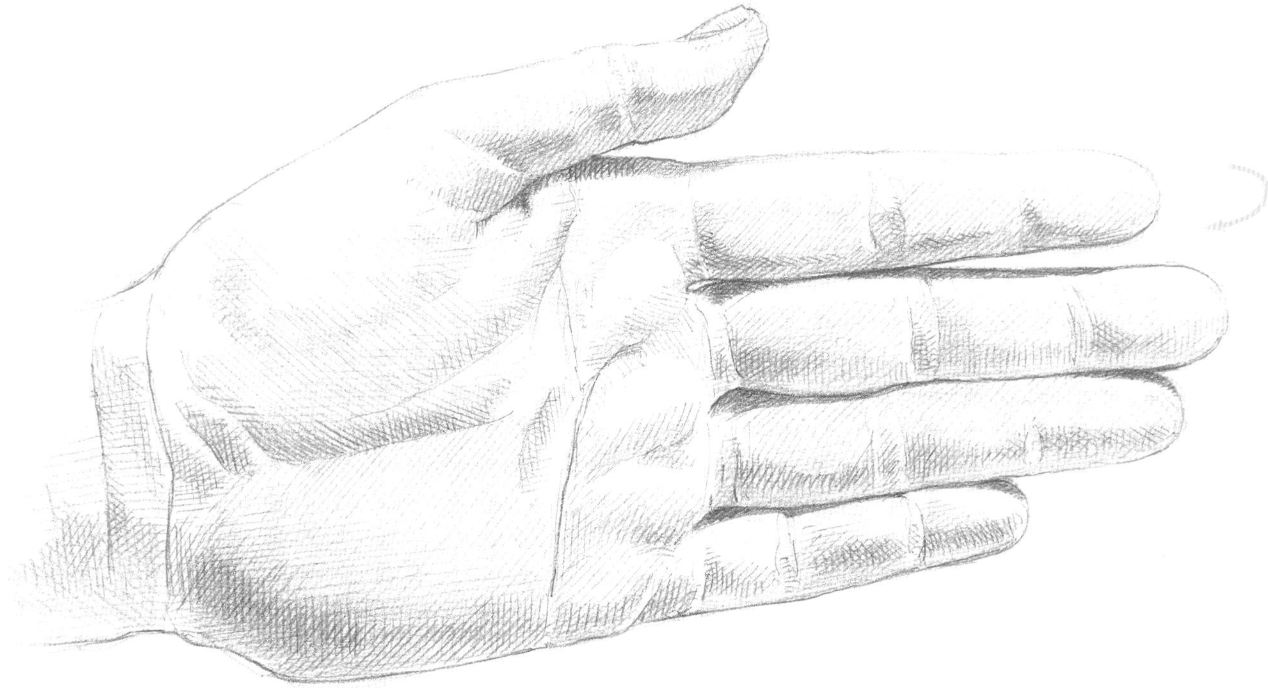

palmar aspect

Fig. 93
Origin and Insertion Surfaces of Muscles in the Carpus and Lower Arm Bones

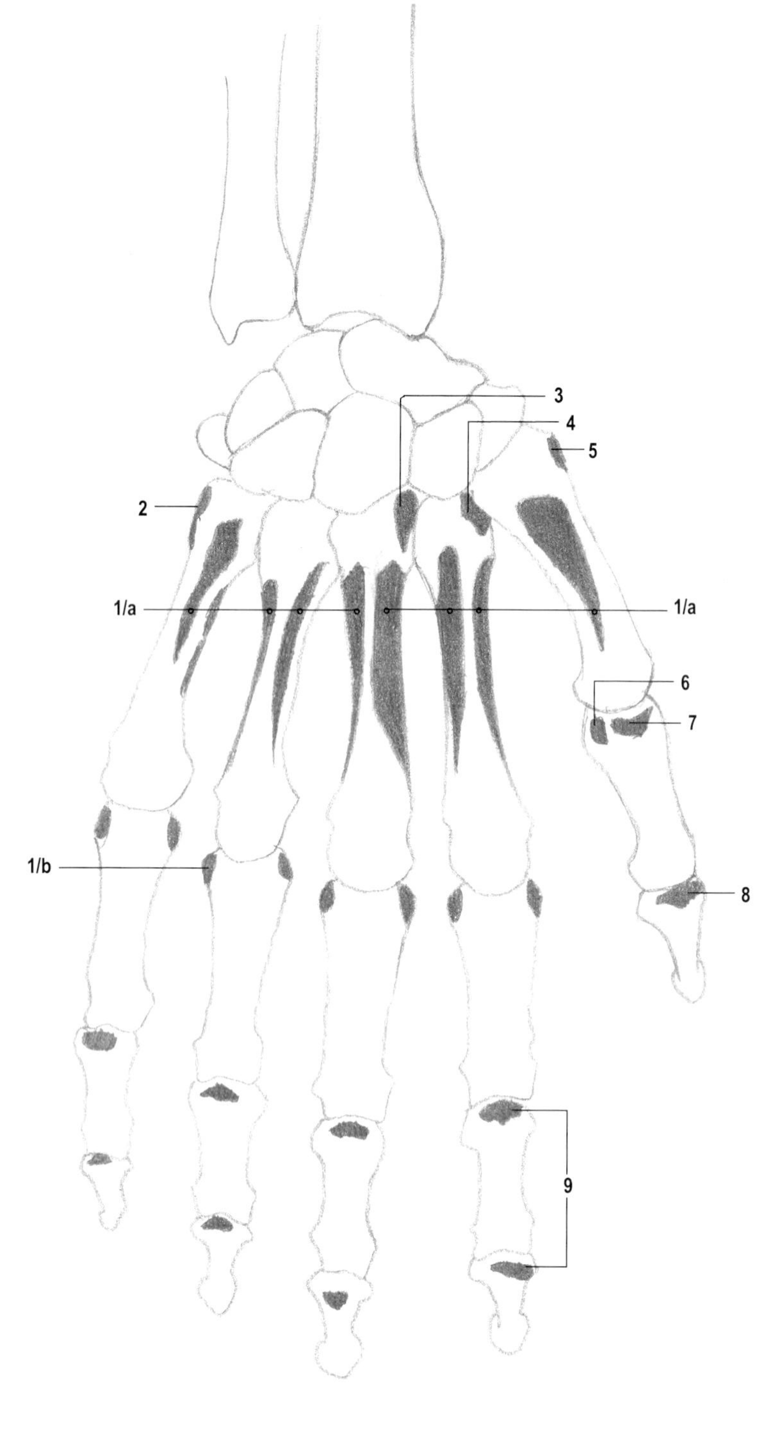

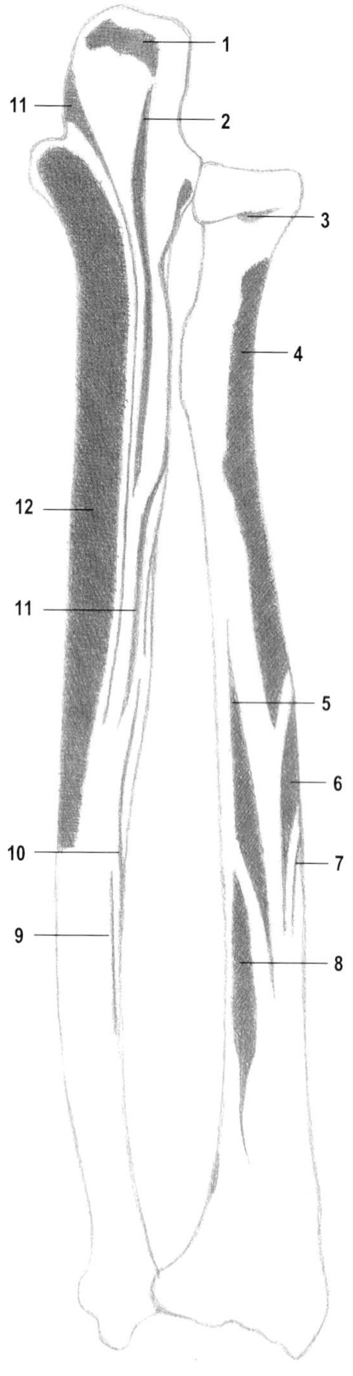

Dorsal Surface

O = Origin
I = Insertion

Dorsal Surface

1 Interosseous muscles, a = O, b = I (*75*)
2 Extensor carpi ulnaris, I (*64*)
3 Extensor carpi radialis brevis, I (*61*)
4 Extensor carpi radialis longus, I (*61*)
5 Abductor pollicis longus, I (*66*)
6 Abductor pollicis brevis, I (*70*)
7 Extensor pollicis brevis, I (*67*)
8 Extensor pollicis longus, I (*68*)
9 Extensor digitorum, I (*62*)

Dorsal Surface

1 Triceps, I (*50*)
2 Anconeus, I (*51*)
3 Extensor carpi ulnaris, O (*64*)
4 Supinator, I (*65*)
5 Abductor pollicis longus, O (*66*)
6 Pronator teres, I (*52*)
7 Flexor digitorum superficialis, O (*56*)
8 Extensor pollicis brevis, O (*67*)
9 Extensor indicis, O (*69*)
10 Flexor pollicis longus, O (*58*)
11 Flexor carpi ulnaris, O (*55*)
12 Flexor digitorum profundus, O (*57*)

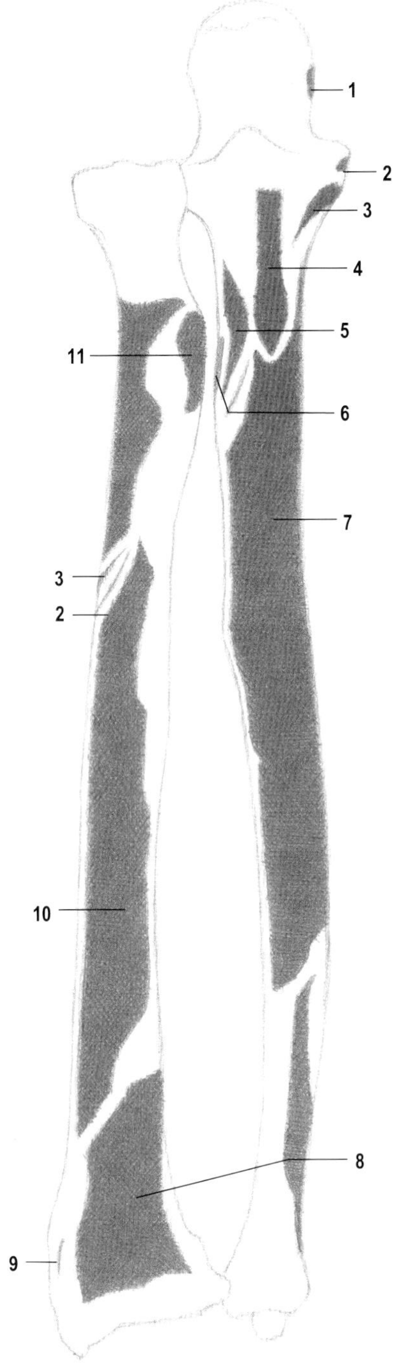

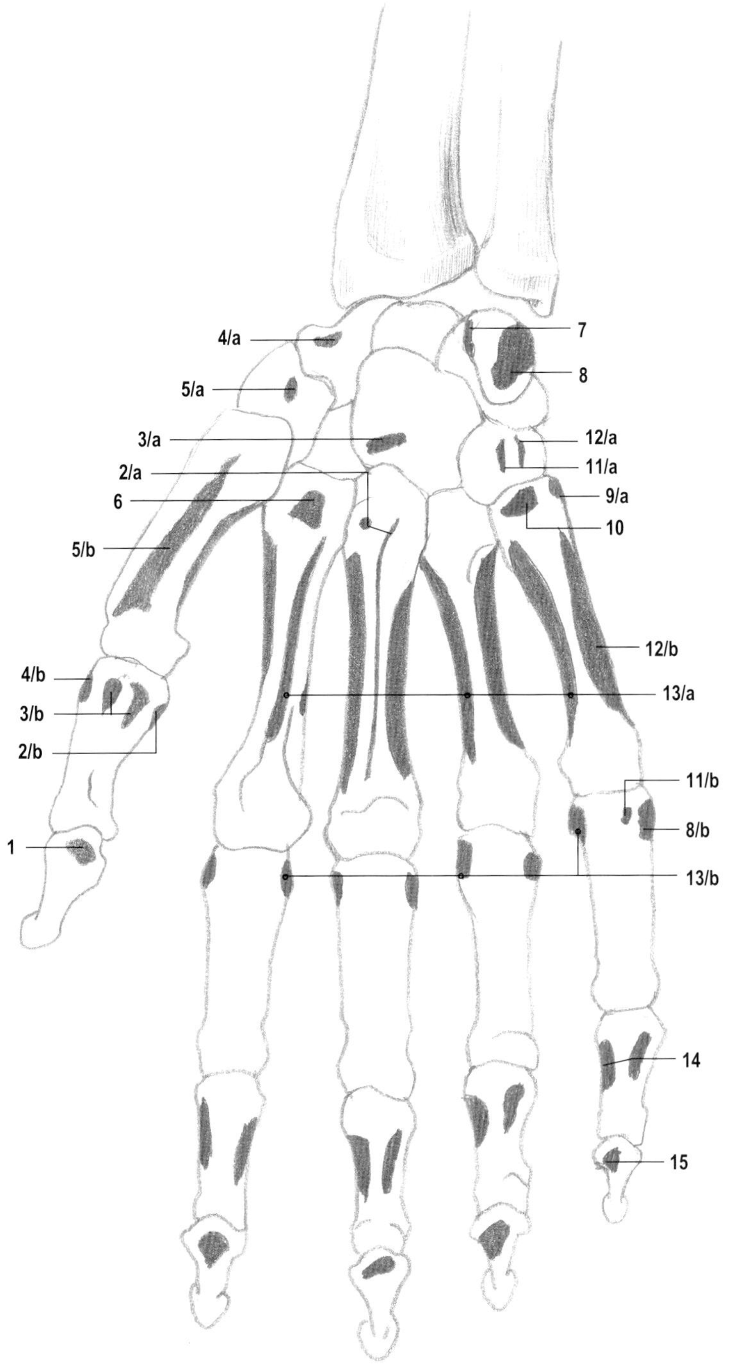

VENTRAL SURFACE

1 Flexor carpi ulnaris, O (*55*)
2 Flexor digitorum superficialis, O (*56*)
3 Pronator teres, I (*52*)
4 Brachialis, I (*49*)
5 Supinator, I (*65*)
6 Extensor carpi ulnaris, O (*64*)
7 Flexor digitorum profundus, O (*57*)
8 Pronator quadratus, I (*59*)
9 Brachioradialis, I (*60*)
10 Flexor pollicis longus, O (*58*)
11 Biceps, I (*47*)

PALMAR SURFACE

1 Flexor pollicis longus, I (*58*)
2 Adductor pollicis, **a** = O, **b** = I (*73*)
3 Flexor pollicis brevis, **a** = O, **b** = I (*71*)
4 Abductor pollicis brevis, **a** = O, **b** = I (*70*)
5 Opponens pollicis, **a** = O, **b** = I (*72*)
6 Flexor carpi radialis, I (*53*)
7 Flexor carpi ulnaris, I (*55*)
8 Abductor digiti minimi, **a** = O, **b** = I (*76*)
9 Extensor carpi ulnaris, I (*64*)
10 Flexor carpi ulnaris, I (*55*)
11 Flexor digiti minimi brevis, **a** = O, **b** = I (*77*)
12 Opponens digiti minimi, **a** = O, **b** = I (*78*)
13 Interosseous muscles, **a** = O, **b** = I (*75*)
14 Flexor digitorum superficialis, I (*56*)
15 Flexor digitorum profundus, I (*57*)

Fig. 94
The Muscles of the Hand

The flat, spindle-shaped muscles of the hand form on both sides a convex surface and in the middle a concave surface. The vault of the thumb muscle is clearly visible.

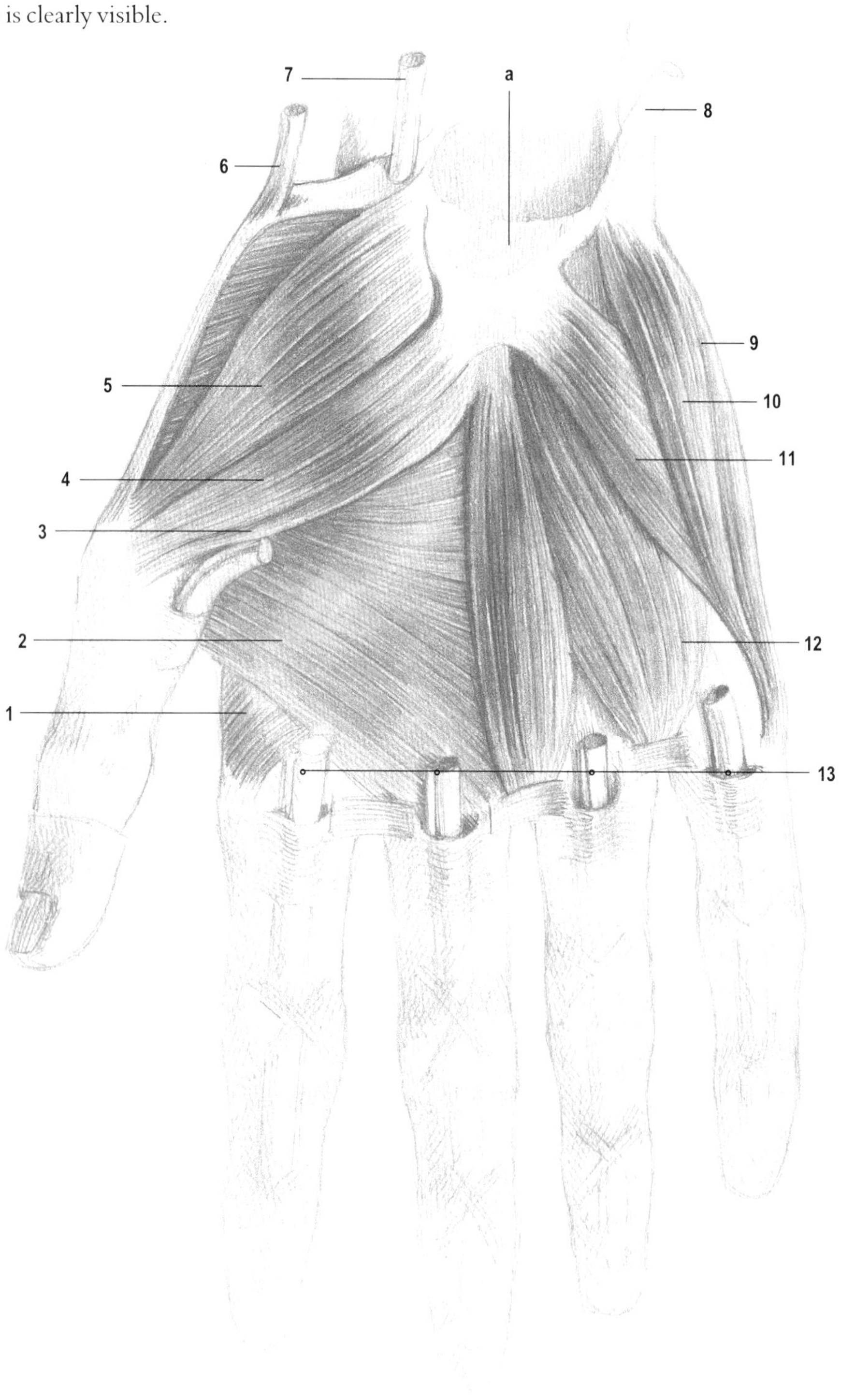

PALMAR SURFACE

1 Interosseous muscles (*75*)
2 Adductor pollicis, diagonal part (*73*)
3 Adductor pollicis, slanted part (*73*)
4 Flexor pollicis brevis (*71*)
5 Abductor pollicis brevis, I (*70*)
6 Abductor pollicis longus (*66*)

7 Flexor carpi radialis (*53*)
8 Tendon of the Flexor carpi ulnaris (*55*)
9 Abductor digiti minimi (*76*)
10 Flexor digiti minimi brevis (*77*)
11 Opponens digiti minimi (*78*)
12 Lumbricals (*74*)

13 Tendons of the Flexor digitorum superficialis and Flexor digitorum profundus (*56, 57*)

a Transverse carpal ligament

The muscles only protrude on the sides of the hand. The dorsum is thin and tendinous; the muscles are located between the bones.

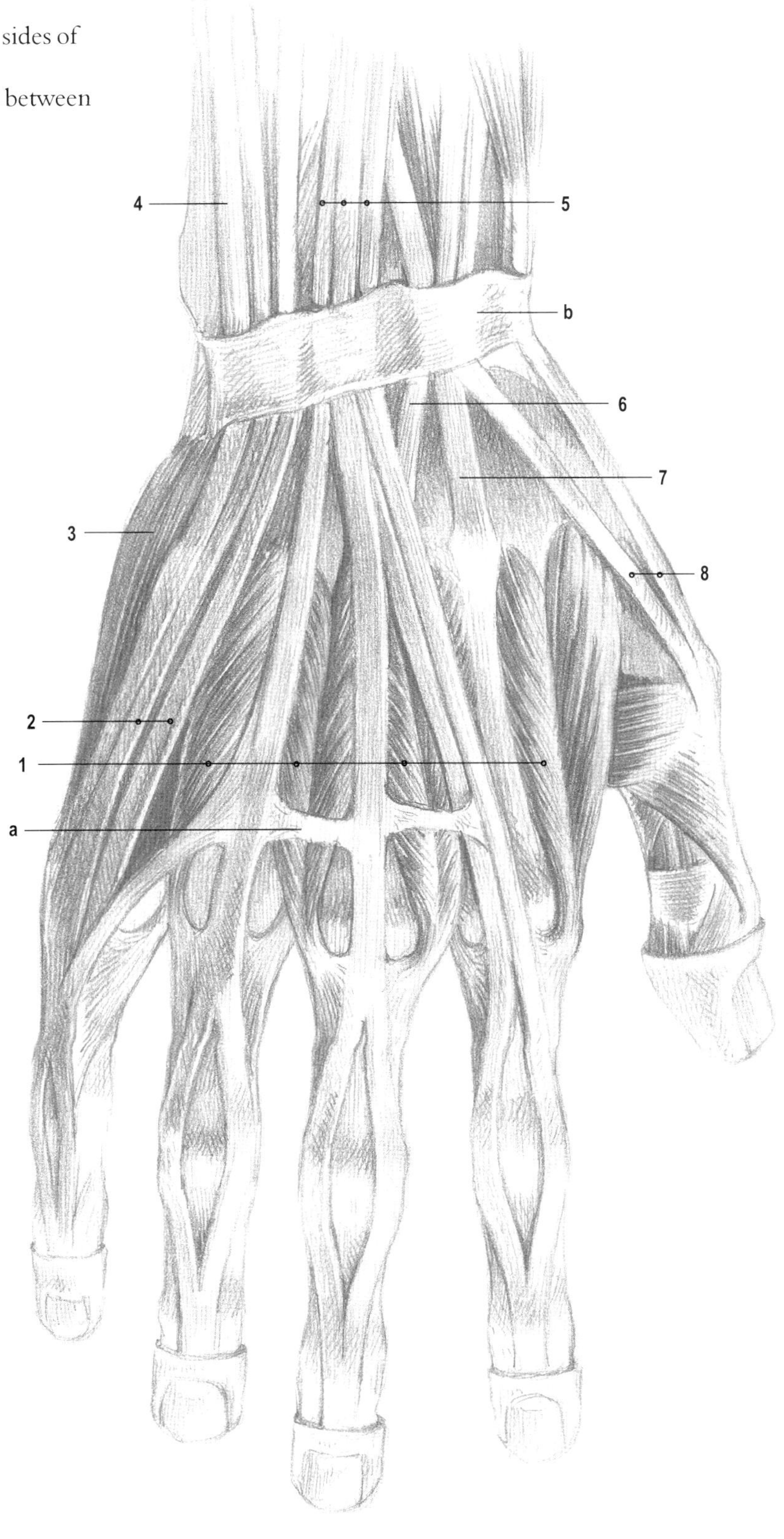

Dorsal Surface

1 Interosseous muscles (*75*)
2 Extensor digiti minimi (*63*)
3 Abductor digiti minimi (*76*)
4 Extensor carpi ulnaris (*64*)
5 Extensor digitorum (*62*)
6 Extensor carpi radialis longus (*61*)
7 Extensor carpi radialis brevis (*61*)
8 Extensor pollicis brevis and Extensor pollicis longus (*67, 68*)

a Dorsal transverse ligament which keeps the tendons together
b Dorsal transverse ligament of the carpus

Fig. 95
The Muscles of the Third Finger

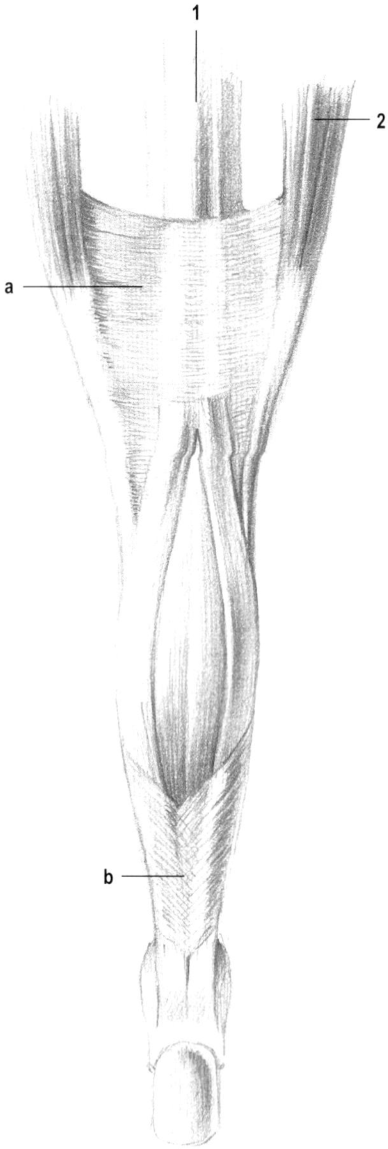

dorsal aspect

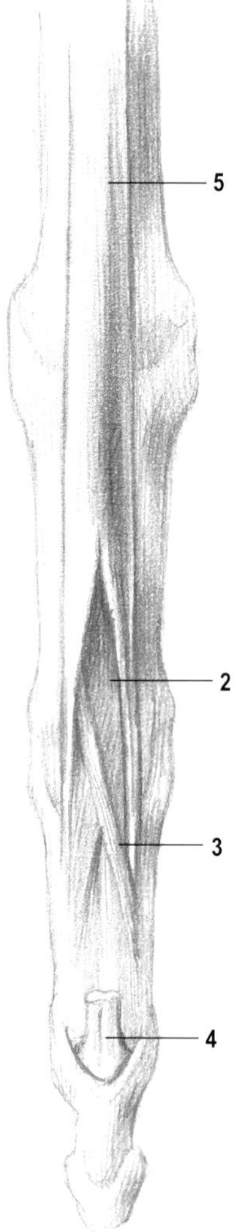

palmar aspect

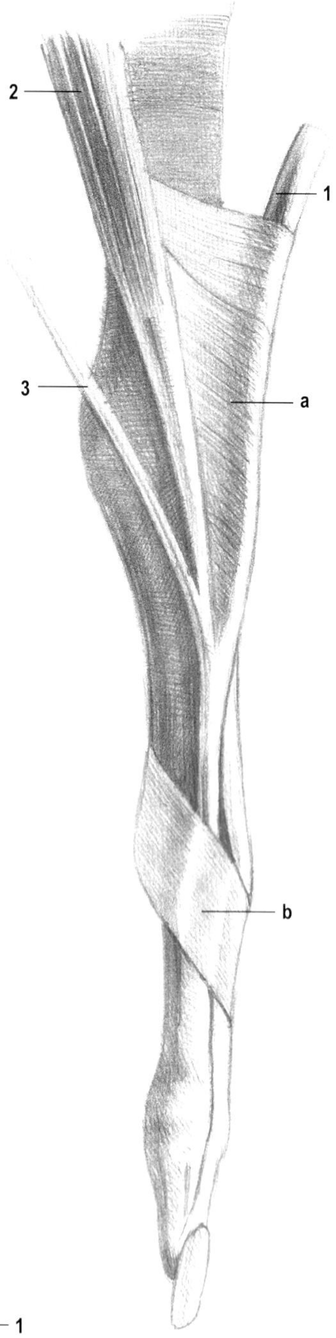

lateral aspect

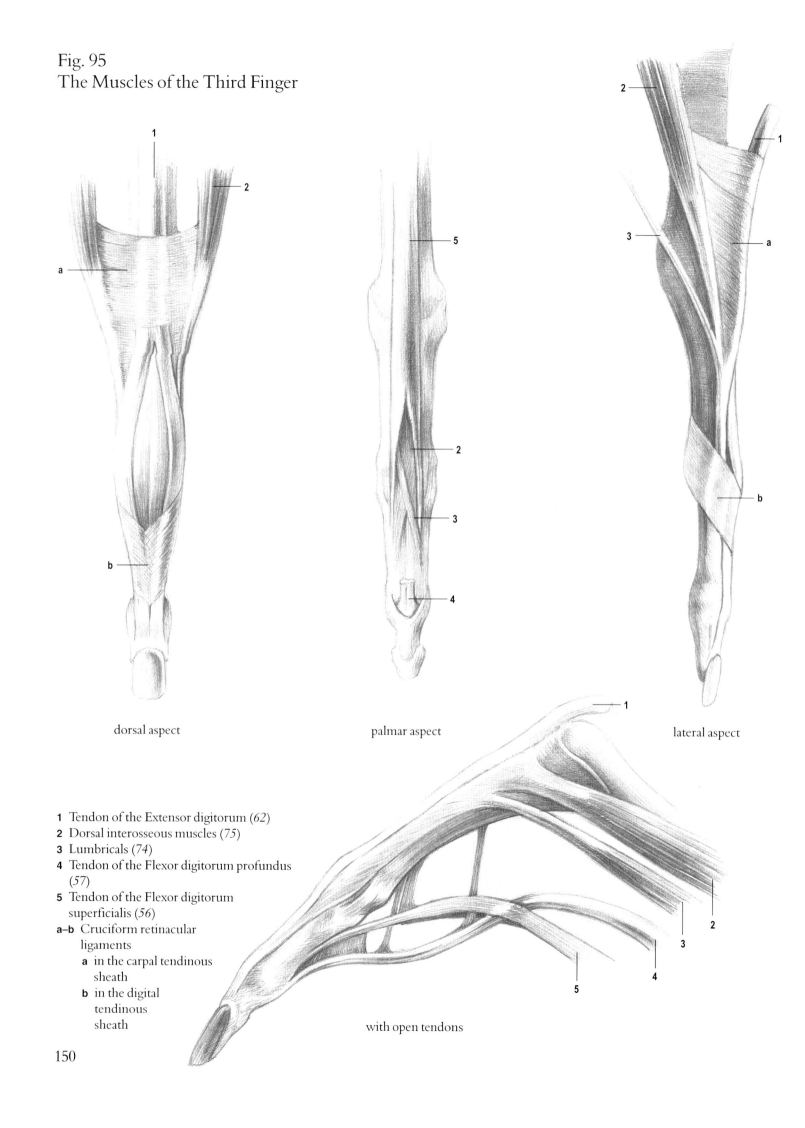

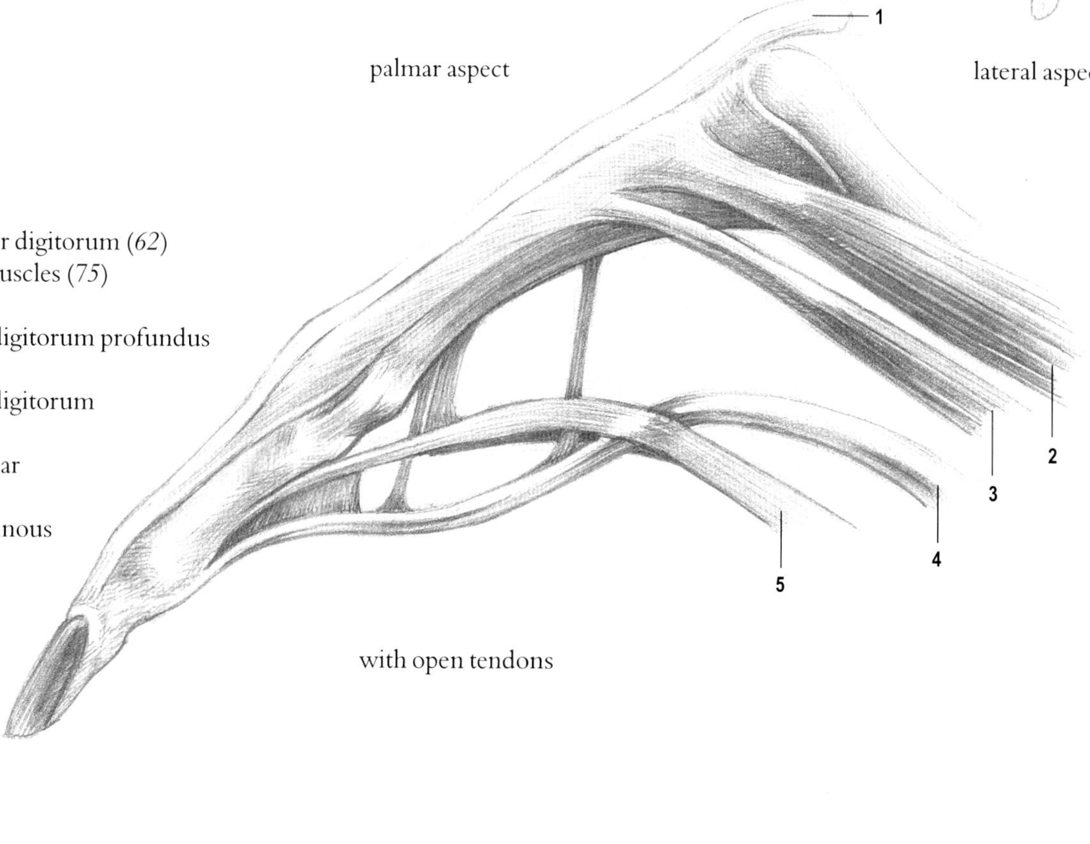

with open tendons

1 Tendon of the Extensor digitorum (*62*)
2 Dorsal interosseous muscles (*75*)
3 Lumbricals (*74*)
4 Tendon of the Flexor digitorum profundus (*57*)
5 Tendon of the Flexor digitorum superficialis (*56*)
a–b Cruciform retinacular ligaments
- **a** in the carpal tendinous sheath
- **b** in the digital tendinous sheath

Fig. 96
The Movement of Fingers

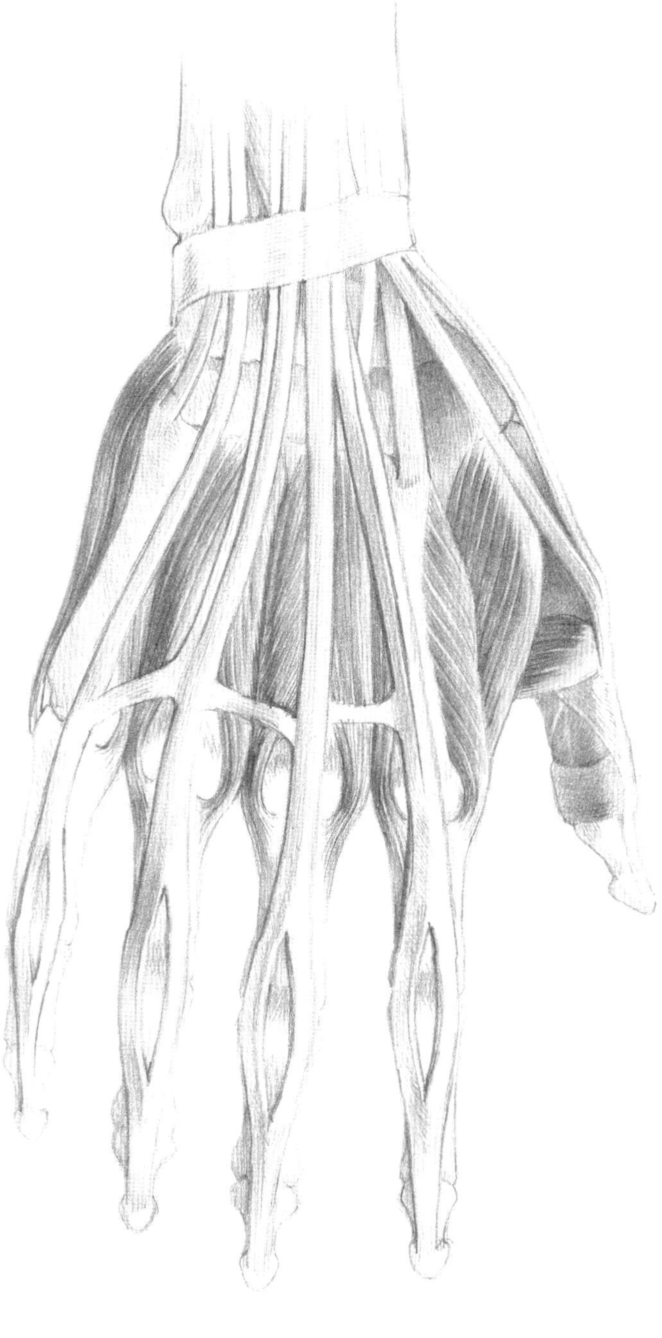

Abduction – Adduction

Fig. 97
The Hand

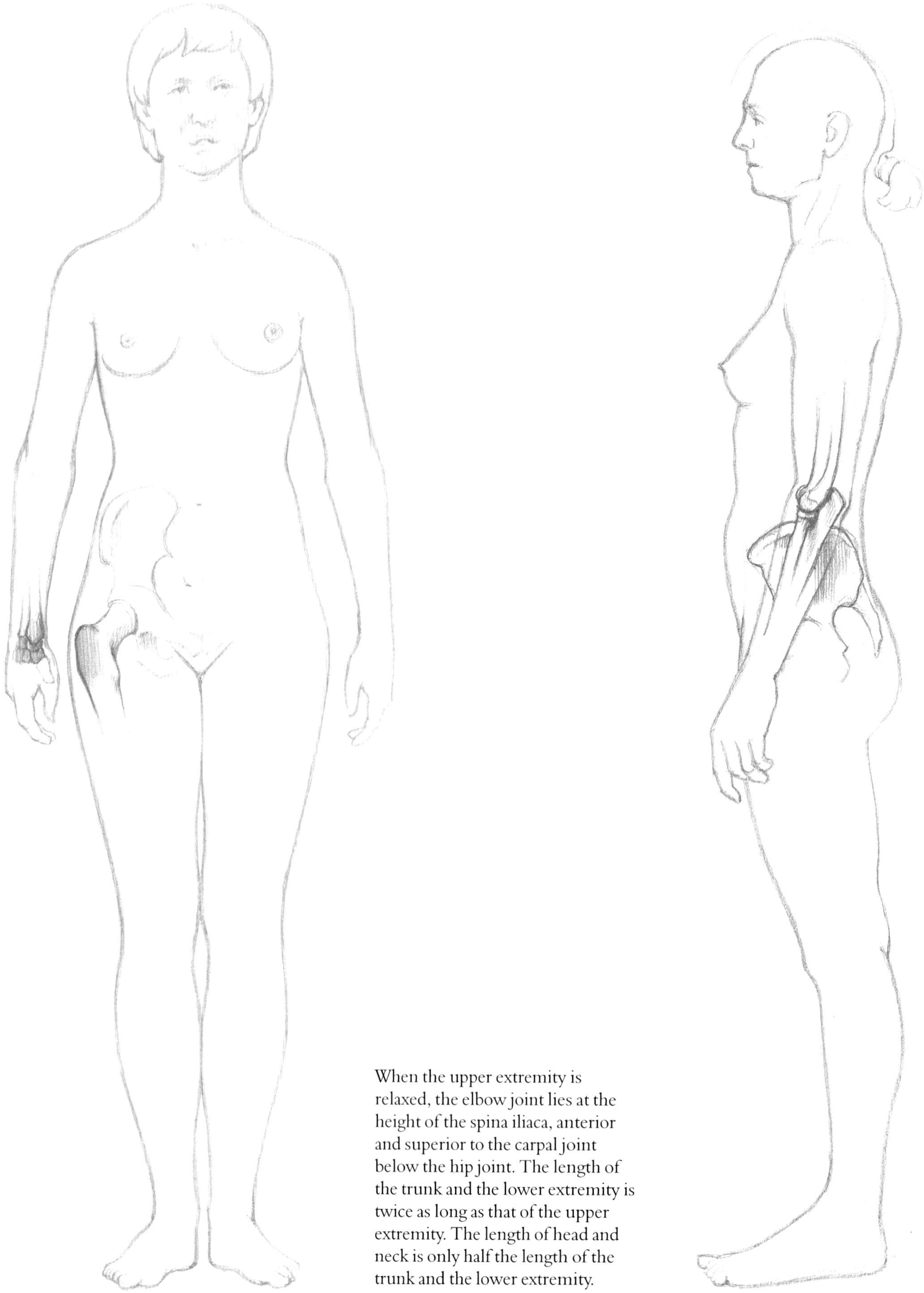

When the upper extremity is relaxed, the elbow joint lies at the height of the spina iliaca, anterior and superior to the carpal joint below the hip joint. The length of the trunk and the lower extremity is twice as long as that of the upper extremity. The length of head and neck is only half the length of the trunk and the lower extremity.

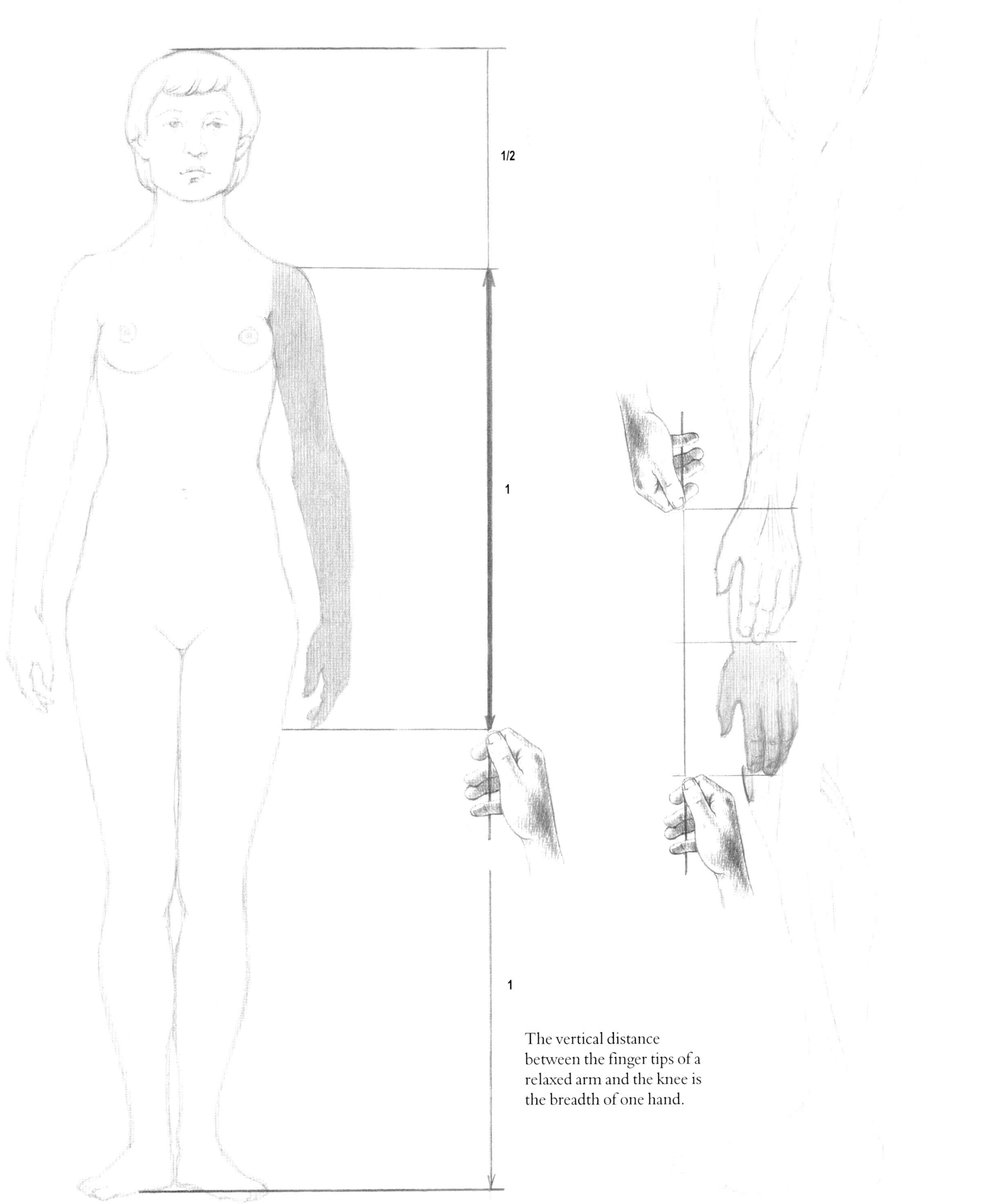

The vertical distance between the finger tips of a relaxed arm and the knee is the breadth of one hand.

Fig. 98
Hand Study

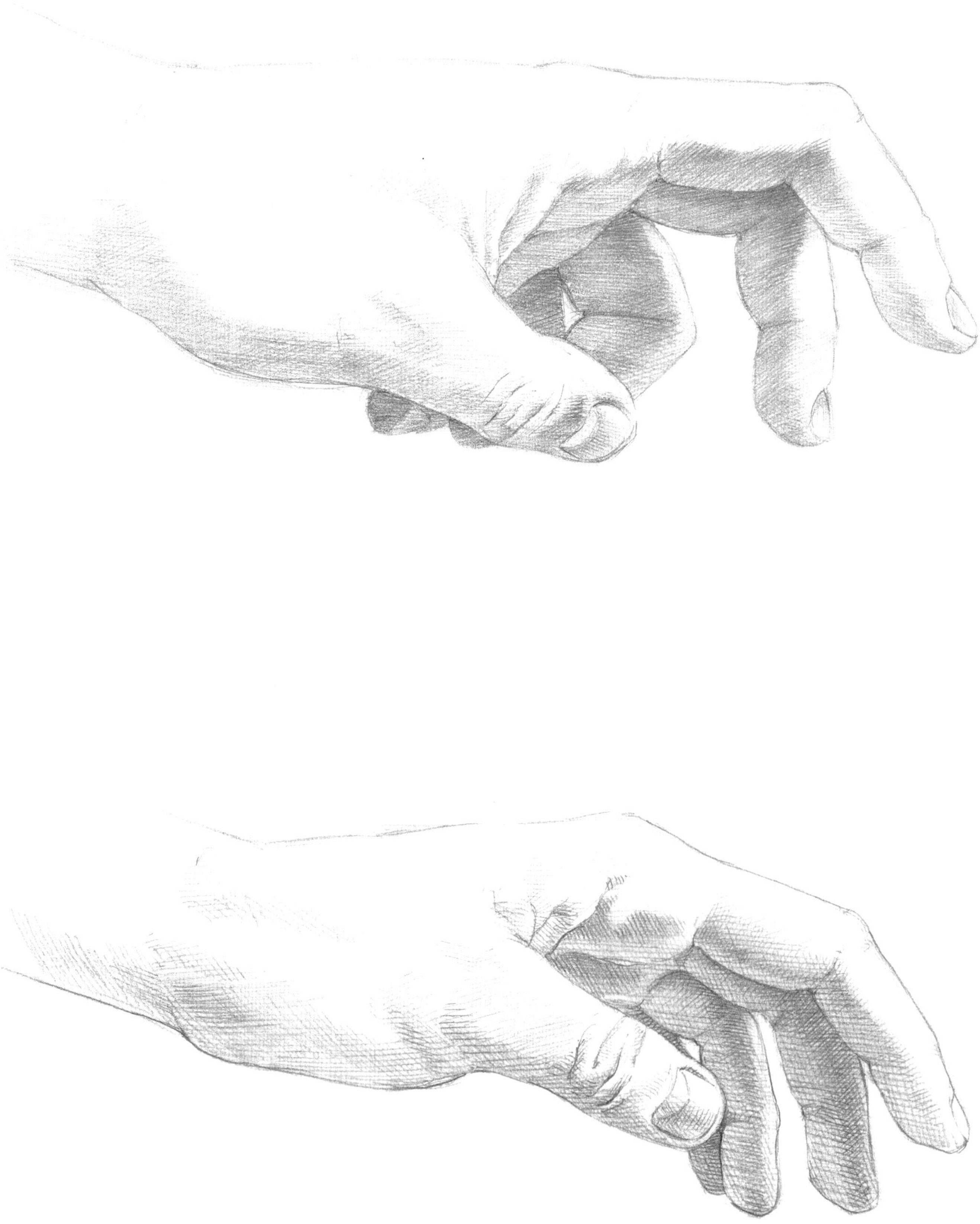

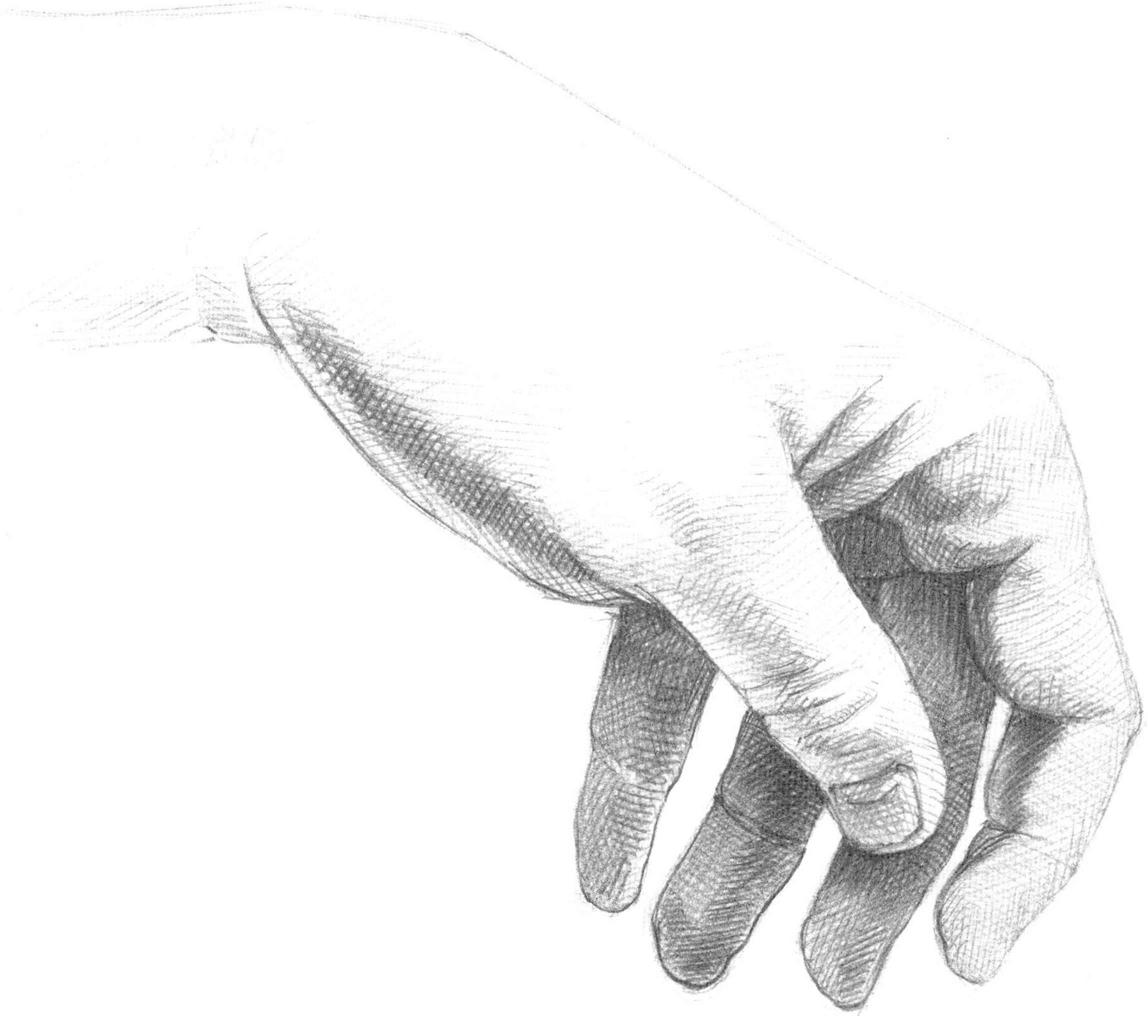

Fig. 99
The Arm at Rest

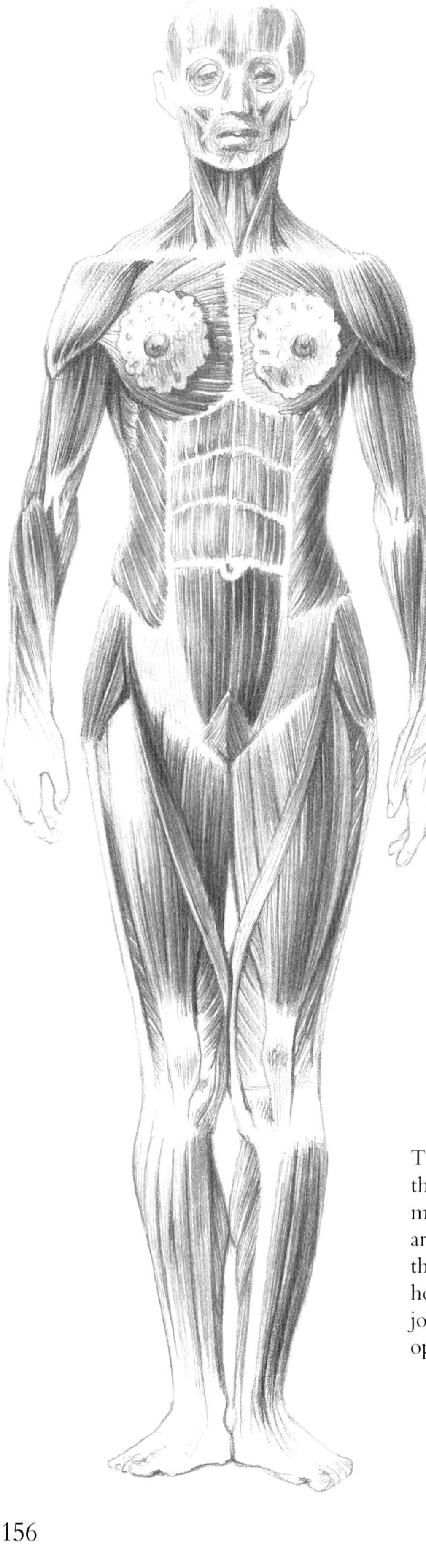

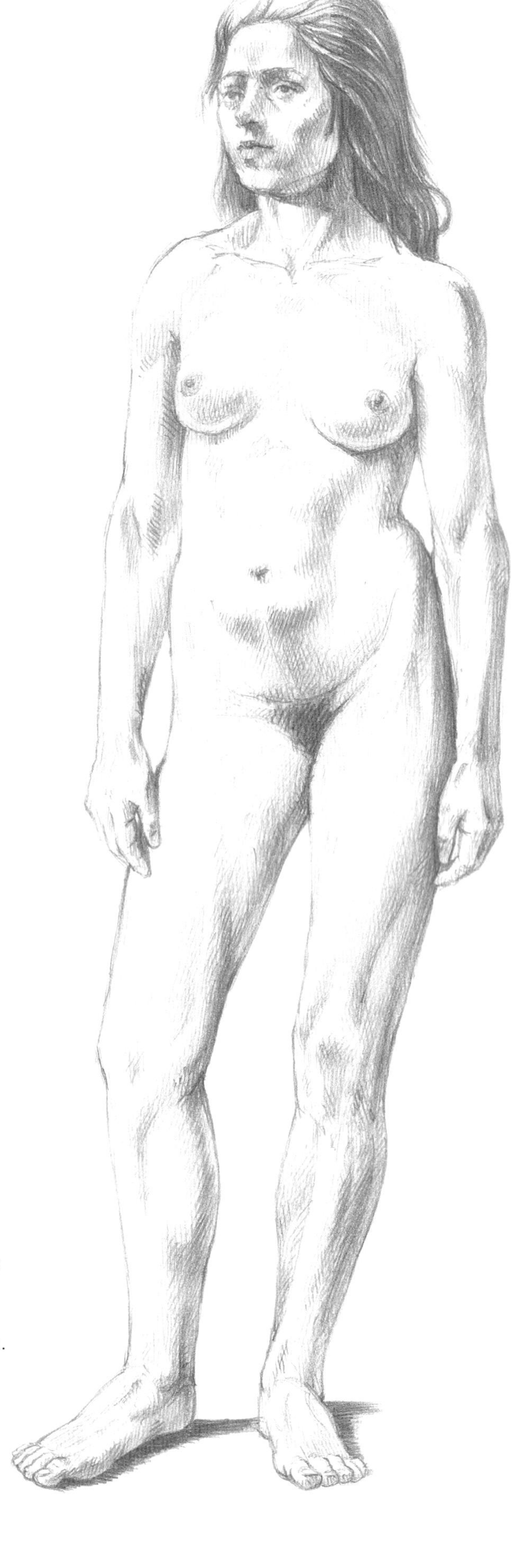

The shoulder joint overhangs the trunk, enabling the arm to move freely. When at rest the arm hangs vertically adjacent to the trunk. The clavicle is almost horizontal, the elbow and hand joints are slightly flexed and the open palm faces the upper thigh.

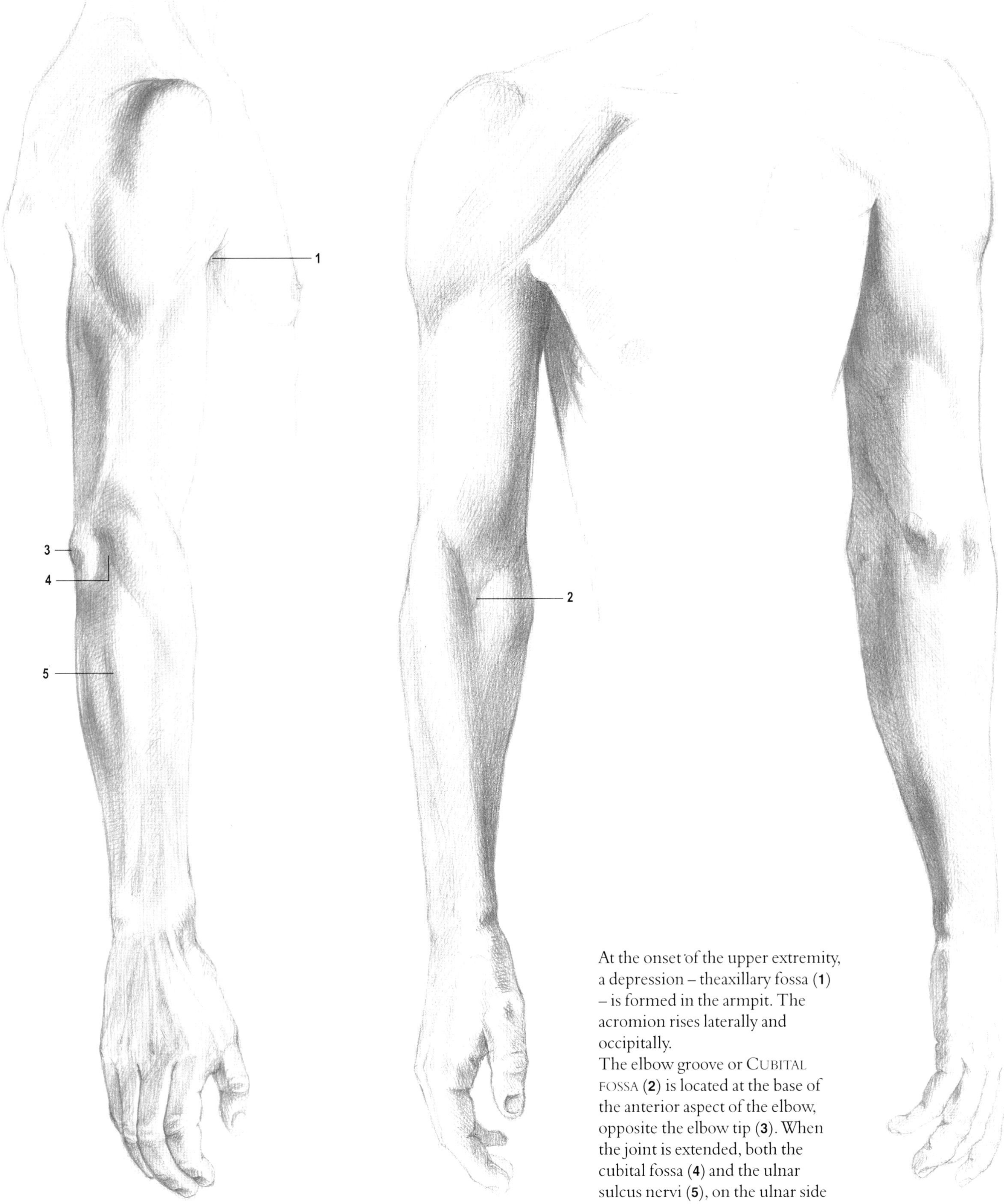

At the onset of the upper extremity, a depression – theaxillary fossa (**1**) – is formed in the armpit. The acromion rises laterally and occipitally.
The elbow groove or CUBITAL FOSSA (**2**) is located at the base of the anterior aspect of the elbow, opposite the elbow tip (**3**). When the joint is extended, both the cubital fossa (**4**) and the ulnar sulcus nervi (**5**), on the ulnar side of the elbow, clearly protrude.

The Arm at Rest
(cont.)

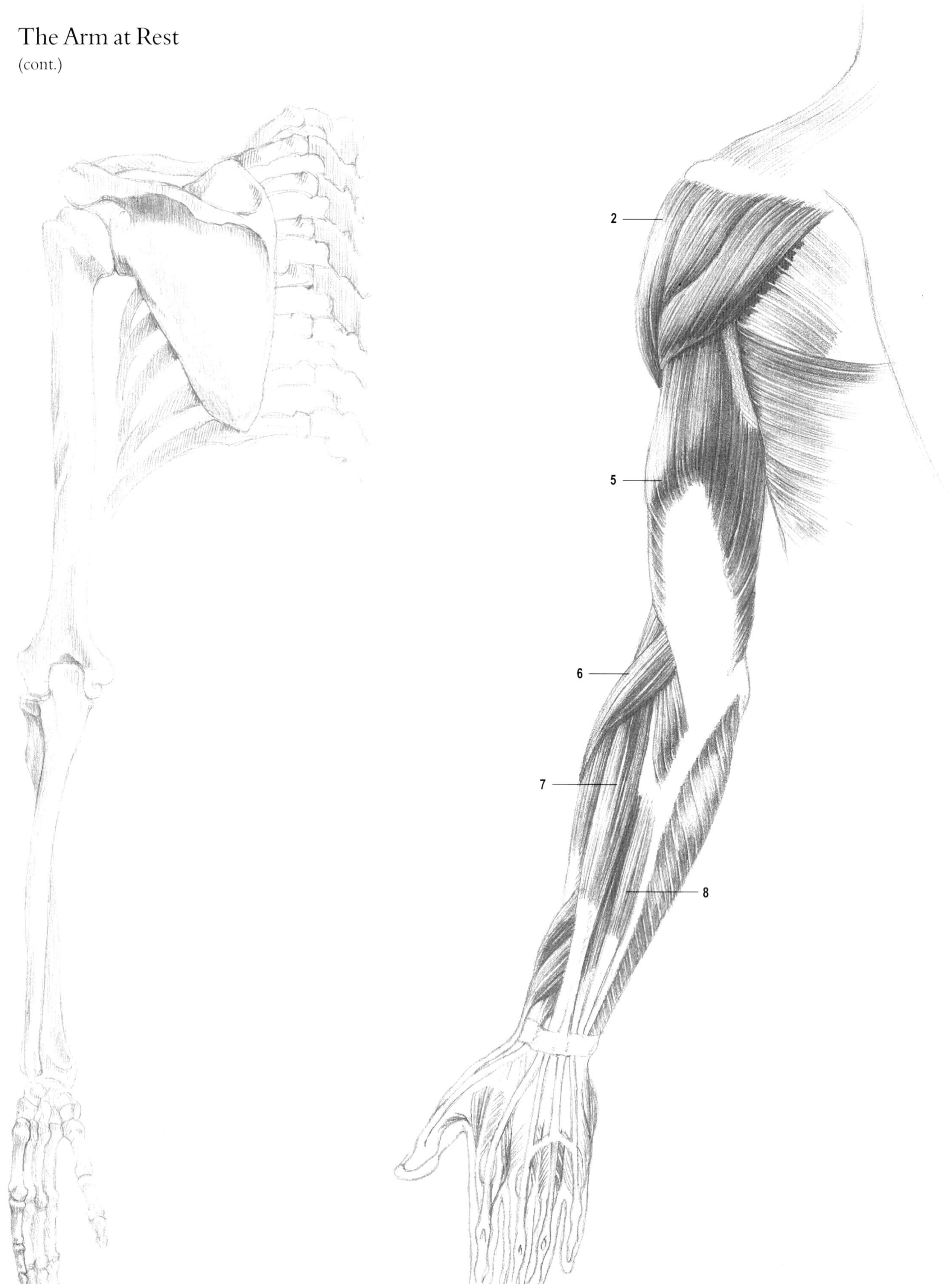

dorsal aspect

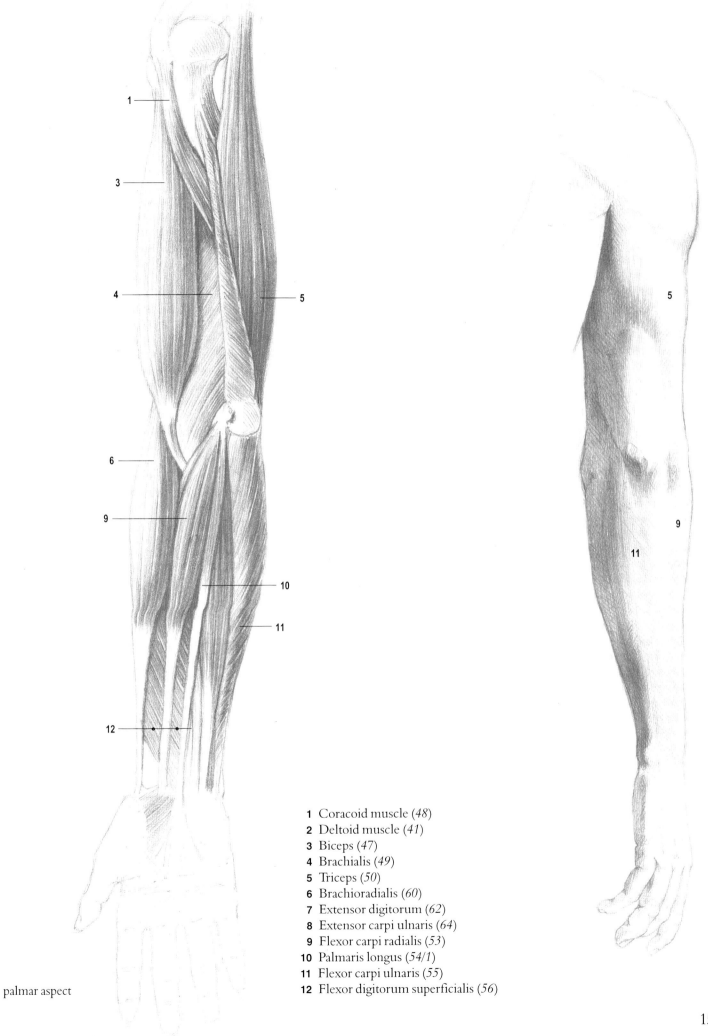

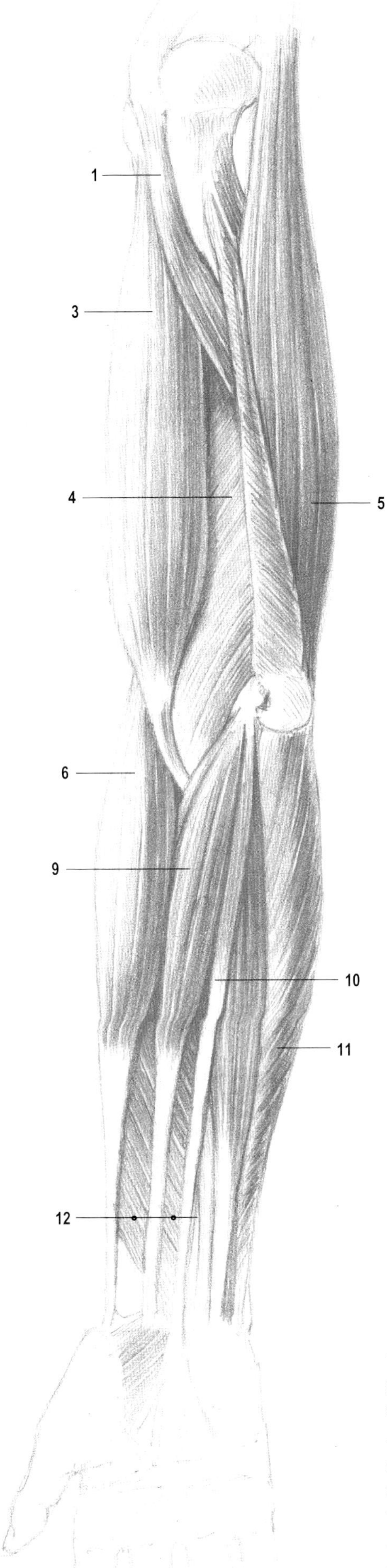

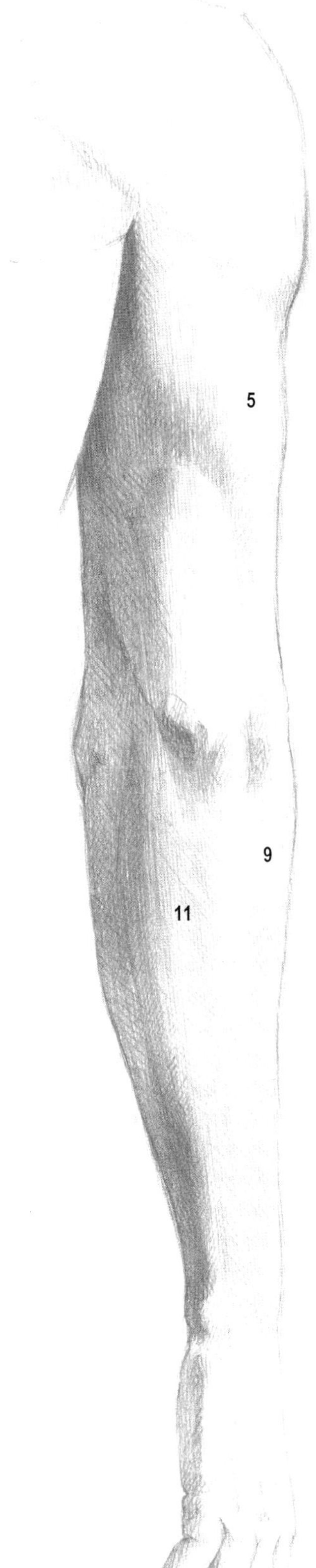

1 Coracoid muscle (*48*)
2 Deltoid muscle (*41*)
3 Biceps (*47*)
4 Brachialis (*49*)
5 Triceps (*50*)
6 Brachioradialis (*60*)
7 Extensor digitorum (*62*)
8 Extensor carpi ulnaris (*64*)
9 Flexor carpi radialis (*53*)
10 Palmaris longus (*54/1*)
11 Flexor carpi ulnaris (*55*)
12 Flexor digitorum superficialis (*56*)

palmar aspect

The Arm at Rest
(cont.)

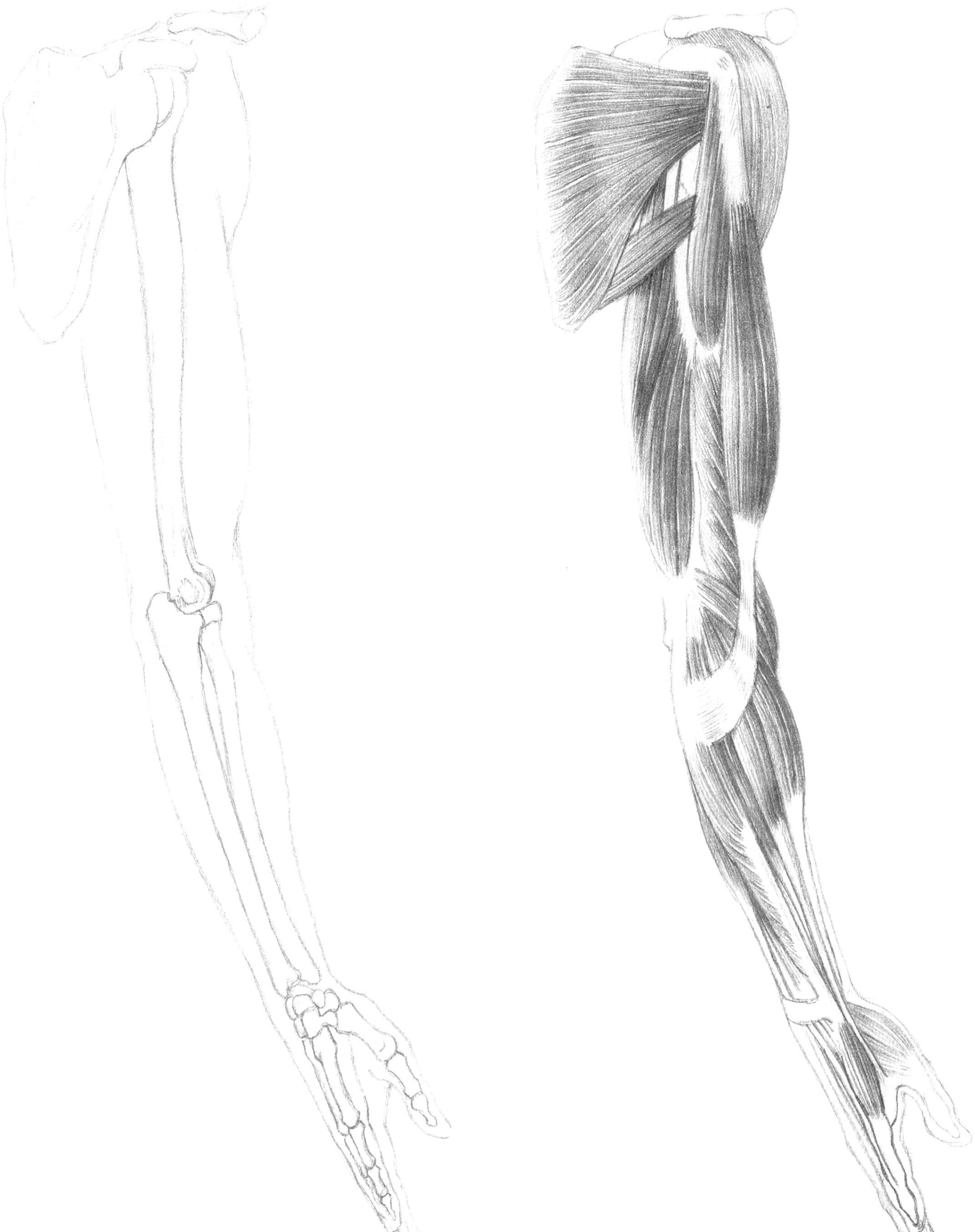

medial aspect

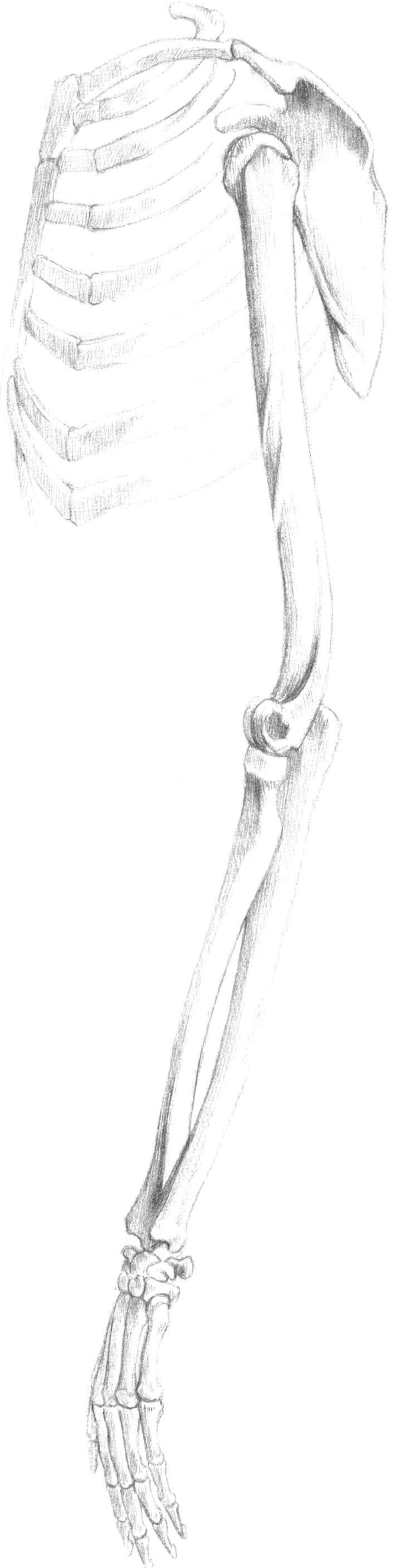

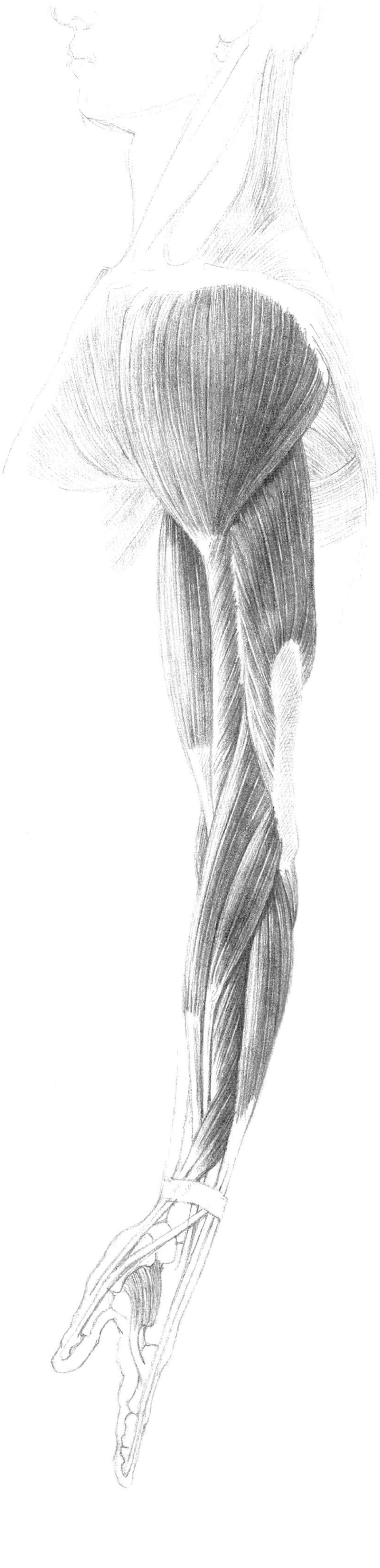

lateral aspect

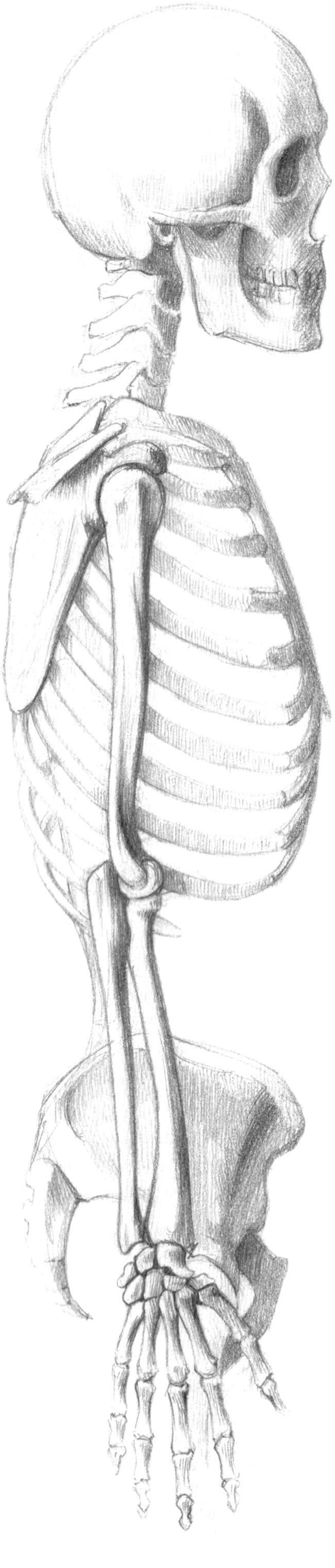

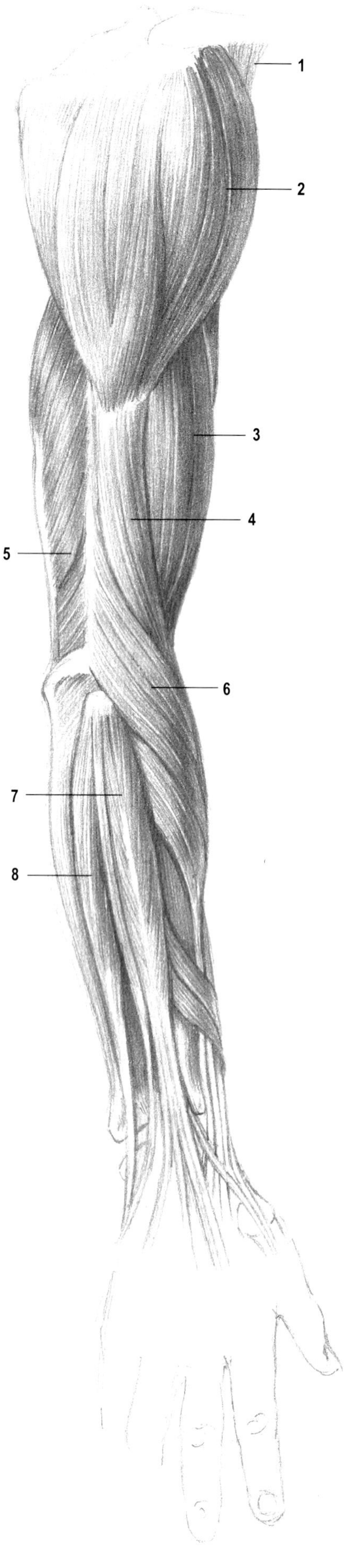

lateral aspect

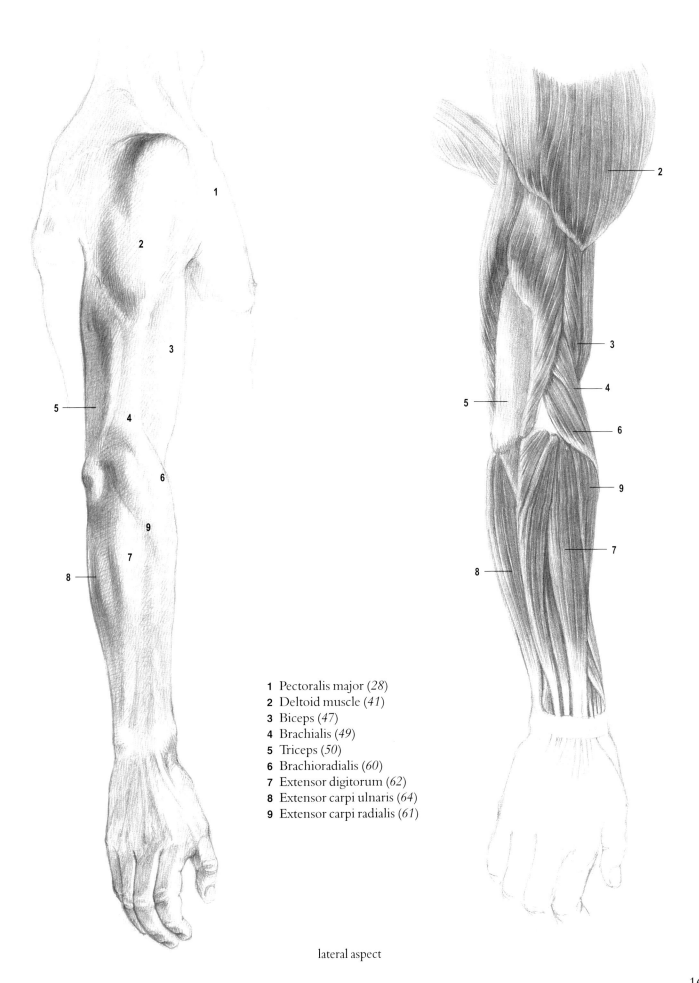

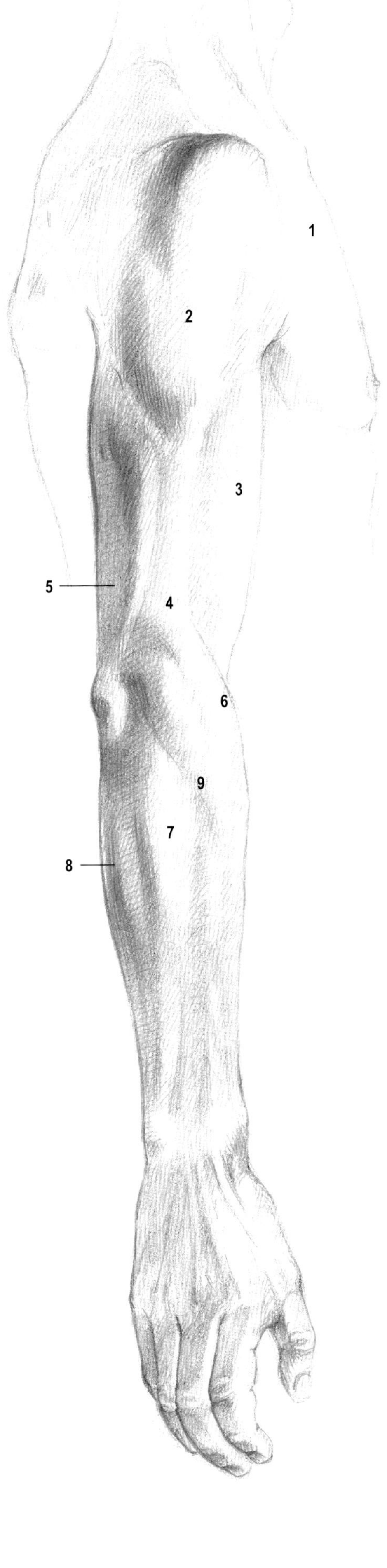

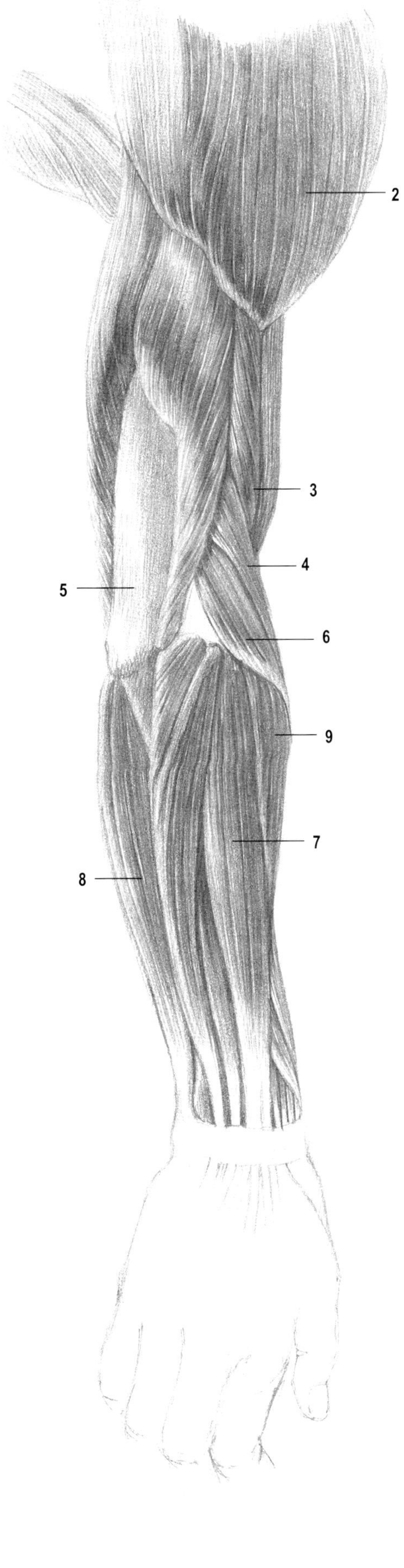

1 Pectoralis major (*28*)
2 Deltoid muscle (*41*)
3 Biceps (*47*)
4 Brachialis (*49*)
5 Triceps (*50*)
6 Brachioradialis (*60*)
7 Extensor digitorum (*62*)
8 Extensor carpi ulnaris (*64*)
9 Extensor carpi radialis (*61*)

lateral aspect

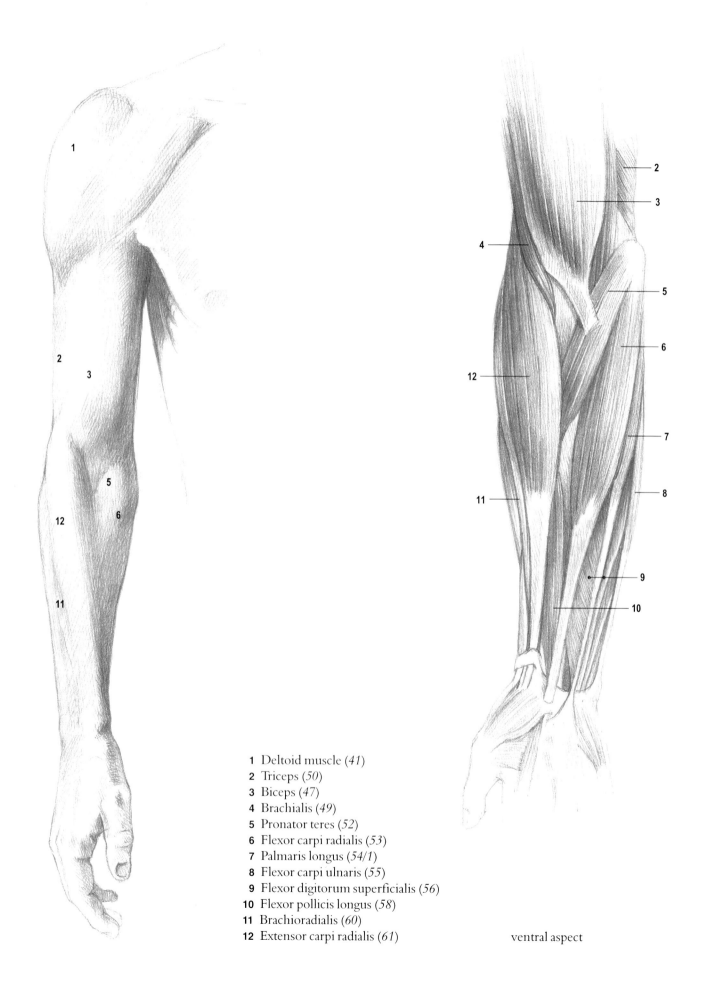

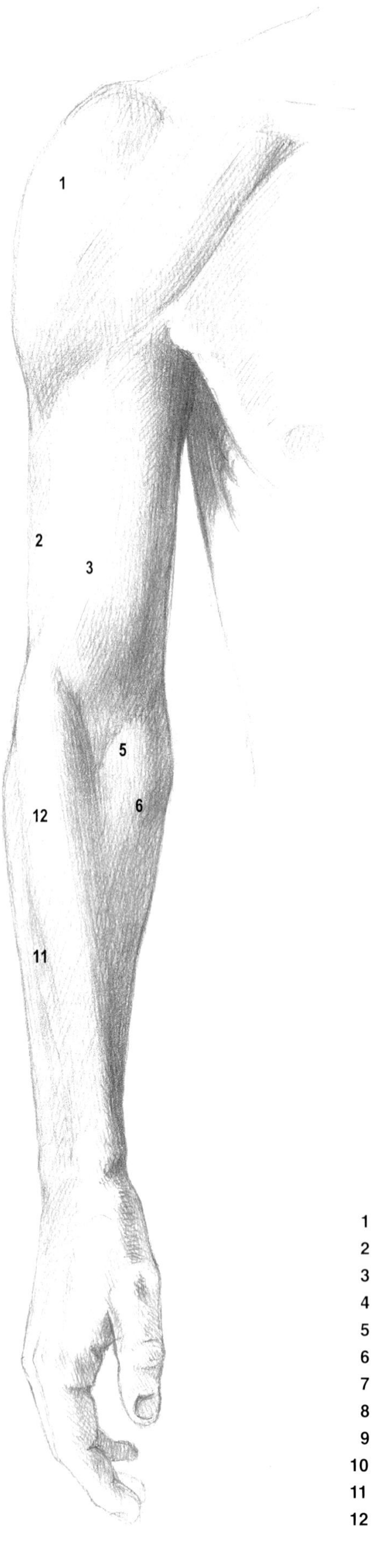

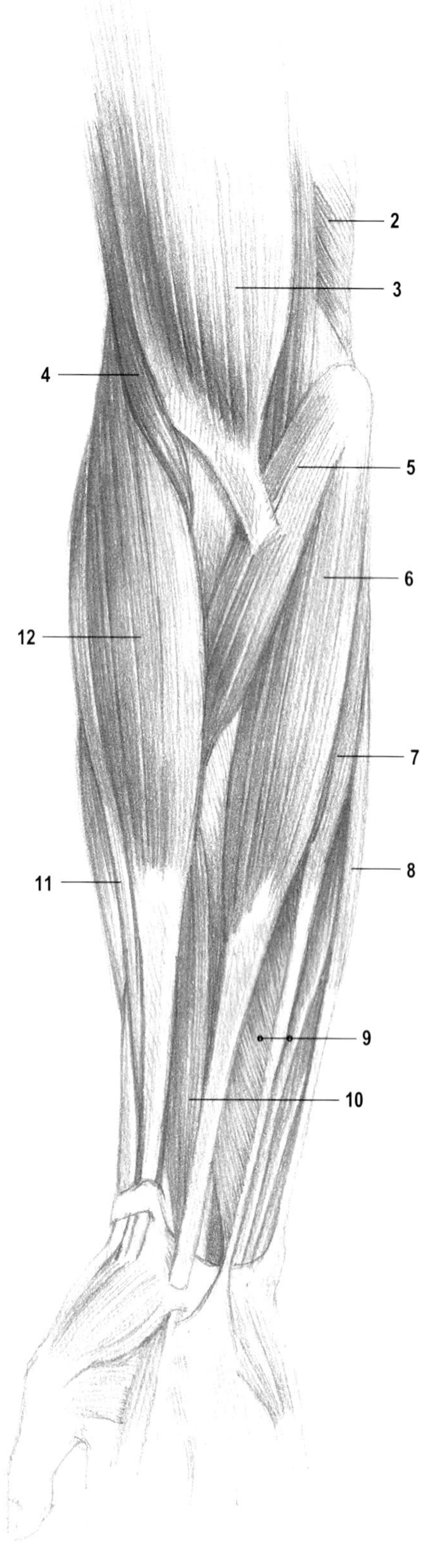

1 Deltoid muscle (*41*)
2 Triceps (*50*)
3 Biceps (*47*)
4 Brachialis (*49*)
5 Pronator teres (*52*)
6 Flexor carpi radialis (*53*)
7 Palmaris longus (*54/1*)
8 Flexor carpi ulnaris (*55*)
9 Flexor digitorum superficialis (*56*)
10 Flexor pollicis longus (*58*)
11 Brachioradialis (*60*)
12 Extensor carpi radialis (*61*)

ventral aspect

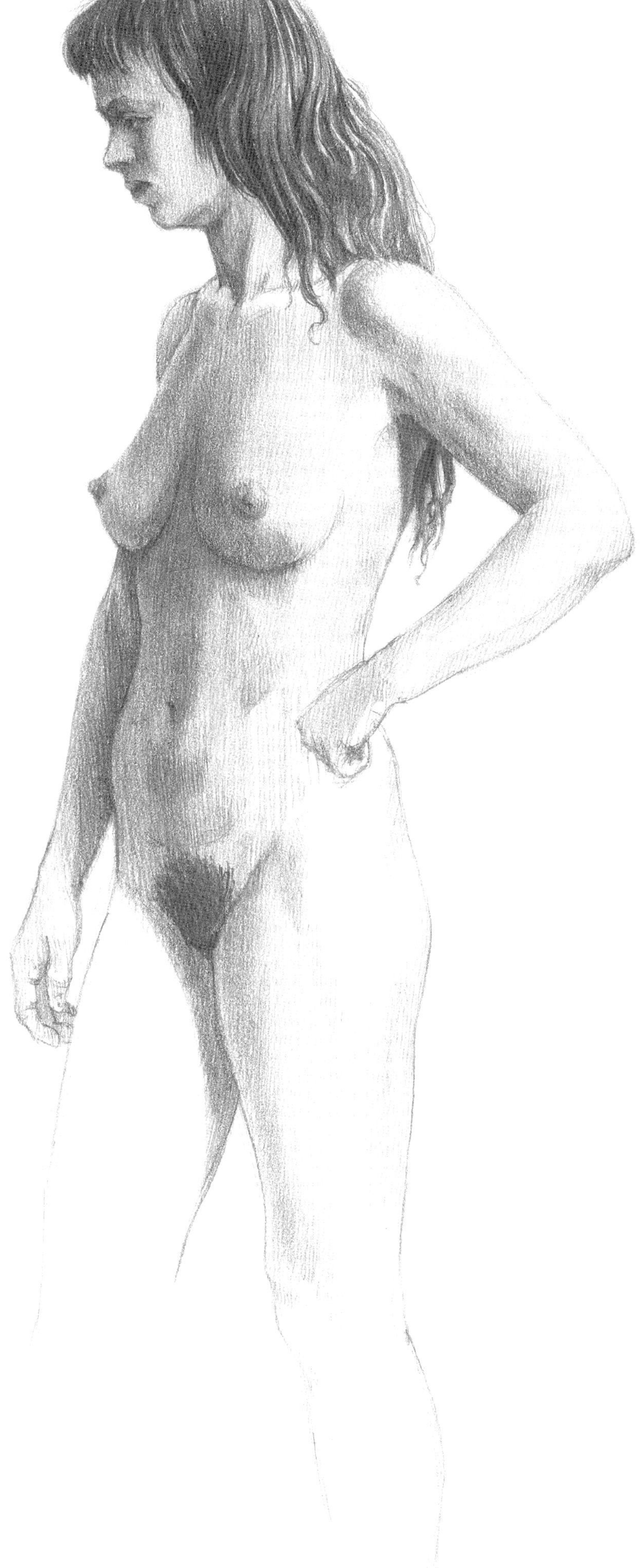

Fig. 100
The Arm in Movement

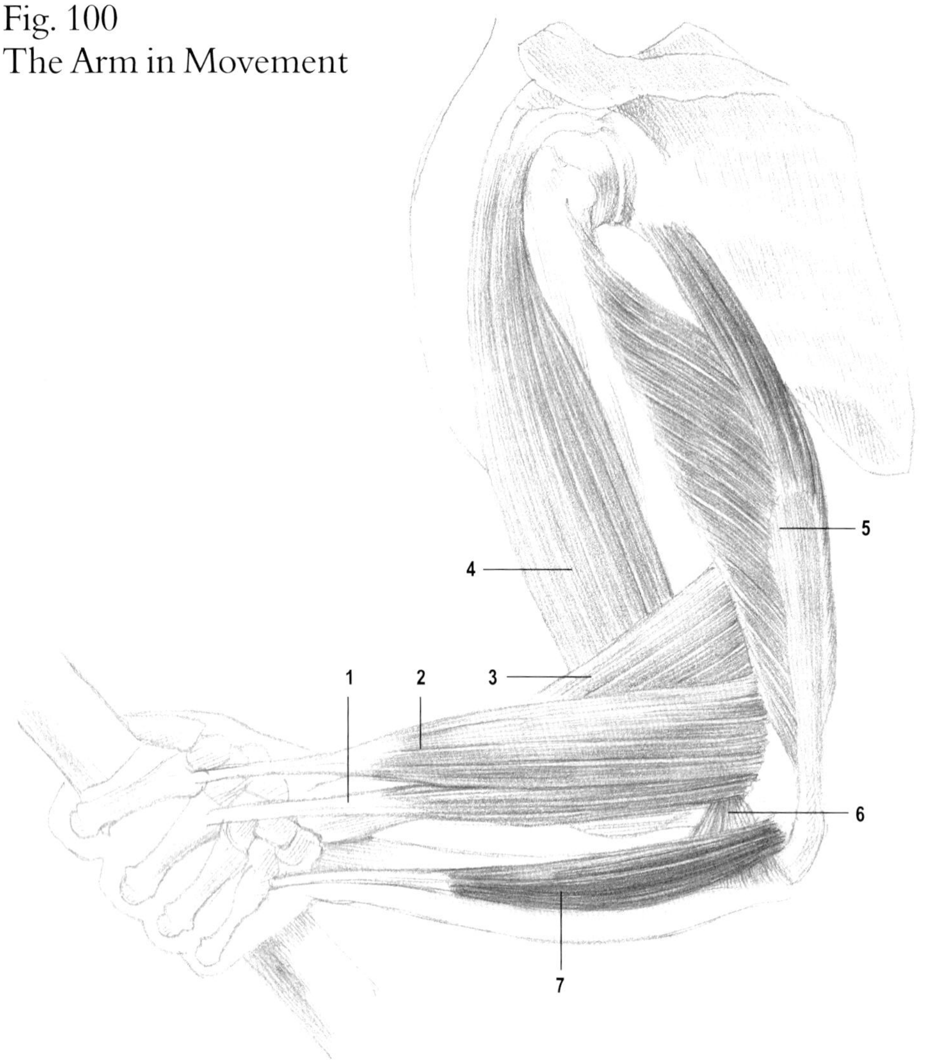

dorsolateral aspect

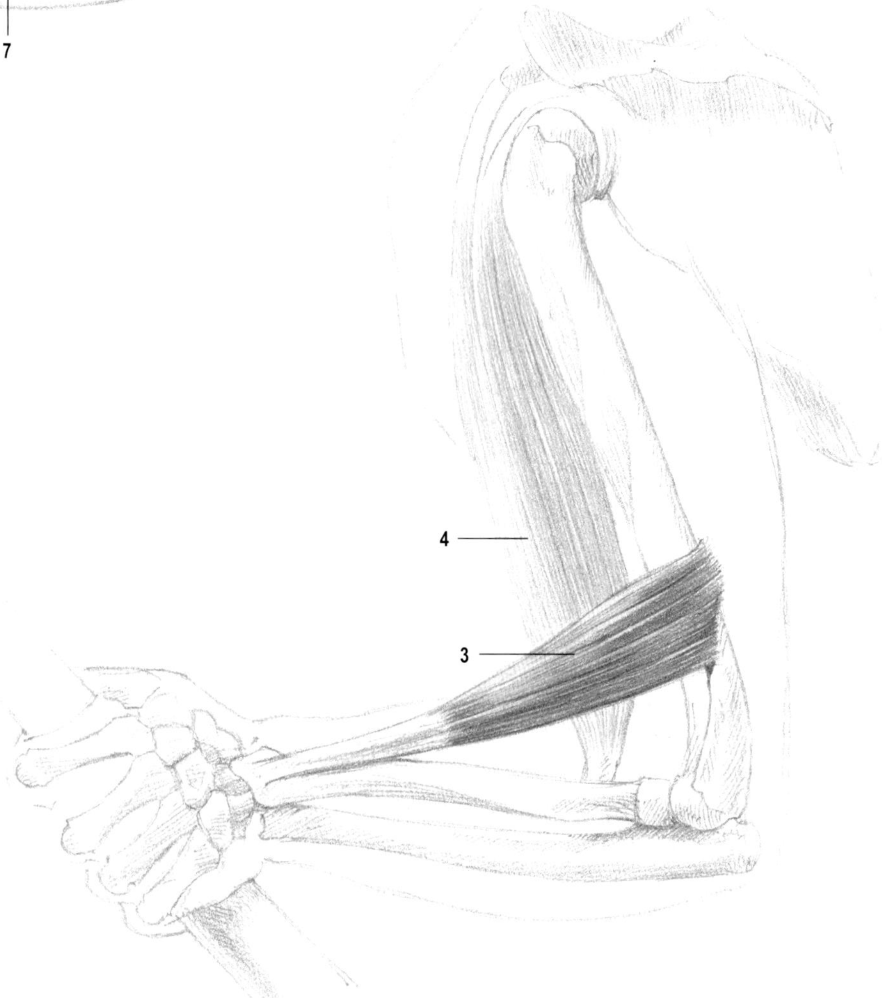

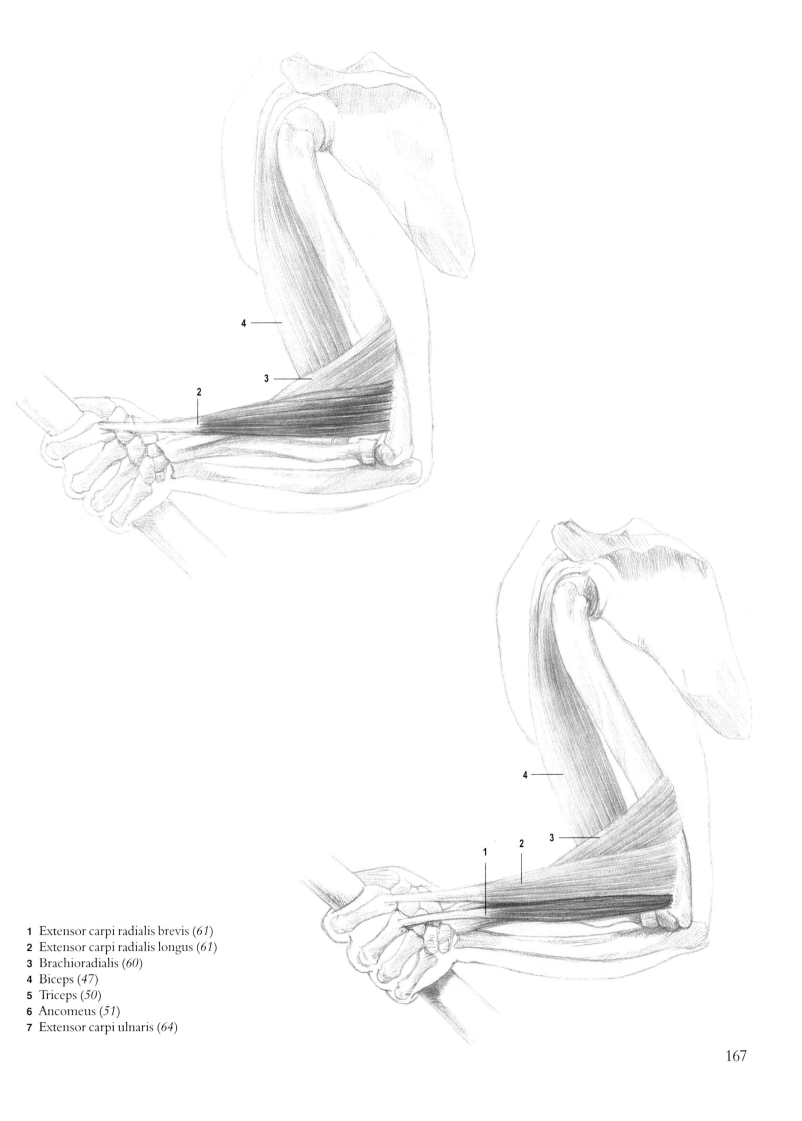

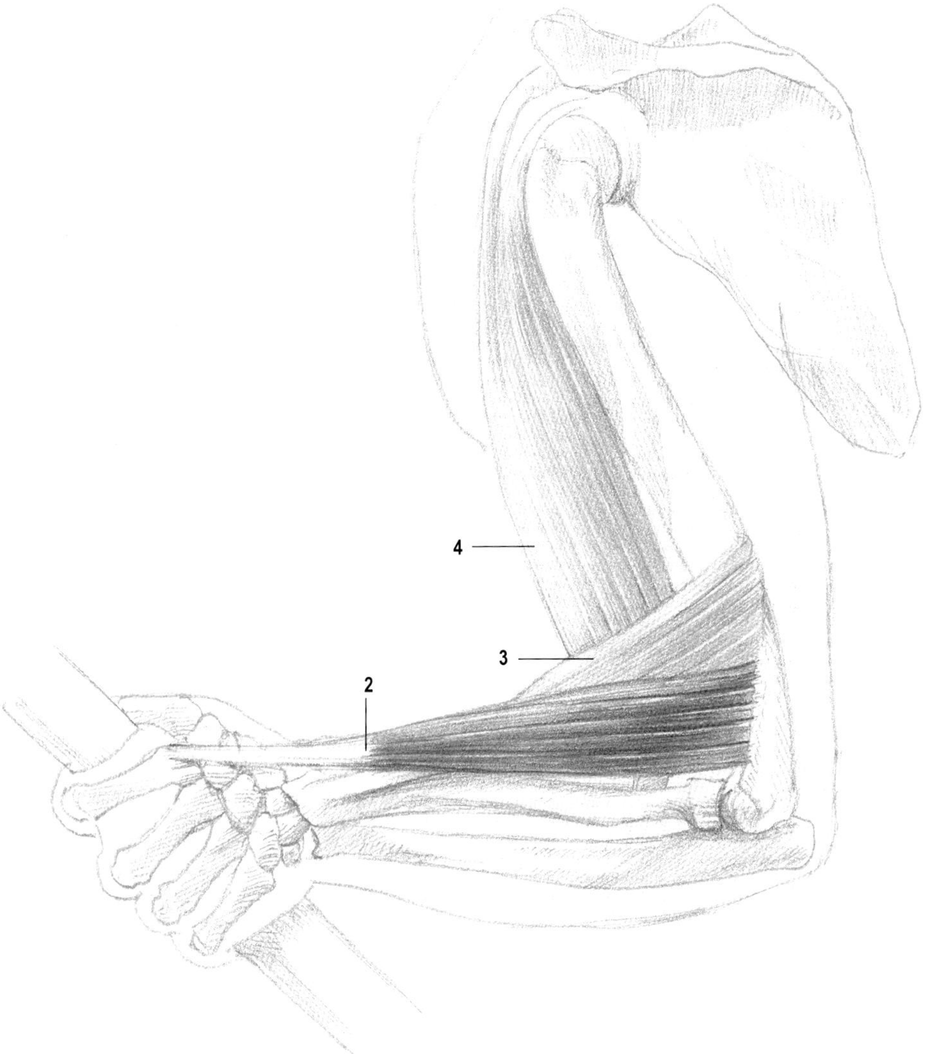

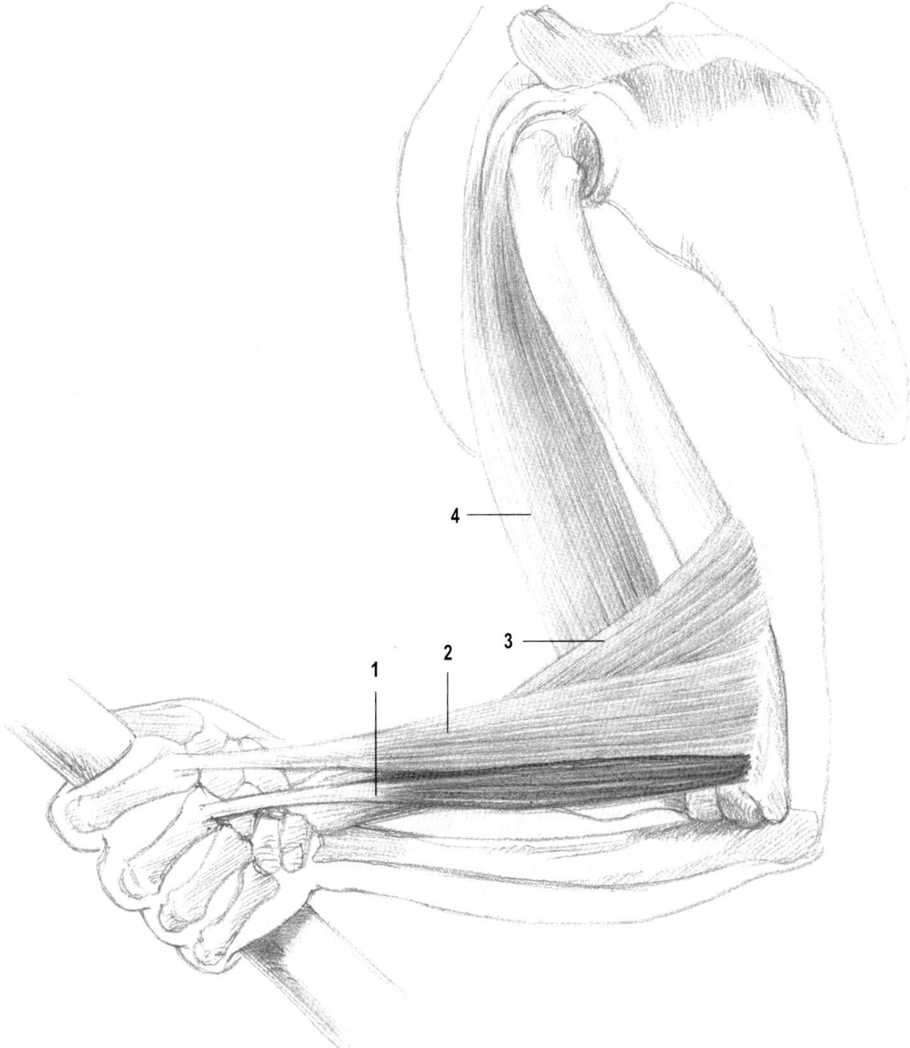

1 Extensor carpi radialis brevis (*61*)
2 Extensor carpi radialis longus (*61*)
3 Brachioradialis (*60*)
4 Biceps (*47*)
5 Triceps (*50*)
6 Ancomeus (*51*)
7 Extensor carpi ulnaris (*64*)

Fig. 101
Muscle Studies

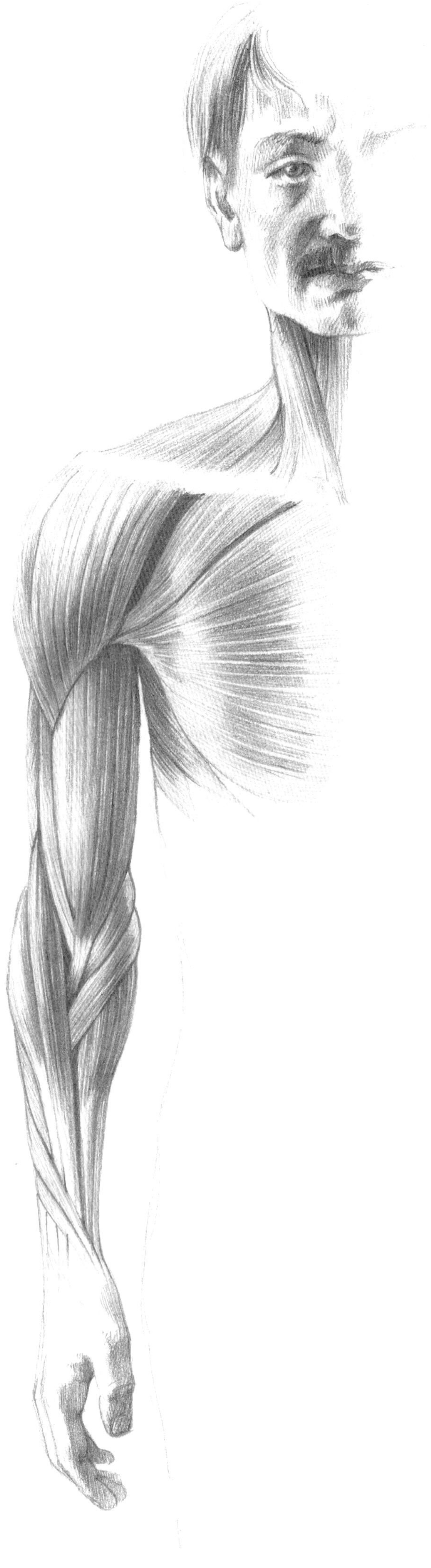

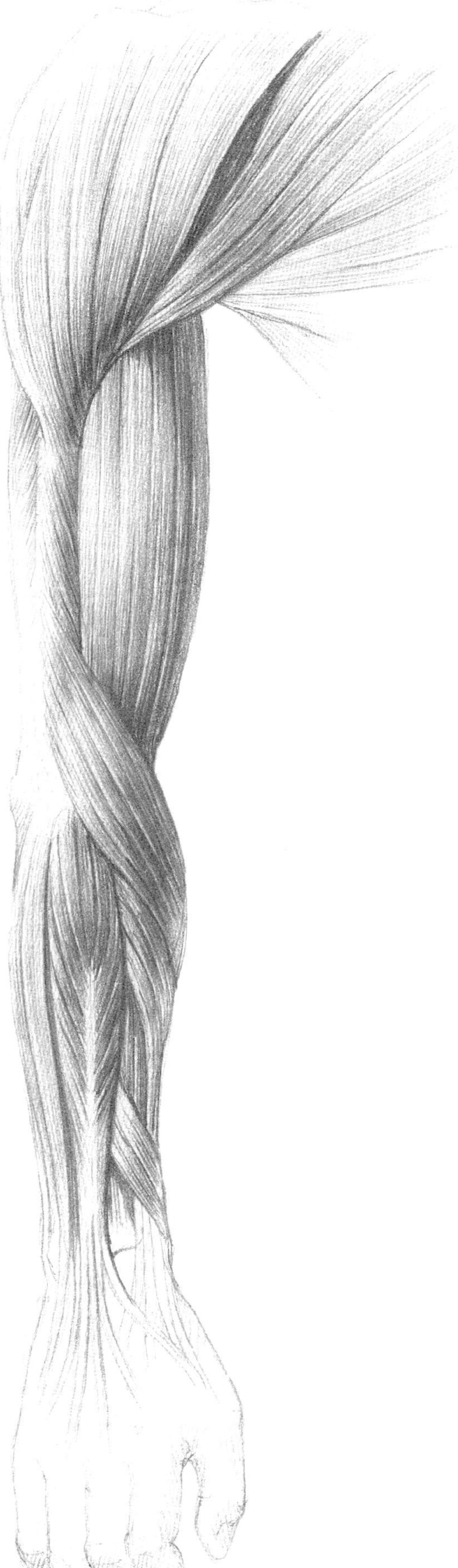

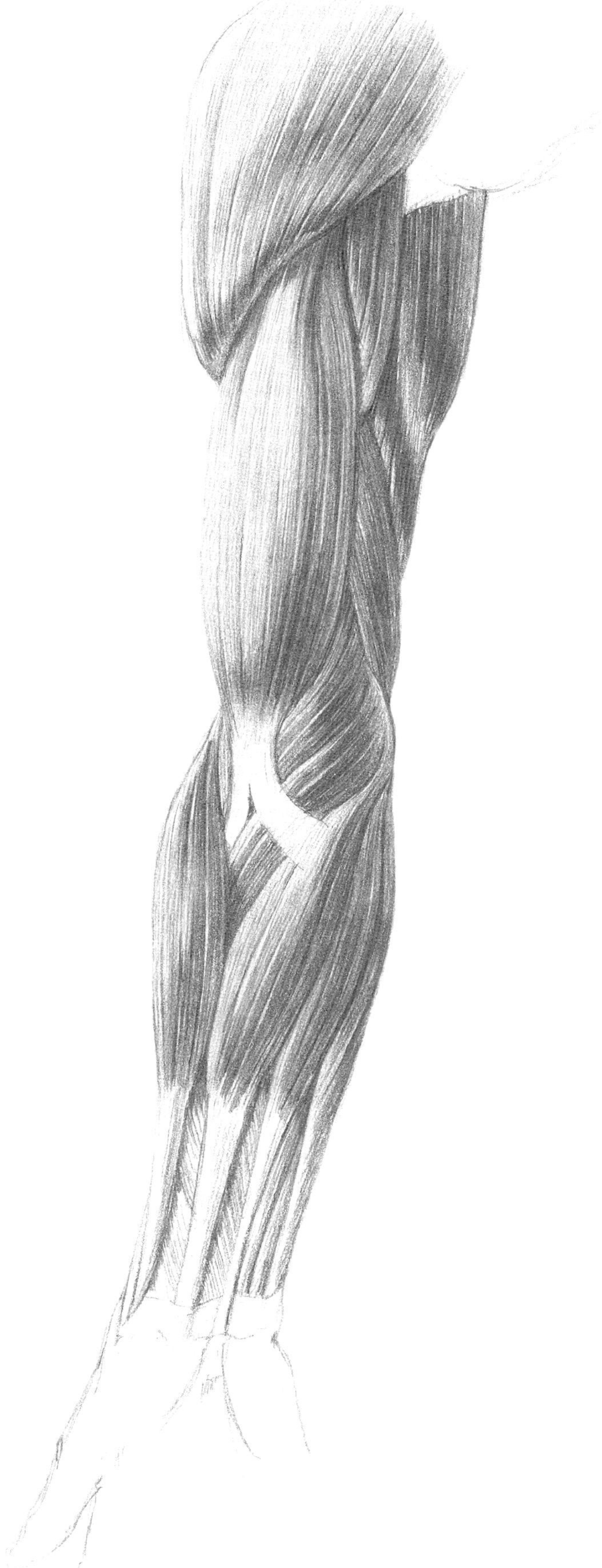

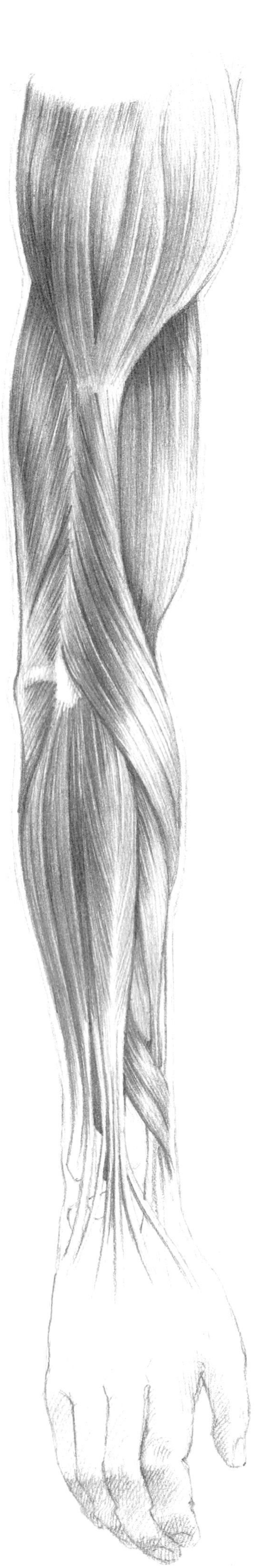

Fig. 102
Studies in Movement

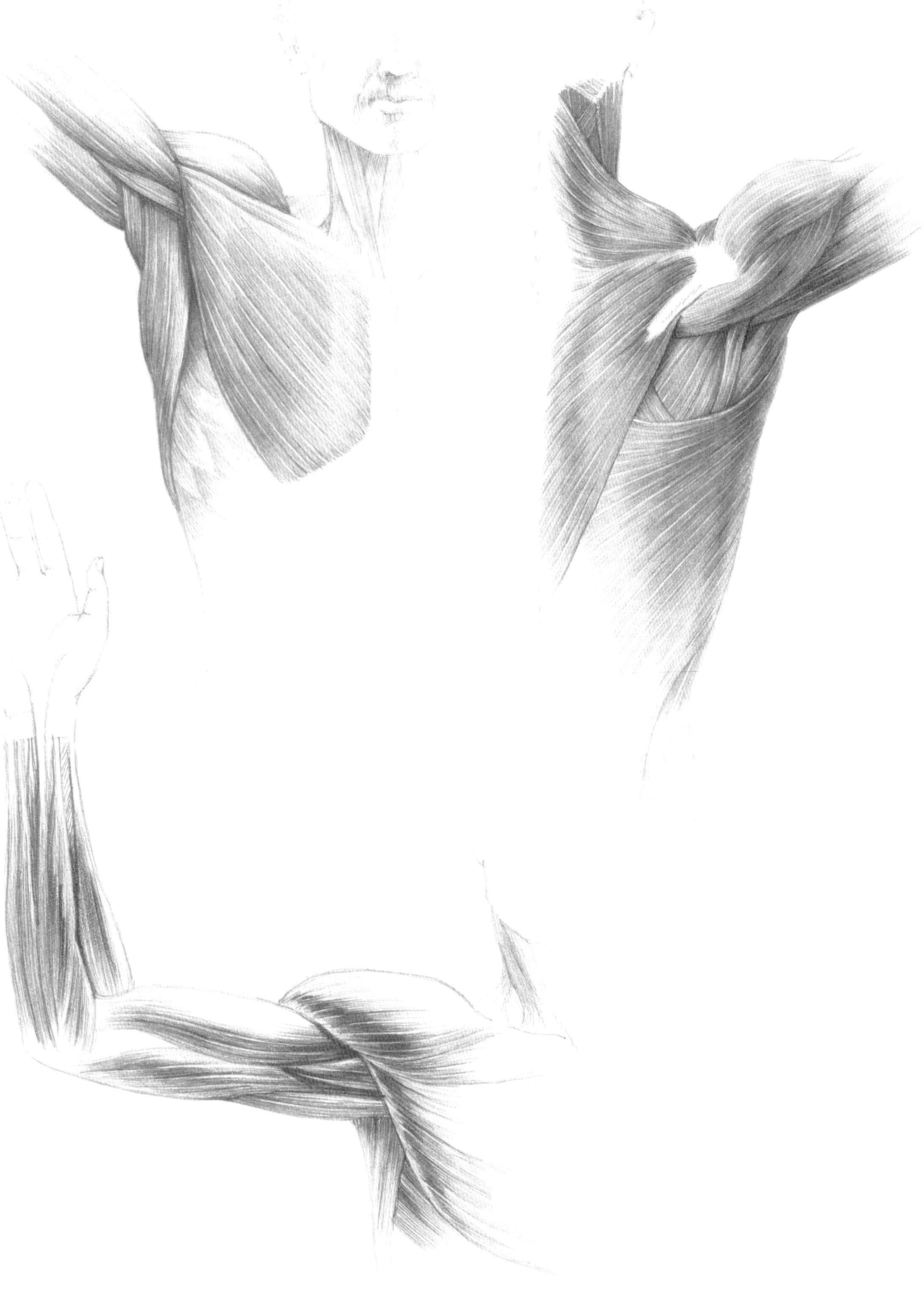

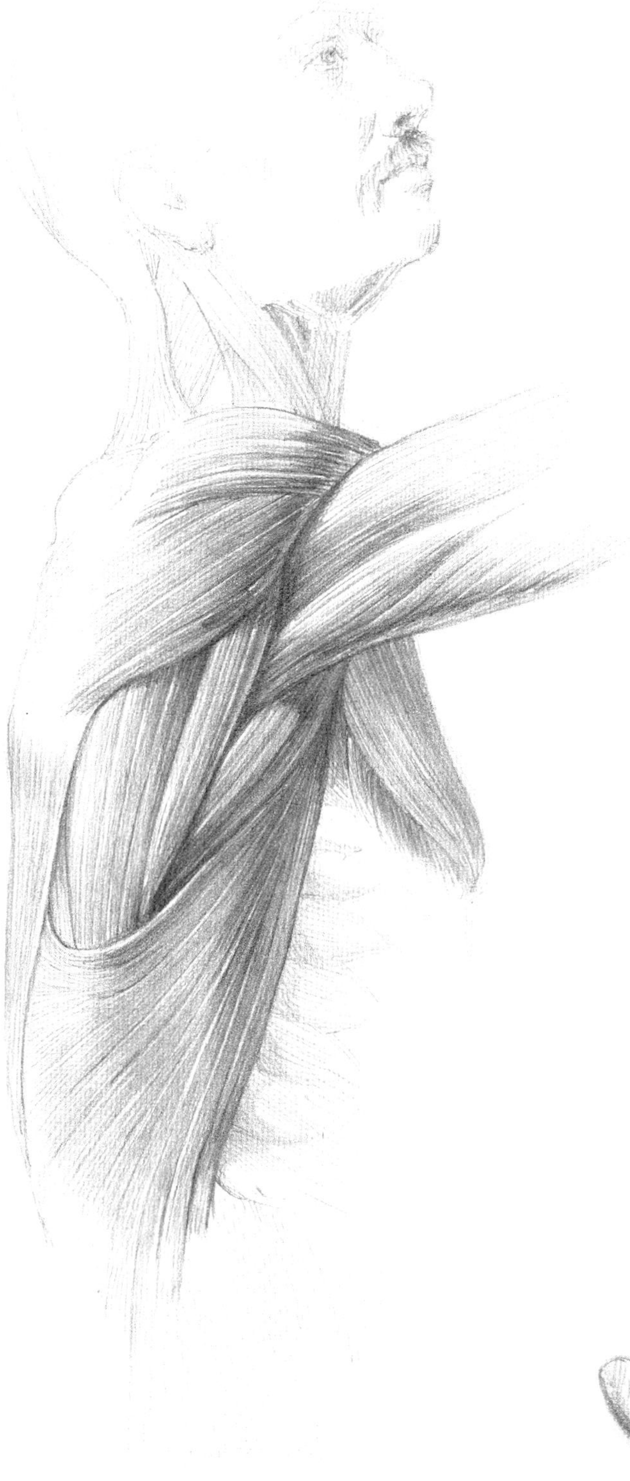

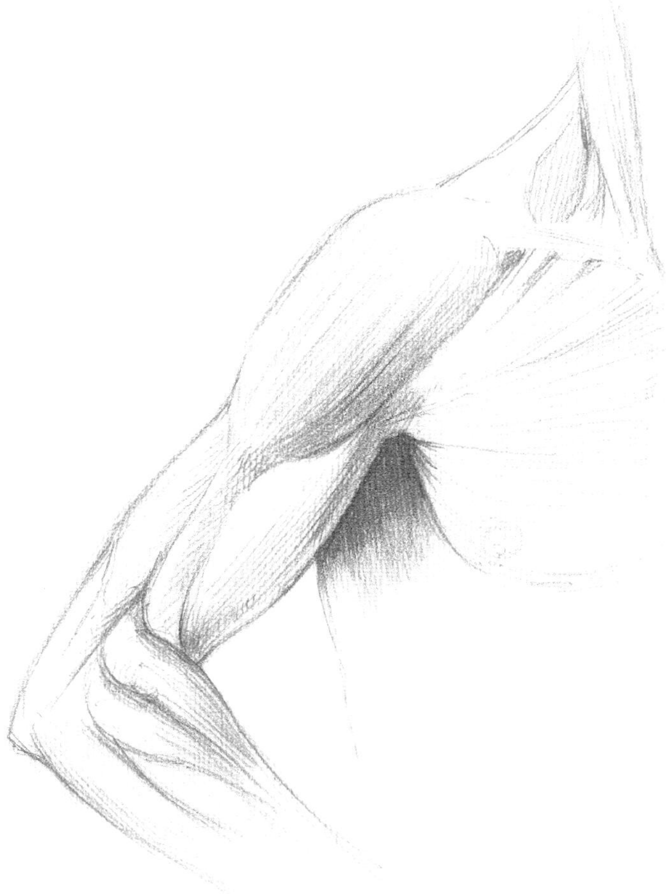

THE LOWER EXTREMITY

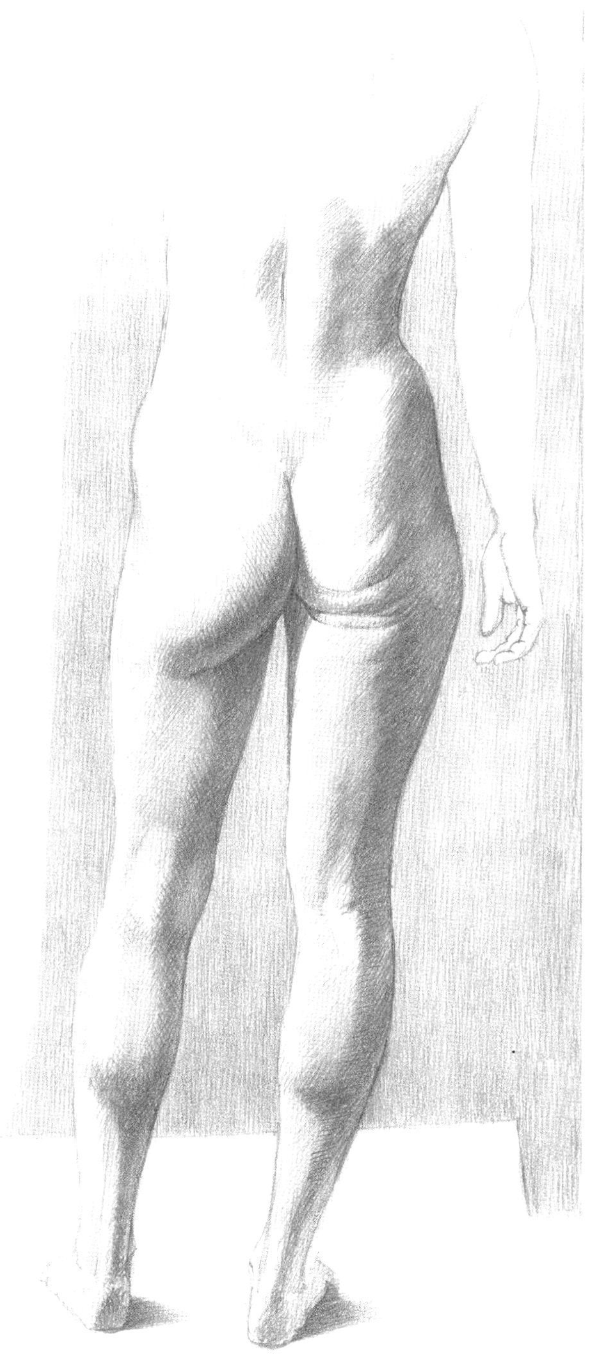

Fig. 103
The Bones
of the Lower Extremity

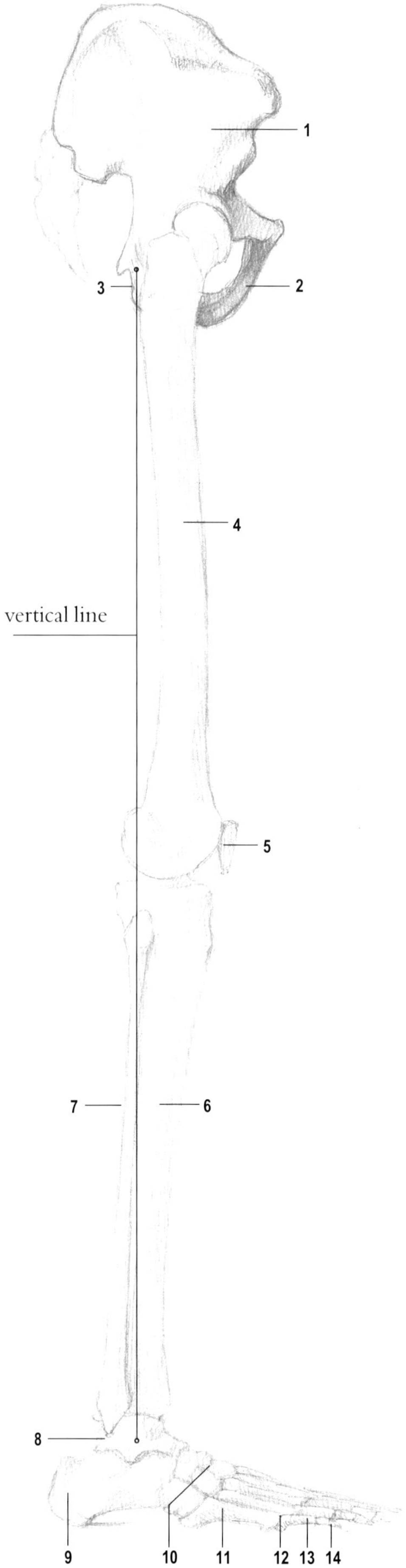

The vertical line begins at the tuberosity of the os sedentarium and divides the ankle bone.

lateral aspect

1 Os ilium
2 Pubic bone
3 Os sedentarium
4 Femur
5 Patella
6 Shinbone
7 Fibula
8 Ankle bone
9 Heelbone
10 Distal row of foot bones
11 Metatarsal bones
12 Proximal pediphalanx
13 Middle pediphalanx
14 End pediphalanx

Fig. 104
The Joints of the Leg

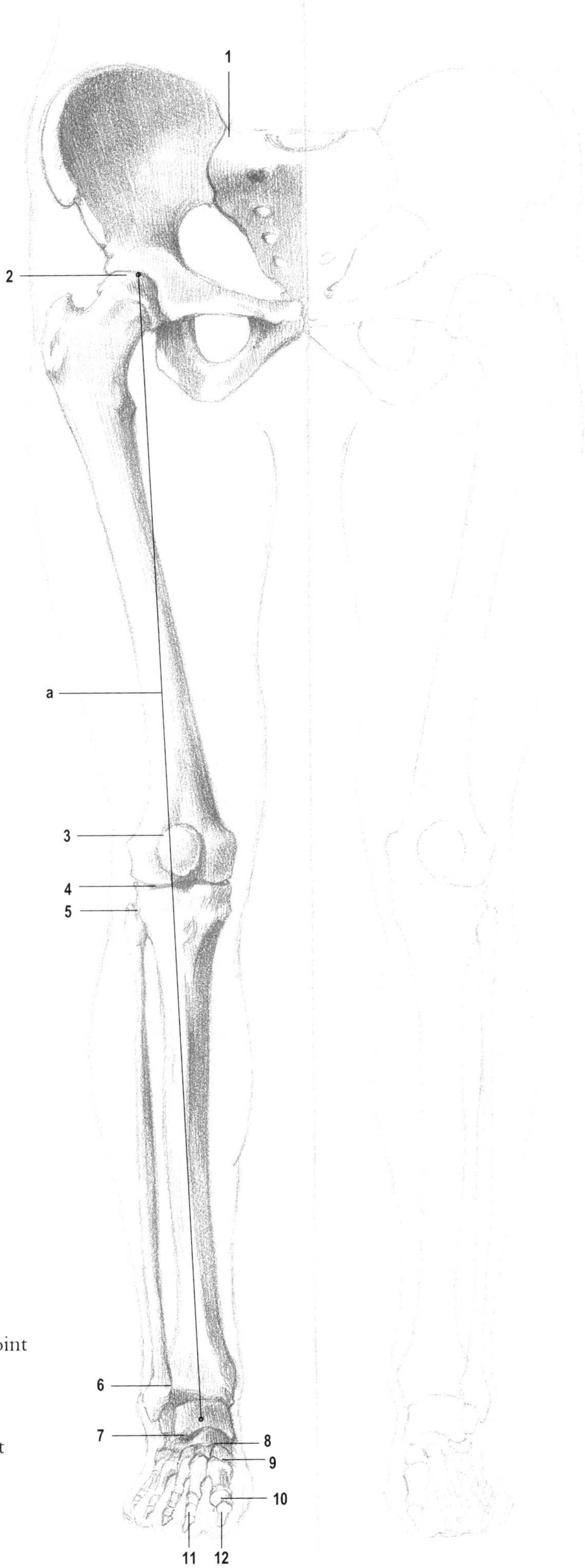

1 Iliosacral joint
2 Hip joint
3 Femoral-patellar joint
4 Femoral-shinbone joint
5 Proximal shinbone-fibular joint
6 Distal shinbone-fibular joint
7 Ankle joint
8 Tarsus pedis joint
9 Tarsus pedis-metatarsus joint
10 Proximal toe joint
11 Middle toe joint
12 Distal toe joint

An imaginary vertical line, beginning at the hip bone, divides the ankle bone.

ventral aspect

Fig. 105
The Male Pelvis

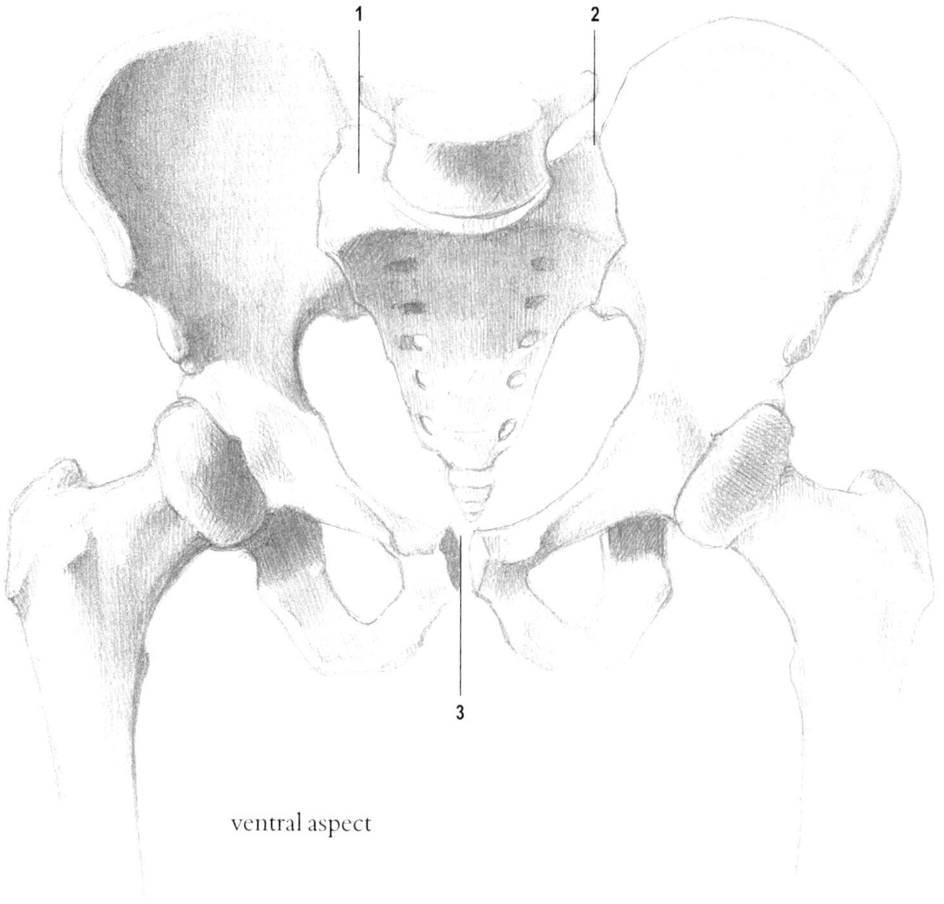

ventral aspect

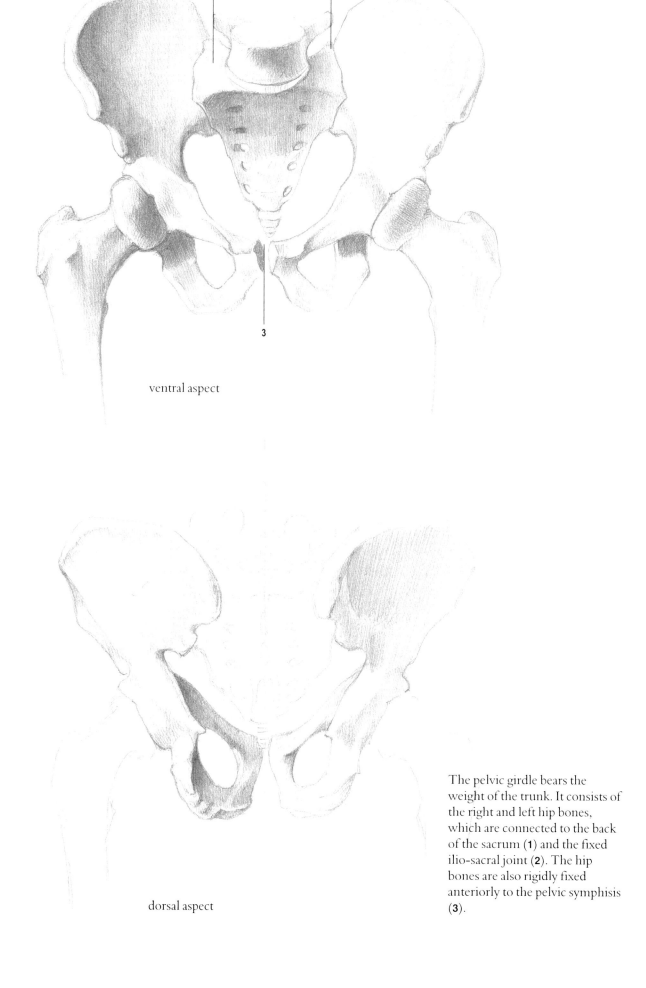

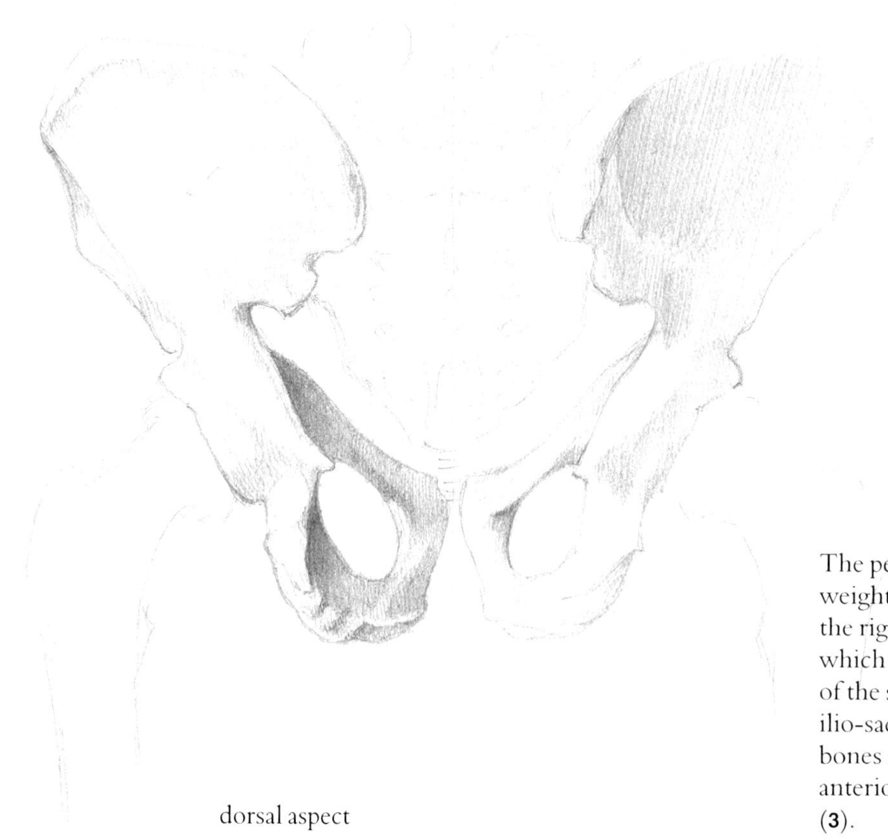

dorsal aspect

The pelvic girdle bears the weight of the trunk. It consists of the right and left hip bones, which are connected to the back of the sacrum (**1**) and the fixed ilio-sacral joint (**2**). The hip bones are also rigidly fixed anteriorly to the pelvic symphisis (**3**).

Fig. 106
The Female Pelvis

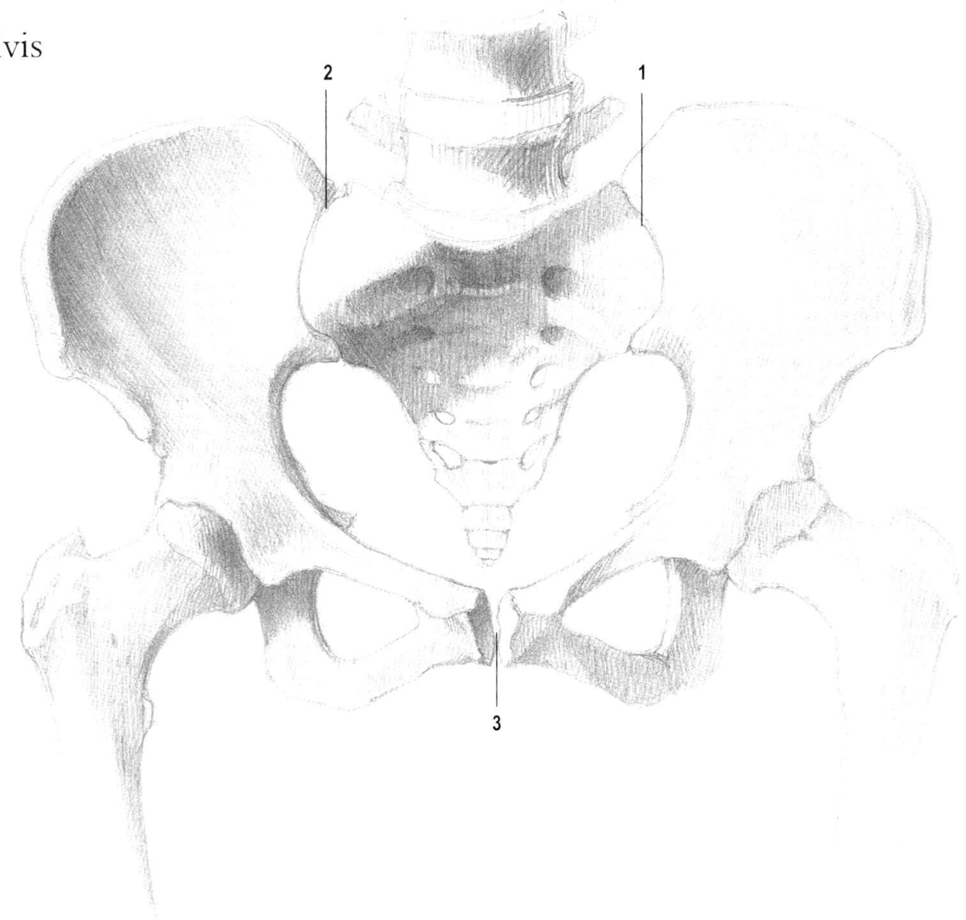

ventral aspect

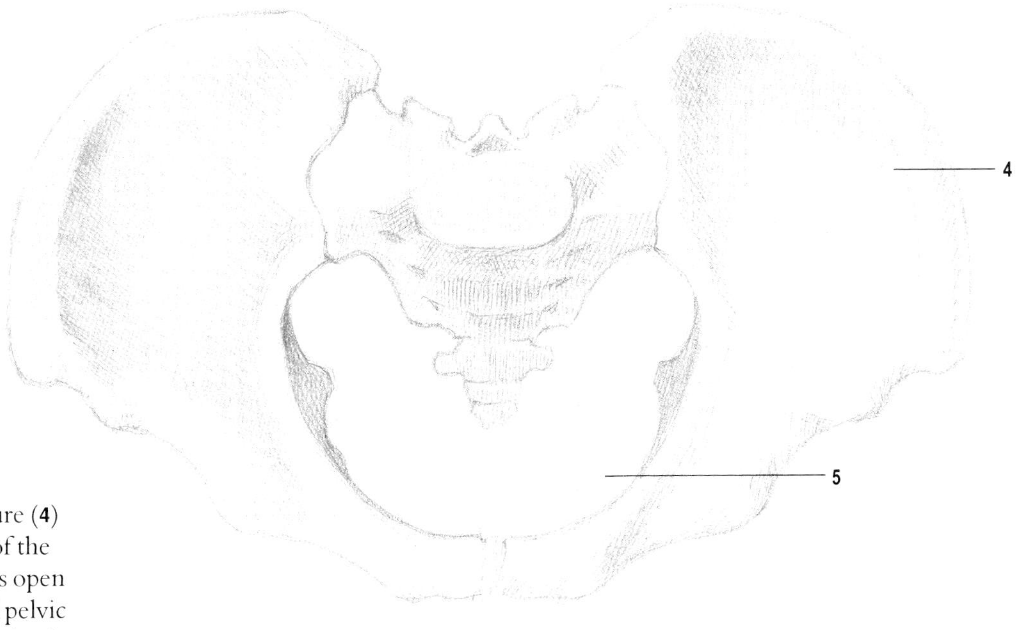

The large pelvic aperture (**4**) forms the lower edge of the abdominal cavity and is open to the front. The small pelvic aperture (**5**) is the bony birth canal. Its entrance is slightly inclined to the front. The female pelvis is wider than that of the male.

superior aspect

Fig. 107
The Hip-Bone

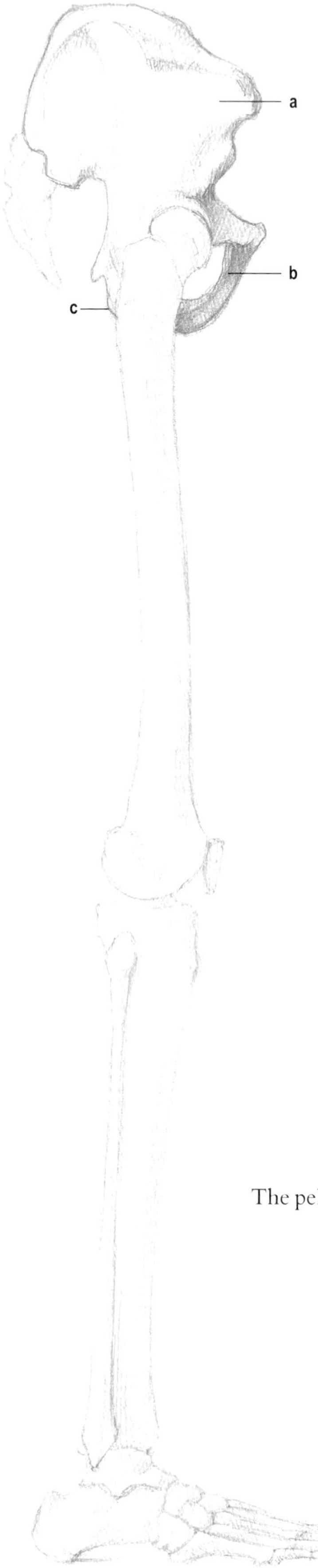

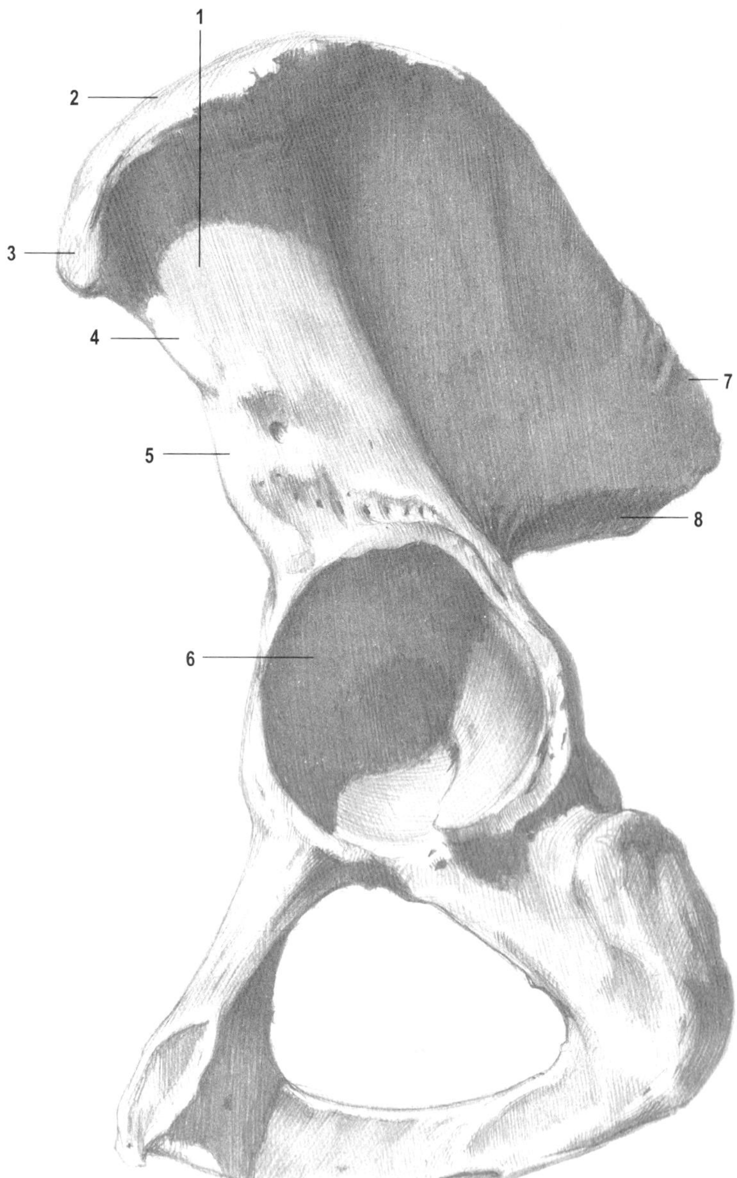

lateral aspect

The hip-bone consists of the os ilium (**a**), pubic bone (**b**) and os sedentarium (**c**).

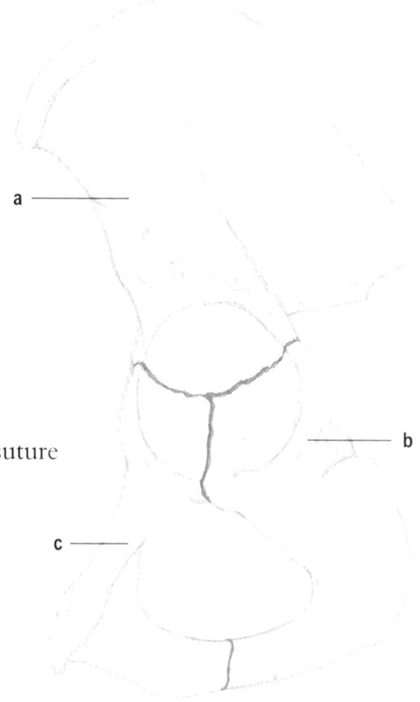

The pelvic suture

The Os Ilium

The os ilium is flat and triangular (**1**), outwardly concave and inwardly convex. The s-shaped os iliumkamm [Ed note; check this] (**2**) forms the upper edge. On the anterior lateral end of the iliac crest the anterior superior iliac spine (**3**) and beneath it, on the body of the ilium is the anterior inferior iliac spine (**4**). The body of theos ilium (**6**) fuses with the pubic bone (**5**) and os sedentarium in the glenoidcavity of the hip bone. On the posterior end of the iliac crest the posterior superior (**7**) and posterior inferior (**8**) iliac spine can be clearly seen.

The Os Sedentarium

The body of the S limits the articular cavity (**1**). The branch (**2**) end in the of the protuberance of the os sedentarium (**3**). The laterally protruding crest is the spinous process of the os sedentarium(**4**).

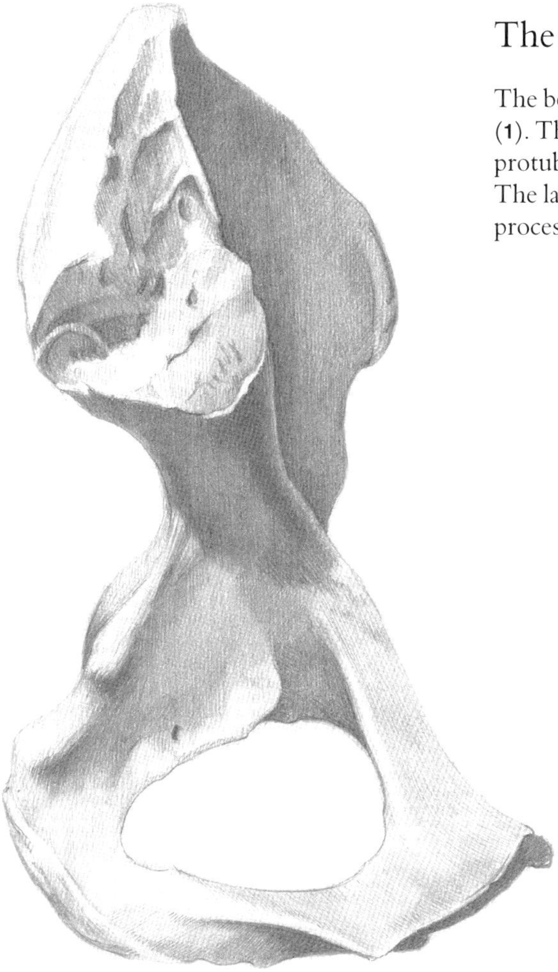

medial aspect

The Pubic Bone

The crest of the pubic bone (**1**) is located on the anterior part of the body of the pubic bone (**2**). The symphisis (**4**), at first cartiligenous and then bony, connects the two branches of the pubic bone. On the superior pubic branch (**5**) the tuberculum pubicum (**6**) and the os pubis (**7**) can be recognised. The pubic bone borders on the foramen obturatum (**8**).

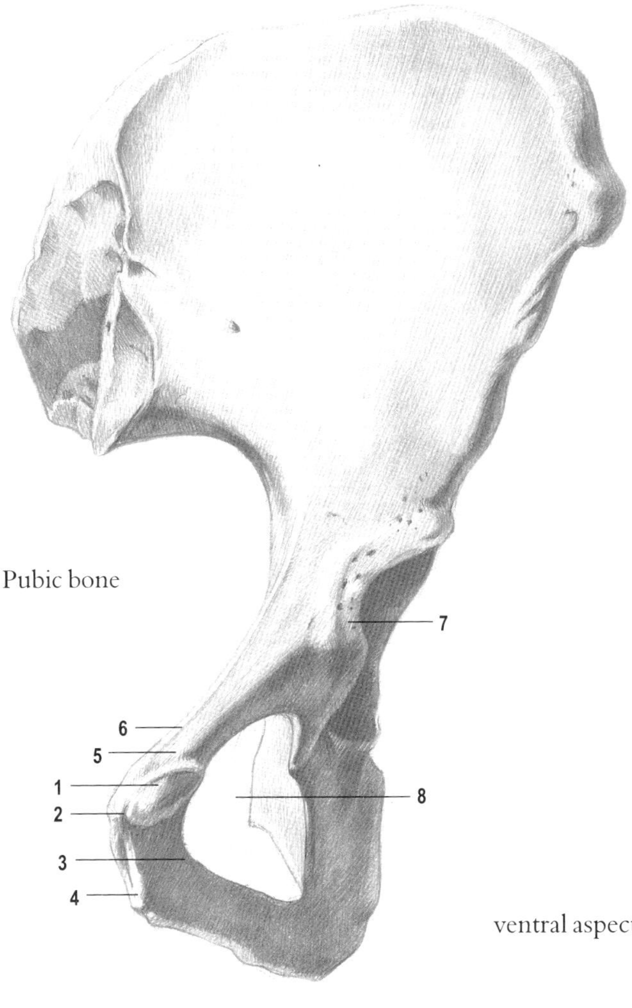

Pubic bone

ventral aspect

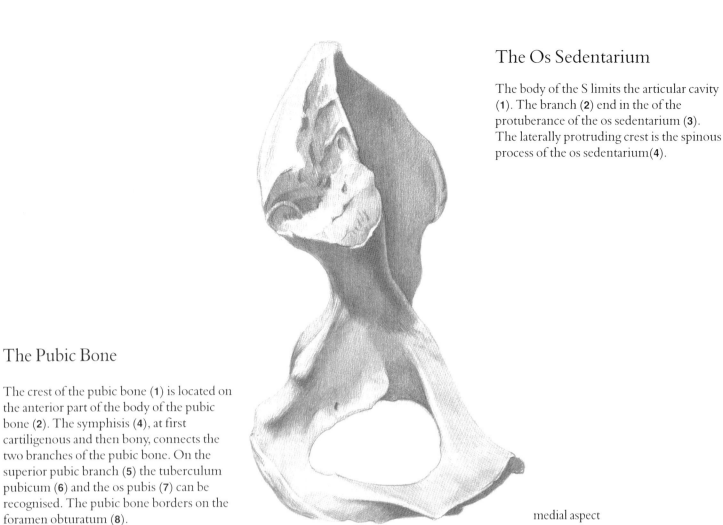

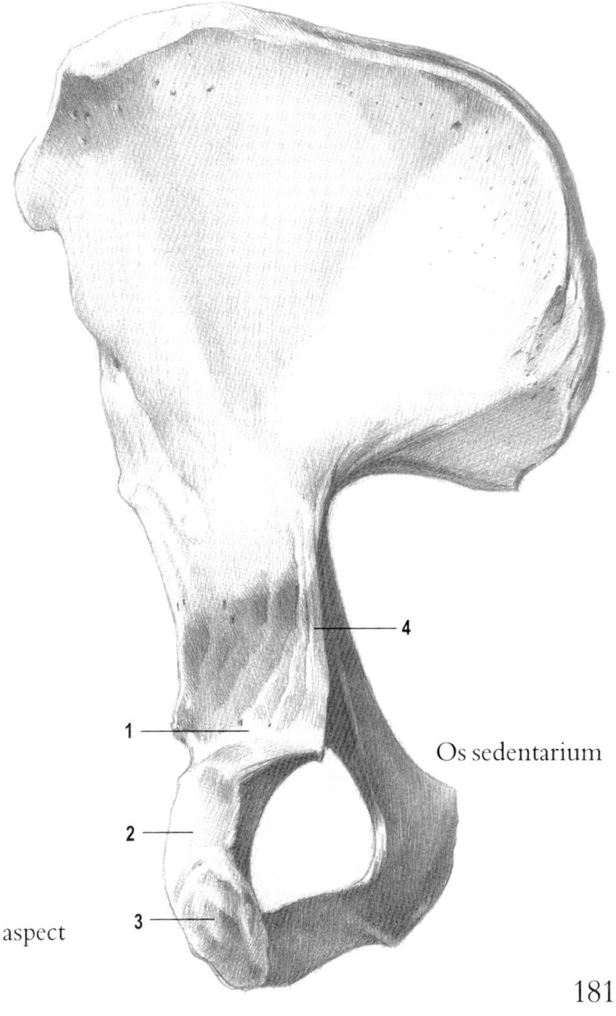

dorsal aspect

Os sedentarium

Fig. 108
The Sacrum

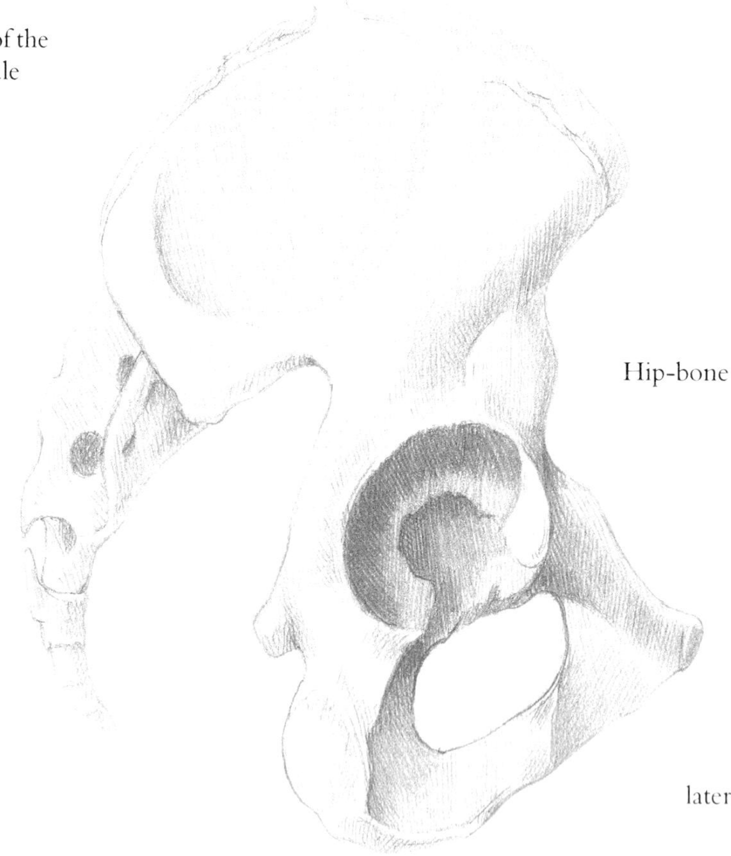

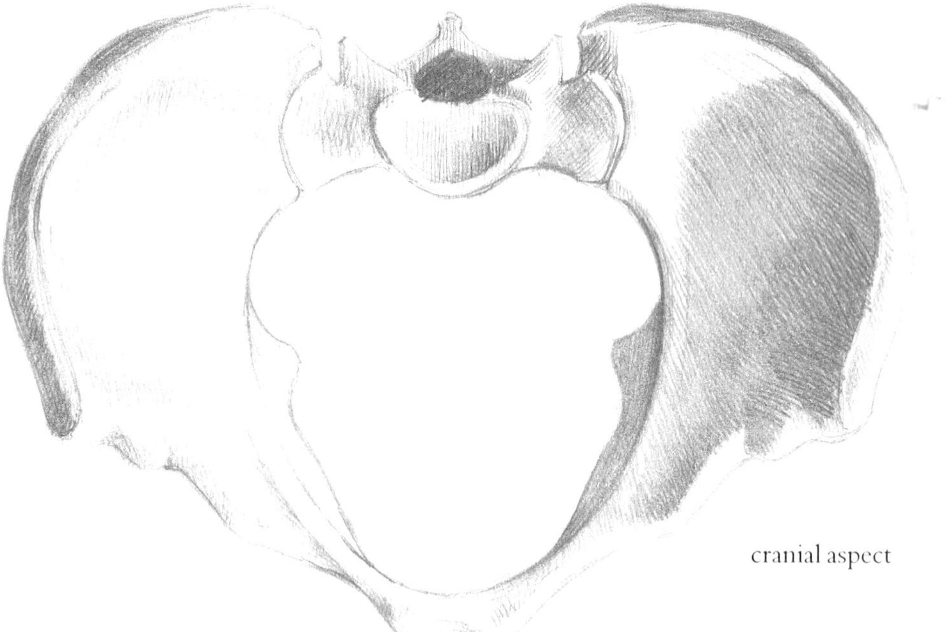

The sacrum is a forwardly concave spade-shaped bone. It forms the posterior part of the pelvic girdle. On the superior edge of the girdle the promontorium of the first sacral vertebra (**1**) is located. The raw articulatory surface (**3**) at its lateral edge forms (**2**) the fixed joint with the ala of the os ilium. On the inferior edge (**4**) the coccyx (**5**) is fixed. On the rear surface the fused spinous process forms the middle crest (**6**) of the sacrum; the capitulum process forms the lateral crest (**7**) and the transverse process forms the sacral crest (**8**). Branches from the sacral nerve, originating in the spinal marrow protrude through the anterior (**9**) and posterior (**10**) aperture. In old age the sacrum fuses with the coccyx.

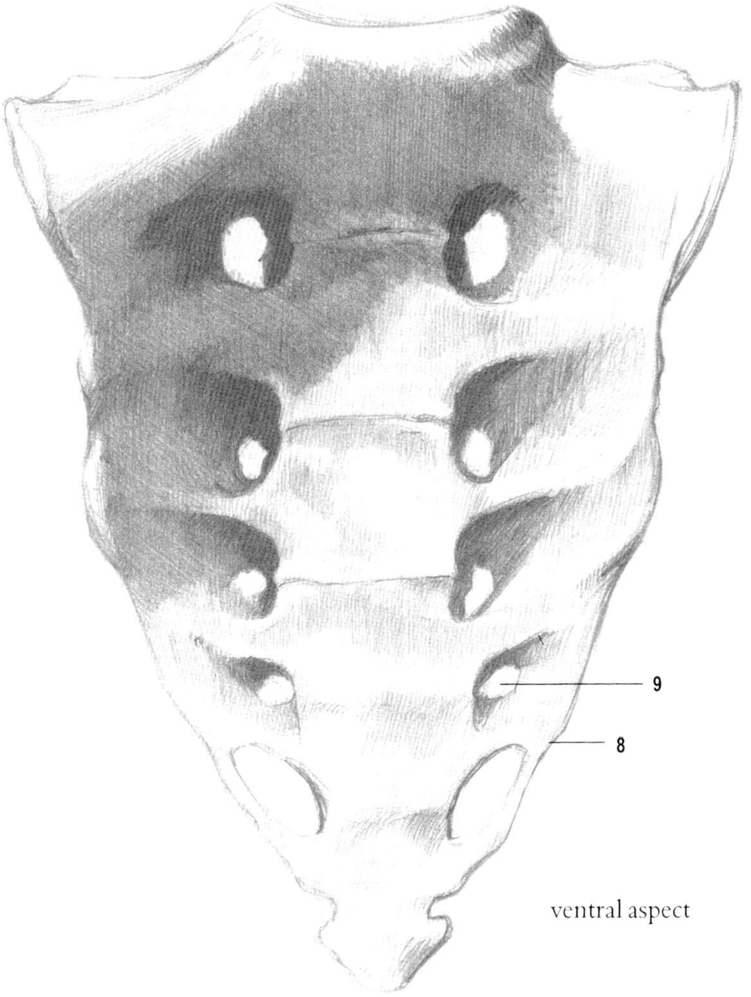

ventral aspect

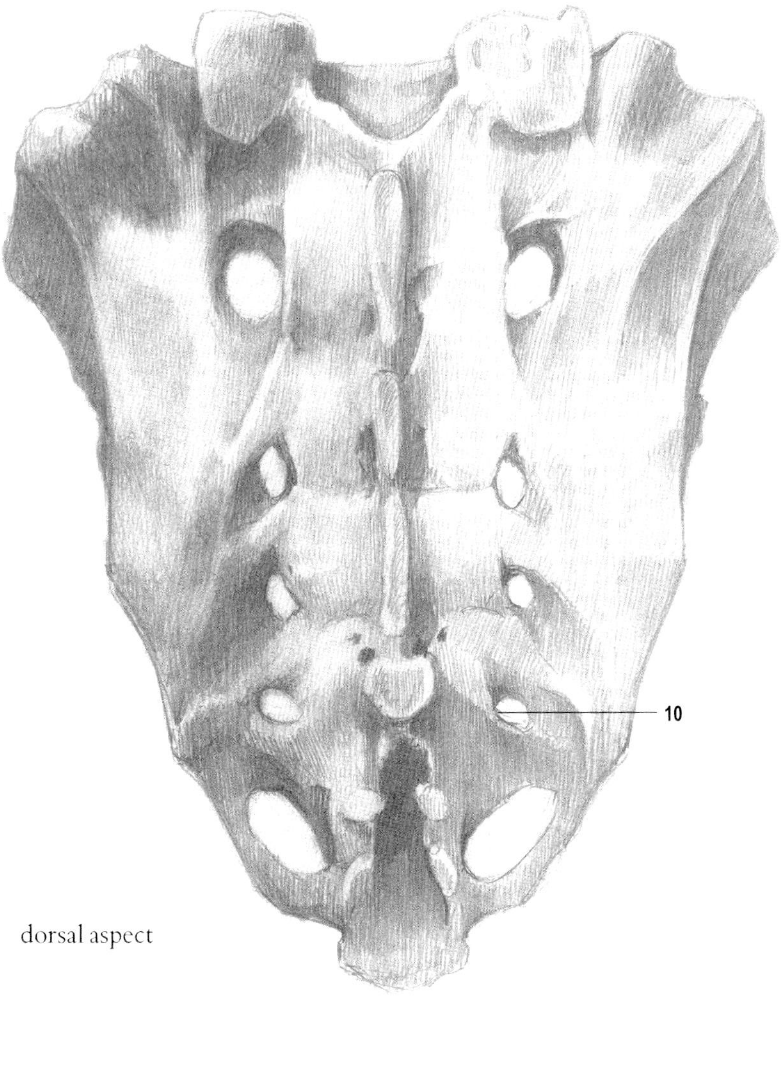

dorsal aspect

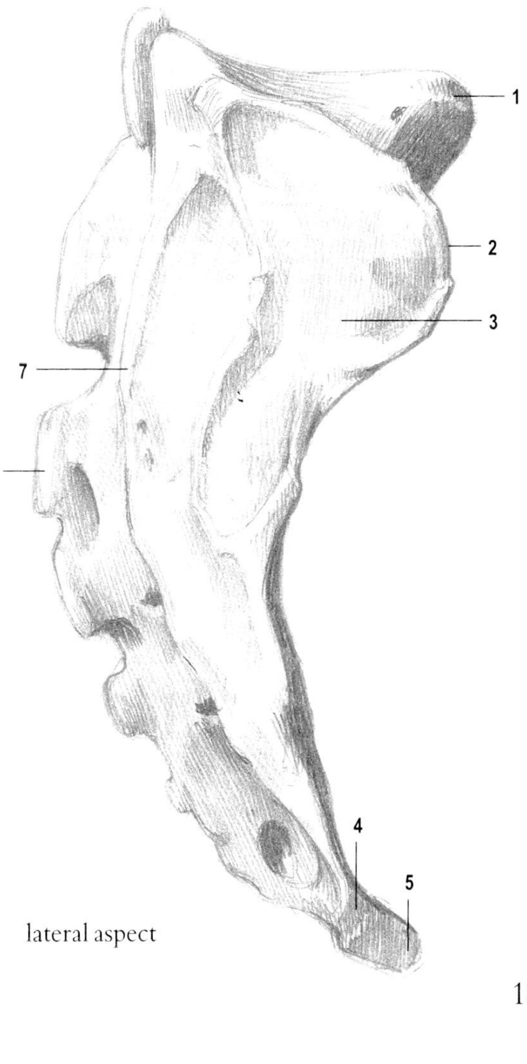

lateral aspect

Fig. 109
The Femur

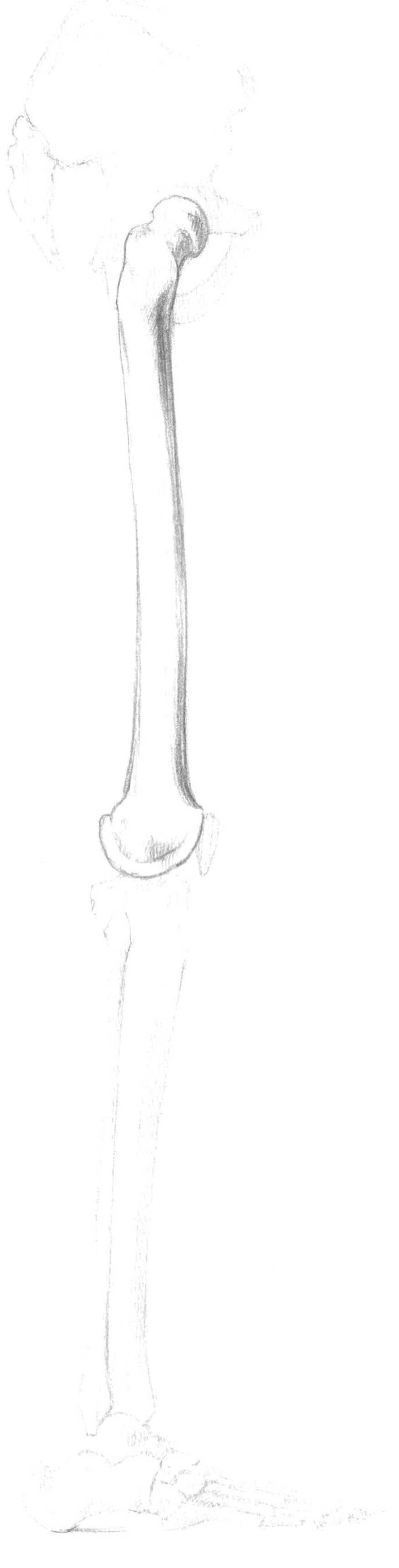

The femur is the longest bone in the body, its axis slightly forward-leaning. The head (**1**) is half-spherical in form; the neck (**2**) forms an angle of 45 degrees with the shaft. Its greater (**3**) and lesser (**4**) trochanter are the insertion surfaces for the gluteal muscles. The shaft (**5**) is slightly backwards-leaning. The medial and lateral condyles on the distal end (**6**) form a joint with the shin bone; the anterior condyle (**7**) forms a joint with the patella. On the posterior surface (**8**), below the greater condyle, the lesser trochanter and beneath it the linea aspera (**9**) are located. The facies poplitea (**10**) is clearly visible above the trochanter. The medial and lateral epicondyles (**11**) can be seen next to the joint head.

lateral aspect

medial aspect

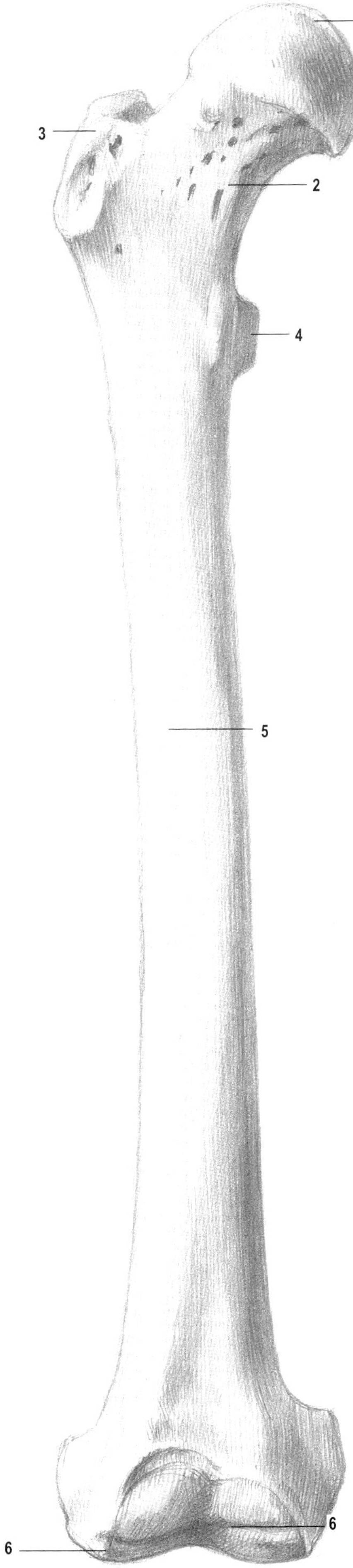

ventral aspect

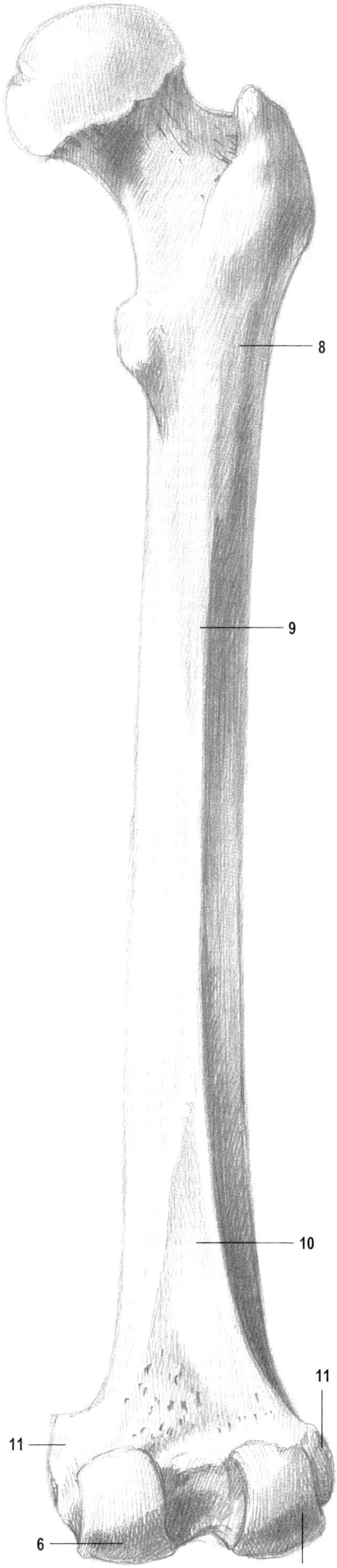

dorsal aspect

Fig. 110
The Shin Bone

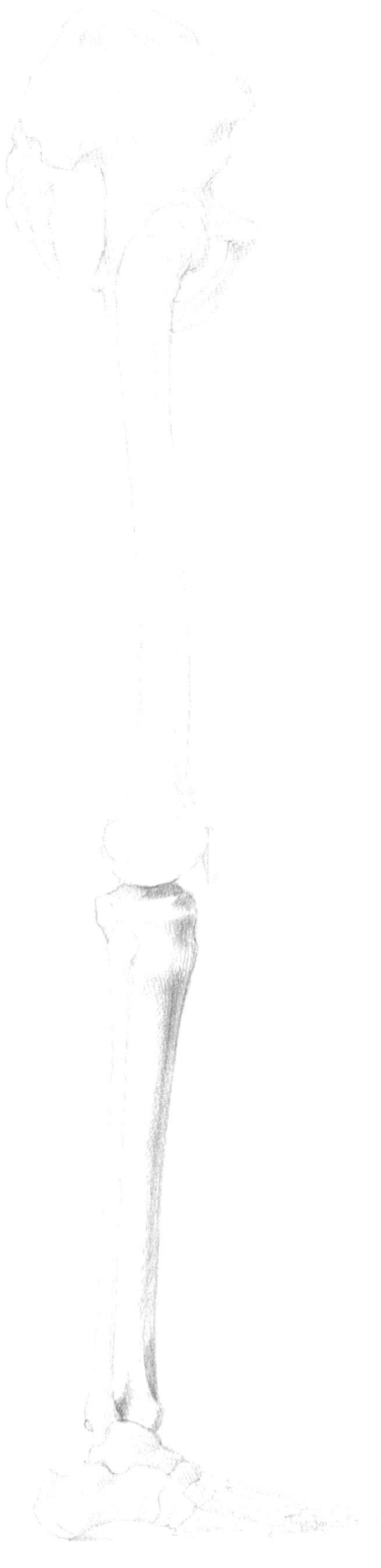

The shin bone has two flat tuberosities at its proximal end (**1**), which form a joint with the condyles of the femur. The joint surfaces of the lateral condyle (**2**) form a joint with the head of the fibula. The insertion (**3**) of the inner ligaments of the knee joint is located between the condyles. From the head on the anterior aspect (**4**) a bony crest runs (**5**) along the shaft of the bone (**6**). Directly beneath the skin (**7**) lies the bone, medially to this crest. The medial protuberance at the distal end of the bone is the condyle of the shin (**8**). The distal end of the fibula forms a joint with the distal end of the shin bone (**9**). The joint surface of the shin bone (**10**) forms a joint with the ankle bone.

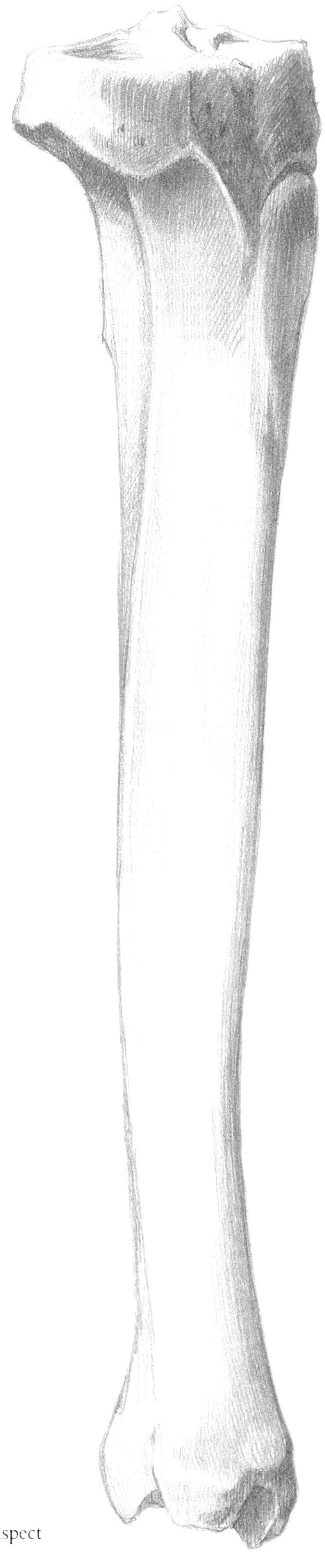

lateral aspect

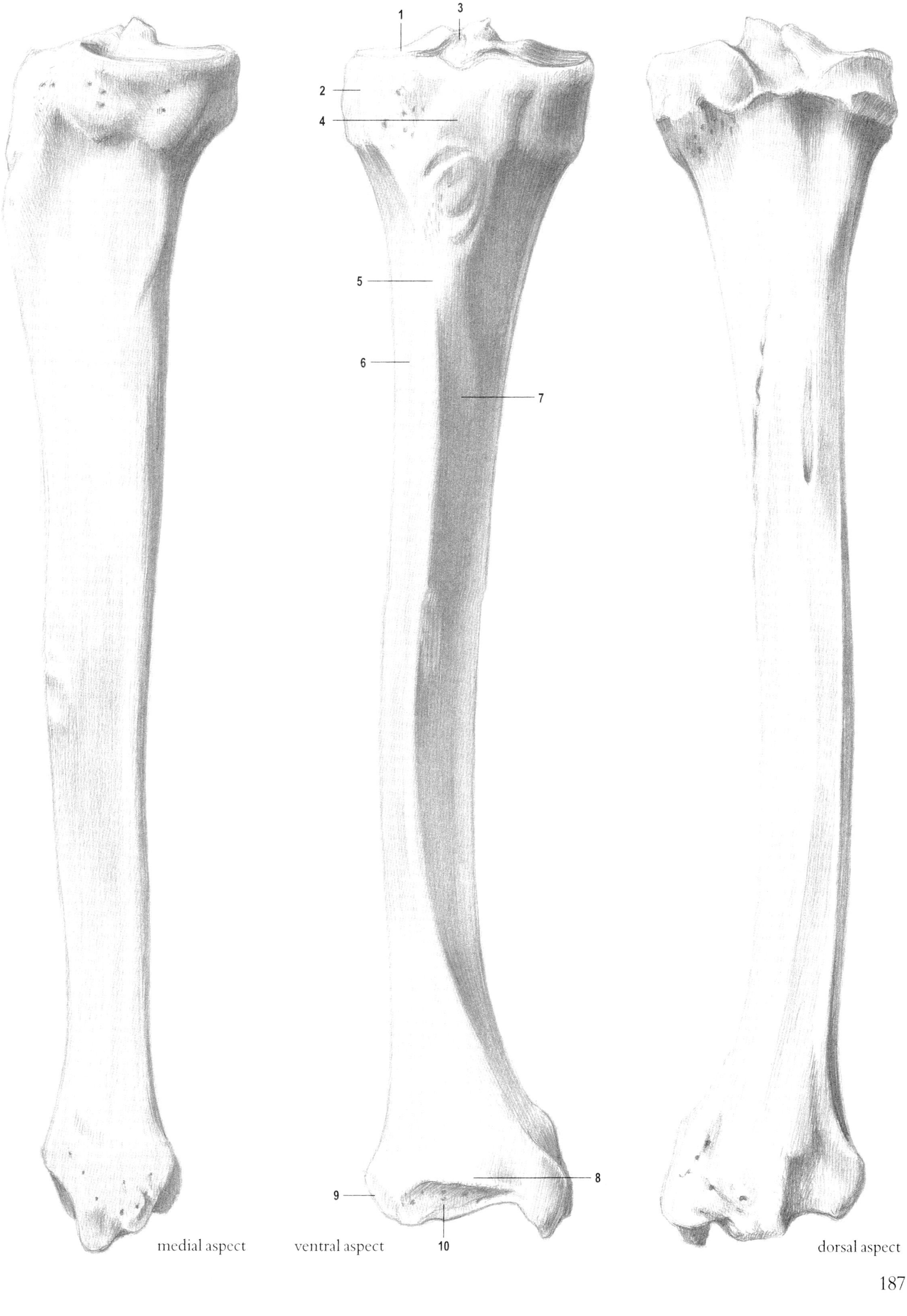

medial aspect ventral aspect dorsal aspect

Fig. 111
The Fibula

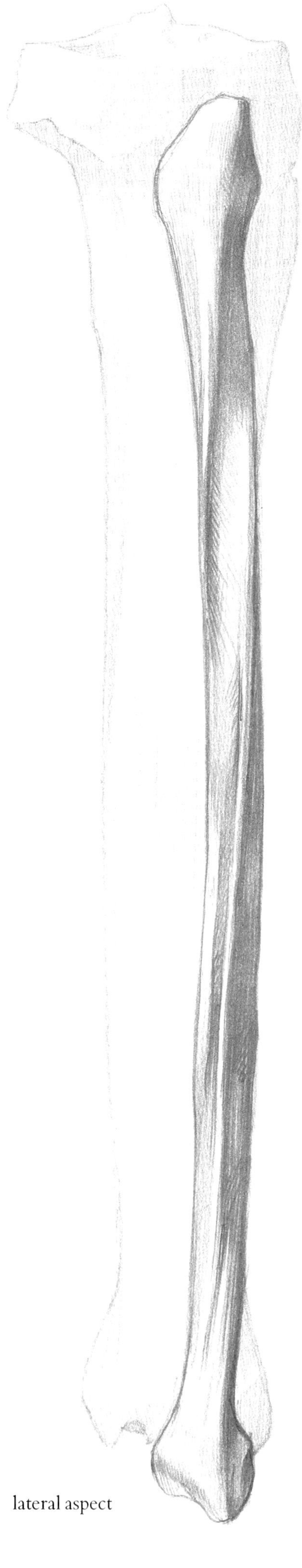

lateral aspect

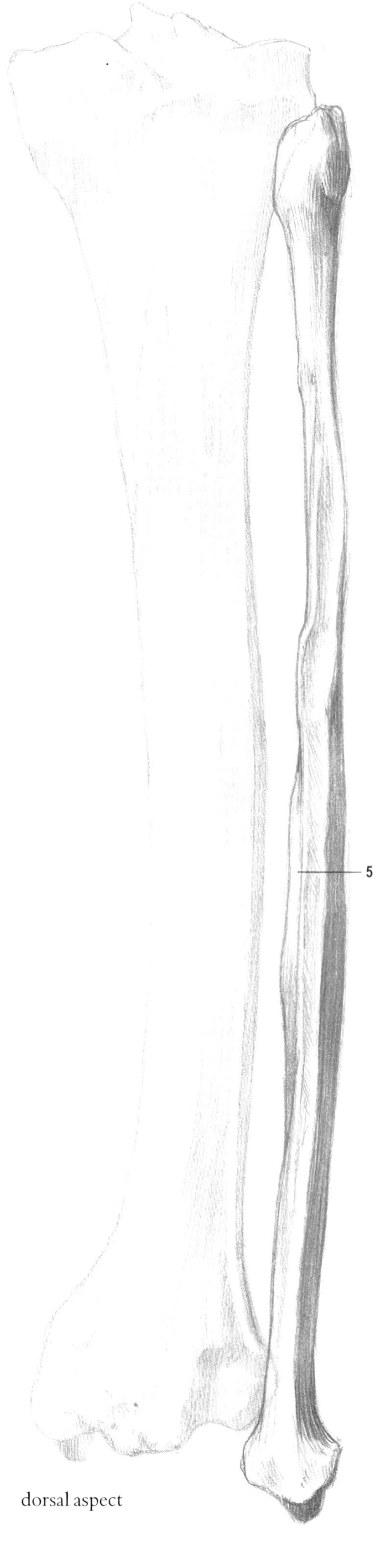

dorsal aspect

Fig. 112
The Ligaments of the Shin-Fibula Joint

medial aspect

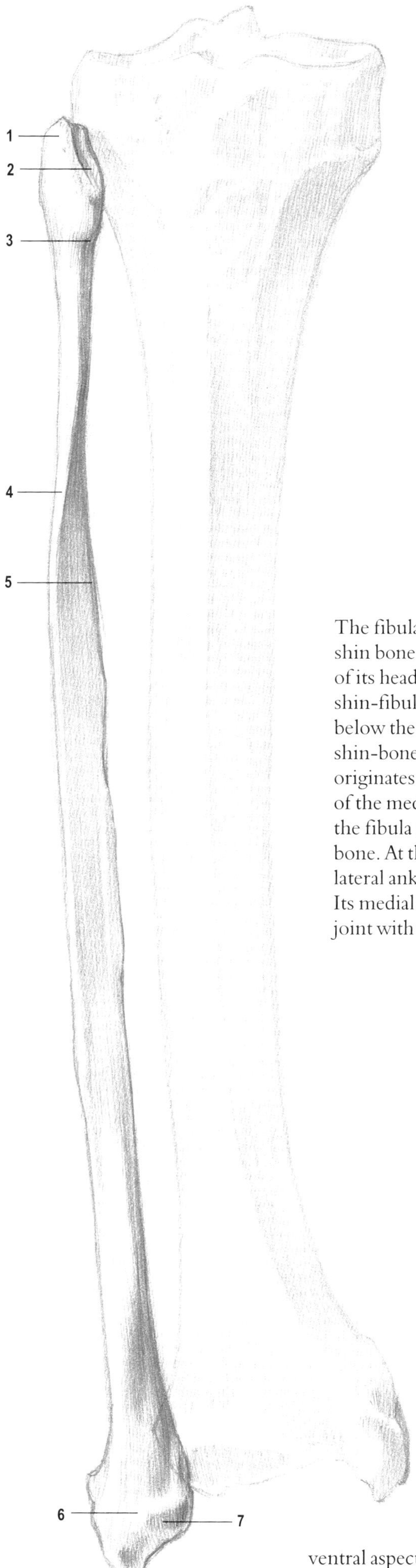

ventral aspect

The fibula is thinner than the shin bone. The joint surface (**2**) of its head forms the superior shin-fibula joint. The shaft (**4**) below the neck (**3**) is thin: The shin-bone-fibula ligament originates at the bony crest (**5**) of the medial surface, and fixes the fibula to the shaft of the shin bone. At the inferior end the lateral ankle (**6**) can be found. Its medial surface (**7**) forms a joint with the ankle bone.

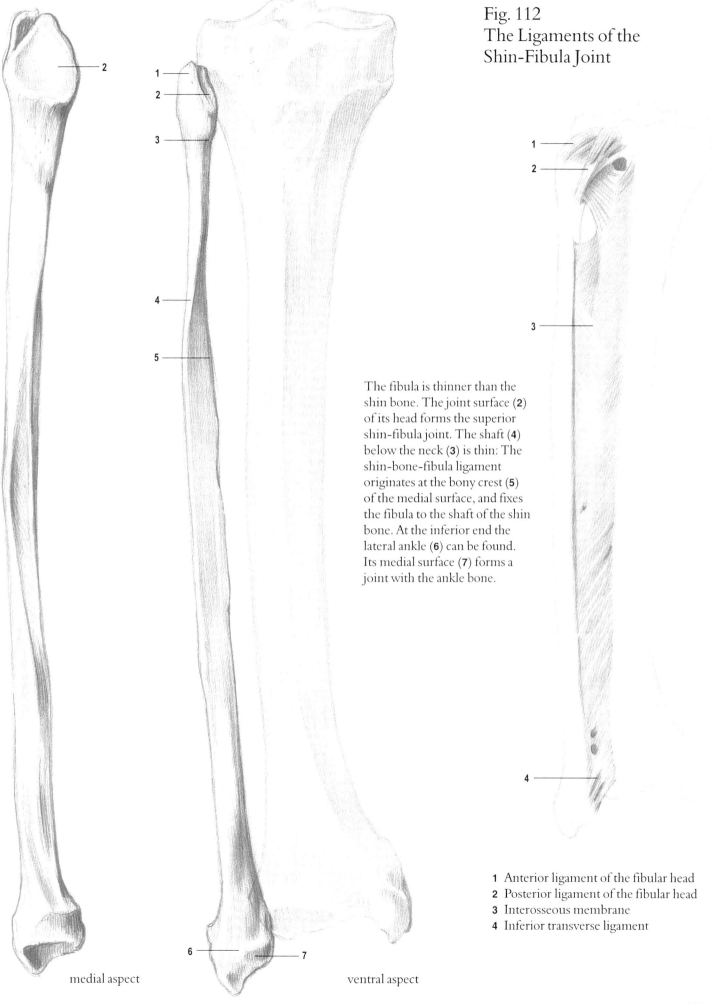

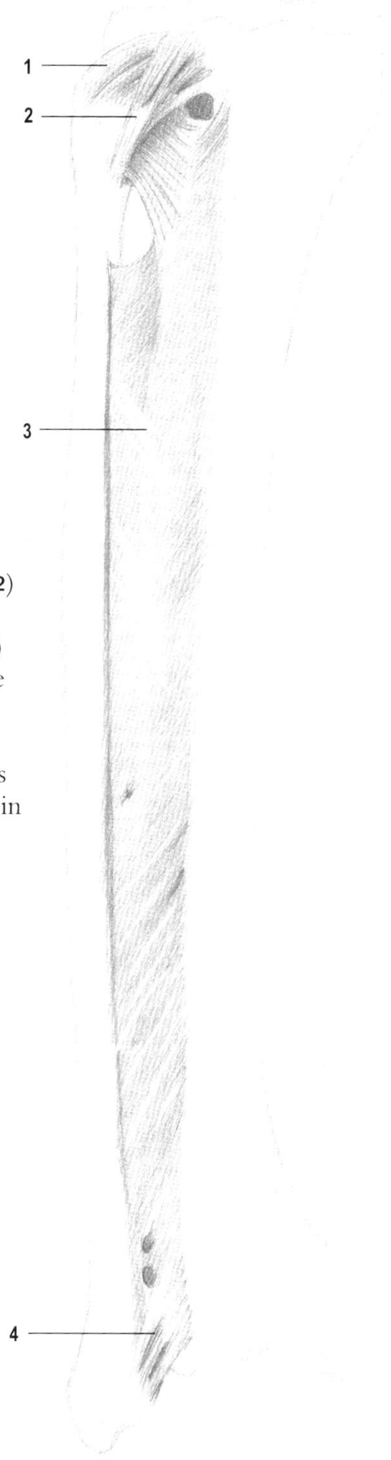

1 Anterior ligament of the fibular head
2 Posterior ligament of the fibular head
3 Interosseous membrane
4 Inferior transverse ligament

Fig. 113
The Patella

lateral aspect

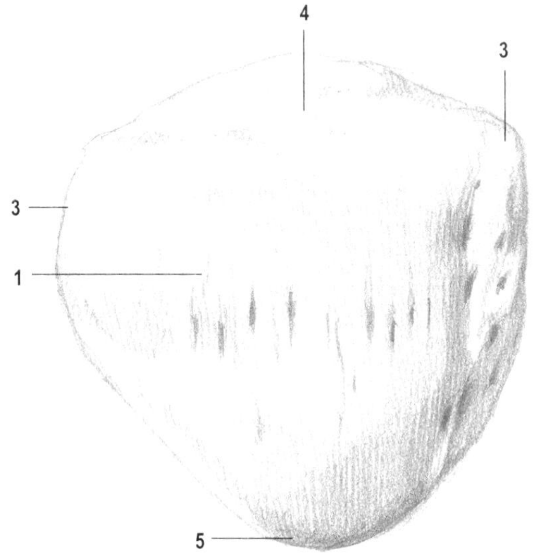

ventral aspect

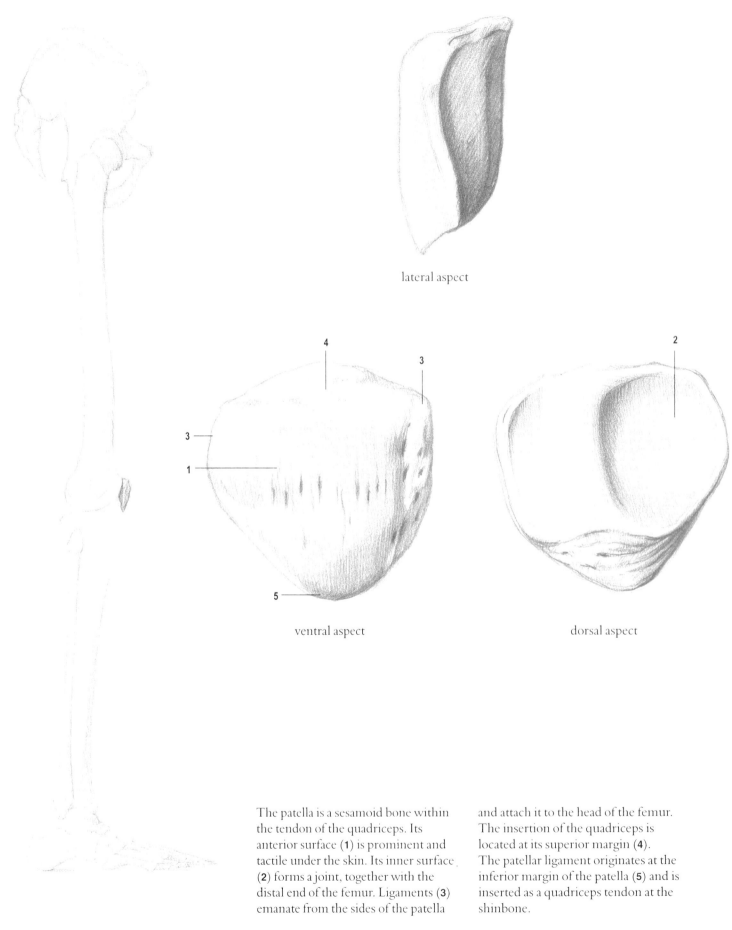

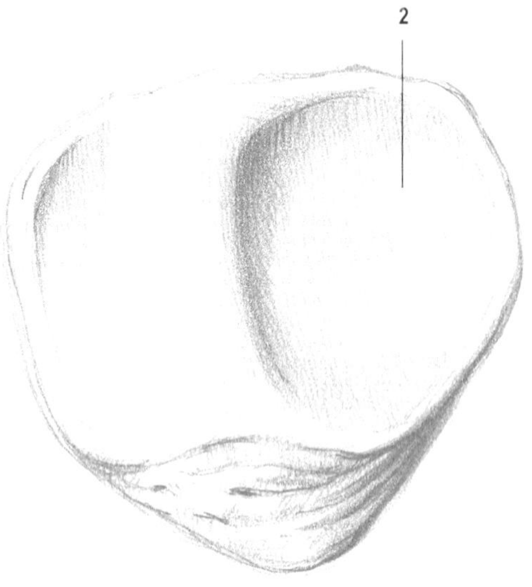

dorsal aspect

The patella is a sesamoid bone within the tendon of the quadriceps. Its anterior surface (**1**) is prominent and tactile under the skin. Its inner surface (**2**) forms a joint, together with the distal end of the femur. Ligaments (**3**) emanate from the sides of the patella and attach it to the head of the femur. The insertion of the quadriceps is located at its superior margin (**4**). The patellar ligament originates at the inferior margin of the patella (**5**) and is inserted as a quadriceps tendon at the shinbone.

Fig. 114
The Bones of the Foot

The foot consists of the bones of the tarsus pedis (**1–7**), the metatarsal bones (**8**) and the phalanges of the toes (**9–11**). Since they bear the weight of the body they are large bones. The big toe is twice as thick as the other toes and consists of only two phalanges.

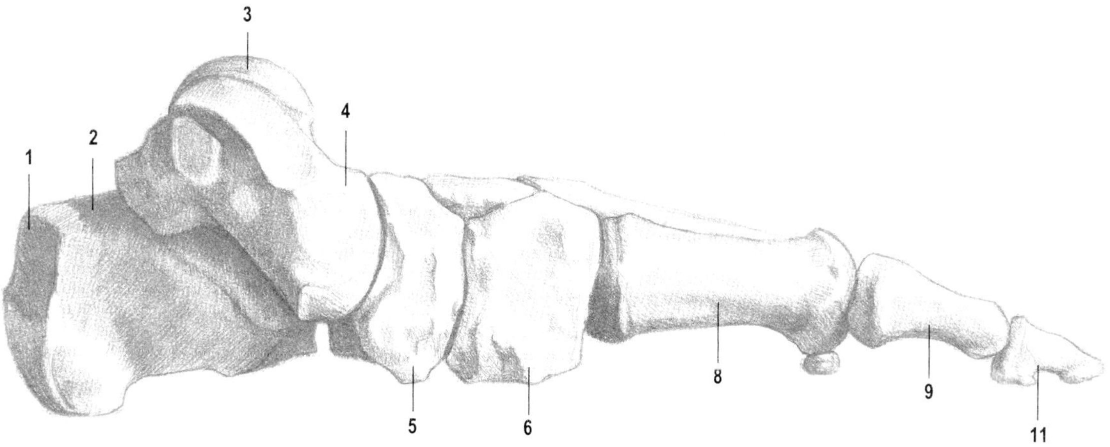

medial aspect

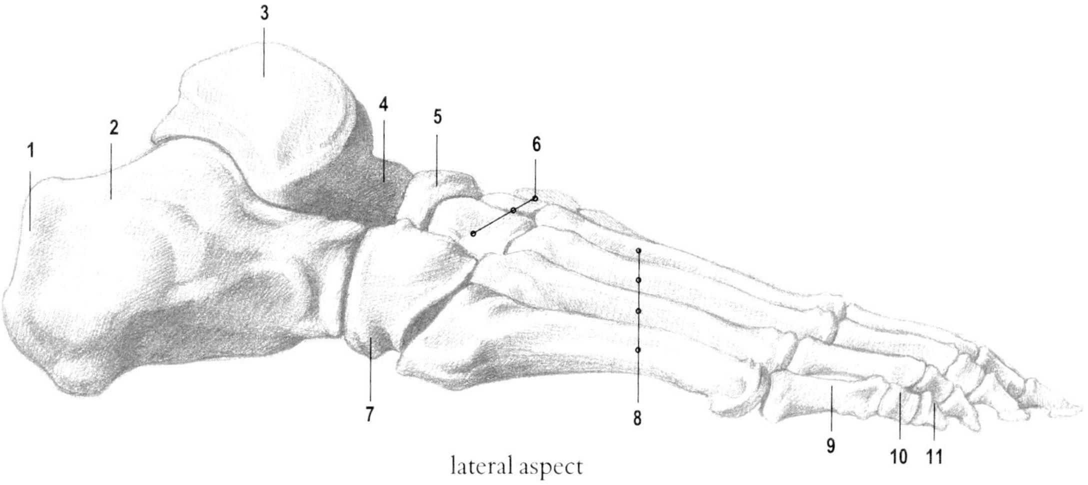

lateral aspect

The bones of the toe phalanges of the second-fifth toes.

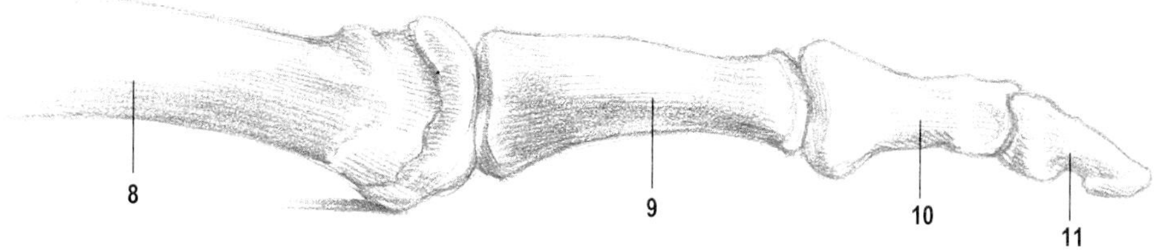

1 Protuberance of the ankle bone
2 Heel bone
3 Articulatory surface of the ankle bone
4 Body of the ankle bone
5 Navicular bone
6 Sphenoidal bones
7 Cuboid bone
8 Metatarsal bones
9 Proximal phalanges
10 Middle phalanax
11 Distal phalanx

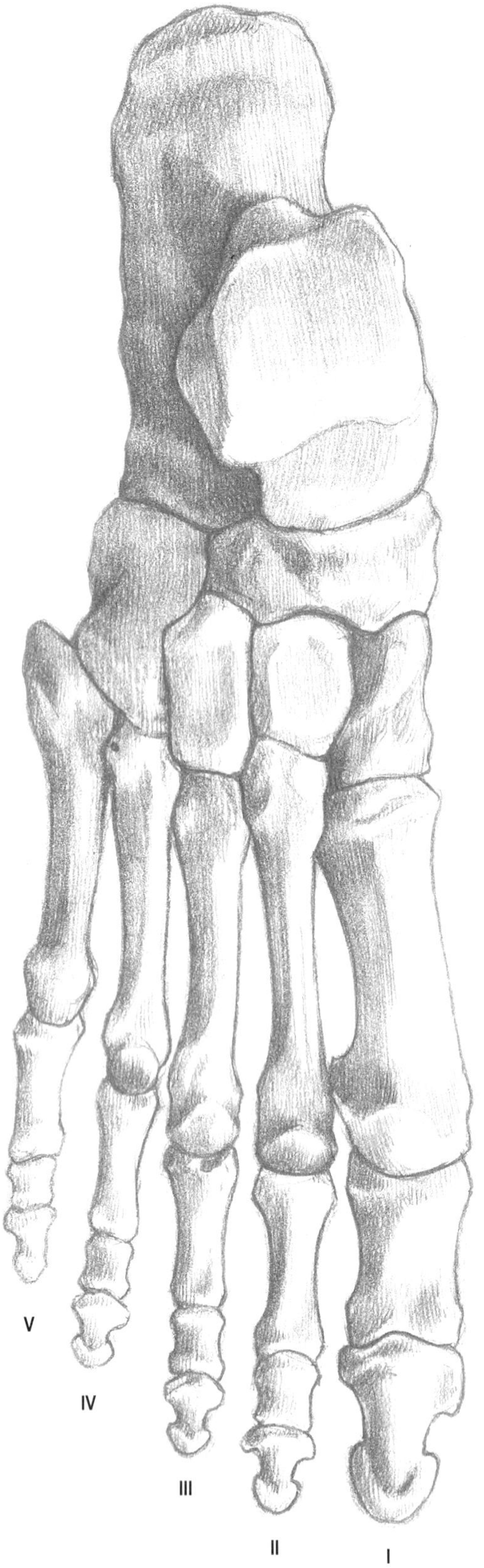

dorsal aspect

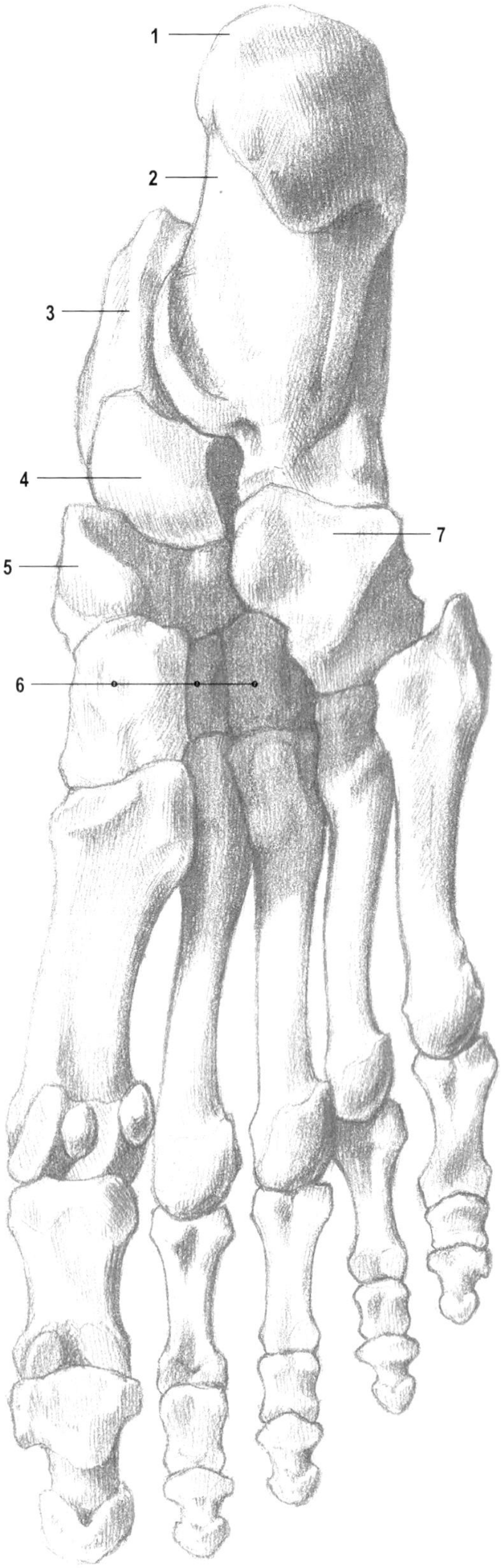

plantar aspect

1 Protuberance of the ankle bone
2 Heel bone
3 Ankle bone
4 Body of the ankle
5 Avicular bone
6 Sphenoidal bones
7 Cuboid bone
I-V: Toes

Fig. 115
Movements of the Foot Joint

Flexion and extension are at their strongest in the superior and inferior ankle joint (**1**) of the tarsus pedis. The joints of the tarsus pedis and the metatarsal bones perform only limited movement. The joints of the pediphalanges can be flexed or extended. The foot touches the ground posteriorly with the two processes of the heel protuberances (**2**) and anteriorly with the root of the big toe (**3**) and small toe (**4**). When the weight of the body is placed on the foot, the arch in the metatasal region (**5**) becomes less pronounced. The foot is like a feather: It becomes flat first and then arches again. It is supported by the anterior shin muscles and the long fibular muscles.

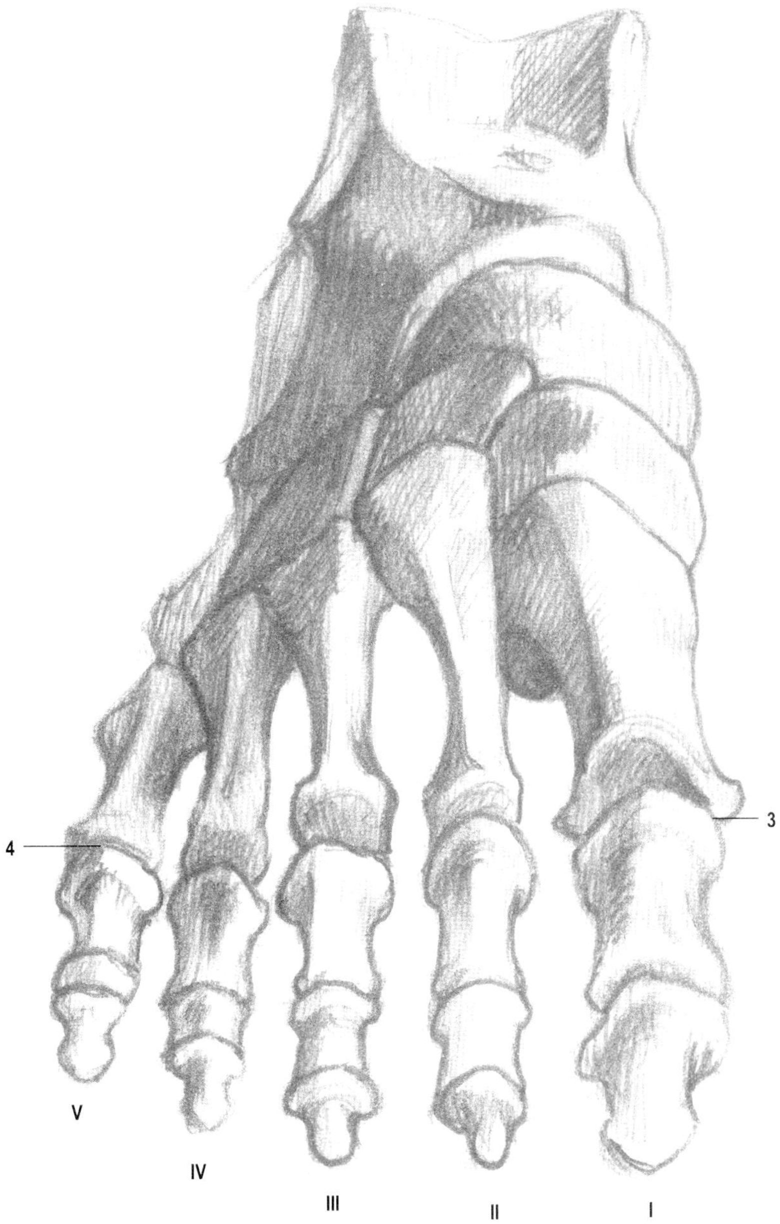

ventral aspect

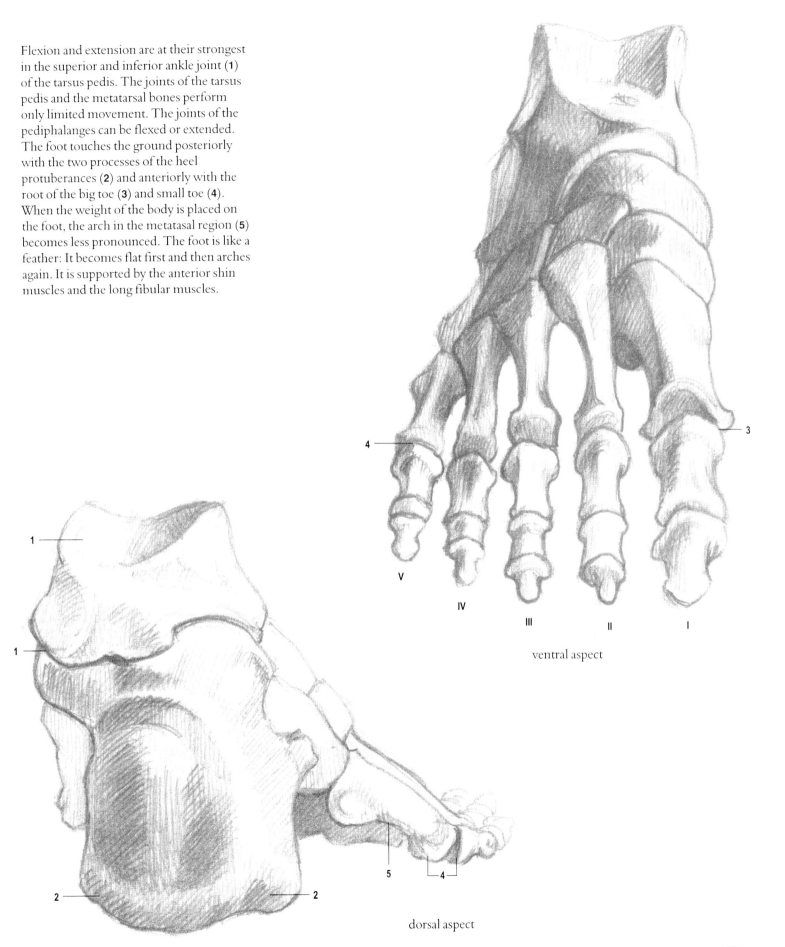

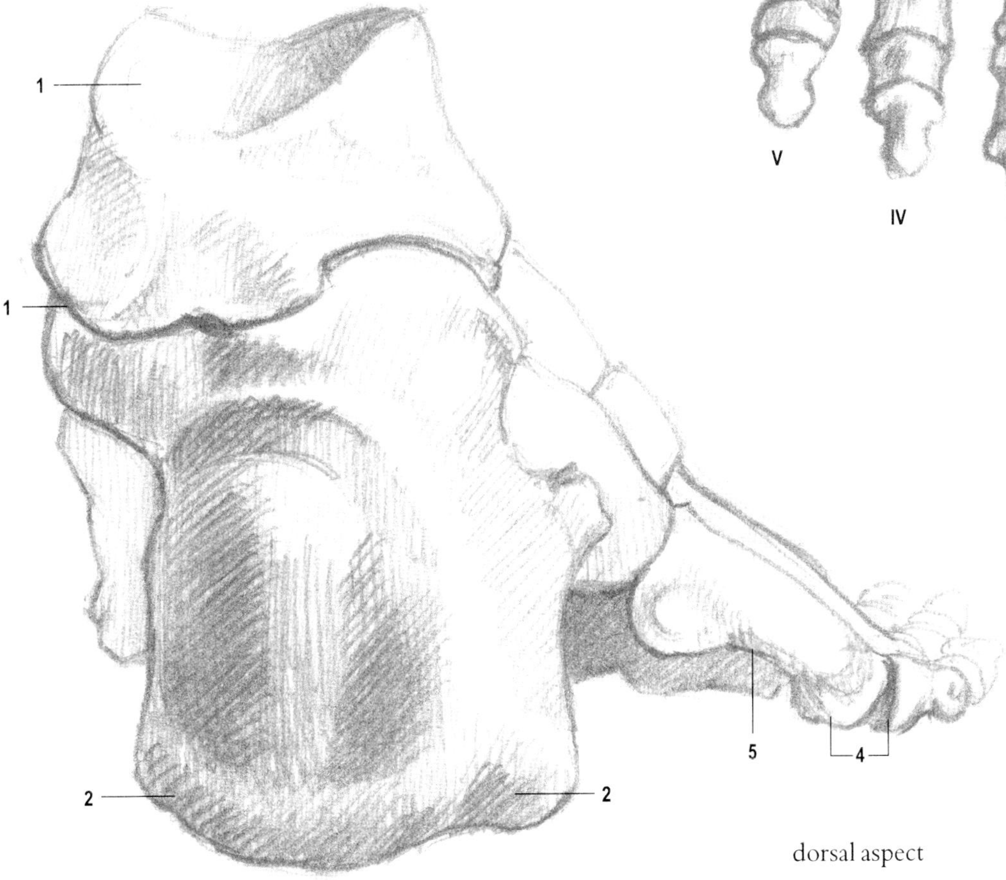

dorsal aspect

Fig. 116
Studies of the Foot

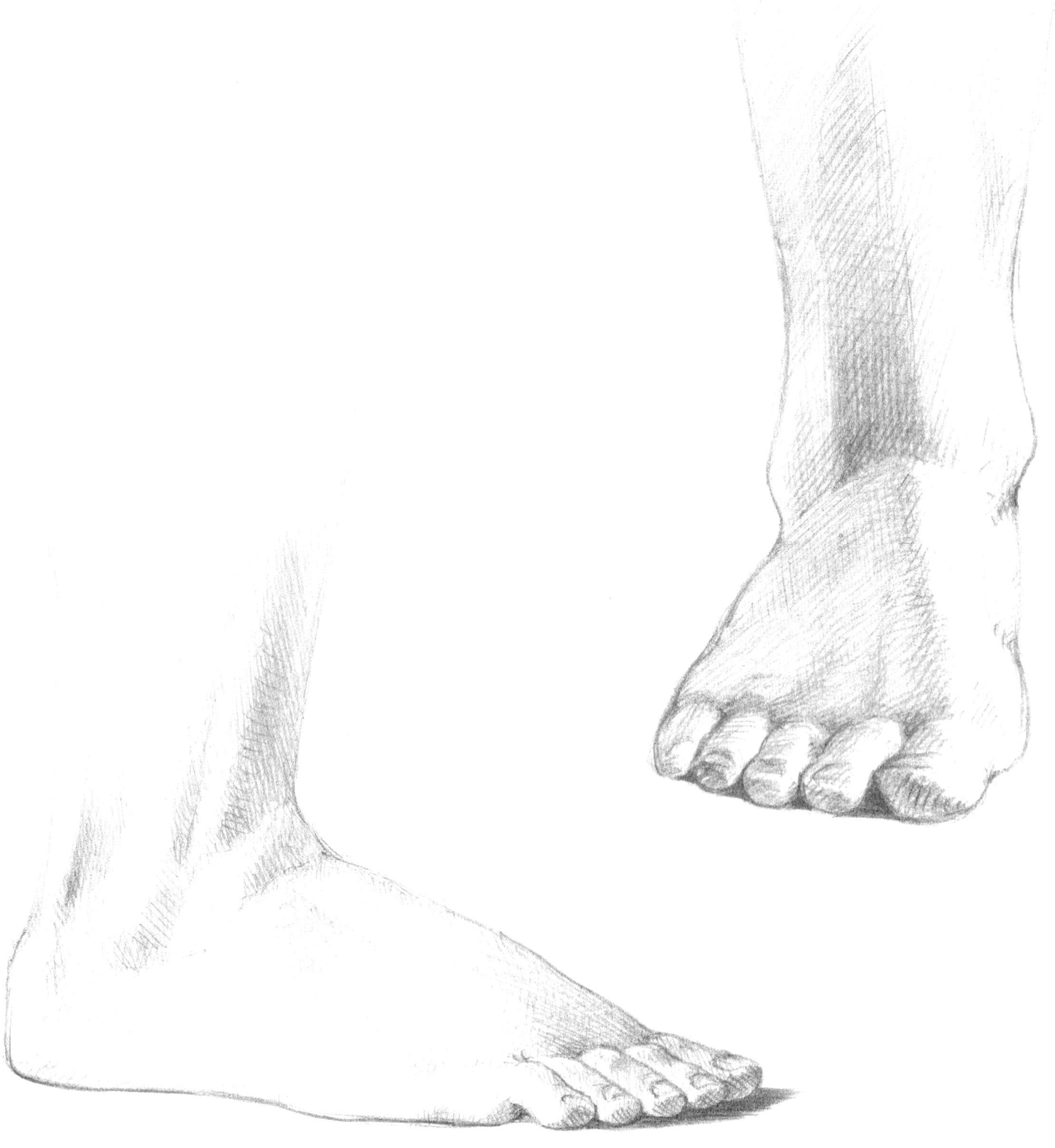

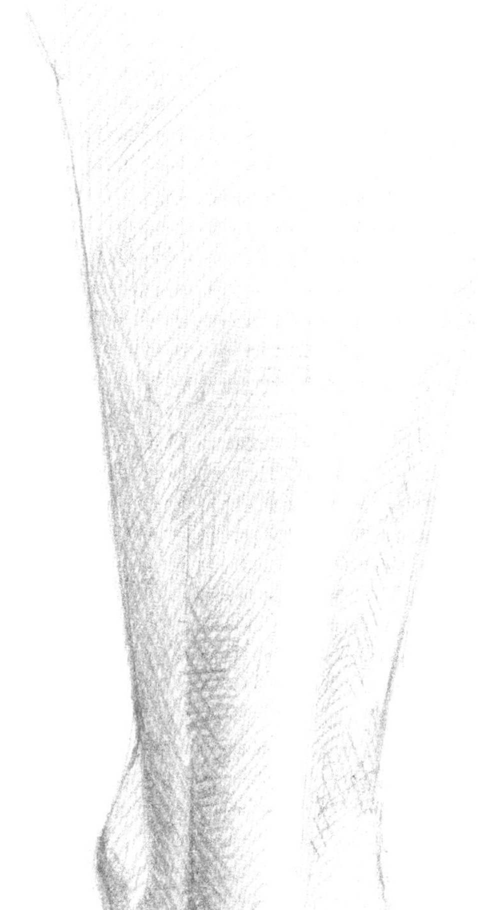

Fig. 117
The Ligaments of the Iliosacral Joint

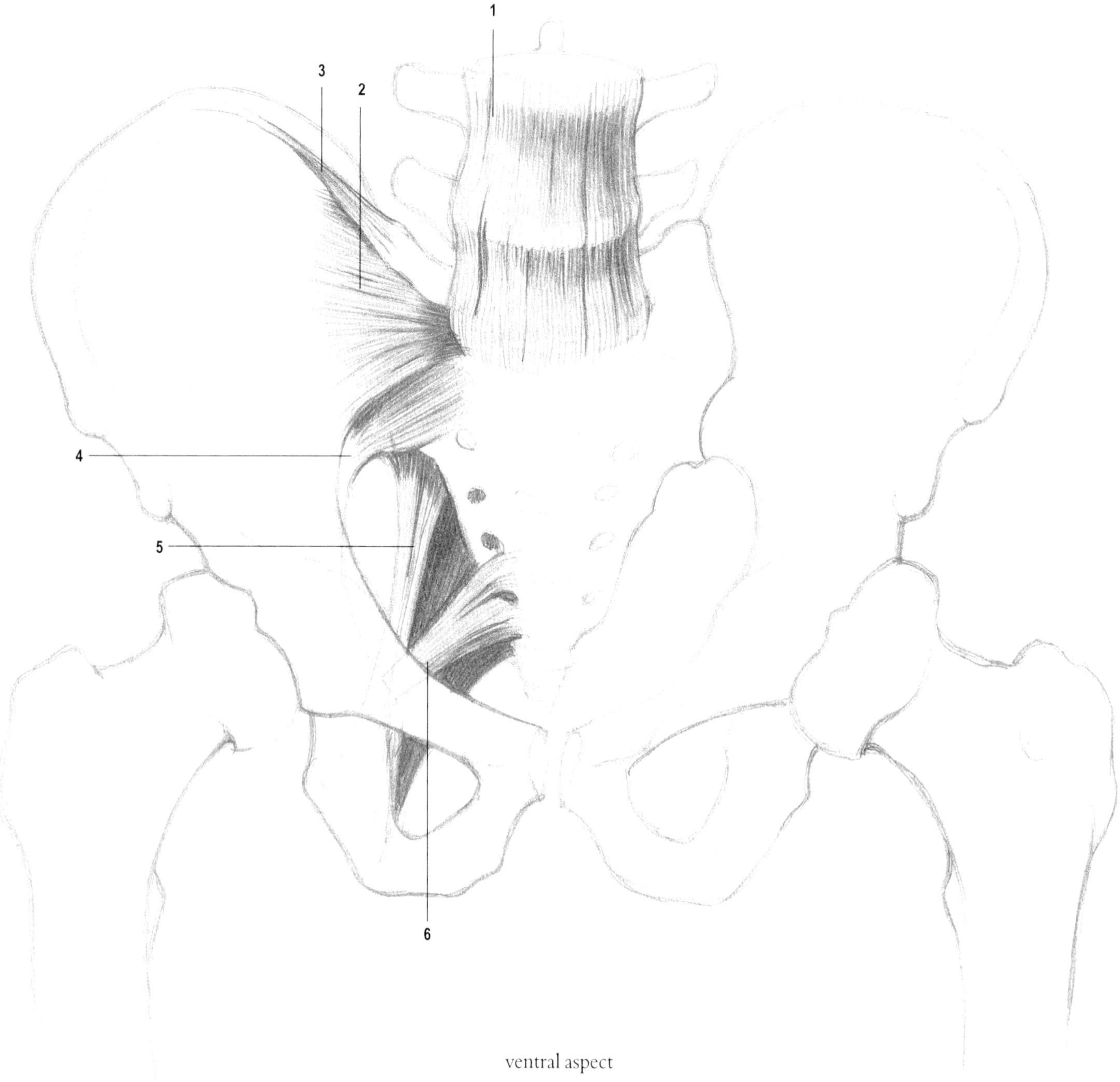

ventral aspect

1 Anterior long spinal ligament
2 Iliac-lumbar ligament
3 Lumbar-iliac ligament
4 Inner iliac-sacral ligament
5 Superior posterior sacrum-os sedentarium ligament
6 Inferior-anterior sacrum-os sedentarium ligament

Fig. 118
The Ligaments of the Hip Joint

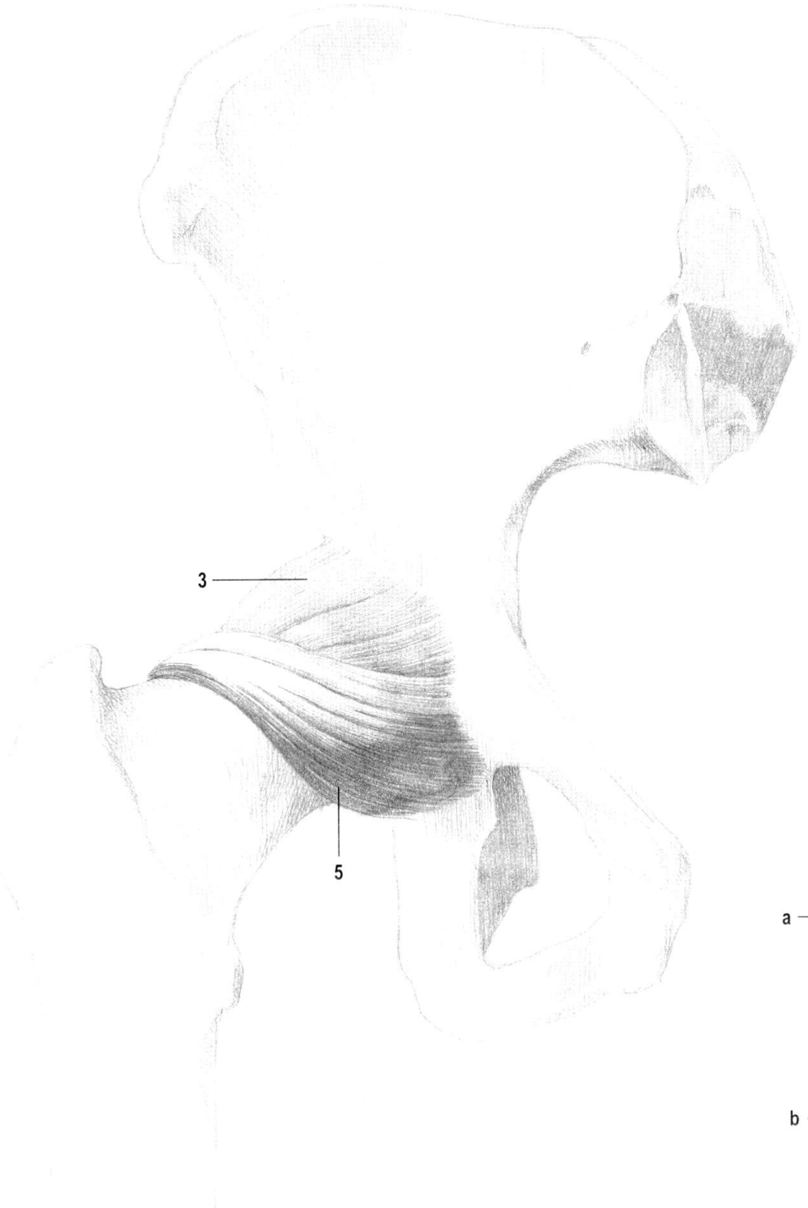

dorsal aspect

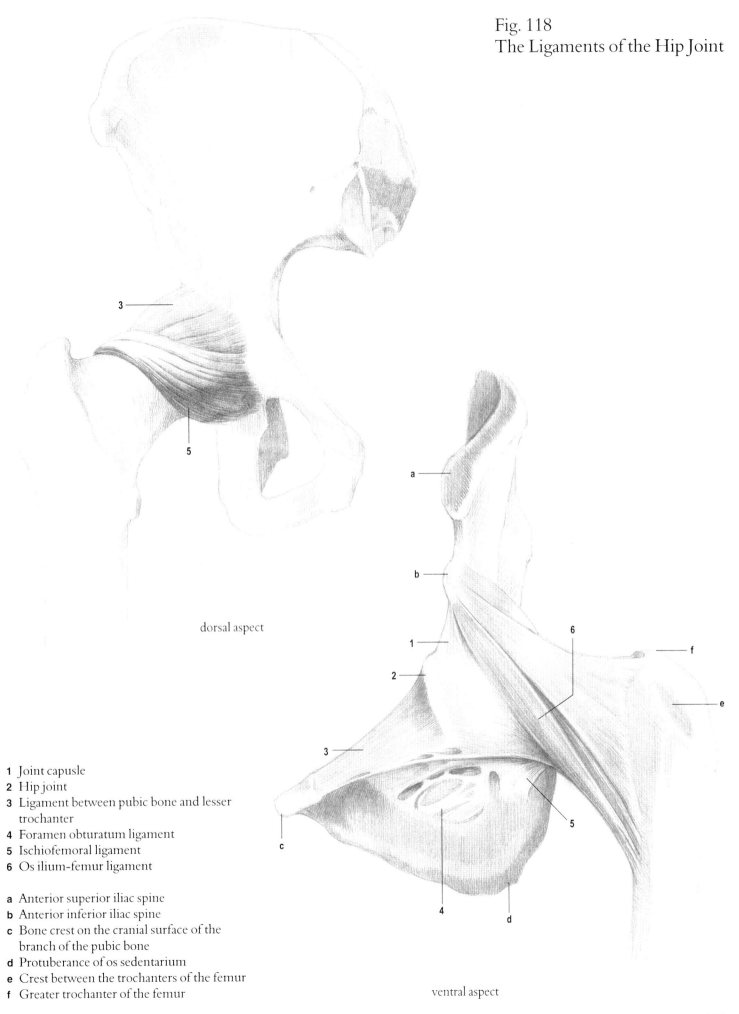

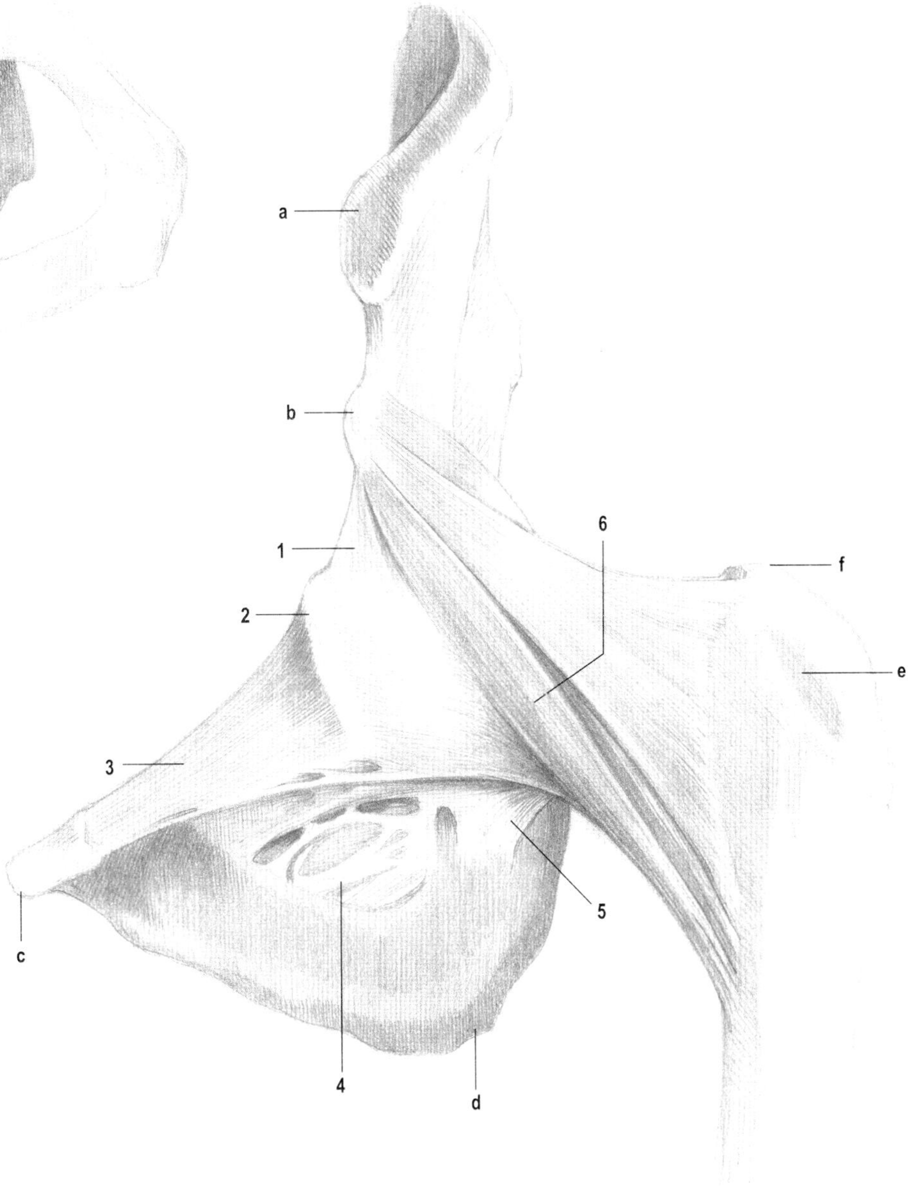

ventral aspect

1 Joint capusle
2 Hip joint
3 Ligament between pubic bone and lesser trochanter
4 Foramen obturatum ligament
5 Ischiofemoral ligament
6 Os ilium-femur ligament

a Anterior superior iliac spine
b Anterior inferior iliac spine
c Bone crest on the cranial surface of the branch of the pubic bone
d Protuberance of os sedentarium
e Crest between the trochanters of the femur
f Greater trochanter of the femur

Fig. 119
The Movement of the Hip Axis

Slow walking involves a series of shifts in body weight. As a result both the body and the hips swing left and right.

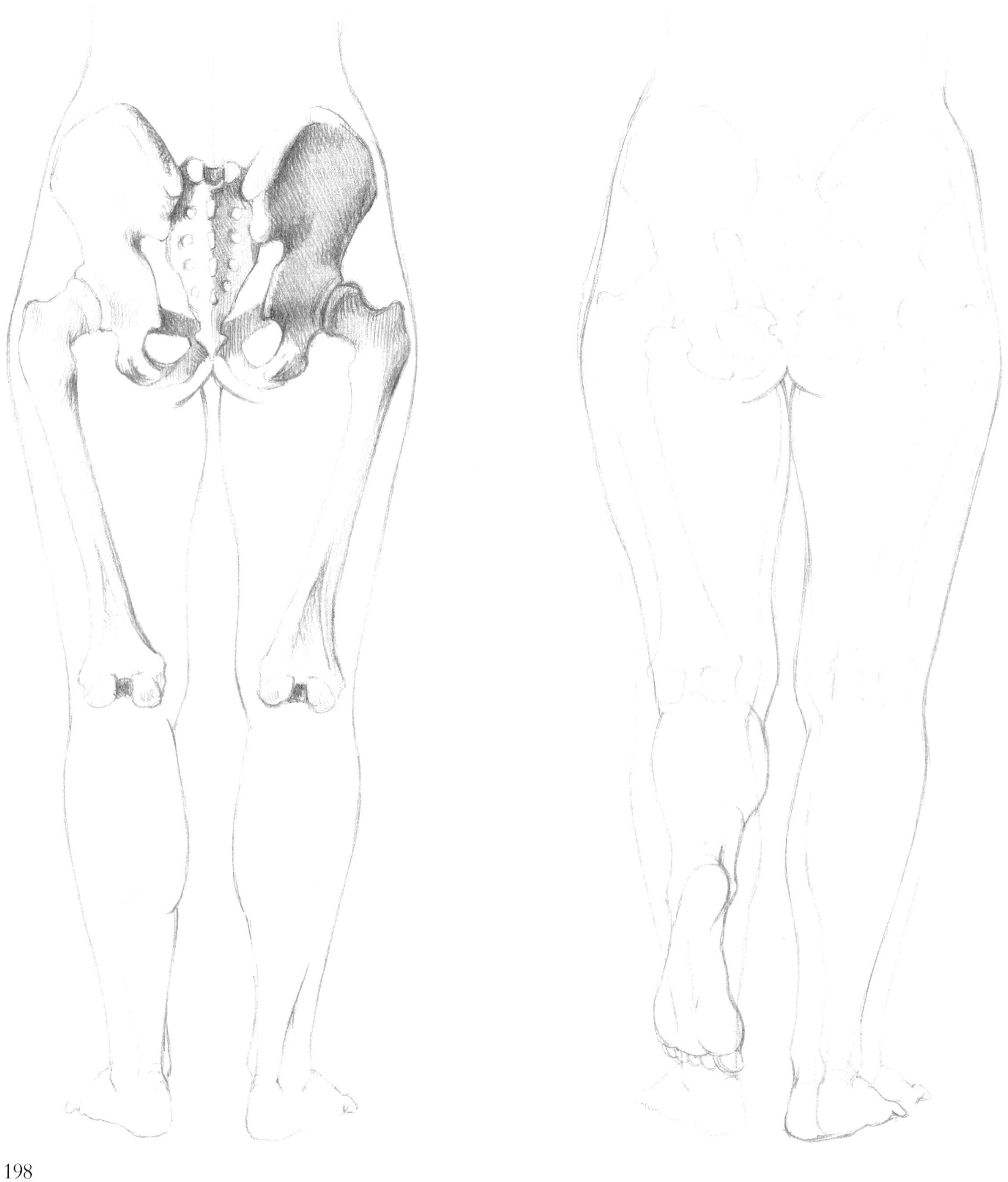

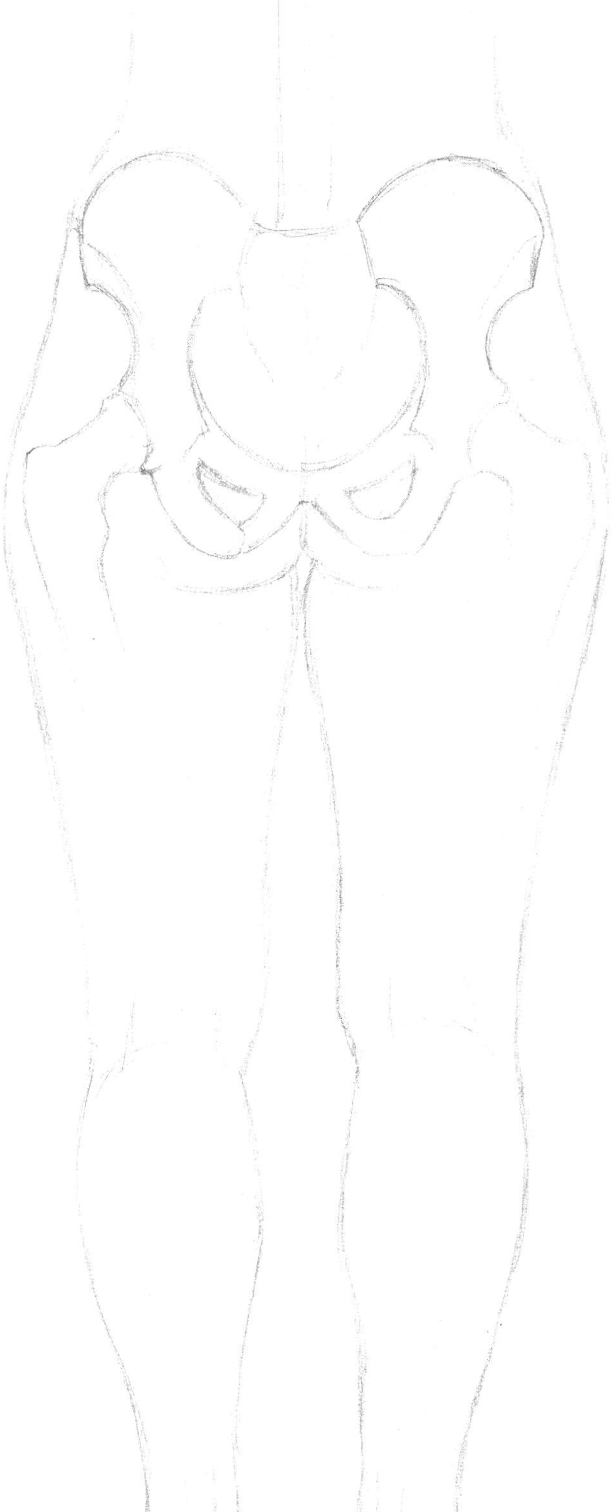

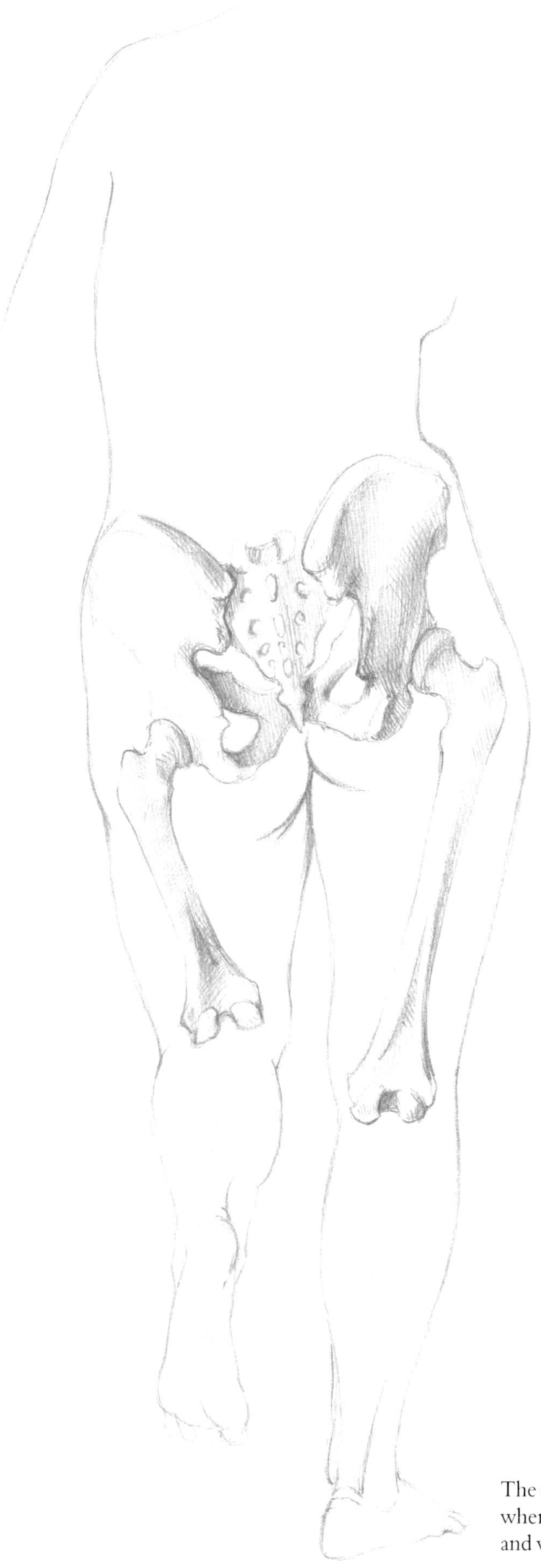

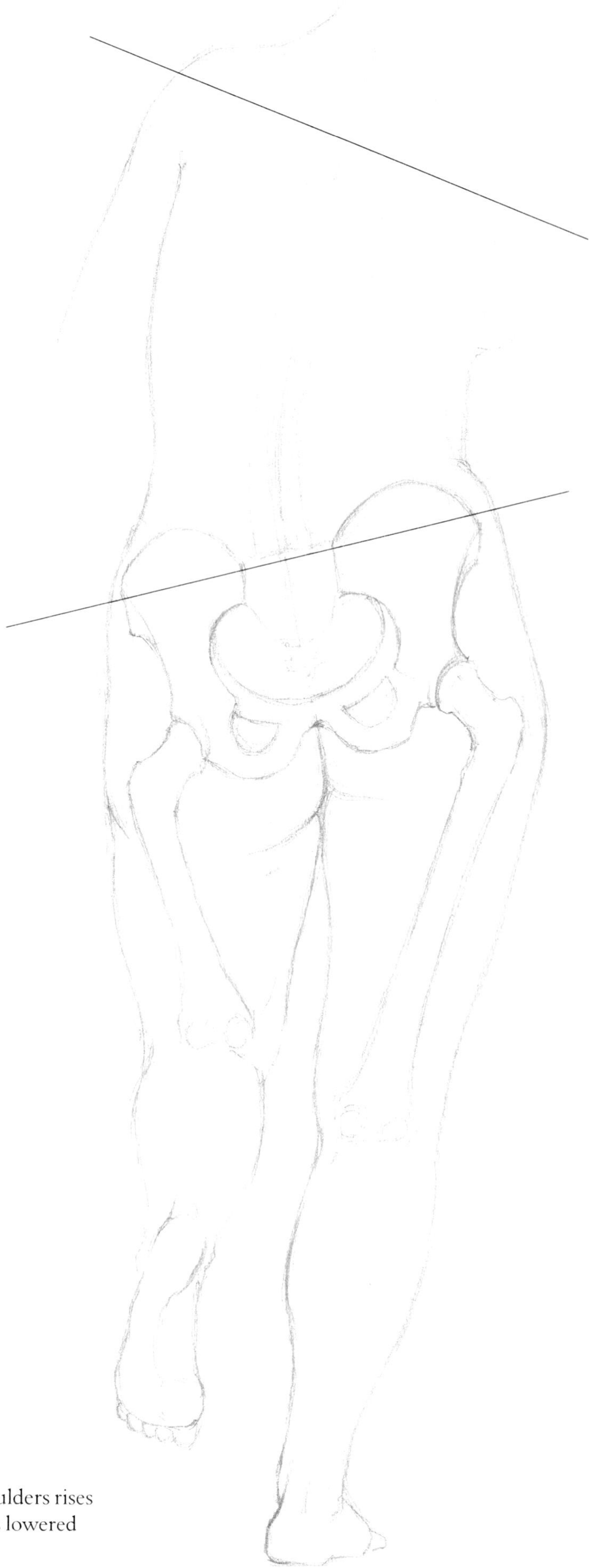

The axis of the shoulders rises when the hip axis is lowered and vice-versa.

Fig. 120
Movements of the Lower Extremity

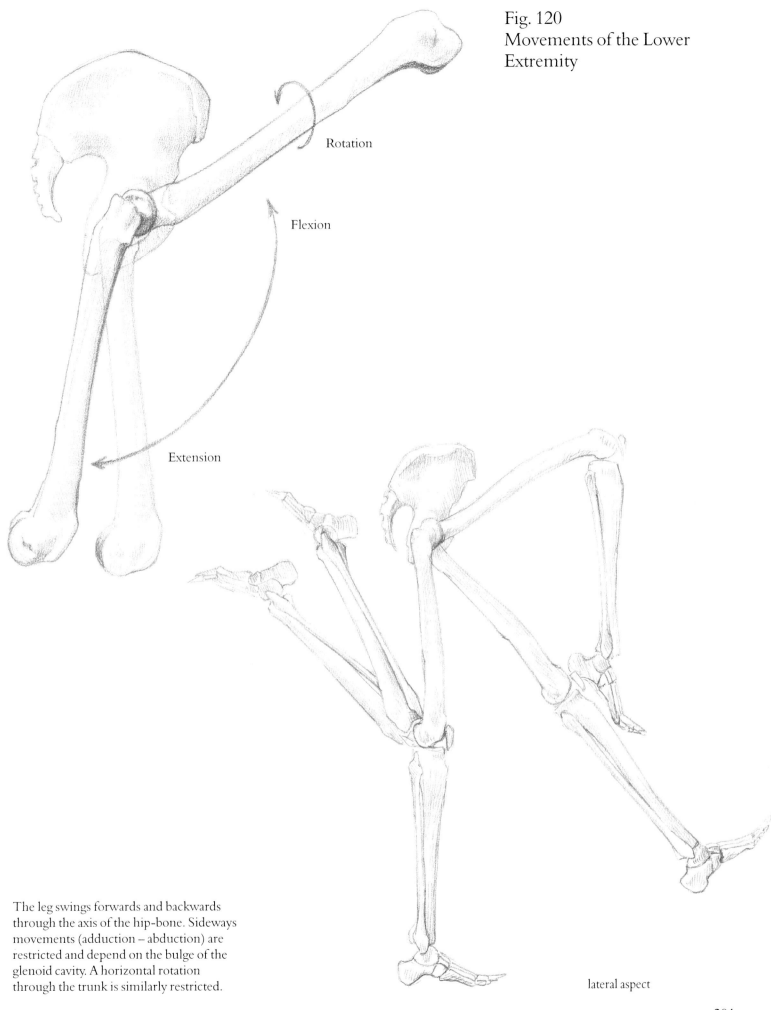

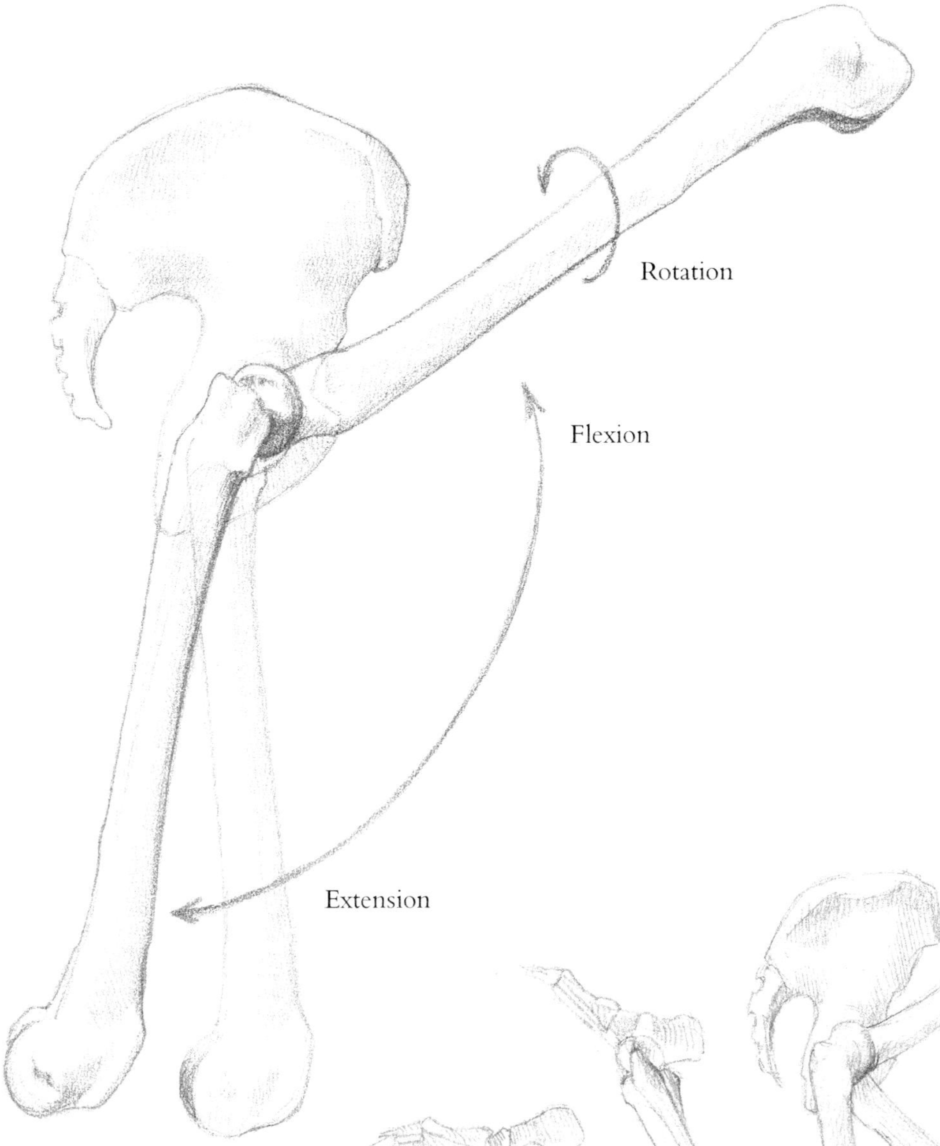

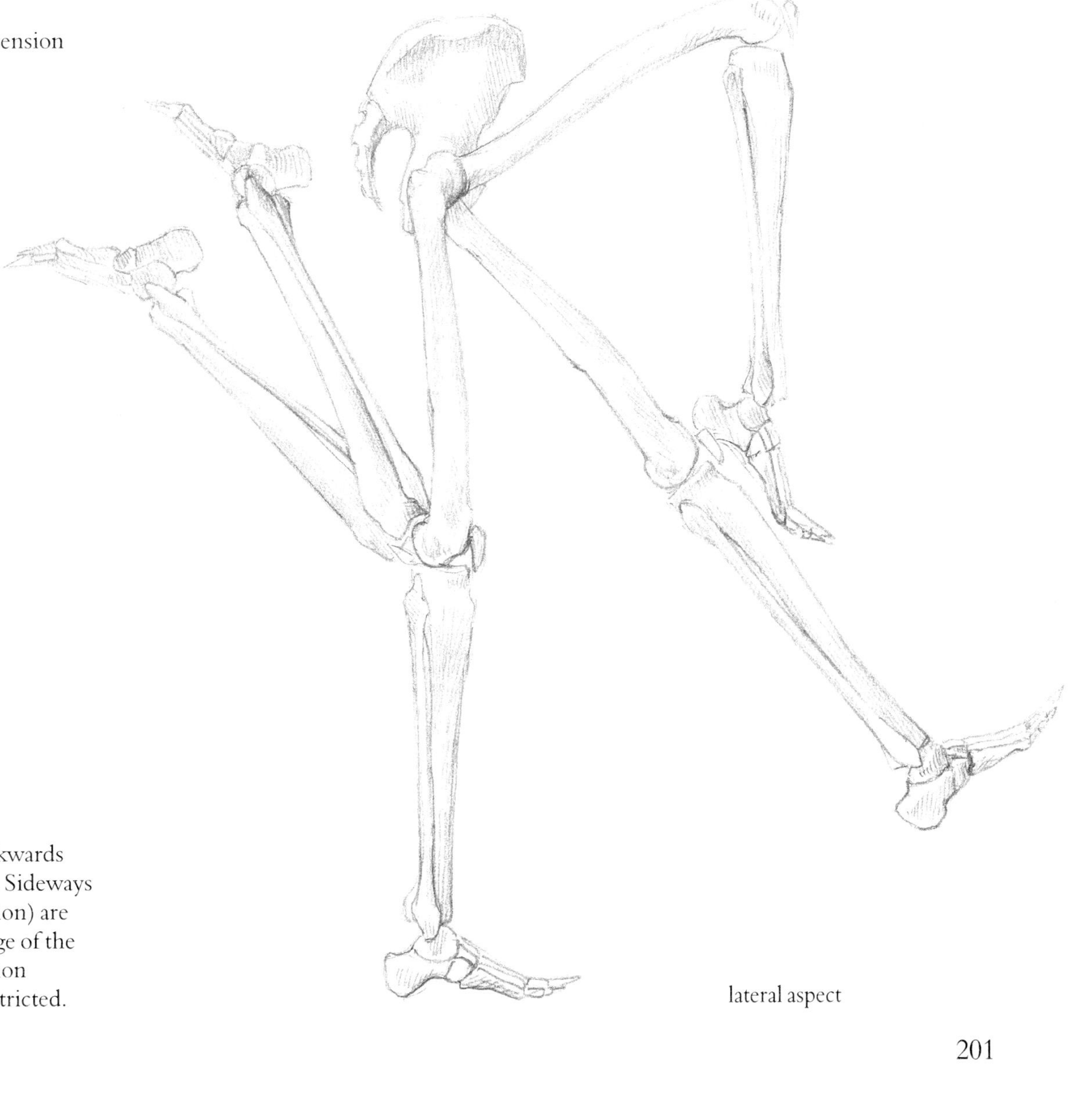

The leg swings forwards and backwards through the axis of the hip-bone. Sideways movements (adduction – abduction) are restricted and depend on the bulge of the glenoid cavity. A horizontal rotation through the trunk is similarly restricted.

Fig. 121
The Movement of the Hip Joint

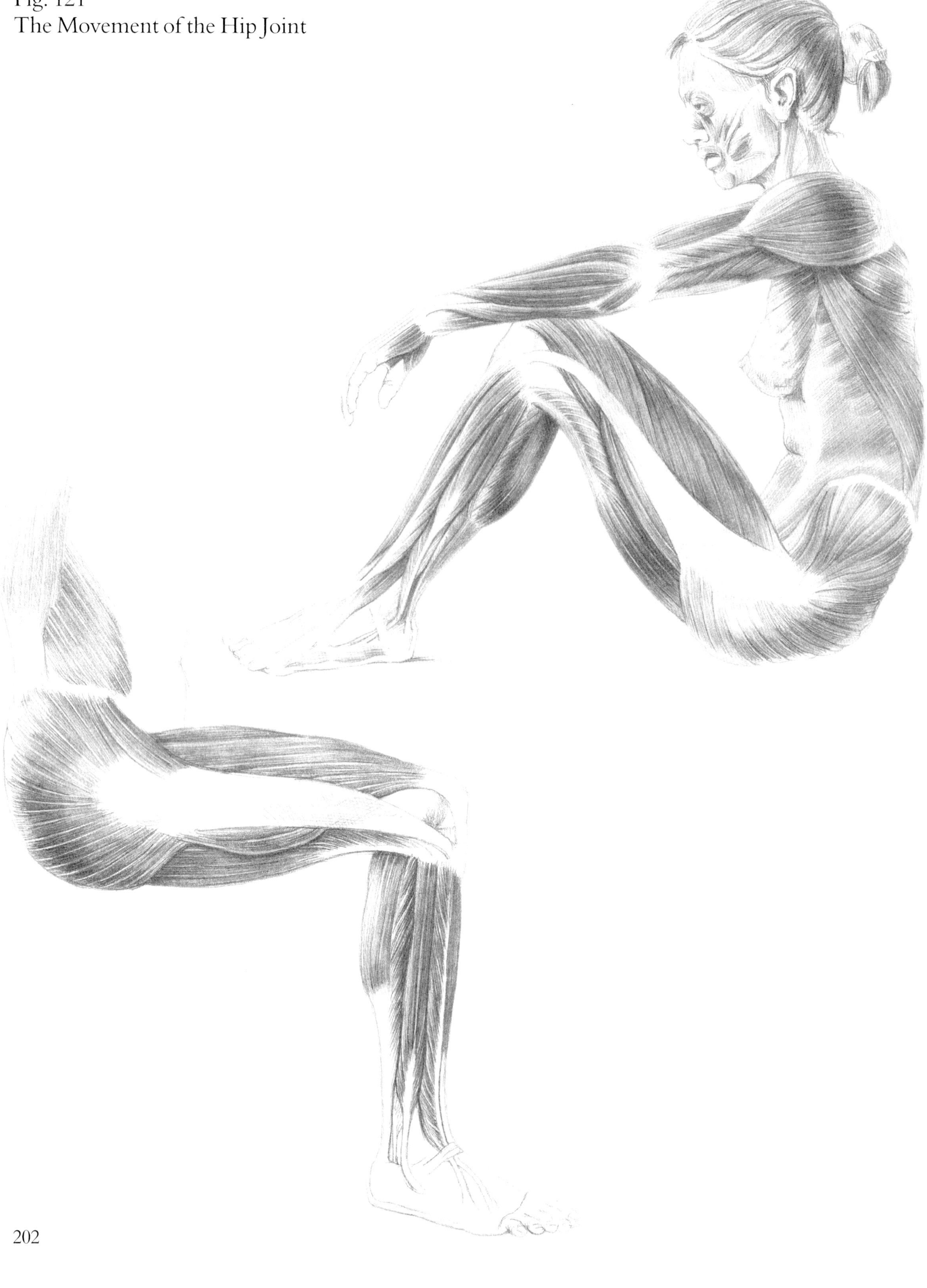

In a kneeling position the hip joint must rest on the ankle joint.

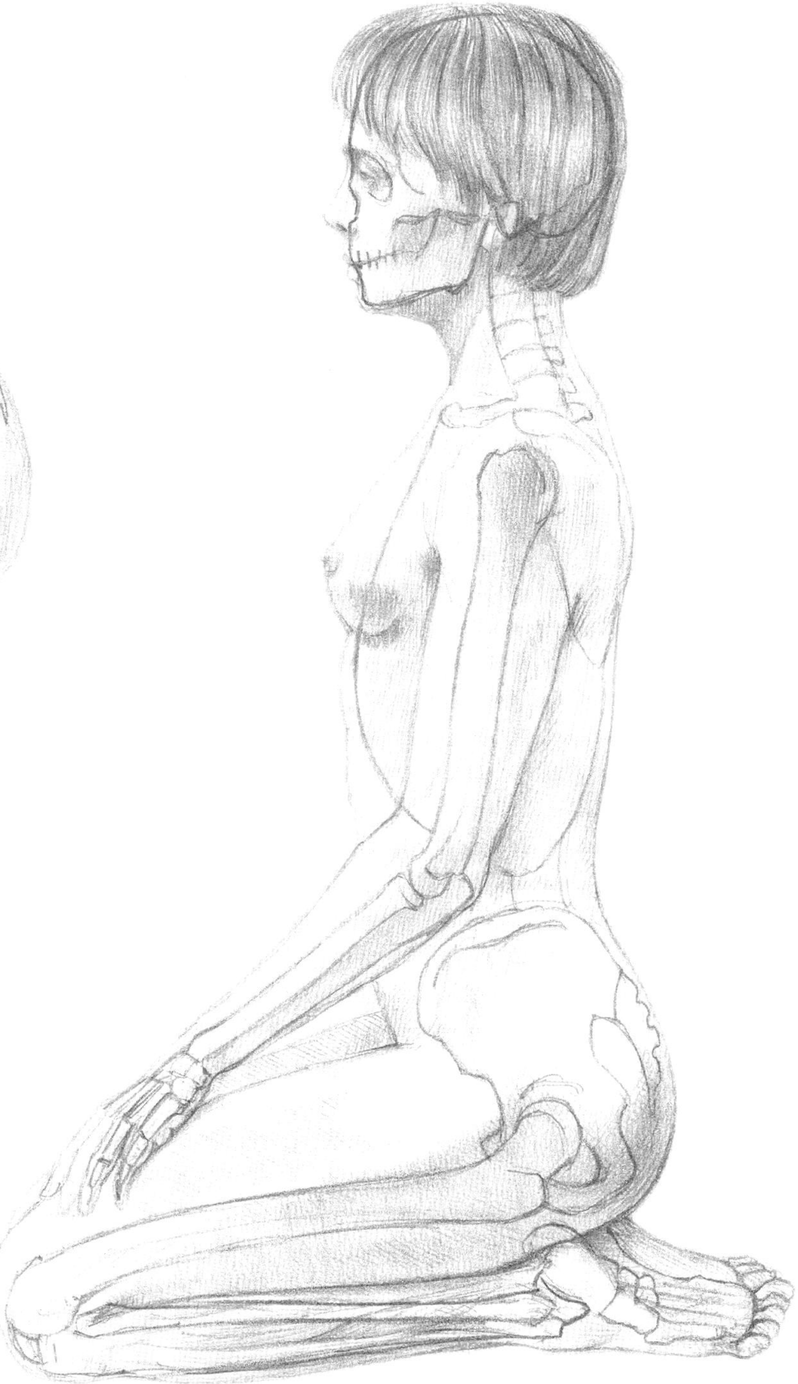

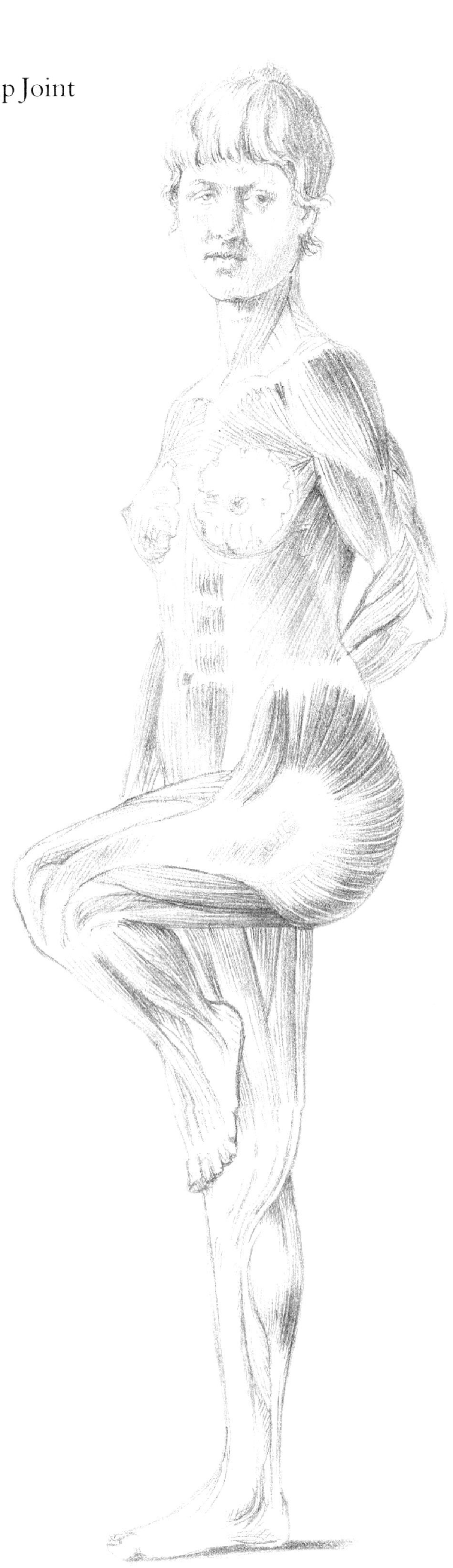

Fig. 122
The Ligaments of the Right Foot

1 Sphenoidal-navicular ligament
2 Ankle bone-navicular ligament
3 Synovial capsule of the ankle-shin-bone ligament
4 Lateral ankle-fibula ligament
5 Long plantar ligament
6 Inner ankle ligament
7 Dorsal navicular-sphenoidal ligaments
8 Plantar ligament between tarsus pedis and metatarsus
9 Plantar and lateral ligaments of the first toe joint

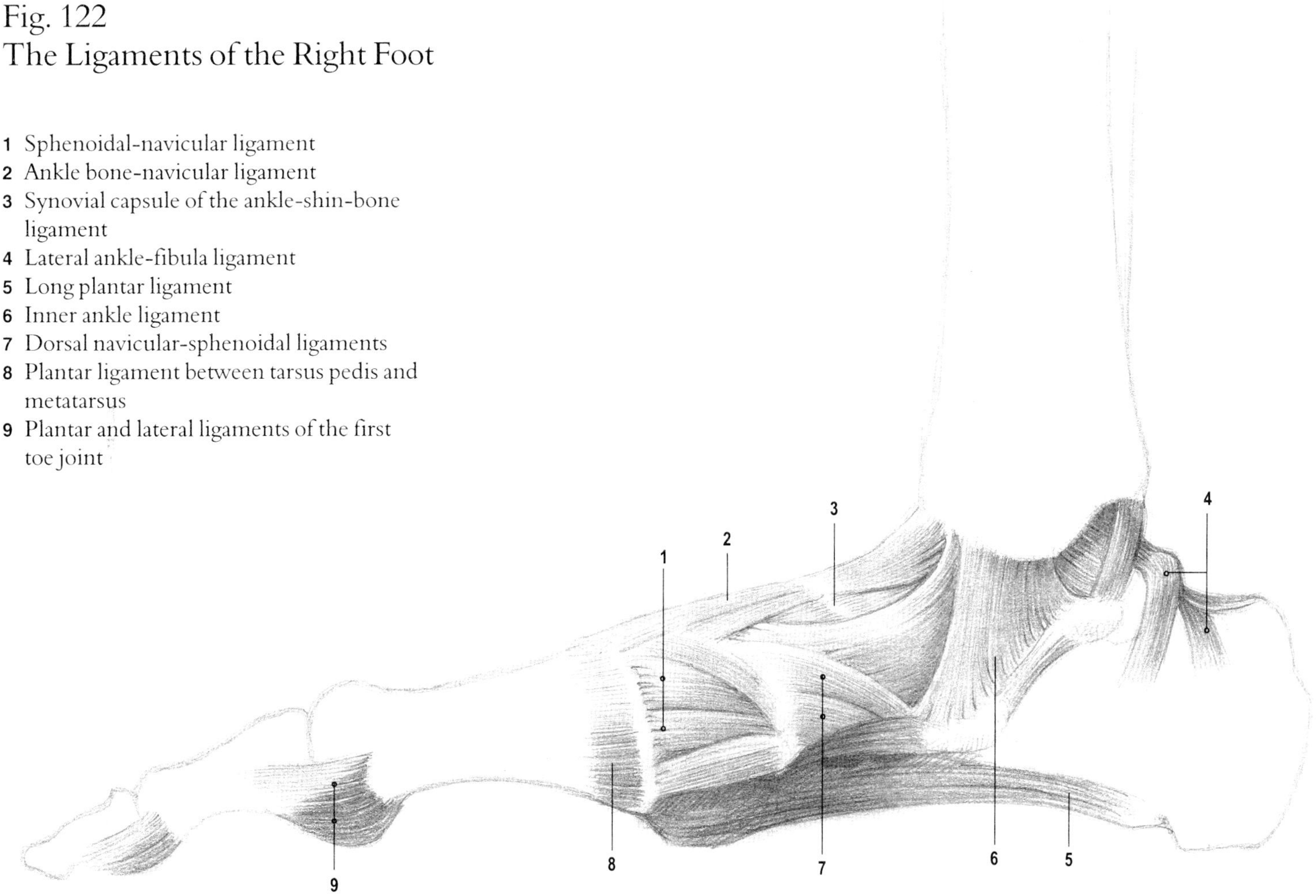

medial aspect

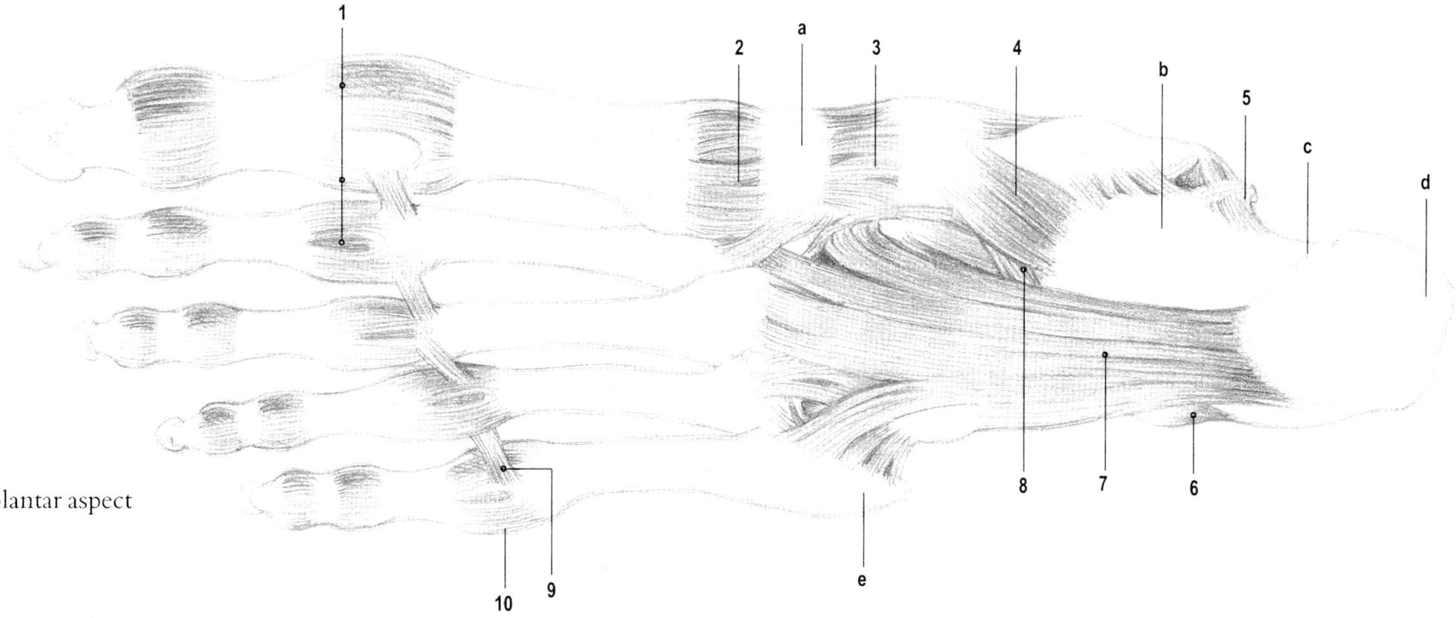

plantar aspect

1 Long plantar ligament
2 Plantar ligaments between tarsus pedis and metatarsus
3 Sphenoidal-navicular ligament
4 Heel bone-navicular ligament
5 Inner lateral ligament
6 Heel bone-shin ligament
7 Plantar ligaments of the second toe
8 Heel scaphoid bone ligament
9 Deep transverse ligament (interphalangeal ligament)
10 Lateral ligaments of the joints of the fifth toe
a Medial sphenoidal bone
b Talus support
c Tendon groove
d Heel bone protuberance
e Metatarsal protuberance of the fifth toe

Fig. 123
The Knee Joint

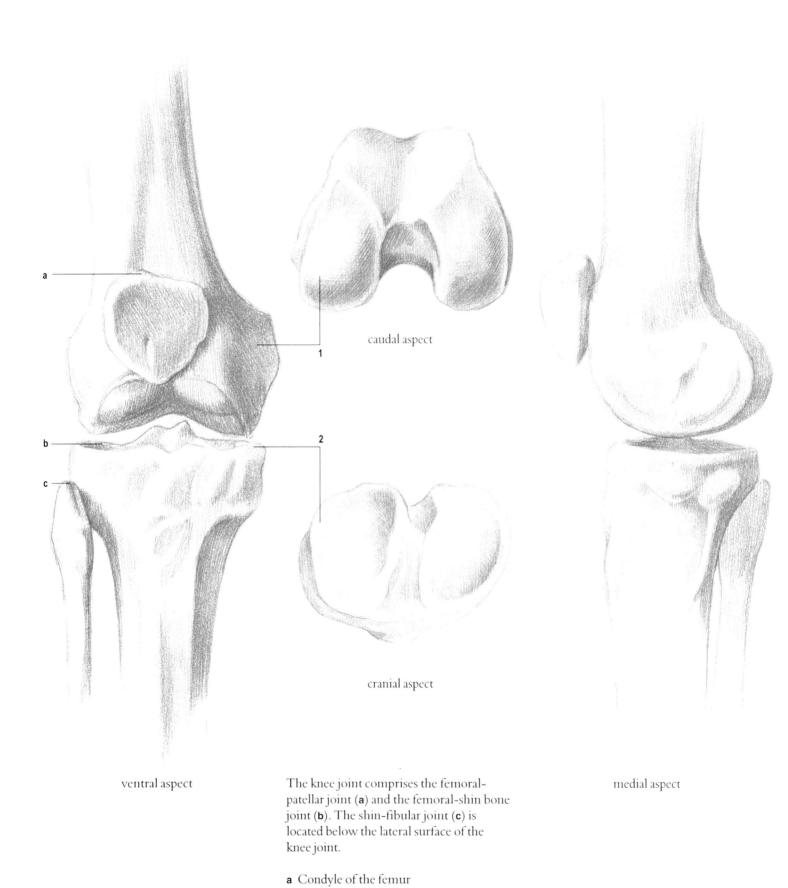

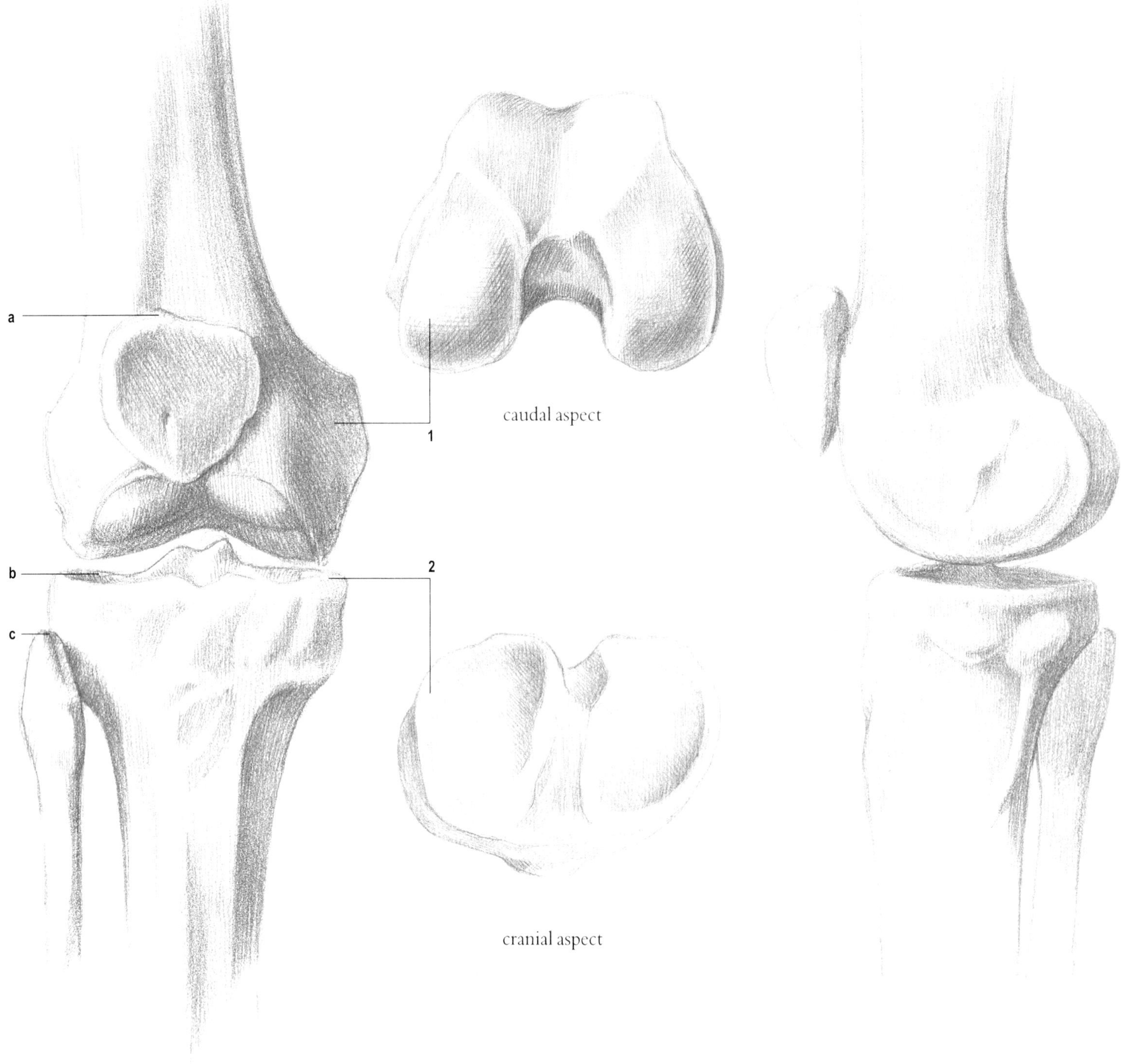

ventral aspect

medial aspect

The knee joint comprises the femoral-patellar joint (**a**) and the femoral-shin bone joint (**b**). The shin-fibular joint (**c**) is located below the lateral surface of the knee joint.

a Condyle of the femur
b Condyle of the shin bone

Fig. 124
The Ligaments of the Knee Joint

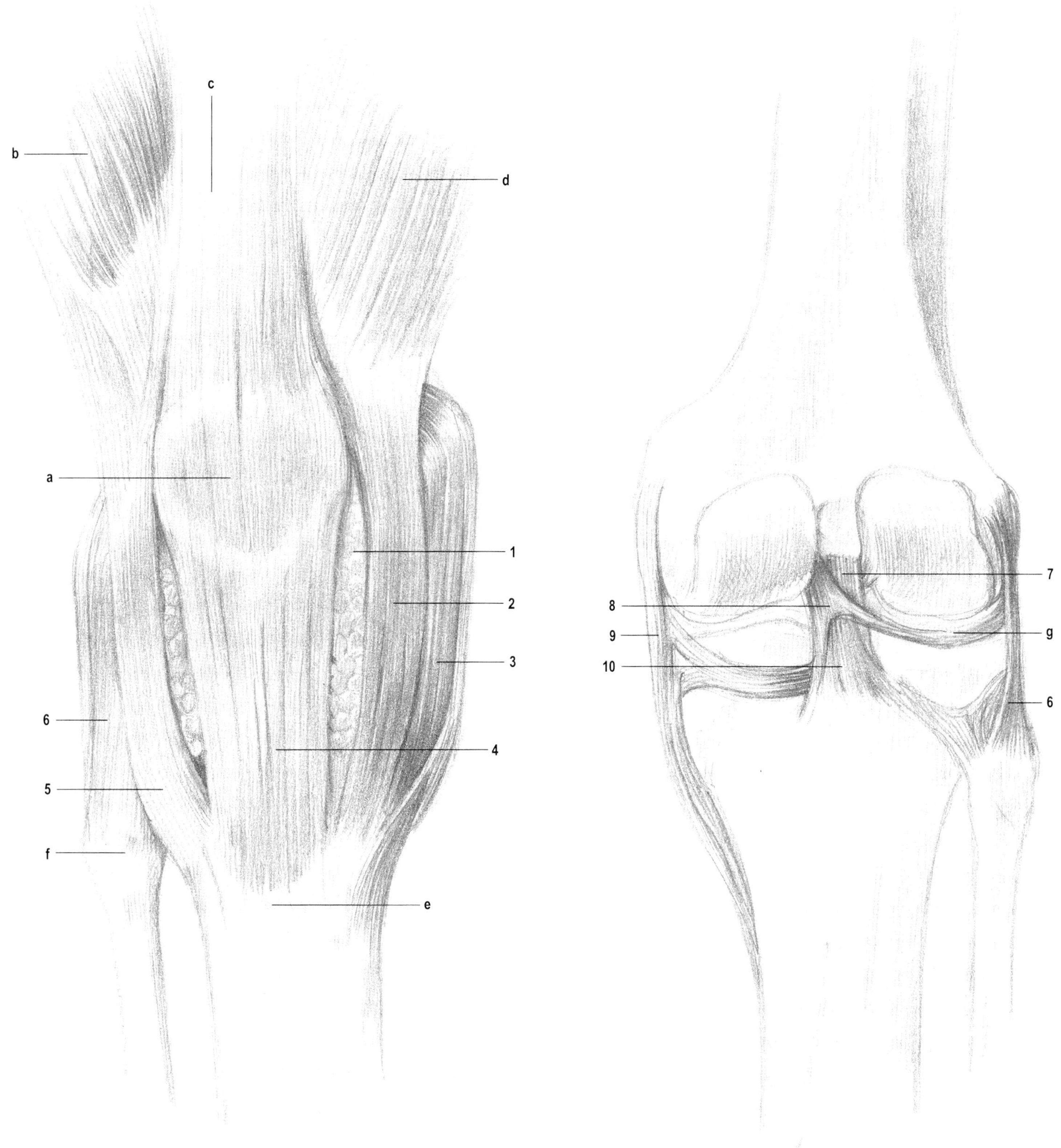

ventral aspect

dorsal aspect

1 Synovial fatty tissue
2 Tendon of the inner thigh muscle
3 Inner lateral ligament
4 Patellar ligament
5 Lateral straight patellar ligament
6 External lateral ligament

7 Anterior transverse ligament
8 Rear meniscus-femoral ligament
9 Internal lateral ligament
10 Posterior transverse ligament

a Patella

b Vastus lateralis (*97/3*)
c Rectus femoris (*97/1*)
d Vastus medialis (*97/2*)
e Protuberance of the shinbone
f Head of fibula
g Lateral meniscus

Fig. 125
Movements of the Knee Joint

Viewed from the side, the upper and lower legs are vertical when the knee is extended. The C-shaped cartilage of the shin bone (menisci) compensates for any unevenness in the articulatory surfaces of the femur and the shin bone. The patella is located above the articulatory condyles of the femur.

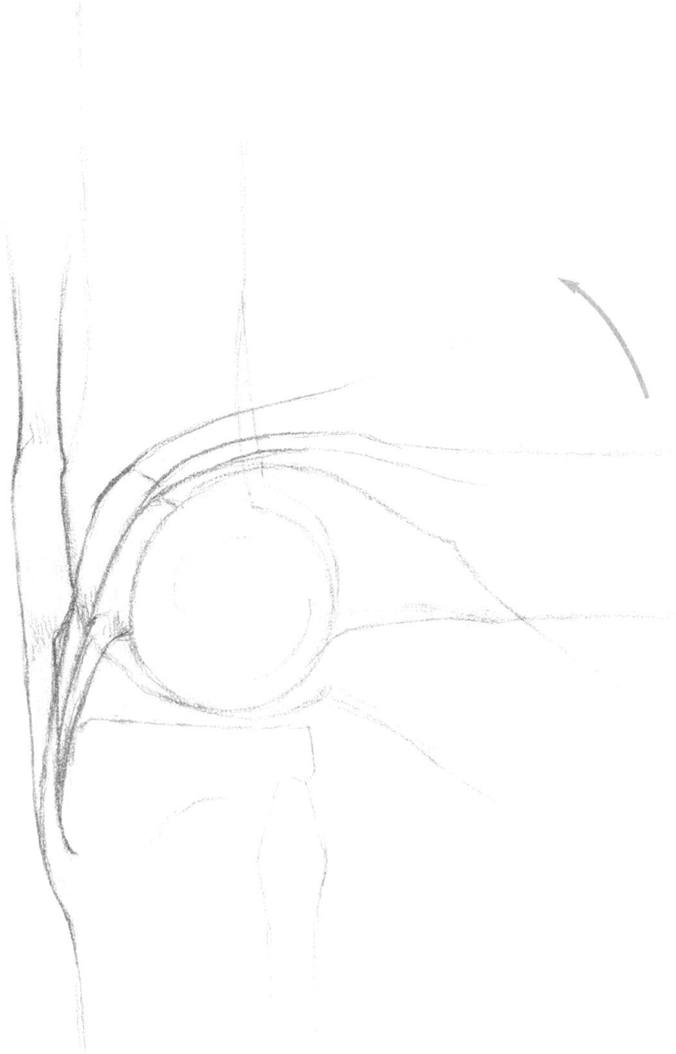

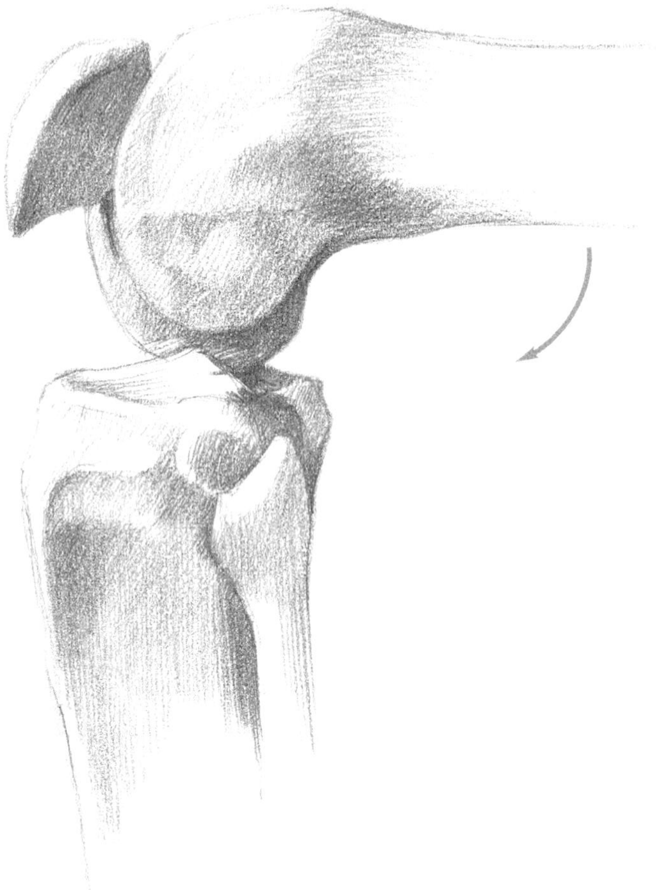

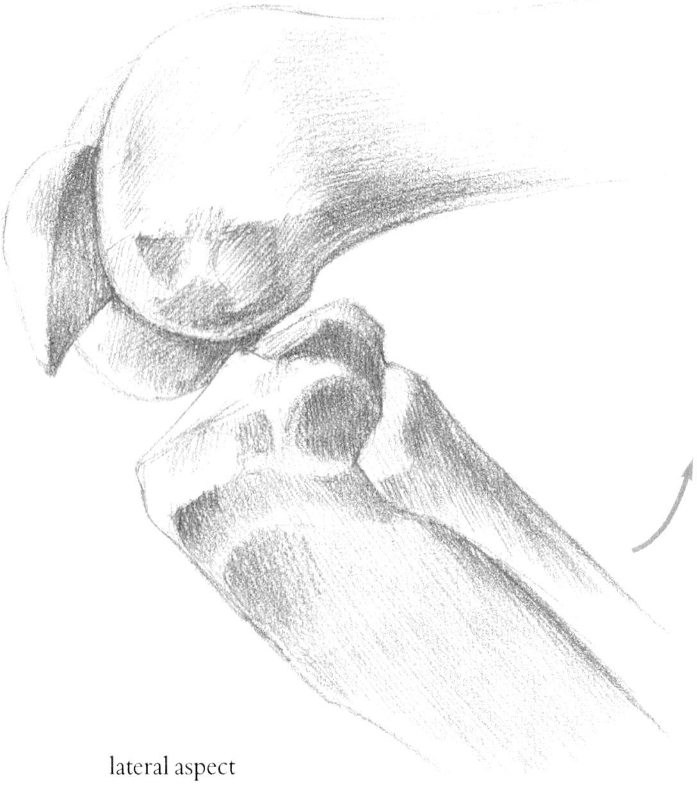

When the knee joint is flexed the patella glides up and down against the articulatory surface of the femur. The distance between the shin bone protuberance and the patella remains constant. The shin bone and fibula do not move relative to each other.

lateral aspect

Fig. 126
The Ligaments of the Knee Joint in Movement

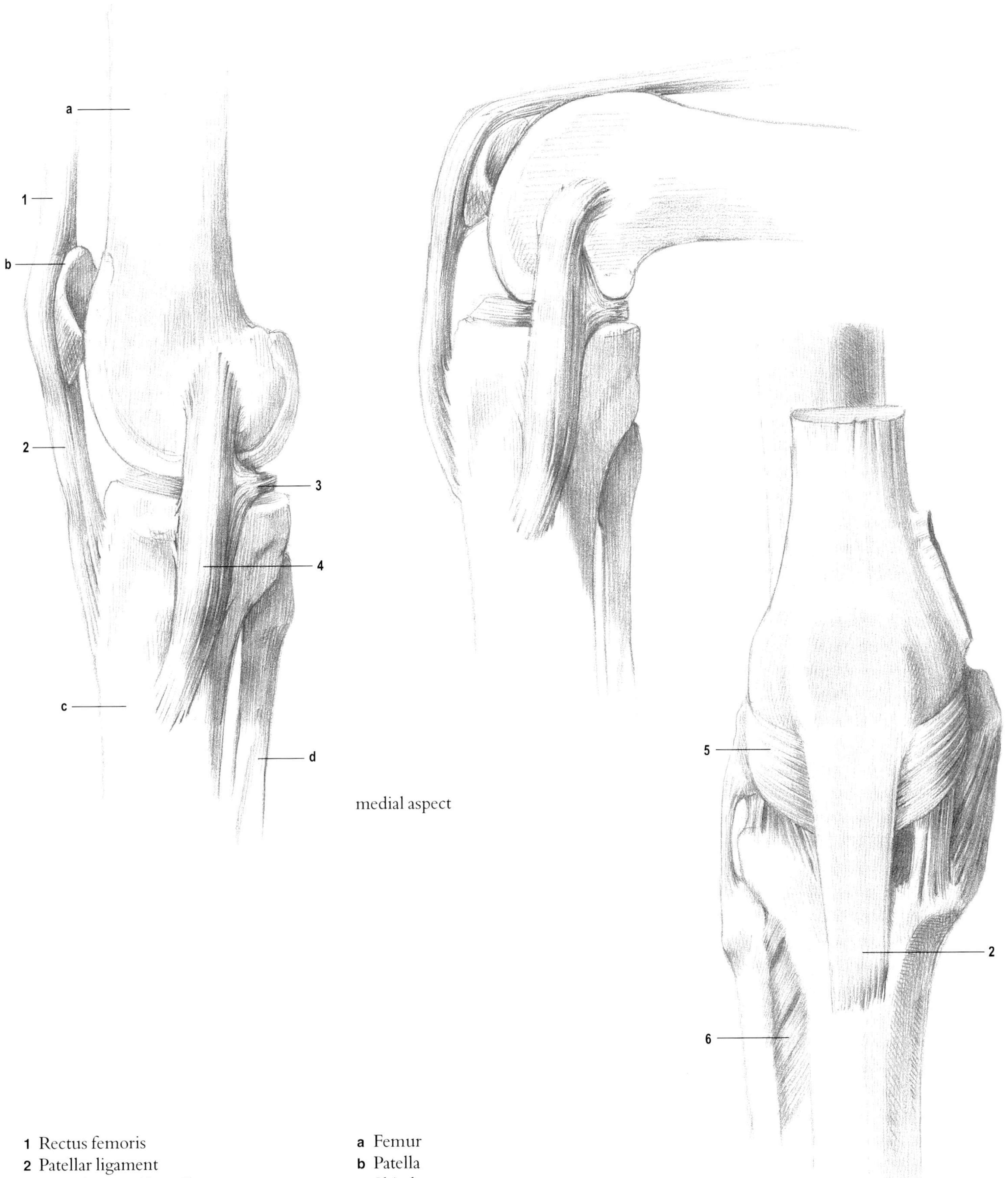

ventral aspect

1 Rectus femoris
2 Patellar ligament
3 Posterior cruciform ligament
4 Inner lateral ligament
5 Anterior transverse ligament
6 Superior interosseous ligament

a Femur
b Patella
c Shin bone
d Fibula

Fig. 127
The Bones and Muscles of the Lower Extremity

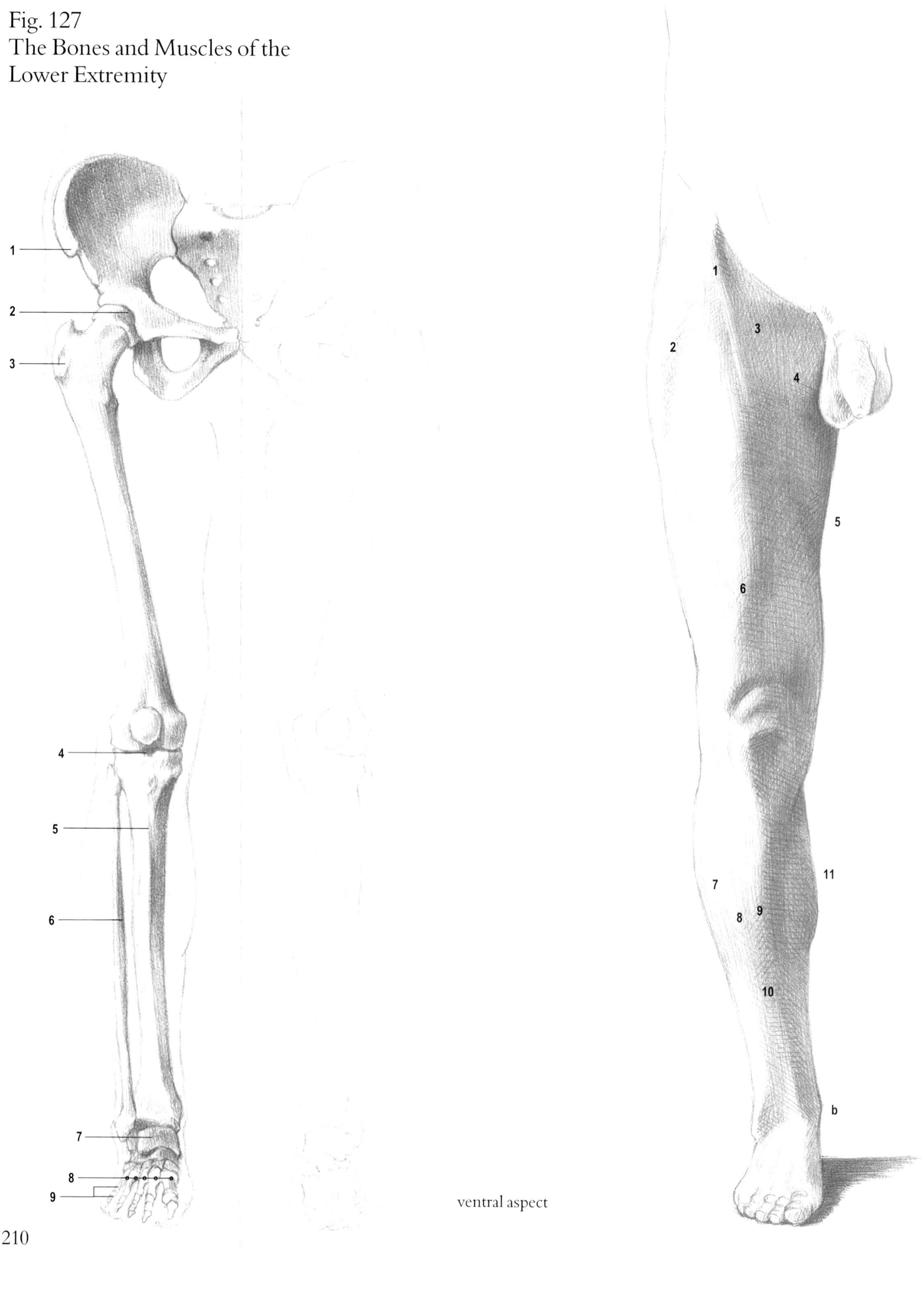

ventral aspect

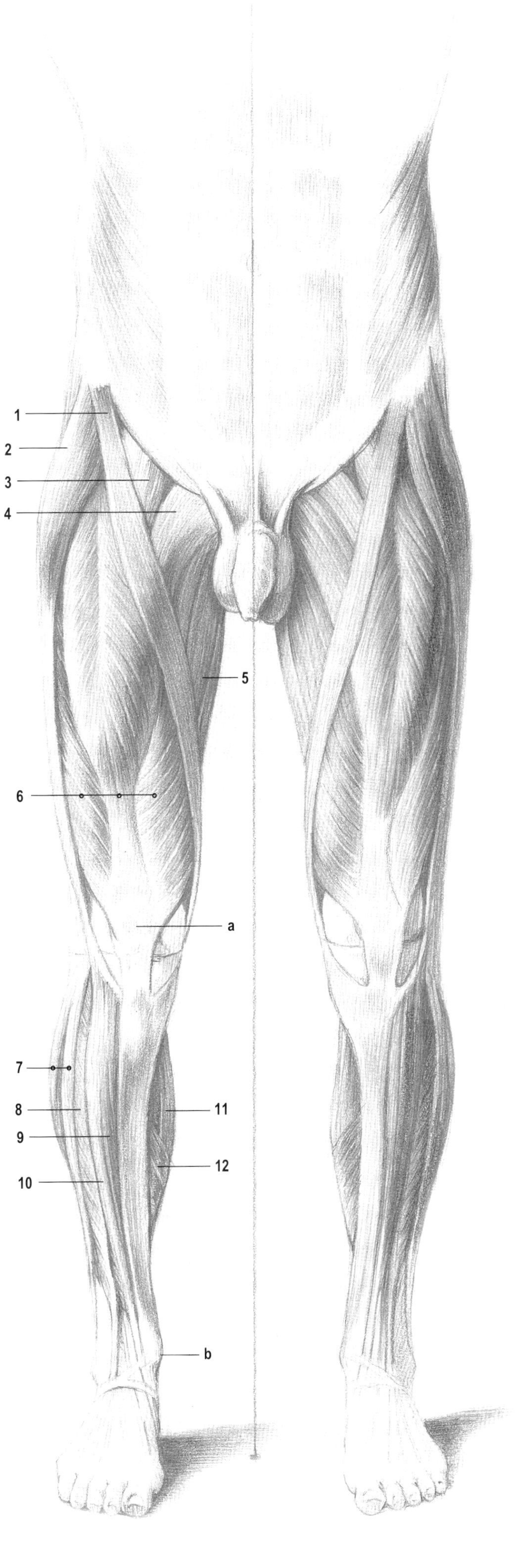

Bones

1 Anterior superior iliac spine
2 Hip joint
3 Greater trochanter
4 Knee joint
5 Shin bone
6 Fibula
7 Ankle bone
8 Metatarsal bones
9 Pediphalanges

Muscles

1 Sartorius (*88*)
2 Tensor faciae latae (*84*)
3 Iliac muscle (*79/1*)
4 Pectineus (*90*)
5 Adductor magnus/brevis (*92*)
6 Quadriceps femoris (*97*)
7 Fibularis longus/brevis (*103, 104*)
8 Extensor digitorum longus (*101*)
9 Tibialis anterior (*99*)
10 Extensor hallucis longus (*100*)
11 Gastrocnemius (*105/1*)
12 Soleus muscle (*105/2*)

a Patella
b Medial ankle bone

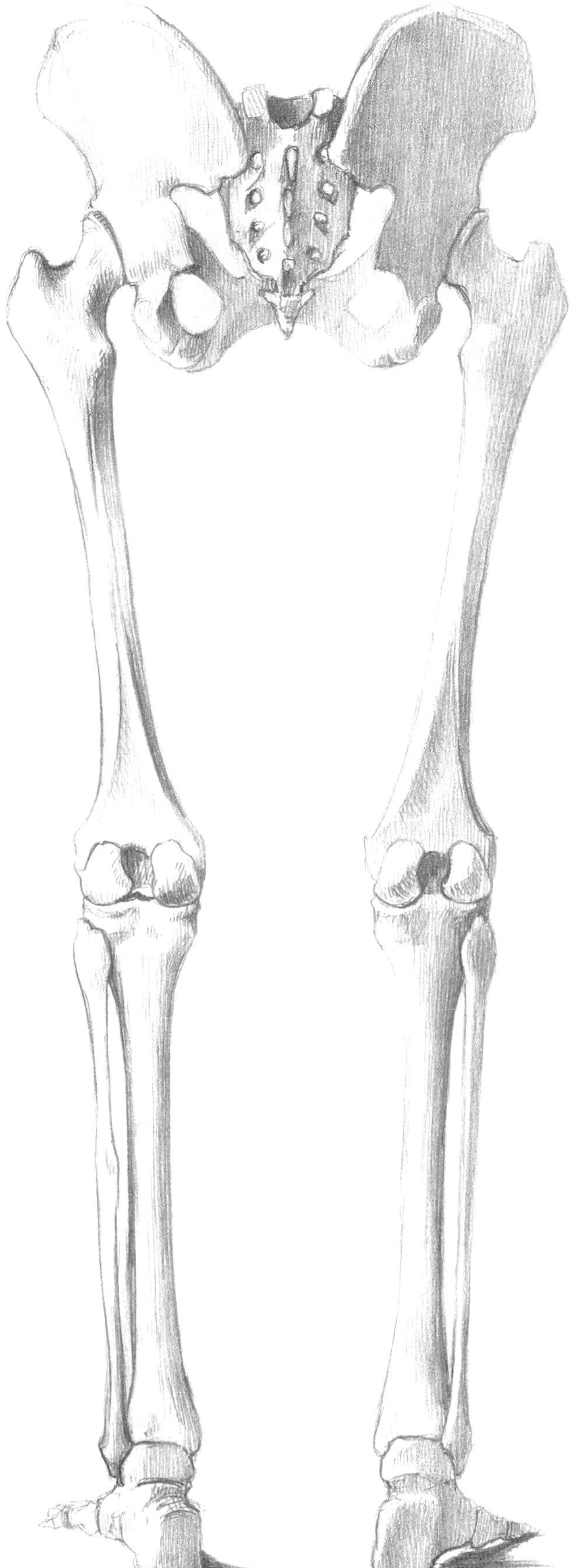

dorsal aspect

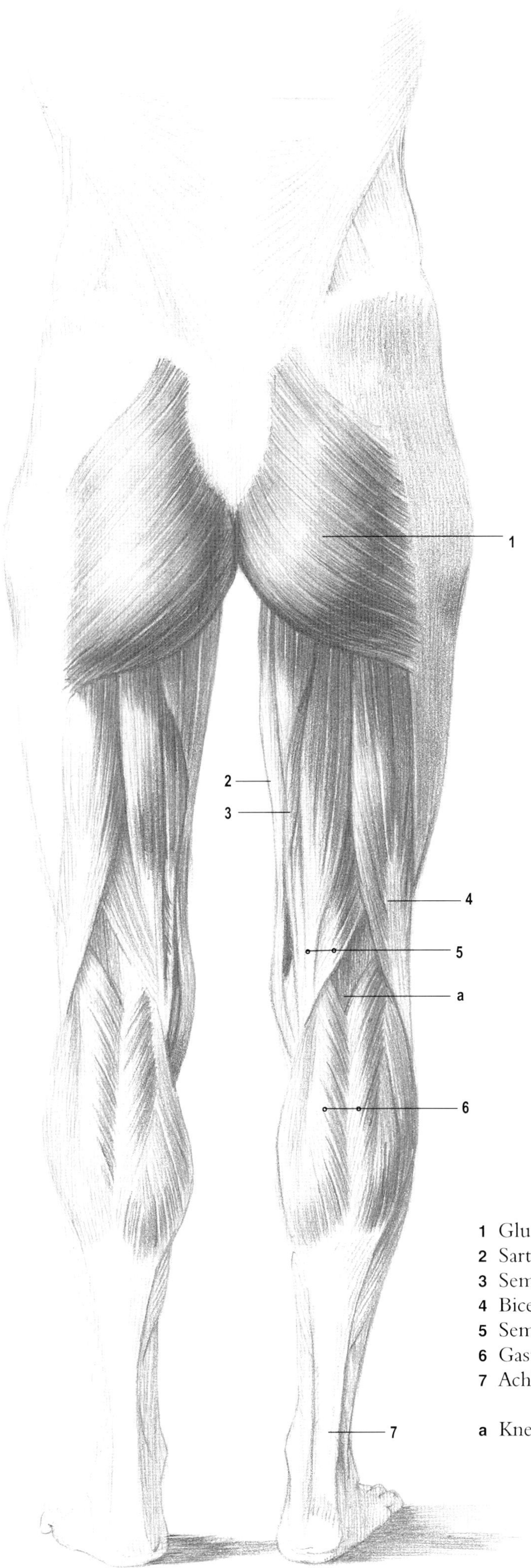

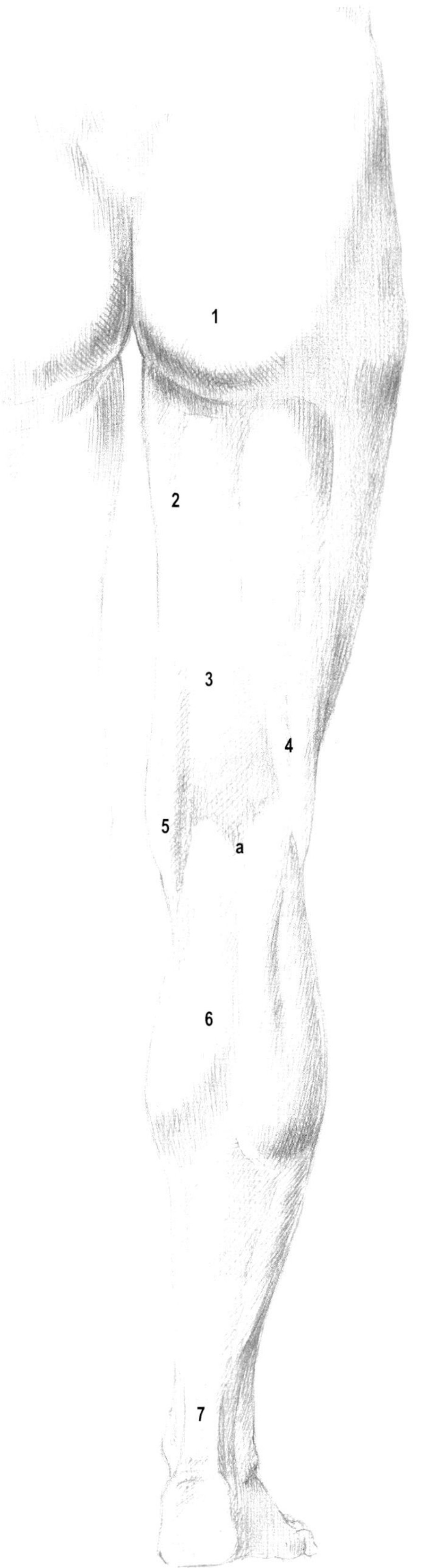

1 Gluteus maximus (*81*)
2 Sartorius (*88*)
3 Semitendinosus (*86*)
4 Biceps femoris (*85*)
5 Semimembranosus (*87*)
6 Gastrocnemius (*105/1*)
7 Achilles tendon

a Knee popliteal space

The Bones and Muscles of the Lower Extremity
(cont.)

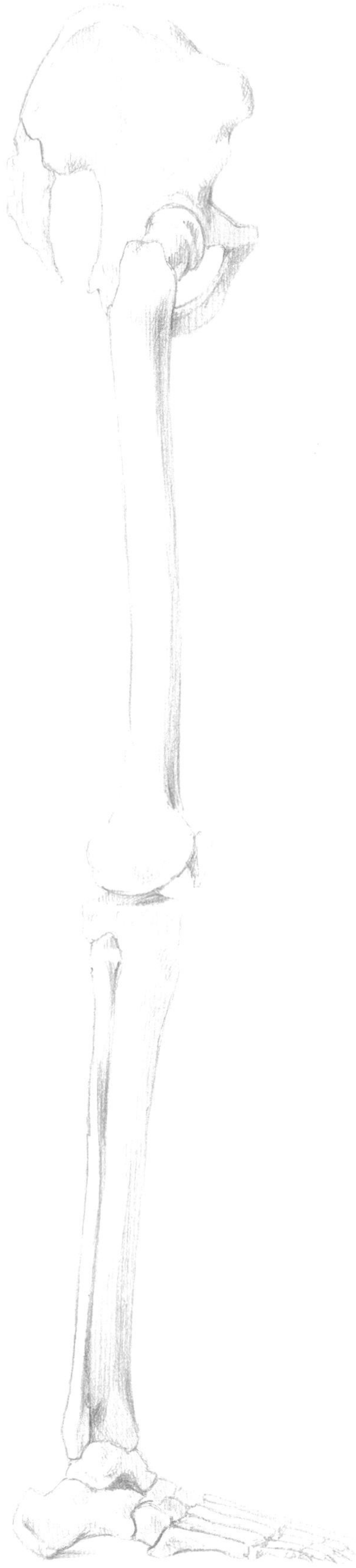

The gluteus maximus forms the buttocks and the lateral contour of the pelvis with the femoral faciae latae tensors. The form of the femur depends on the quadriceps femoris anteriorly and posteriorly on the gluteal muscles. The medial surface of the shin bone is covered only by skin. The anterior and lateral arch of the lower leg is formed by the foot extensors and toe extensors, whereas the foot flexors and toe flexors form the posterior curve.

1 Tensor faciae latae (*84*)
2 Sartorius (*88*)
3 Quadriceps femoris (*97*)
4 Extensor digitorum longus (*101*)
5 Extensor digitorum brevis (*111*)
6 Fibularis longus and Fibularis brevis (*103, 104*)
7 Gastrocnemius (*105/1*)
8 Semimembranosus (*87*)
9 Semitendinosus (*86*)
10 Biceps femoris (*85*)
11 Gluteus maximus (*81*)

a Patellar joint
b Knee joint
c Achilles tendon

lateral aspect

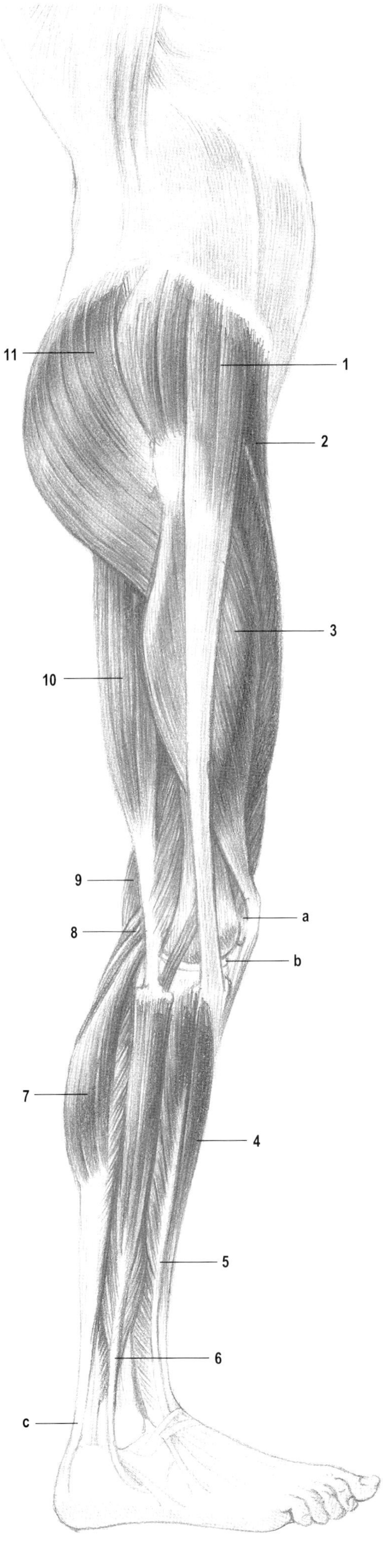
11
1
2
3
10
9
a
8
b
7
4
5
6
c

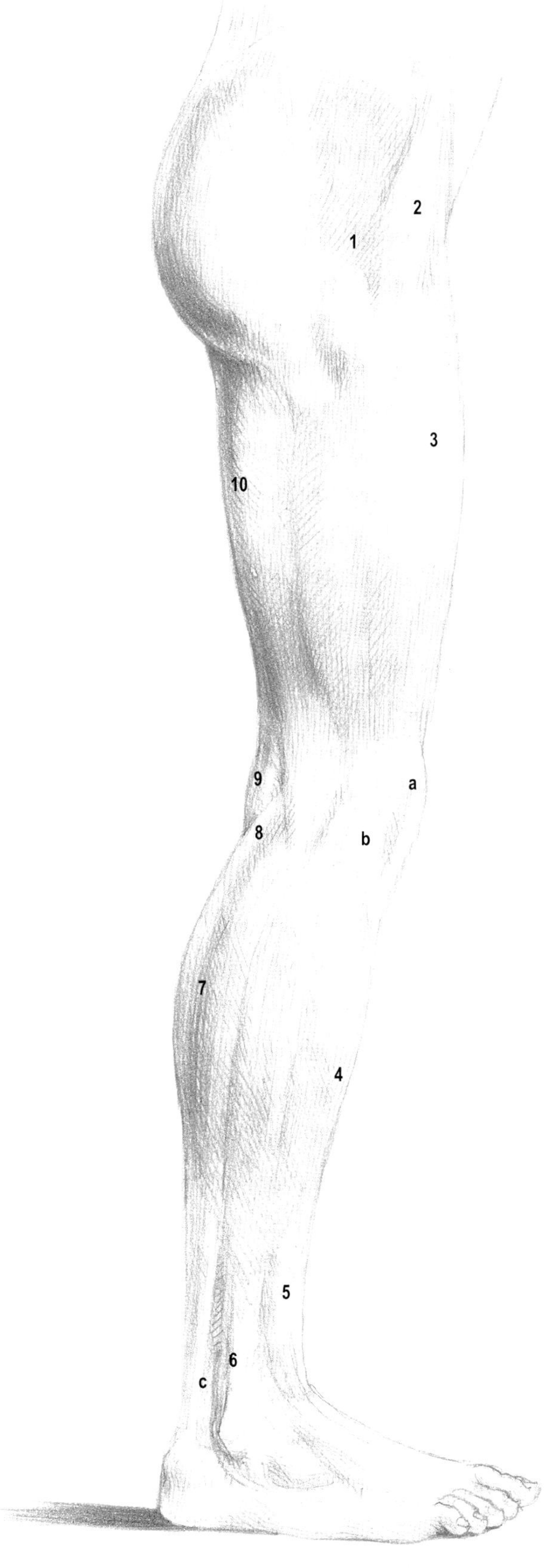
2
1
3
10
9
a
8
b
7
4
5
6
c

Fig. 128
The Greater Psoas Muscle
(M. psoas major, *79/2*)

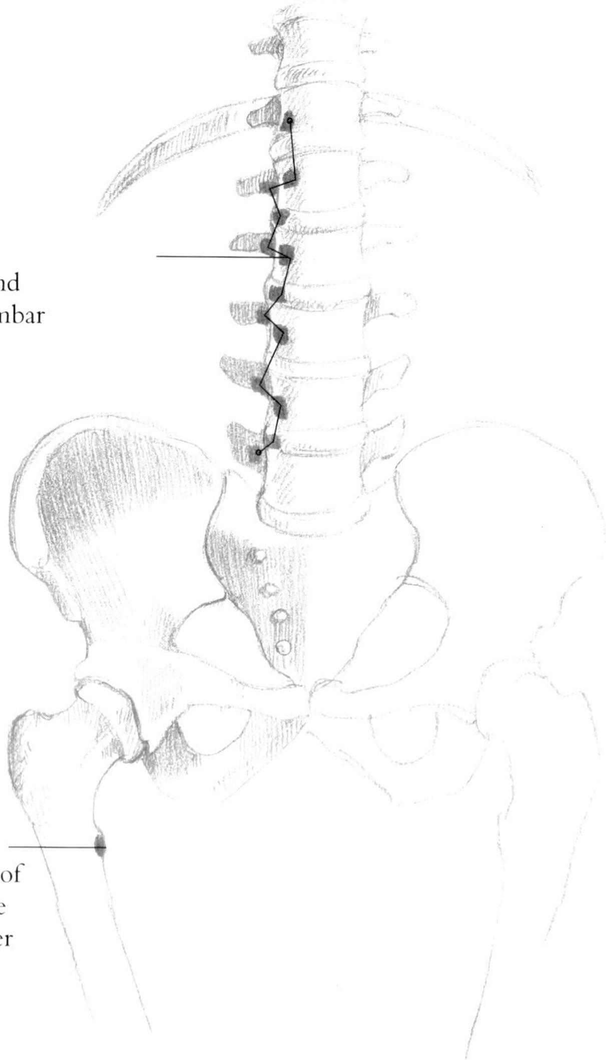

ORIGIN:
Last thoracic and first to fifth lumbar vertebrae

INSERTION:
Medial surface of the femur at the lesser trochanter

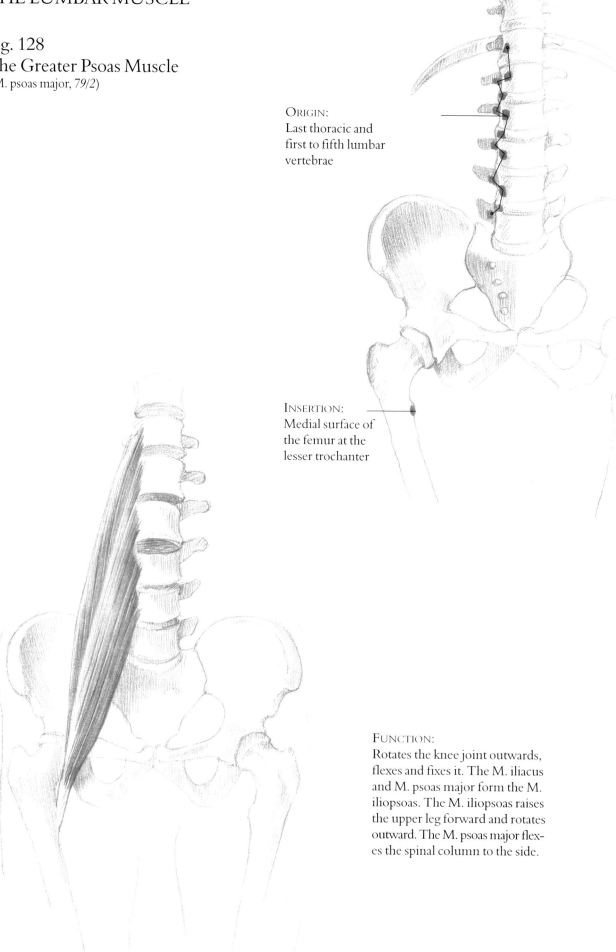

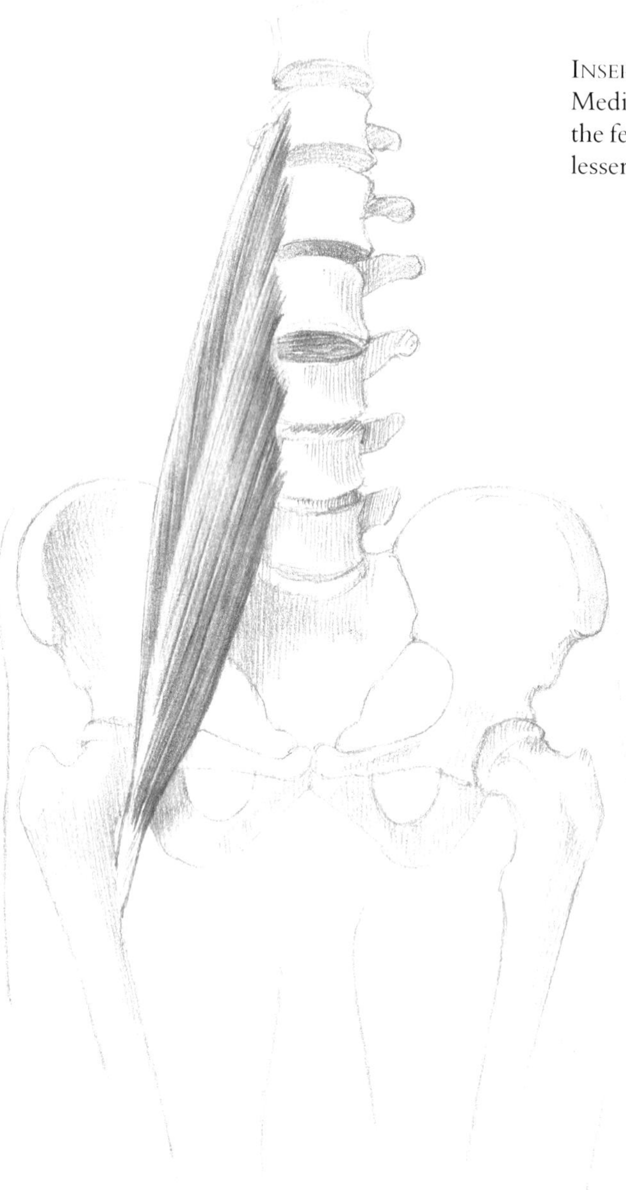

FUNCTION:
Rotates the knee joint outwards, flexes and fixes it. The M. iliacus and M. psoas major form the M. iliopsoas. The M. iliopsoas raises the upper leg forward and rotates outward. The M. psoas major flexes the spinal column to the side.

ventral aspect

Fig. 129
The Iliac Muscle
(Iliacus, *79/1*)

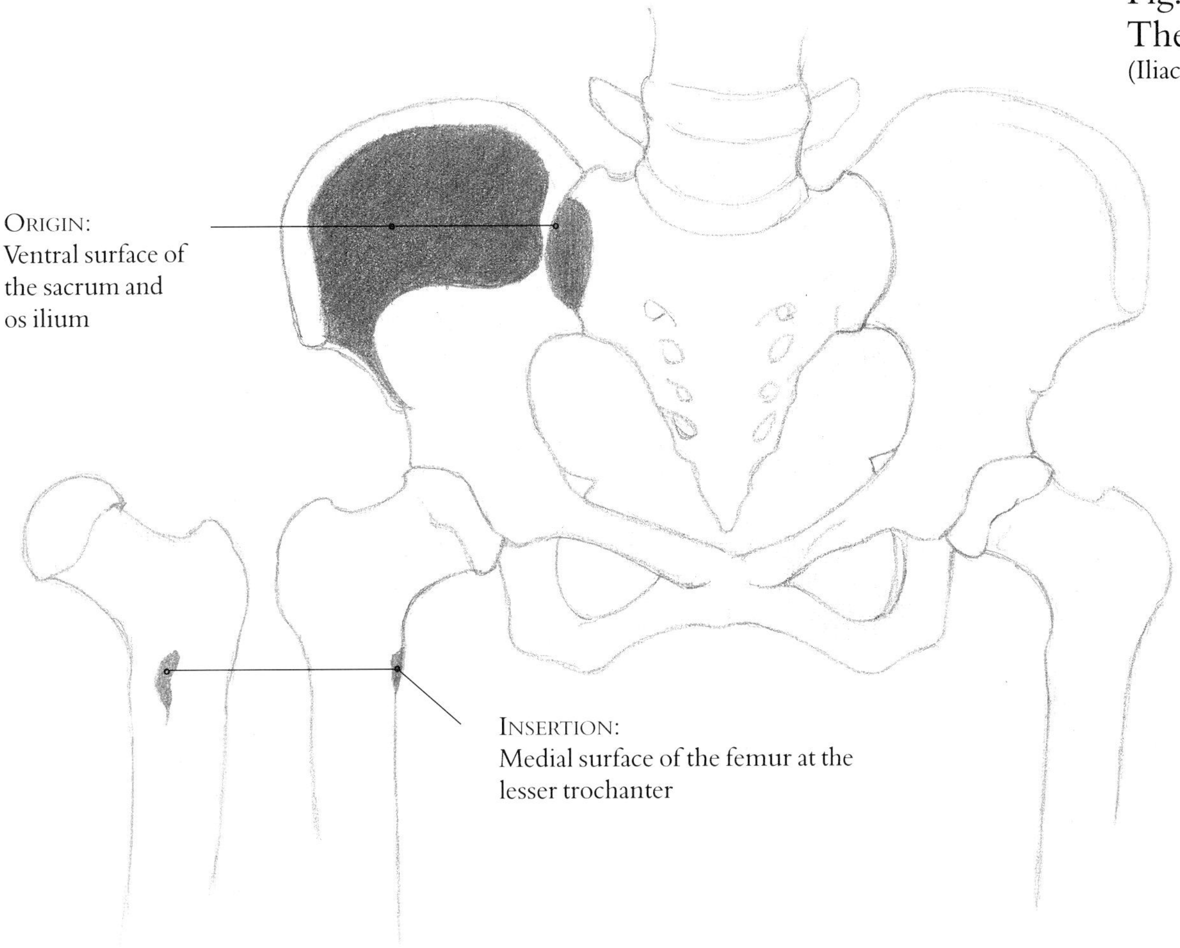
ORIGIN:
Ventral surface of the sacrum and os ilium
INSERTION:
Medial surface of the femur at the lesser trochanter

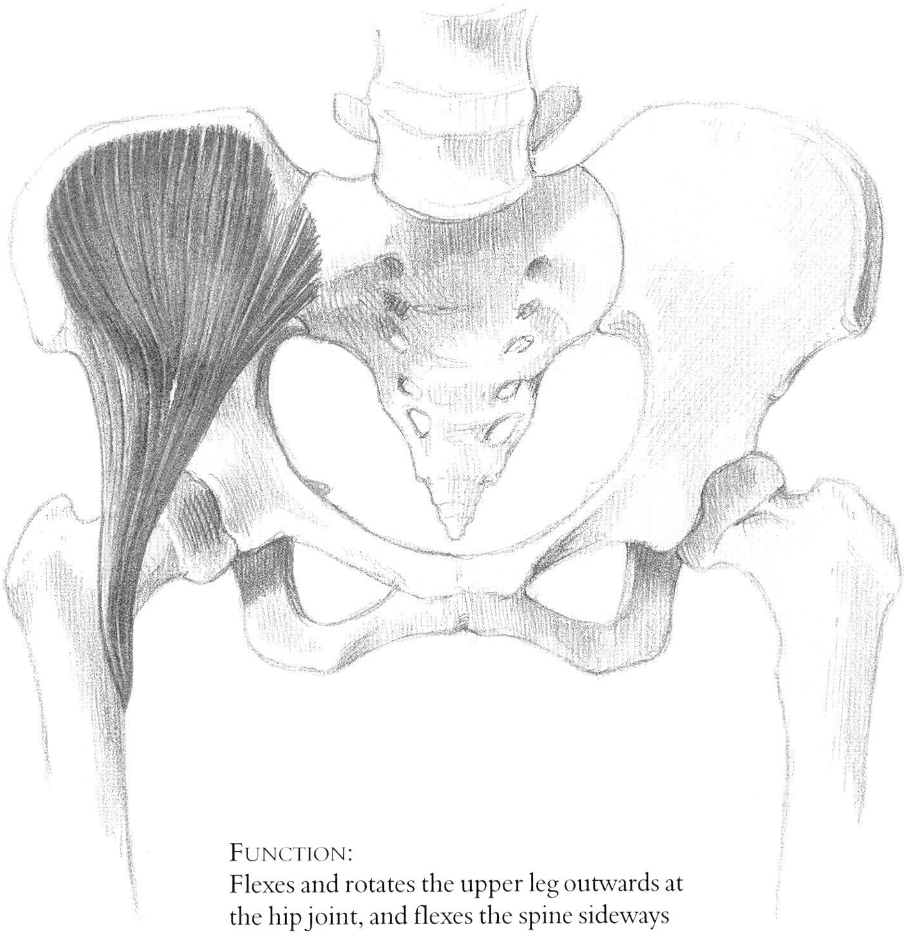
FUNCTION:
Flexes and rotates the upper leg outwards at the hip joint, and flexes the spine sideways

ventral aspect

Fig. 130
The Quadrate Muscle of the Thigh
(M. quadratus femoris, *96*)

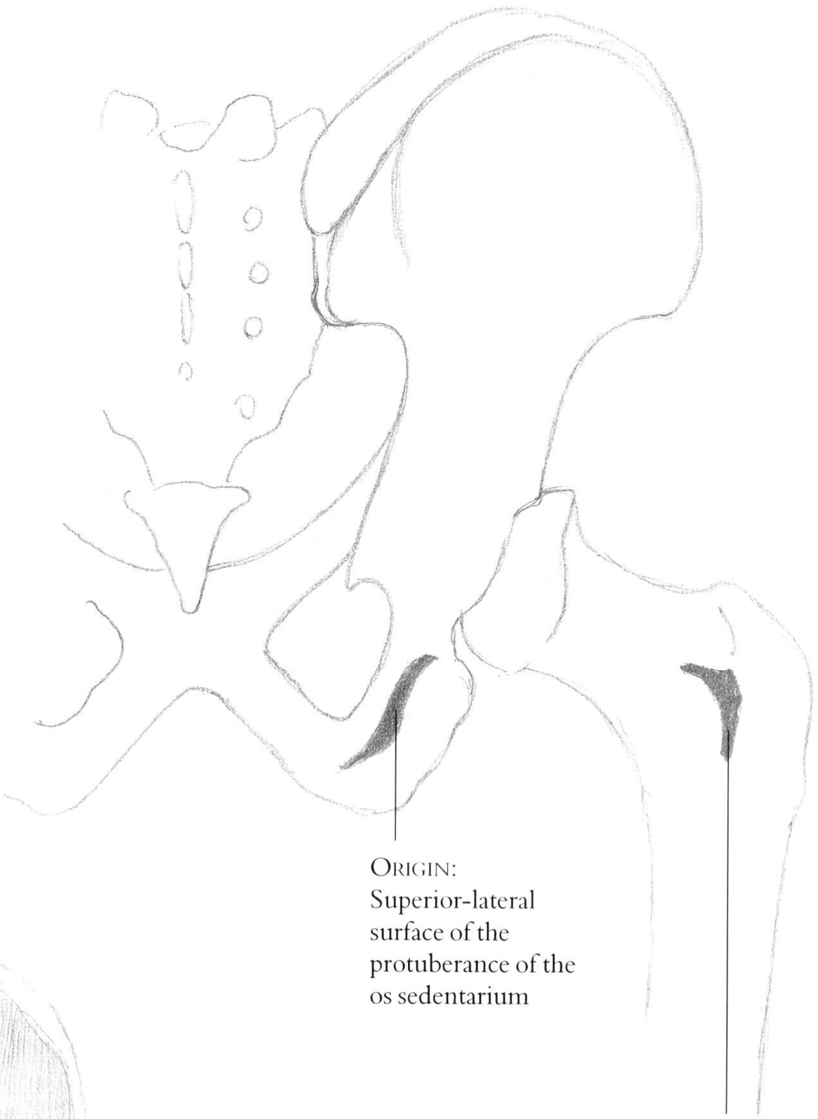

ORIGIN:
Superior-lateral surface of the protuberance of the os sedentarium

INSERTION:
Bony crest between the femoral trochanter

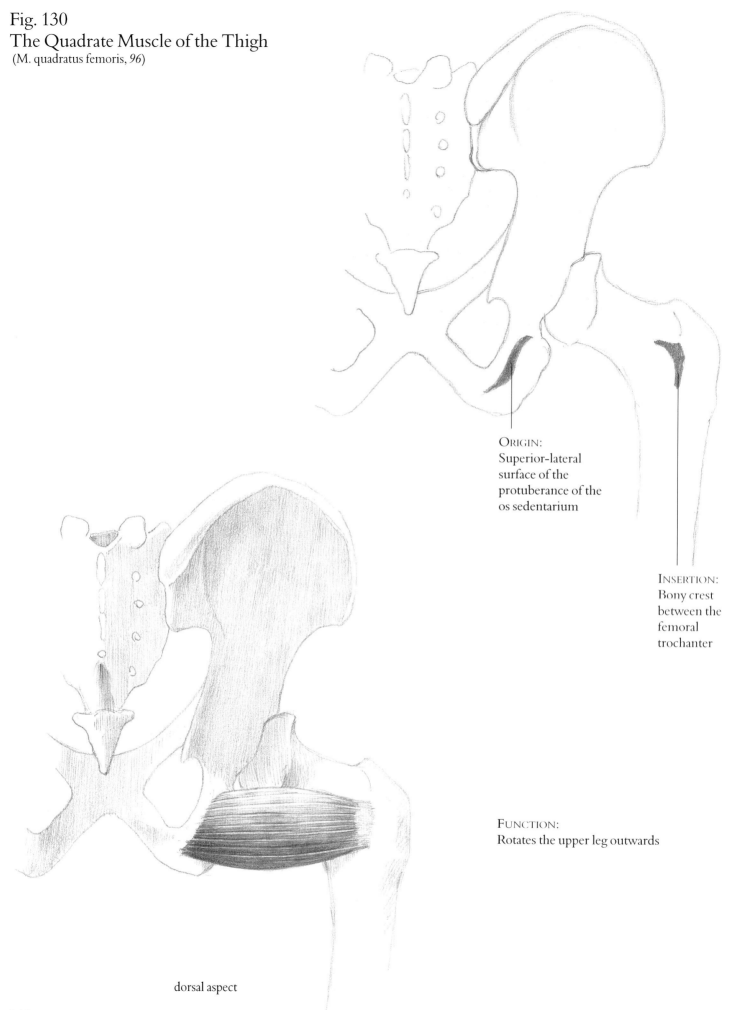

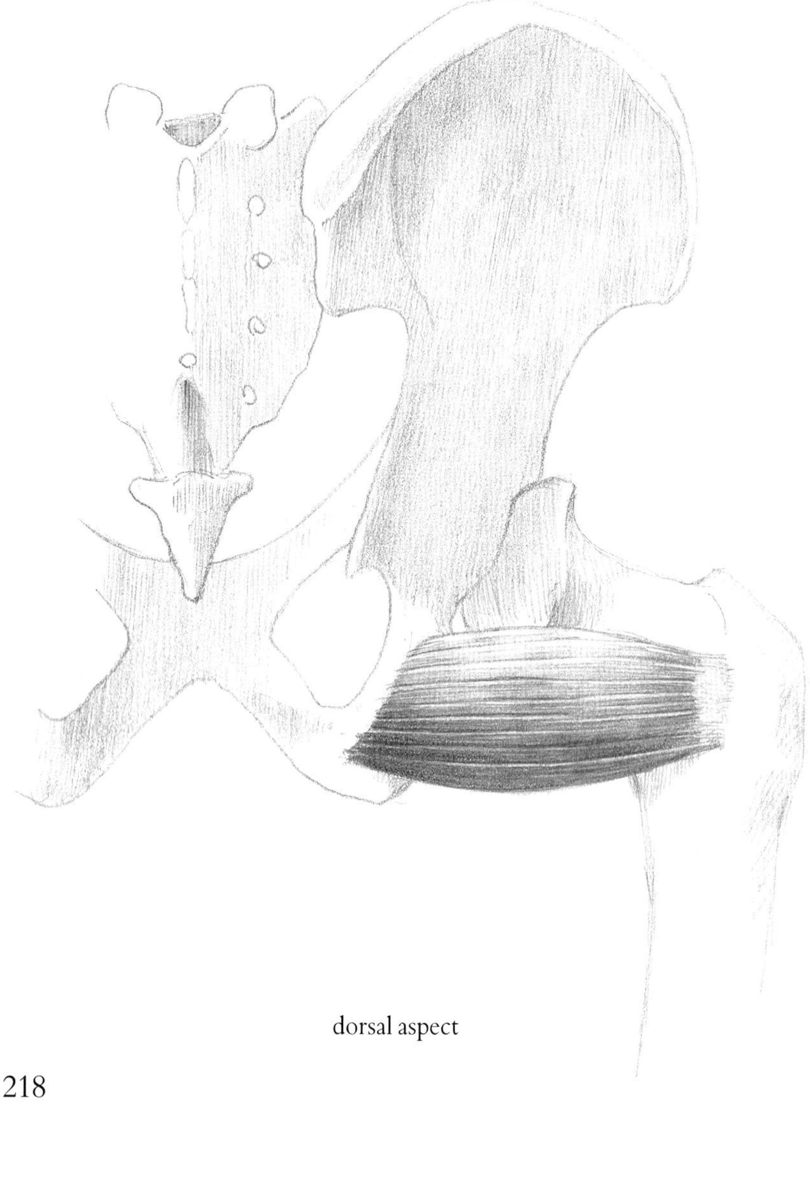

FUNCTION:
Rotates the upper leg outwards

dorsal aspect

Fig. 131
The Gemelli, *95*

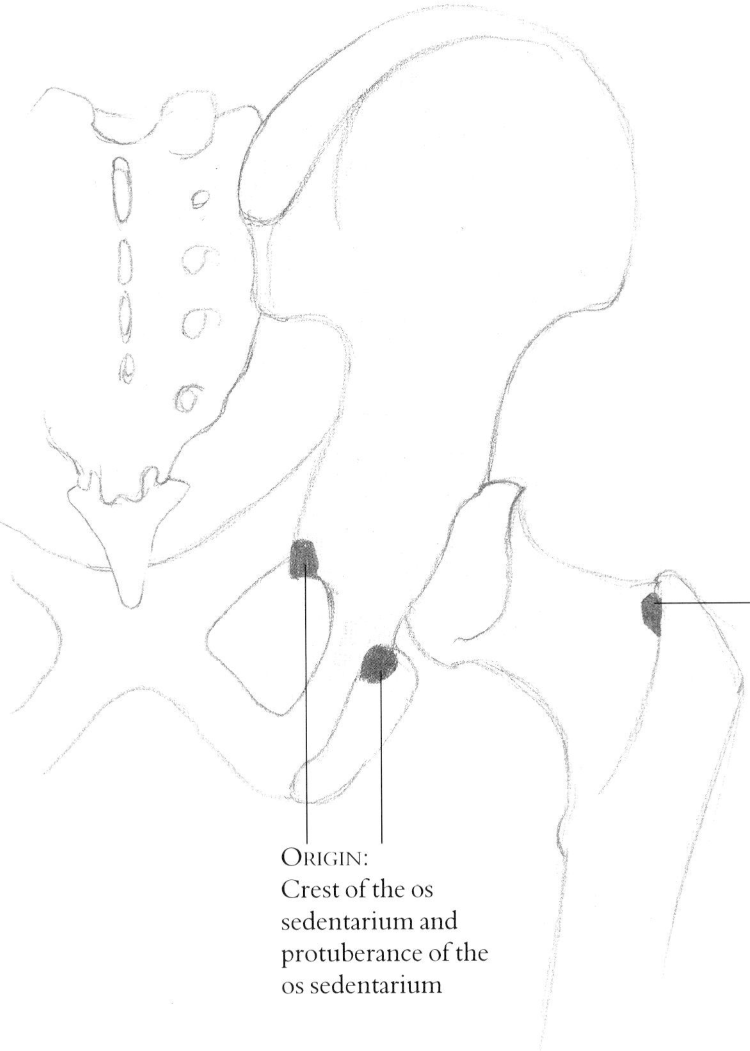

INSERTION:
Below the groove next to the femoral trochanter

ORIGIN:
Crest of the os sedentarium and protuberance of the os sedentarium

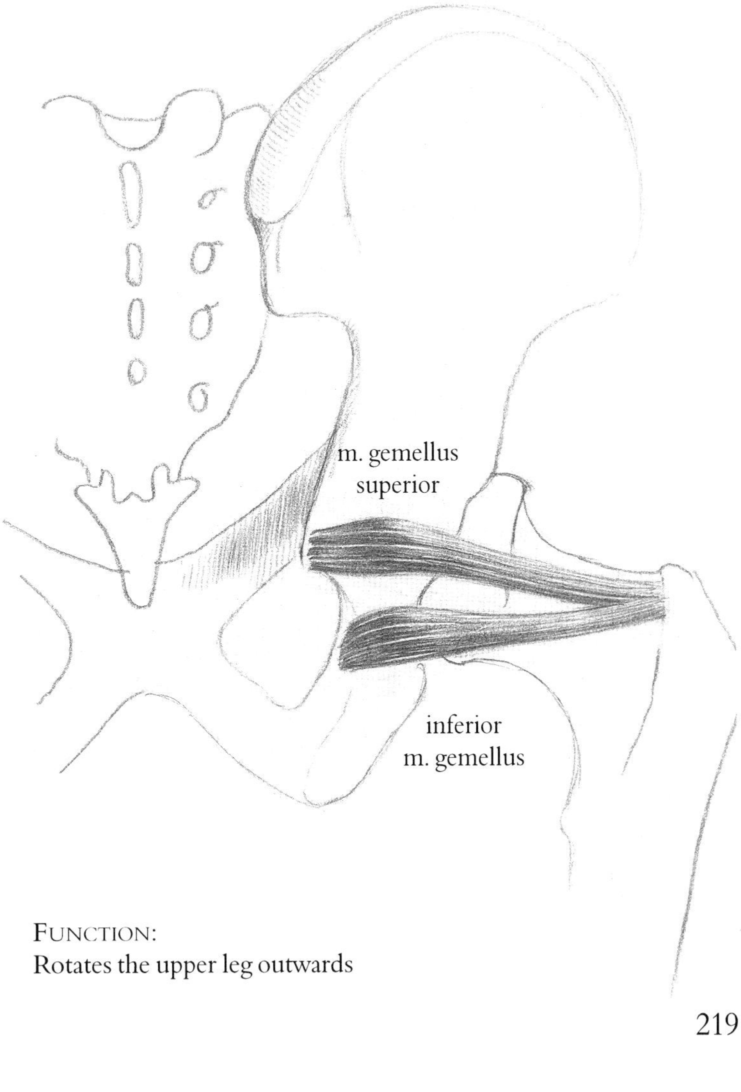

dorsal aspect

FUNCTION:
Rotates the upper leg outwards

Fig. 132
The Piriform Muscle
(M. piriformis, *91*)

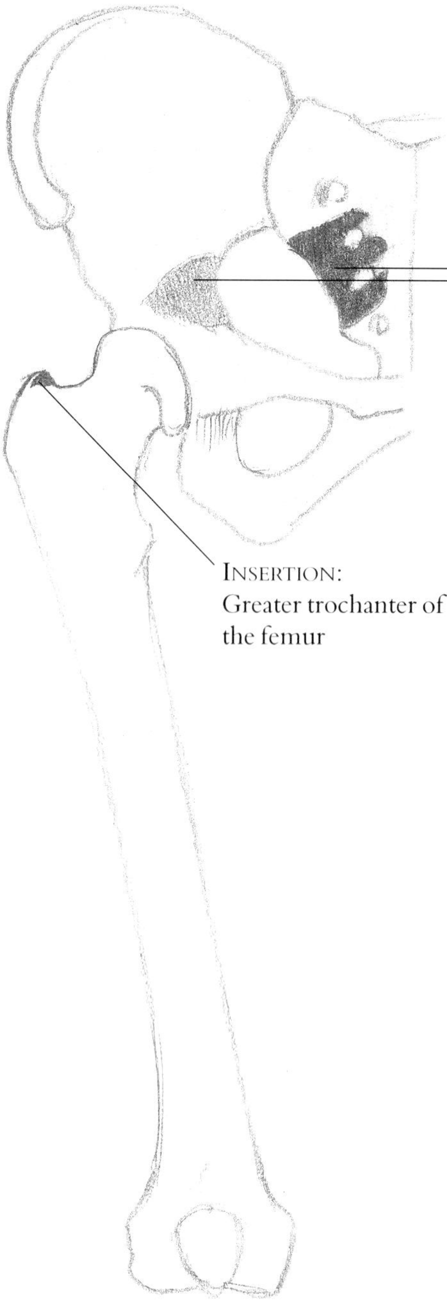

ORIGIN:
Ventral surface of
the sacrum
INSERTION:
Greater trochanter of
the femur

ventral aspect

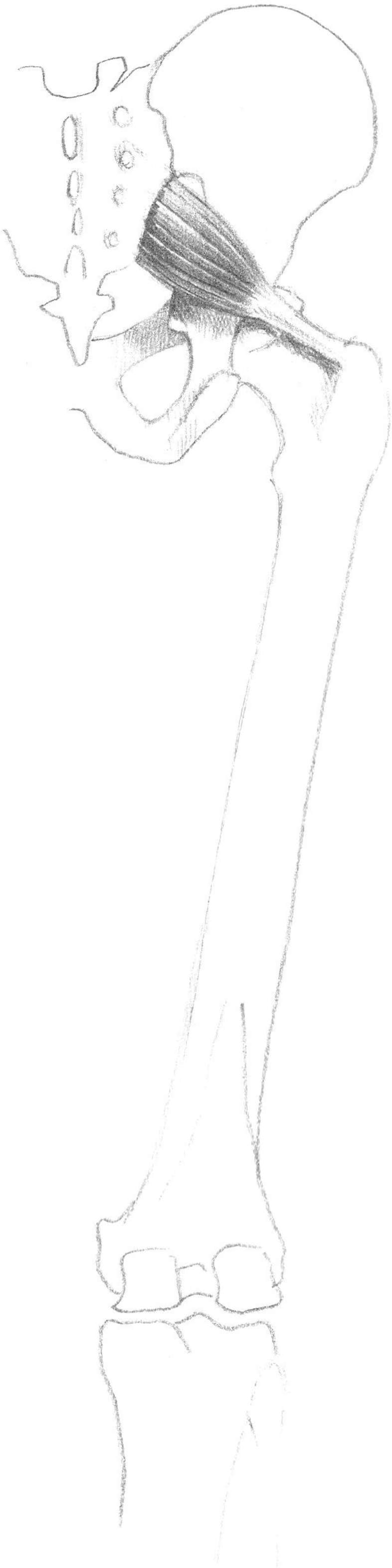

dorsal aspect

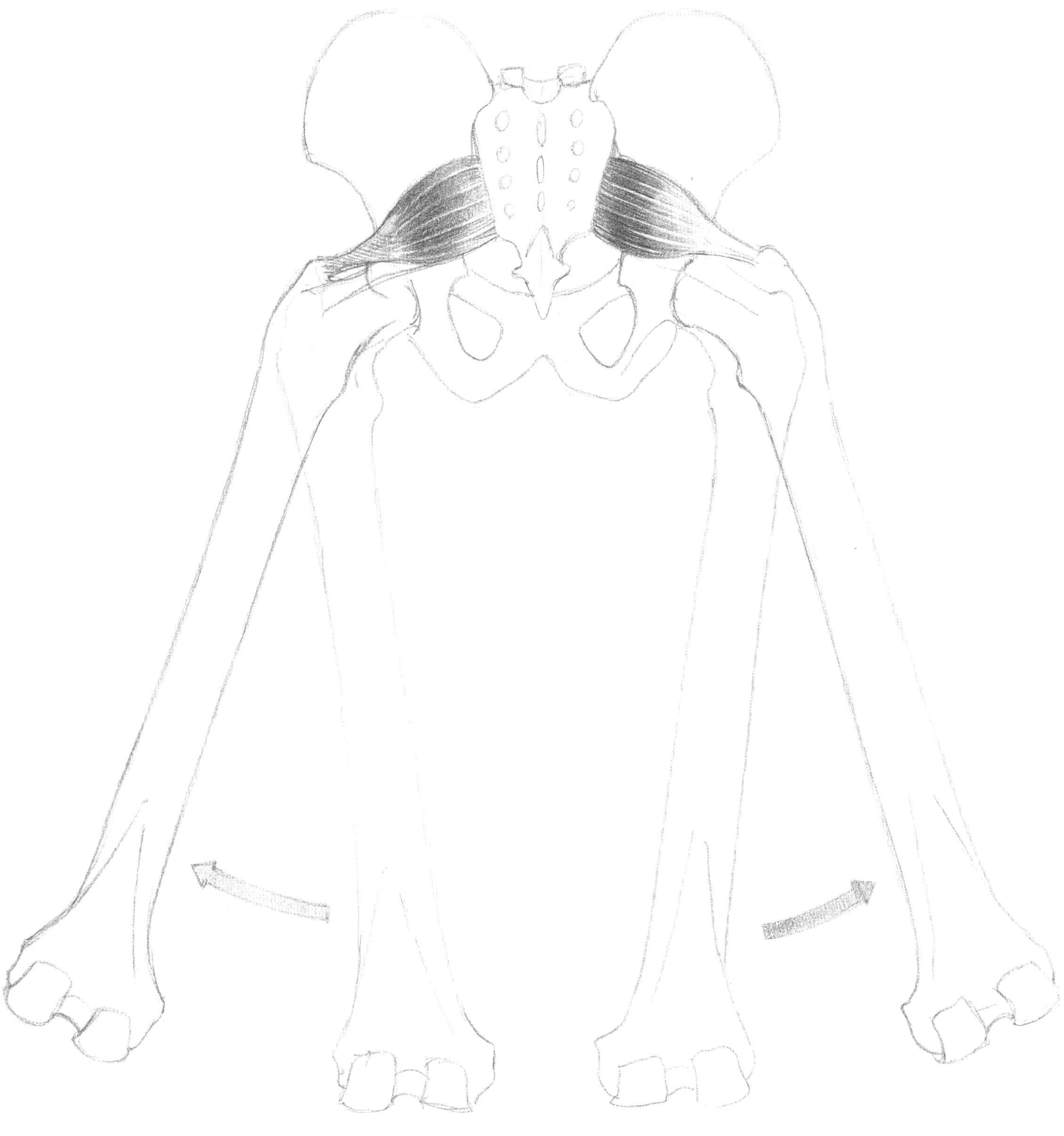

FUNCTION:
Abducts the upper leg and rotates it outwards

Fig. 133
The Gluteus Maximus, *81*

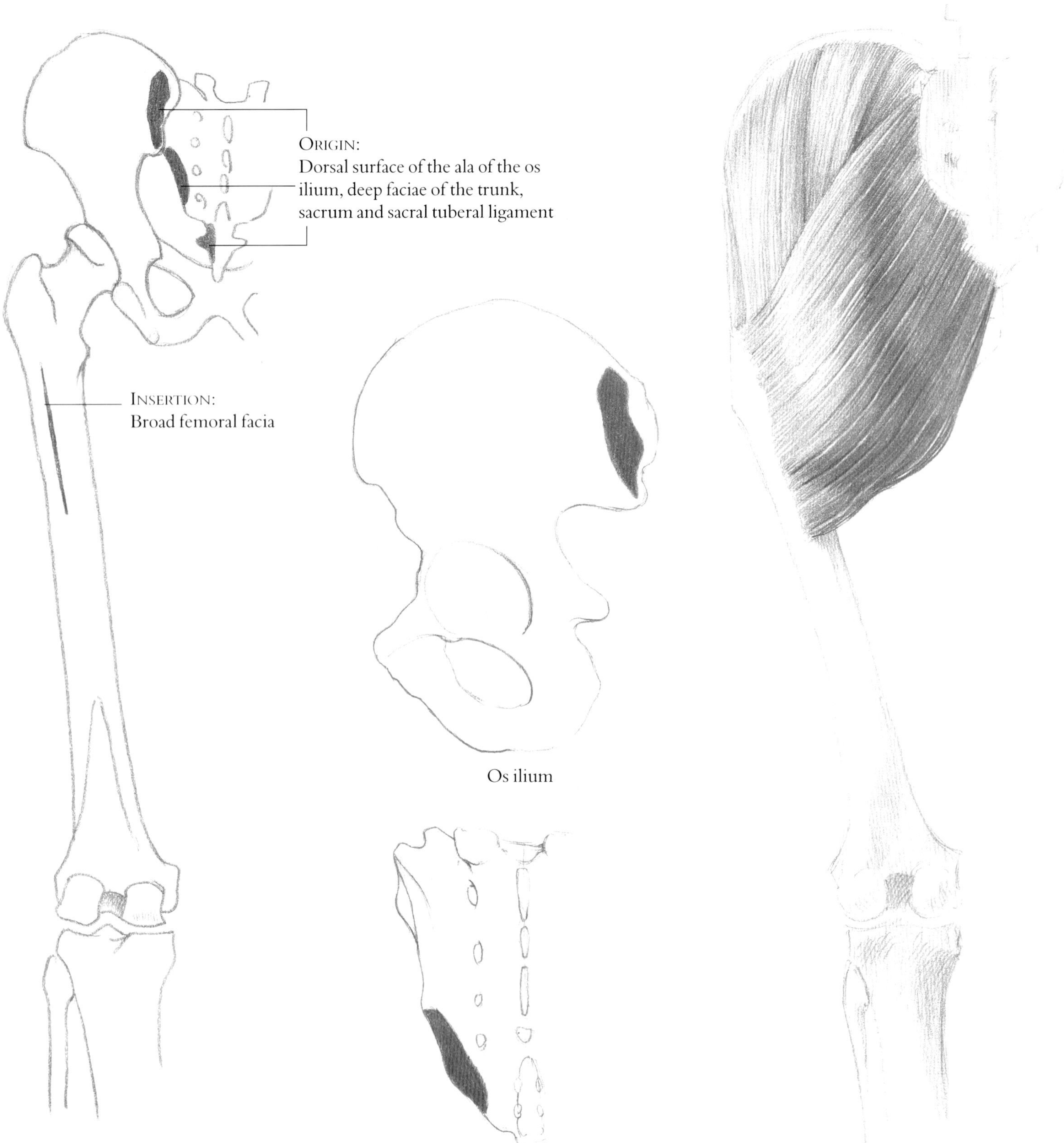
ORIGIN:
Dorsal surface of the ala of the os ilium, deep faciae of the trunk, sacrum and sacral tuberal ligament
INSERTION:
Broad femoral facia
Os ilium

dorsal aspect

sacrum

FUNCTION:
Extends, spreads and rotates the leg from the hip joint and fixes the knee joint

Fig. 134
The Gluteus Medius, *82*

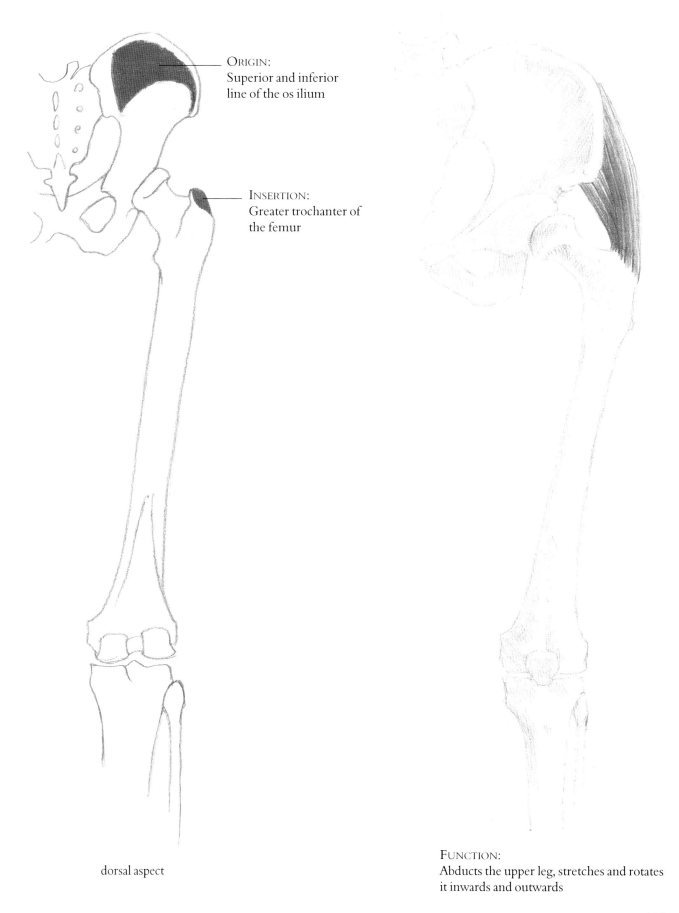

ORIGIN:
Superior and inferior line of the os ilium
INSERTION:
Greater trochanter of the femur

dorsal aspect

FUNCTION:
Abducts the upper leg, stretches and rotates it inwards and outwards

Fig. 135
The Gluteus Minimus, *83*

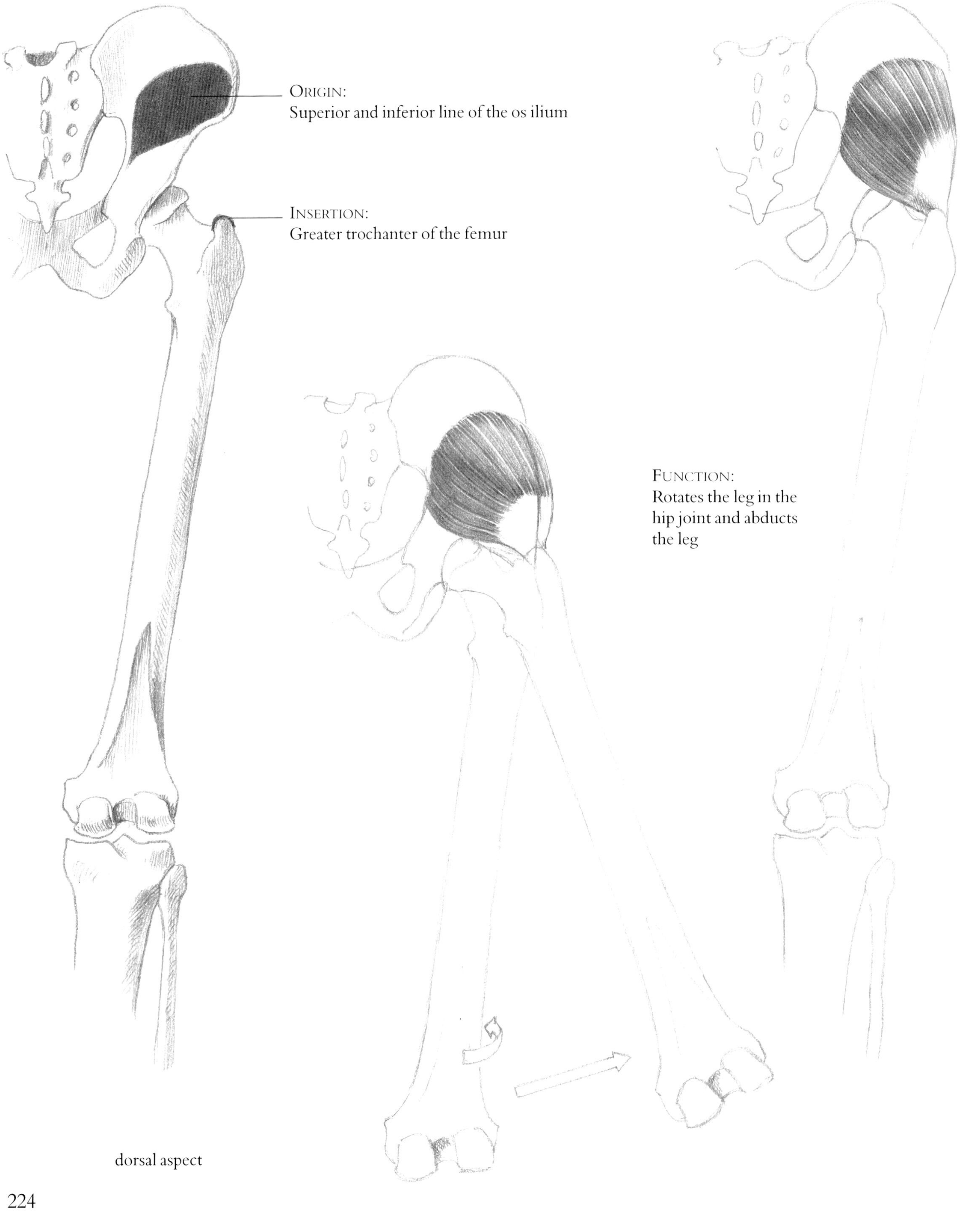
ORIGIN:
Superior and inferior line of the os ilium
INSERTION:
Greater trochanter of the femur
FUNCTION:
Rotates the leg in the hip joint and abducts the leg
dorsal aspect

Fig. 136
The Obturator Interius Muscle
(M. obturatorius internus, *93*)

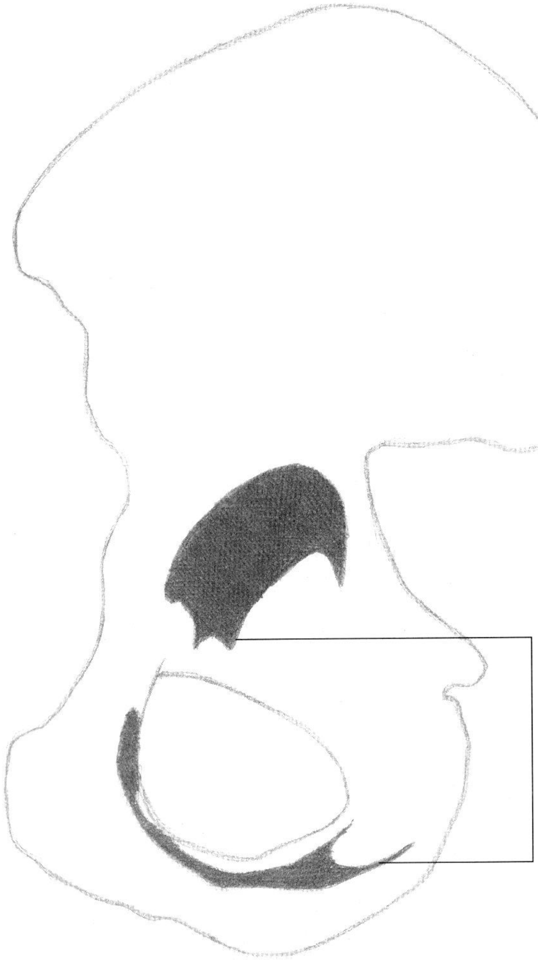

ORIGIN:
Ventral edge and
ligament of the
hip cavity

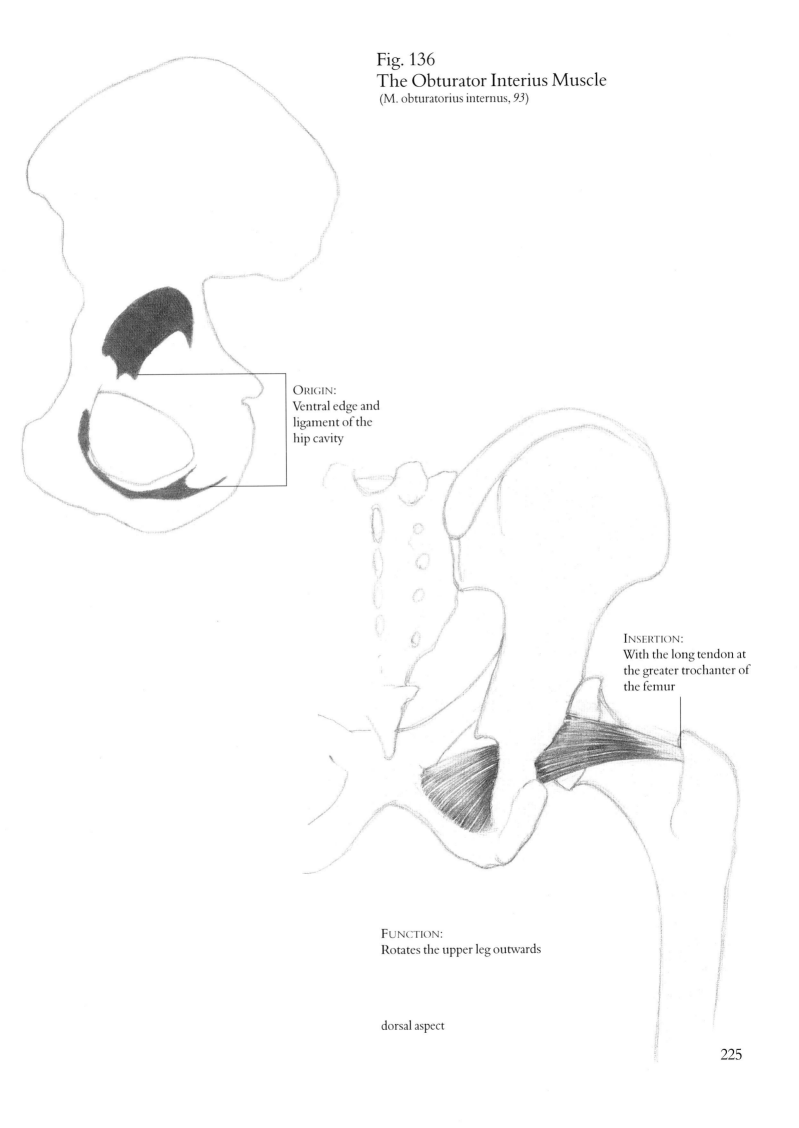

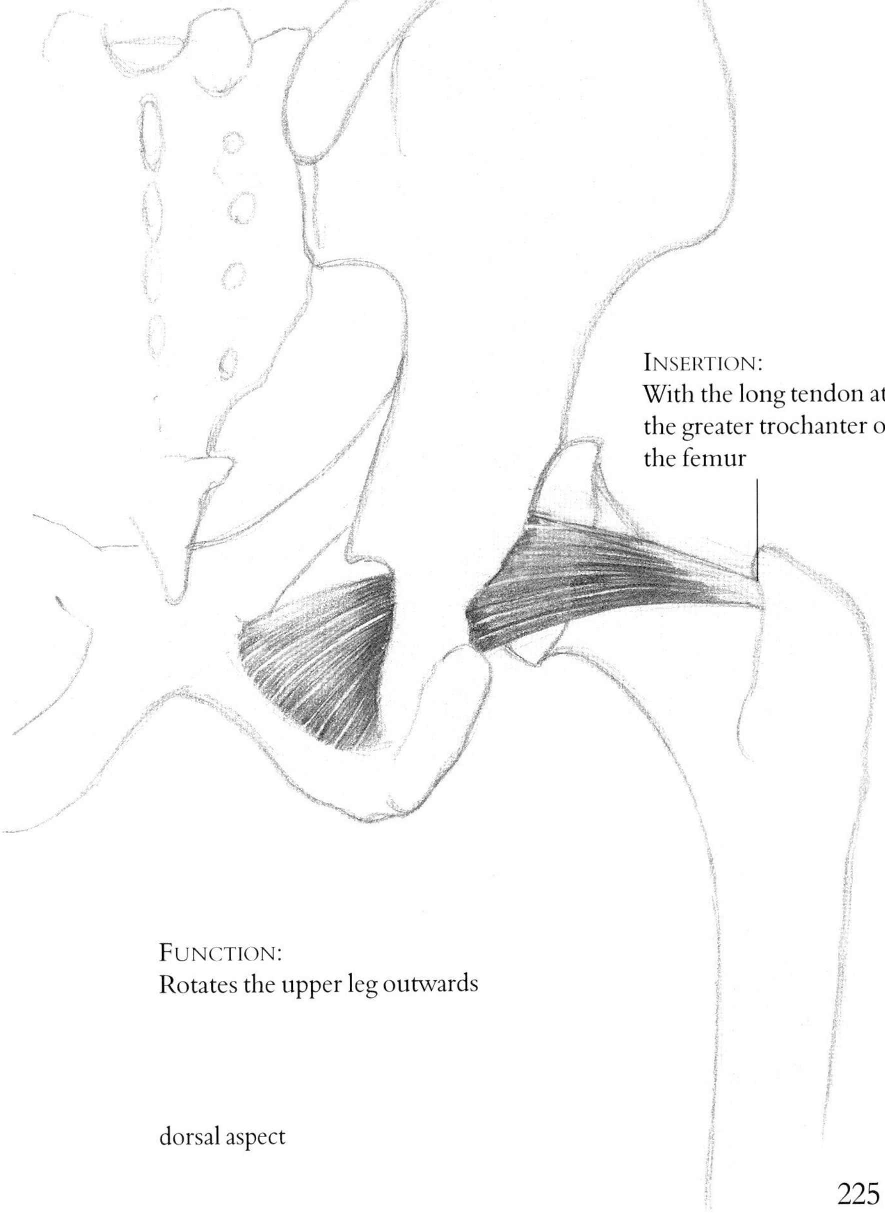

INSERTION:
With the long tendon at
the greater trochanter of
the femur

FUNCTION:
Rotates the upper leg outwards

dorsal aspect

Fig. 137
The Femur Muscle

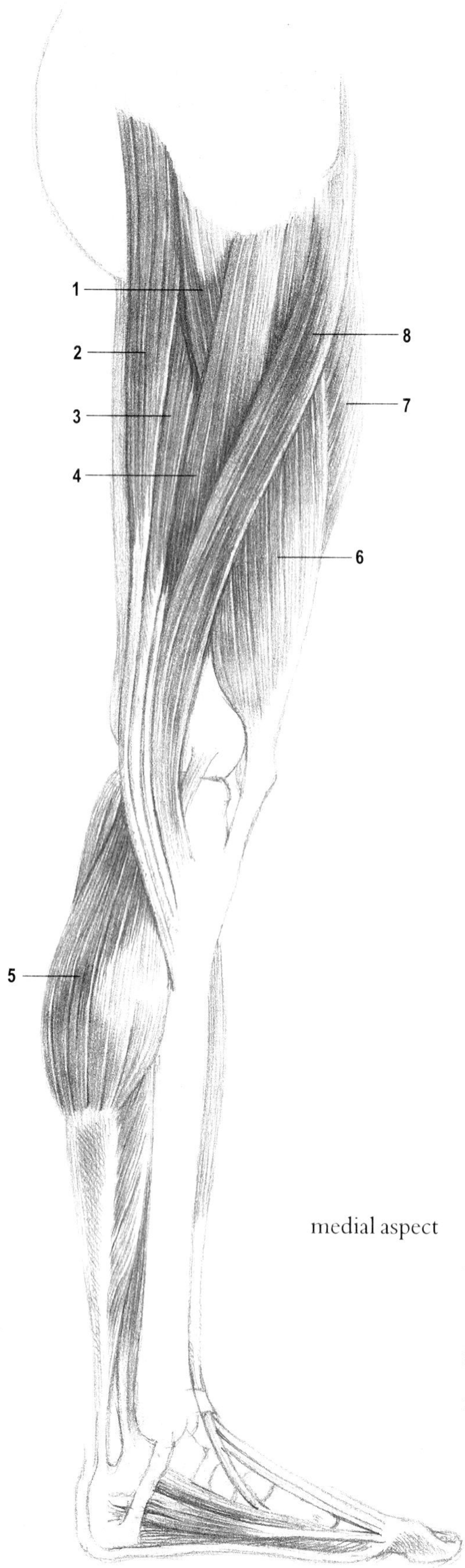

medial aspect

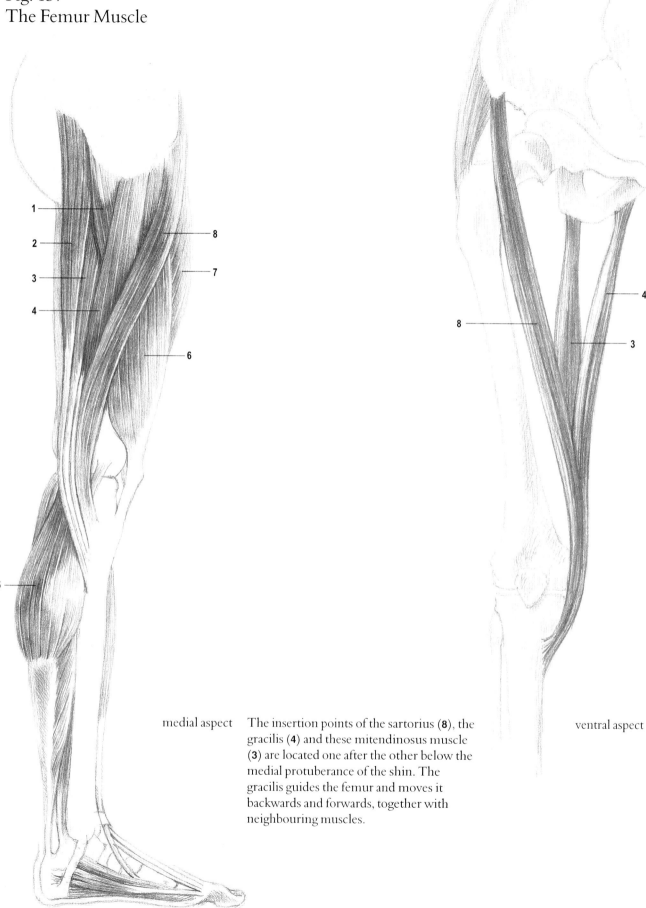

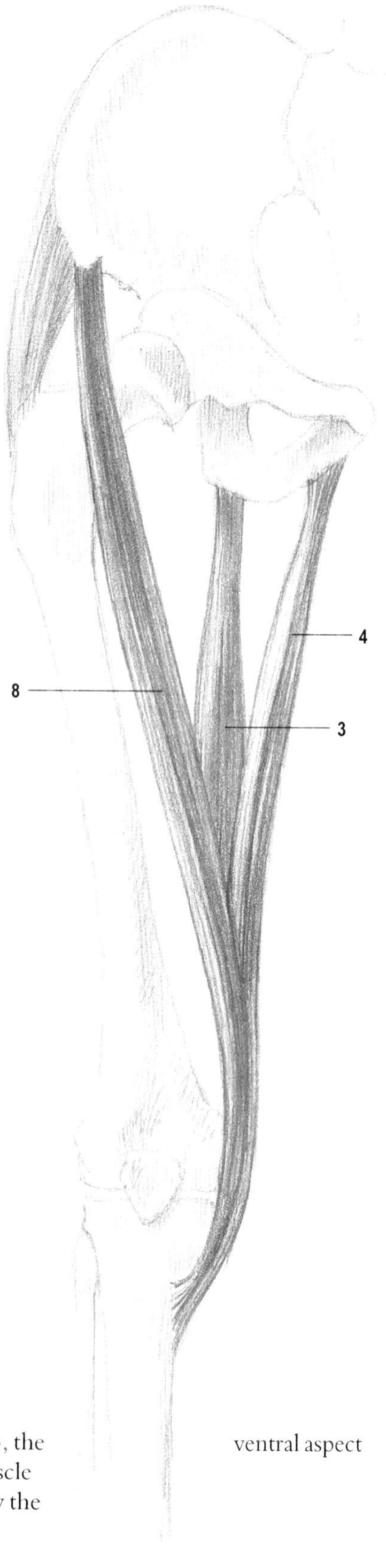

ventral aspect

The insertion points of the sartorius (**8**), the gracilis (**4**) and these mitendinosus muscle (**3**) are located one after the other below the medial protuberance of the shin. The gracilis guides the femur and moves it backwards and forwards, together with neighbouring muscles.

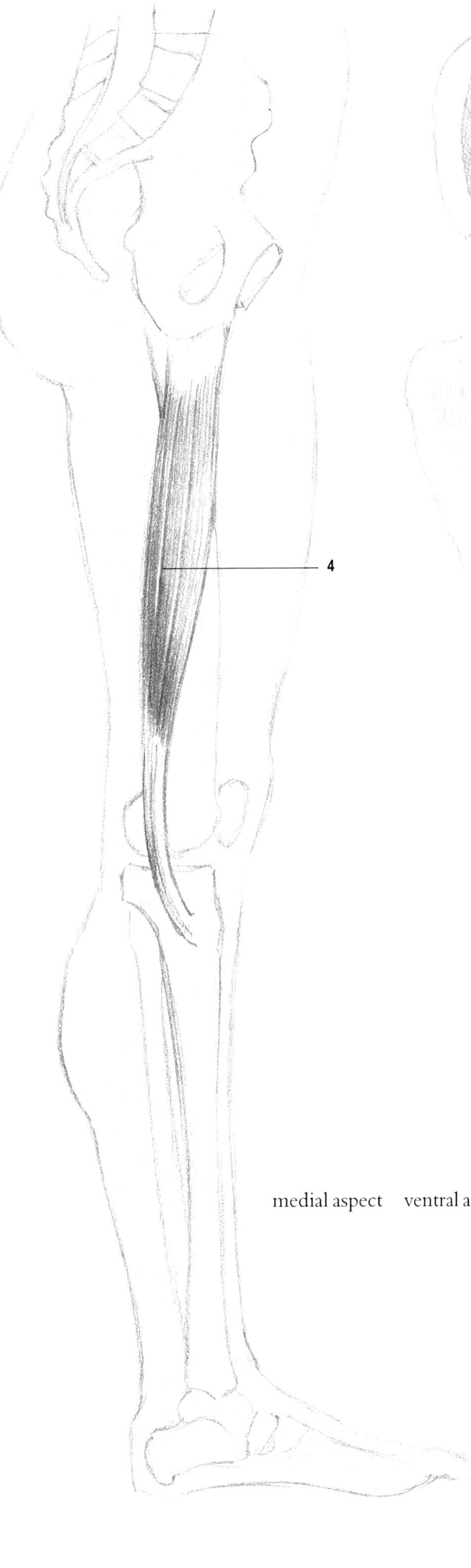

medial aspect

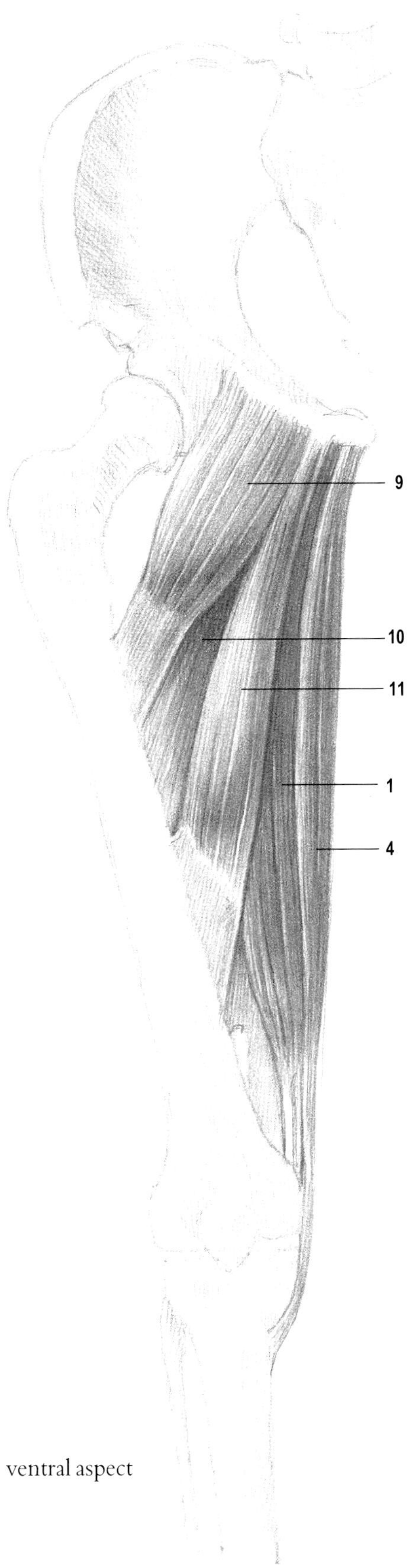

ventral aspect

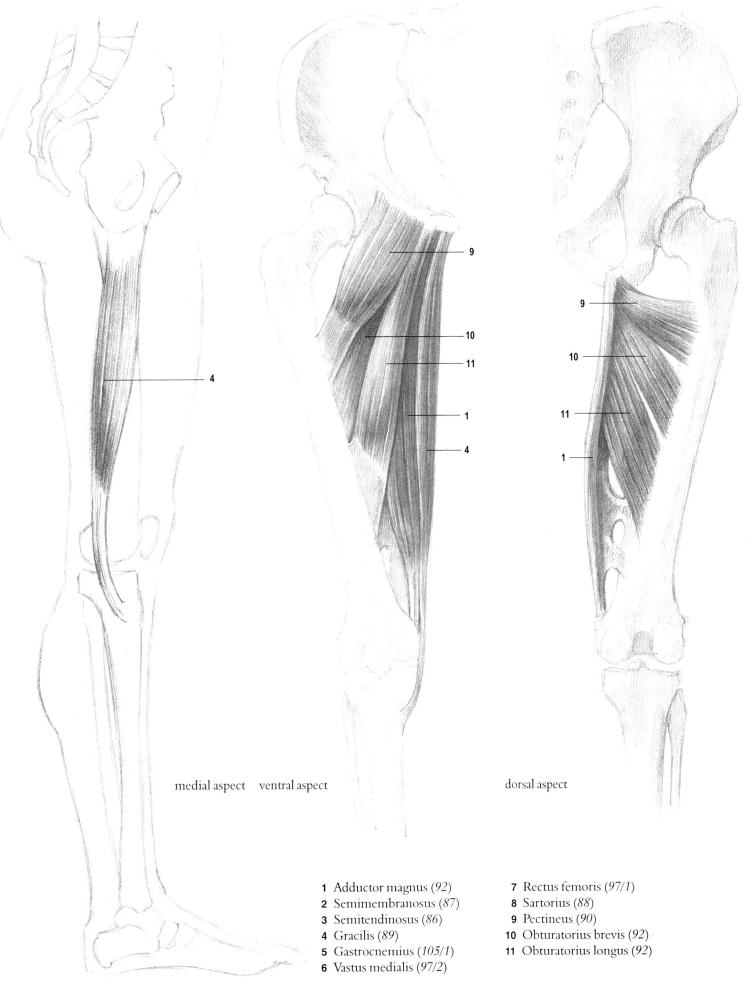

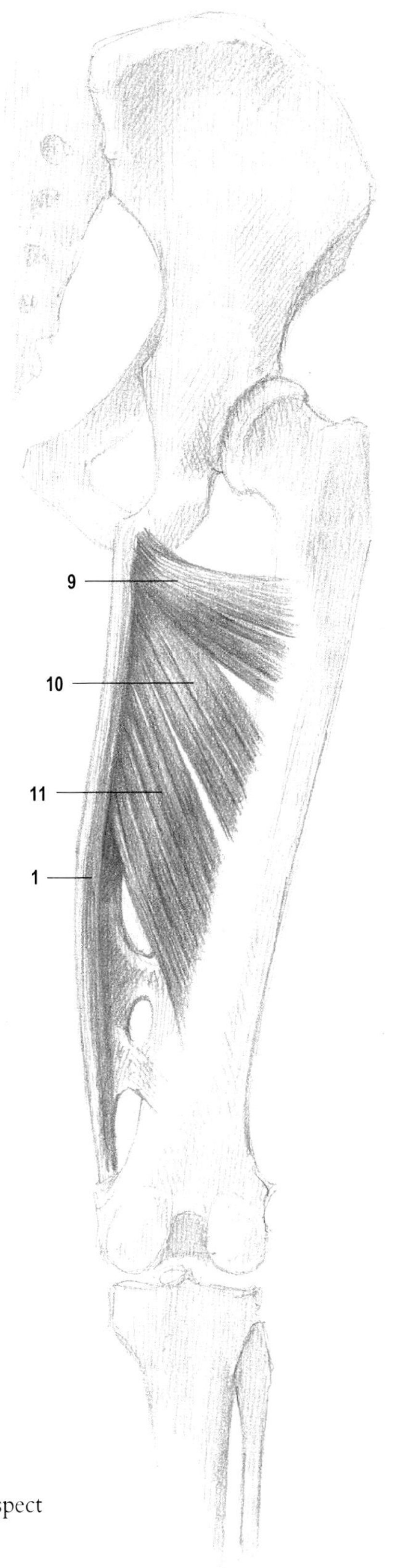

dorsal aspect

1 Adductor magnus (*92*)
2 Semimembranosus (*87*)
3 Semitendinosus (*86*)
4 Gracilis (*89*)
5 Gastrocnemius (*105/1*)
6 Vastus medialis (*97/2*)
7 Rectus femoris (*97/1*)
8 Sartorius (*88*)
9 Pectineus (*90*)
10 Obturatorius brevis (*92*)
11 Obturatorius longus (*92*)

Fig. 138
The Tensor Fasciae Latae, *84*

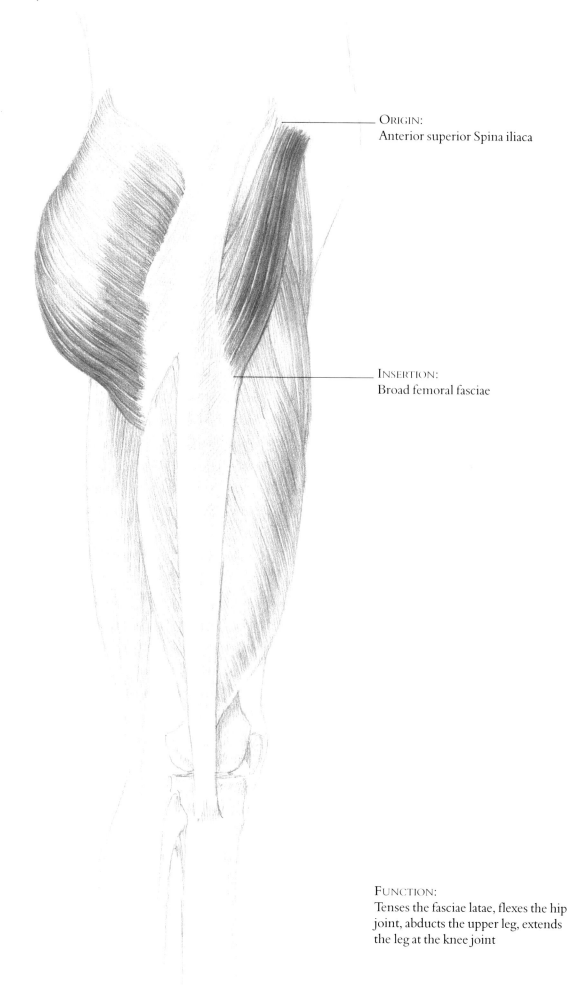

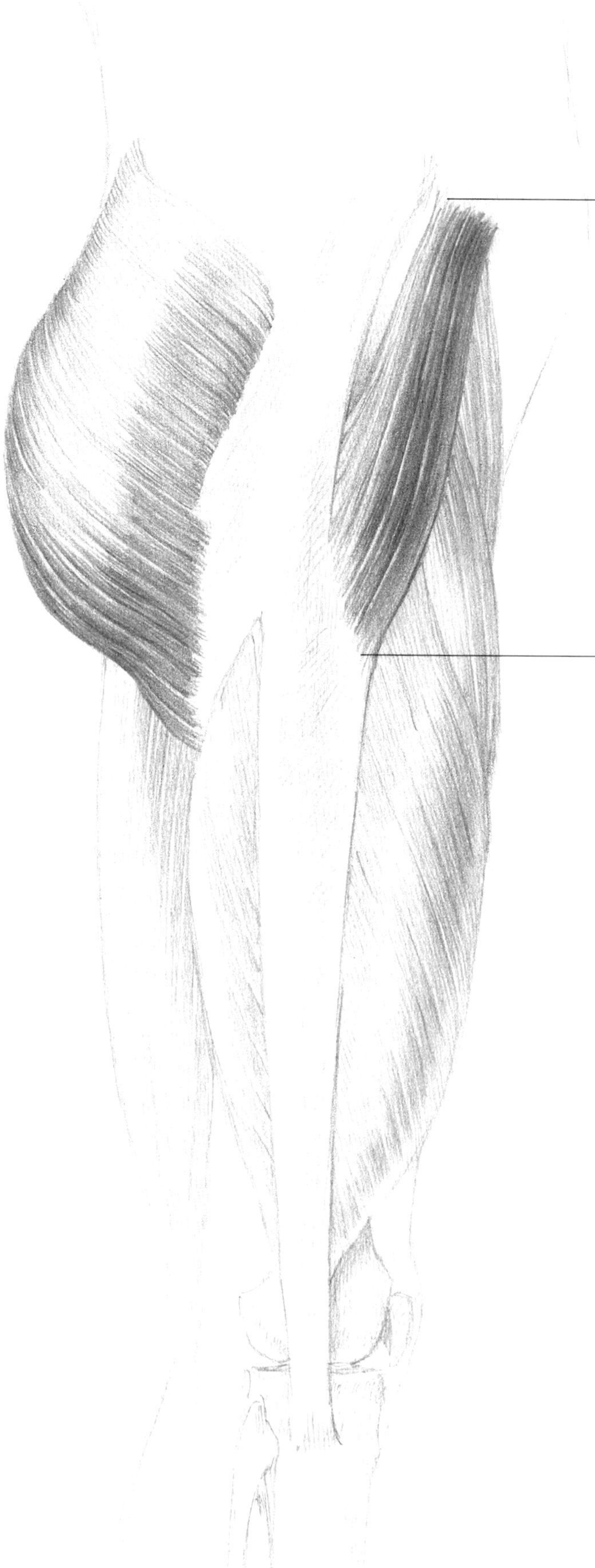

ORIGIN:
Anterior superior Spina iliaca

INSERTION:
Broad femoral fasciae

FUNCTION:
Tenses the fasciae latae, flexes the hip joint, abducts the upper leg, extends the leg at the knee joint

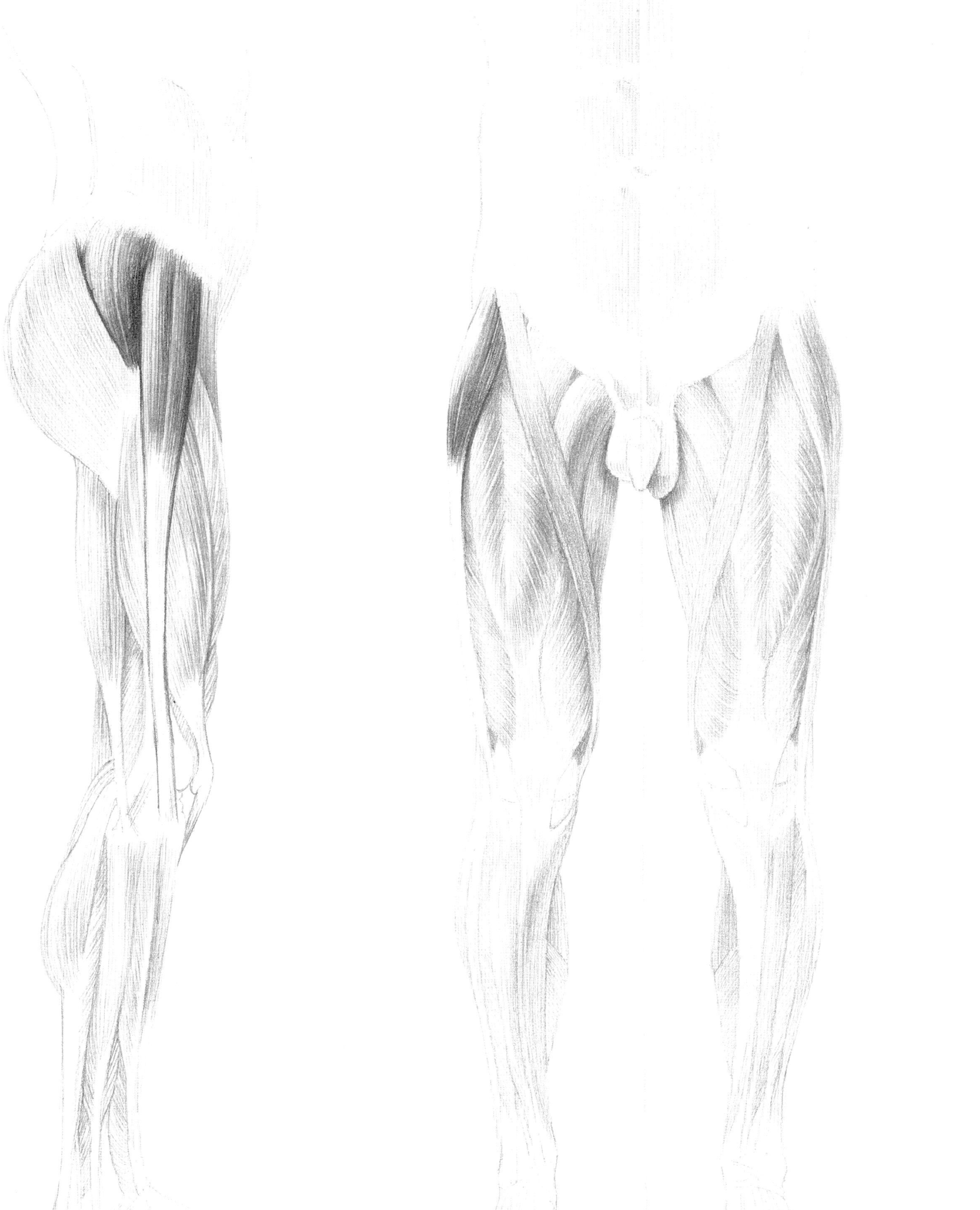

Fig. 139
The Quadriceps Femoris, *97*

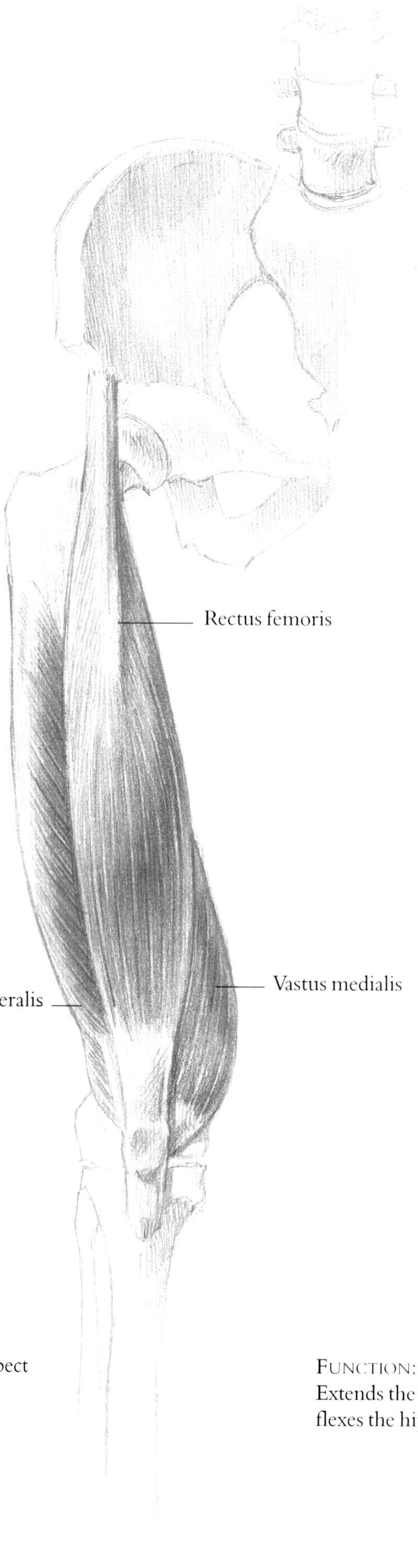

ventral aspect

FUNCTION:
Extends the lower leg at the knee joint, flexes the hip joint

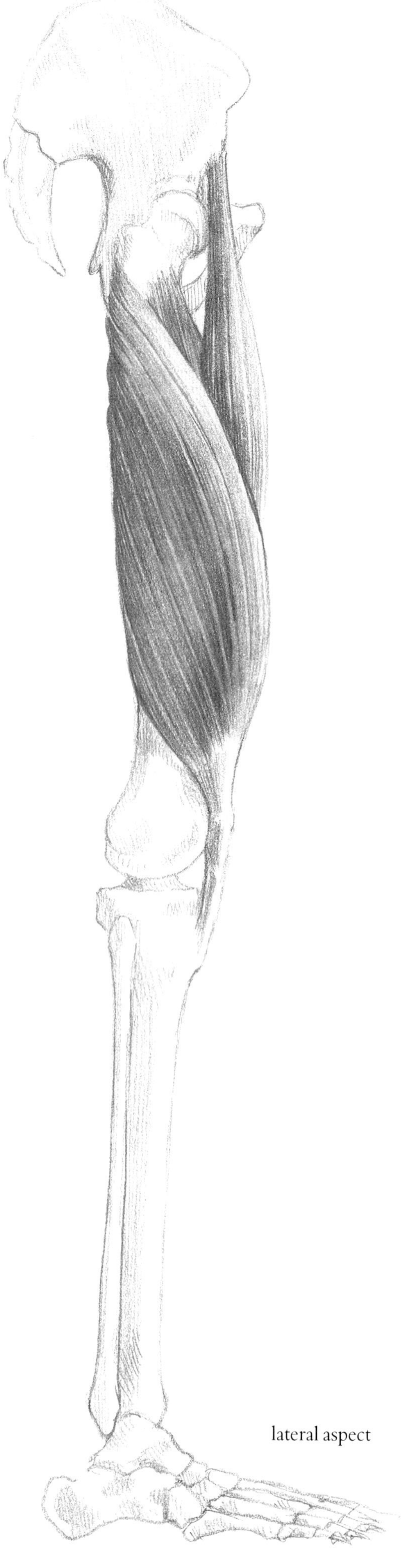

lateral aspect

Fig. 139/a
The Vastus Lateralis, *97/3*

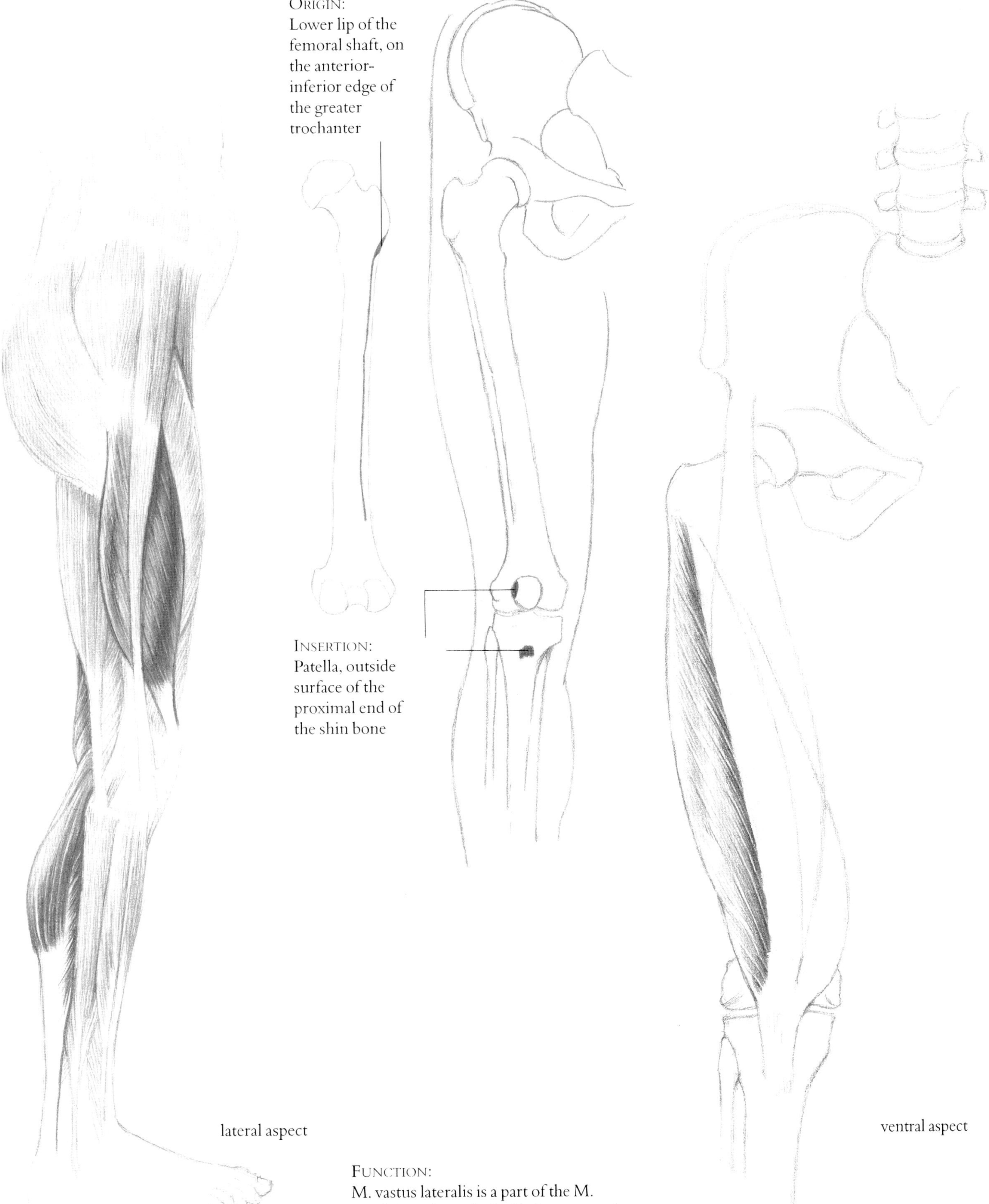

ORIGIN:
Lower lip of the femoral shaft, on the anterior-inferior edge of the greater trochanter

INSERTION:
Patella, outside surface of the proximal end of the shin bone

FUNCTION:
M. vastus lateralis is a part of the M. quedriceps femoris. It aids knee extension.

Fig. 139/b
The Vastus Medialis, *97/2*

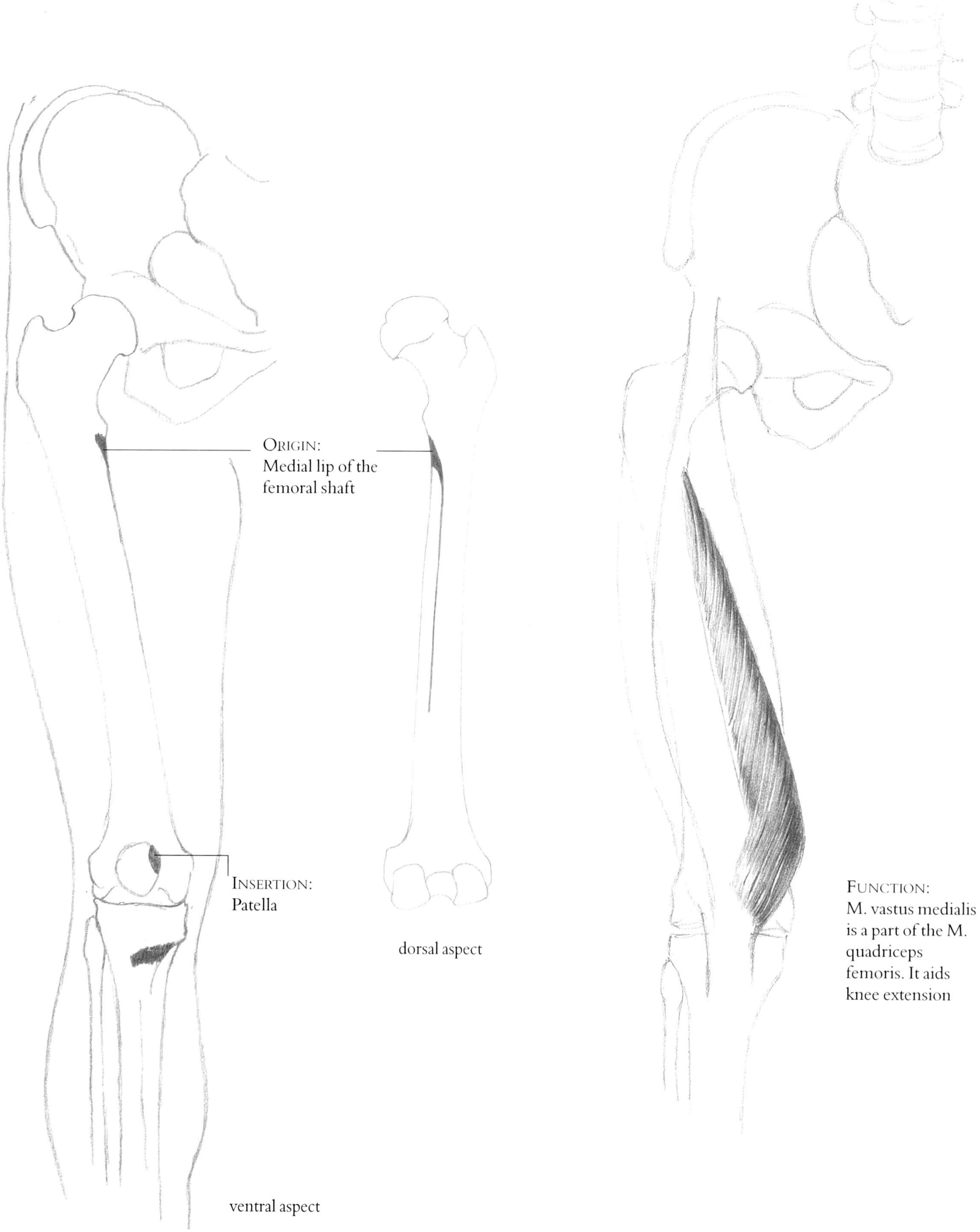
ORIGIN:
Medial lip of the
femoral shaft
INSERTION:
Patella
dorsal aspect
FUNCTION:
M. vastus medialis
is a part of the M.
quadriceps
femoris. It aids
knee extension
ventral aspect

Fig. 139/c
The Vastus Intermedius, *97/4*

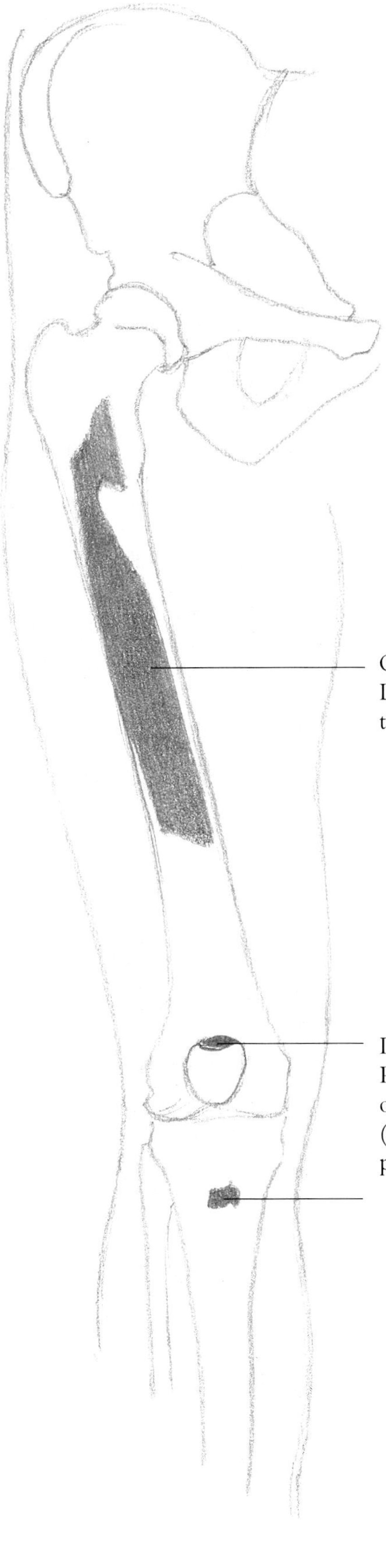

ORIGIN:
Lateral surface on lower third of femur
INSERTION:
Patella, at the crest of the shin bone (with straight patellar ligament)

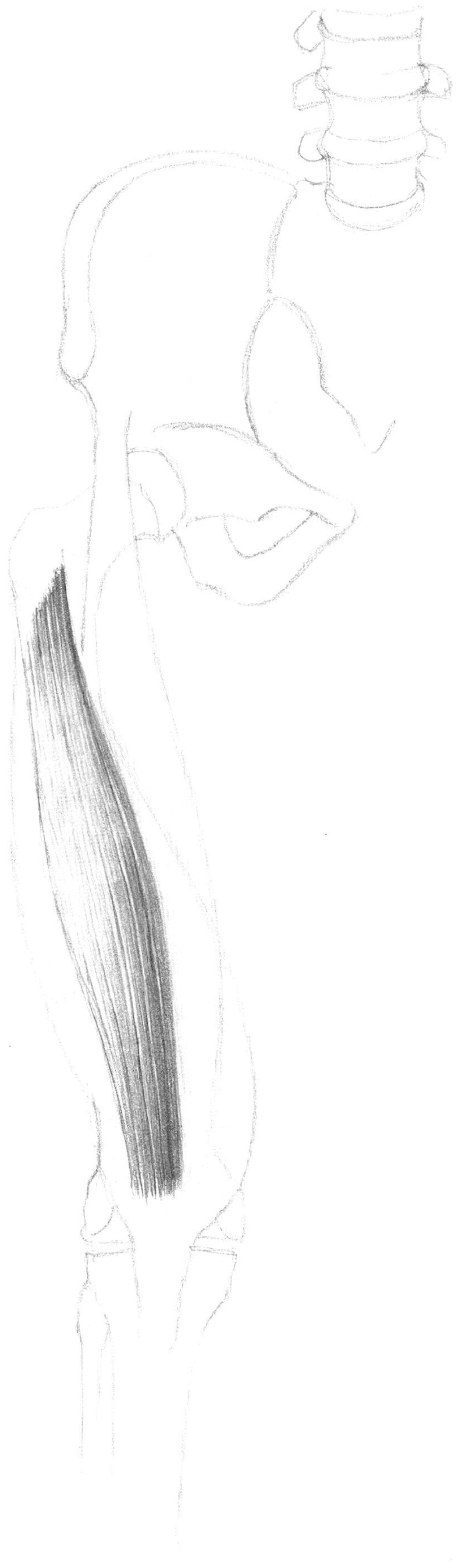

ventral aspect

Fig. 139/d
The Rectus Femoris, *97/1*

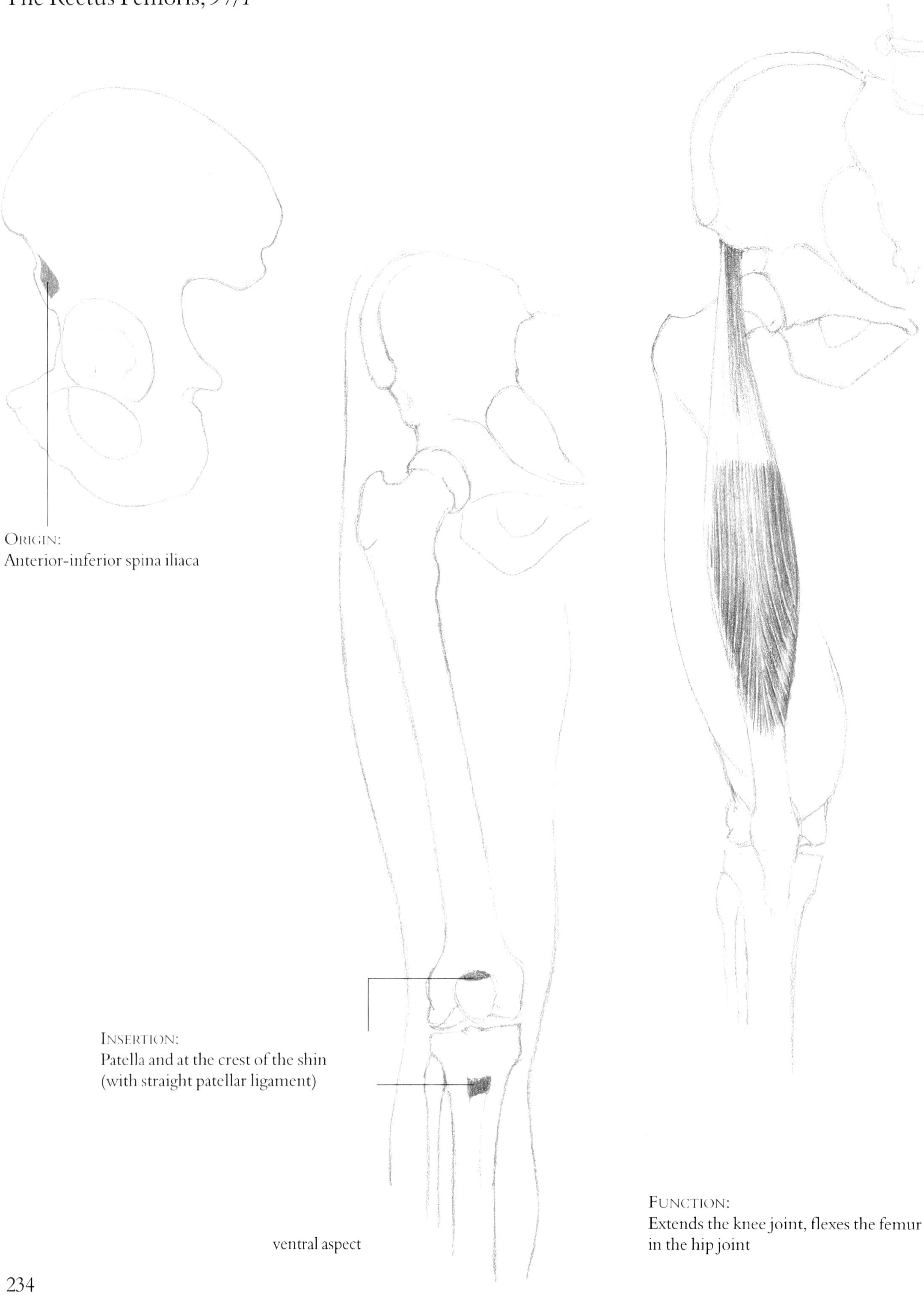

ORIGIN:
Anterior-inferior spina iliaca

INSERTION:
Patella and at the crest of the shin
(with straight patellar ligament)

ventral aspect

FUNCTION:
Extends the knee joint, flexes the femur
in the hip joint

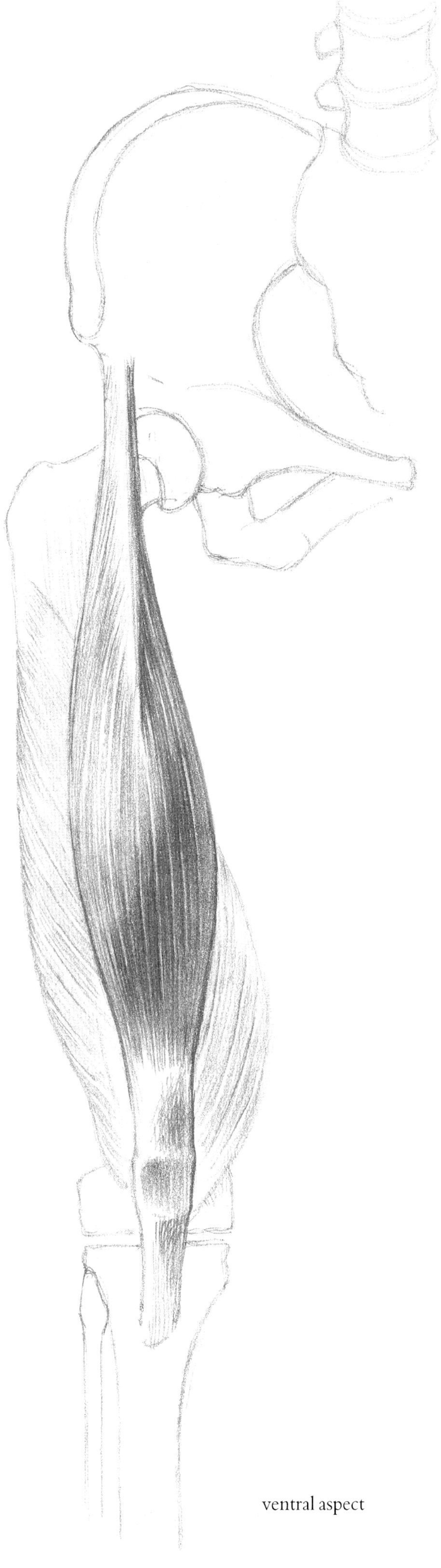

ventral aspect

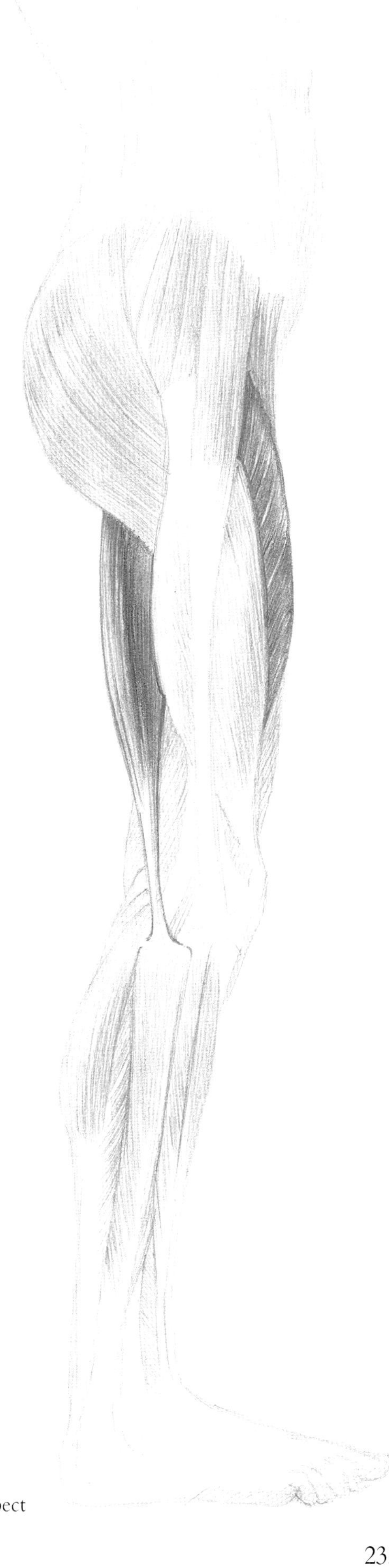

lateral aspect

Fig. 140
The Muscles of the Knee Region

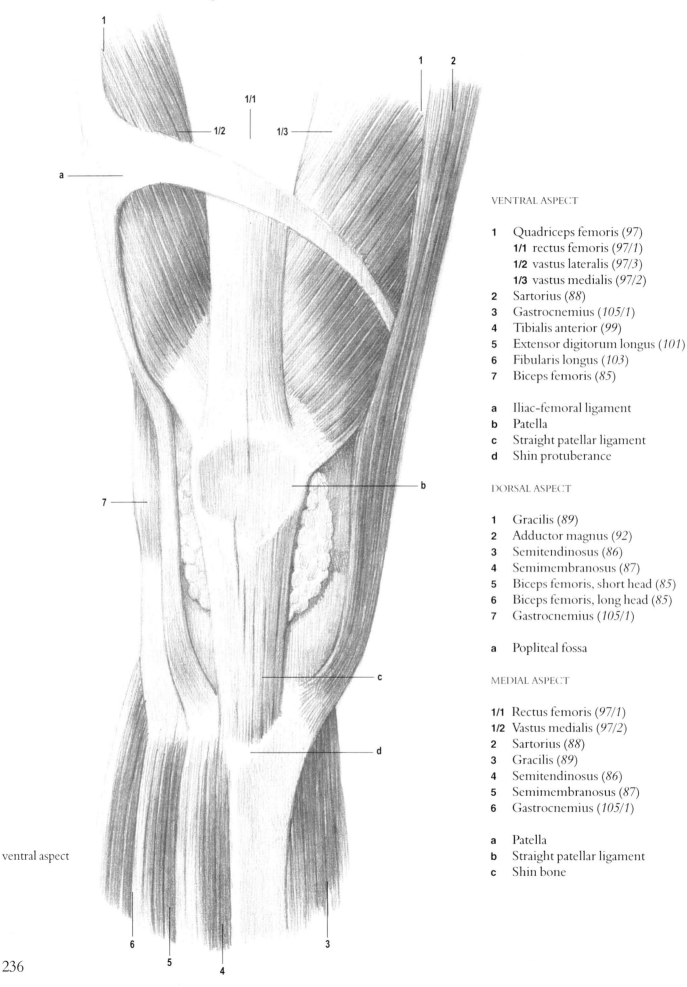

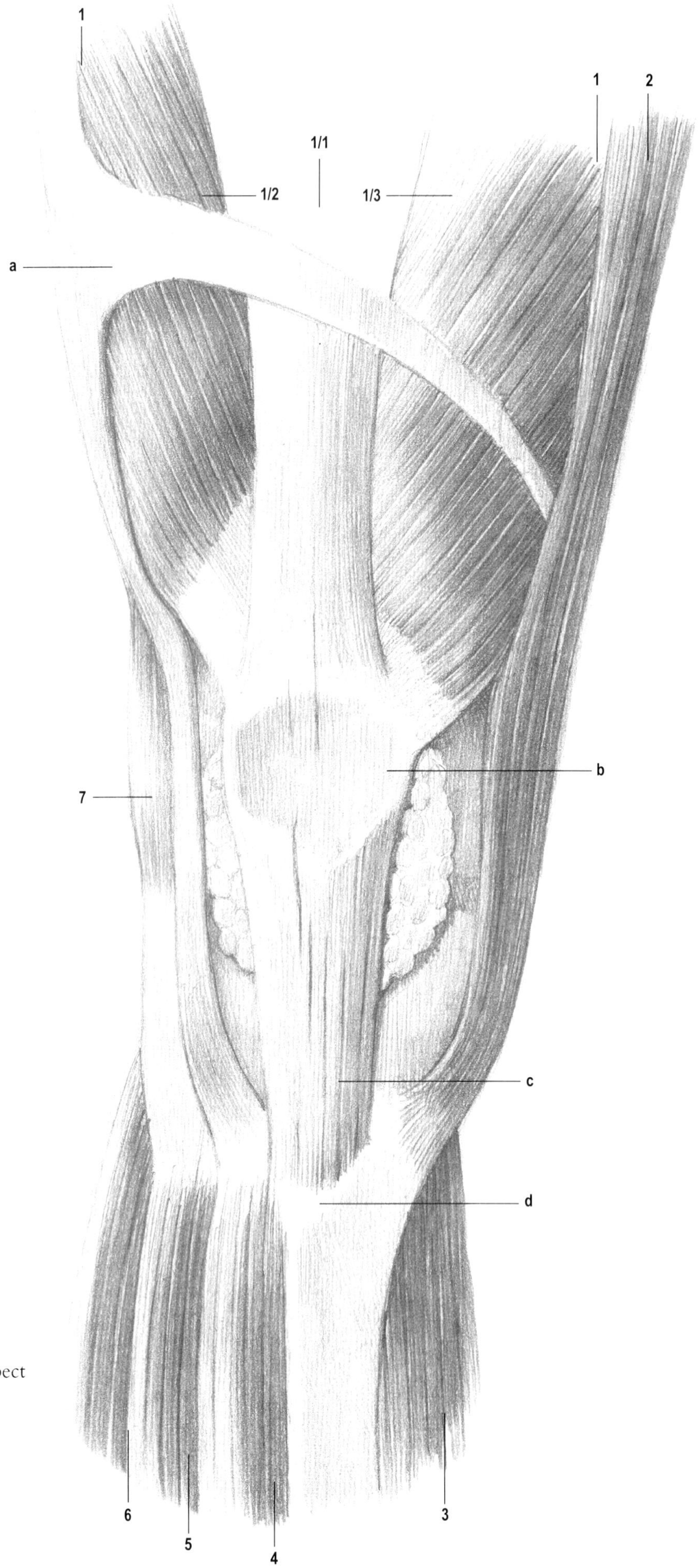

ventral aspect

VENTRAL ASPECT

1 Quadriceps femoris (*97*)
1/1 rectus femoris (*97/1*)
1/2 vastus lateralis (*97/3*)
1/3 vastus medialis (*97/2*)
2 Sartorius (*88*)
3 Gastrocnemius (*105/1*)
4 Tibialis anterior (*99*)
5 Extensor digitorum longus (*101*)
6 Fibularis longus (*103*)
7 Biceps femoris (*85*)

a Iliac-femoral ligament
b Patella
c Straight patellar ligament
d Shin protuberance

DORSAL ASPECT

1 Gracilis (*89*)
2 Adductor magnus (*92*)
3 Semitendinosus (*86*)
4 Semimembranosus (*87*)
5 Biceps femoris, short head (*85*)
6 Biceps femoris, long head (*85*)
7 Gastrocnemius (*105/1*)

a Popliteal fossa

MEDIAL ASPECT

1/1 Rectus femoris (*97/1*)
1/2 Vastus medialis (*97/2*)
2 Sartorius (*88*)
3 Gracilis (*89*)
4 Semitendinosus (*86*)
5 Semimembranosus (*87*)
6 Gastrocnemius (*105/1*)

a Patella
b Straight patellar ligament
c Shin bone

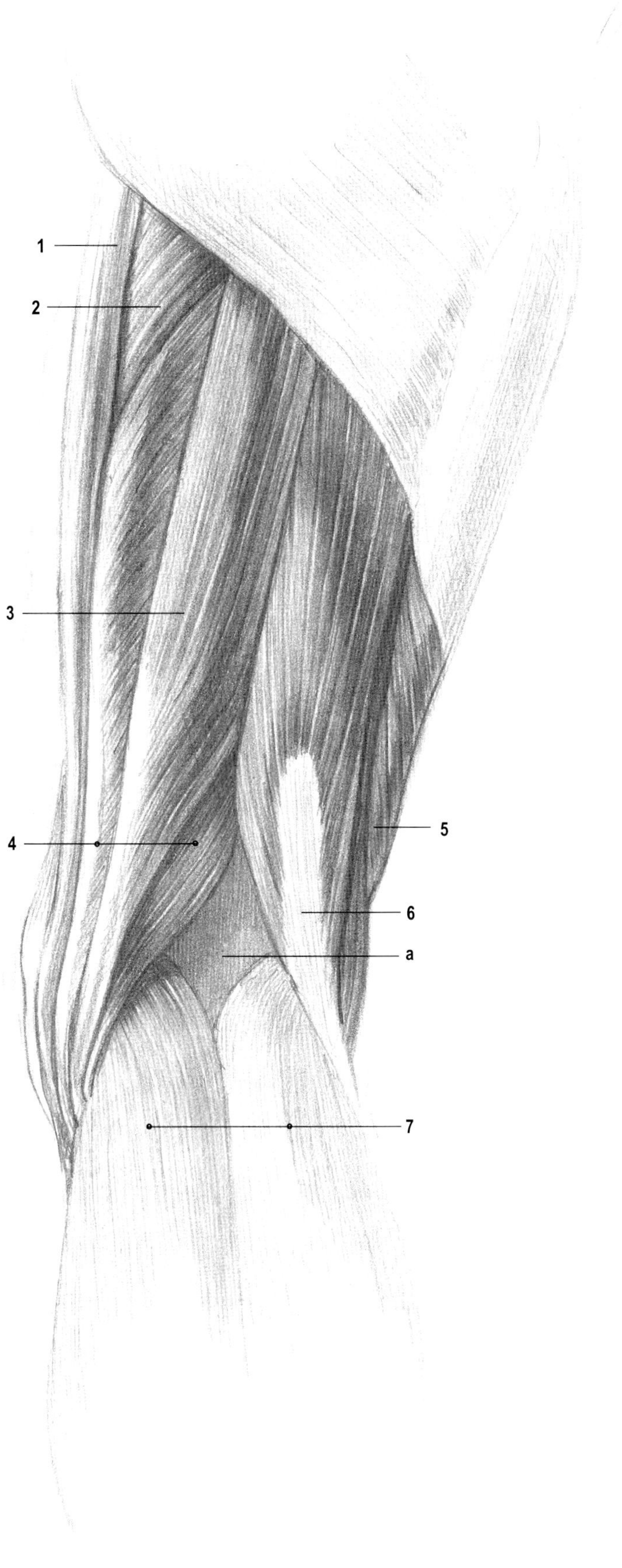

dorsal aspect

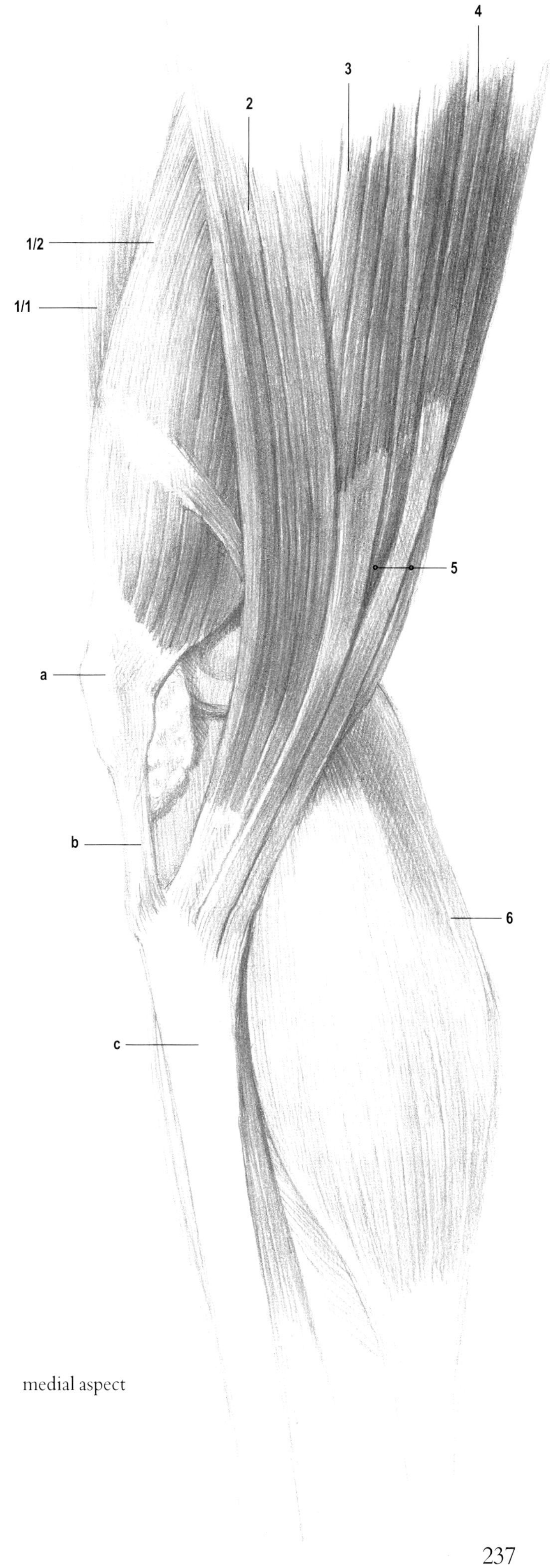

medial aspect

Fig. 141
The Posteria Femoral Muscles

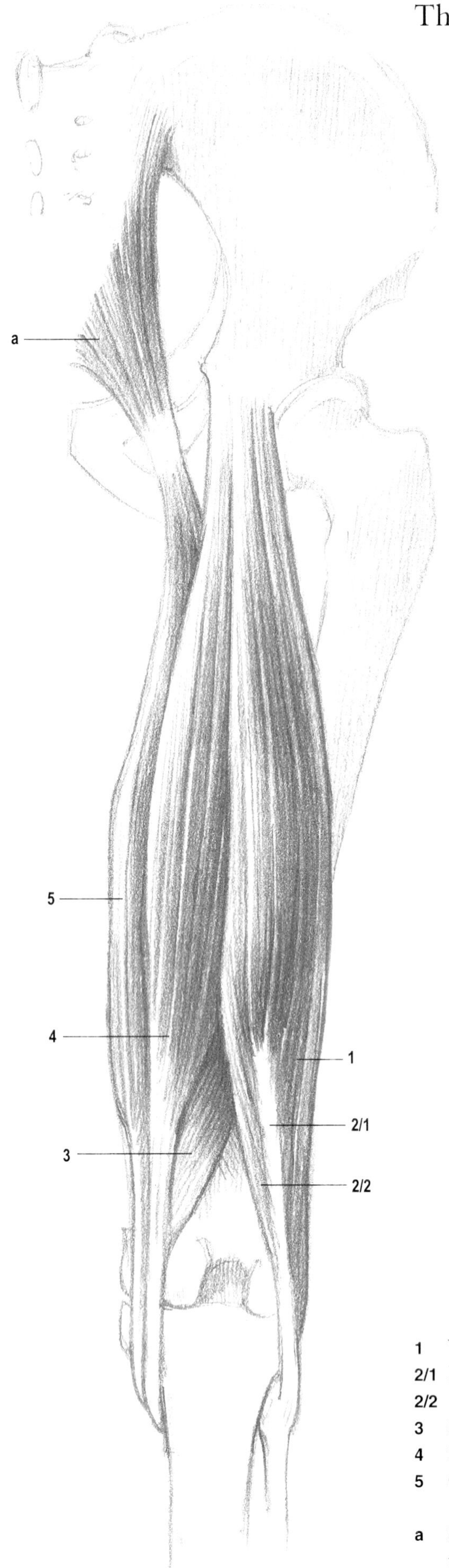

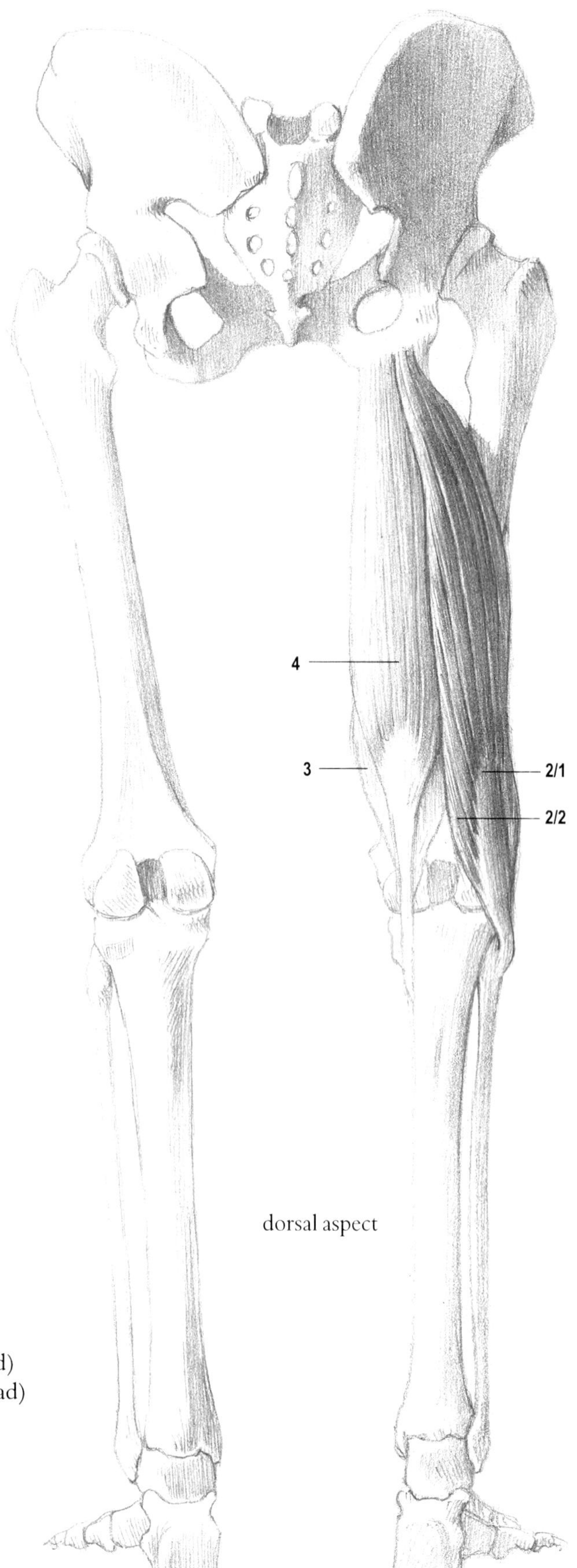

1 Vastus lateralis
2/1 Biceps femoris (long head)
2/2 Biceps femoris (short head)
3 Semimembranosus
4 Semitendinosus
5 Gracilis

a Sacrum-sedentarium protuberance ligament

Fig. 142
The Biceps Muscle of the Thigh
(M. biceps femoris, *85*)

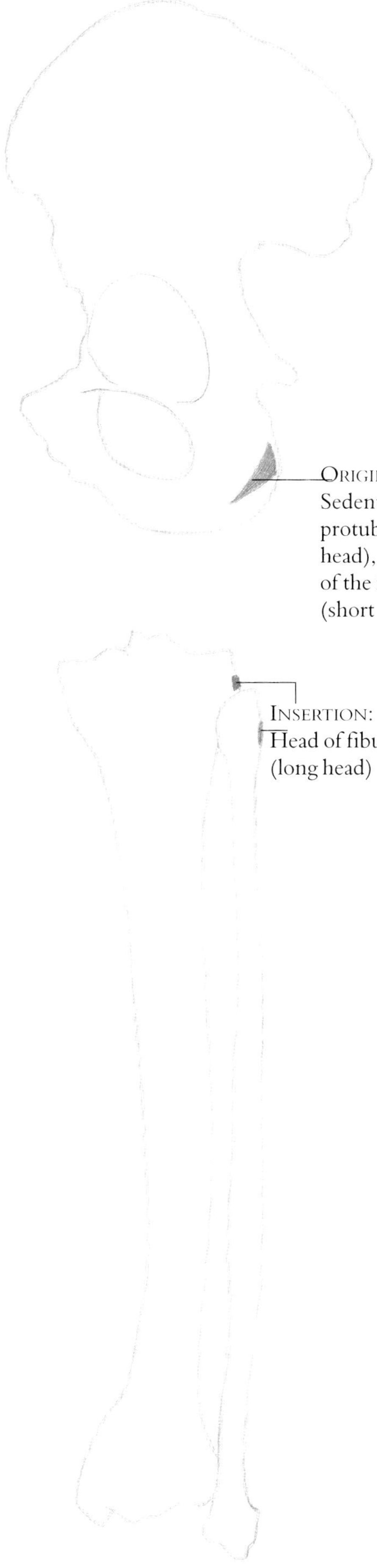

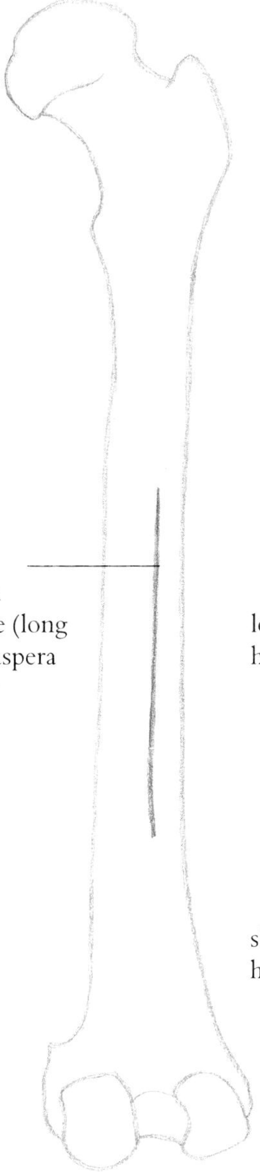

ORIGIN:
Sedentarium protuberance (long head), linea aspera of the femur (short head)

INSERTION:
Head of fibula (long head)

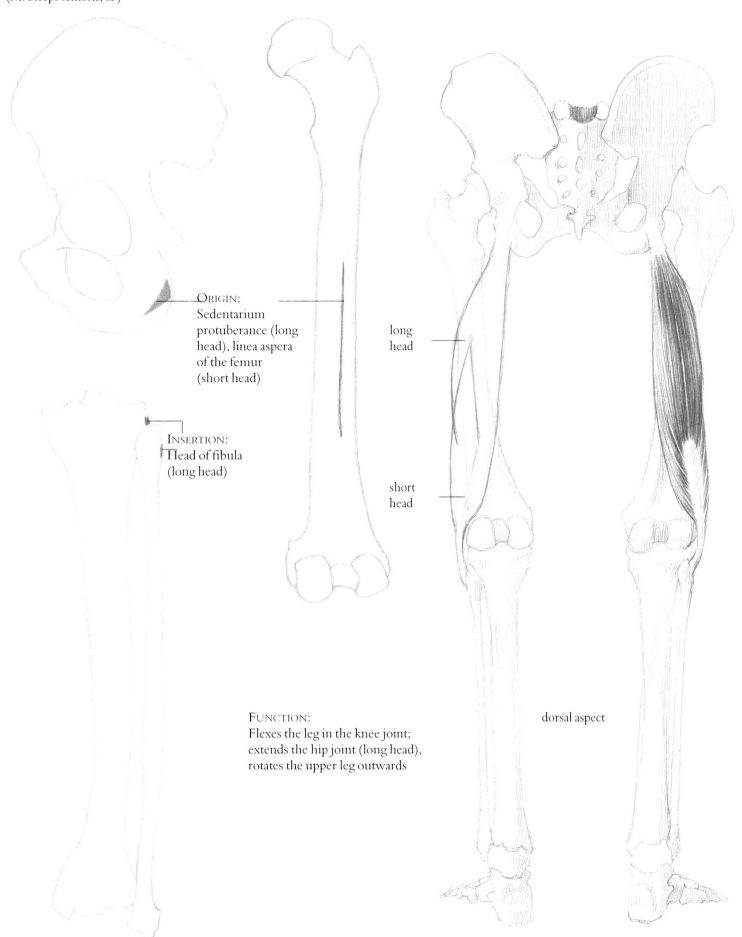

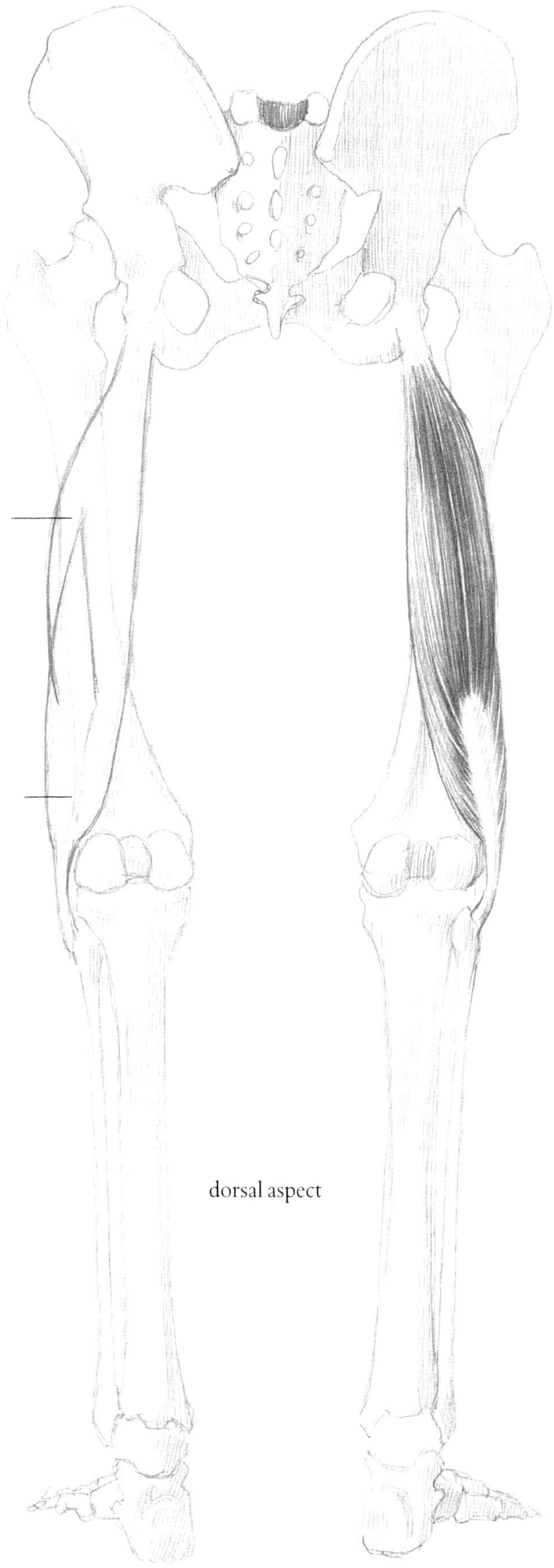

dorsal aspect

FUNCTION:
Flexes the leg in the knee joint; extends the hip joint (long head), rotates the upper leg outwards

Fig. 143
The Semitendinous Muscle
(M. semitendinosus, *86*)

ORIGIN:
Protuberance of os sedentarium, fused with biceps femoris

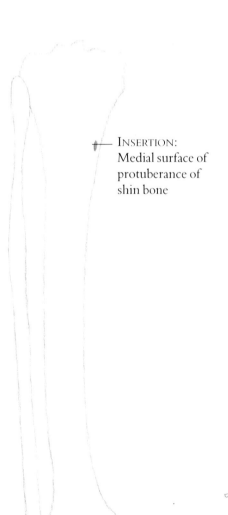

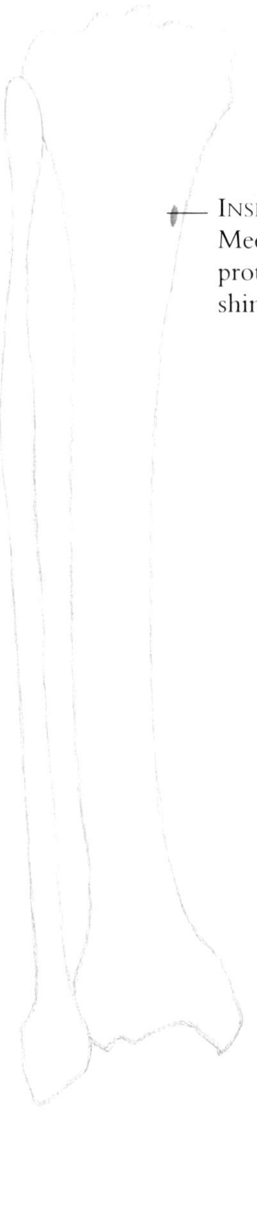

INSERTION:
Medial surface of protuberance of shin bone

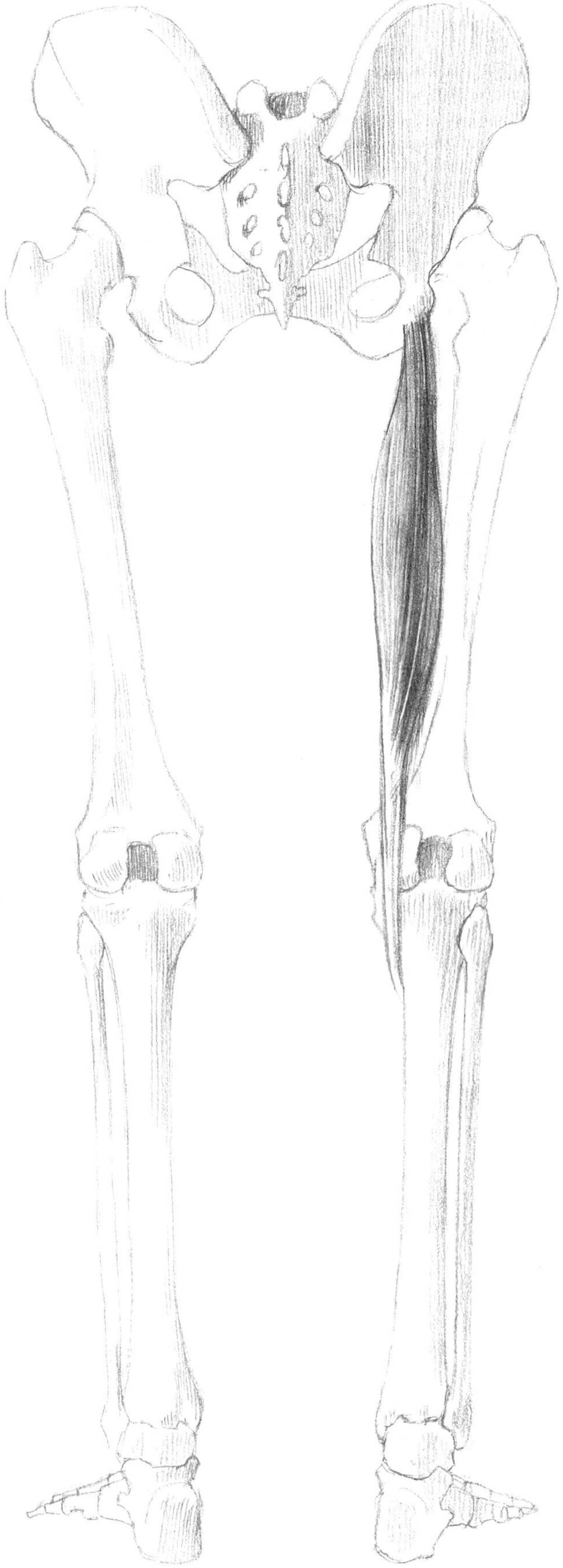

dorsal aspect

FUNCTION:
Rotates the lower leg inwards and flexes the knee joint, extends the hip joint

Fig. 144
The Semimembraneous Muscle

(M. semimembranosus, *87*)

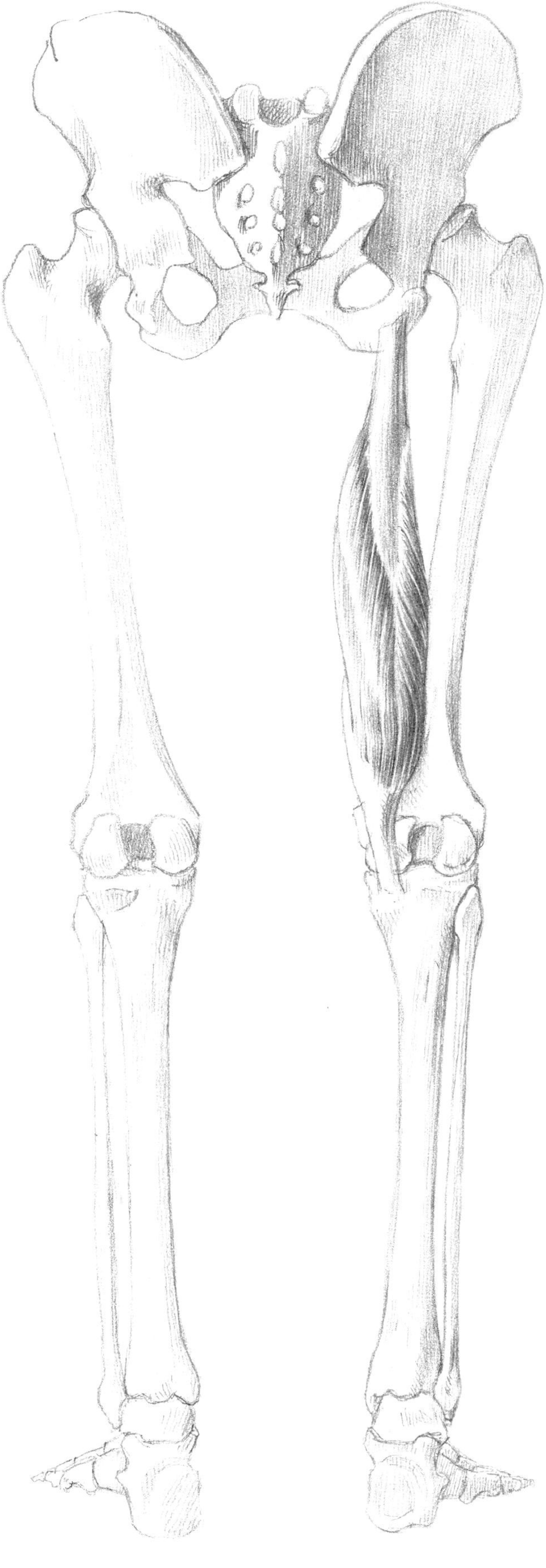

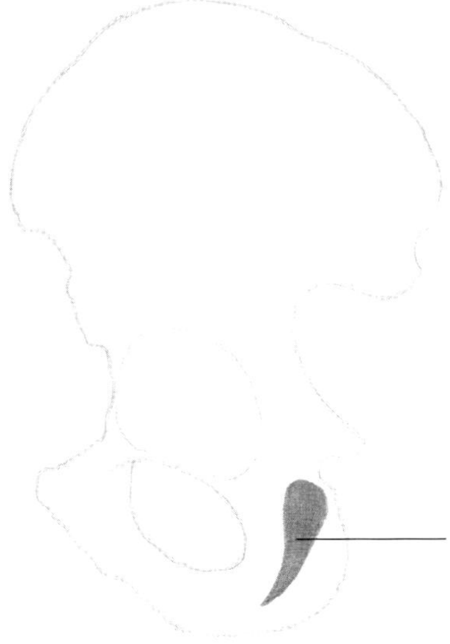

ORIGIN:
Protuberance of os sedentarium

INSERTION:
Medial shin tuberosity and at popliteal ligament

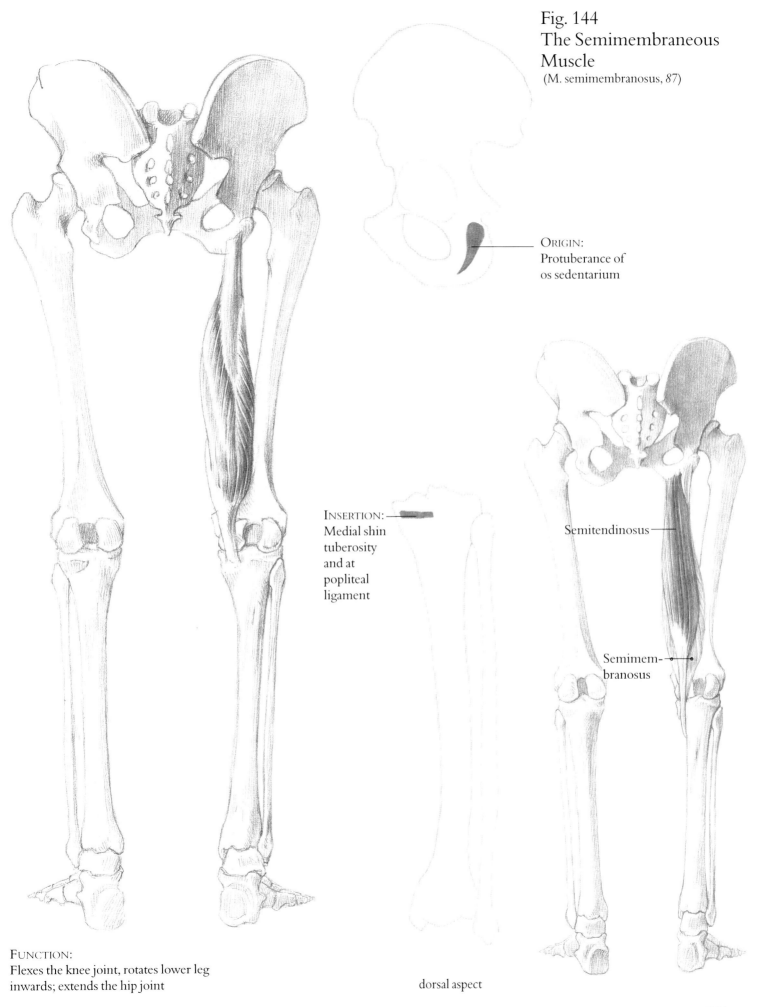

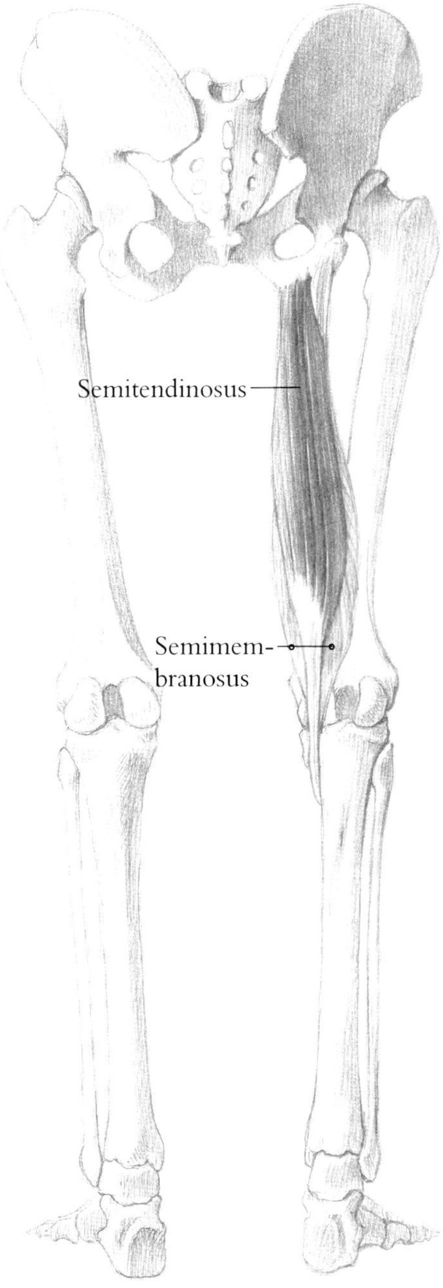

FUNCTION:
Flexes the knee joint, rotates lower leg inwards; extends the hip joint

dorsal aspect

Fig. 145

The Great Adductor Muscle

(M. adductor magnus, *92*)

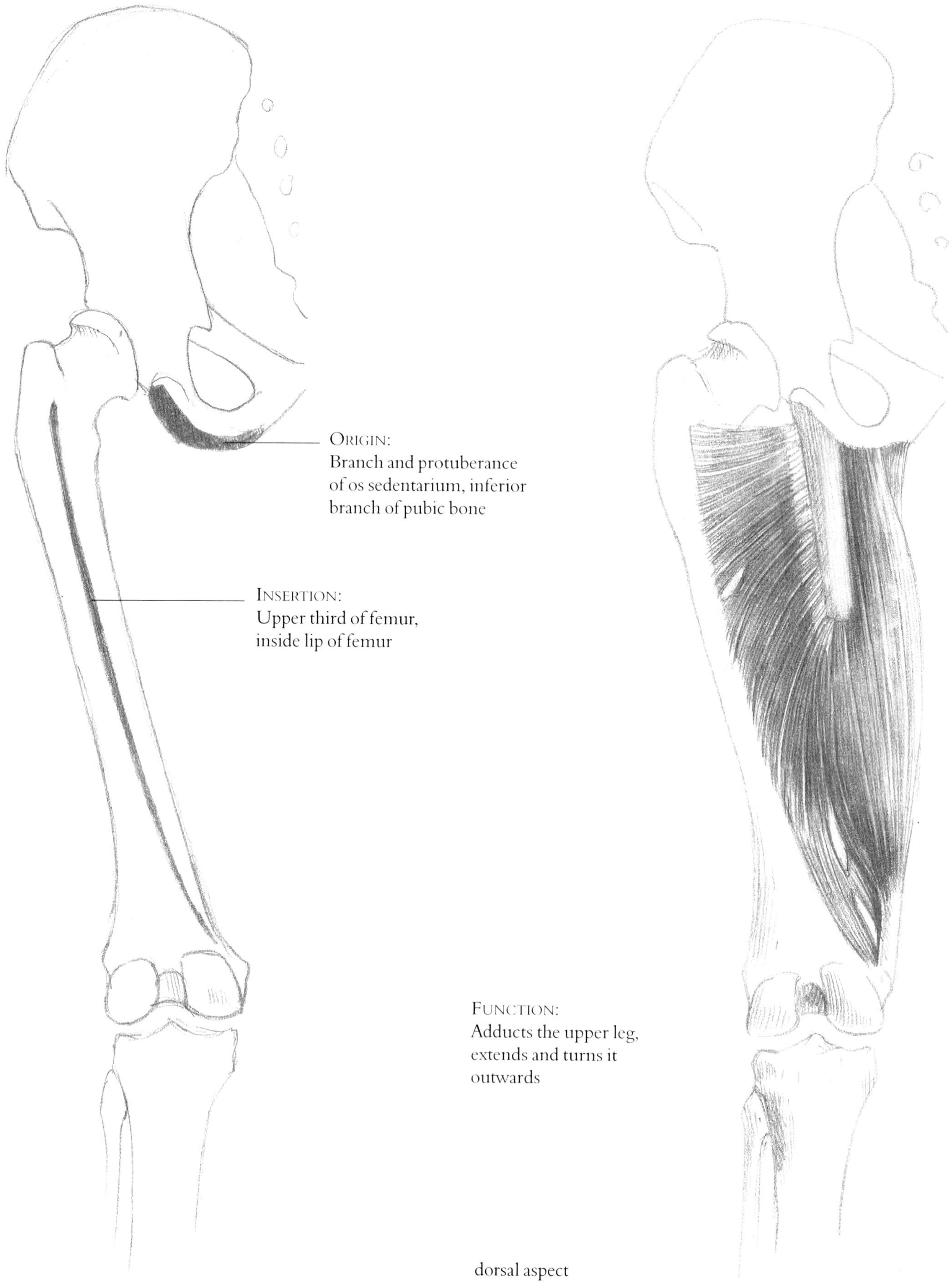
ORIGIN:
Branch and protuberance of os sedentarium, inferior branch of pubic bone
INSERTION:
Upper third of femur, inside lip of femur
FUNCTION:
Adducts the upper leg, extends and turns it outwards

dorsal aspect

Fig. 146
The Short Adductor Muscle
(M. adductor brevis, *92*)

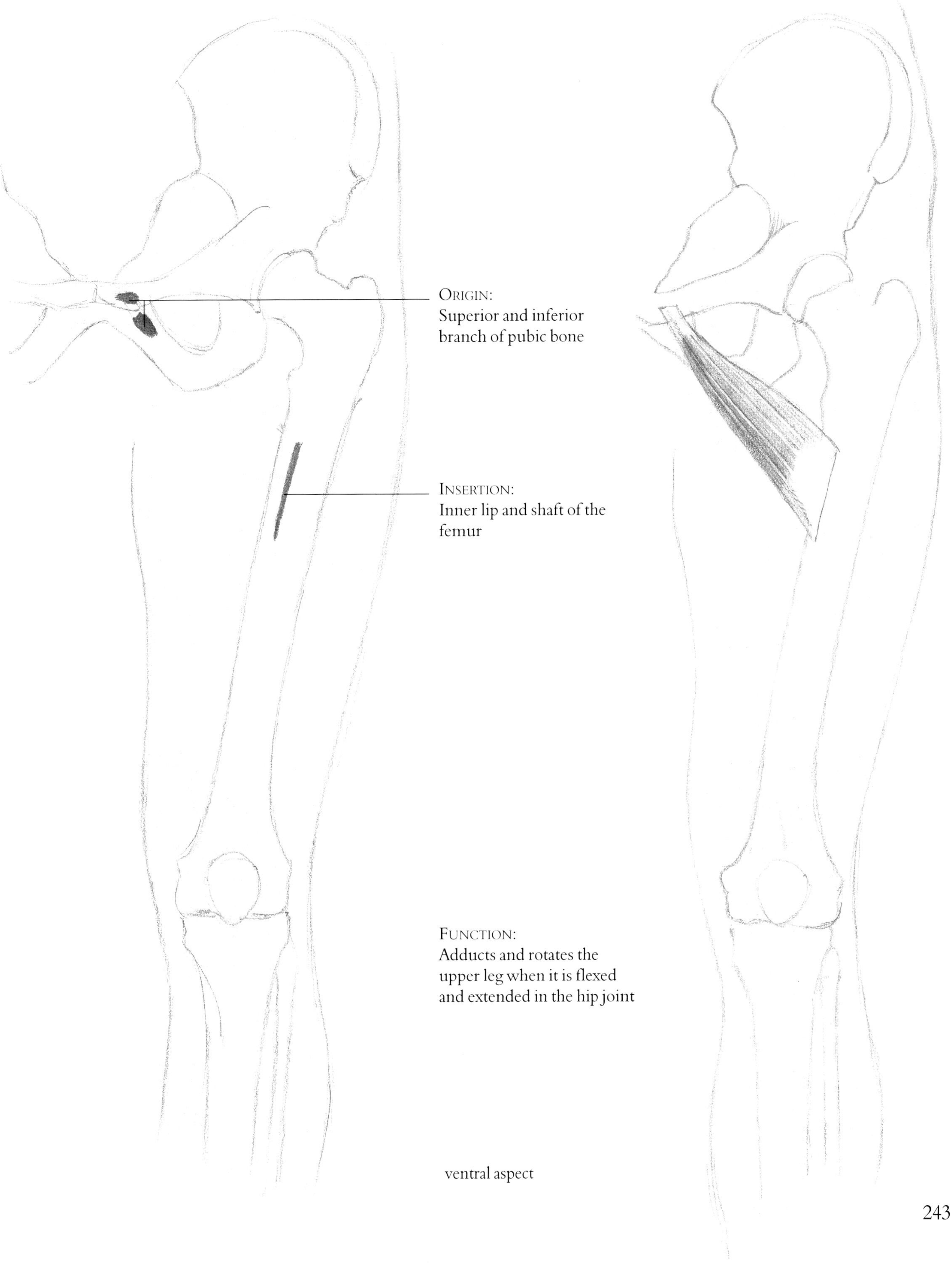
ORIGIN:
Superior and inferior branch of pubic bone
INSERTION:
Inner lip and shaft of the femur
FUNCTION:
Adducts and rotates the upper leg when it is flexed and extended in the hip joint

ventral aspect

Fig. 147

The Long Adductor Muscle

(M. adductor longus, *92*)

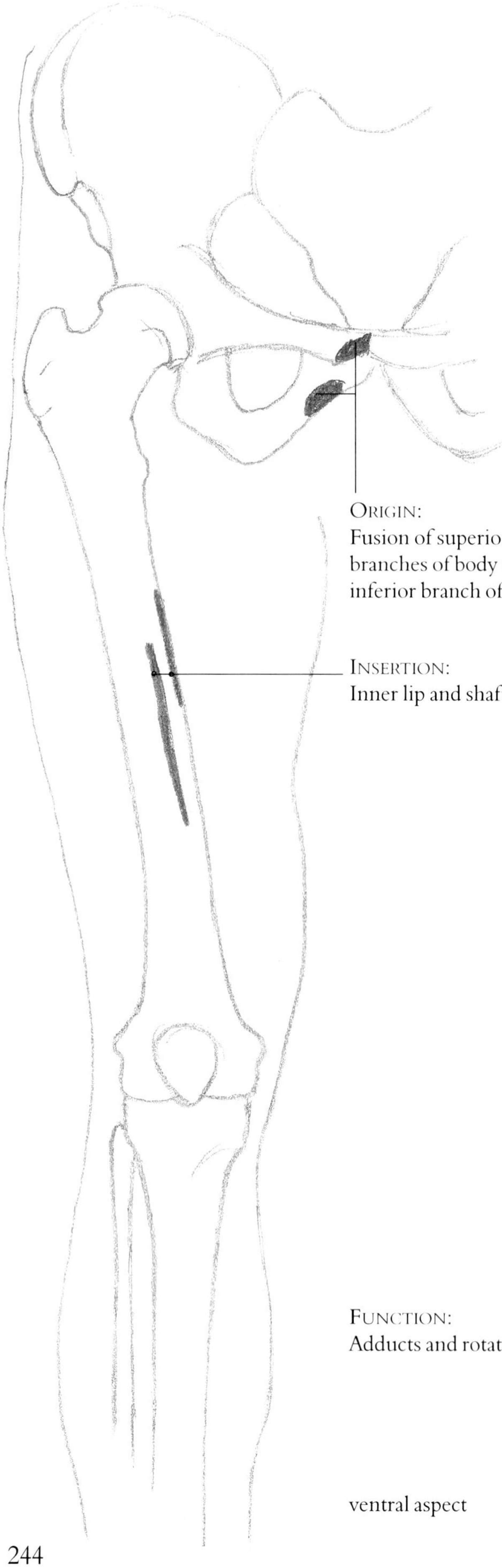

ORIGIN:
Fusion of superior and inferior branches of body of pubic bone and inferior branch of the pubic bone

INSERTION:
Inner lip and shaft of the femur

FUNCTION:
Adducts and rotates the upper leg

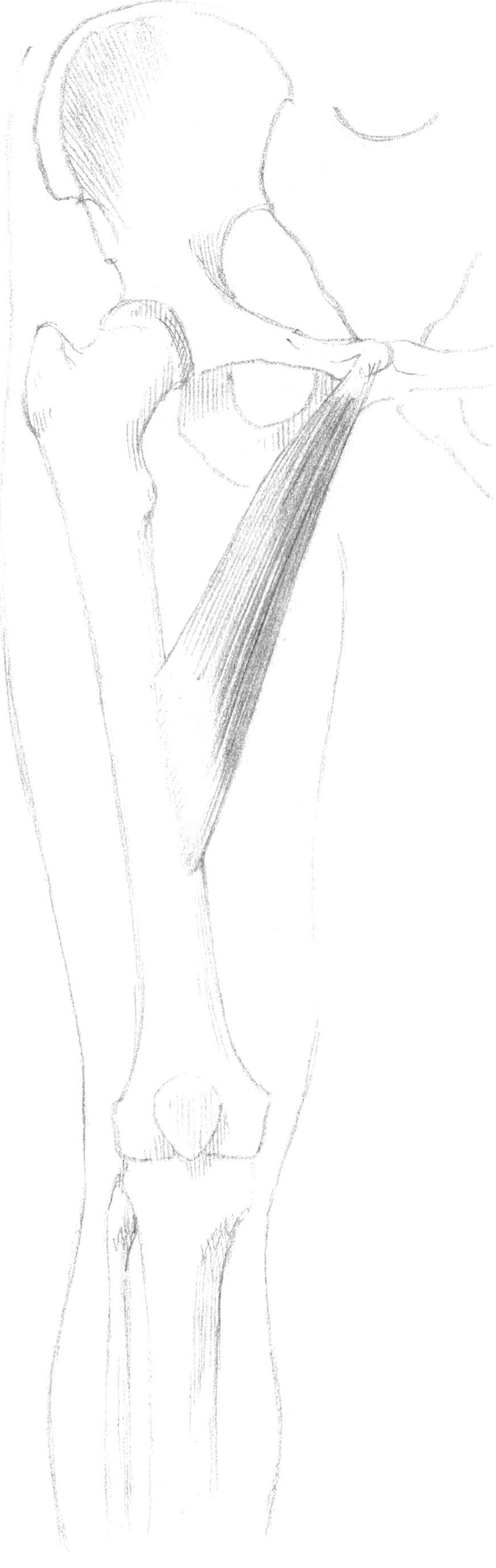

ventral aspect

Fig. 148
The Pectineal Muscle
(M. pectineus, *90*)

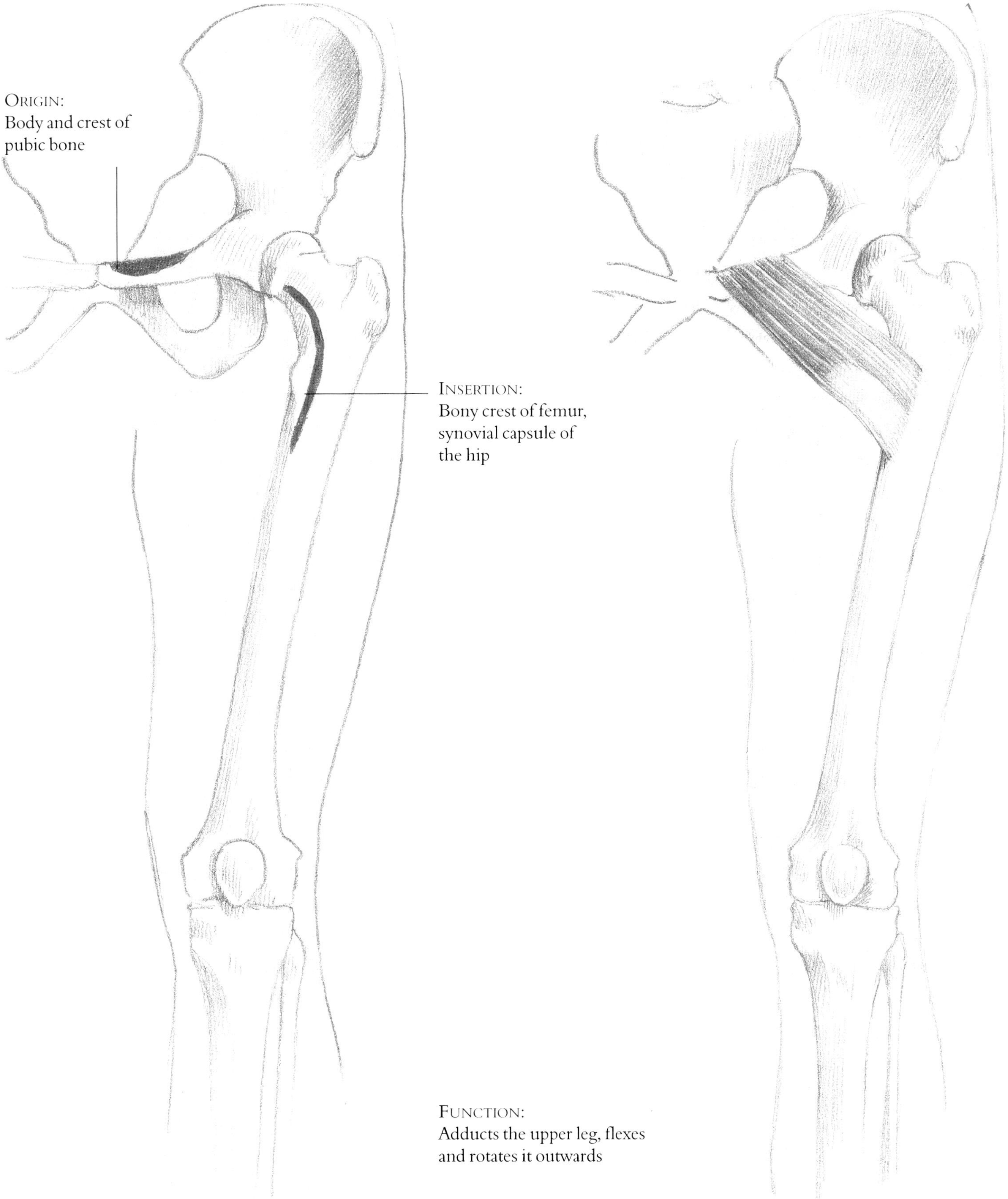

FUNCTION:
Adducts the upper leg, flexes
and rotates it outwards

ventral aspect

Fig. 149
The Tailor's Muscle
(M. sartorius, *88*)

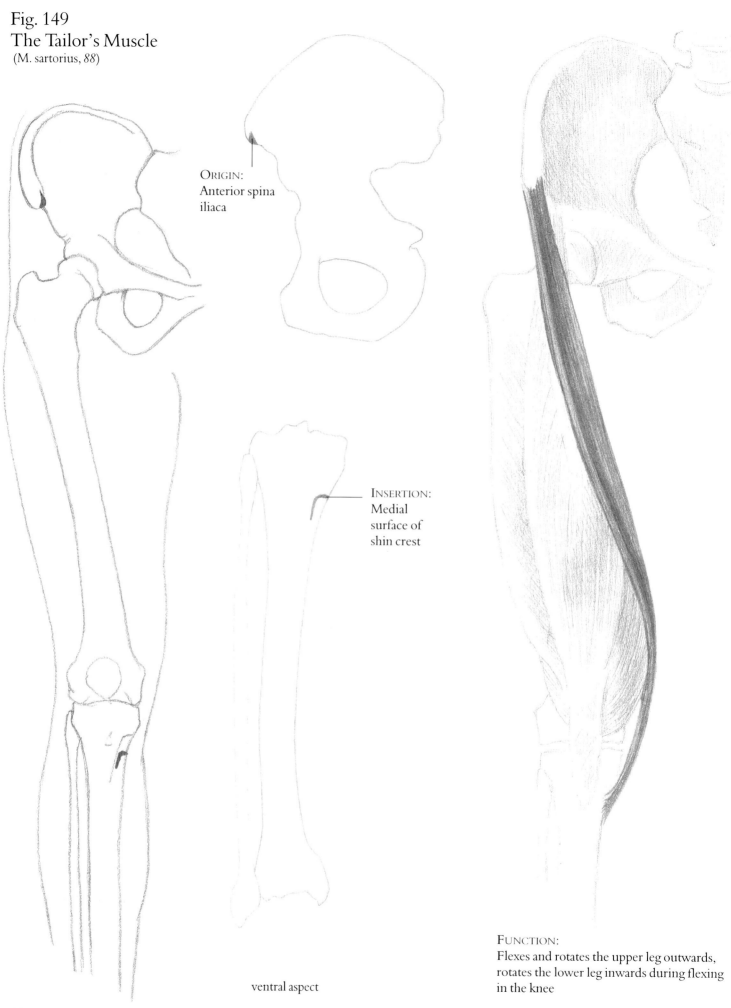

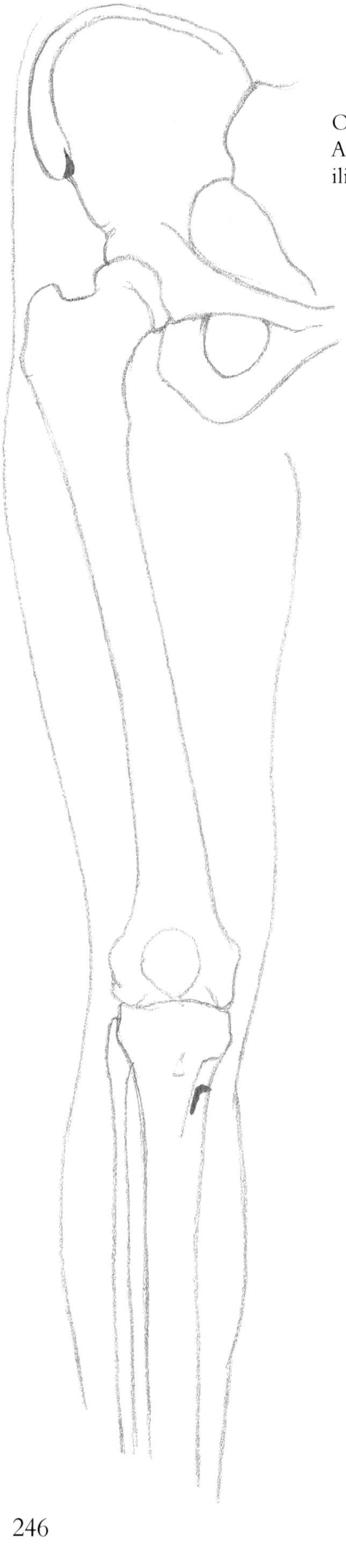

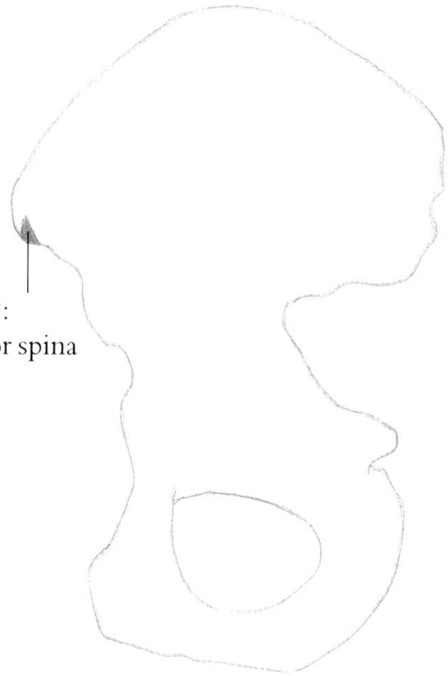

ORIGIN:
Anterior spina iliaca

INSERTION:
Medial surface of shin crest

FUNCTION:
Flexes and rotates the upper leg outwards, rotates the lower leg inwards during flexing in the knee

ventral aspect

Fig. 150
The Gracilis, *89*

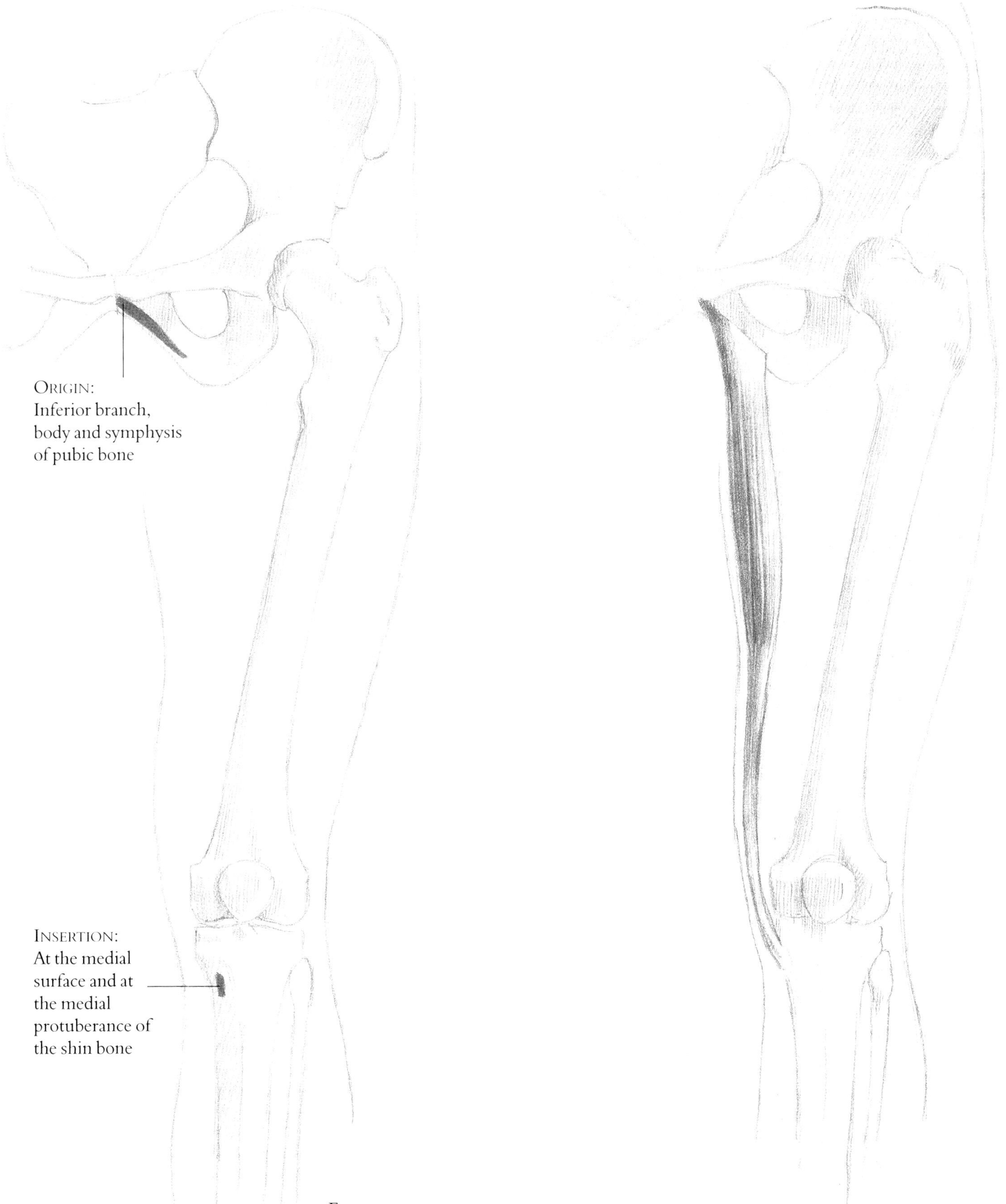

ORIGIN:
Inferior branch,
body and symphysis
of pubic bone

INSERTION:
At the medial
surface and at
the medial
protuberance of
the shin bone

FUNCTION:
Adducts the upper leg, flexes the knee, pulls the
lower leg towards the center

ventral aspect

Fig. 151
Origins and Insertion Points of Muscles

0 = Origin
I = Insertion

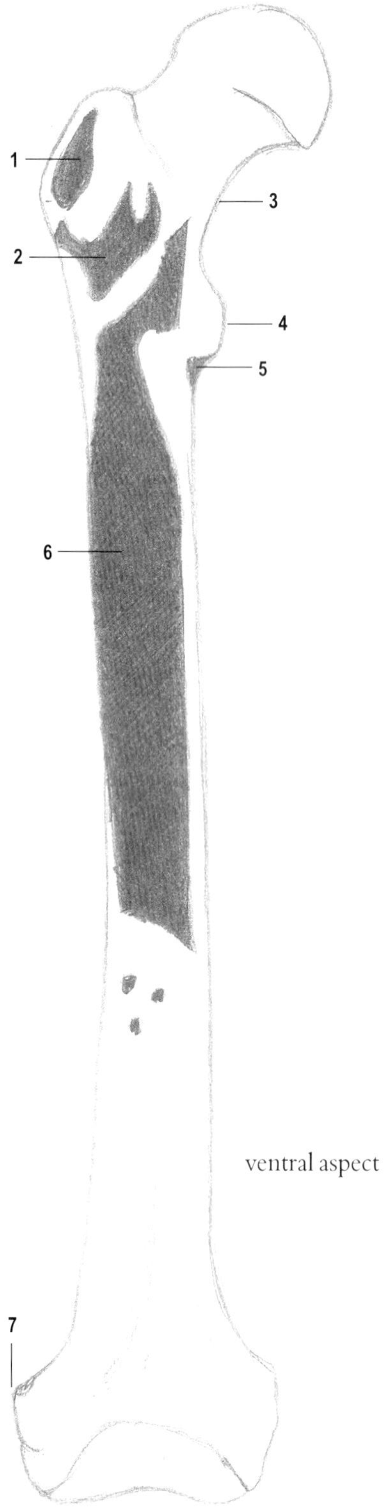

ventral aspect

Fig. 151/a
Origins and Points of Insertion of Muscles in the Sacrum Region

1 Iliacus, O (*79/1*)
2 Priform muscle, O (*91*)
3 Tail muscle, O I
4 Semispinalis, I (*26/4*)
5 Glutaeus maximus, O (*81*)

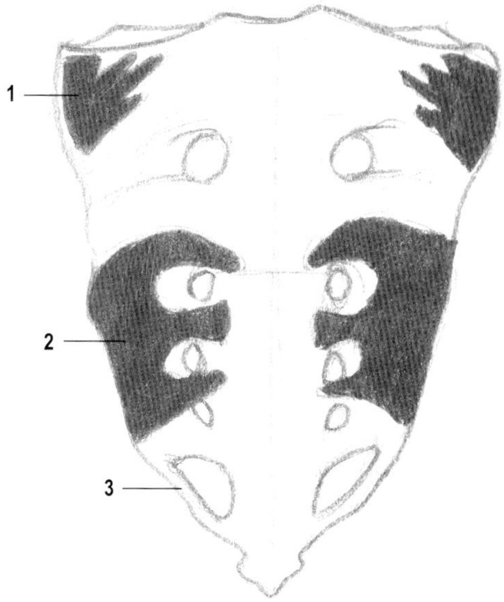

ventral surface

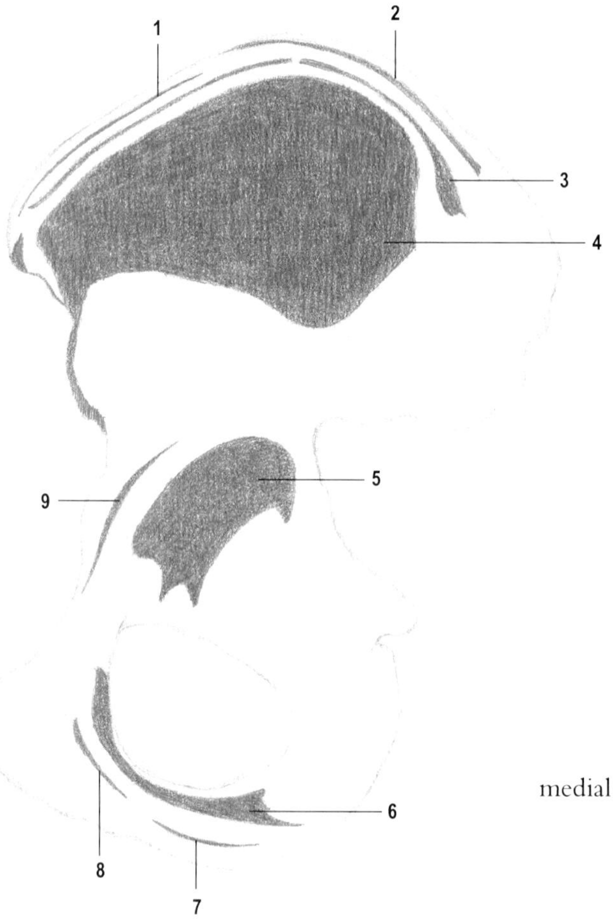

medial surface

Fig. 151/b
Origins and Points of Insertion of Muscles in the Region of the Pelvic Bone

1 Tensor fasciae latae, O (*84*)
2 Quadratus lumbar muscle, O (*80*)
3 Inner loxic abdominis, O (*38*)
4 Iliacus, I (*79/1*)
5 Obturatorius internus, O (*93*)
6 Transverse perinei
7 Levantor ani
8 Pectineus, O (*90*)
9 Transversus abdominis, I (*39*)
10 Outer loxic abdominis, I (*37*)
11 Glutaeus medius, O (*82*)
12 Glutaeus minimus, O (*83*)

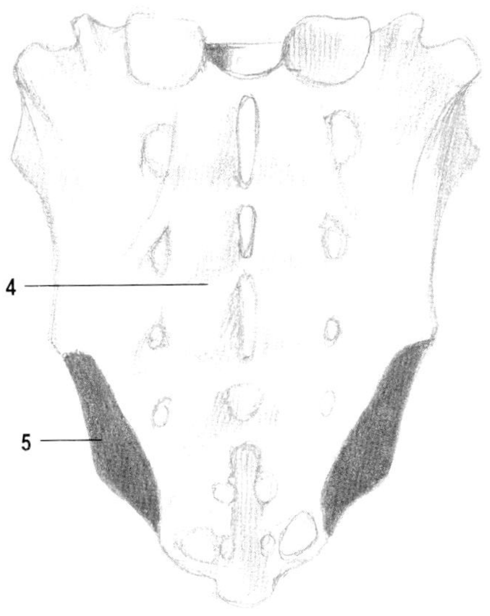

dorsal aspect

Fig. 151/c
Origins and Points of Insertion of the Muscles in the Region of the Femur

1 Glutaeus minimus, I (*83*)
2 Vastus lateralis, O (*97/3*)
3 Pectineus, I (*90*)
4 Psoas major, I (*79/2*)
5 Vastus medialis, O (*97/2*)
6 Vastus intermedialis, O (*97/4*)
7 Gastrocnemius, O (*105/1*)
8 Obturatorius externus, I (*94*)
9 Glutaeus medius, I (*82*)
10 Quadratus femoris, I (*96*)
11 Iliacus, (*79/1*)
12 Glutaeus maximus, I (*81*)
13 Adductor magnus, I (*92*)
14 Adductor longus, I (*92*)
15 Biceps femoris, O (*85*)
16 Adductor brevis, I (*92*)
17 Popliteus, O (*107*)
18 Plantaris, O (*106*)

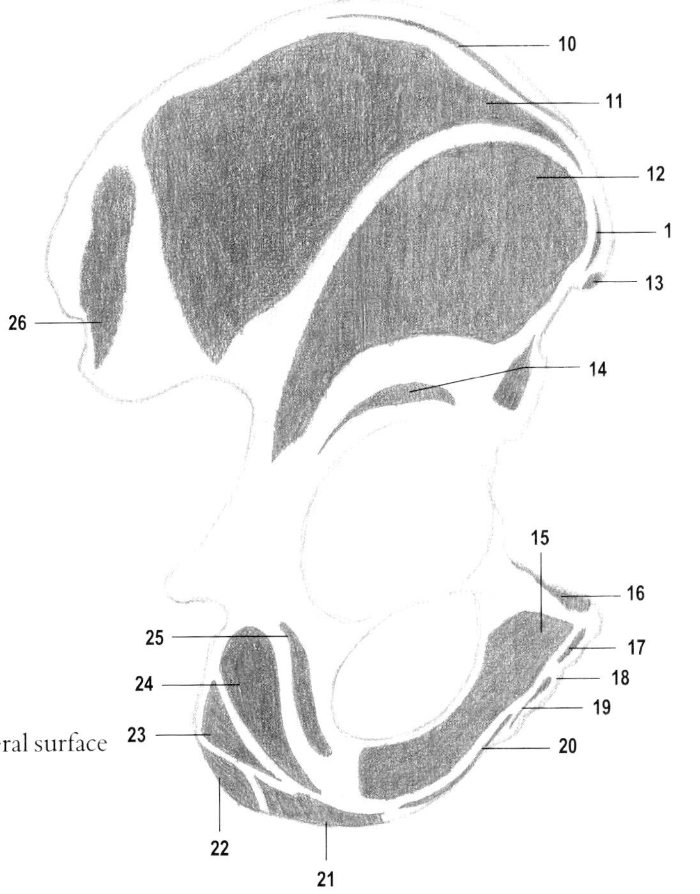

dorsolateral surface

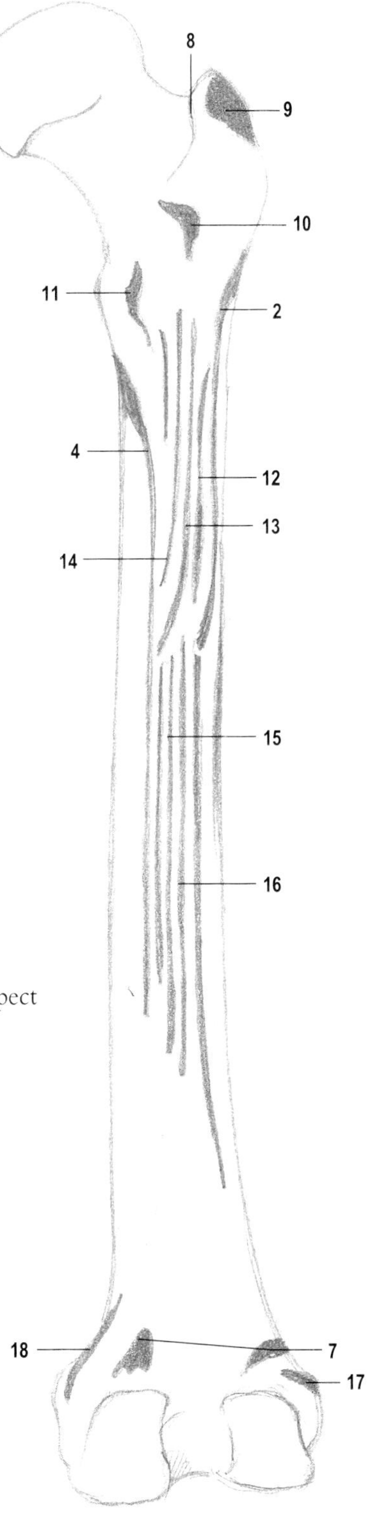

dorsal aspect

13 Sartorius, O (*88*)
14 Rectus femoris, O (*97/1*)
15 Adductor longus, O (*92*)
16 Pectineus, O (*90*)
17 Rectus abdominis, I (*35*)
18 Pyramid muscle, I (*36*)
19 Adductor brevis, O (*92*)
20 Adductor minimus, O (*92*)
21 Adductor magnus, O (*92*)
22 Semitendinosus, O (*86*)
23 Biceps femoris, O (*85*)
24 Semimembranosus, O (*87*)
25 Quadratus femoris, O (*96*)
26 Glutaeus maximus, O (*81*)

Fig. 152
The Triceps Surae, *105*

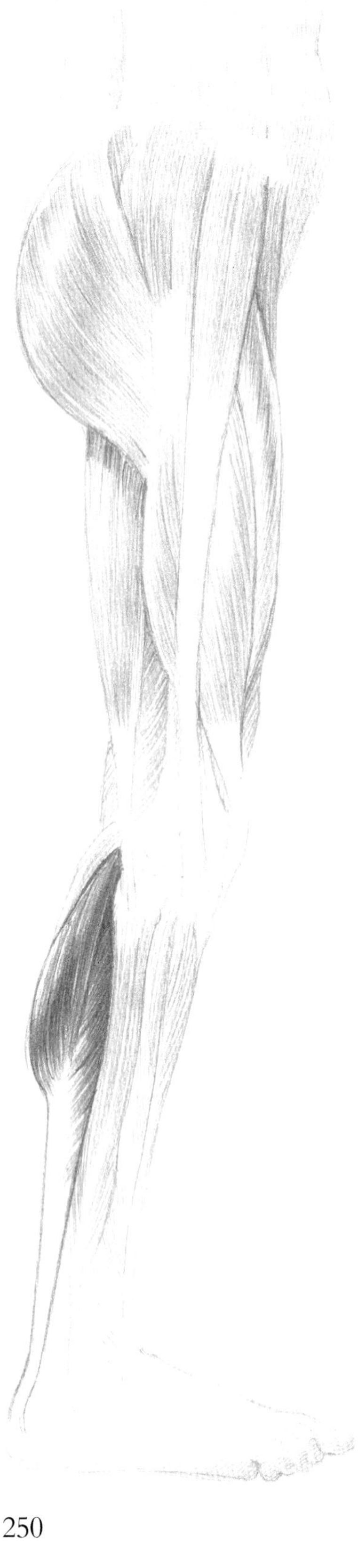

The two heads (**a**) of the triceps surae originate at the inner and outer head of the femur (**1**) and at the synovial capsule of the knee. They cover the third head (soleus muscle, *105/2*) (**b**), which originates at the shin and fibula. The tendon of the three muscles is the Achilles tendon (**2**), which is inserted at the protuberance of the heel (**3**).

FUNCTION:
Flexion of the foot (plantar flexion), outwards rotation of the foot

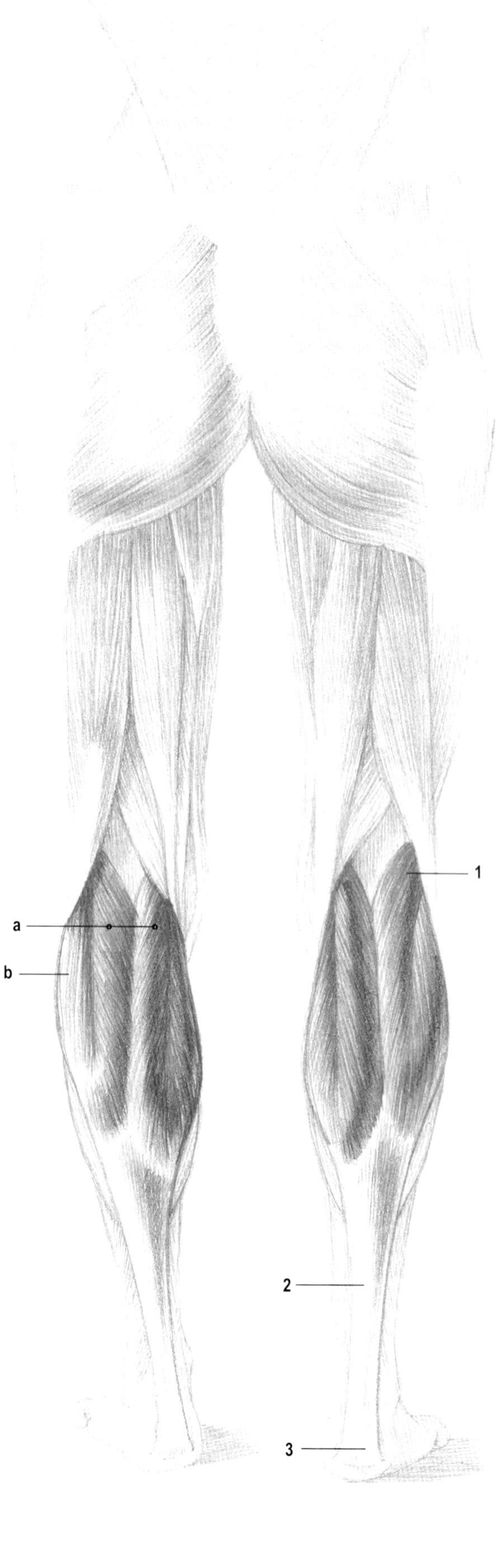

Fig. 153
The Gastrocnemius, *105/1*

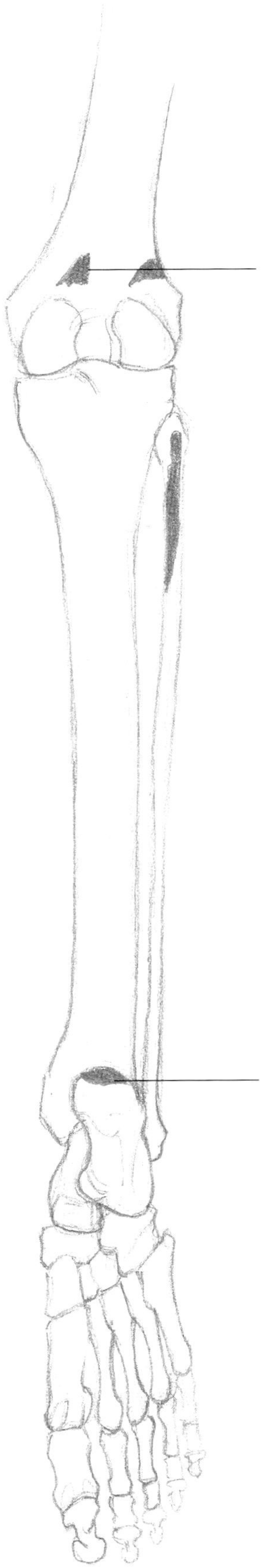

Origin:
At the medial (inner head) and lateral (outer head) articula rtuberosity of the femur

Insertion:
With the Achilles tendon at the heel bone

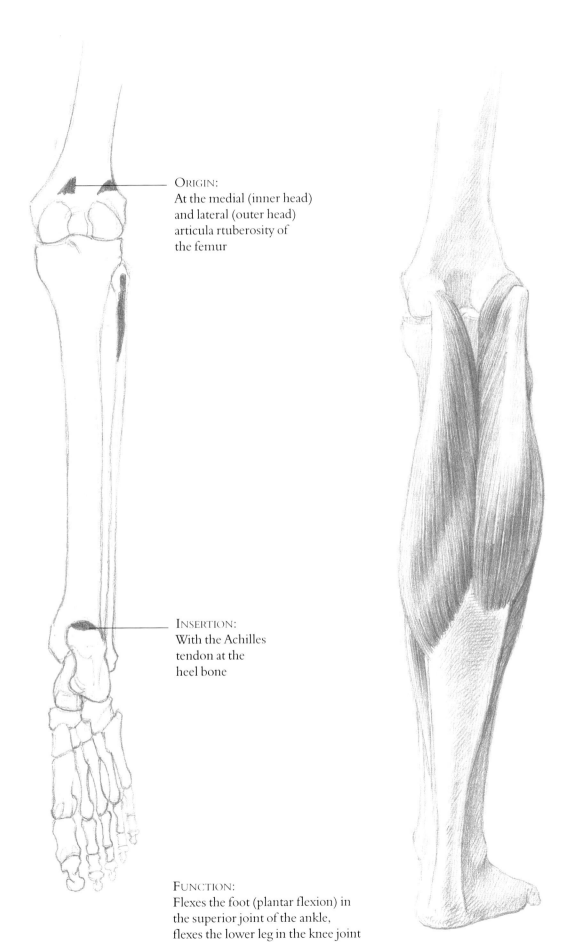

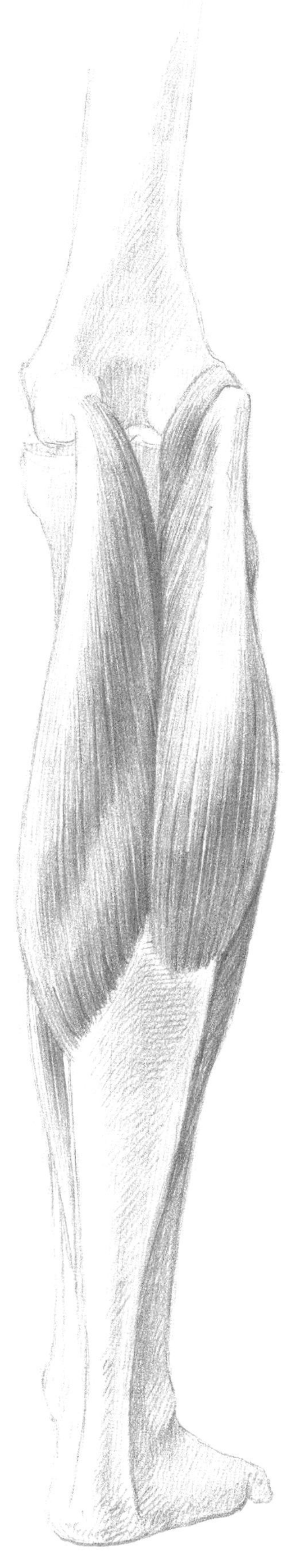

Function:
Flexes the foot (plantar flexion) in the superior joint of the ankle, flexes the lower leg in the knee joint

Fig. 154
The Popliteus, *107*

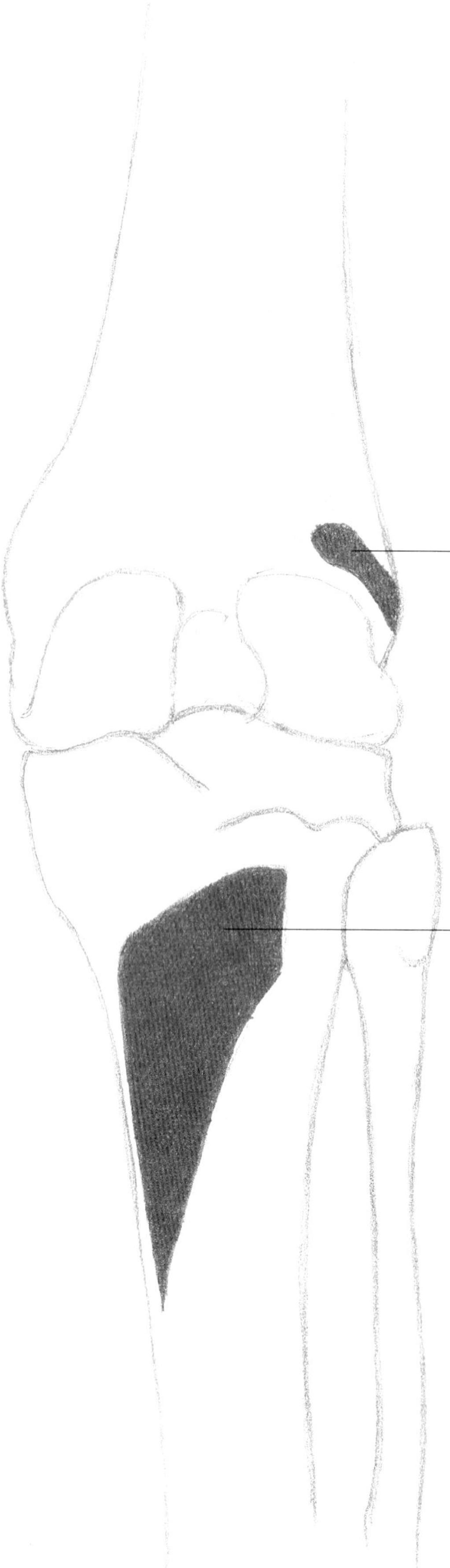

ORIGIN:
Lateral ligament protuberance of the femoral condyle and at the synovial capsules

INSERTIONS:
Inferior medial surface of the proximal of the tuberosity of the shin bone

FUNCTION:
Flexes the knee, rotates the lower leg inwards when knee is bent

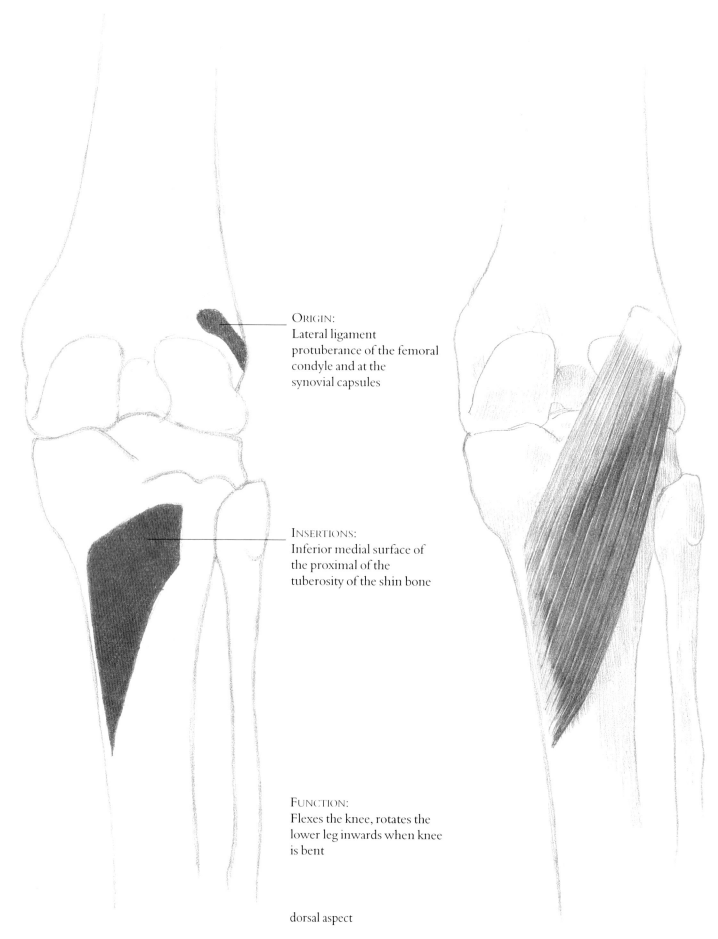

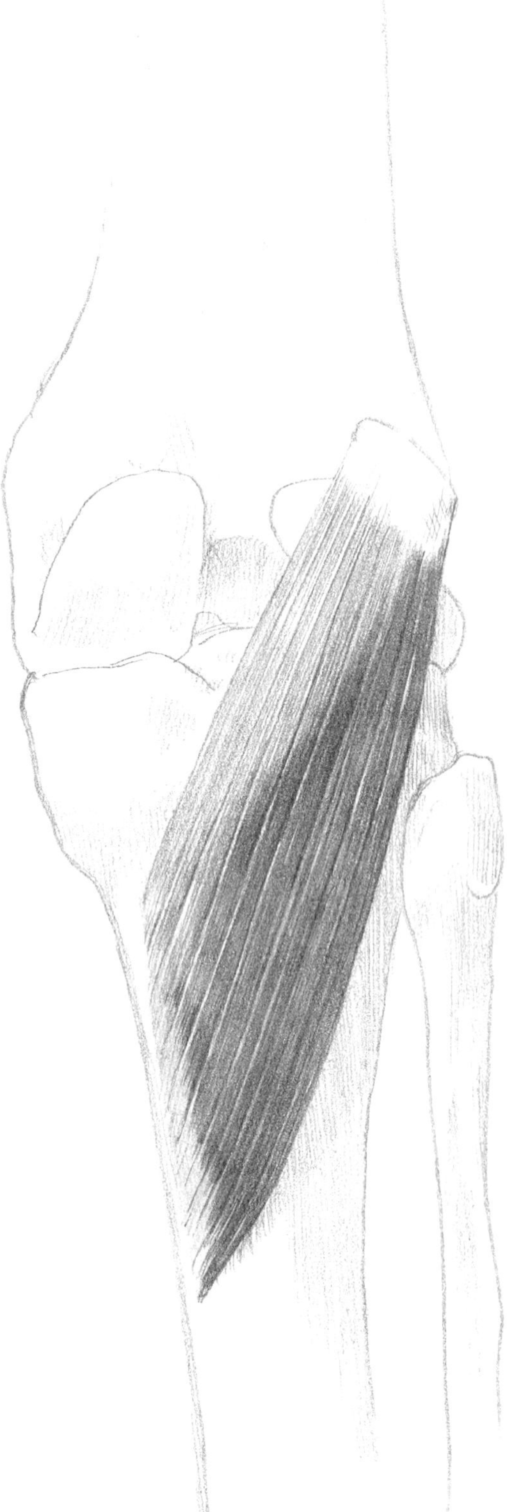

dorsal aspect

Fig. 155
The Tibialis Anterior, *99*

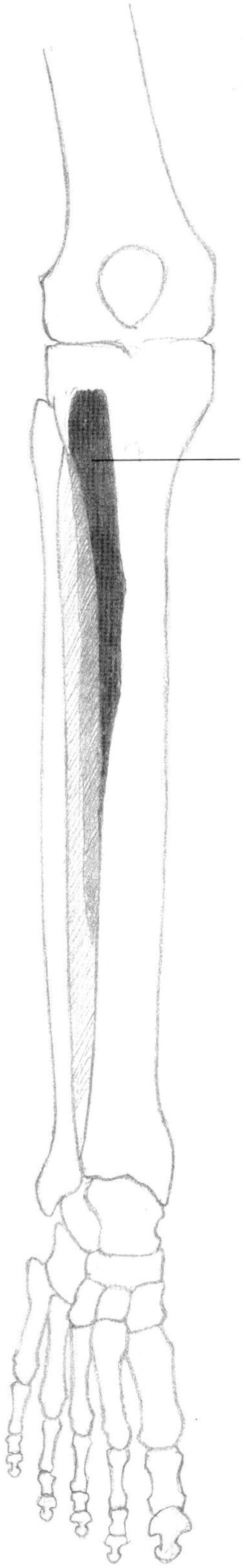

ORIGIN:
Lateral surface and tuberosity of shin bone, interosseous ligament

ventral aspect

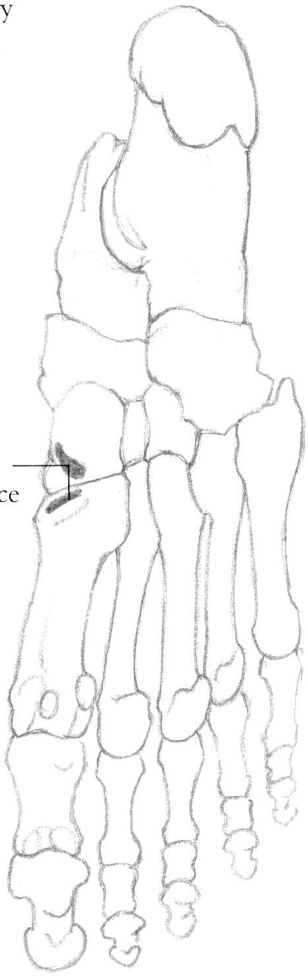

INSERTION:
Medial surface of the sphenoid bone; first metatarsal bone

plantar aspect

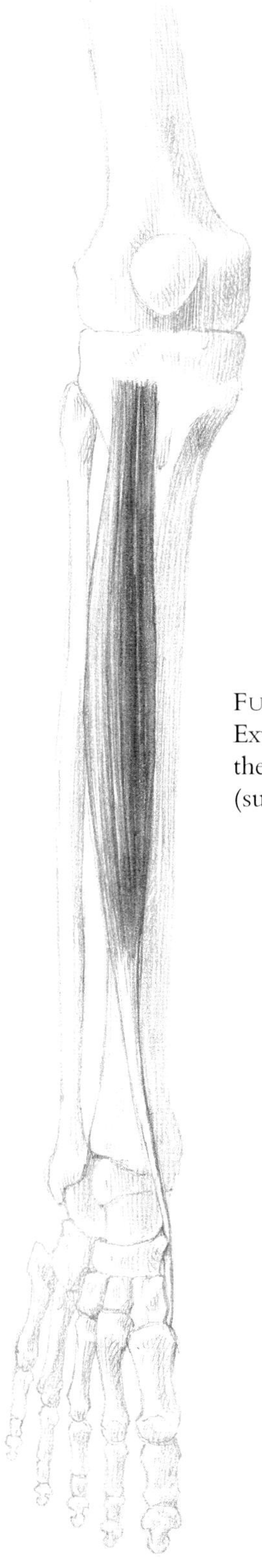

FUNCTION:
Extends (dorsal flexion) the foot in the ankle joint and rotates it outwards (supination)

ventral aspect

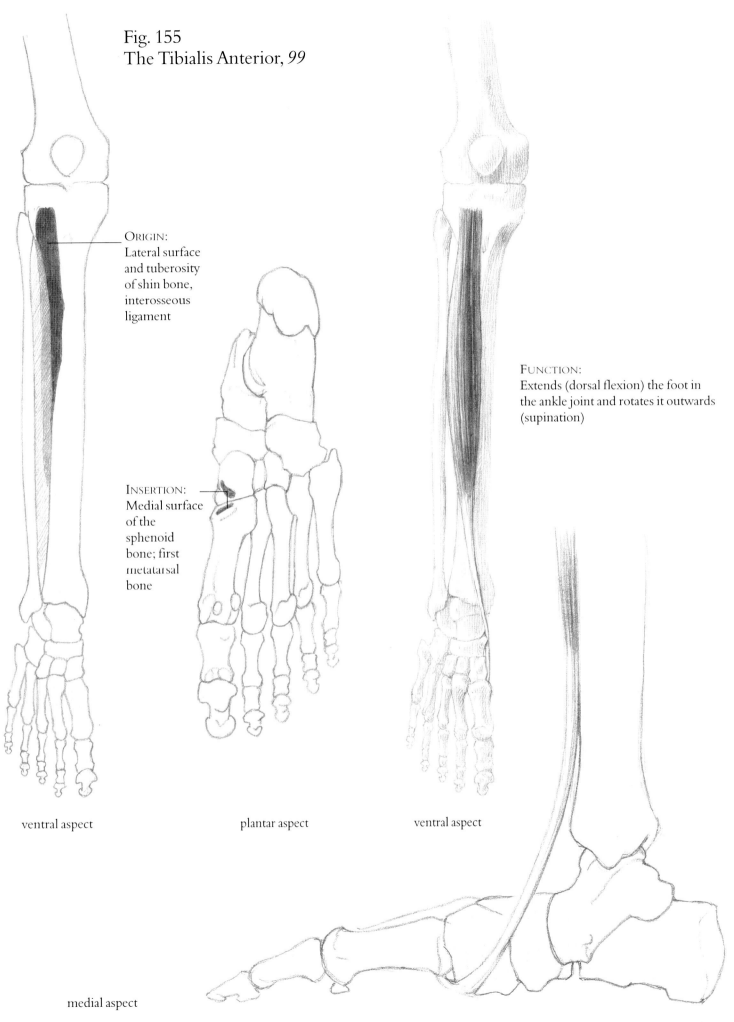

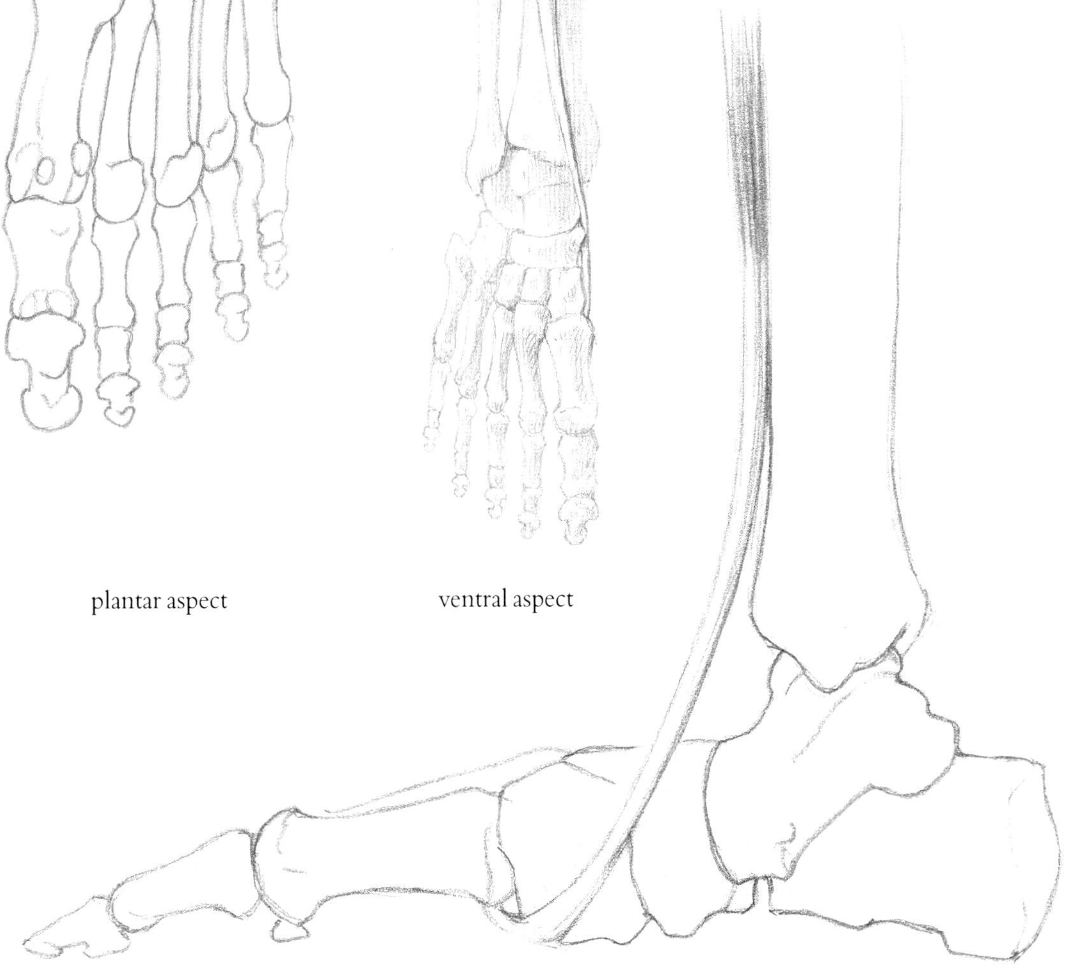

medial aspect

Fig. 156
The Extensor Digitorum Longus, *101*

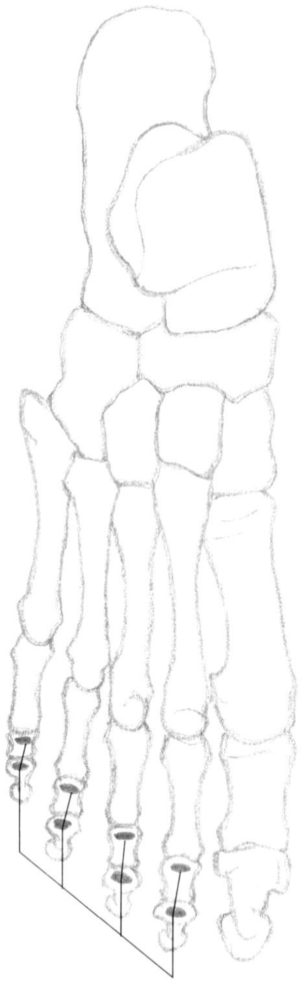

Origin:
At the proximal condyles of the shin bone, ventral surface of the fibula and interosseous ligament

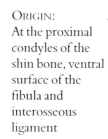

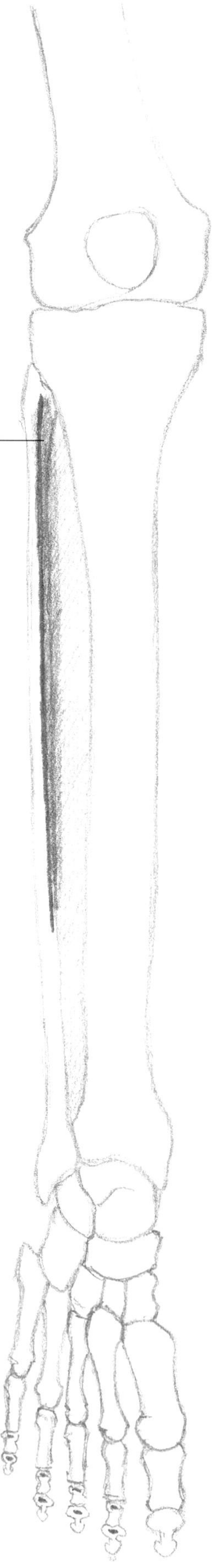

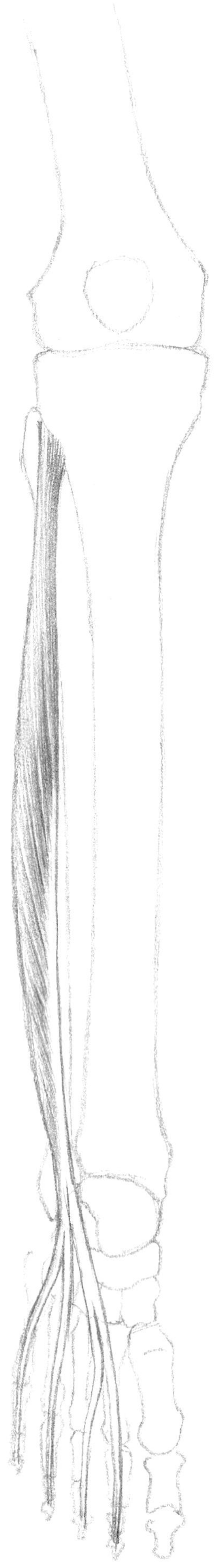

Insertion
Distal pediphalanx of the second to fifth toes

Function:
Extends the second to the fifth toes, helps dorsal flexion of the foot

ventral aspect

Fig. 157
The Extensor Hallucis Longus, *100*

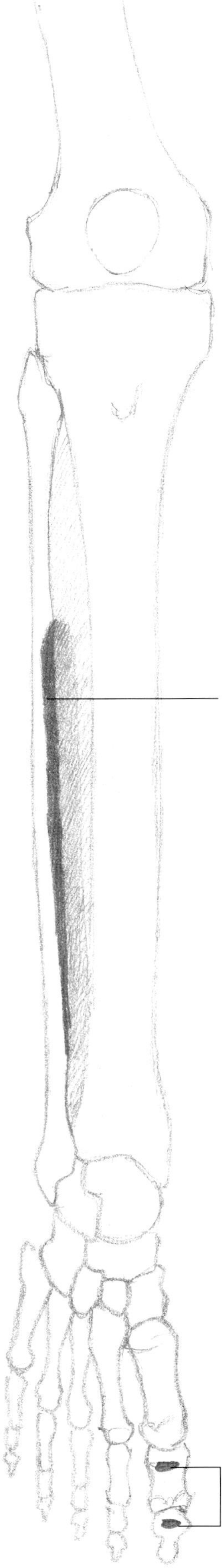

Origin:
Medial surface of fibula and interosseous ligament

Insertion:
Distal pediphalanx of the first toe

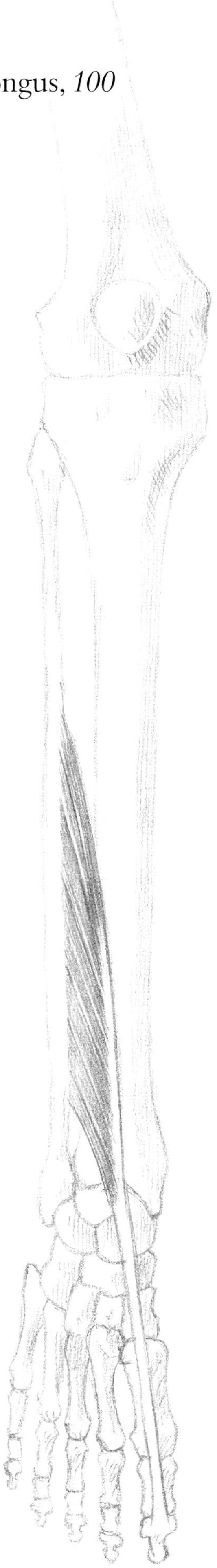

Function:
Extends and adducts the first toe

ventral aspect

Fig. 158
The Tibialis Posterior, *108*

ORIGIN:
Dorsal surface of the shin bone, medial surface of the fibula and interosseous ligament

INSERTION:
Tuberosity of the navicular bone, inner and outer sphenoid bone, and second to fifth metatarsal bones

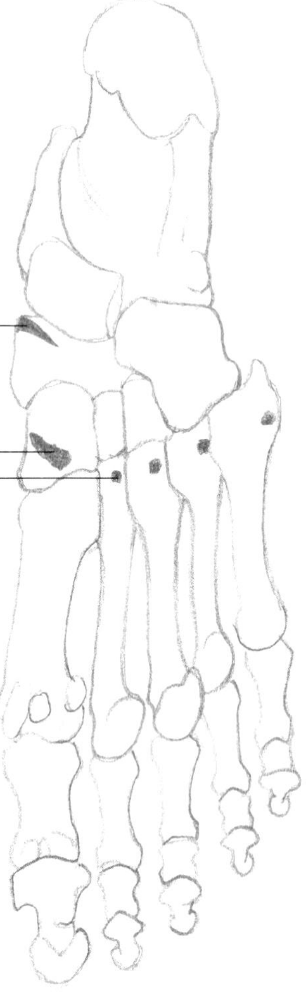

plantar aspect

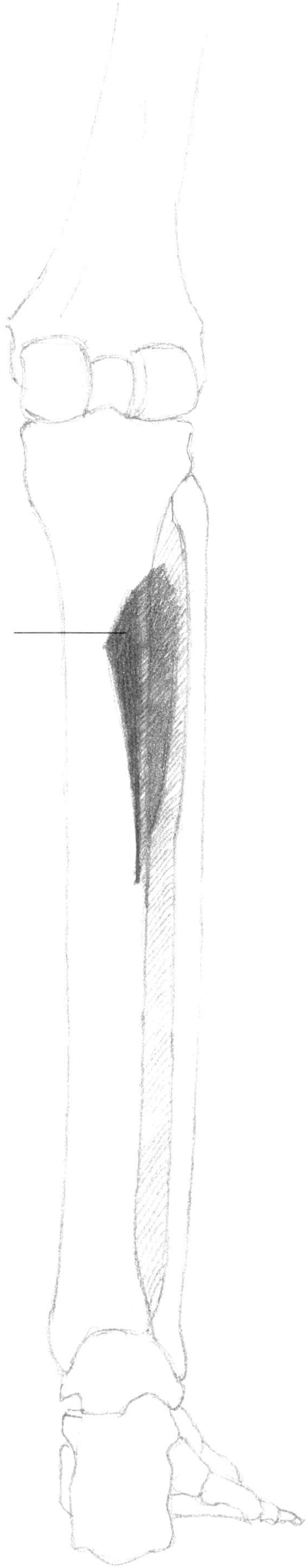

dorsal aspect

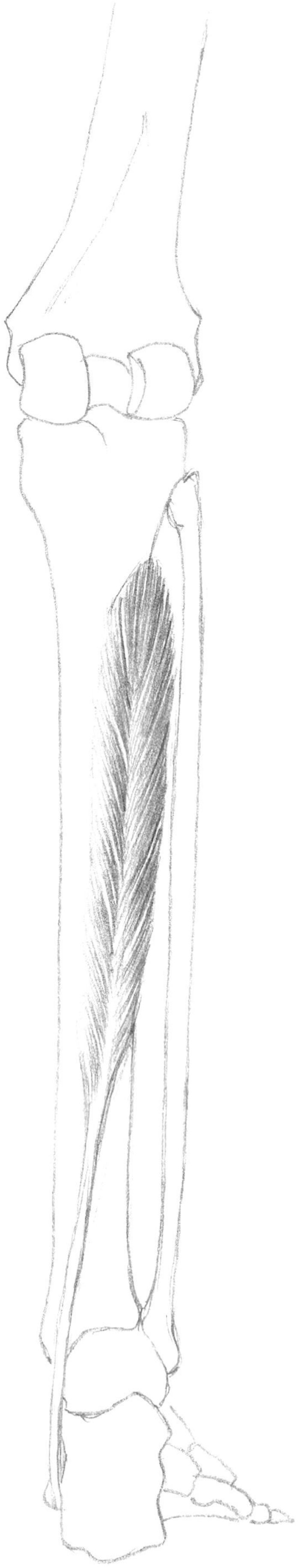

FUNCTION:
Flexes the foot in the ankle joint (plantar flexion) and lifts the medial edge of the foot

Fig. 159
The Flexor
Hallucis Longus,
110

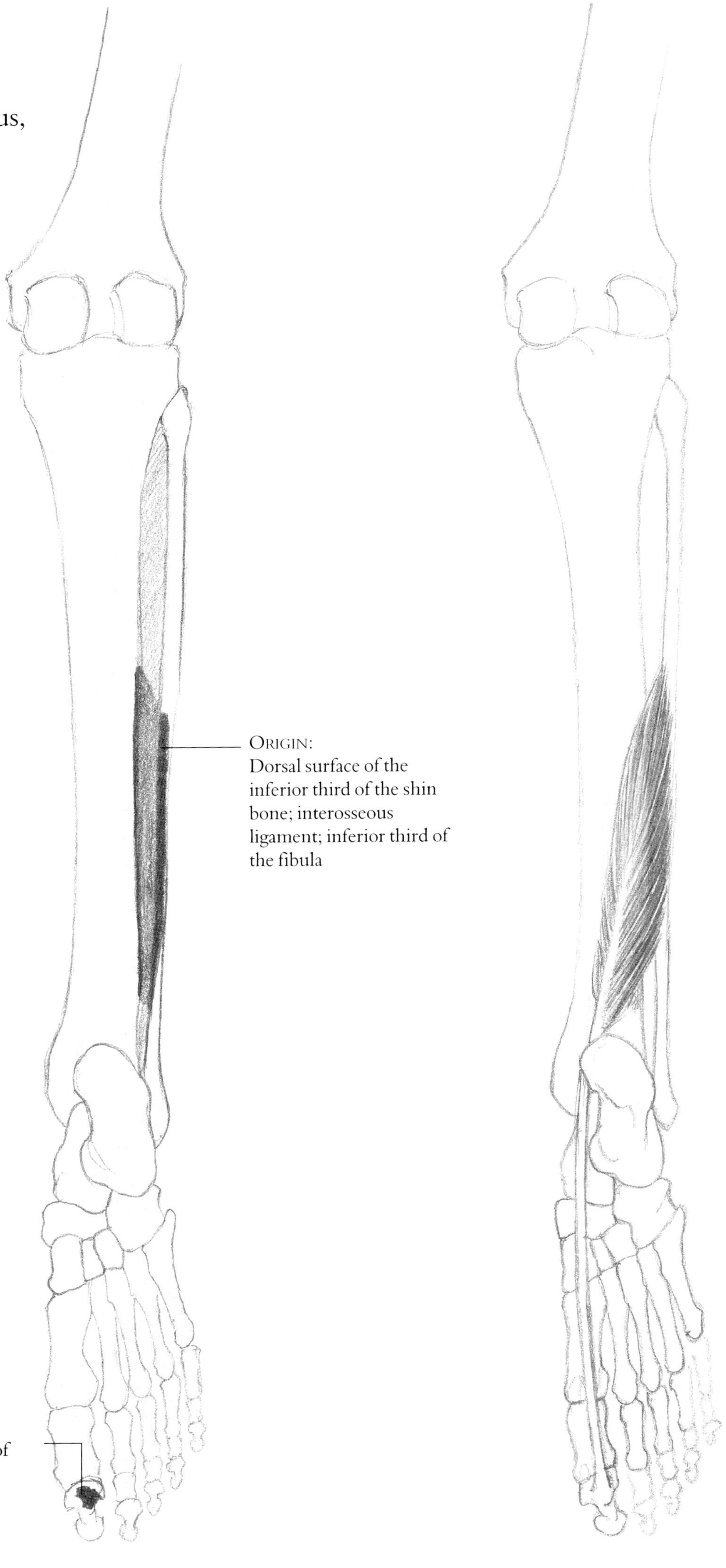

ORIGIN:
Dorsal surface of the inferior third of the shin bone; interosseous ligament; inferior third of the fibula
INSERTION:
Distal pediphalanx of first toe
FUNCTION:
Flexes the great toe and the foot and supinates the foot
dorsal aspect

Fig. 160
The Fibularis Brevis, *104*

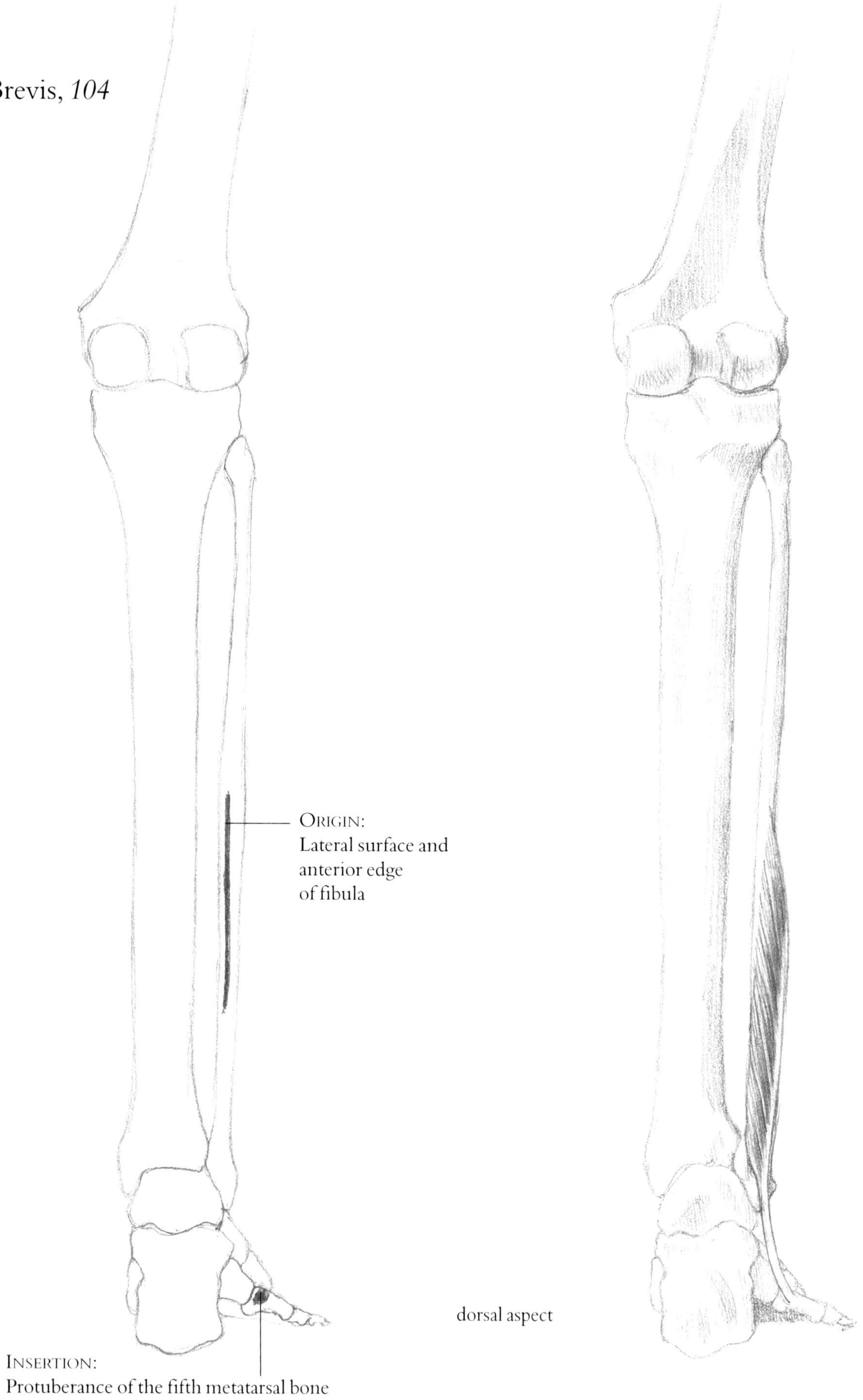

FUNCTION:
Flexes the foot; raises the lateral and lowers the medial edge of the foot (pronation)

Fig. 161
The Fibularis Longus, *103*

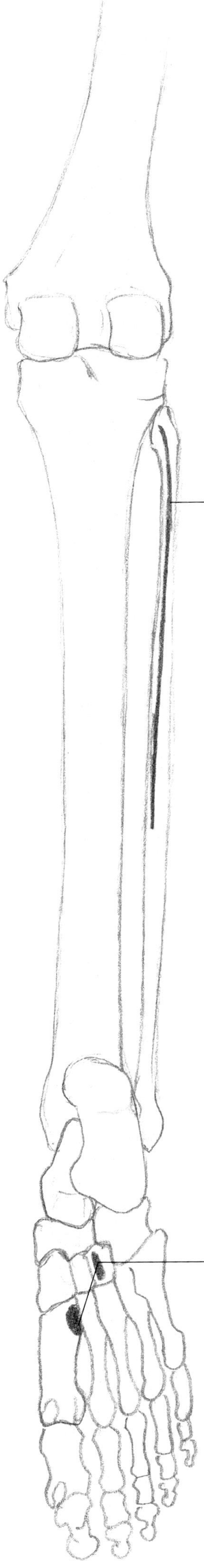

Origin:
Head of fibula and superior two-thirds of fibular shaft

Insertion:
With the long tendon at the first (second) metatarsal bone and medial protuberance of the sphenoid bone

dorsal aspect

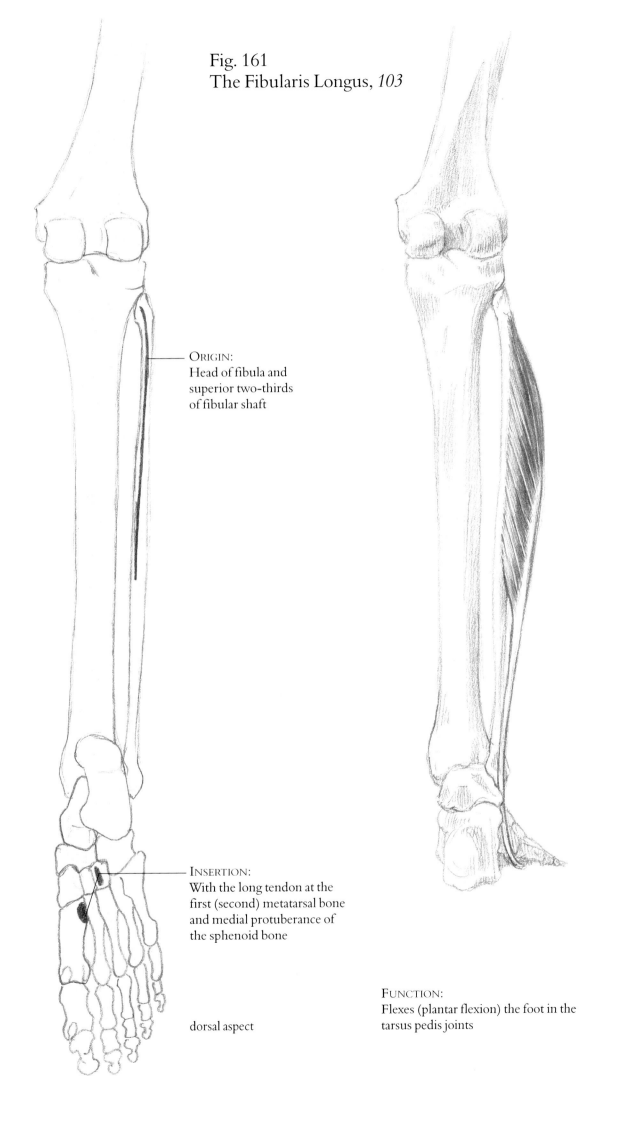

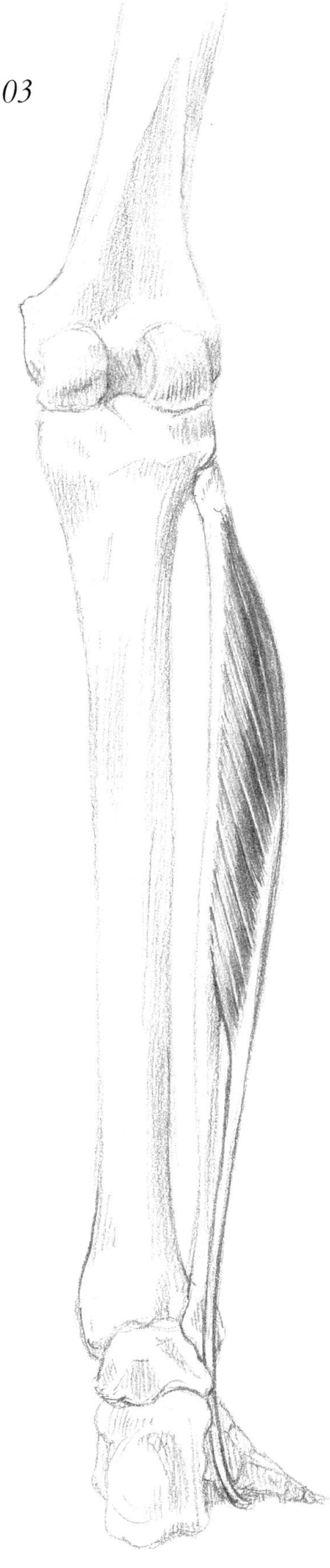

Function:
Flexes (plantar flexion) the foot in the tarsus pedis joints

Fig. 162
The Flexor Digitorum Longus, *109*

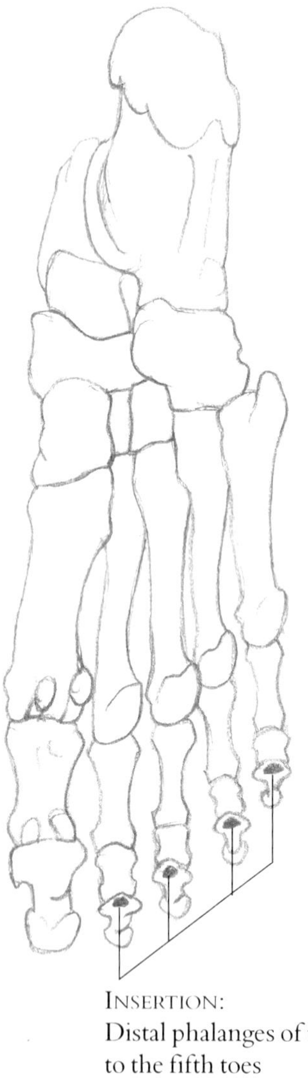

plantar aspect

INSERTION:
Distal phalanges of the second to the fifth toes

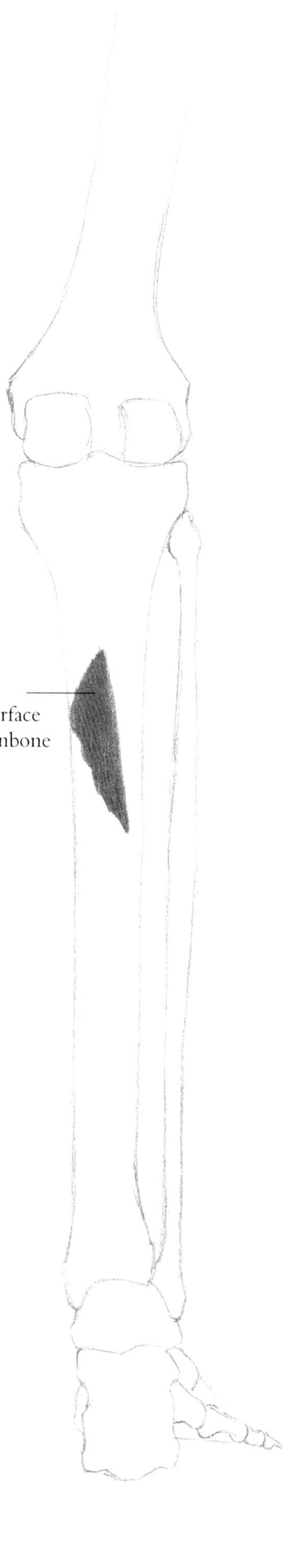

ORIGIN:
Dorsal surface of the shinbone

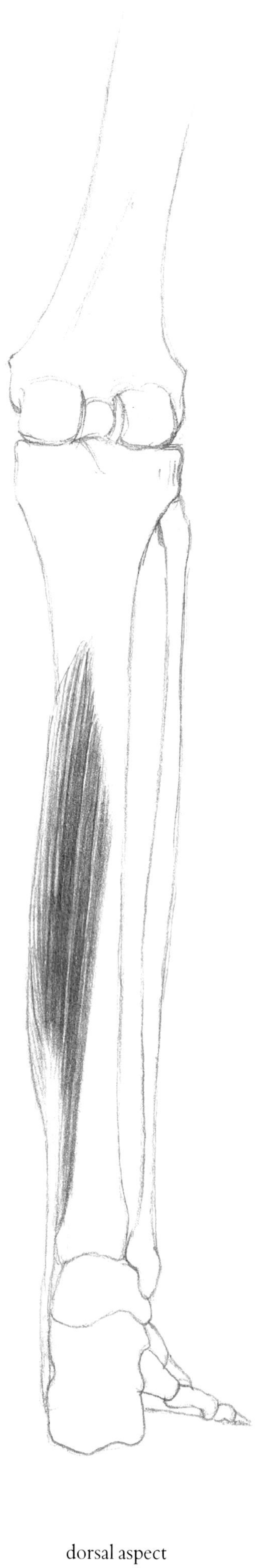

dorsal aspect

Function:
Flexes the end phalanges of the toes, assists in plantar flexion

Fig. 163
Origin and Insertion Surfaces
of the Lower Leg Muscles

O = Origin
I = Insertion

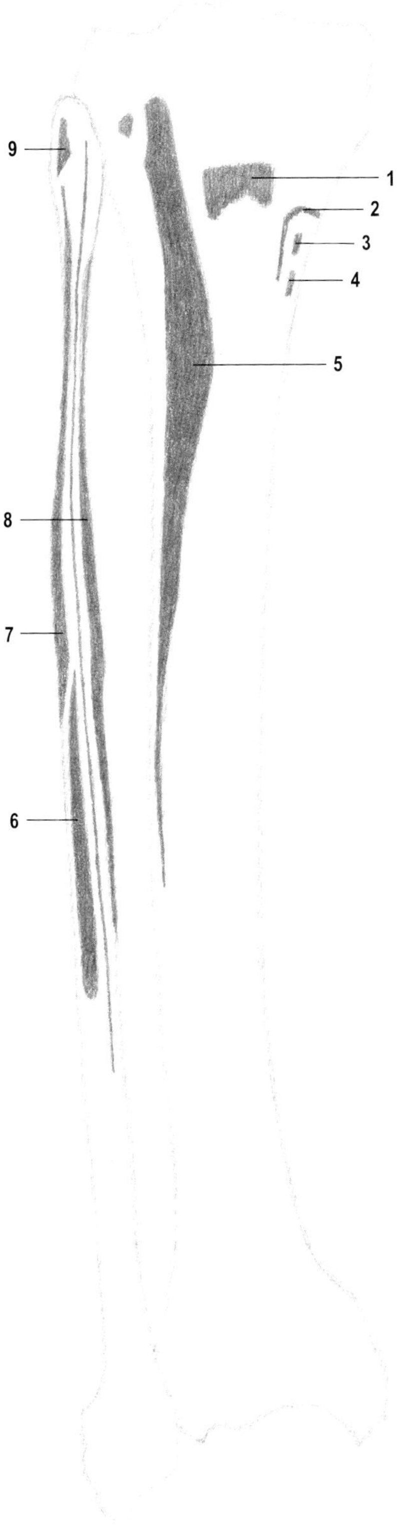

1 Quadriceps femoris, I (*97*)
2 Sartorius, I (*88*)
3 Gracilis, I (*89*)
4 Semitendinosus, I (*86*)
5 Tibilais anterior, O (*99*)
6 Fibularis brevis, O (*104*)
7 Extensor digitorum longus, O (*101*)
8 Tibialis posterior, O (*108*)
9 Fibularis longus, O (*103*)
10 Biceps femoris, I (*85*)
11 Soleus, O (*105/2*)
12 Flexor hallucis longus, O (*110*)
13 Flexor digitorum longus, O (*109*)
14 Popliteus, I (*107*)
15 Semimembranosus, I (*87*)

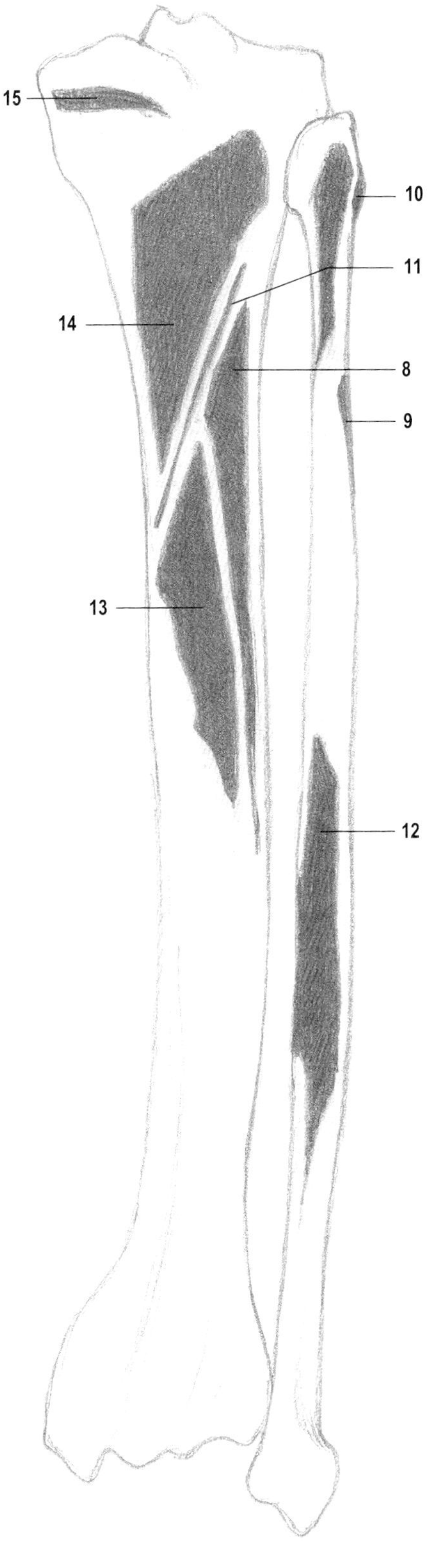

ventral surface

dorsal surface

Fig. 164
The Extensor Digitorum Brevis, *111*

The Extensor Hallucis Brevis, *112*

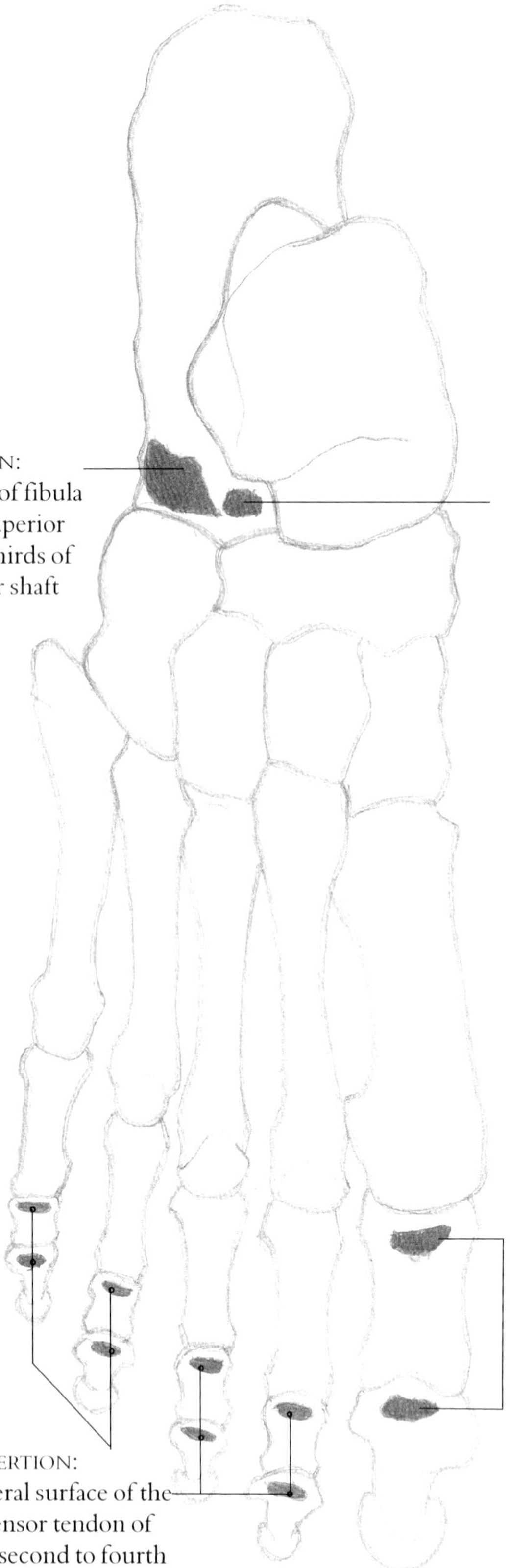

ORIGIN:
Head of fibula and superior two-thirds of fibular shaft (*111*)

INSERTION:
Lateral surface of the extensor tendon of the second to fourth toes (*111*)

ORIGIN:
Dorsal surface of heel bone (*112*)

INSERTION:
Proximal pediphalanx of the first toe (*112*)

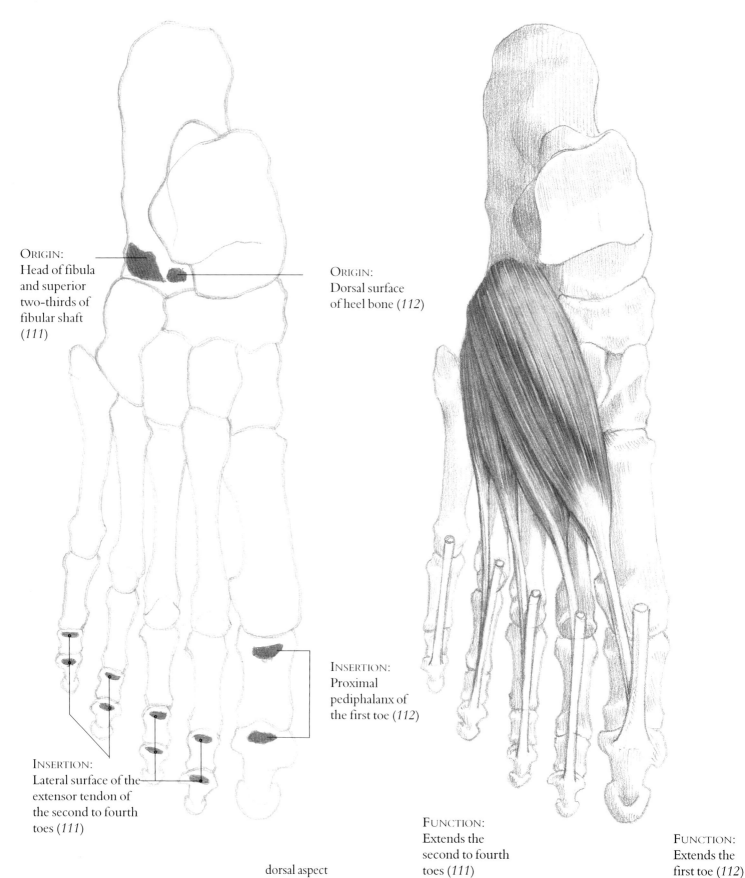

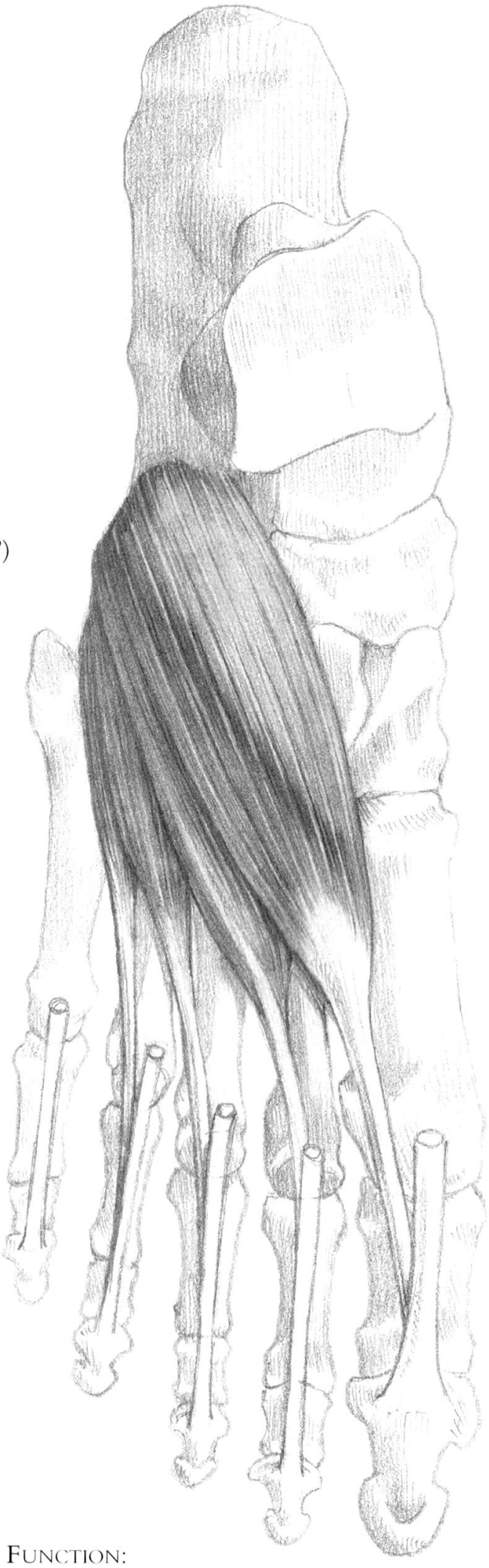

dorsal aspect

FUNCTION:
Extends the second to fourth toes (*111*)

FUNCTION:
Extends the first toe (*112*)

Fig. 165
The Muscles of the Foot

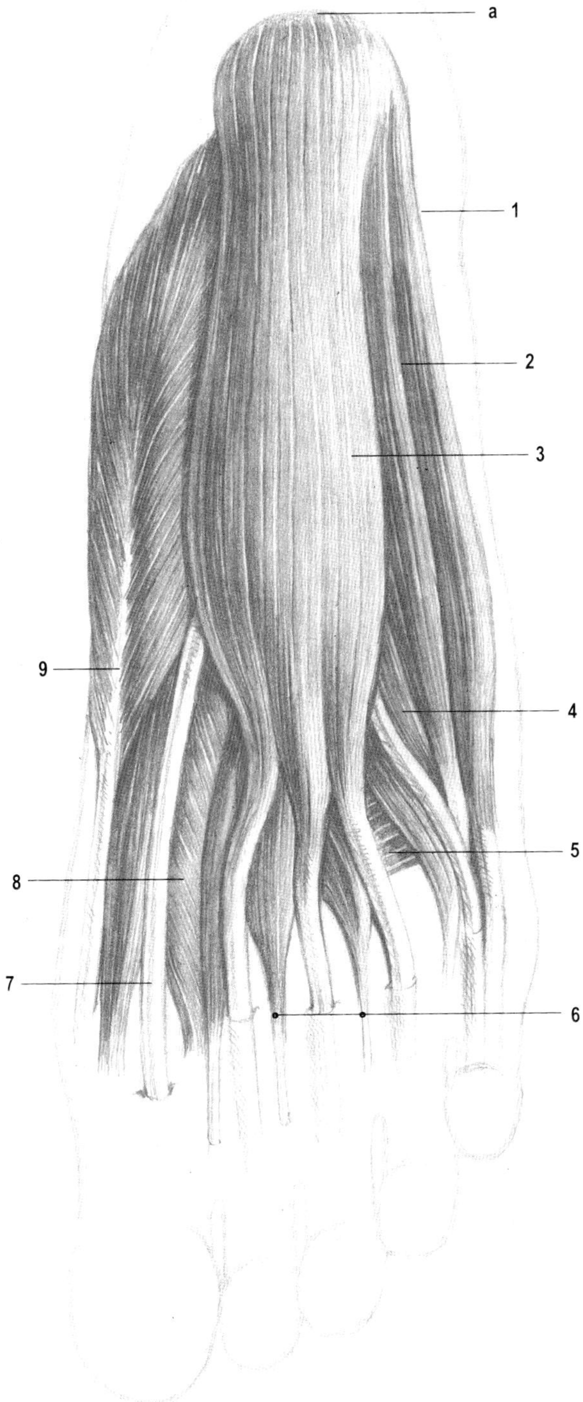

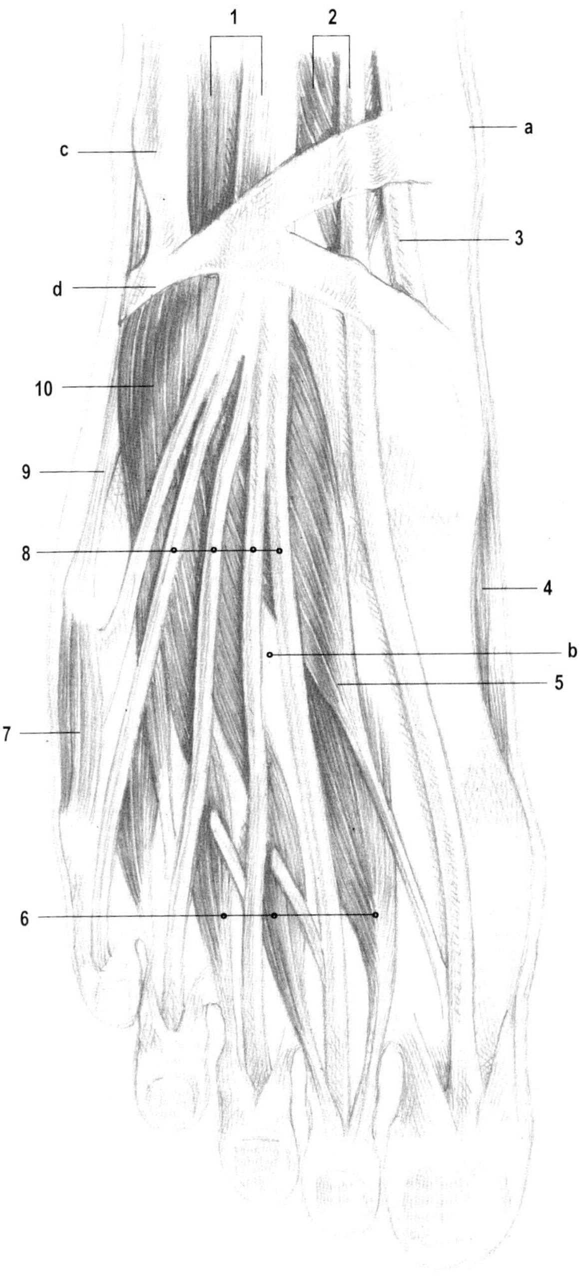

THE MUSCLES OF THE SOLE

1 Abductor digiti minimi (*119*)
2 Flexor digiti minimi brevis (*120*)
3 Flexor digitorum brevis (*122*)
4 Plantar interossei dorsales (*115*)
5 Adductor hallucis (transverse head) (*118*)
6 Lumbricals of the second to fifth toes (*113*)
7 Tendon of the flexor hallucis longus (*110*)
8 Flexor hallucis brevis (*117*)
9 Abductor hallucis (*116*)

a Heel bone

DORSAL ASPECT

1 Extensor digitorum longus (*101*)
2 Extensor hallucis longus (*100*)
3 Tendon of tibialis anterior (*99*)
4 Abductor hallucis (*116*)
5 Extensor hallucis brevis (*112*)
6 Dorsal interosseous muscles (*114*)
7 Abductor digiti minimi (*119*)
8 Tendons of the extensor digitorum longus (*101*)
9 Third fibular muscle (*102*)
10 Extensor digitorum brevis (*111*)

a Medial ankle bone
b Arch of proximal pediphalanx of the big toe
c Lateral heel bone protuberance
d Retinaculum musculorum extensorum superius

Fig. 166
The Interossei Dorsales, *114*

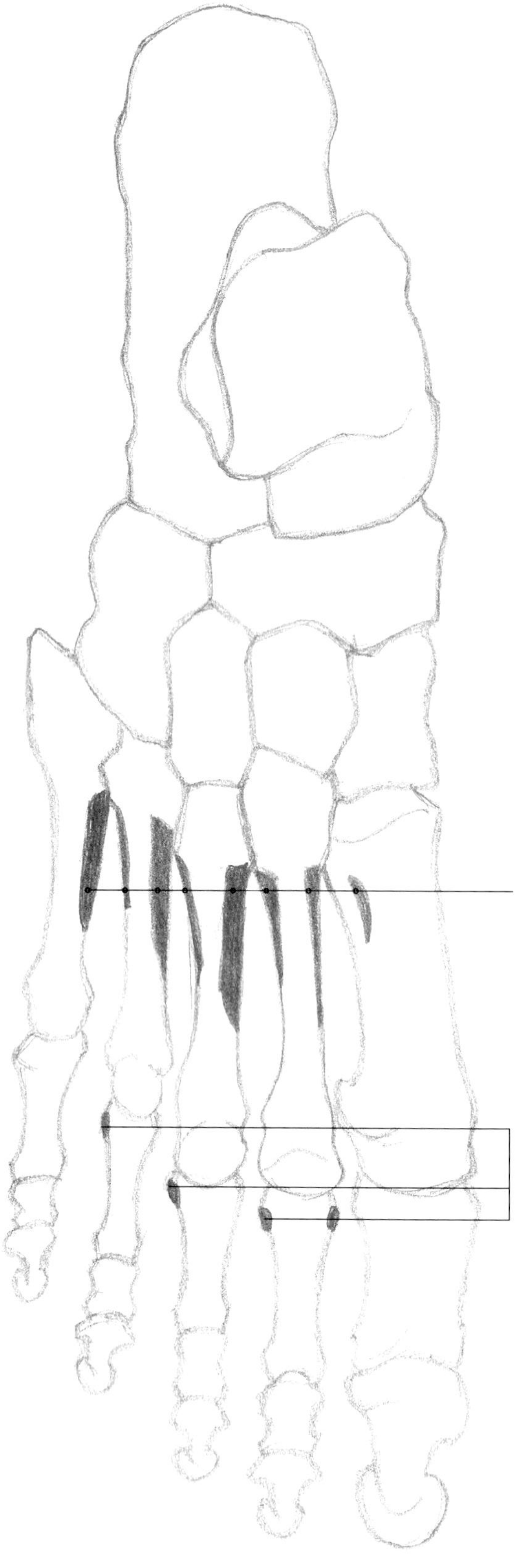

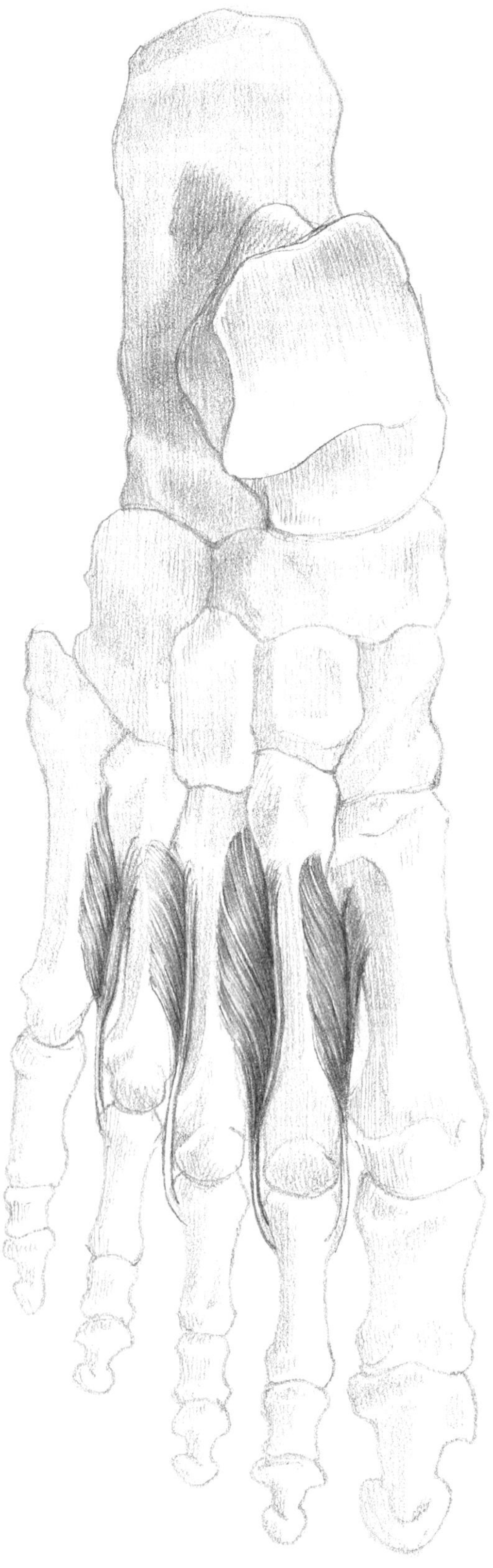

dorsal aspect

Fig. 167
The Interossei Plantares, *115*

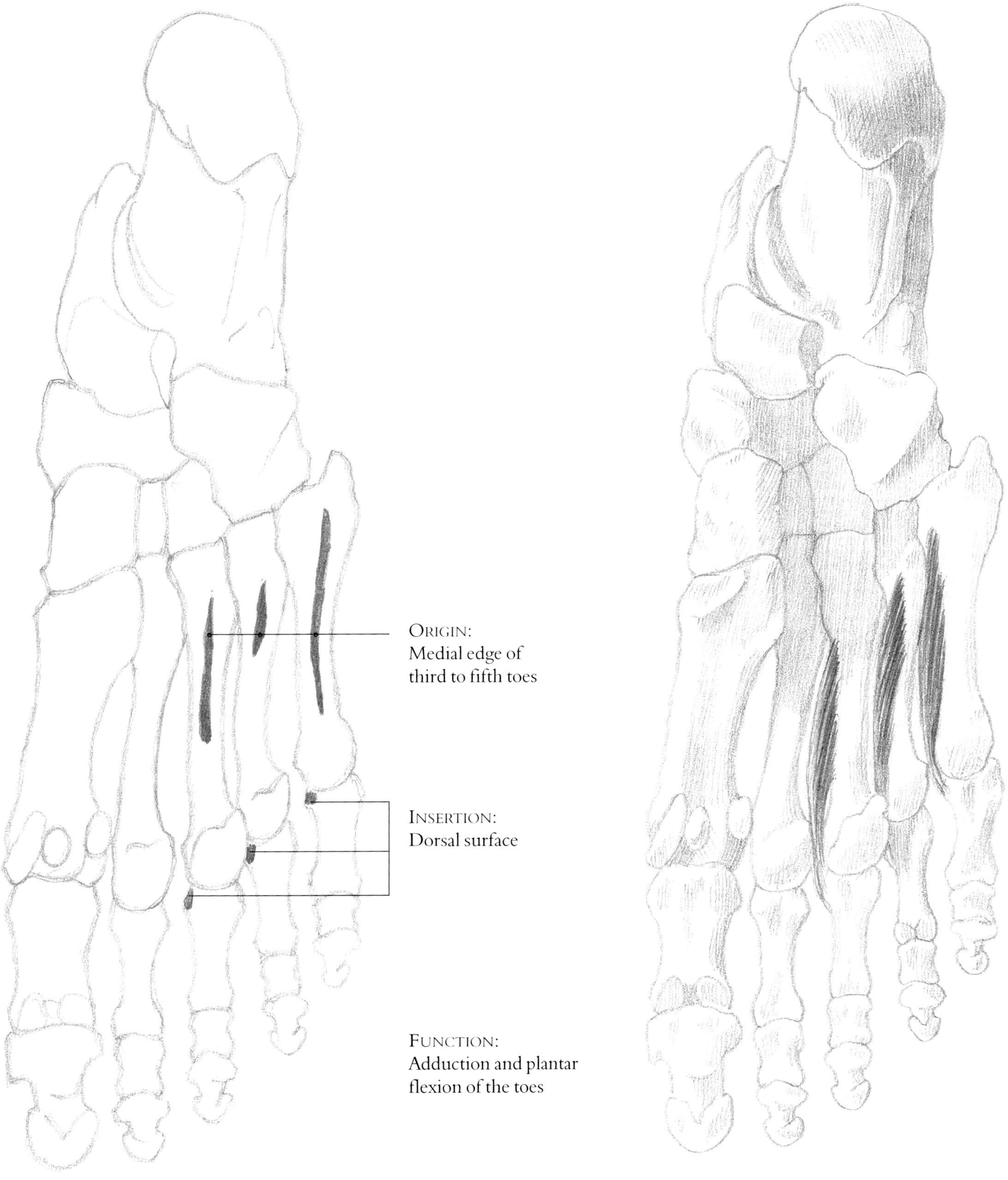

plantar aspect

Fig.168
The Lumbricals, *113*

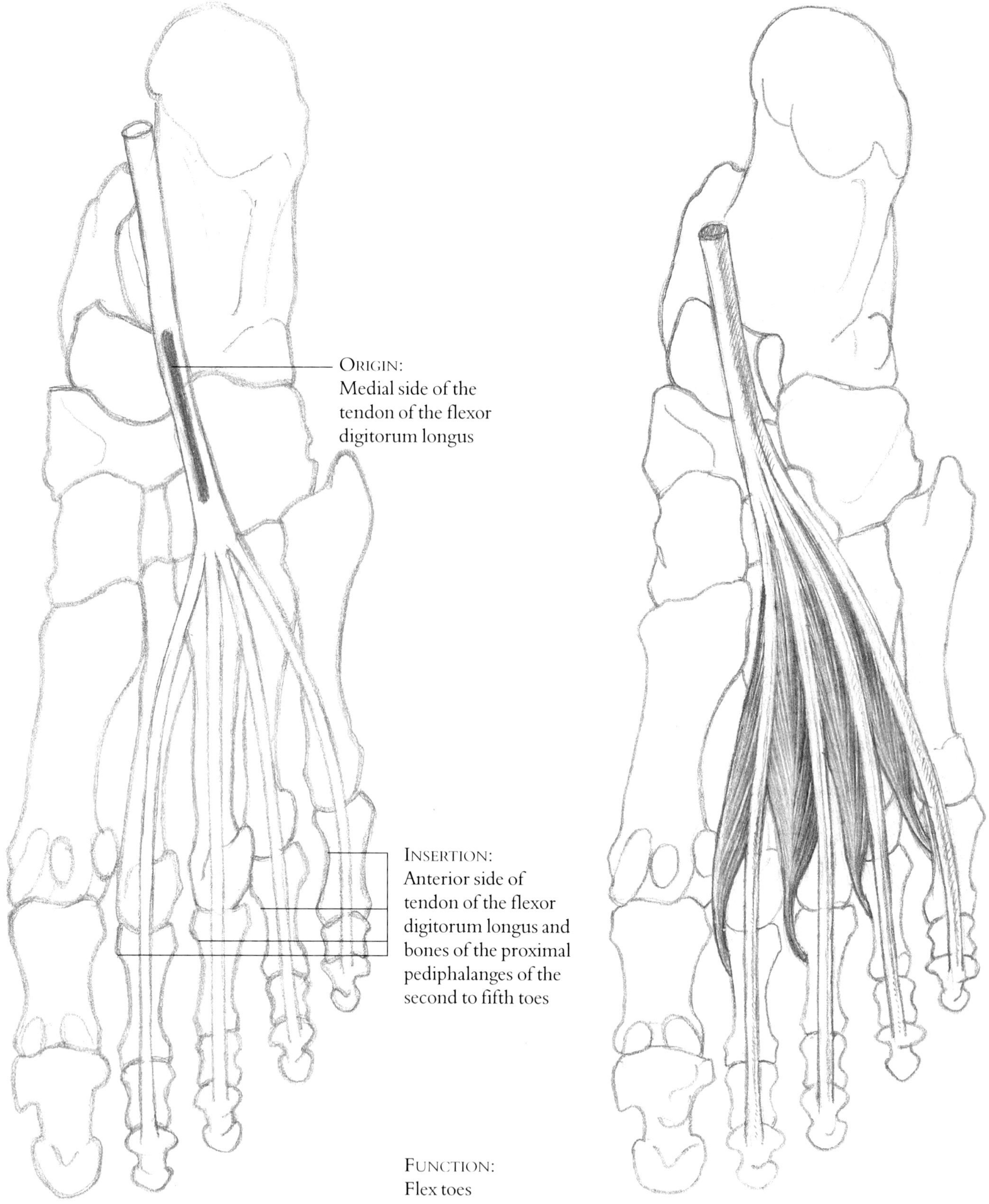
ORIGIN:
Medial side of the tendon of the flexor digitorum longus
INSERTION:
Anterior side of tendon of the flexor digitorum longus and bones of the proximal pediphalanges of the second to fifth toes
FUNCTION:
Flex toes

plantar aspect

Fig. 169
The Quadratus Plantae, *123*

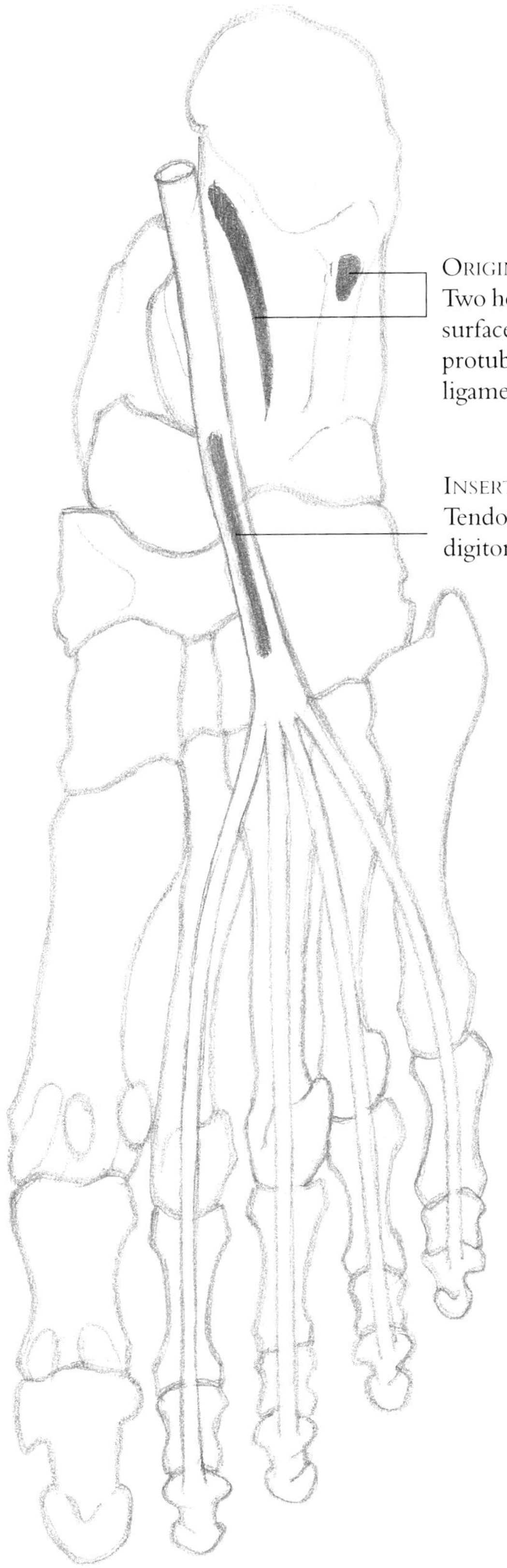

Origin:
Two heads at the posterior surface of the heel bone protuberance and on the ligament of the plantar surface

Insertion:
Tendons of the flexor digitorum longus

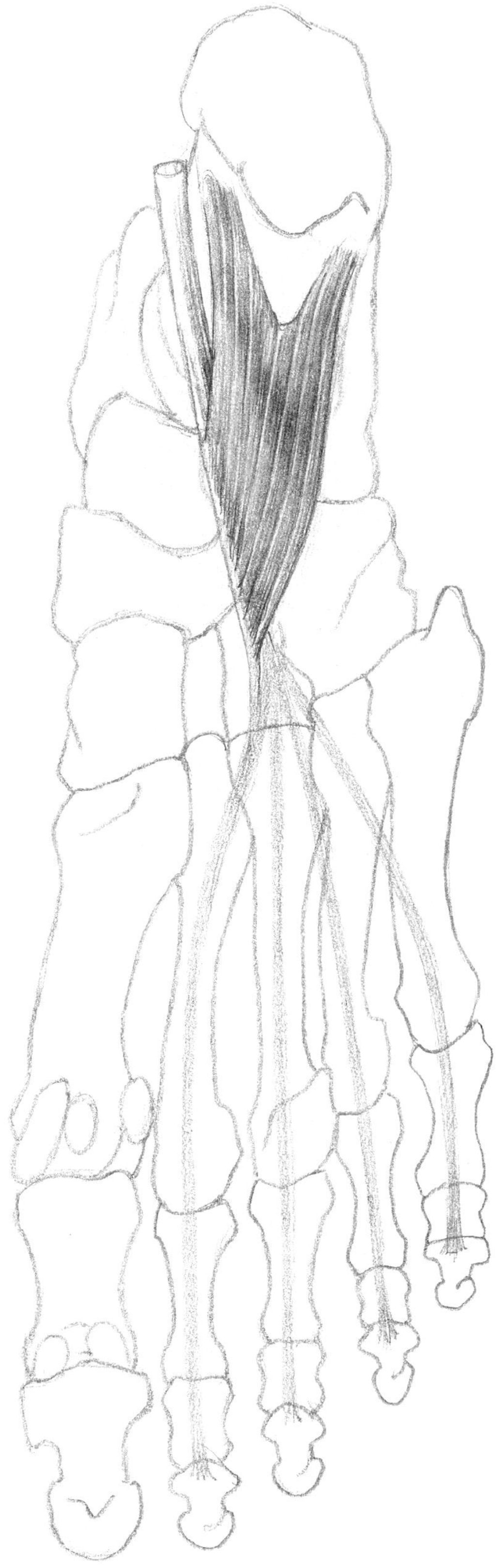

Function:
Supports muscle *109* in plantar flexion of the distal phalanges

plantar aspect

Fig. 170
The Flexor Digiti Minimi Brevis, *120*

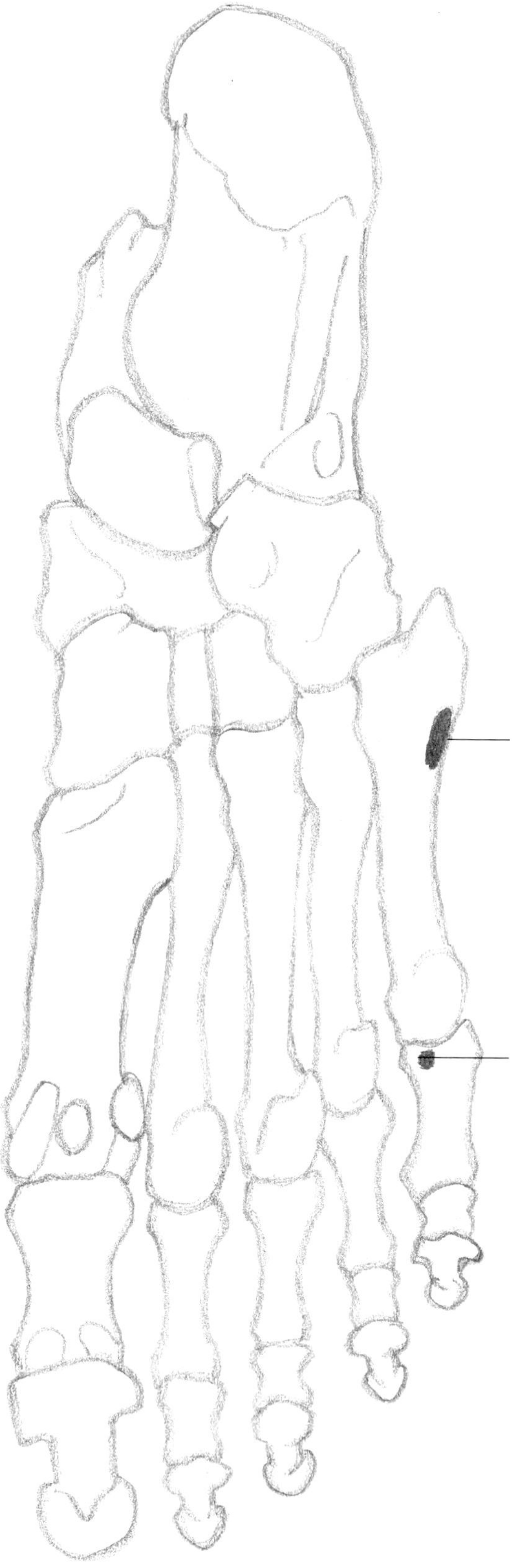

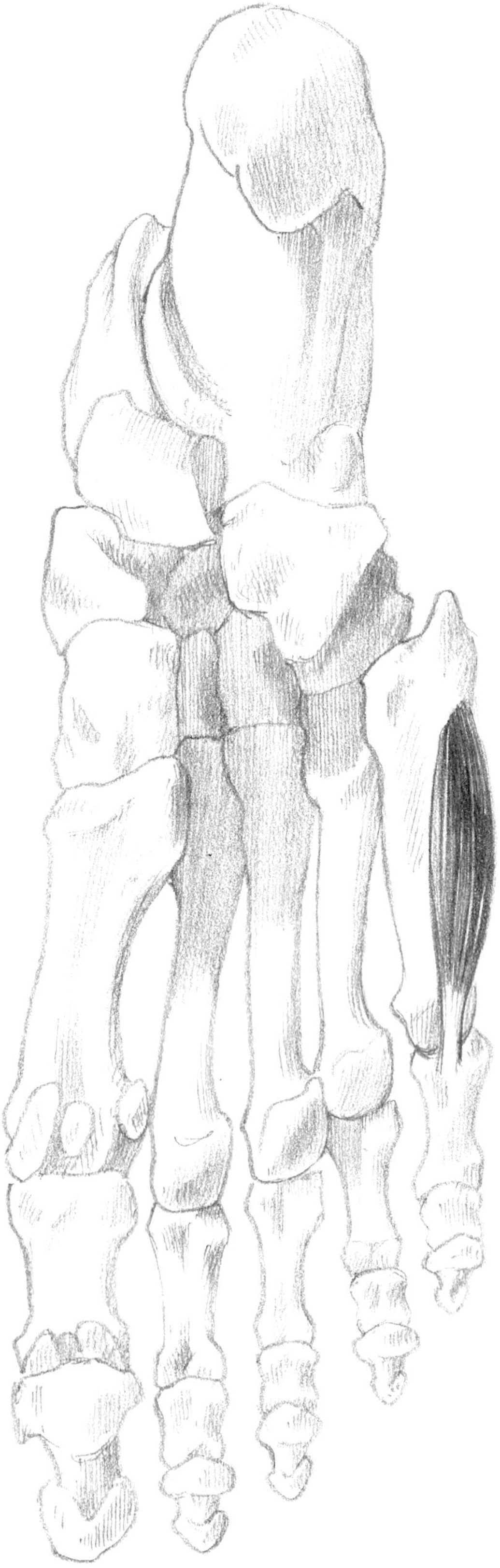

plantar aspect

Fig. 171
The Opponens Digiti Minimi, *121*

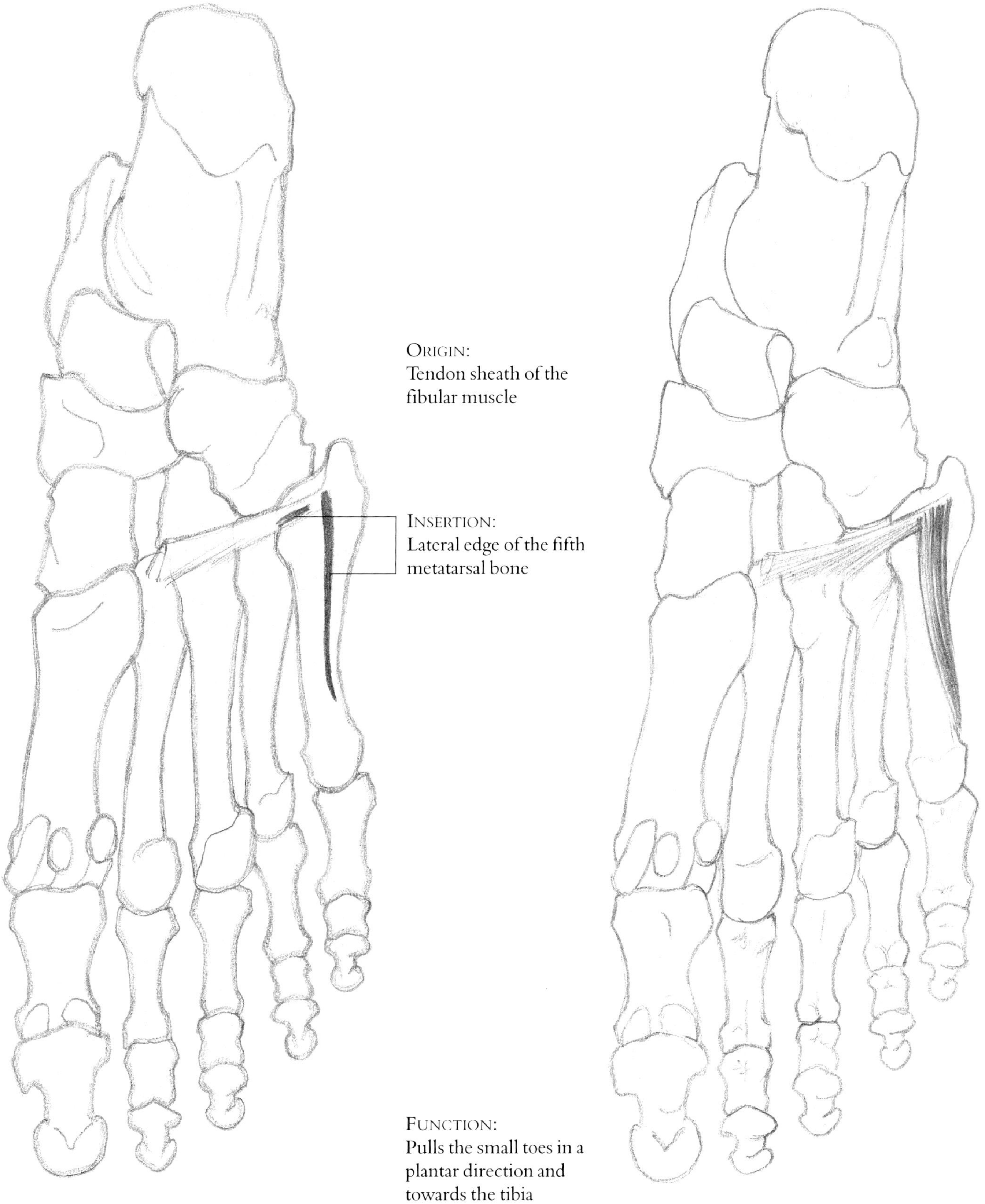

plantar aspect

Fig. 172
The Flexor Digitorum Brevis, *122*

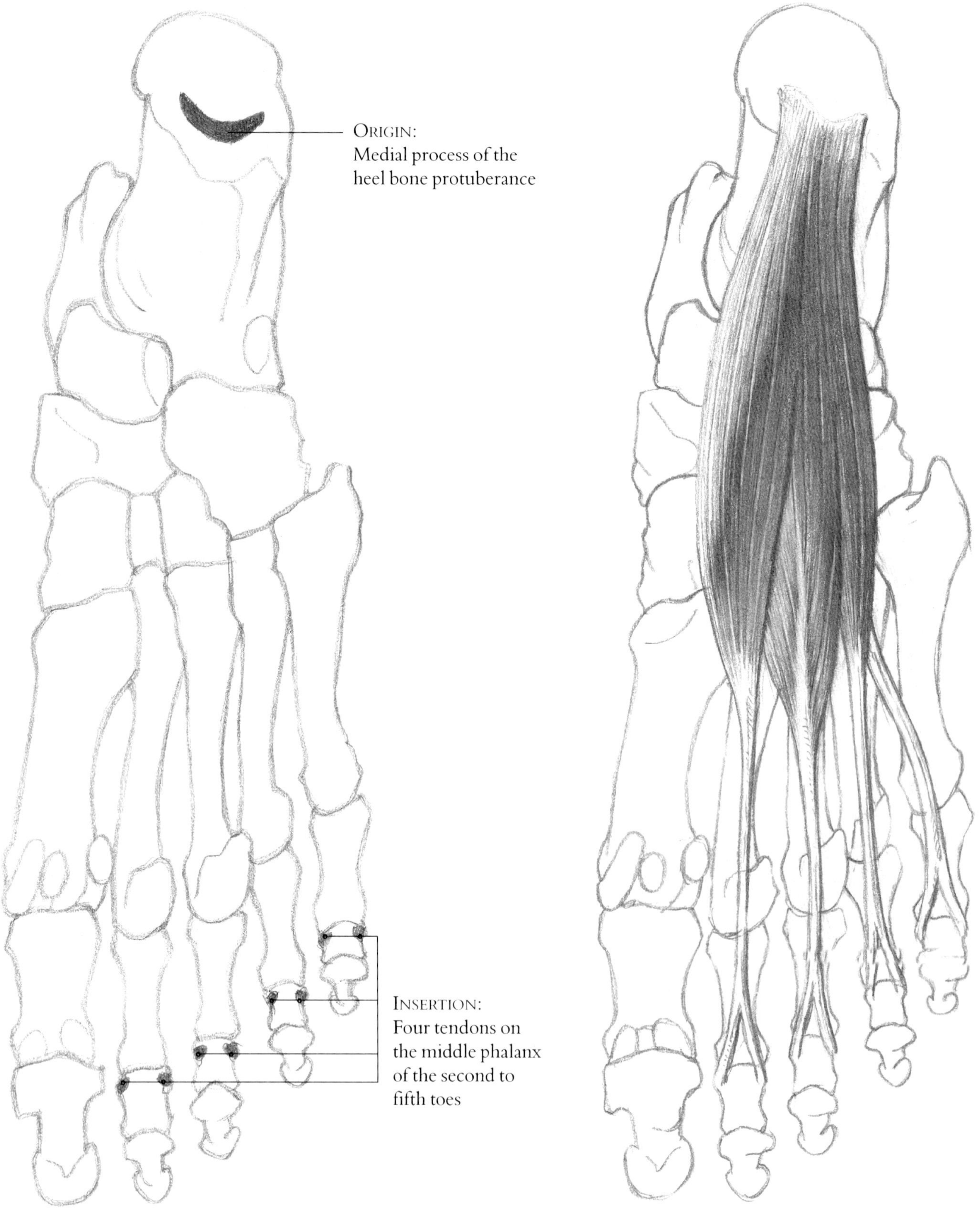

FUNCTION:
flexes the middle
phalanx of the
second to fifth toes

plantar aspect

Fig. 173/a
The Abductor Muscle of the
Great Toe
(M. abductor hallucis, *116*)

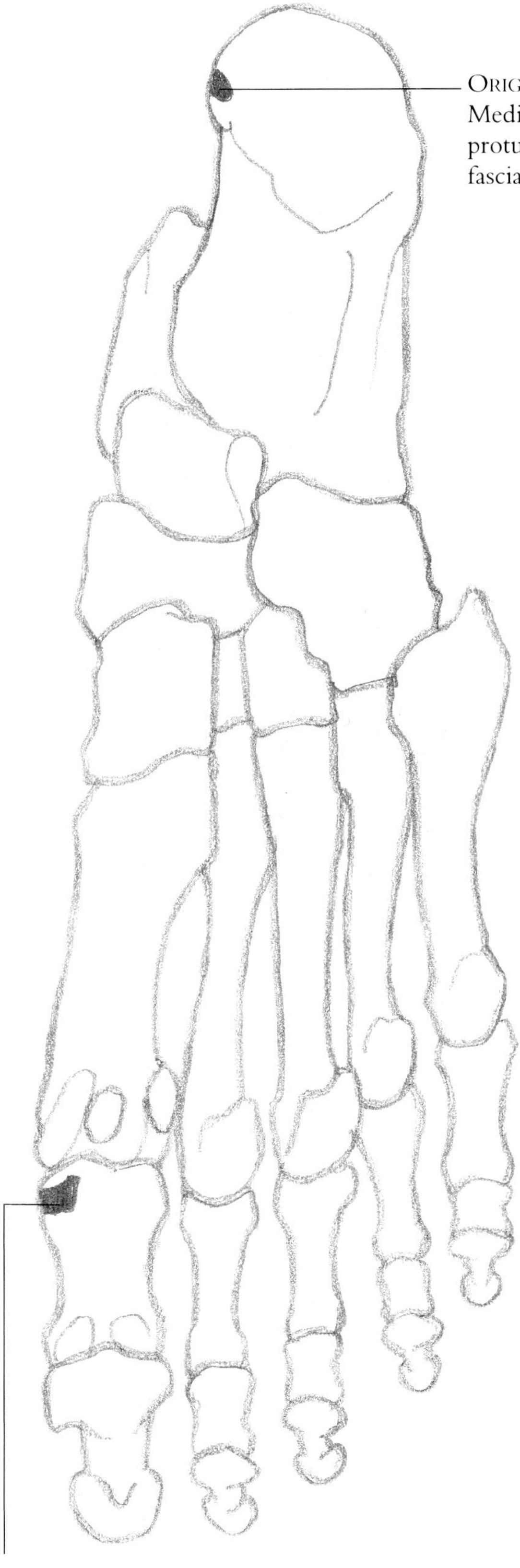

ORIGIN:
Medial protuberance of heel bone protuberance (medial head); deep plantar fascia (lateral head)

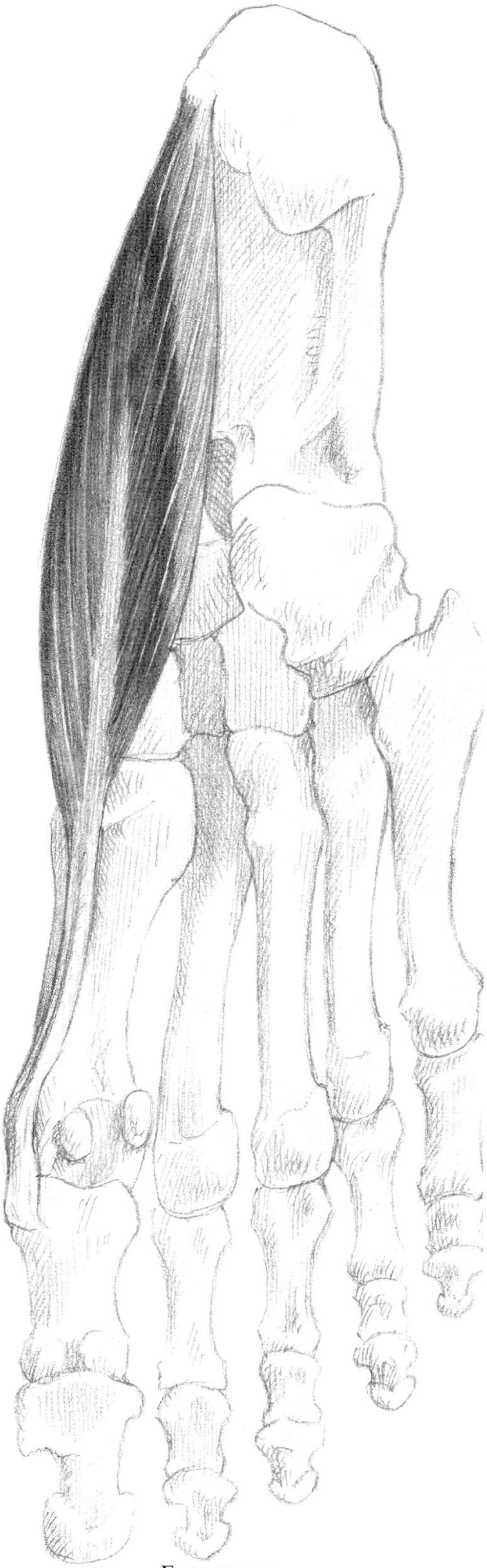

INSERTION:
Proximal pediphalanx of the first toe

FUNCTION:
Flexes and abducts the first toe

plantar aspect

Fig. 173/b
The Abductor Muscle of the Great Toe
(M. adductor hallucis, *118*)

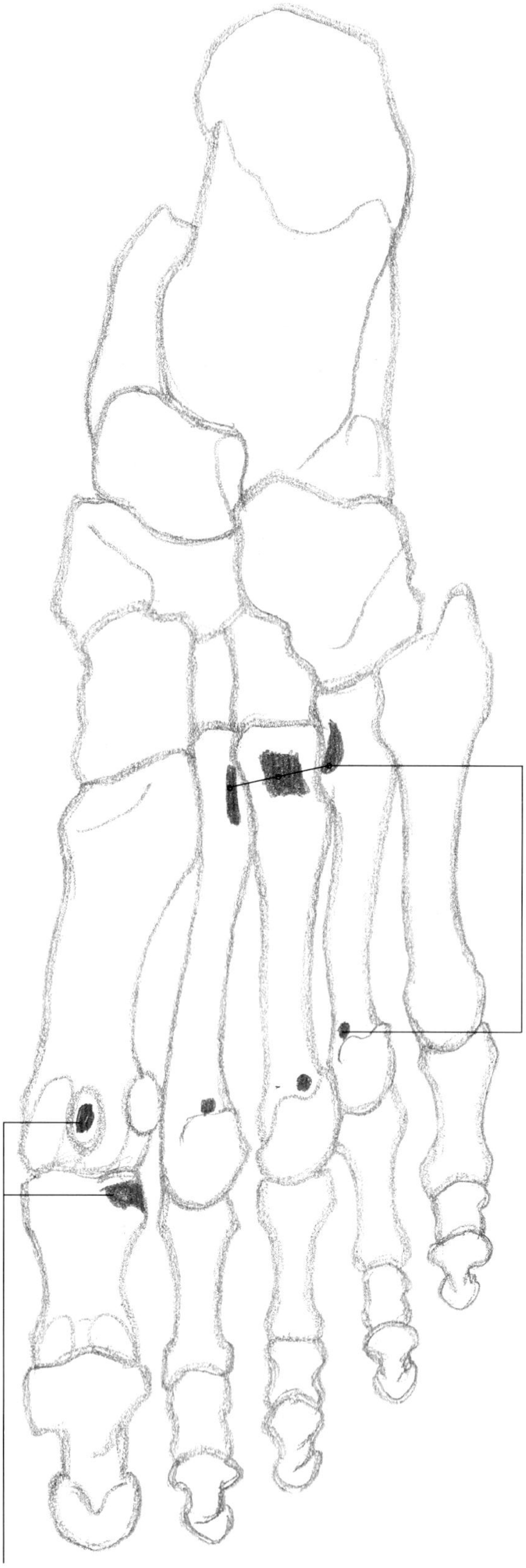

ORIGIN:
Ligament between tarsus pedis and metatarsal bones and long palmar ligament (transverse head); proximal end of first to fourth metatarsal bones (diagonal head)

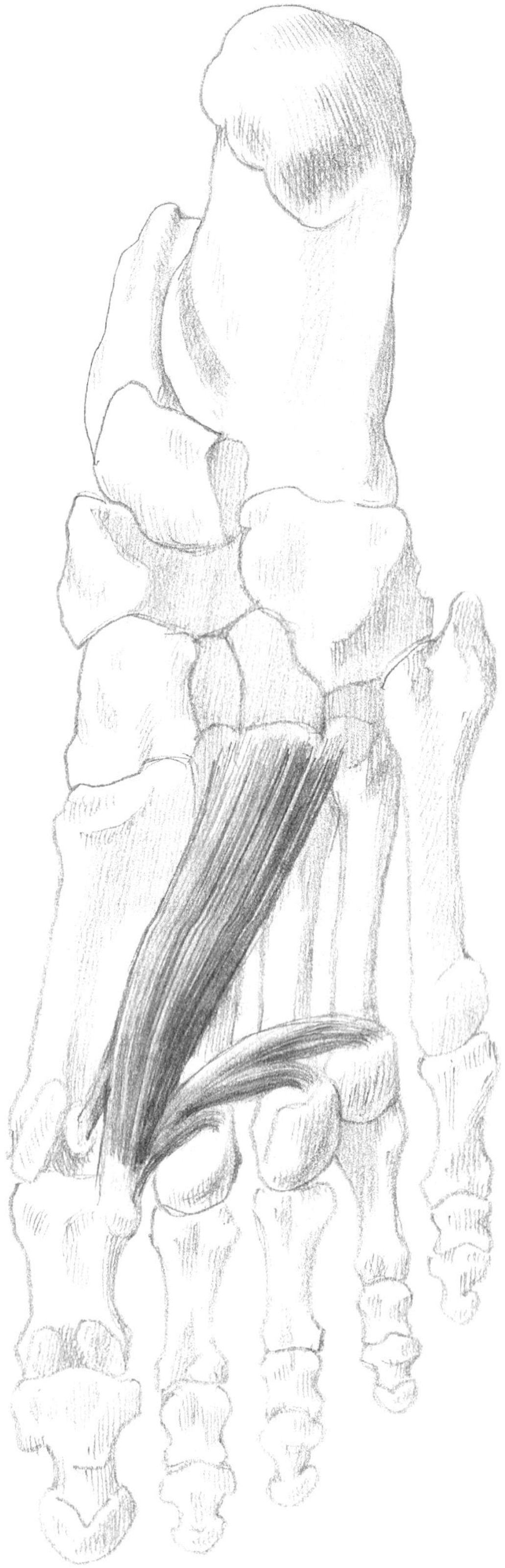

FUNCTION:
Adducts and opposes the big toe, fixes the foot arch

INSERTION:
Lateral sesamoid bone and proximal phalanx of first toe

plantar aspect

Fig. 174
The Flexor Hallucis Brevis, *117*

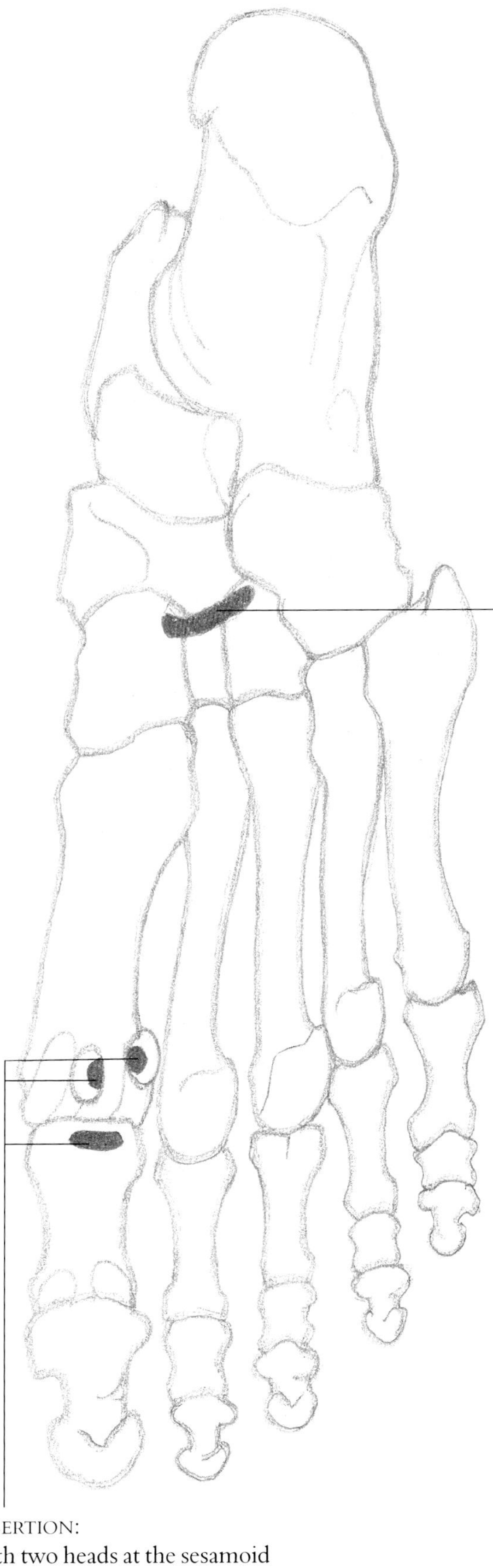

ORIGIN:
Plantar surface of the three lateral sphenoid bones, long plantar fascia

INSERTION:
With two heads at the sesamoid bone and at the proximal phalanx of the first toe

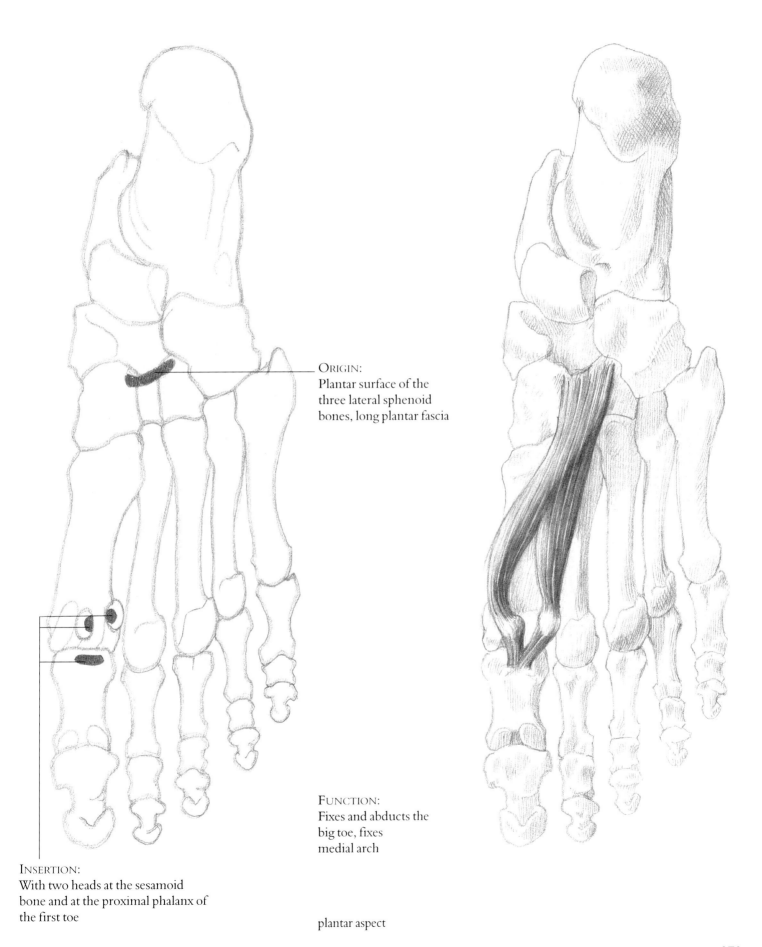

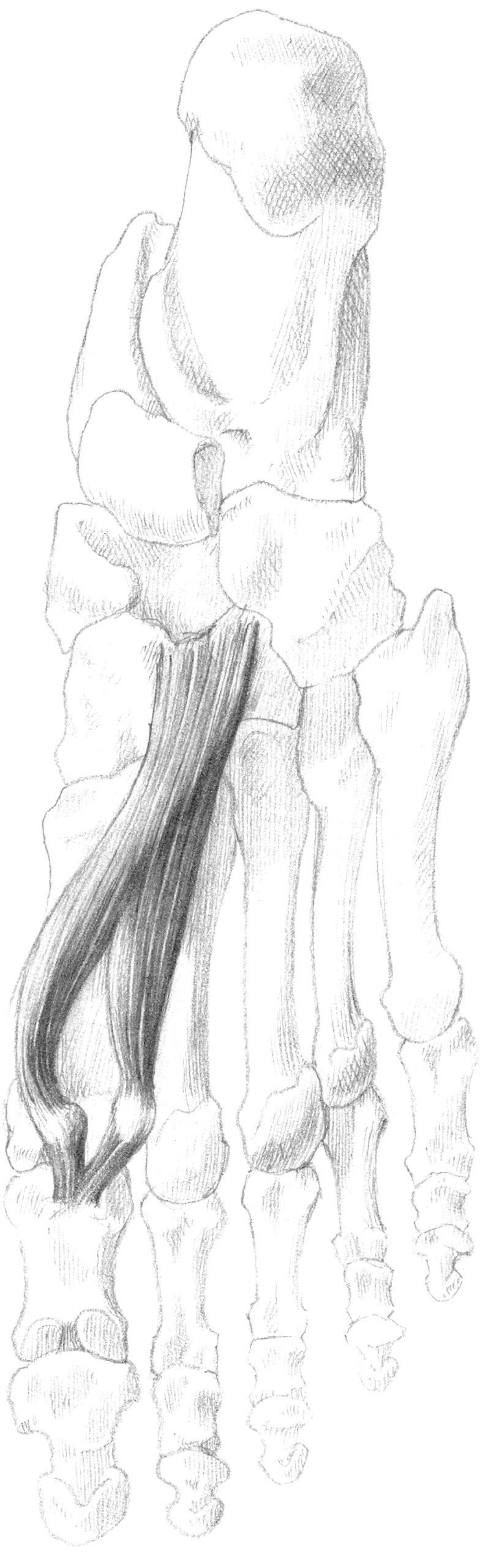

FUNCTION:
Fixes and abducts the big toe, fixes medial arch

plantar aspect

Fig. 175
The Abductor Muscle of the Little Toe

(M. abductor digiti minimi pedis, *119*)

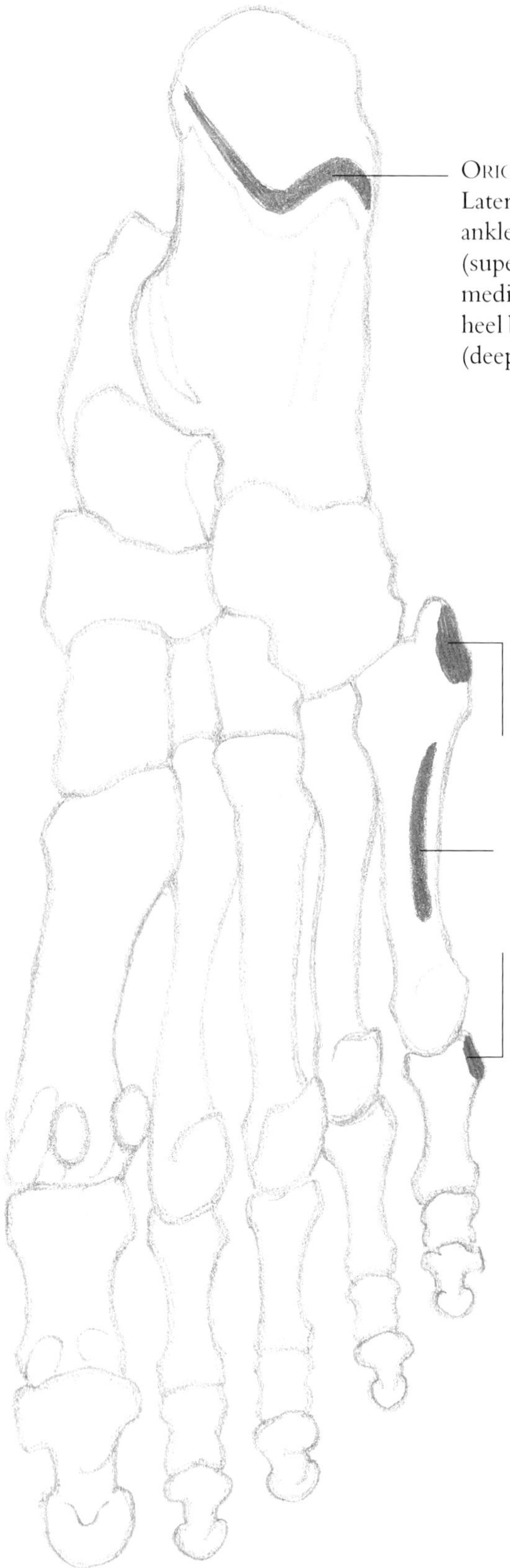

ORIGIN:
Lateral process of the ankle protuberance (superficial head); medial process of the heel bone protuberance (deep head)

INSERTION:
Lateral edge of the proximal phalanx of the fifth toe; posterior surface of the fifth metatarsal bone

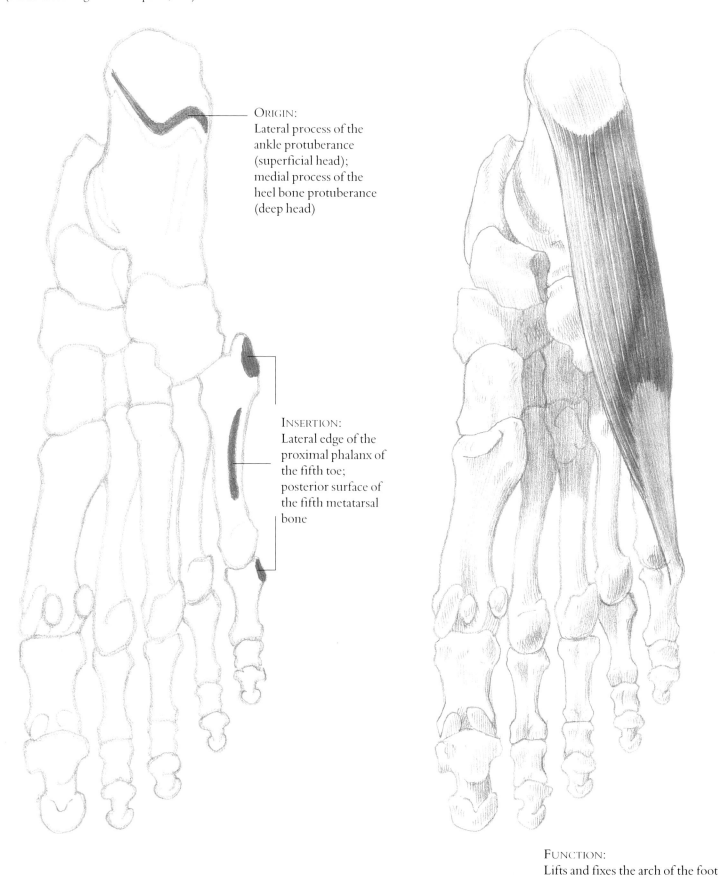

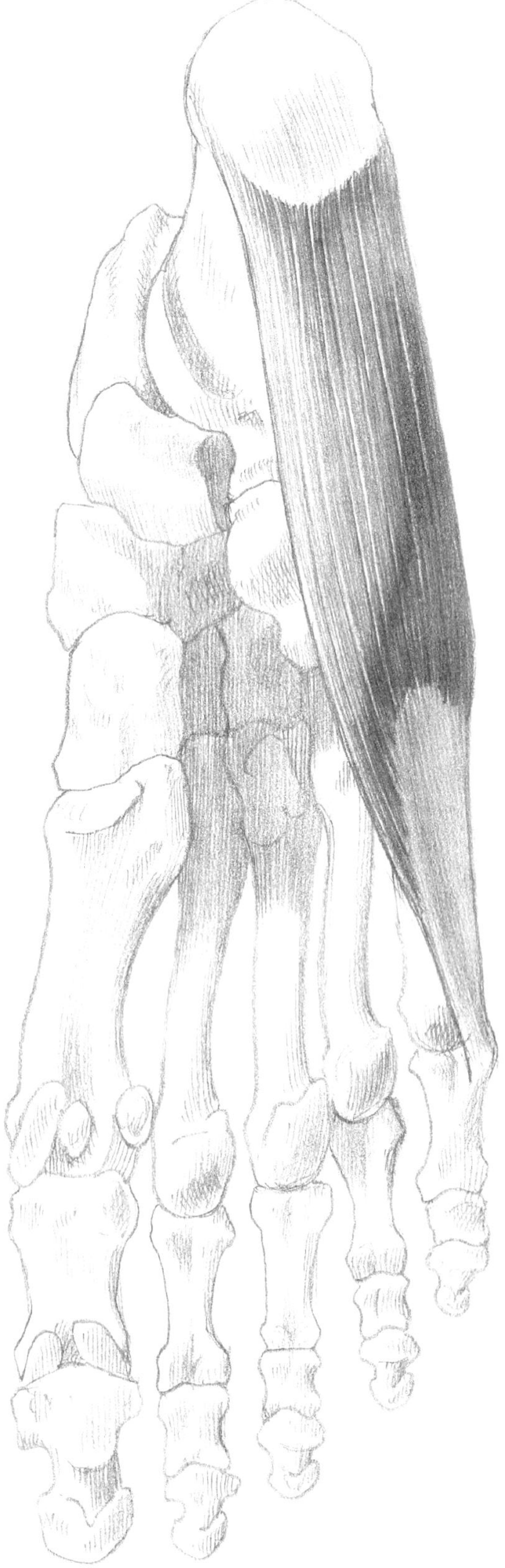

FUNCTION:
Lifts and fixes the arch of the foot

plantar aspect

Fig. 176
Origins and Points of Insertion of the Muscles in the Foot Region

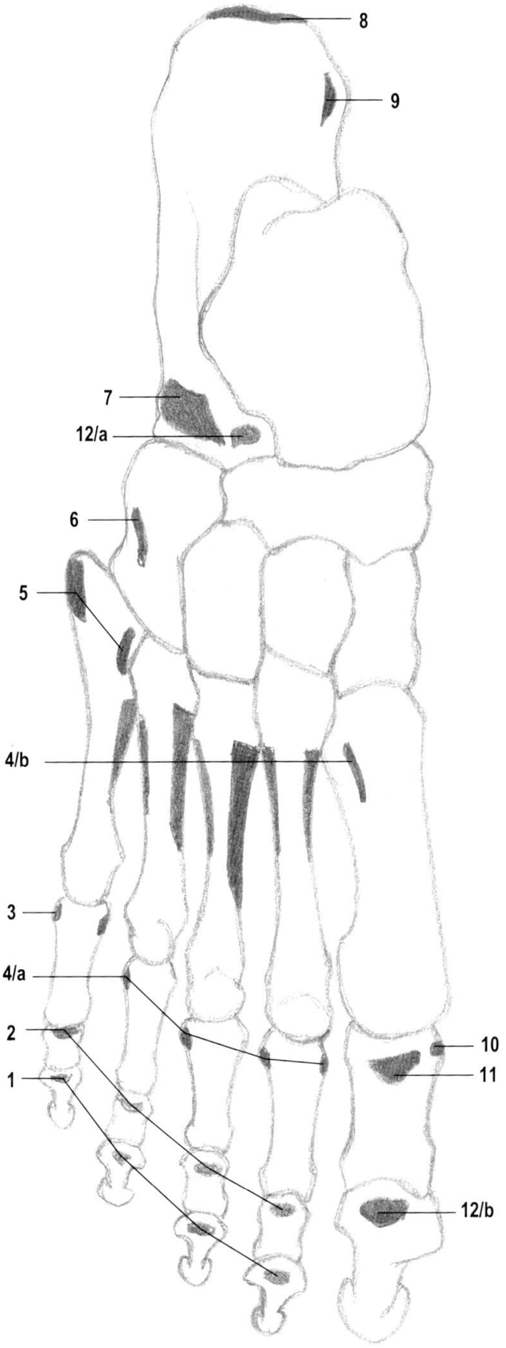

dorsal surface

Dorsal Surface

1 Extensor digitorum longus, I (*101*)
2 Extensor digitorum brevis, I (*111*)
3 Abductor digiti minimi, I (*119*)
4 Dorsal interosseous muscles, **a**=O, **b**=I (*114*)
5 third fibularis, I (*102*)
6 Fibularis brevis, I (*104*)
7 Extensor digitorum brevis, I (*111*)
8 Achilles tendon, I (*105*)
9 Quadratus plantae, O (*123*)
10 Abductor hallucis, I (*116*)
11 Extensor hallucis longus, I (*100*)
12 Extensor hallucis brevis, **a**=O, **b**=I (*112*)

Plantar Surface

1 Flexor hallucis longus, I (*110*)
2 Abductor hallucis, **a**=O, **b**=I (*116*)
3 Abductor hallucis (A: transverse head, B: diagonal head), **a**=O, **b**=I (*118*)
4 Flexor hallucis brevis, **a**=O, **b**=I (*117*)
5 Fibularis longus, I (*103*)
6 Tibialis anterior, I (*99*)
7 Tibialis posterior, I (*108*)
8 Quadratus plantae, **a**=O, **b**=I (*123*)
9 Flexor digitorum brevis, **a**=O, **b**=I (*122*)
10 Abductor digiti minimi, **a**=O, **b**=I (*119*)
11 Flexor digiti minimi brevis, **a**=O, **b**=I (*120*)
12 Opponens digiti minimi, **a**=O, **b**=I (*121*)
13 Plantar interosseous muscles, **a**=O, **b**=I (*115*)
14 Flexor hallucis longus, I (*109*)

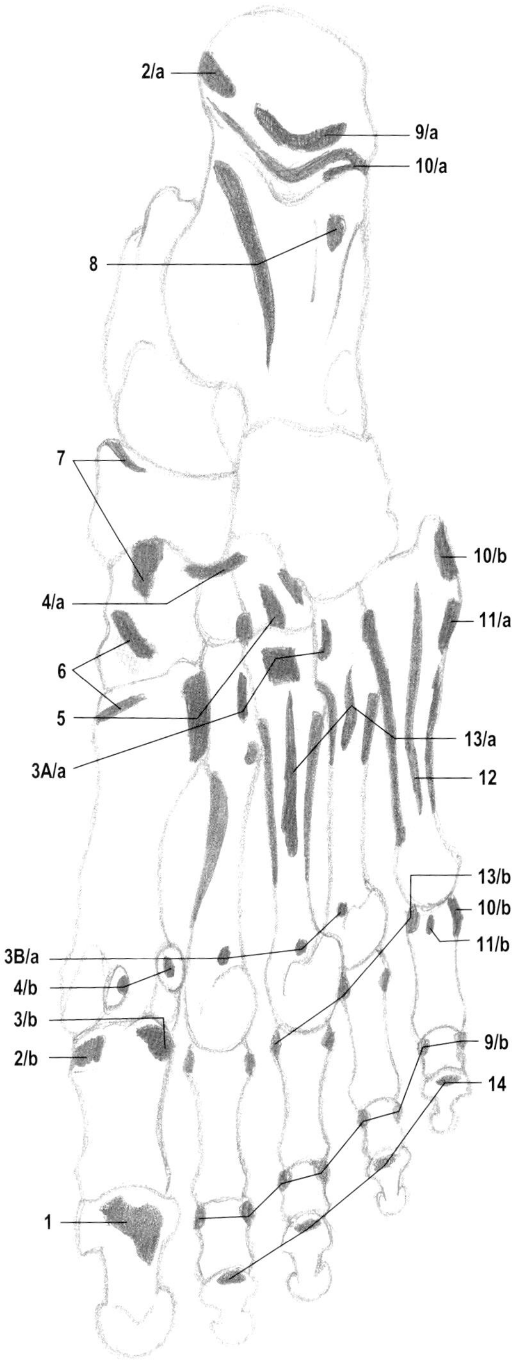

plantar surface

Fig. 177
The Right Foot

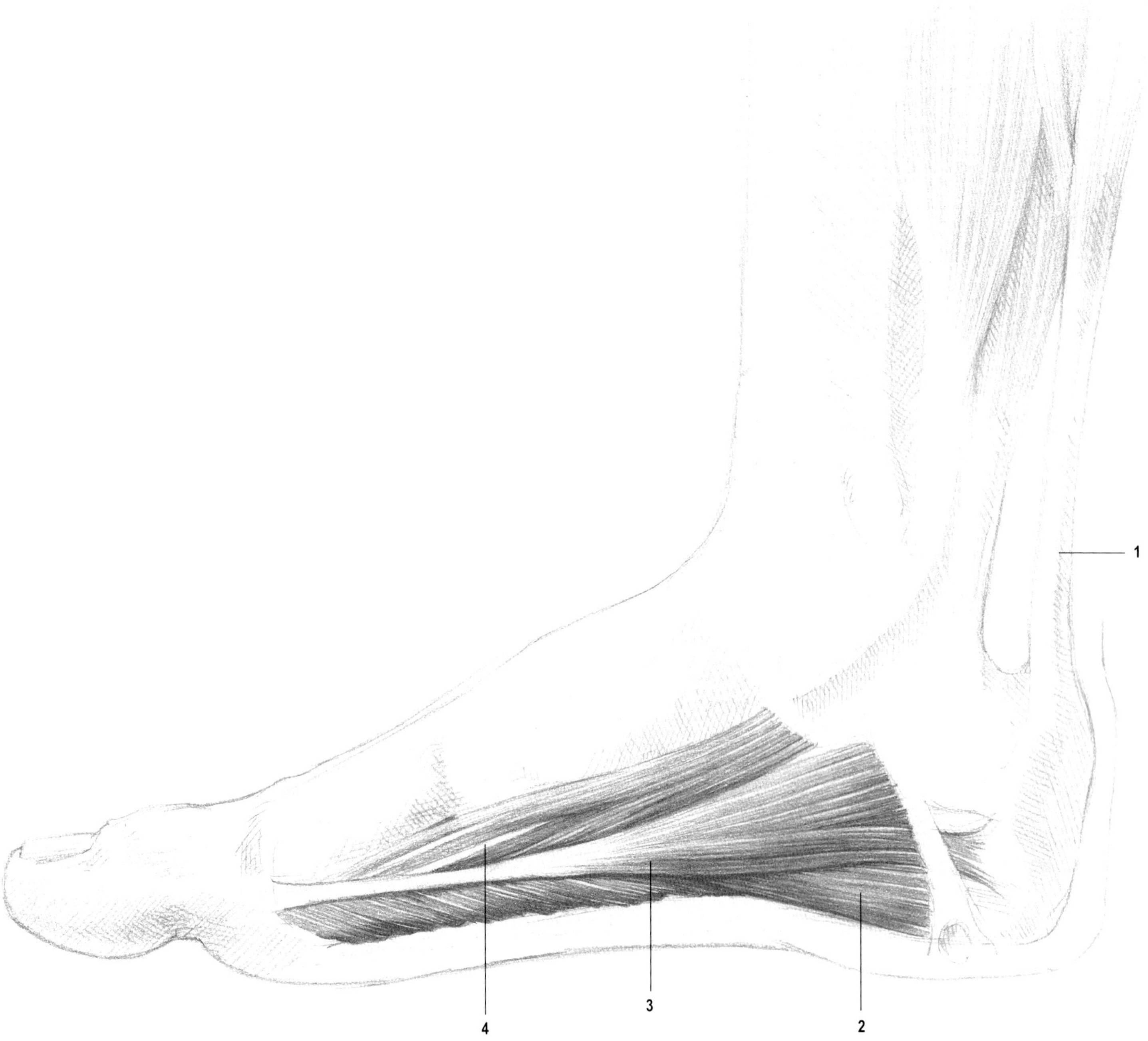

medial aspect

1 Achilles tendon (*105/1*)
2 Flexor digitorum brevis (*122*)
3 Abductor hallucis (*116*)
4 Flexor hallucis brevis (*117*)

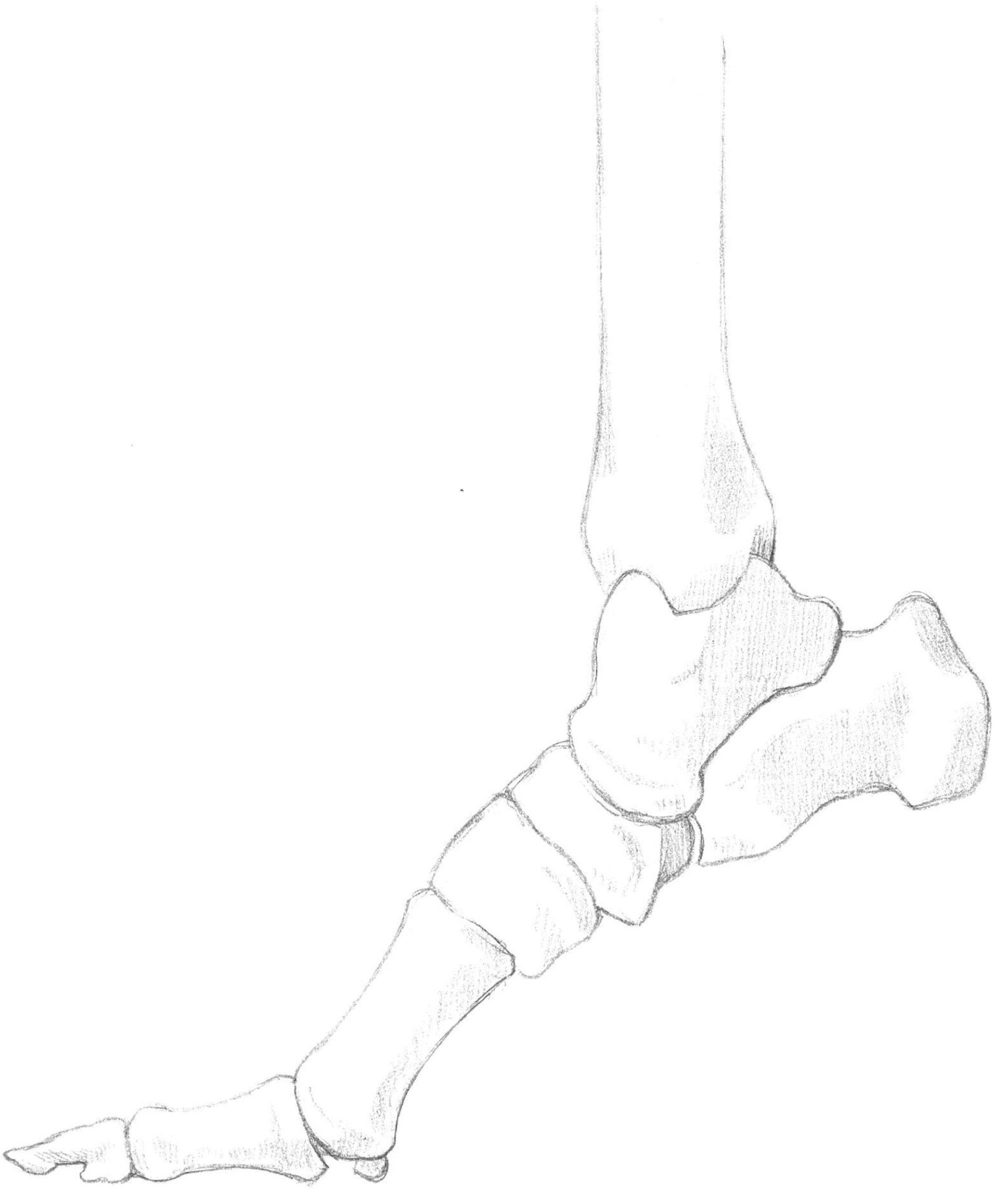

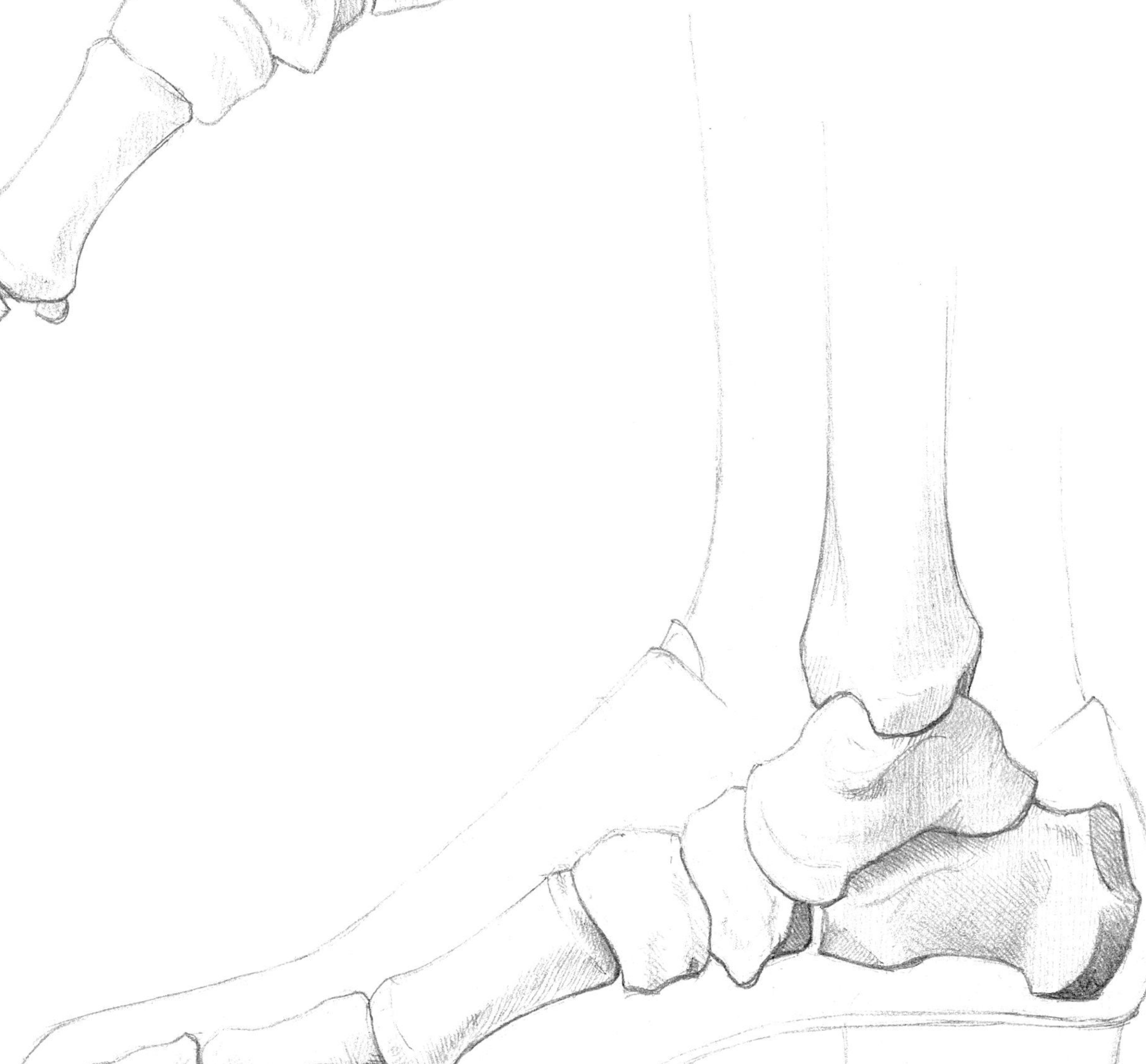

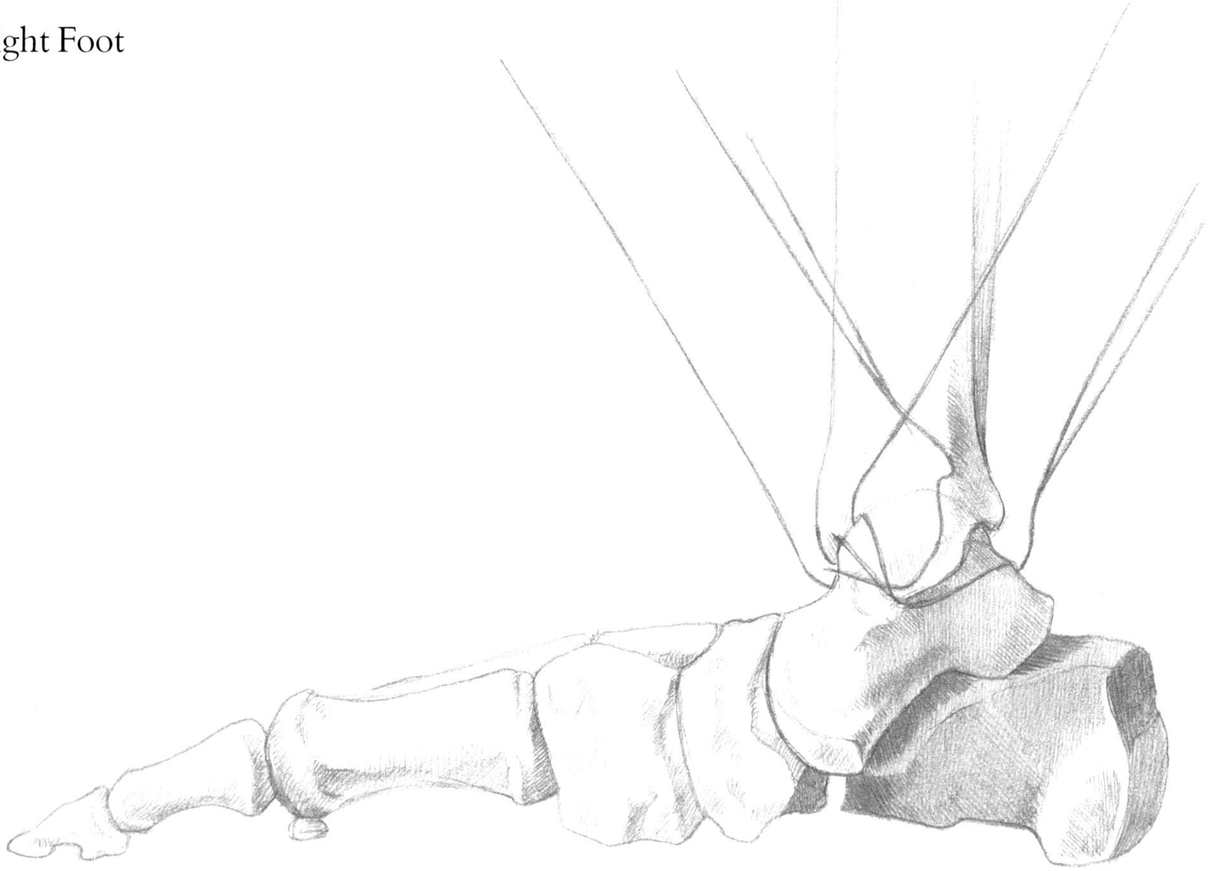

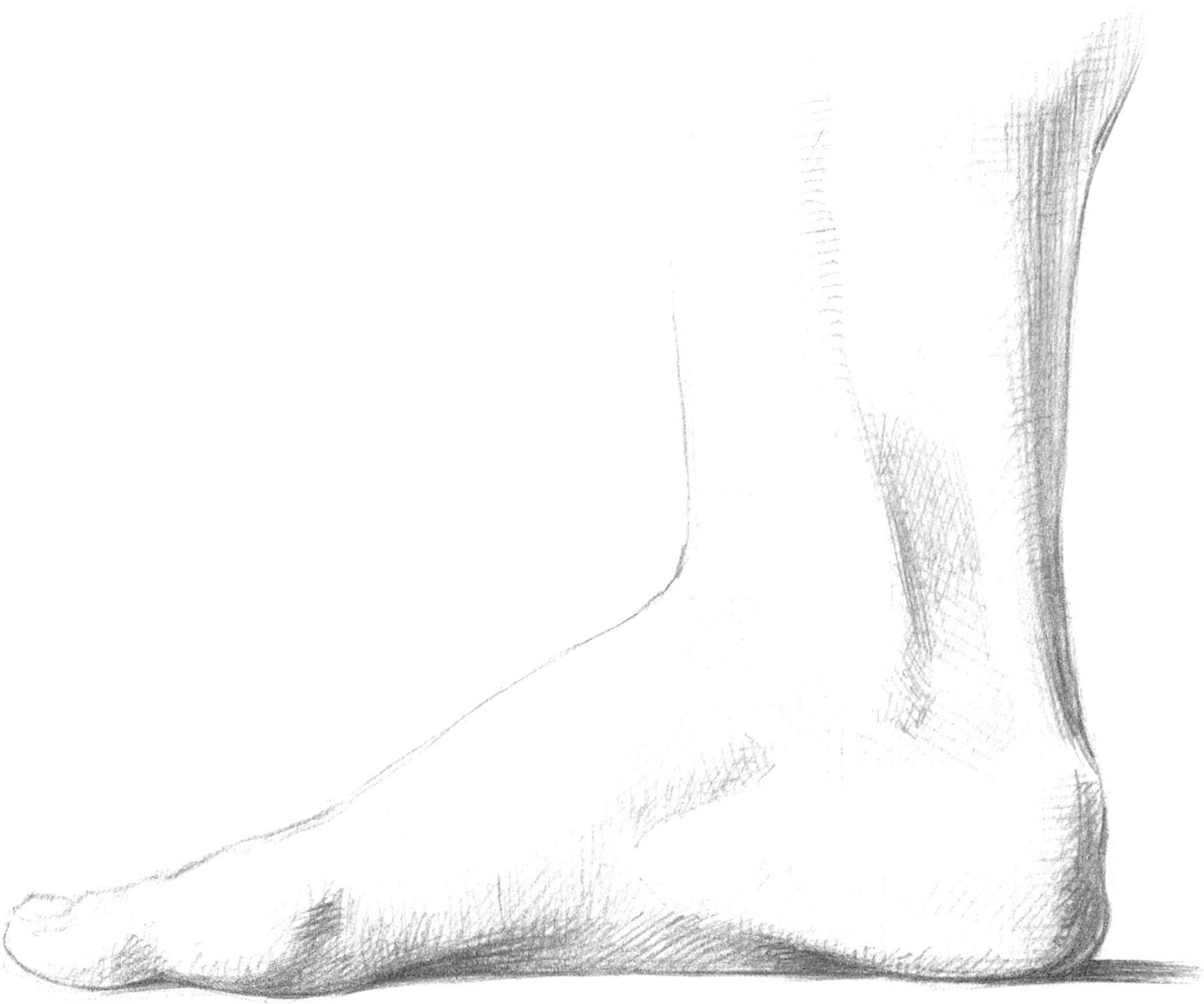

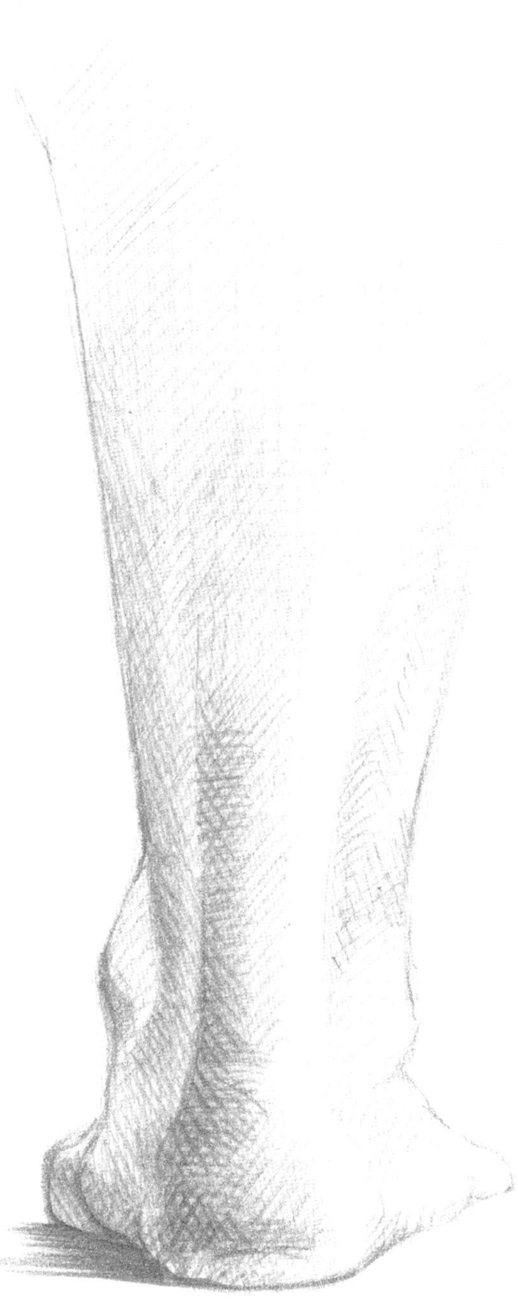

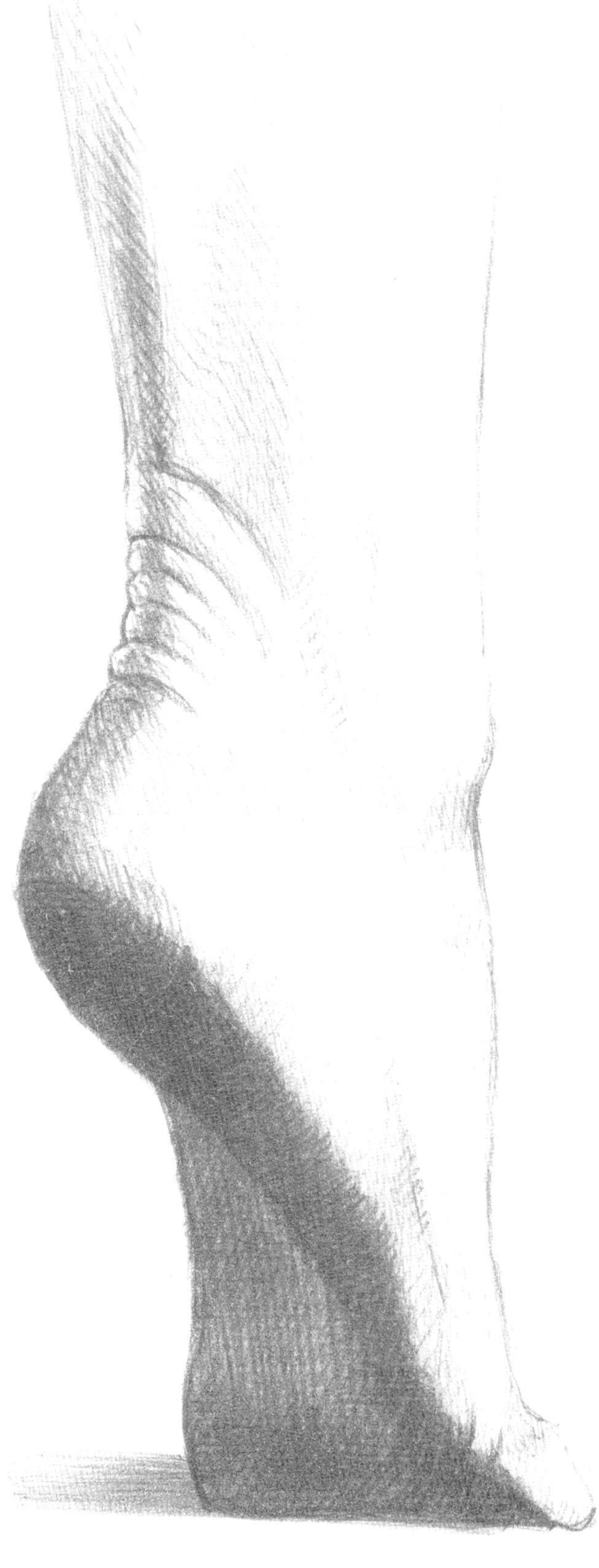

The Right Foot
(cont.)

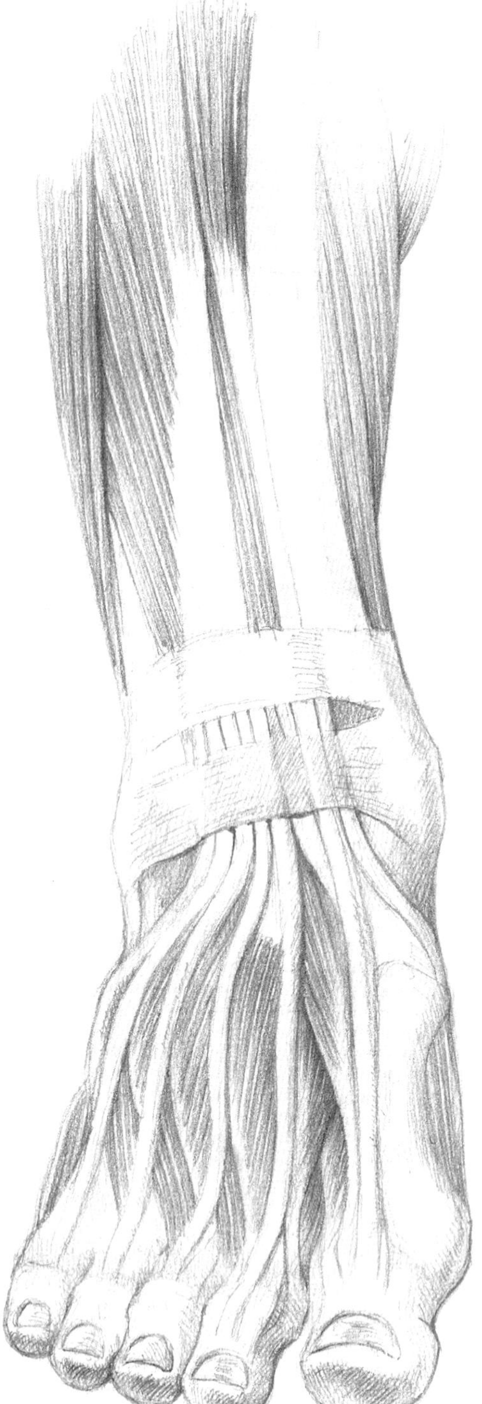

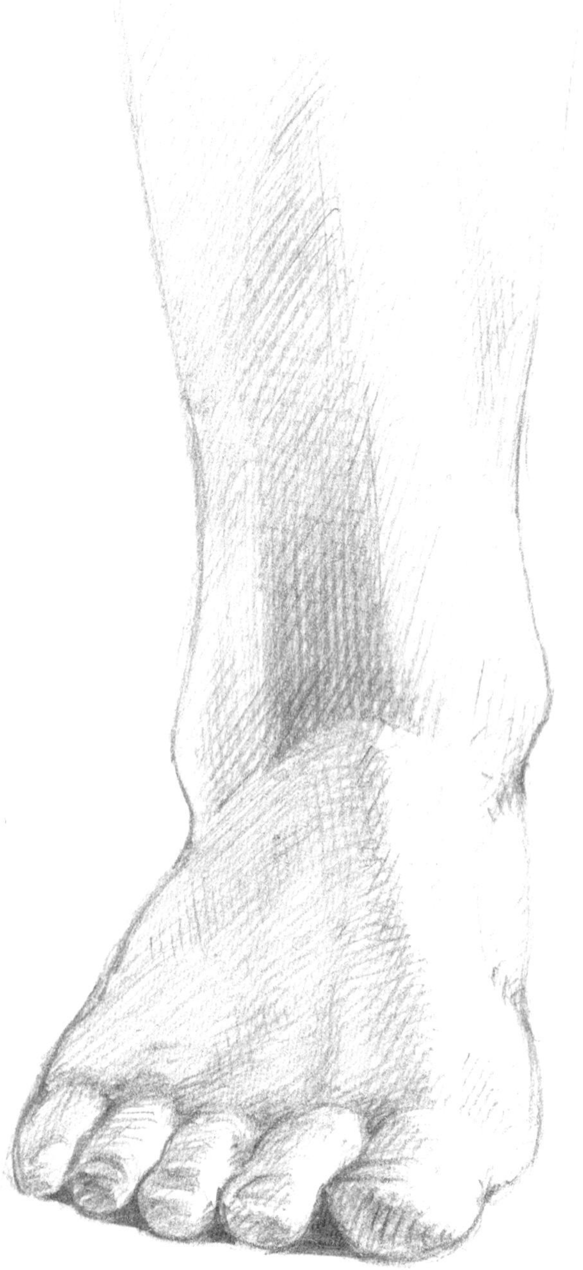

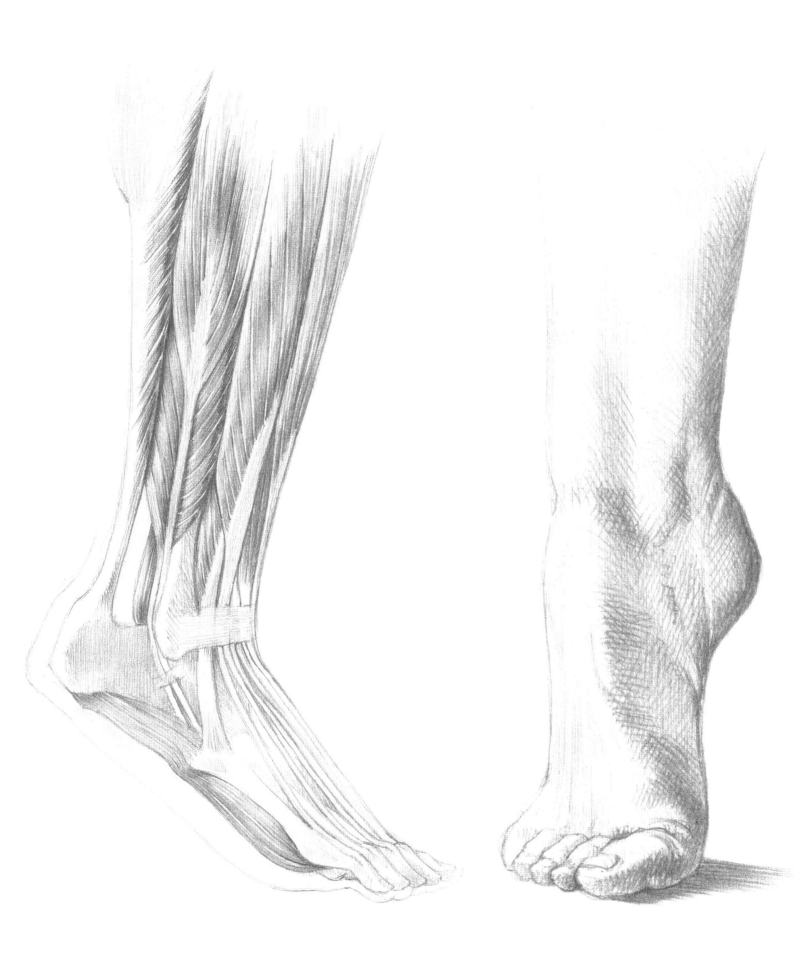

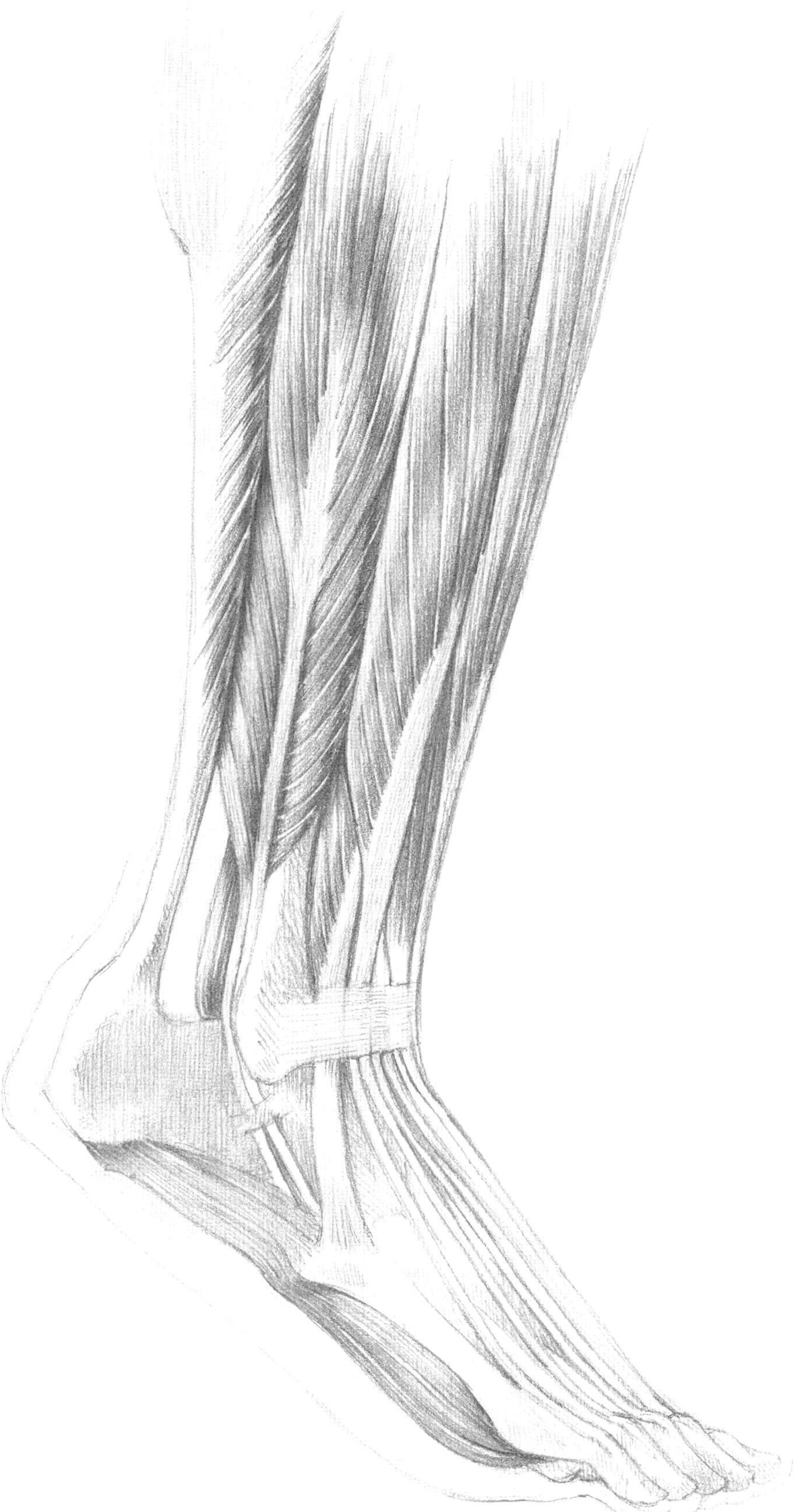
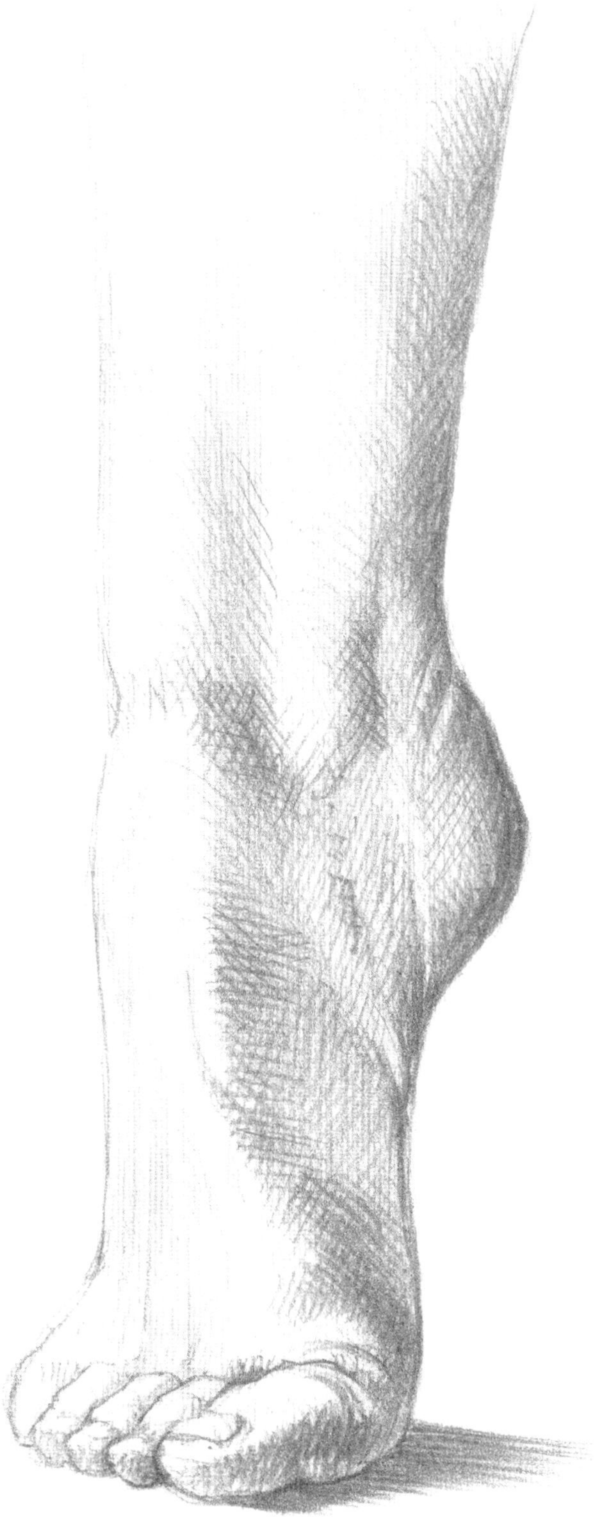

Fig. 178
The Lower Extremity
in Movement

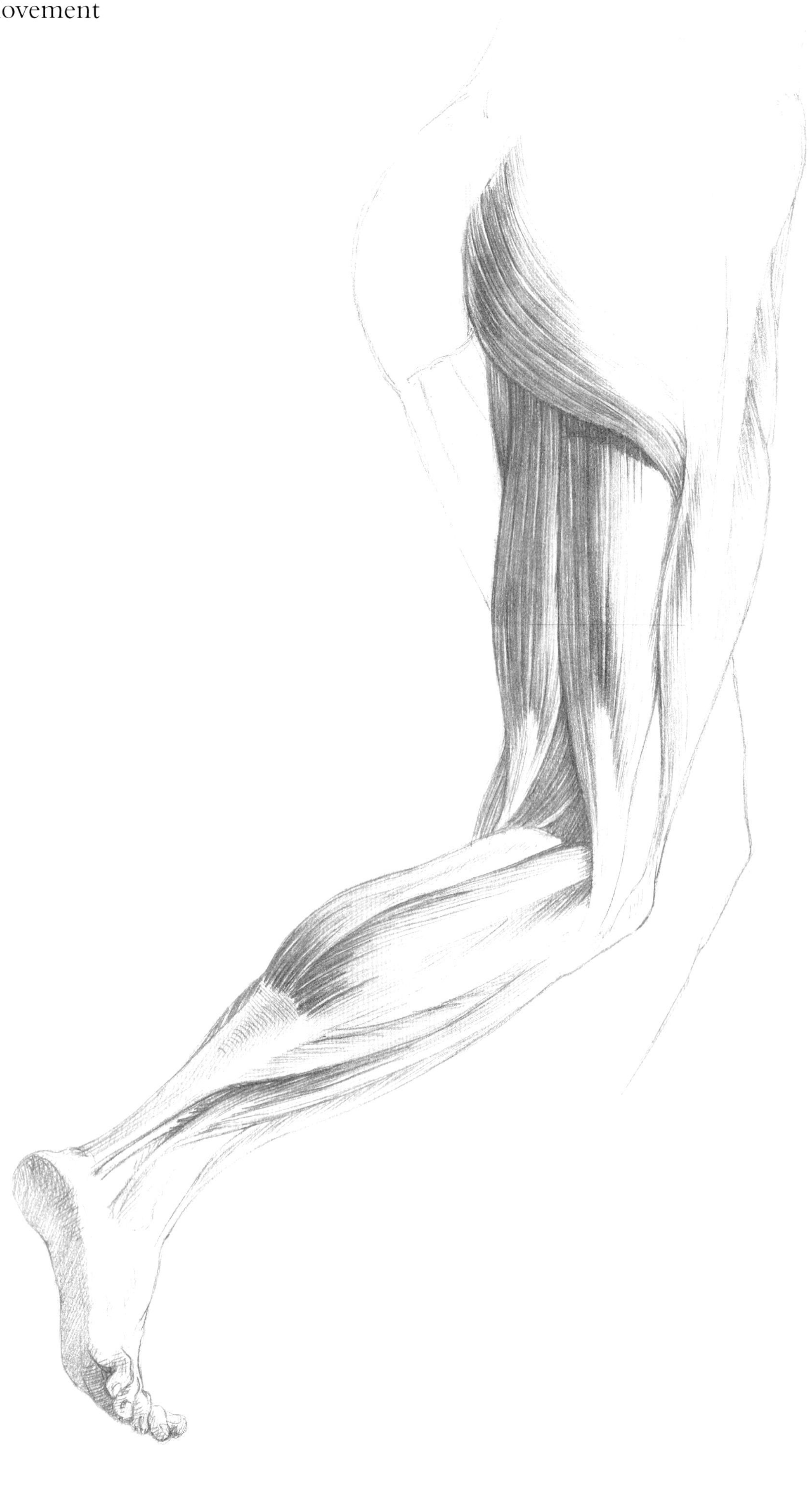

Fig. 179
Contours of the Lower Extremity

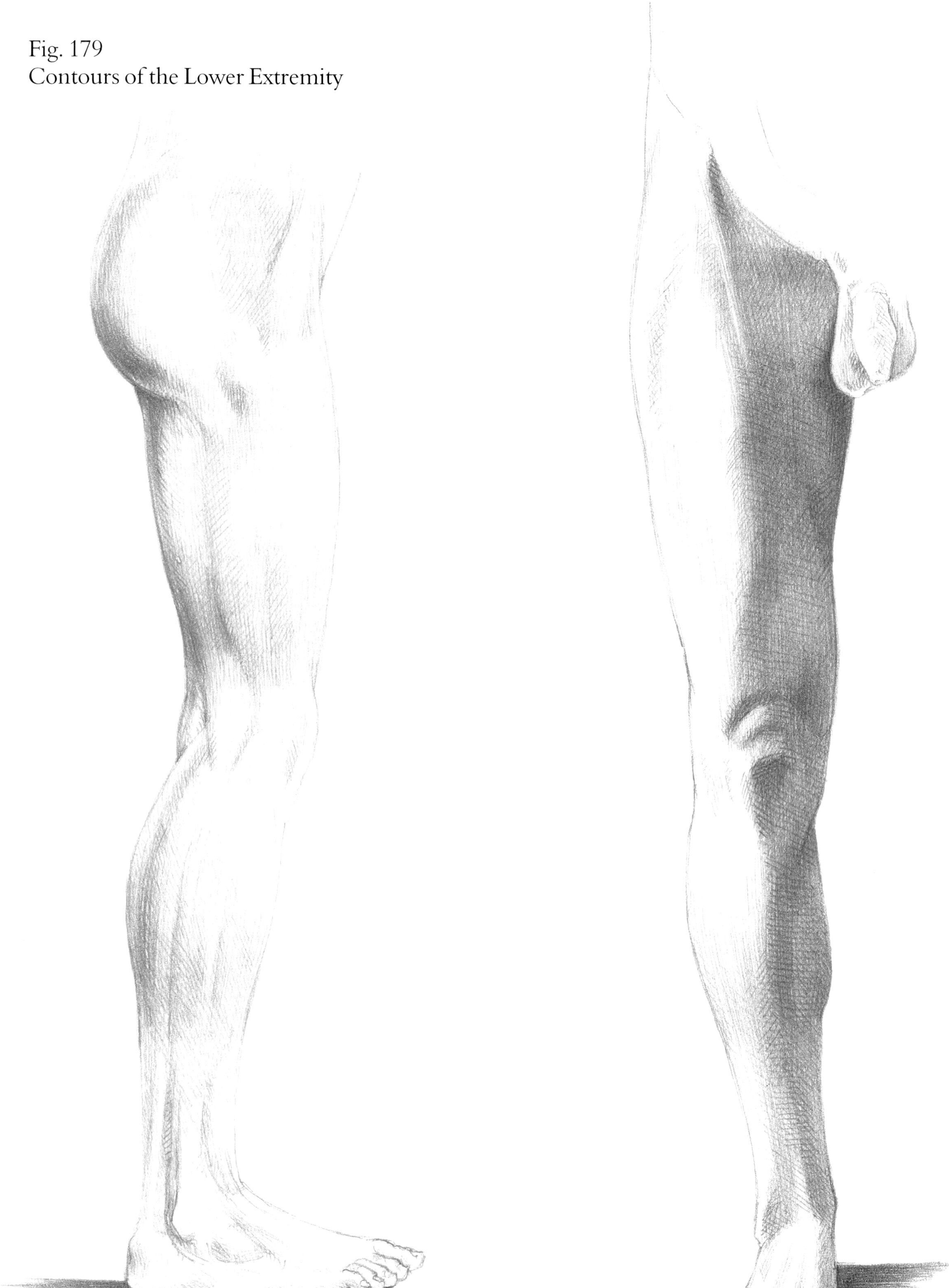

The Femoral Triangle (TRIGONUM FEMORALE) is located on the inferior limb, on the anterior inner side of the femur and the hip bend, below the inguinal bend between the pronator and the sartorius. It continues in the direction of the knee in the hip groove known as the SULCUS FEMORALIS. The anterior longitudinal fossa of the lateral hip muscle can be seen on the lateral surface of the femur. Behind this lies a groove which is formed by the septum between the lateral muscles. The tendons of the quadriceps femoris, the patella and, bilaterally, the epicondyles of the knee joint protrude above the knee joint. In small children the medial epicondyle of the femur protrudes more clearly.

The shin bone can be felt under the skin at the front and inside of the lower leg; anteriorly and laterally the muscle bellies of the ankle joint and toe extensors protrude. The posterior contour of the fibula is formed by the gemelli muscle of the fibula. In the anterior and medial part of the foot, the head of the ankle can be seen; in the medial part the tuberosity of the navicular bone can be seen; when the foot is extended the muscle tendons protrude. There are three tuberosities on the sole of the foot: the heel tuberosity and those of the first and fifth metatarsal bones. There are also two constant grooves in the sole known as the SULCUS PLANTARIS LATERALIS and the SULCUS PLANTARIS MEDIALIS. The great toe is straight while the other toes are bent towards the sole.

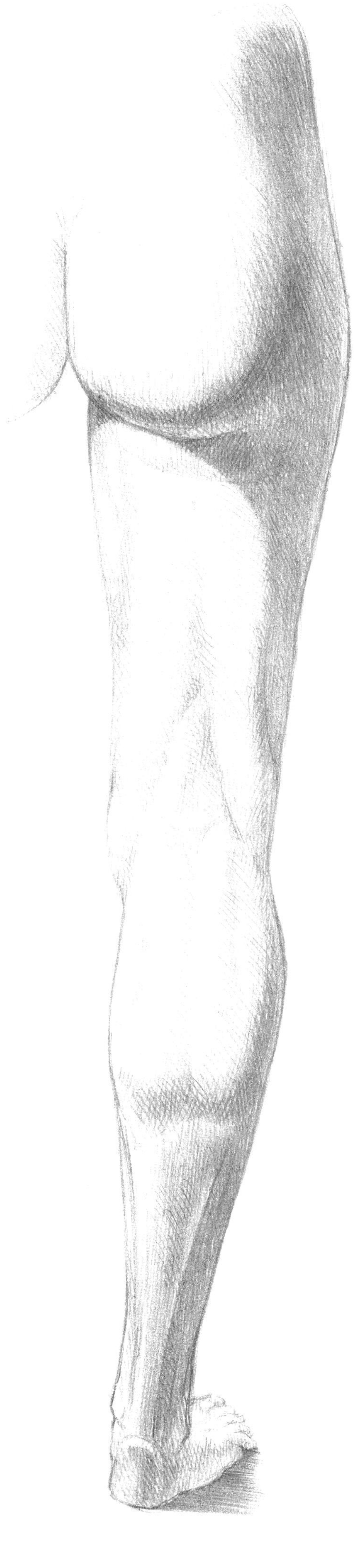

THE TRUNK

Fig 180
The Bones of the Trunk

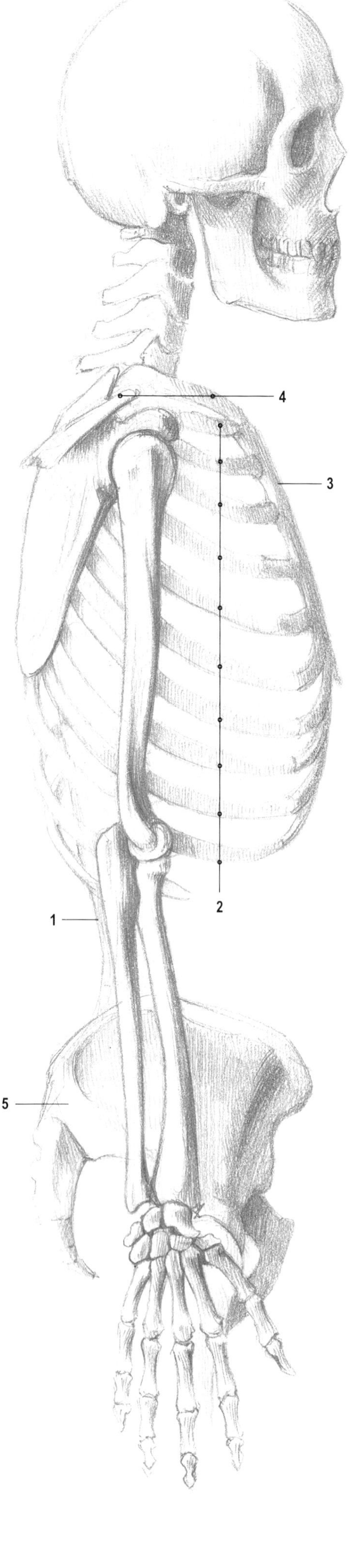

lateral aspect

The skeleton of the trunk comprises the spine (**1**), ribs (**2**) and sternum (**3**). These are complemented superiorly and inferiorly by the shoulder bones (**4**) and the pelvic girdle (**5**). Between the closed chest and the bony pelvis the abdominal region possesses bones only in the back; to the sides and front it has only a muscular wall. From the front, through the skin, it is possible to see the clavicle, mastoid process of the sternum, rib cage, deltoid muscle, chest muscles, anterior sartorius, the edges of the rectus abdominis and the anterior iliac spines. The posterior part of the trunk is formed by the shoulder blades, trapezium, spinous processes of the thoracic and lumbar vertebrae, and the arched gluteal muscles.

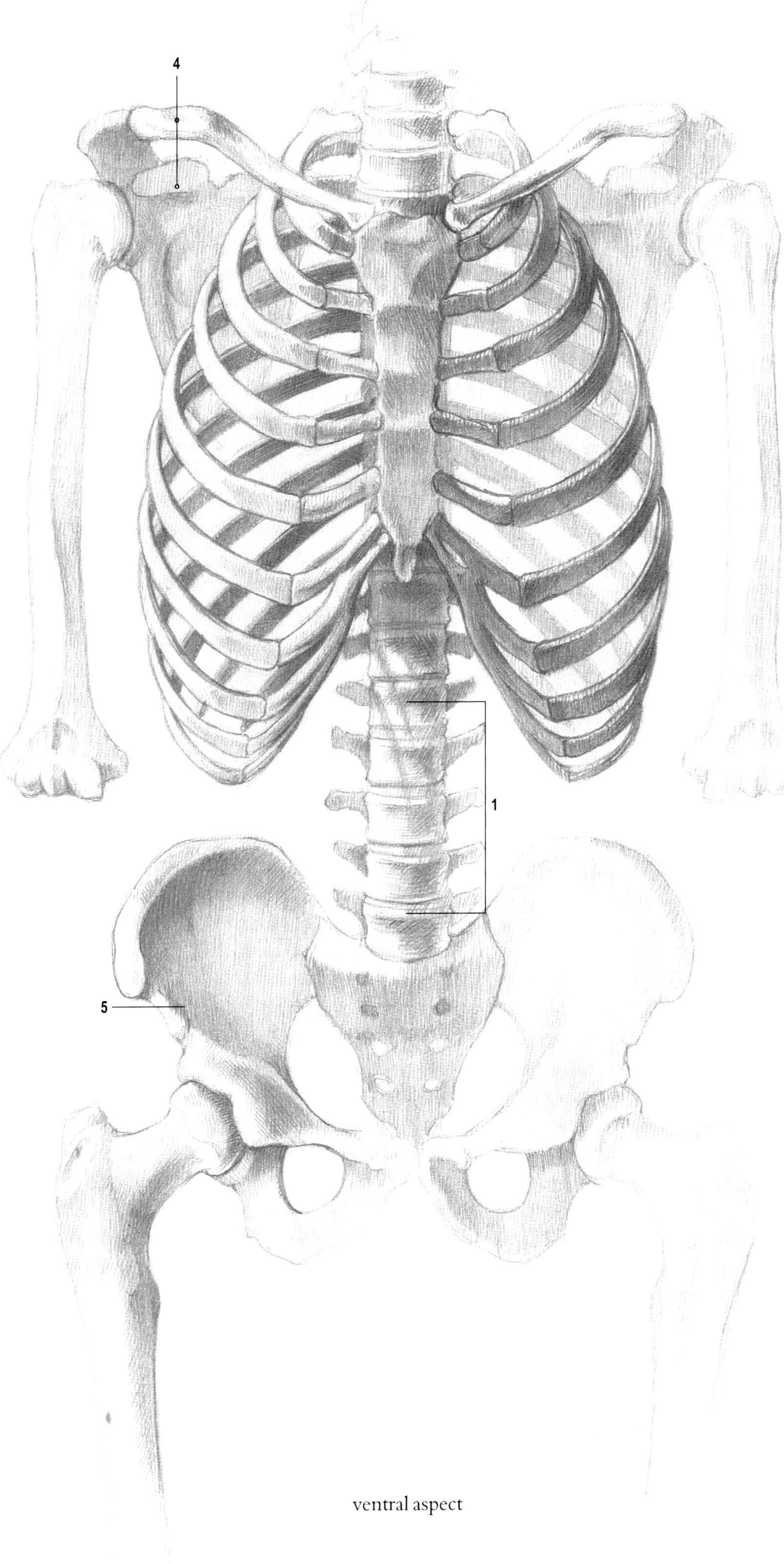

ventral aspect

Fig. 181
The Spine
(M. columna vertebralis)

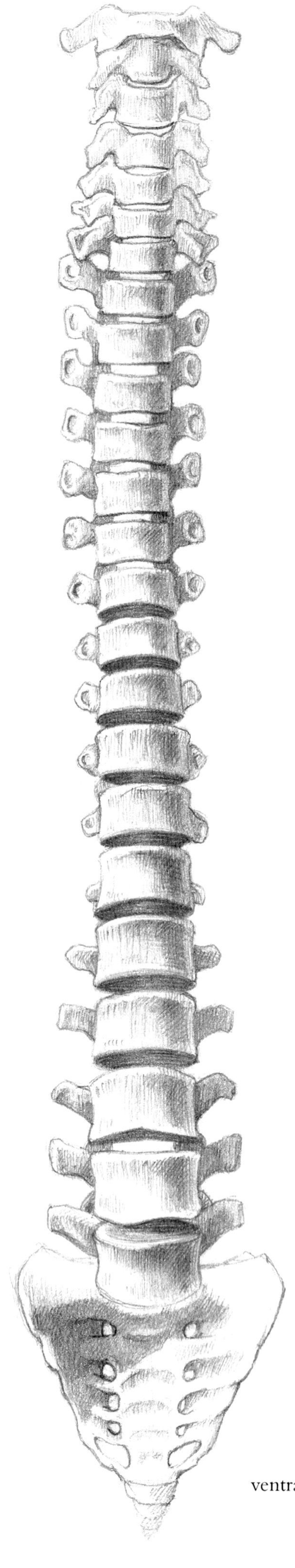

ventral aspect

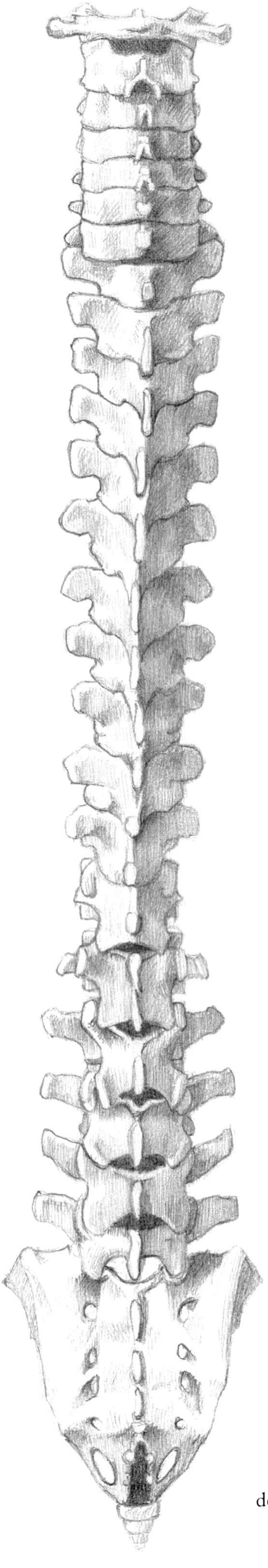

dorsal aspect

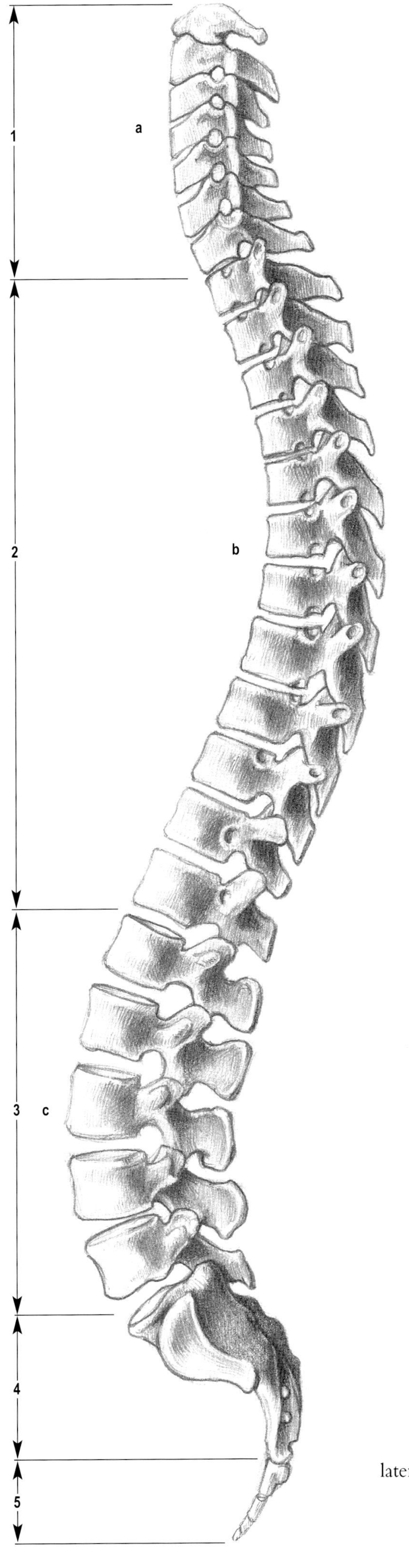

lateral aspect

The spine is located in the median plane of the body and comprises 32–33 vertebrae (7 cervical, 12 thoracic, 5 lumbar, 5 sacral and 3–4 coccyx vertebrae). The vertebrae account for three-quarters of the length of the spine and the intervertebral disks the remaining quarter. The spine is straight when viewed from the front or back; however, from the sides three bends are visible: a neck lordosis (**a**) and lumbar lordosis (**c**) and a thoracic kyphosis (**b**). The spine is divided into several regions: the neck (**1**), thorax (**2**), the lumbar region (**3**) and the sacrum (**4**) and coccyx (**5**).

Fig. 182
The First Cervical Vertebra or Atlas

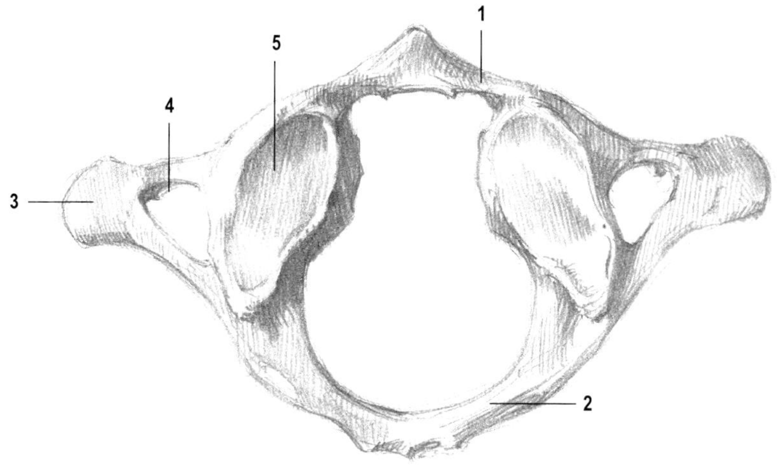

cranial aspect

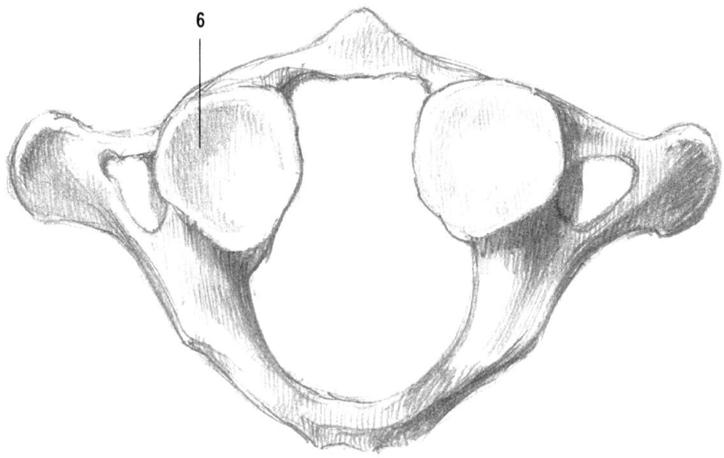

caudal aspect

dorsal aspect

1 Posterior arch
2 Anterior arch
3 Transverse process
4 Aperture of the transverse process
5 Superior articular groove
6 Inferior articular groove

Fig. 183
The Second Cervical Vertebra or Axis

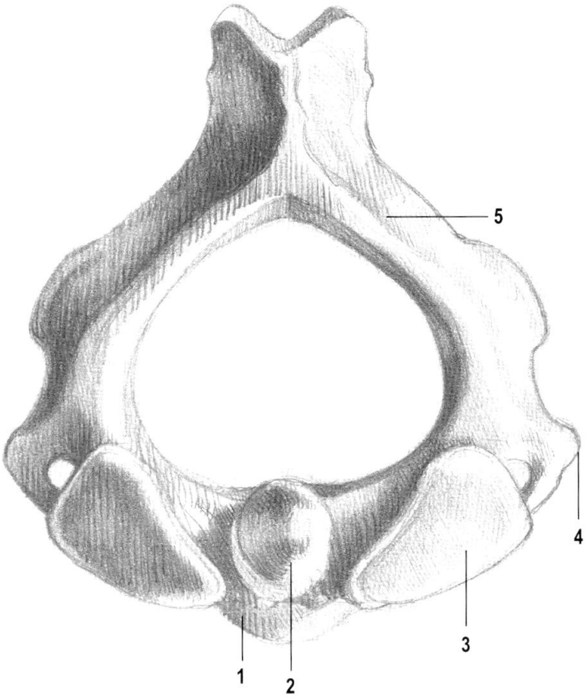

cranial aspect

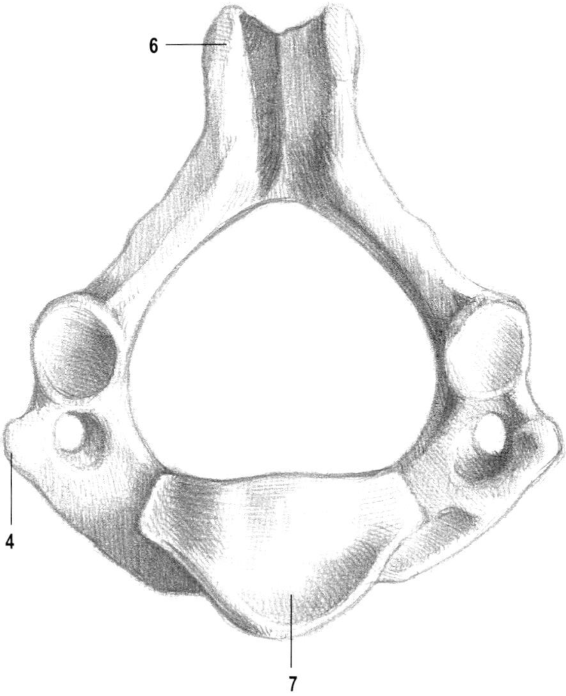

caudal aspect

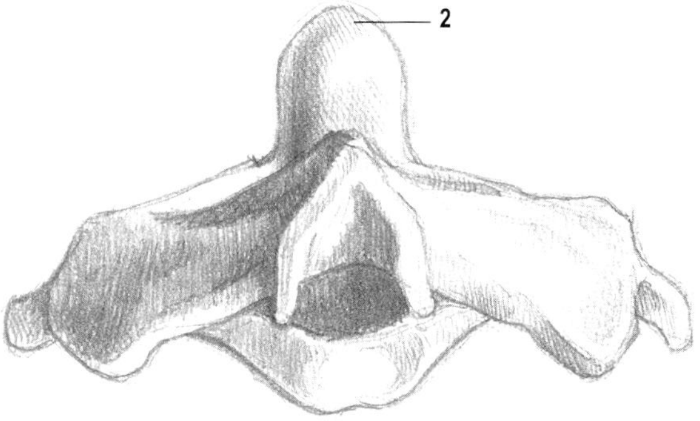

dorsal aspect

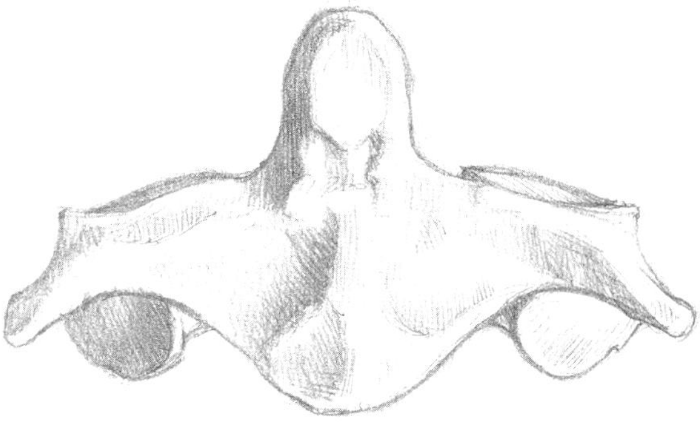

ventral aspect

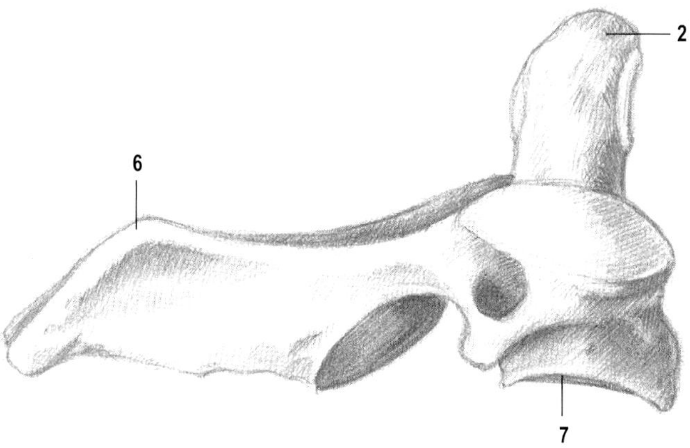

lateral aspect

1 Body
2 Dens (occipitally directed process)
3 Superior articulatory surface
4 Transverse process
5 Arch
6 Spinous process
7 Articular groove

Fig. 184
The Joint Connection between
First and Second Cervical Vertebrae

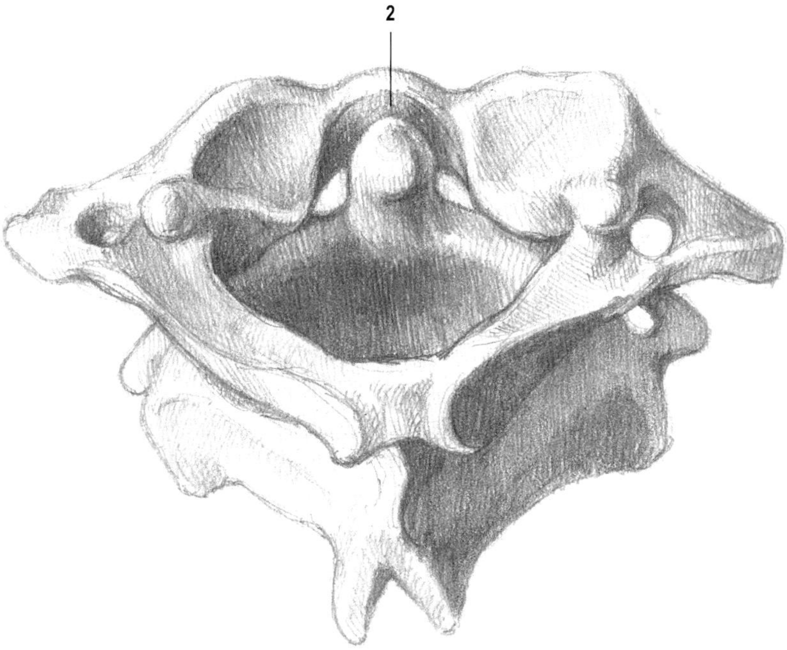

superior-dorsal aspect

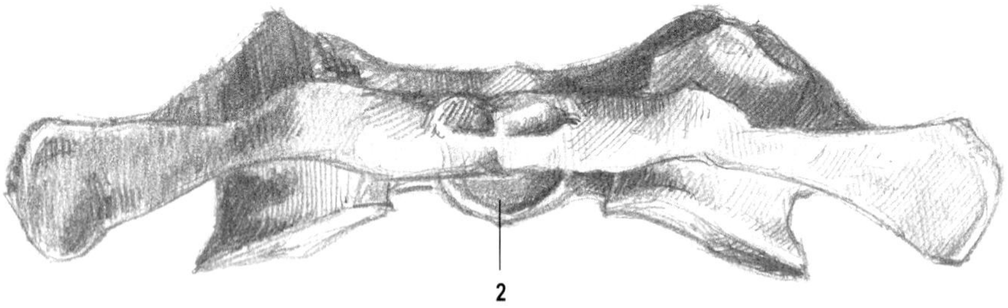

dorsal aspect

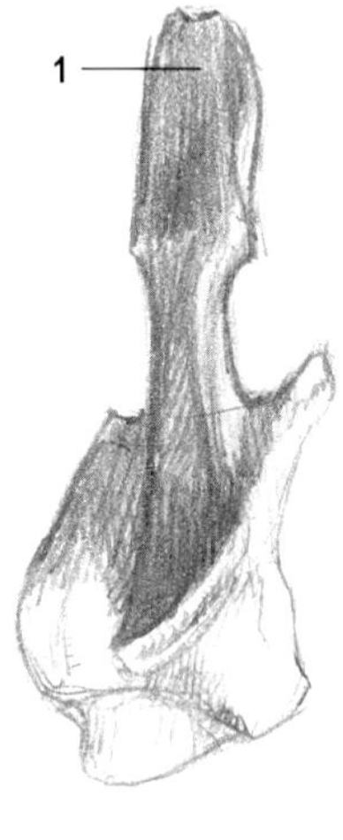

lateral aspect

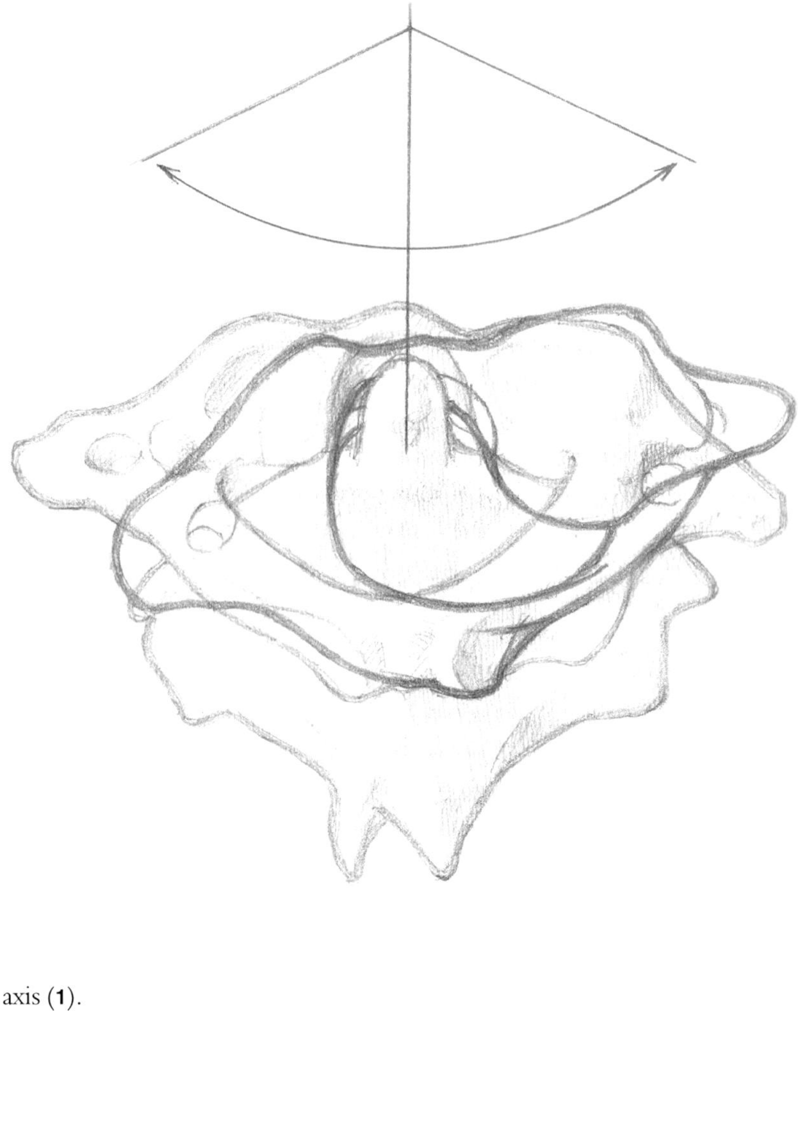

The articulatory groove of the atlas (**2**) rotates in the dens of the axis (**1**).

Fig. 185
The Last or Seventh Cervical Vertebra

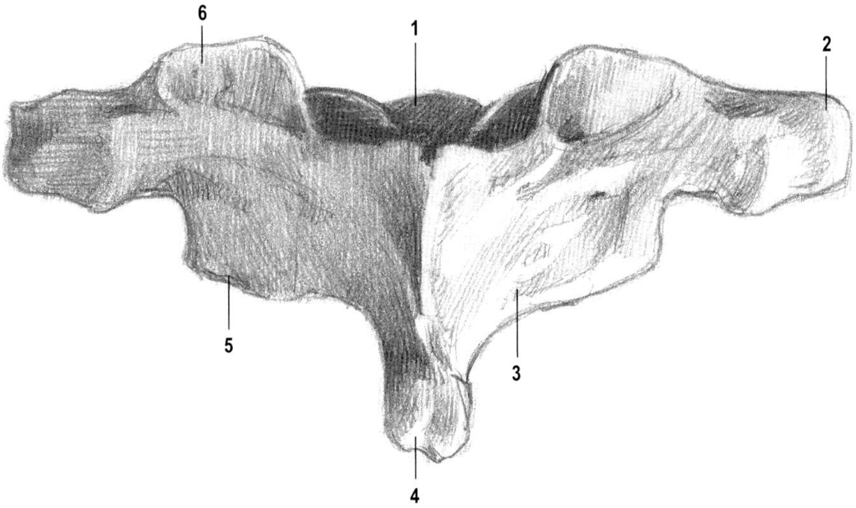

dorsal aspect

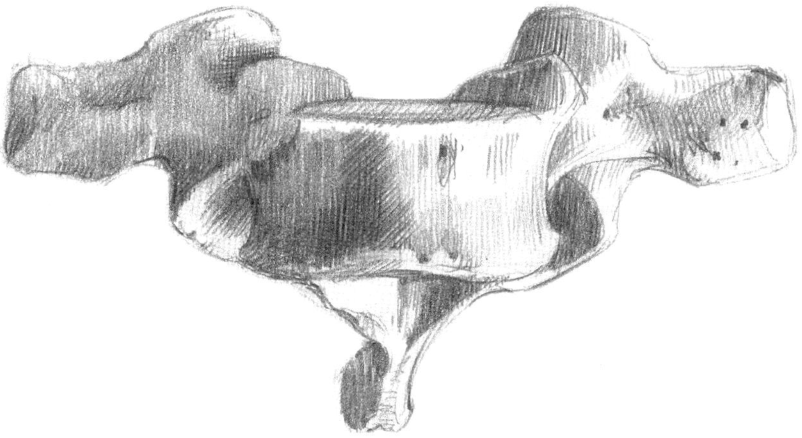

ventral aspect

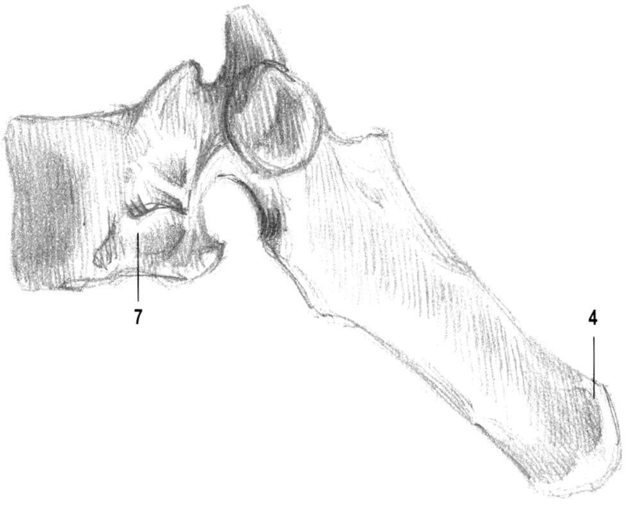

lateral aspect

1 Body
2 Transverse process
3 Arch
4 Spinous process
5 Inferior capitulum
6 Inferior capitulum
7 Inferior articulatory facet

Fig. 186
The Sixth Thoracic Vertebra

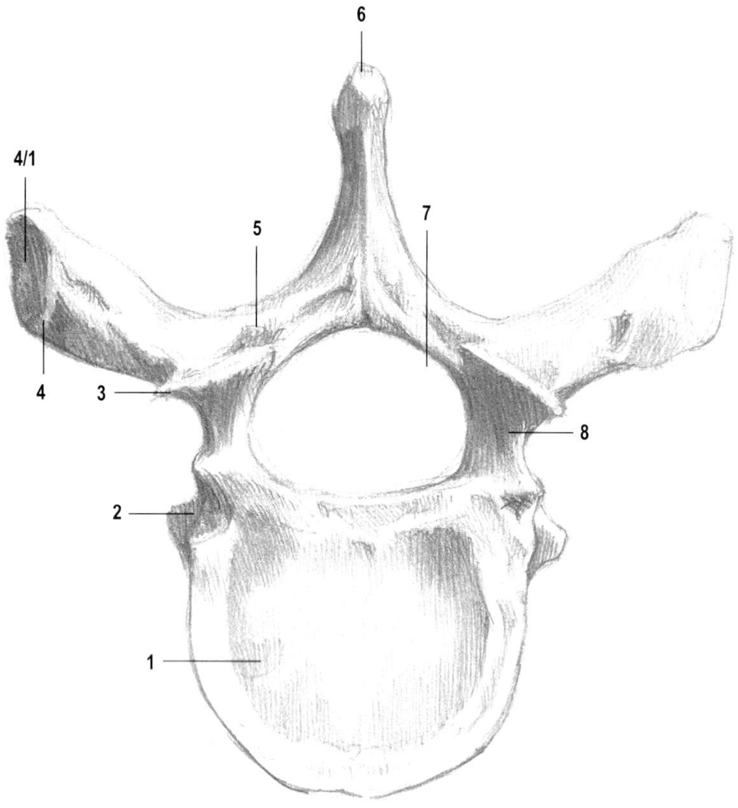

cranial aspect

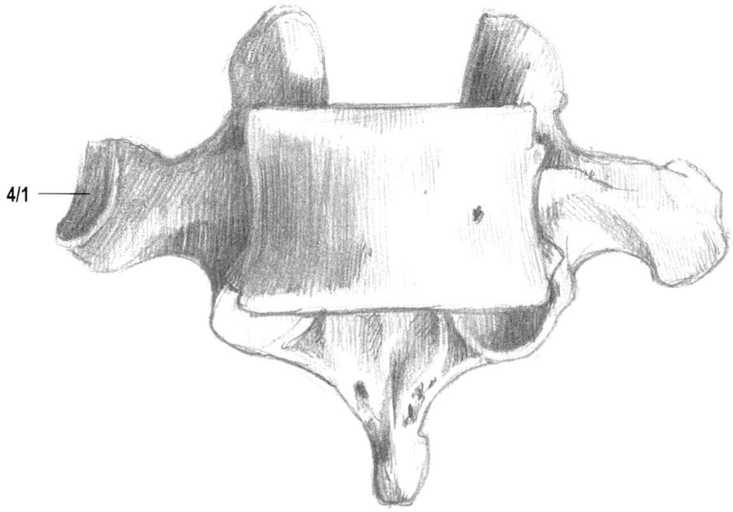

ventral aspect

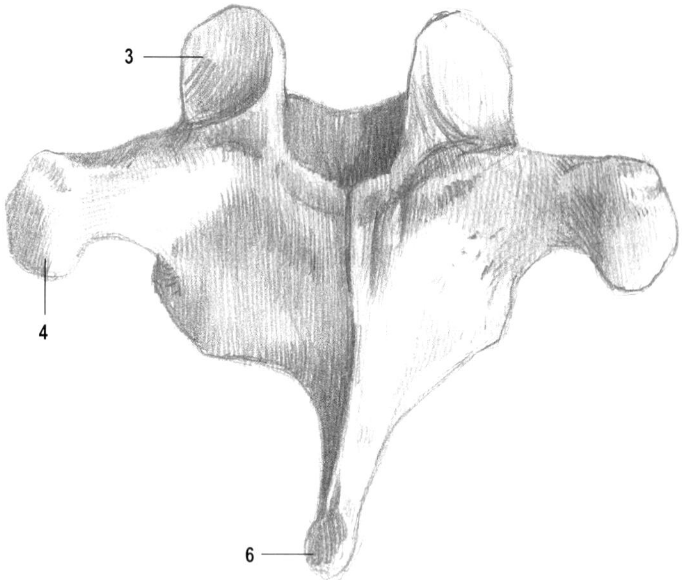

dorsal aspect

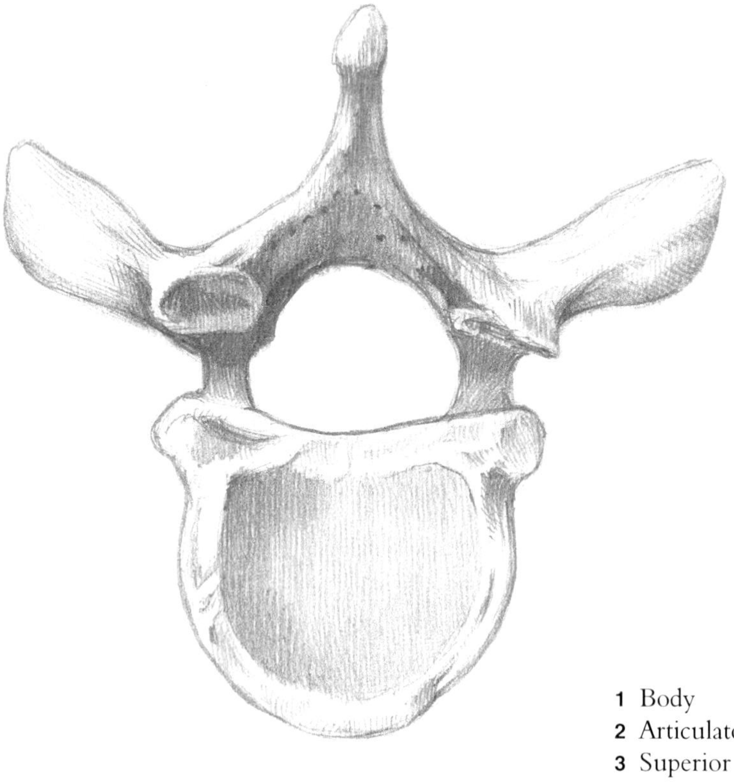

caudal aspect

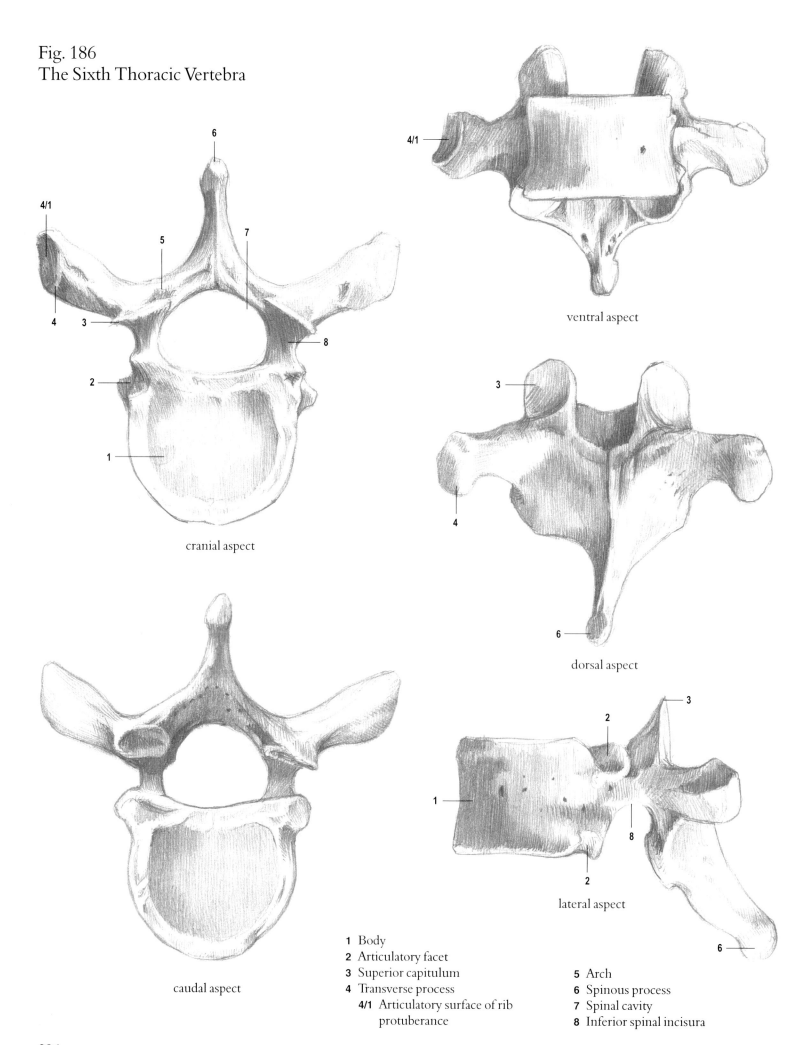

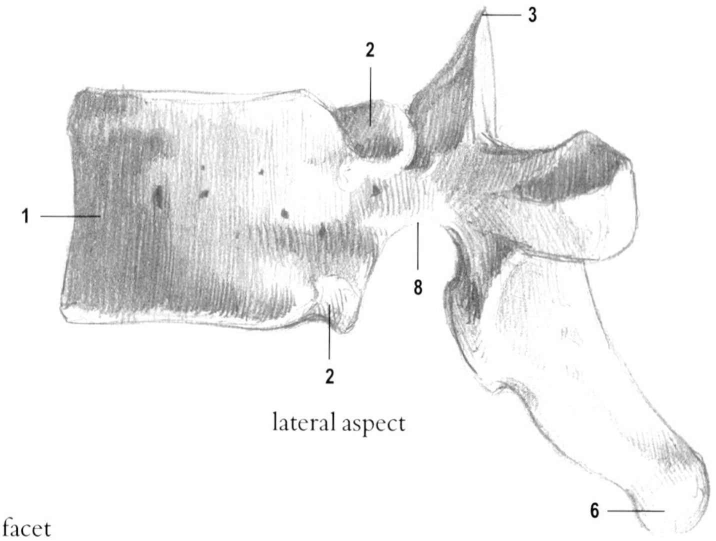

lateral aspect

1 Body
2 Articulatory facet
3 Superior capitulum
4 Transverse process
  4/1 Articulatory surface of rib protuberance
5 Arch
6 Spinous process
7 Spinal cavity
8 Inferior spinal incisura

Fig. 187
The Second Lumbar Vertebra

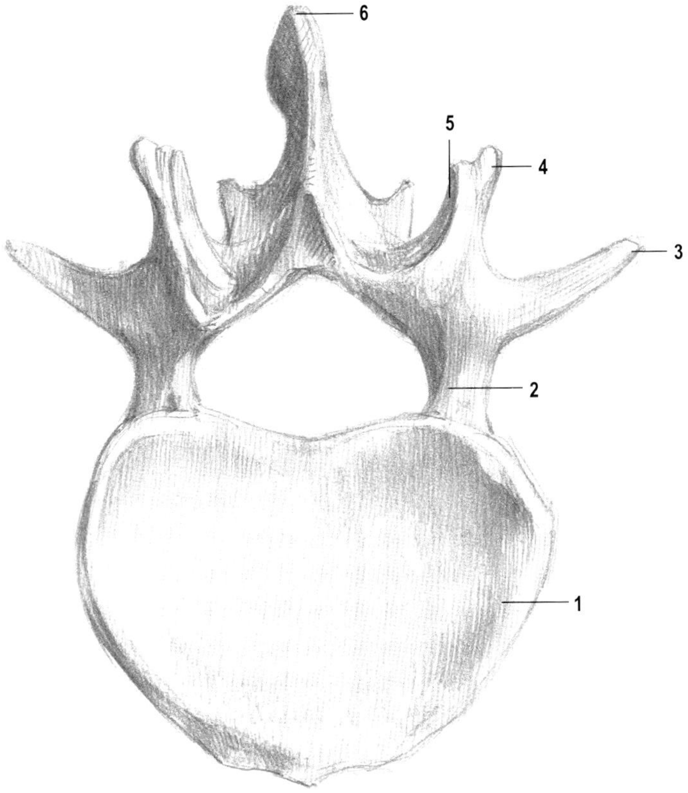

superior aspect

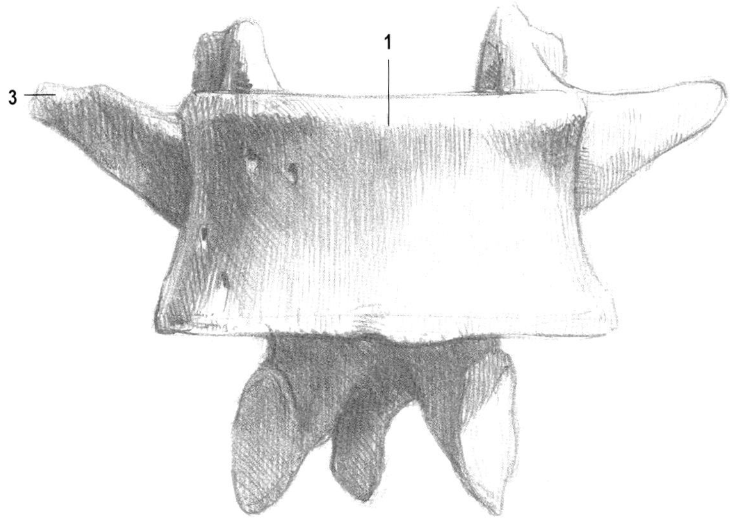

ventral aspect

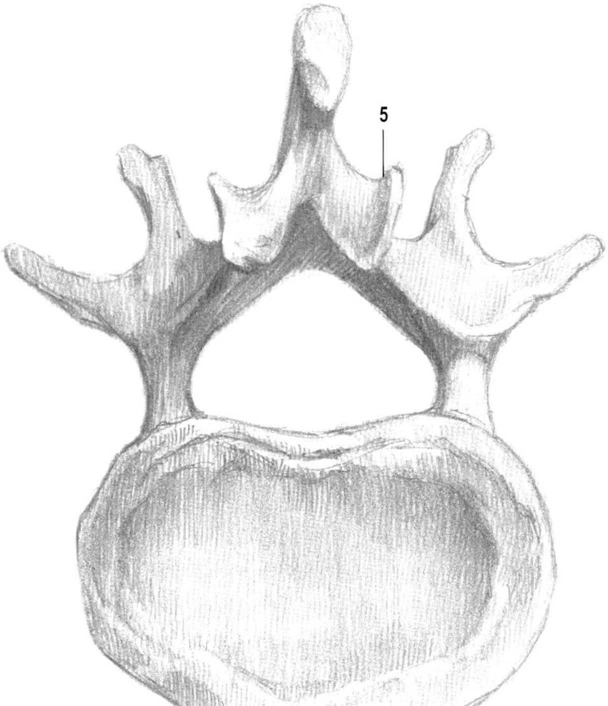

inferior aspect

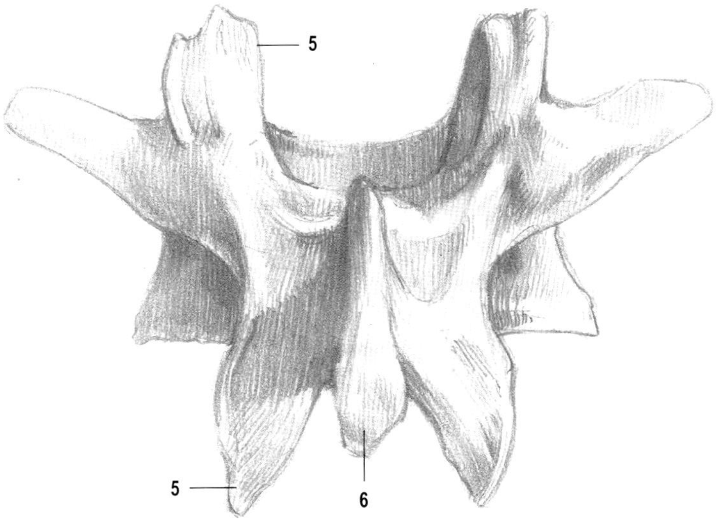

dorsal aspect

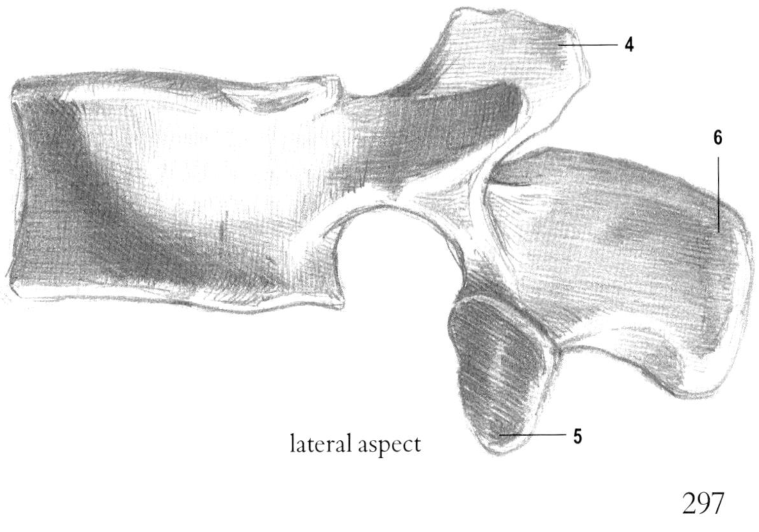

lateral aspect

1 Body
2 Arch
3 Transverse process
4 Mastoid process
5 Inferior and superior capitulum
6 Spinous process

Fig. 188
The Intervertebral Joints

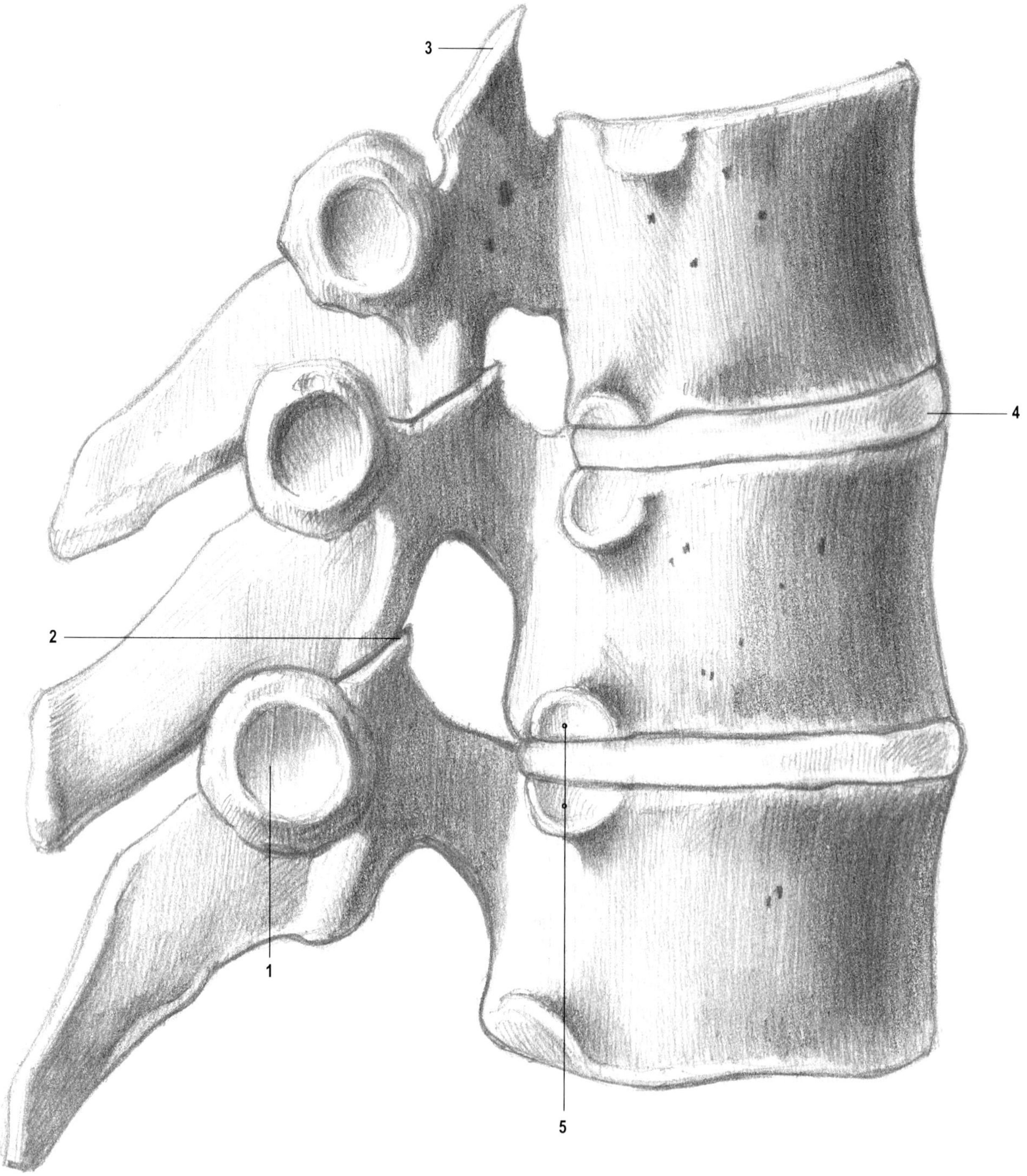

The body of the spine is connected by discs of ligament (**4**), which are composed of an outer fibrous cartilage ring, known as the anulus fibrosus, and a soft core (nucleus pulposus). Apart from the joints (**2**) which are formed by the capitula (**3**), the spine is stabilised by long and short ligaments. The heads of the ribs form joints with the rib facets of the body of the spine (**5**); their protuberances form joints with the transverse rib facets (**1**) of the transverse processes.

Fig. 189
The Spine-Rib Joints

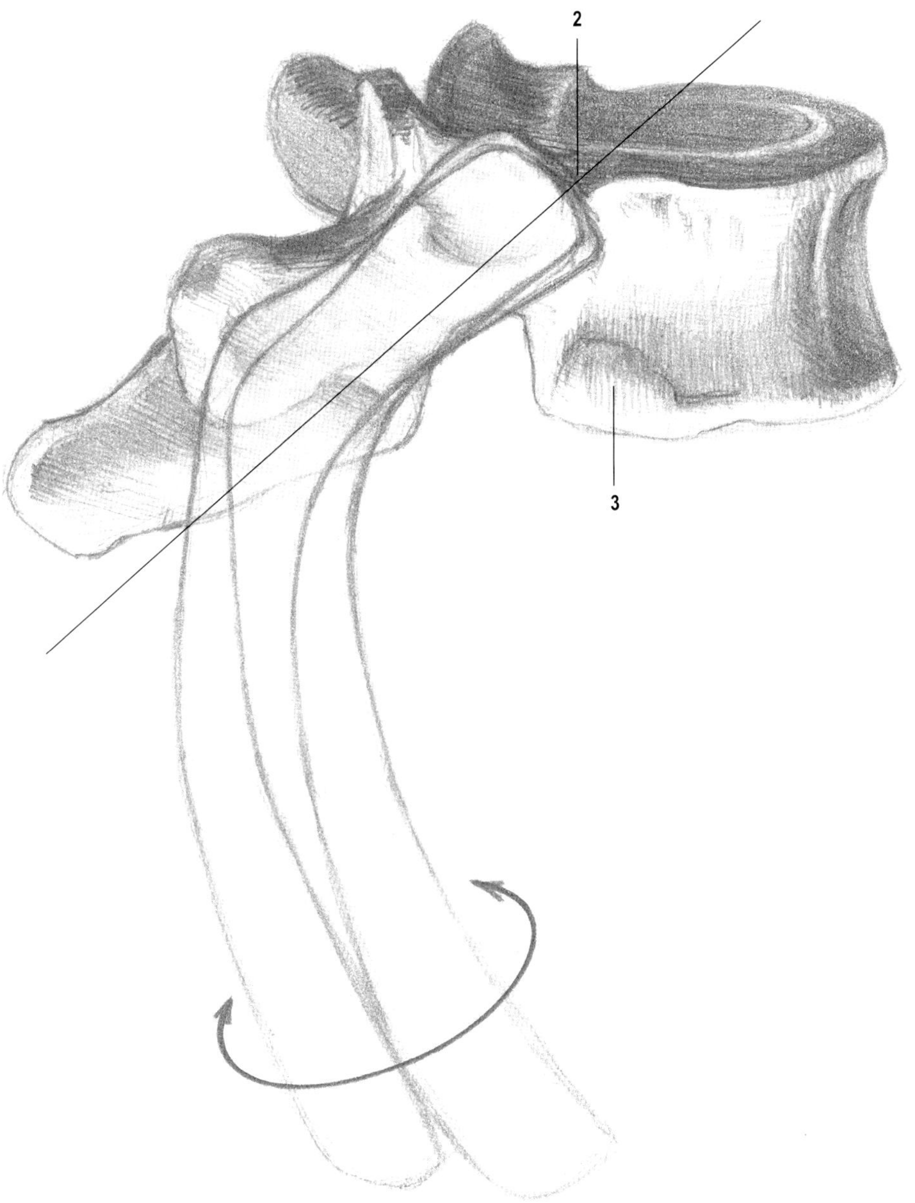

lateral aspect

The rib heads (**1**) form a joint (**4**) with the superior (**2**) and inferior(**3**) capitula of the body of the spine. The rib protuberance (**6**) forms a joint with the transverse rib facet (**5**) of the corresponding vertebra. The axis of rotation of the joint (**7**) is horizontal.

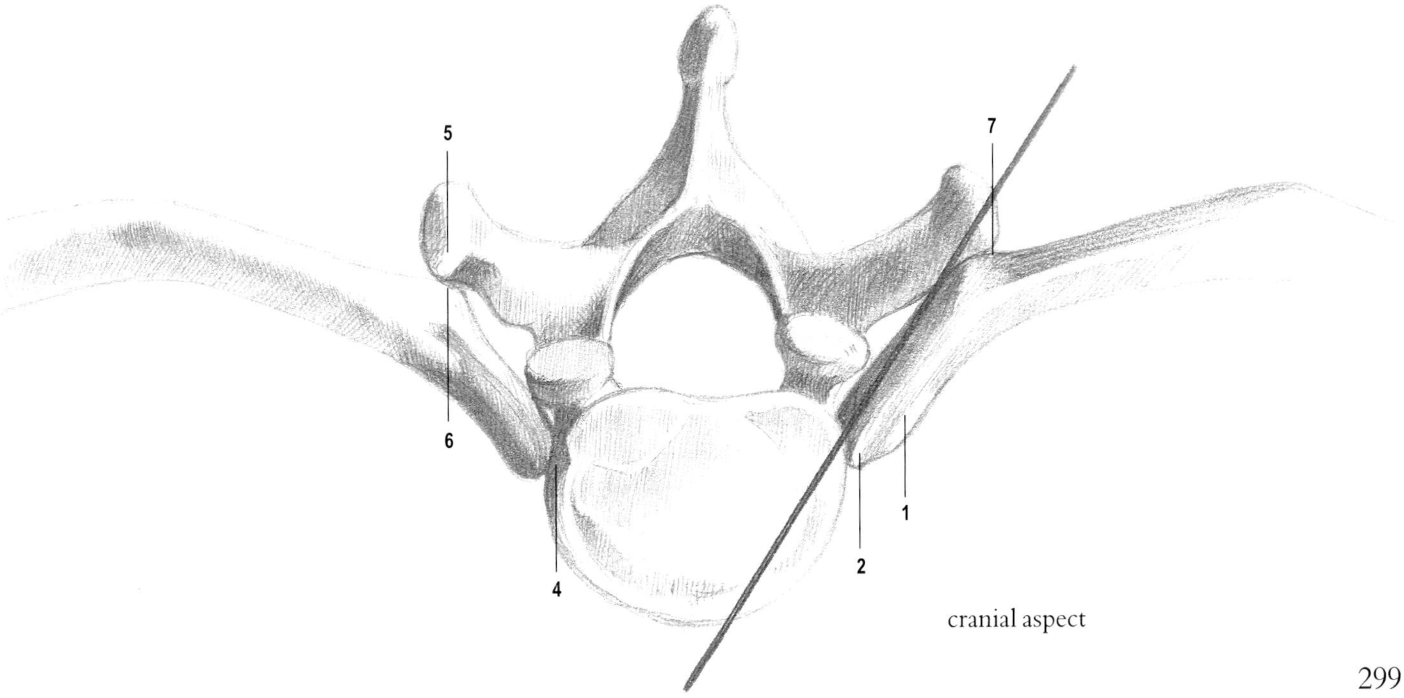

cranial aspect

Fig. 190
The Rib

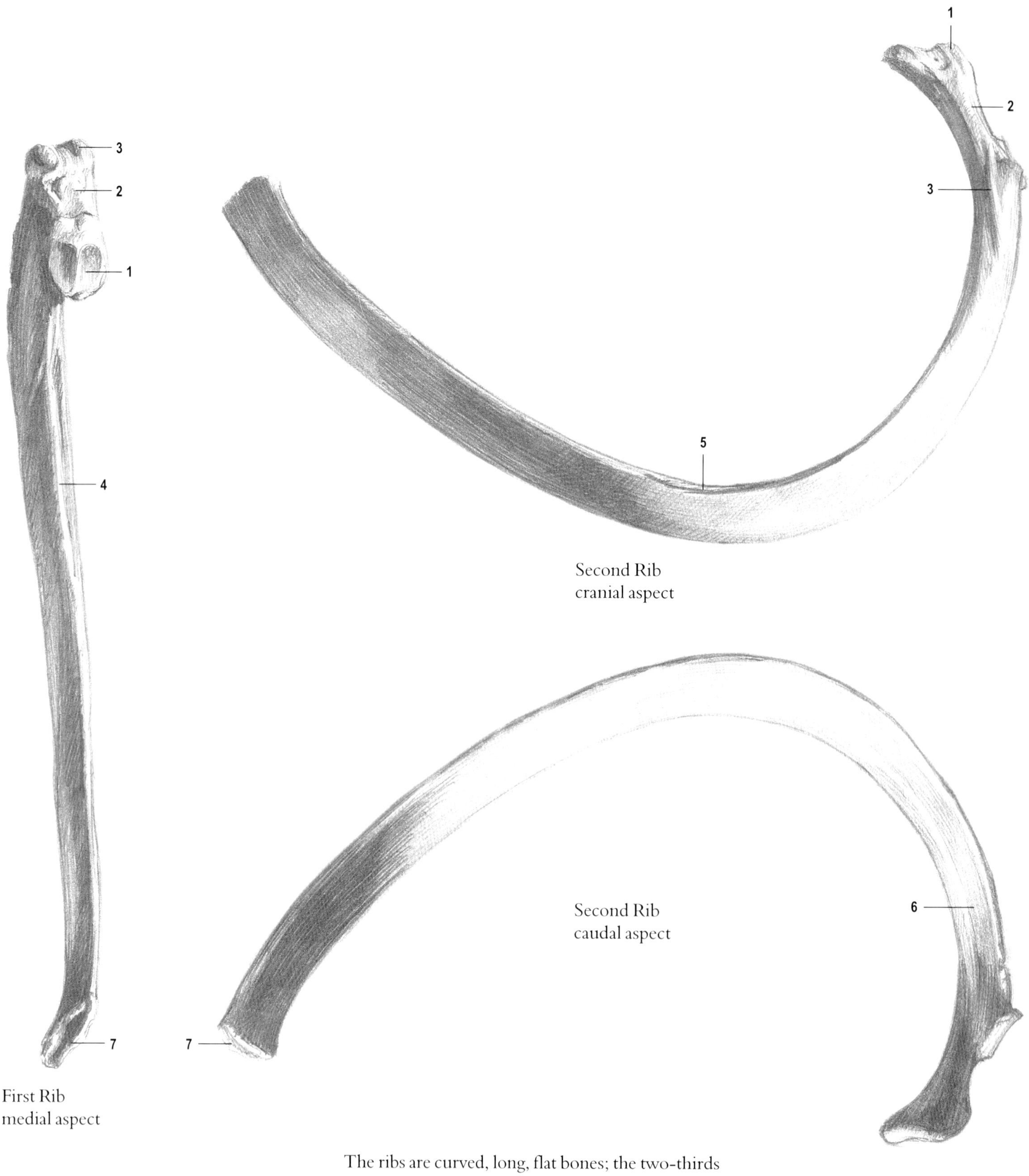

First Rib
medial aspect

The ribs are curved, long, flat bones; the two-thirds located towards the spine are bones whereas the third nearer the chest are composed of rib cartilage. The superior seven ribs – connected with the sternum – are known as true ribs. The cartilage of the eighth to tenth ribs constitute the rib arch. The last two ribs have an open end in the abdominal wall.

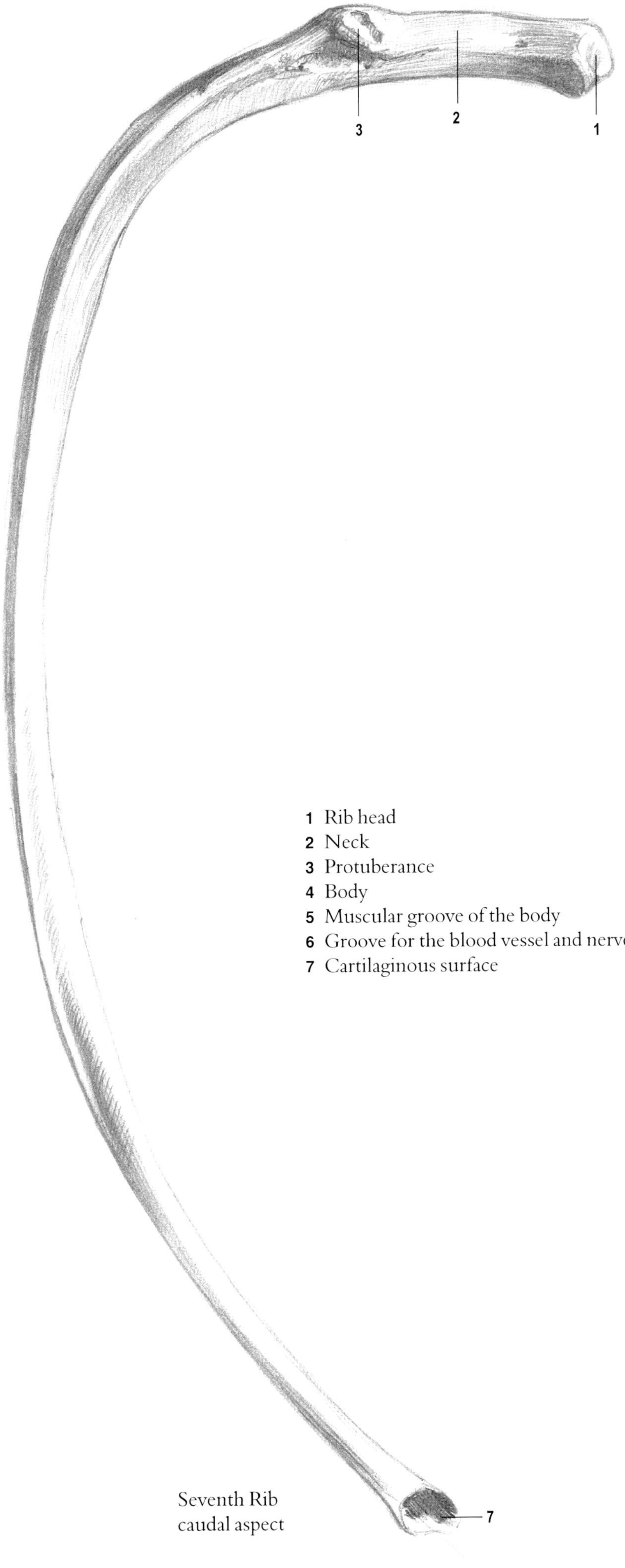

Seventh Rib
caudal aspect

1 Rib head
2 Neck
3 Protuberance
4 Body
5 Muscular groove of the body
6 Groove for the blood vessel and nerve
7 Cartilaginous surface

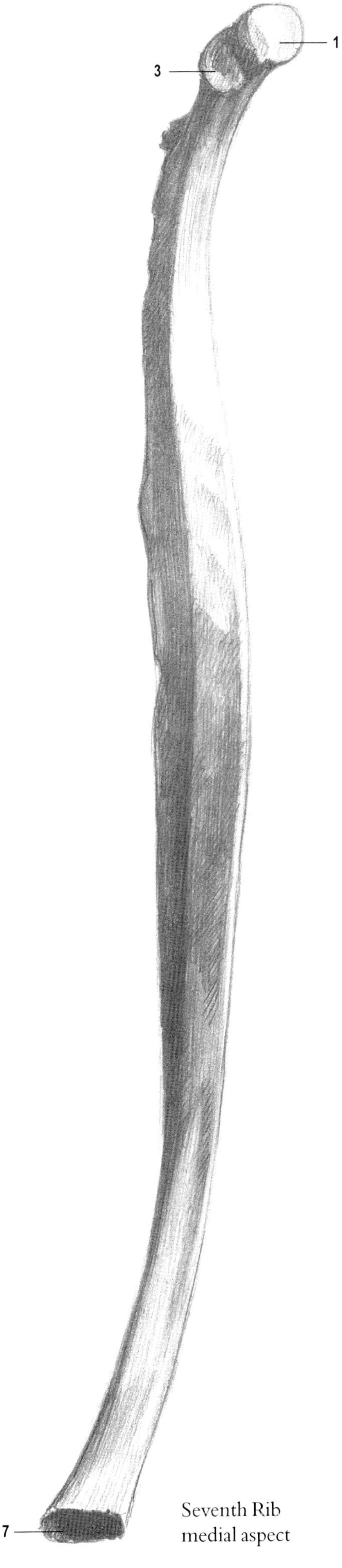

Seventh Rib
medial aspect

Fig. 191
The Sternum

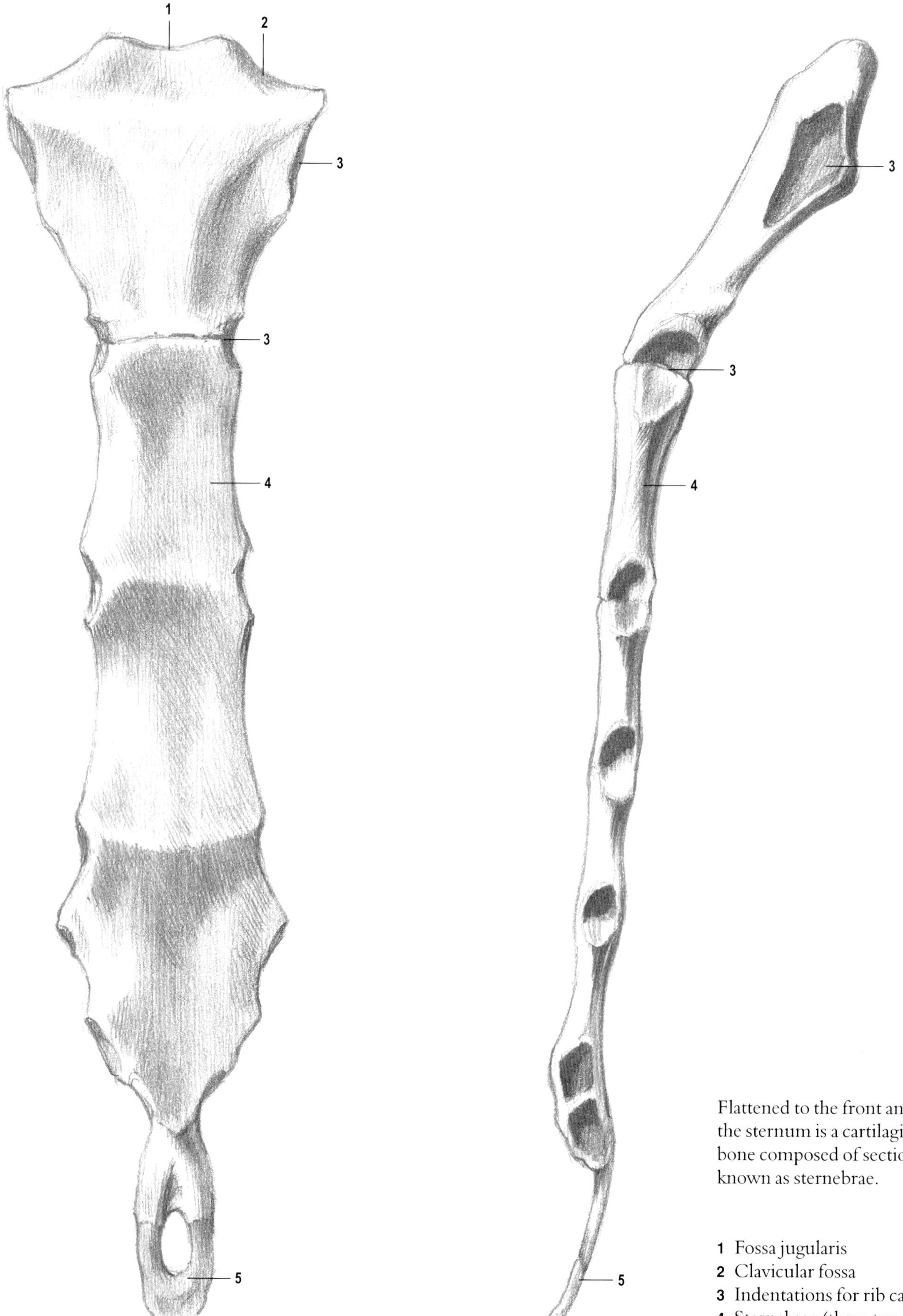

ventral aspect

lateral aspect

Flattened to the front and back, the sternum is a cartilaginous bone composed of sections known as sternebrae.

1 Fossa jugularis
2 Clavicular fossa
3 Indentations for rib cartilage
4 Sternebrae (three transverse crests)
5 Mastoid process

Fig. 192
The Joints of the Sternum

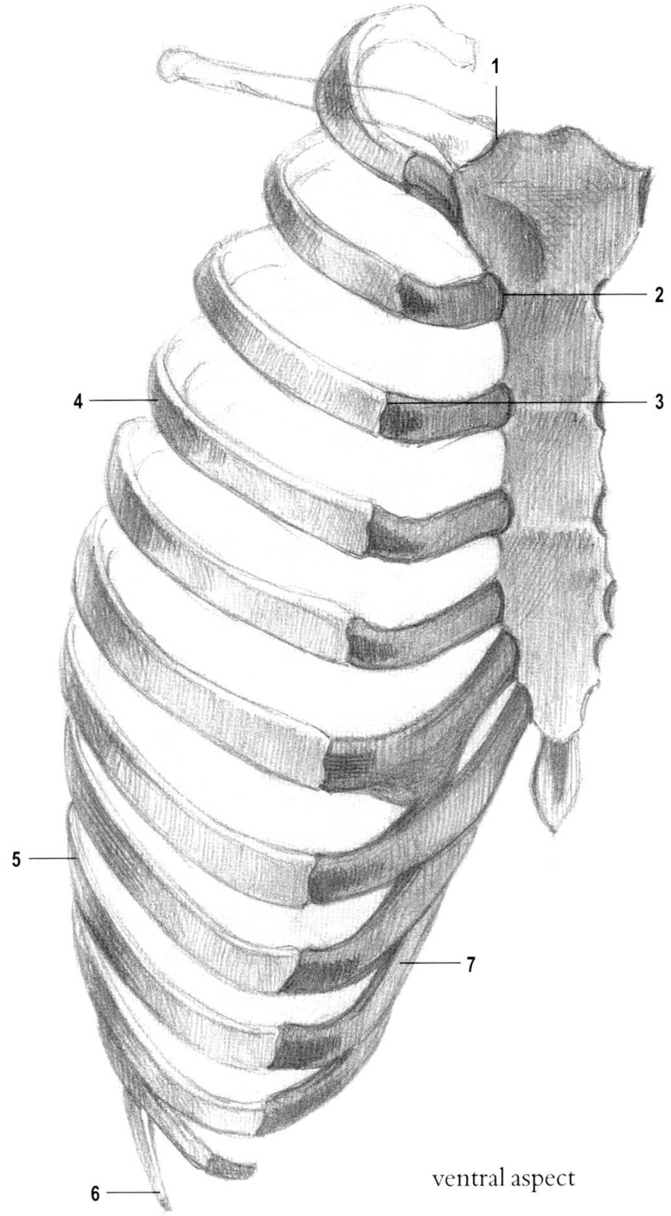

ventral aspect

1 Internal clavicular joint
2 Sternum-rib cartilage joint
3 Joint composed of cartilaginous sections
4 True rib
5 False rib
6 Free rib
7 Rib arch

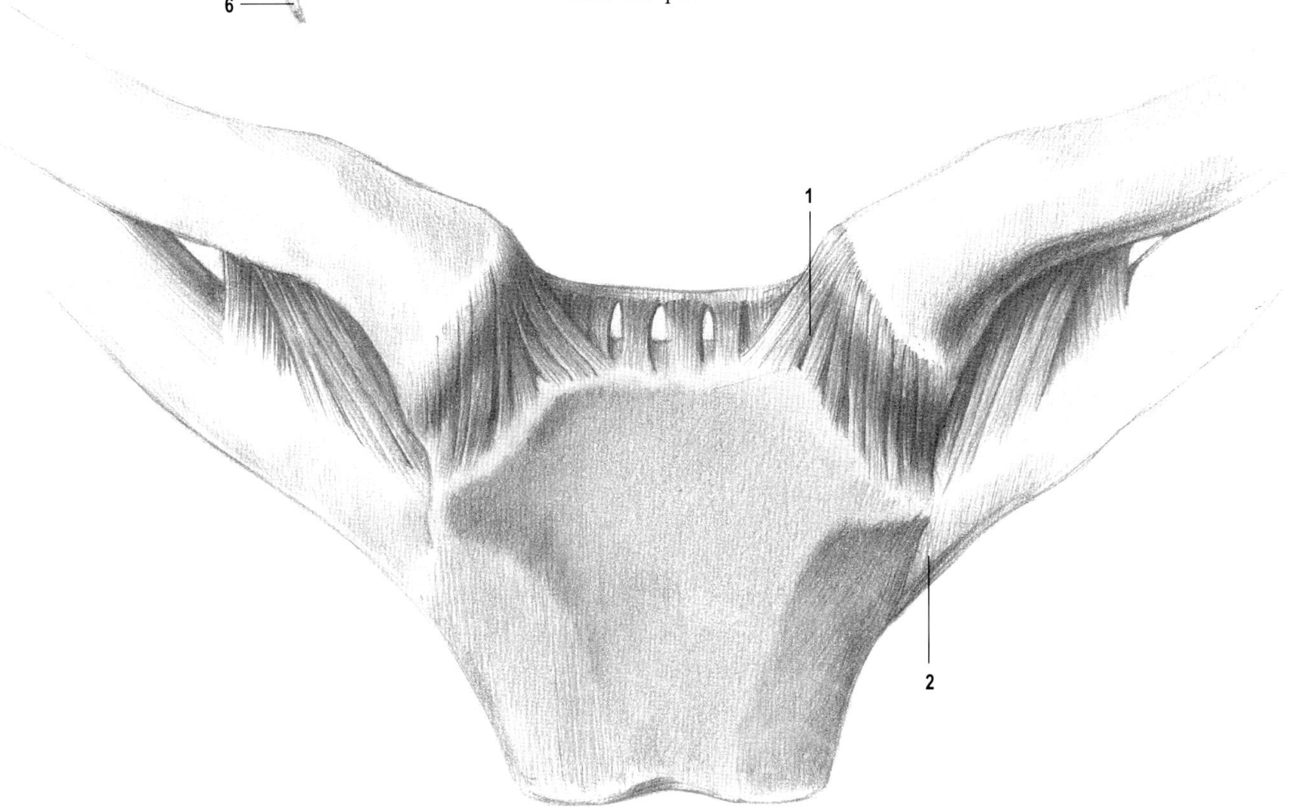

Fig. 193
The Thorax

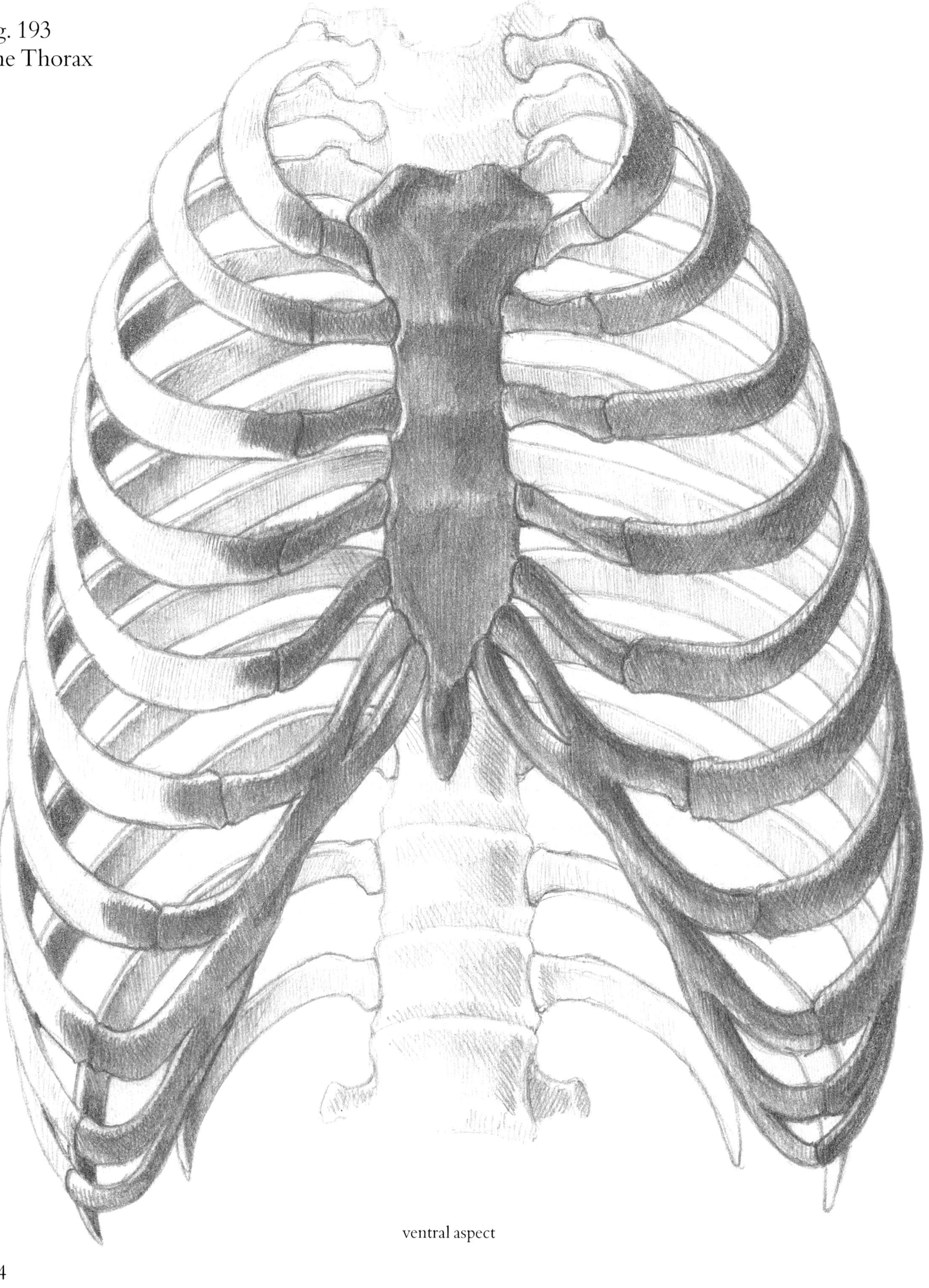

ventral aspect

The thorax broadens towards the pelvis; it possesses an oval cross-section. The inferior end (last rib and costal arch) lifts and expands in inhalation; during exhalation it descends and approaches the spine.

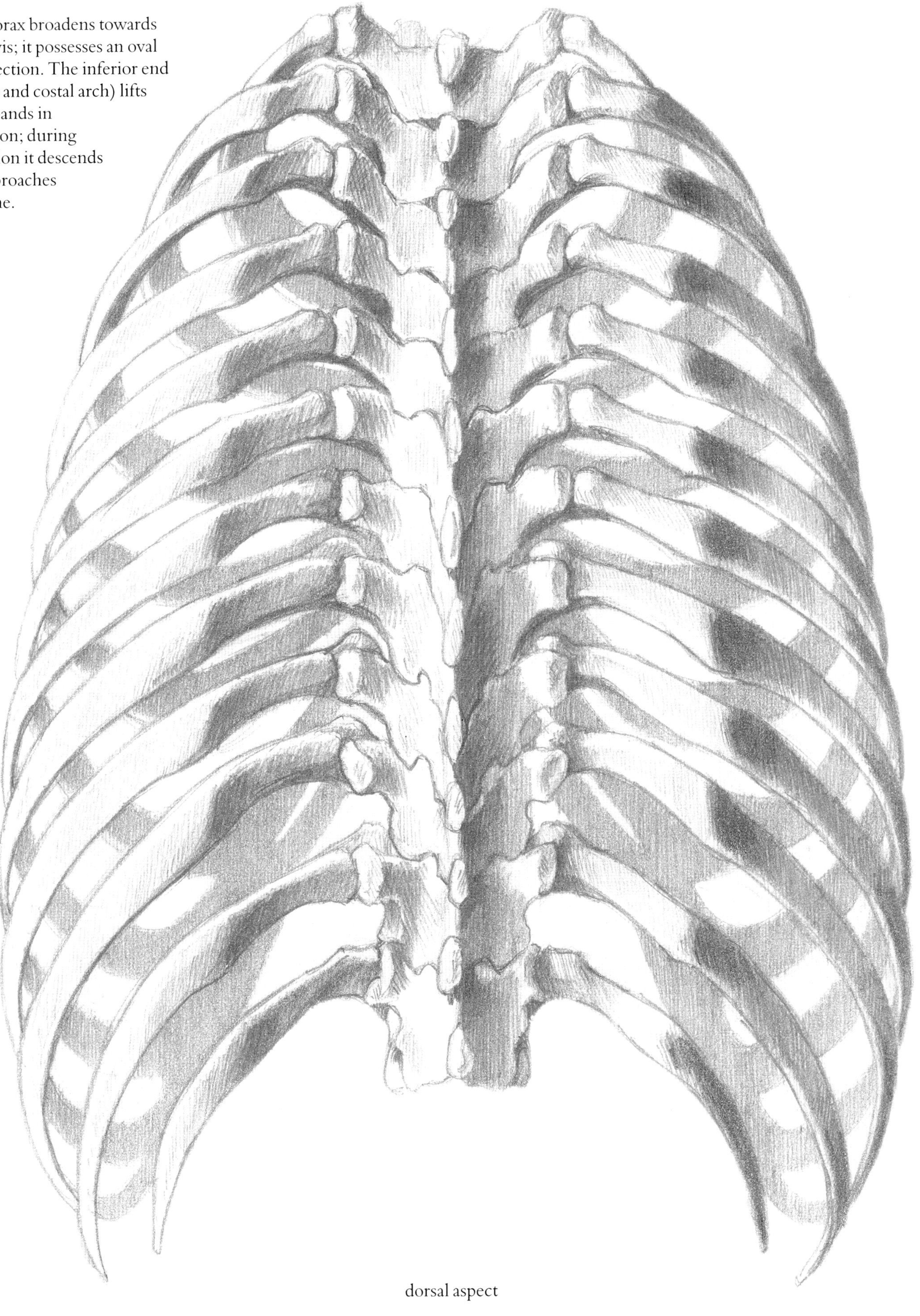

dorsal aspect

The Thorax
(cont.)

The thorax is flat anteriorly and posteriorly. Viewed from the side it is convex.

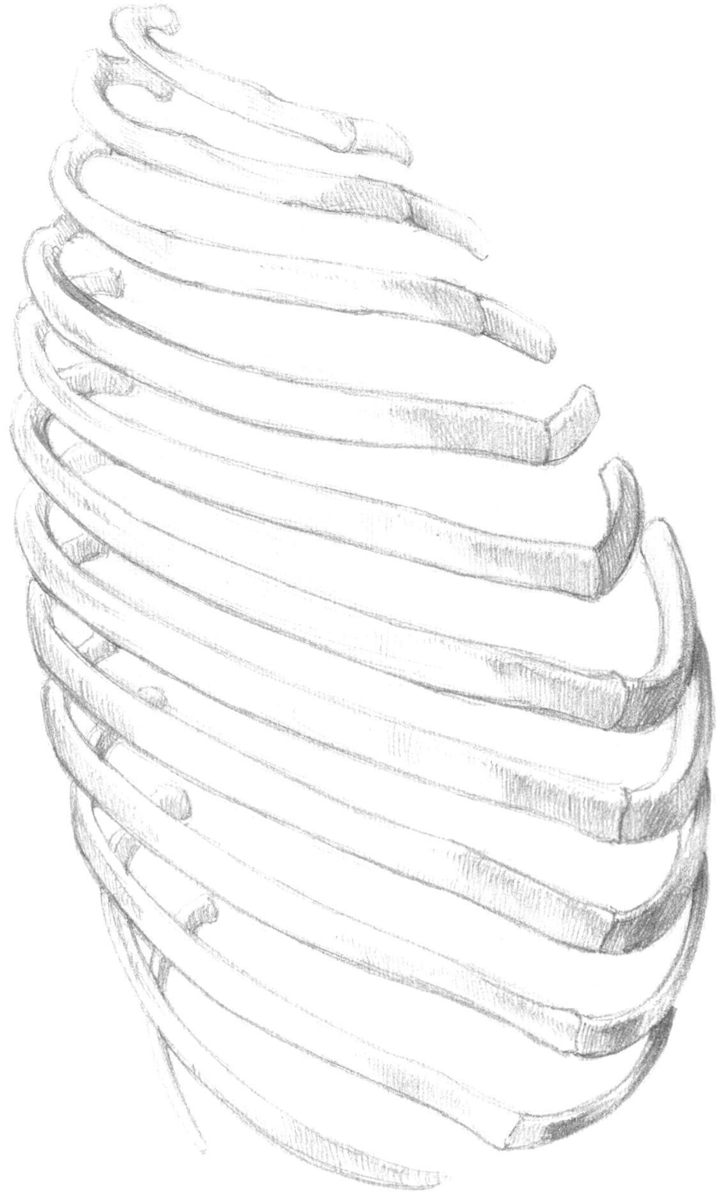

lateral aspect

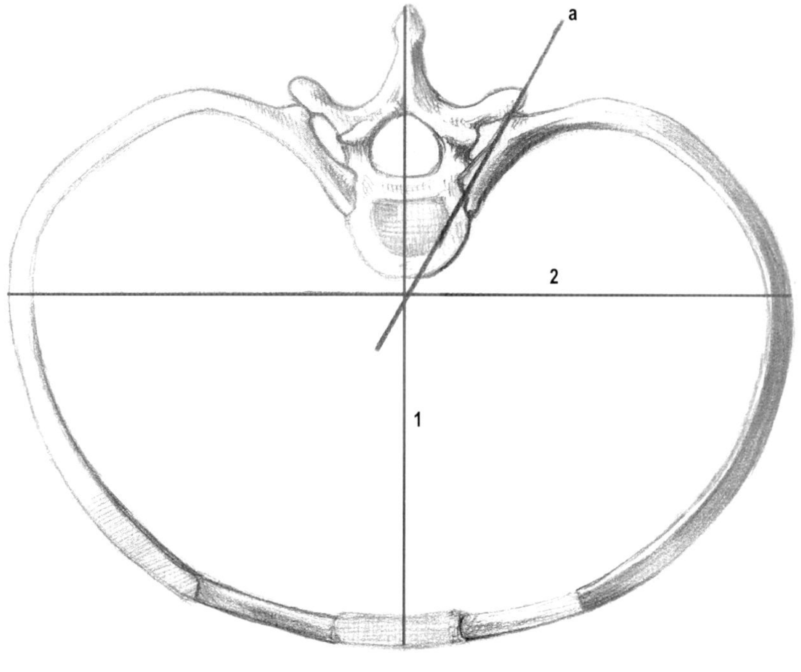

Diameter of the Thorax

1 Medial diameter
2 Transverse diameter

a Rotational axis of the rib-spine joints

Fig. 194
The Movements of the Thorax in Breathing

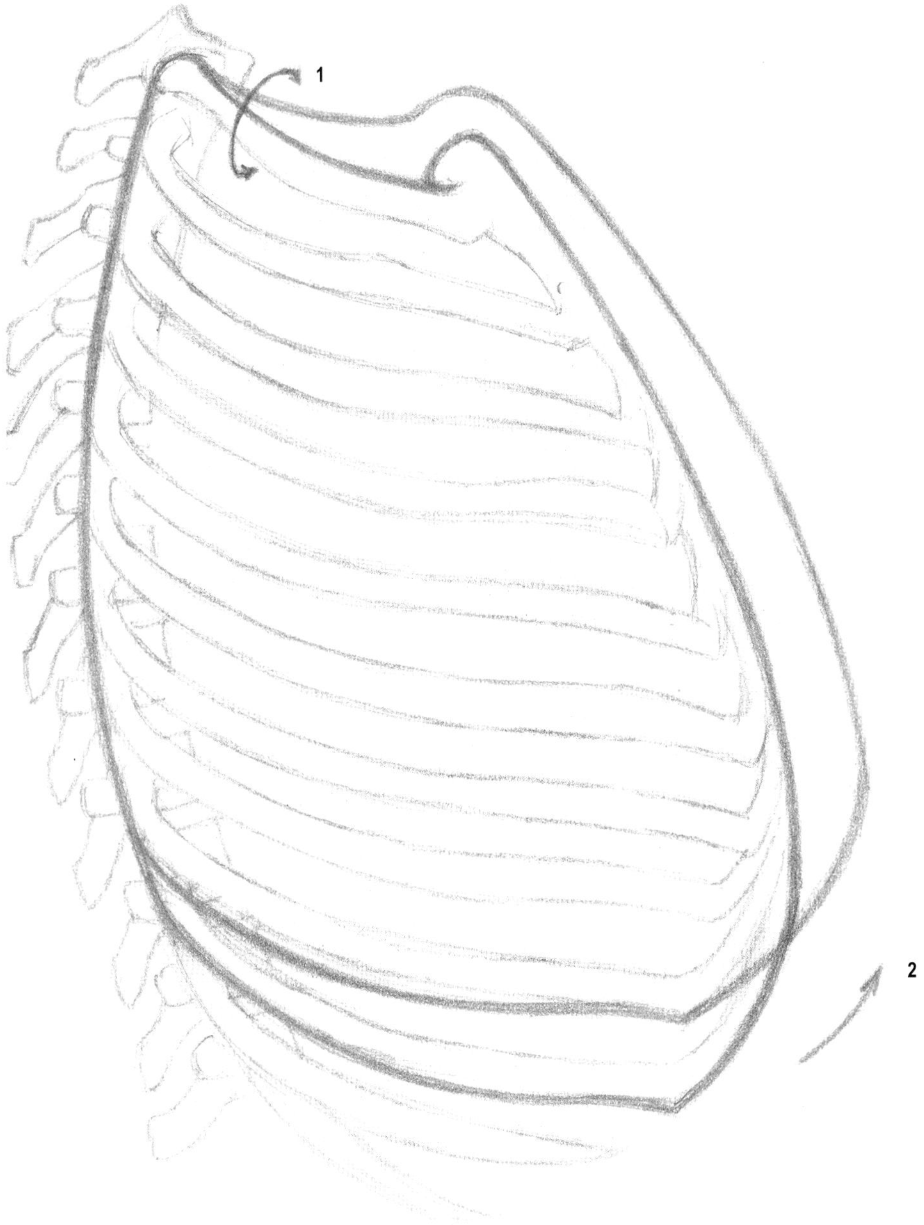

lateral aspect

In inhalation the ribs rotate through the rib-spine joints (**1**) and the thorax expands (**2**) laterally (**3**).

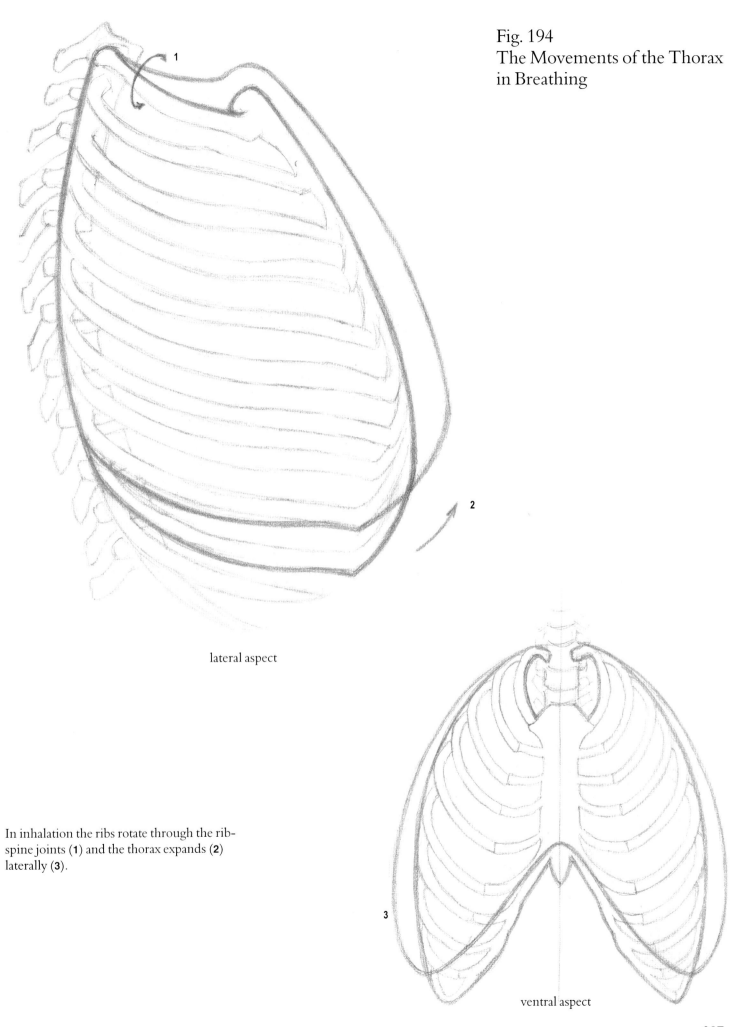

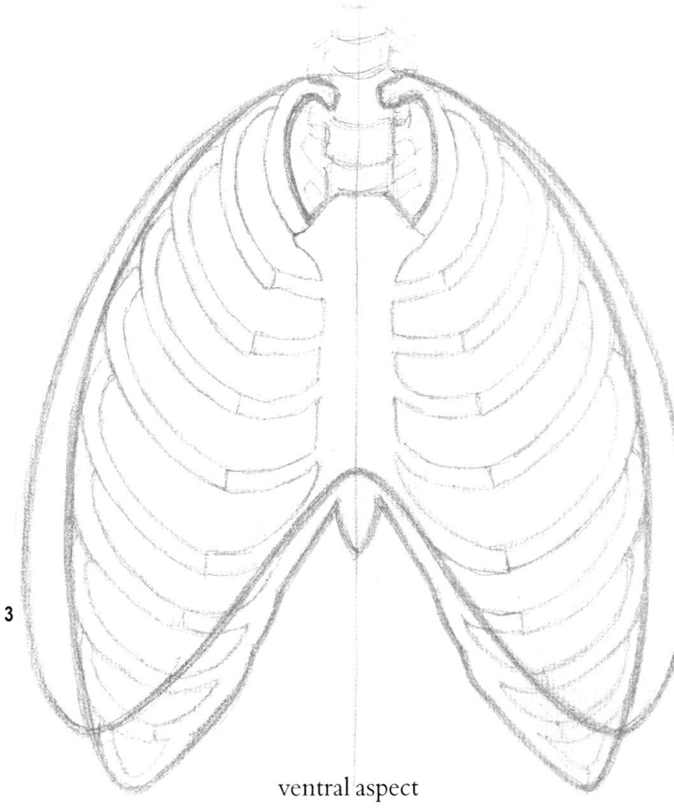

ventral aspect

Fig. 195
The Bones and Superficial
Muscles of the Back

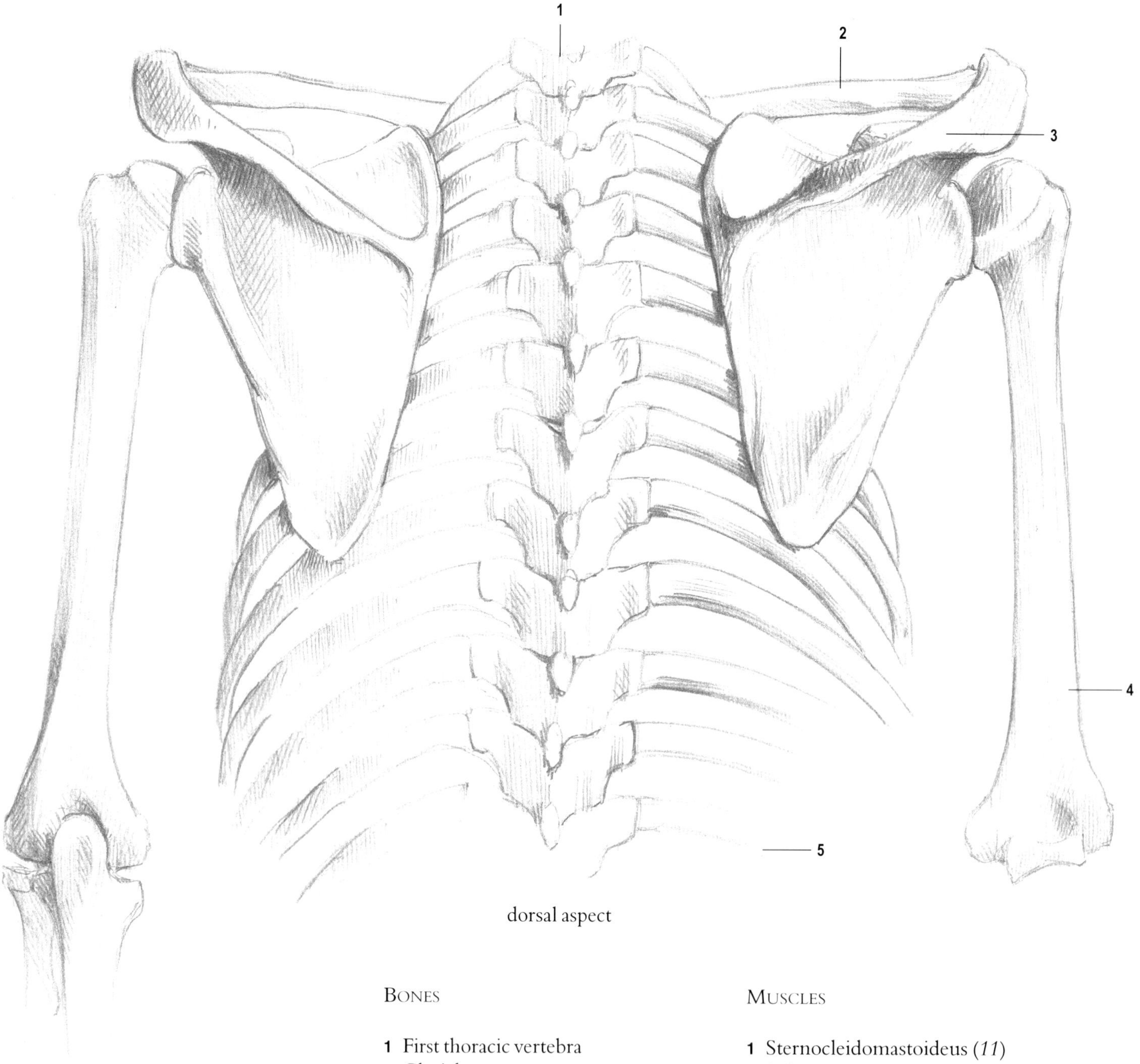

dorsal aspect

BONES

1 First thoracic vertebra
2 Clavicle
3 Shoulder blades
4 Humerus
5 Last (free) rib

MUSCLES

1 Sternocleidomastoideus (*11*)
2 Splenius (*5*)
3 Trapezius (*20*)
4 Two heads of the Deltoid muscle (*41*)
5 Teres minor, teres major (*44, 45*)
6 Triceps (*50*)
7 Latissimus dorsi (*21*)
8 Anterior Serratus (*31*)
9 Obliquus externus abdominis (*37*)

a Tendinous plate above the shoulder blades
b Dorsolumar fasciae

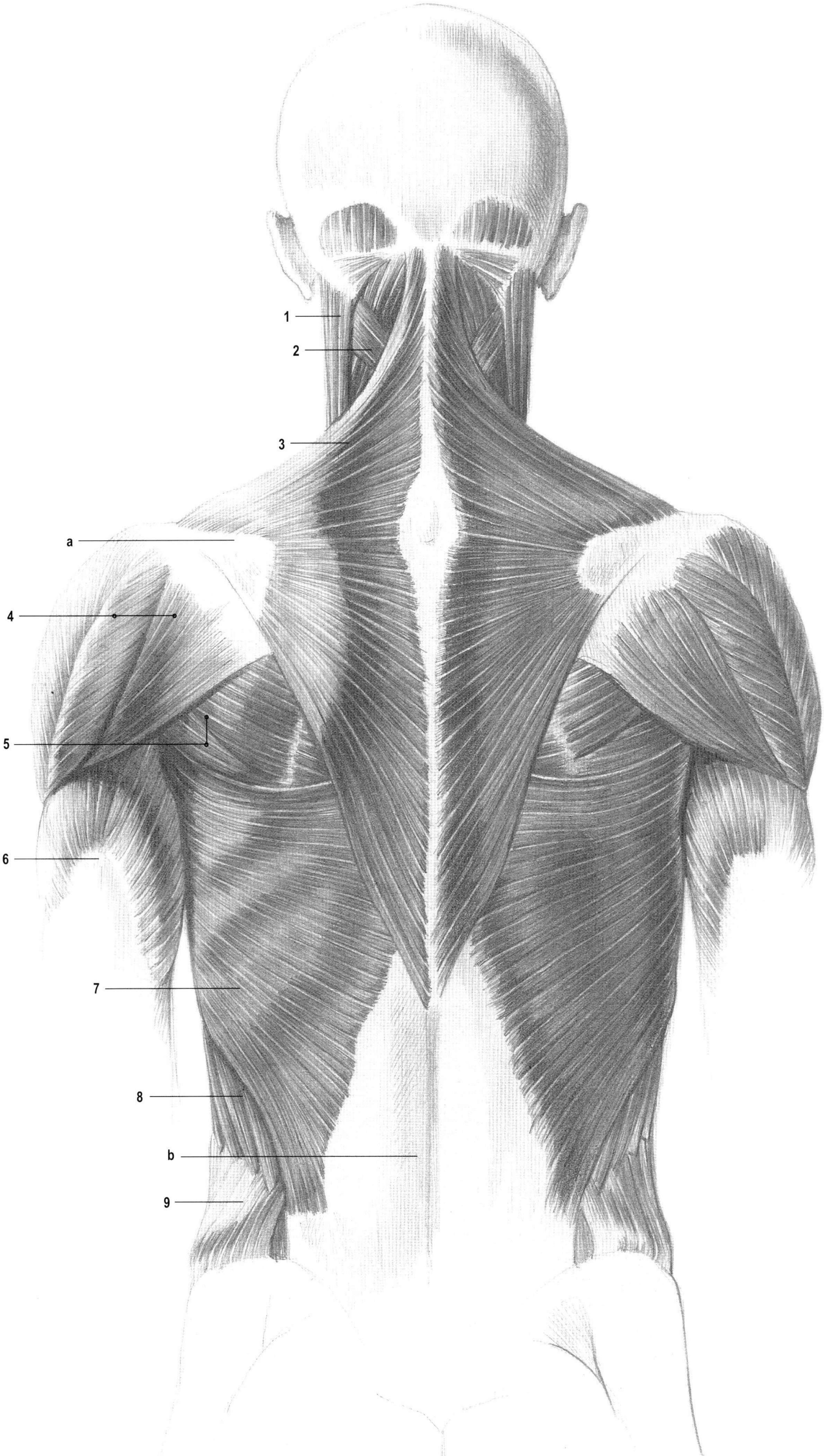

1
2
3
a
4
5
6
7
8
b
9

Fig. 196
The Trapezius, *20*

The trapezius is a flat, triangular muscle.

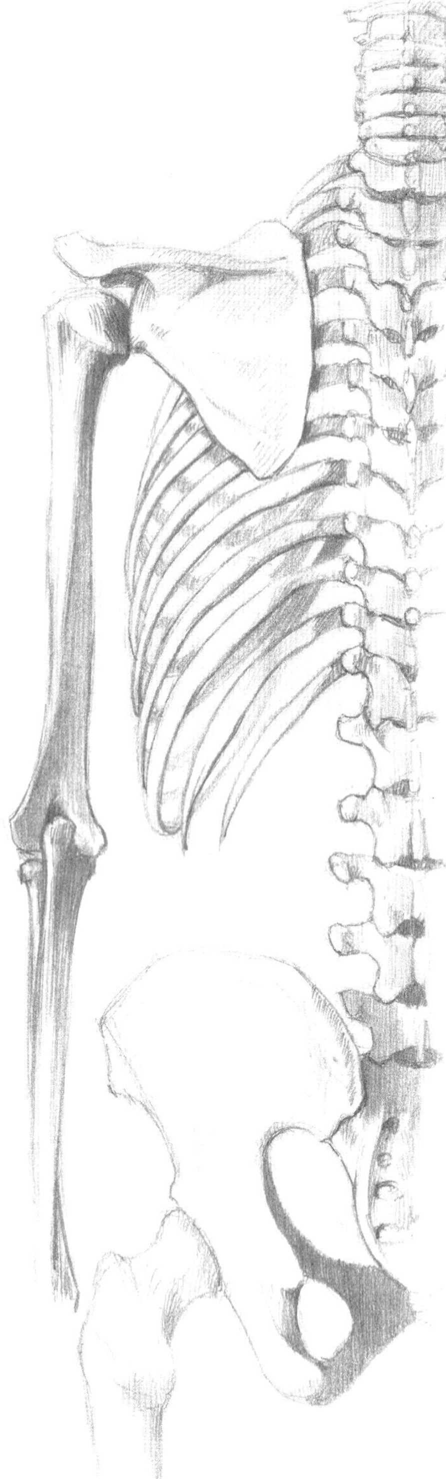

dorsal aspect

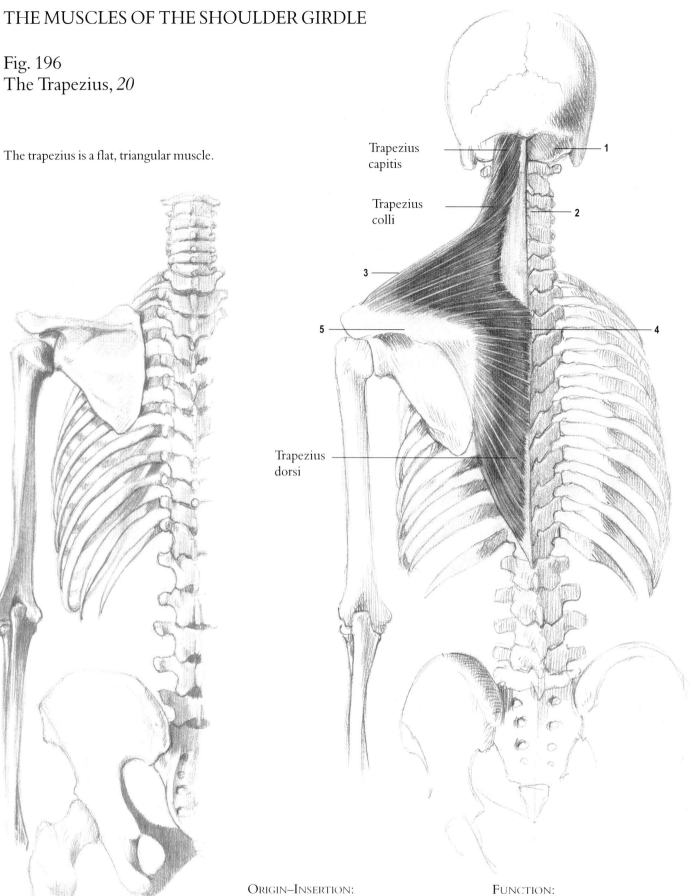

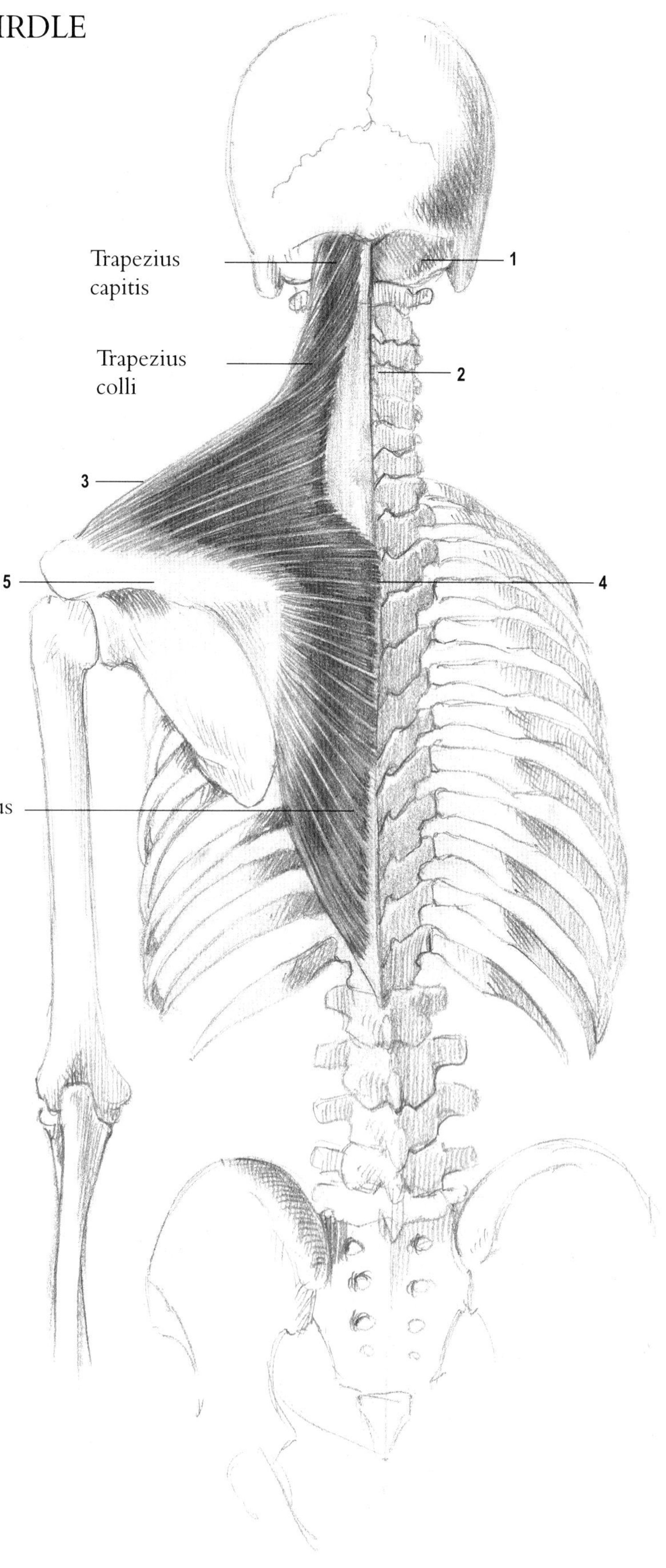

ORIGIN–INSERTION:
The trapezius capitis originates at the rear of the skull (**1**) and at the neck ligament (**2**) and is inserted at the clavicle (**3**); the trapezius dorsi originates at the spinous processes of the spine (**4**) and is inserted at the shoulder blades (**5**)

FUNCTION:
Pulls the shoulder joint and the clavicle upwards; the trapezius colli lifts the shoulder and the trapezius dorsi pulls the shoulder blade to the centre

Fig. 197
The Latissimus Dorsi, *21*

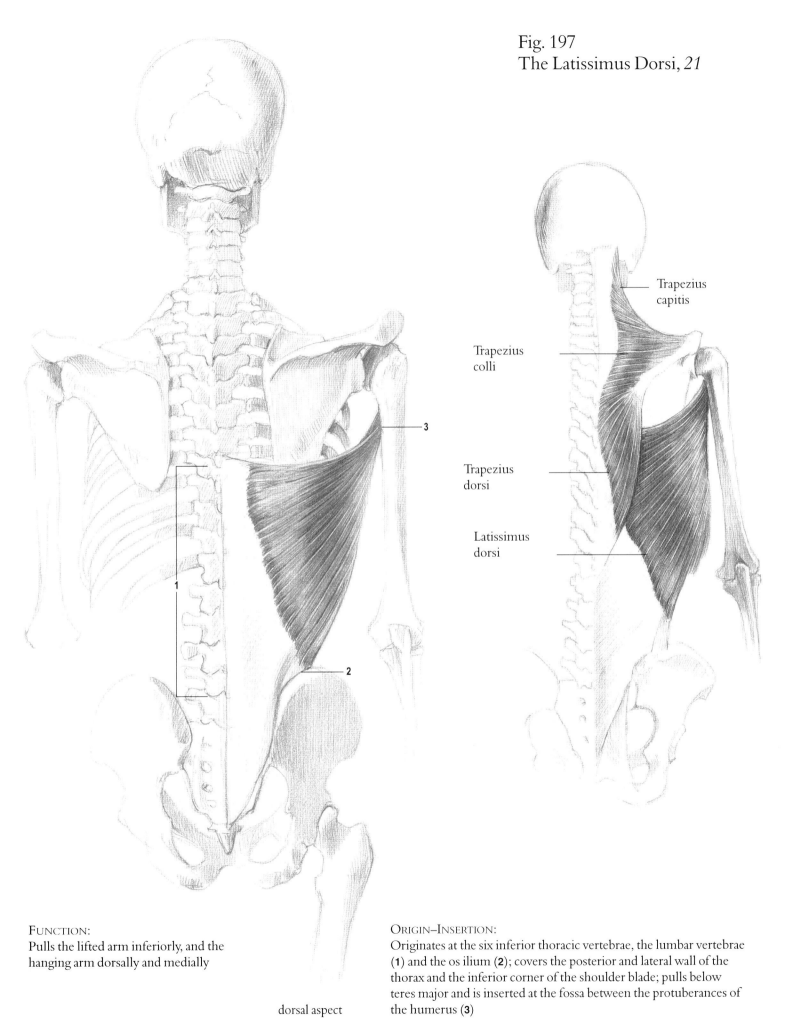

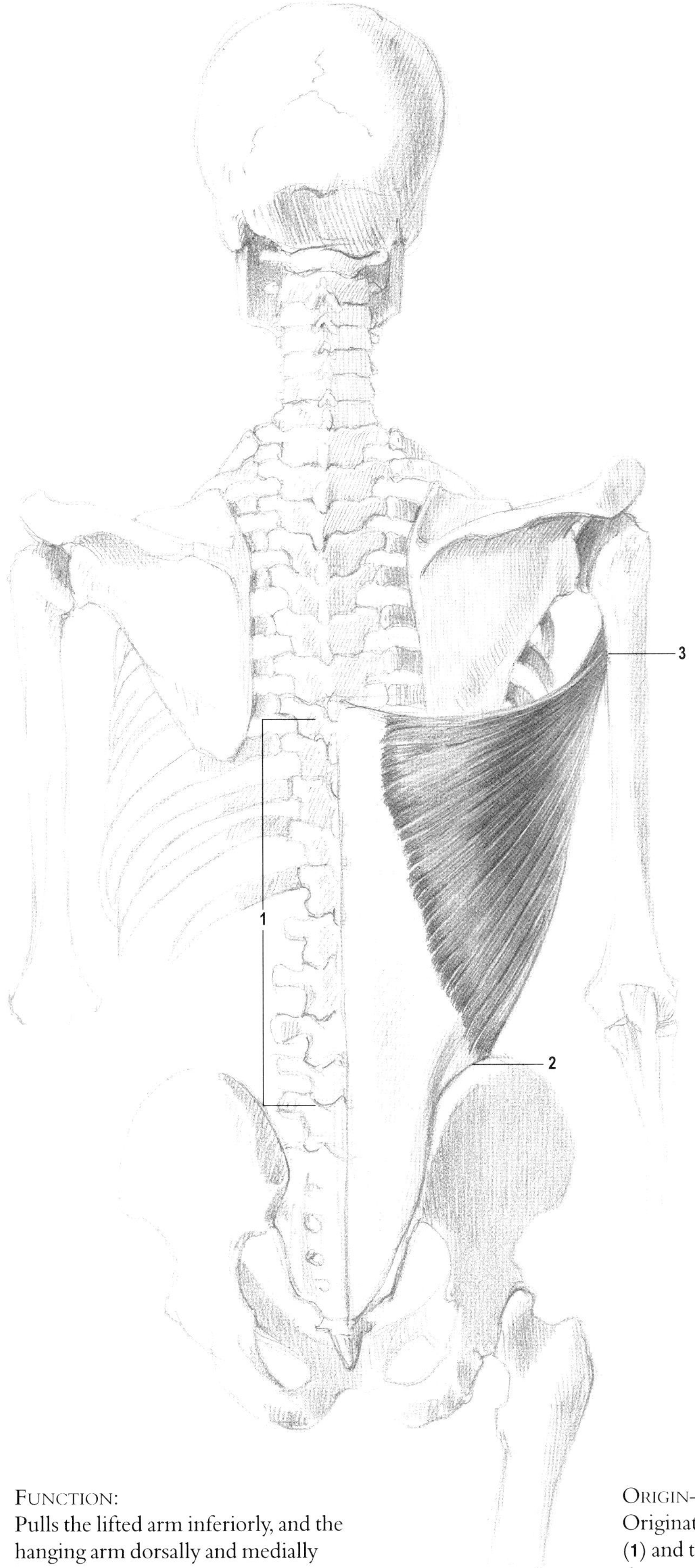

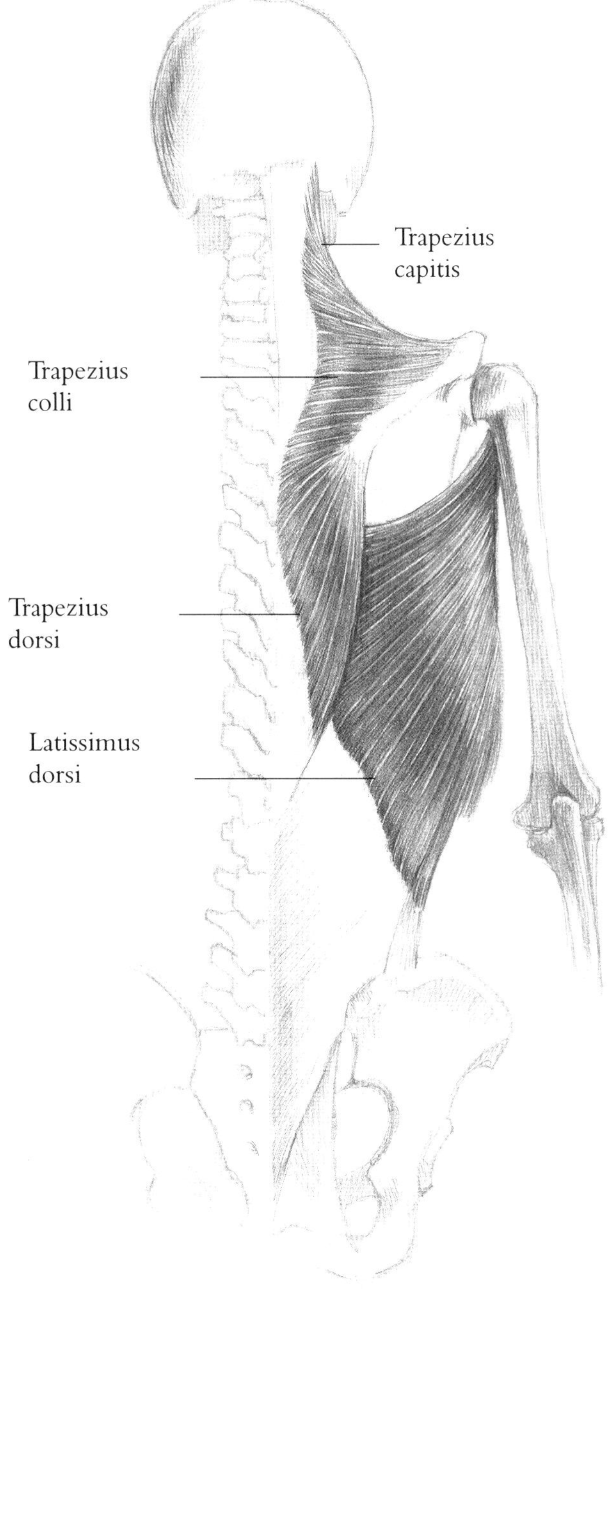

FUNCTION:
Pulls the lifted arm inferiorly, and the hanging arm dorsally and medially

dorsal aspect

ORIGIN–INSERTION:
Originates at the six inferior thoracic vertebrae, the lumbar vertebrae (**1**) and the os ilium (**2**); covers the posterior and lateral wall of the thorax and the inferior corner of the shoulder blade; pulls below teres major and is inserted at the fossa between the protuberances of the humerus (**3**)

Fig. 198 The Greater Rhomboid Muscle

(M. rhomboideus, *22*, *23*)

Rhomboideus major

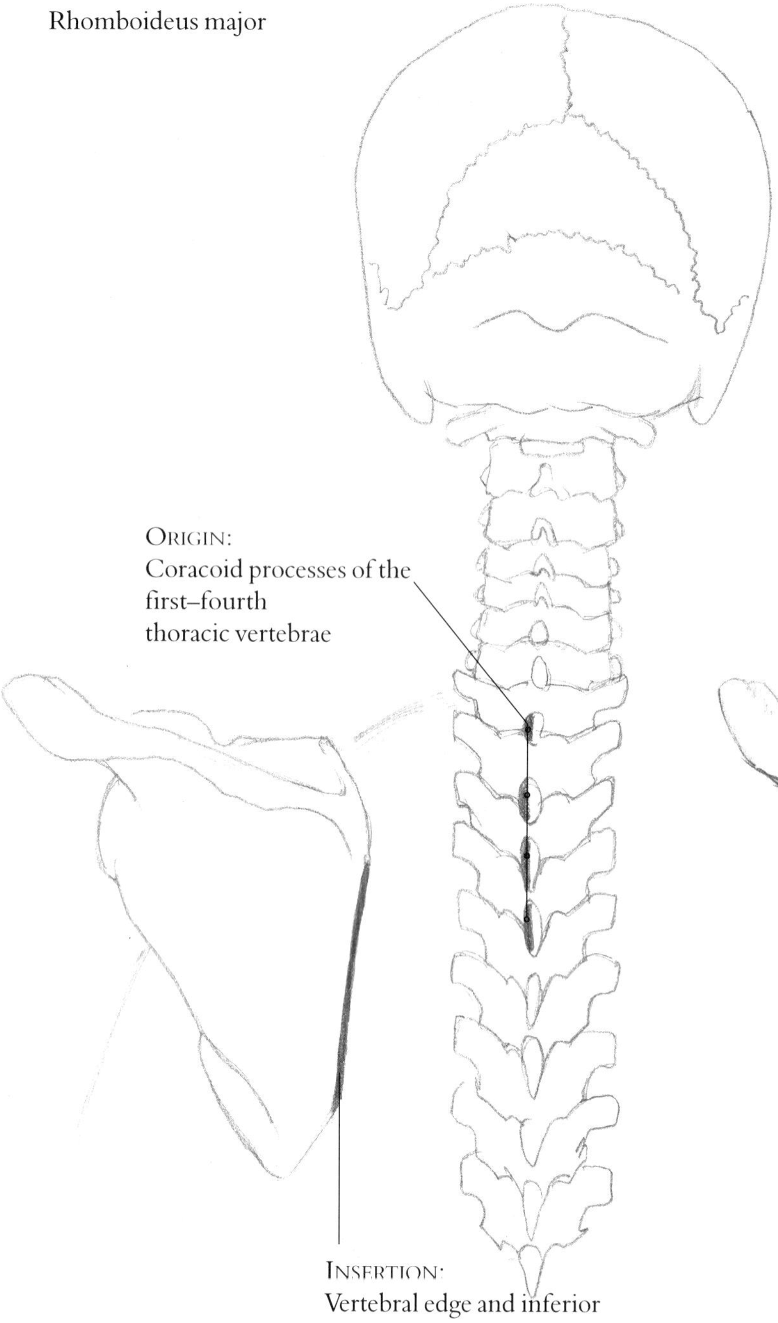

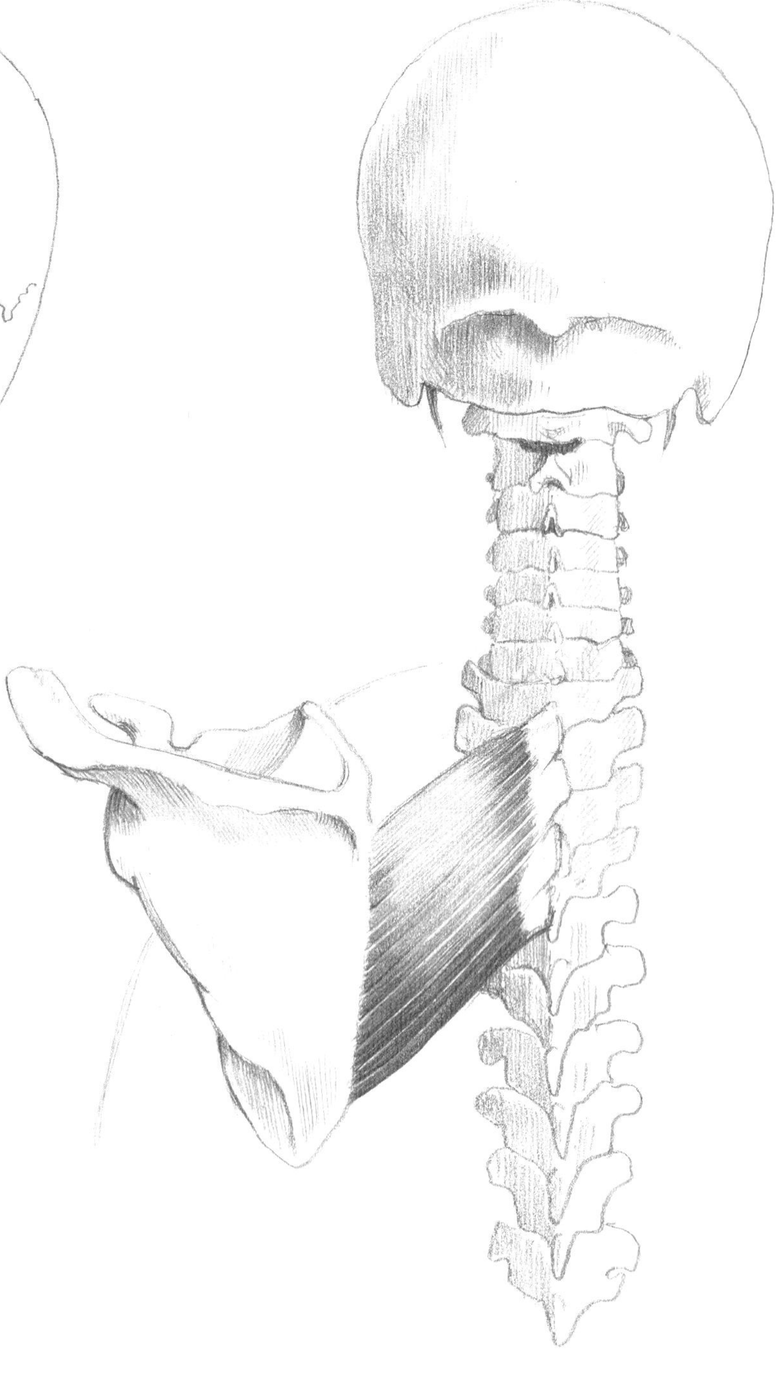

FUNCTION:
Pulls the shoulder blade towards the spine and head

dorsal aspect

Rhomboideus minor

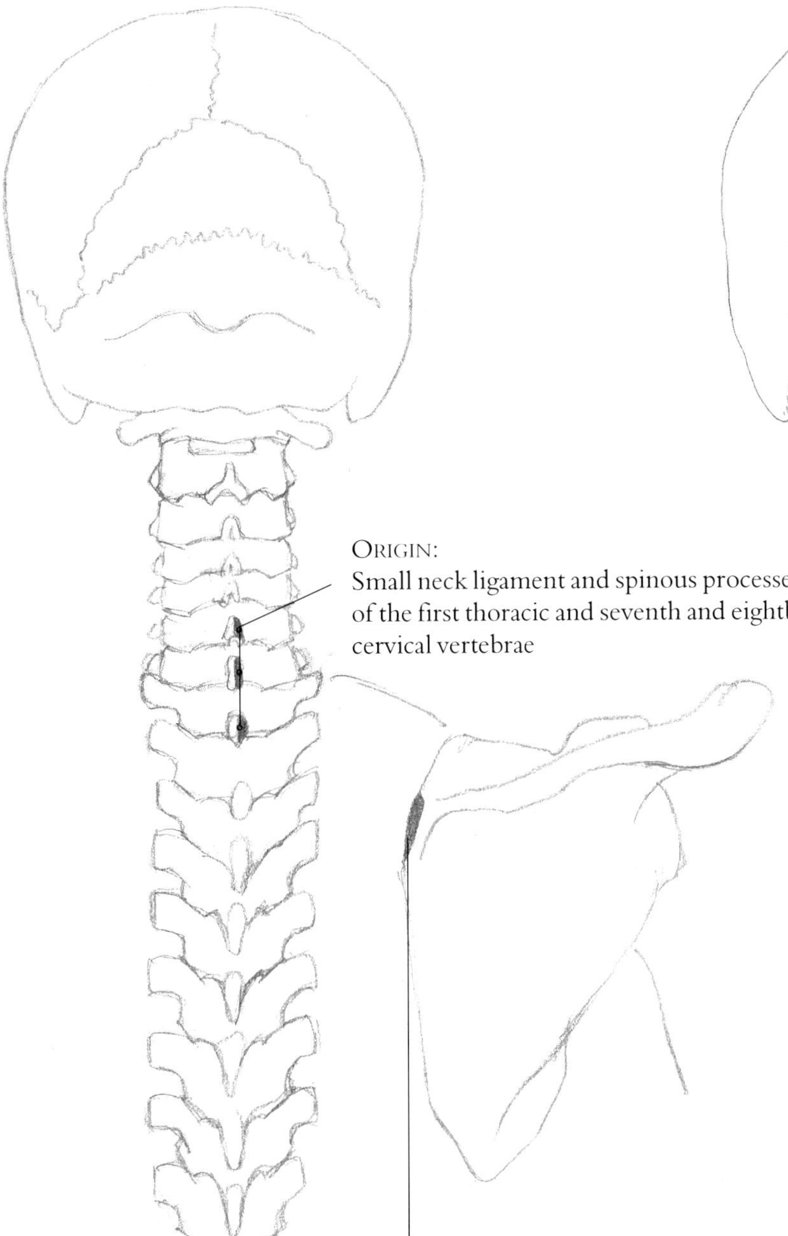

ORIGIN:
Small neck ligament and spinous processes of the first thoracic and seventh and eighth cervical vertebrae

INSERTION:
Medial edge of the shoulder blade

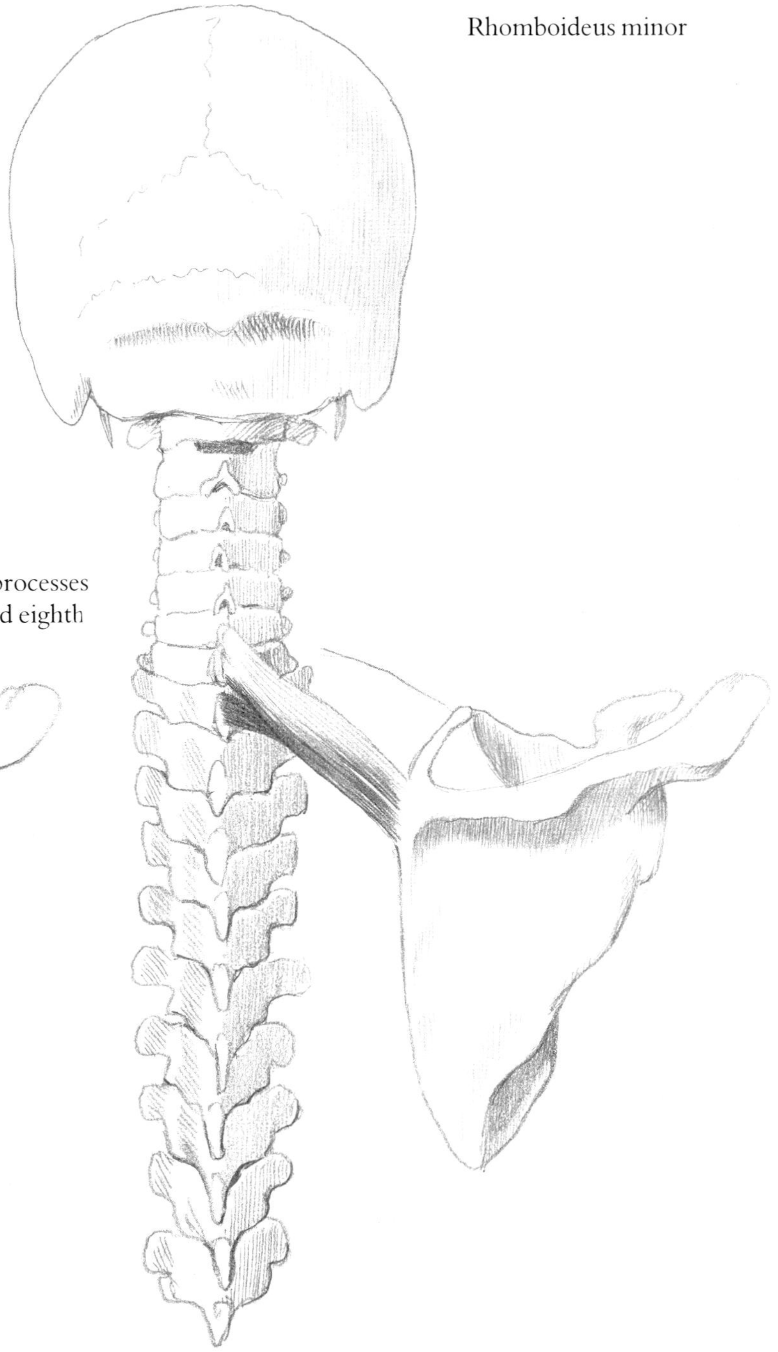

FUNCTION:
Pulls the shoulder blade towards the spiral column and towards the head

dorsal aspect

The Greater Rhomboid Muscle

(cont.)

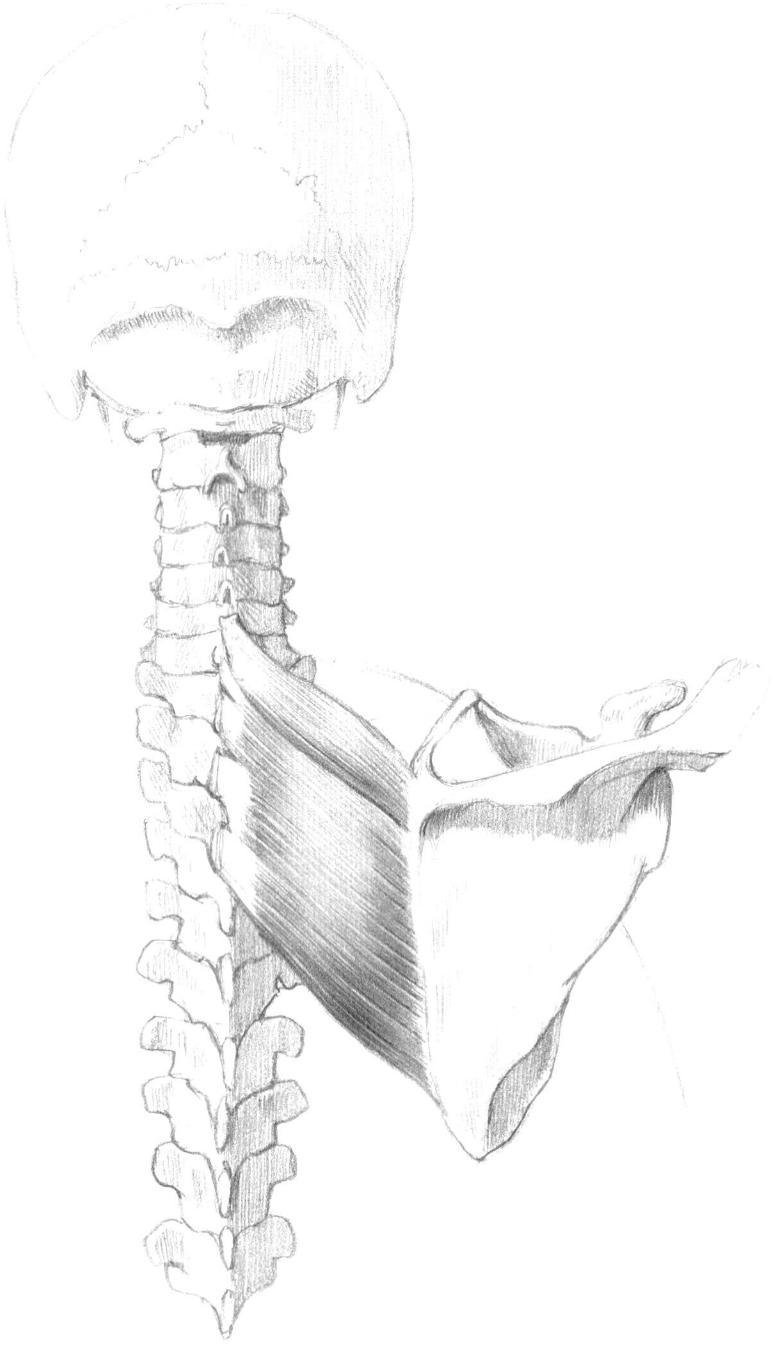

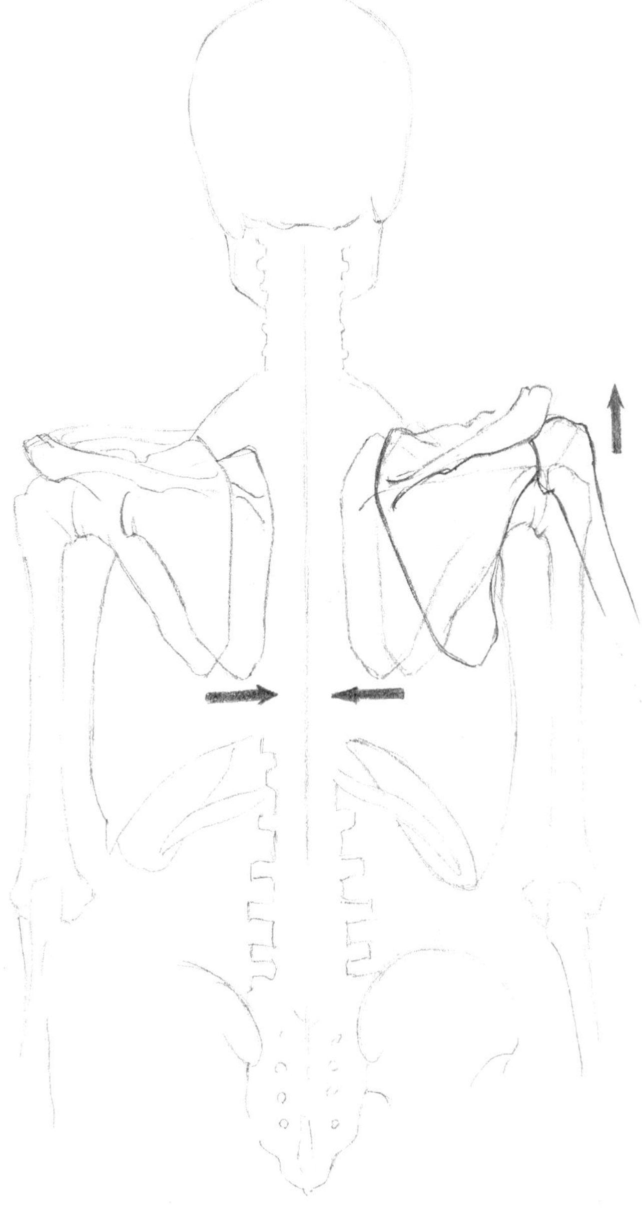

The rhomboideus originates at the coracoid processes of the last two cervical and first four thoracic vertebrae, and is inserted at the medial edge of the shoulder blade.

FUNCTION:
Pulls the shoulder blade towards the spine and head, fixes the shoulder blade

dorsal aspect

Fig. 199
The Muscles of the Shoulder

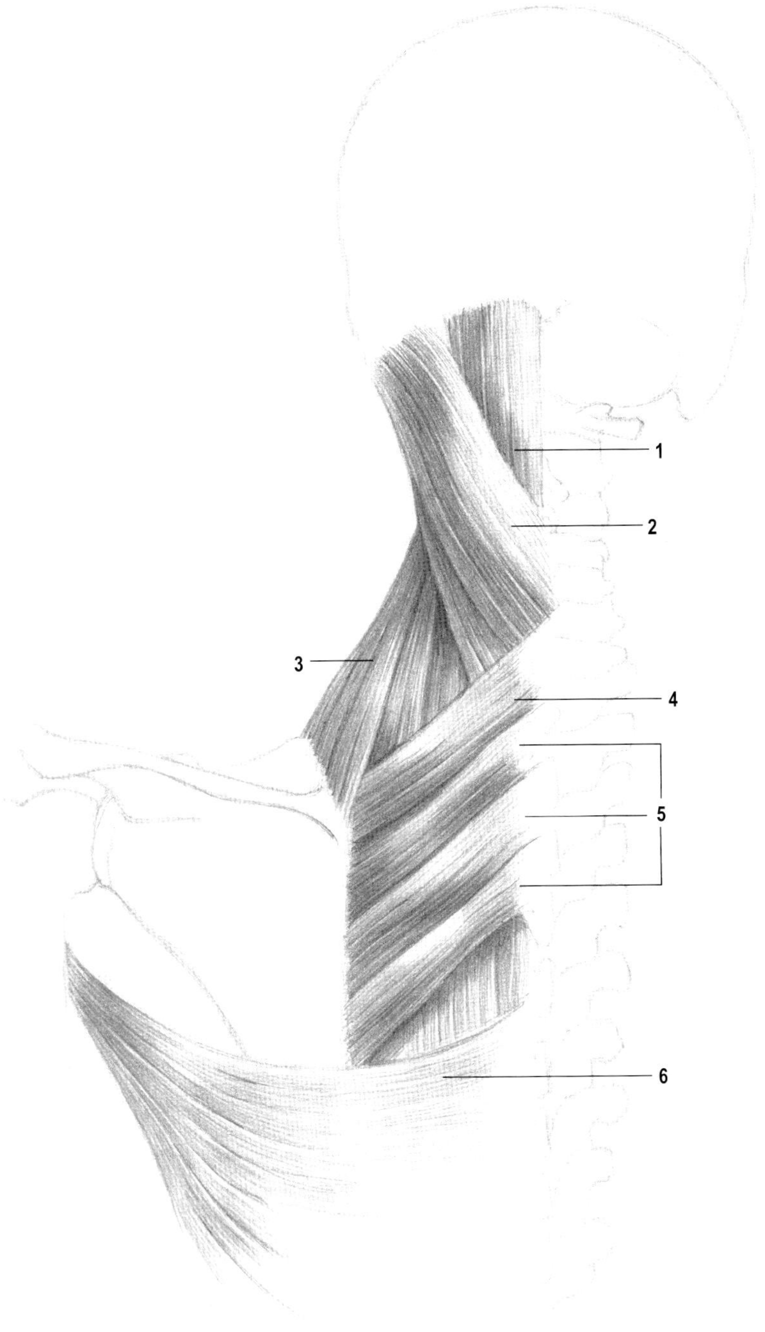

dorsal aspect

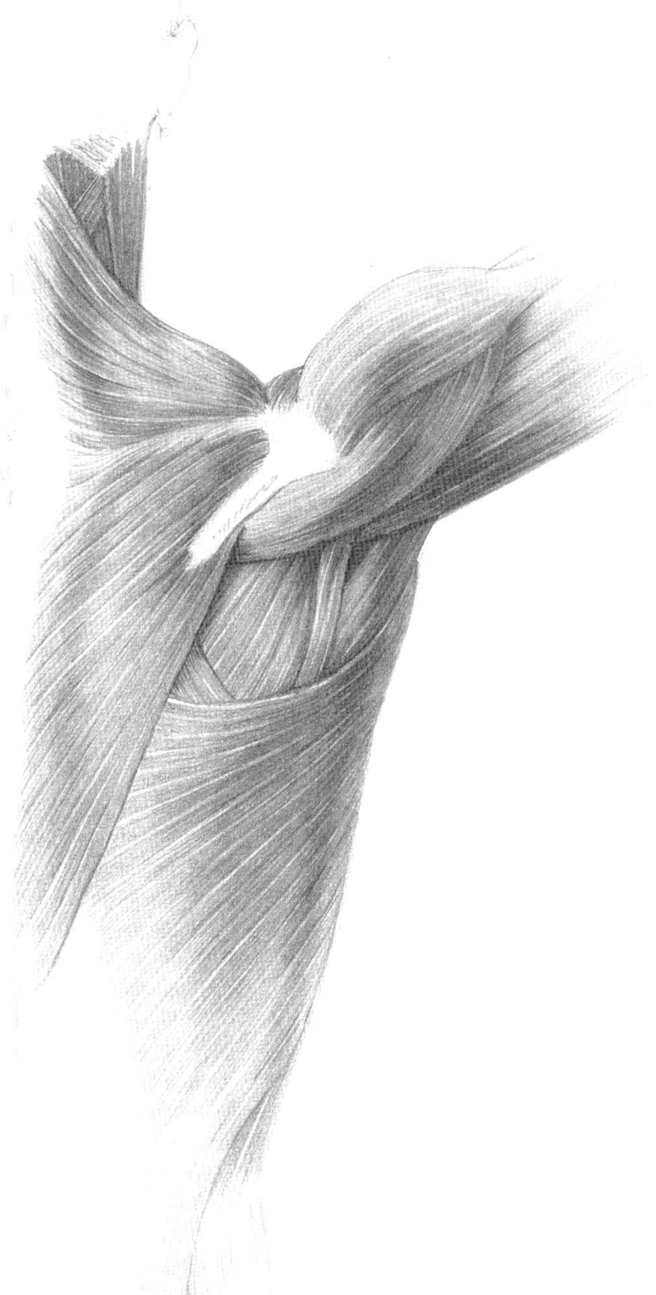

1 Semispinalis (*26/4*)
2 Splenius (*5*)
3 Levator scapulae (*24*)
4 Rhomboideus minor (*23*)
5 Rhomboideus major (*22*)
6 Latissimus dorsi (*21*)

Fig. 200
The Levator Scapulae, *24*

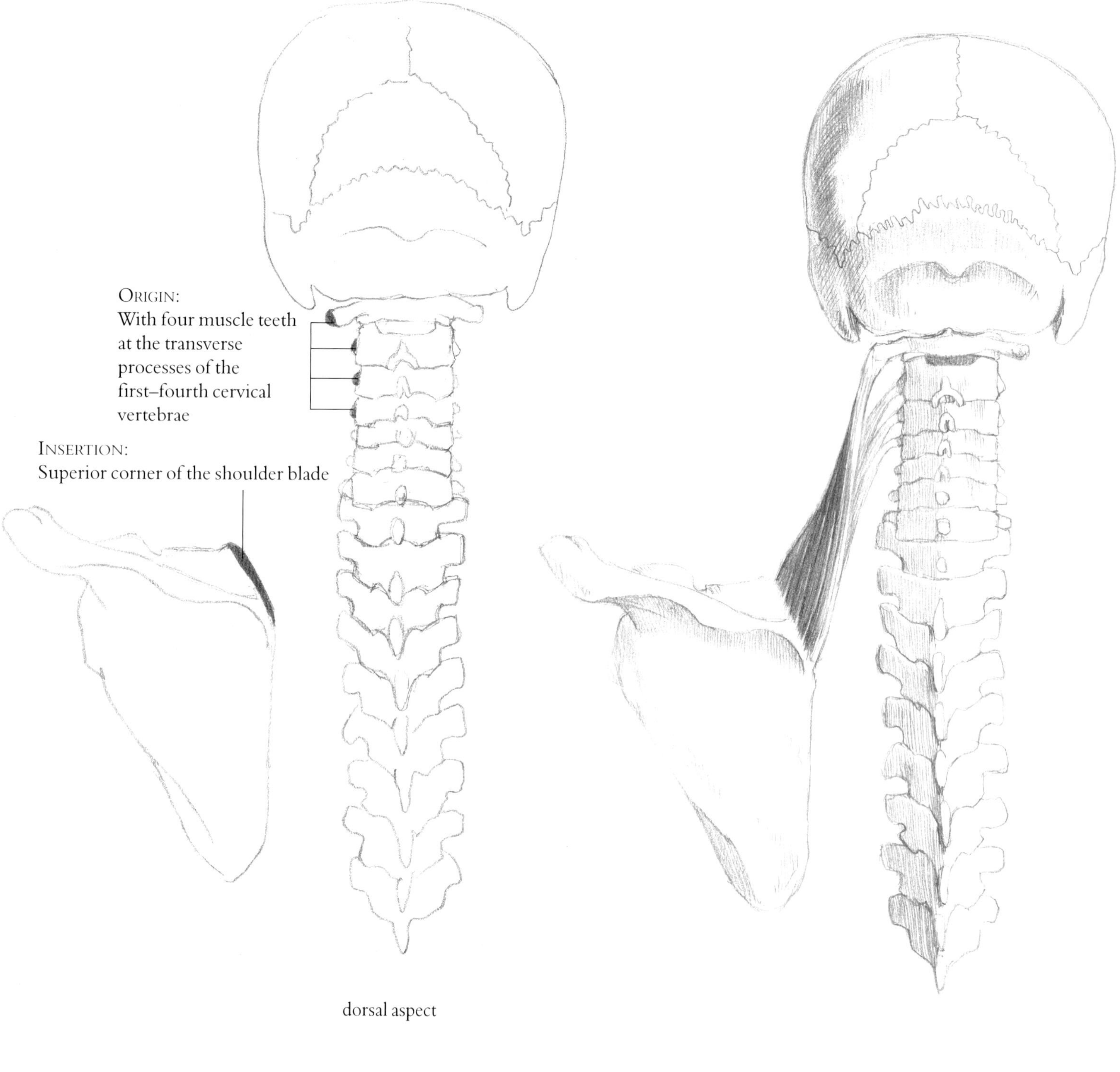

FUNCTION:
Lifts and rotates the shoulder blade

Fig. 201
The Subclavius, *30*

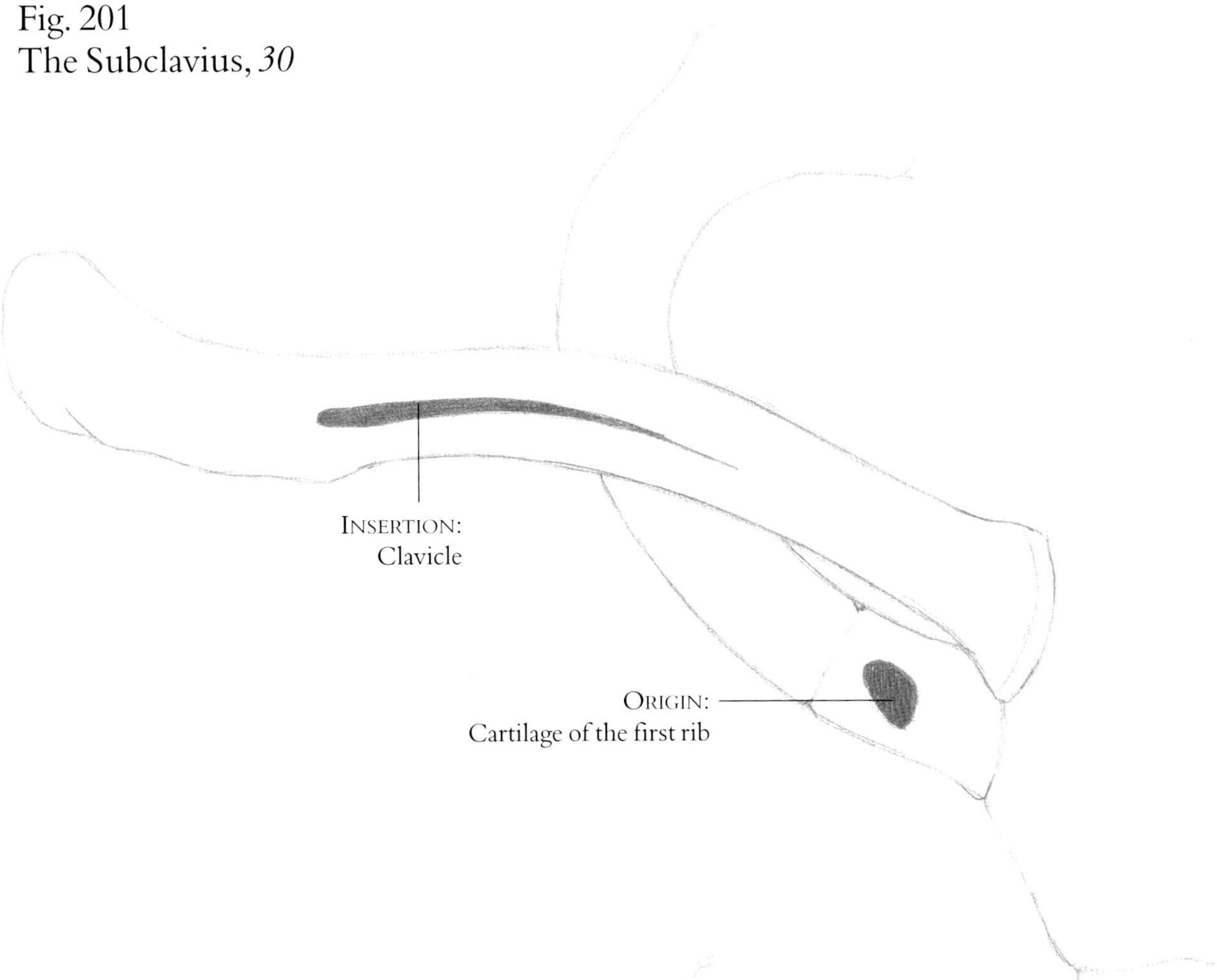

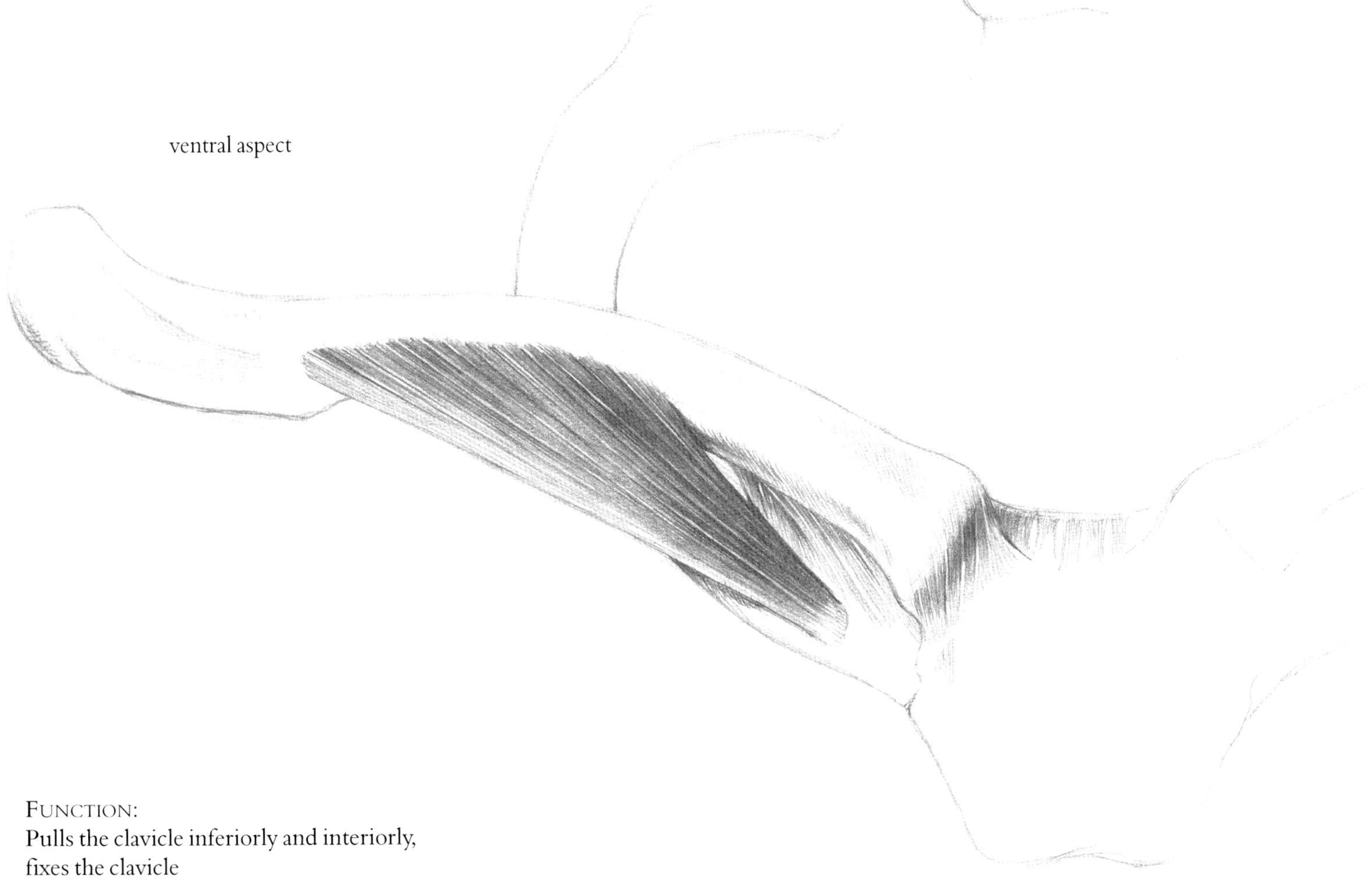

FUNCTION:
Pulls the clavicle inferiorly and interiorly,
fixes the clavicle

Fig. 202
The Back Muscles

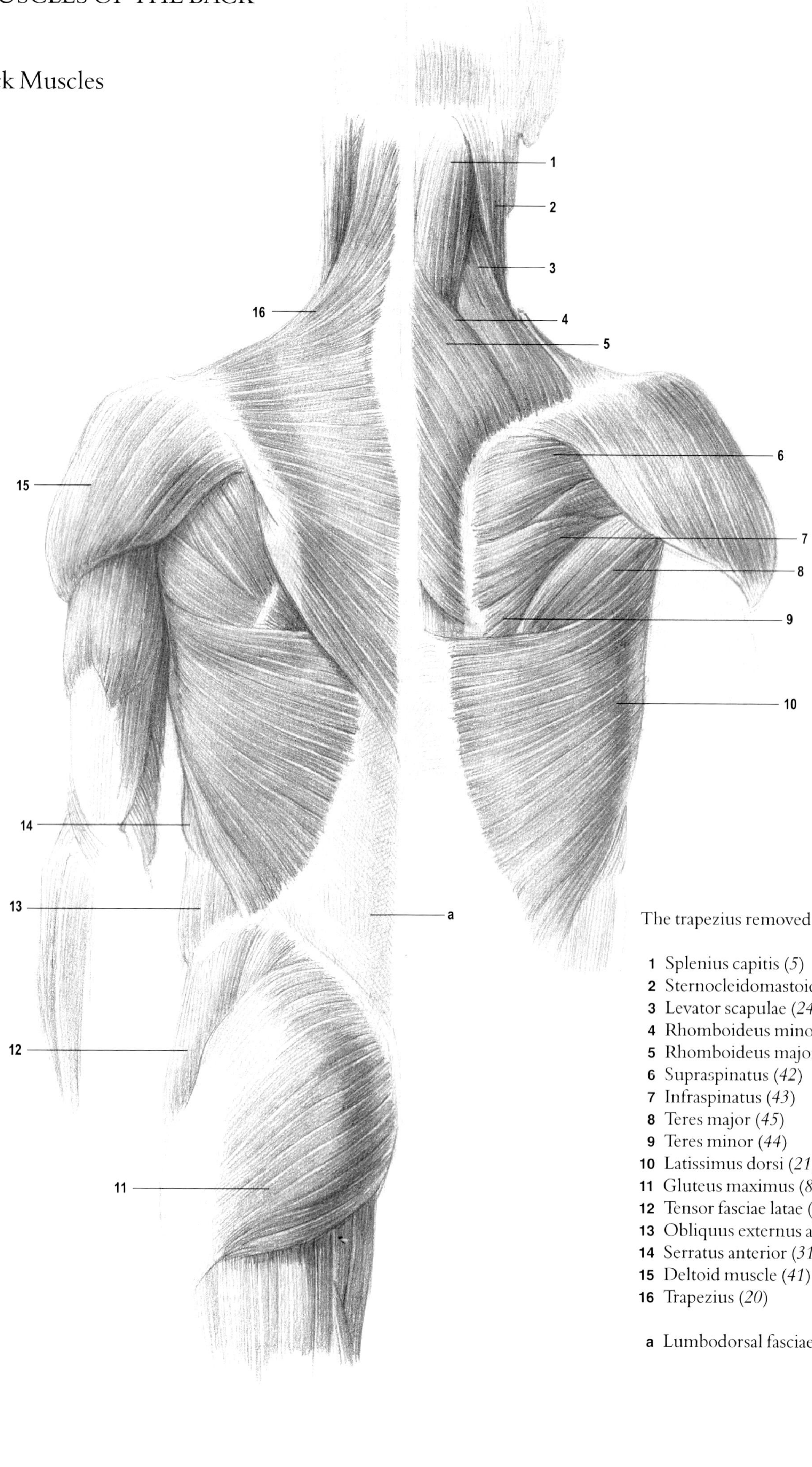

The trapezius removed on the right.

1 Splenius capitis (*5*)
2 Sternocleidomastoideus (*11*)
3 Levator scapulae (*24*)
4 Rhomboideus minor (*23*)
5 Rhomboideus major (*22*)
6 Supraspinatus (*42*)
7 Infraspinatus (*43*)
8 Teres major (*45*)
9 Teres minor (*44*)
10 Latissimus dorsi (*21*)
11 Gluteus maximus (*81*)
12 Tensor fasciae latae (*84*)
13 Obliquus externus abdominis (*37*)
14 Serratus anterior (*31*)
15 Deltoid muscle (*41*)
16 Trapezius (*20*)

a Lumbodorsal fasciae

Fig. 203
The Muscles of the Neck and Back

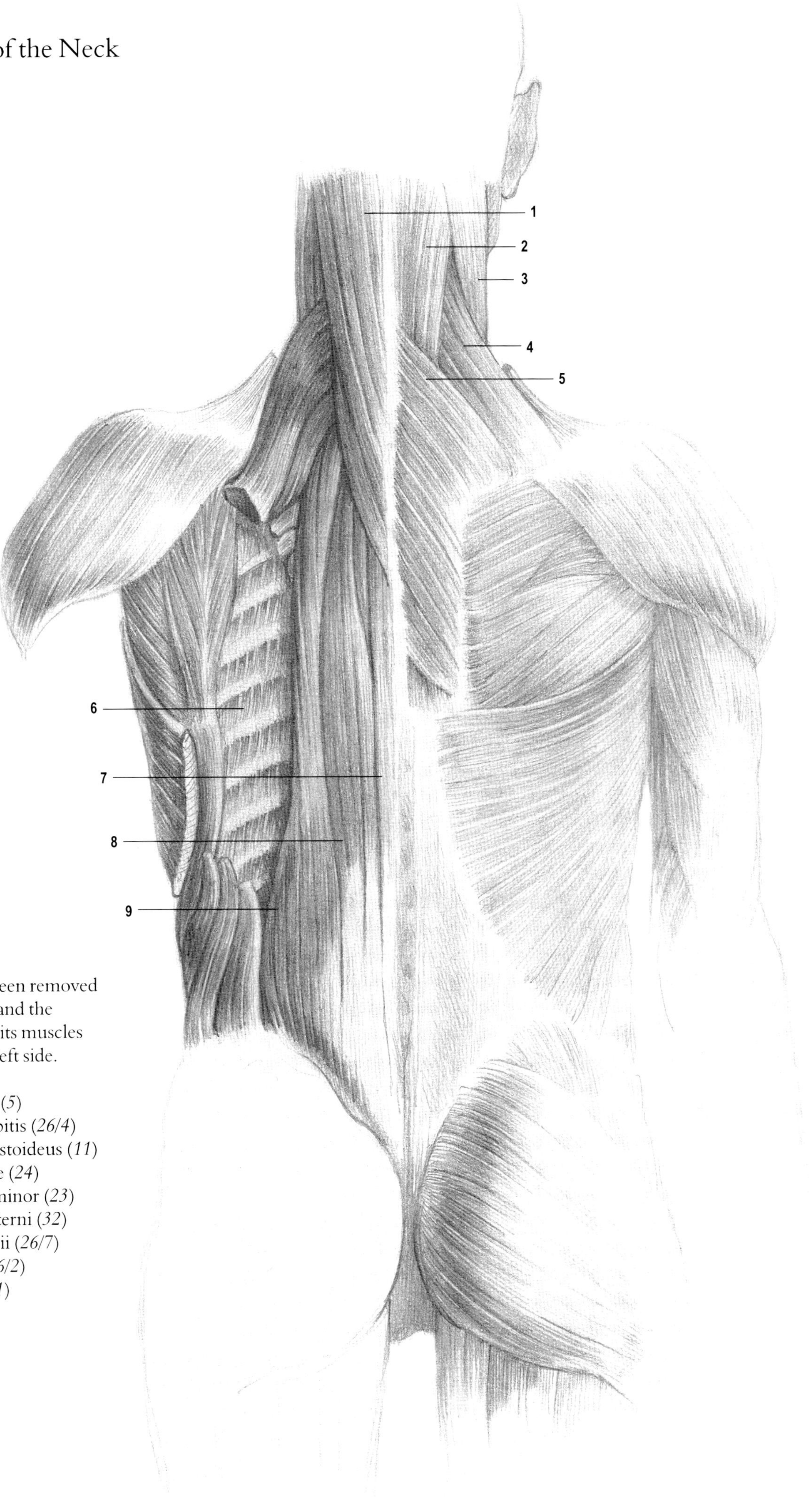

The trapezius has been removed from the right side and the shoulder blade and its muscles removed from the left side.

1 Splenius capitis (*5*)
2 Semispinalis capitis (*26/4*)
3 Sternocleidomastoideus (*11*)
4 Levator scapulae (*24*)
5 Rhomboideus minor (*23*)
6 Intercostales externi (*32*)
7 Intertransversarii (*26/7*)
8 Longissimus (*26/2*)
9 Iliocostalis (*26/1*)

Fig. 204
The Splenius, *5*

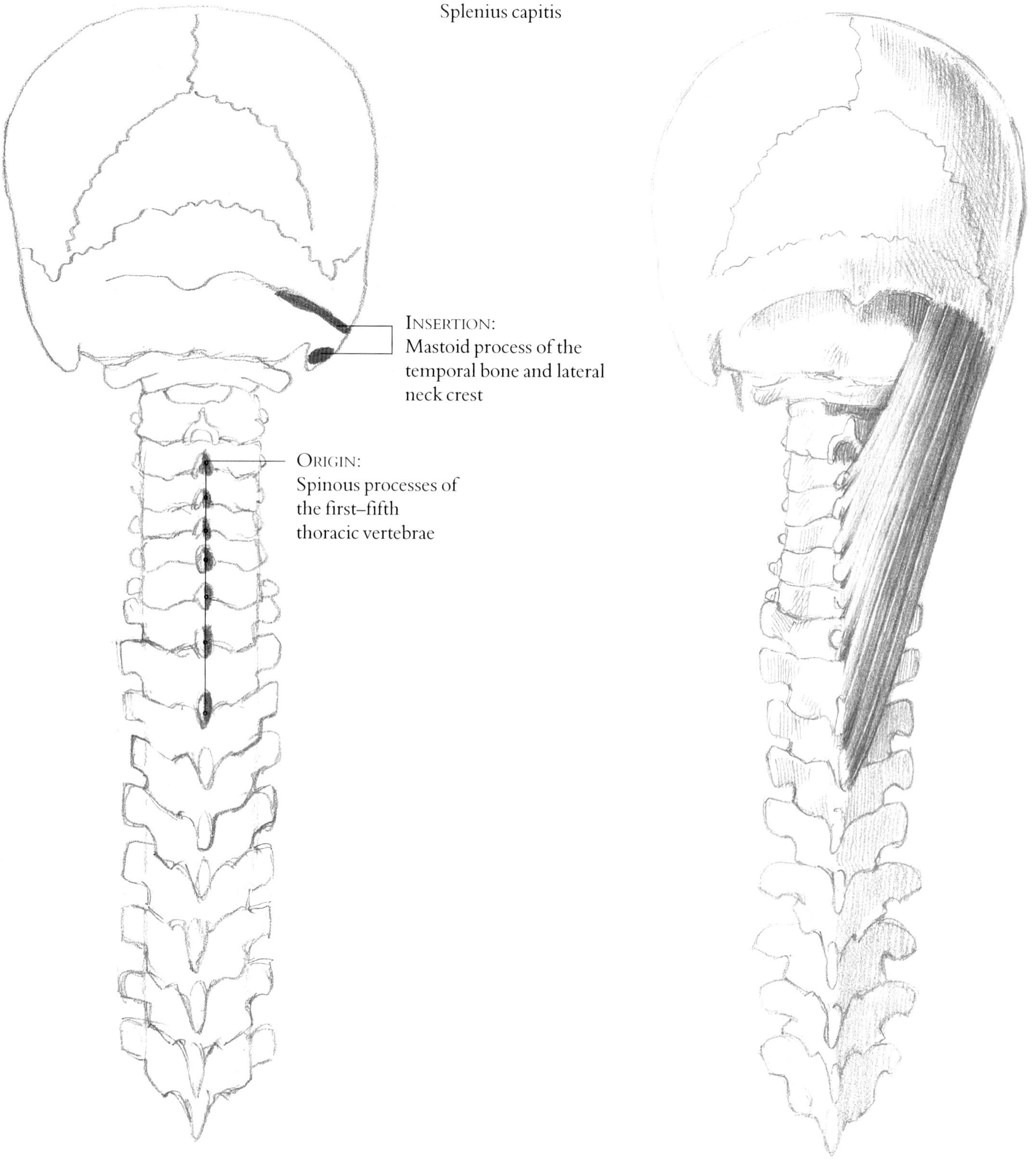

dorsal aspect

FUNCTION:
Pulls the neck laterally and rotates the head

Splenius cervicis

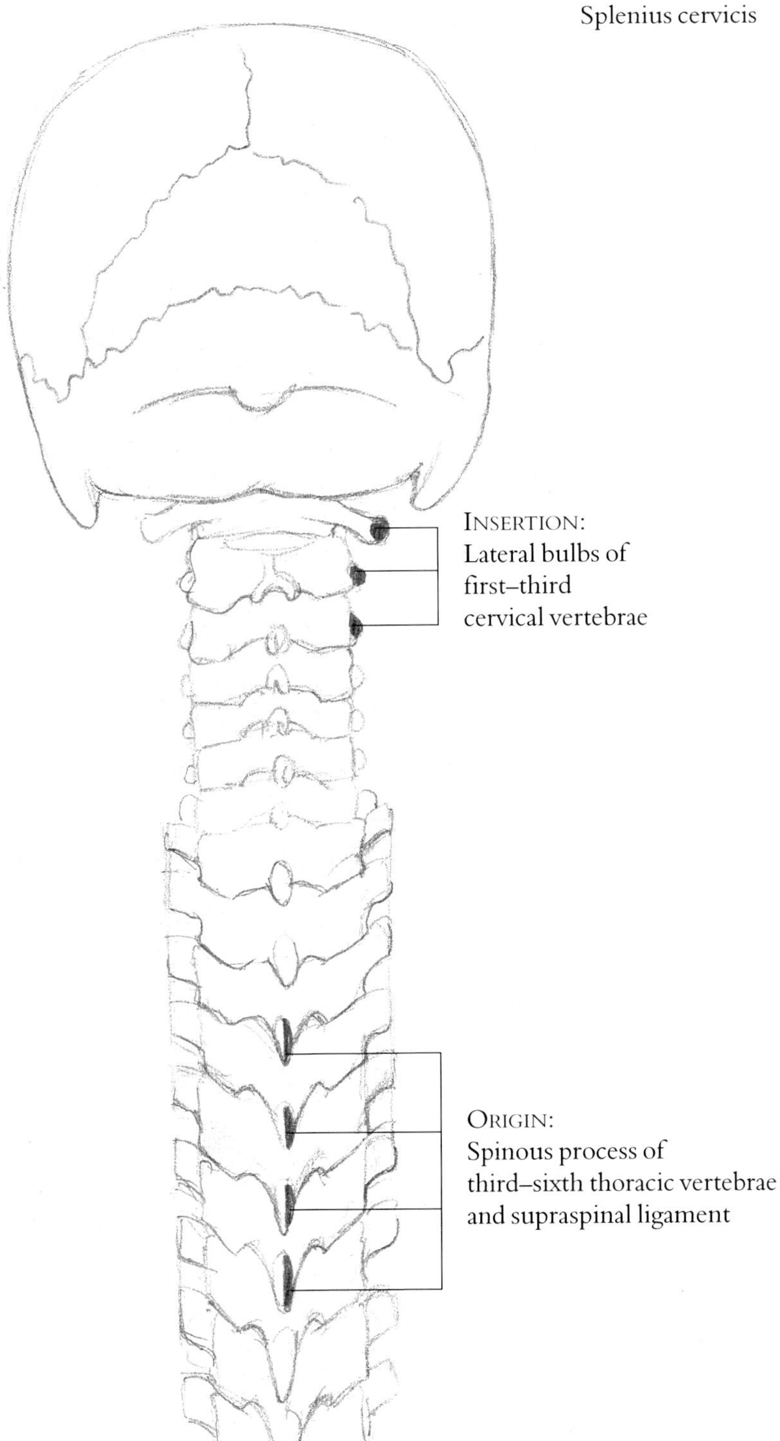

dorsal aspect

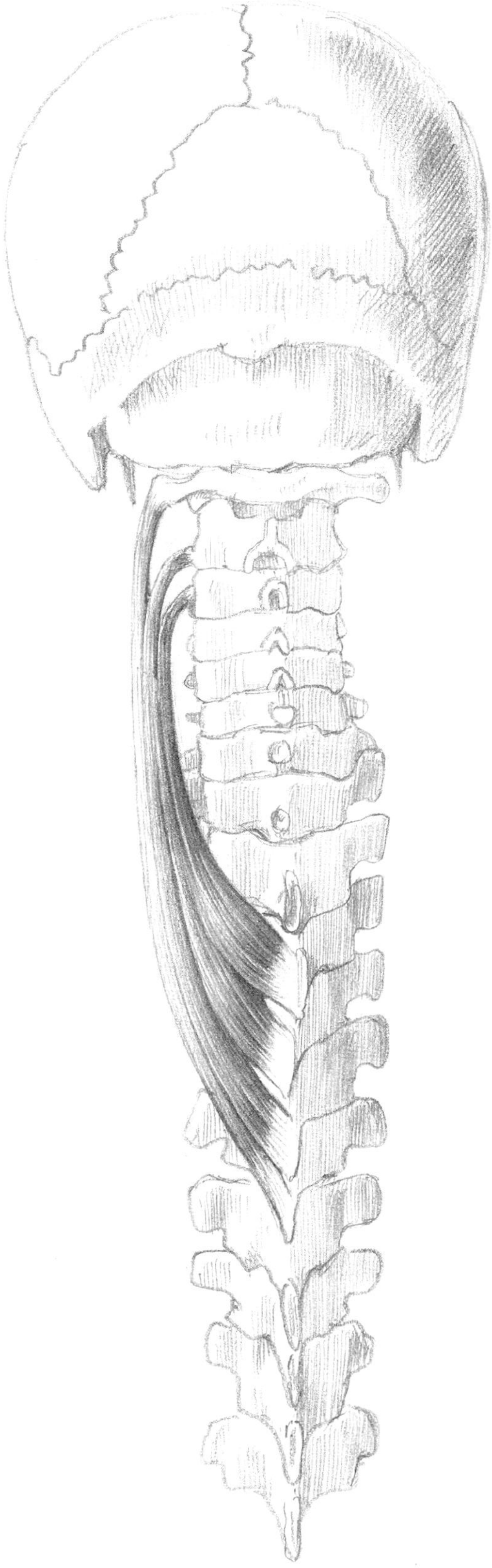

FUNCTION:
Extends the head and pulls the neck laterally

Fig. 205
The Semispinal Muscle of the Head and Neck
(M. semispinalis capitis cervicis, *26/4*)

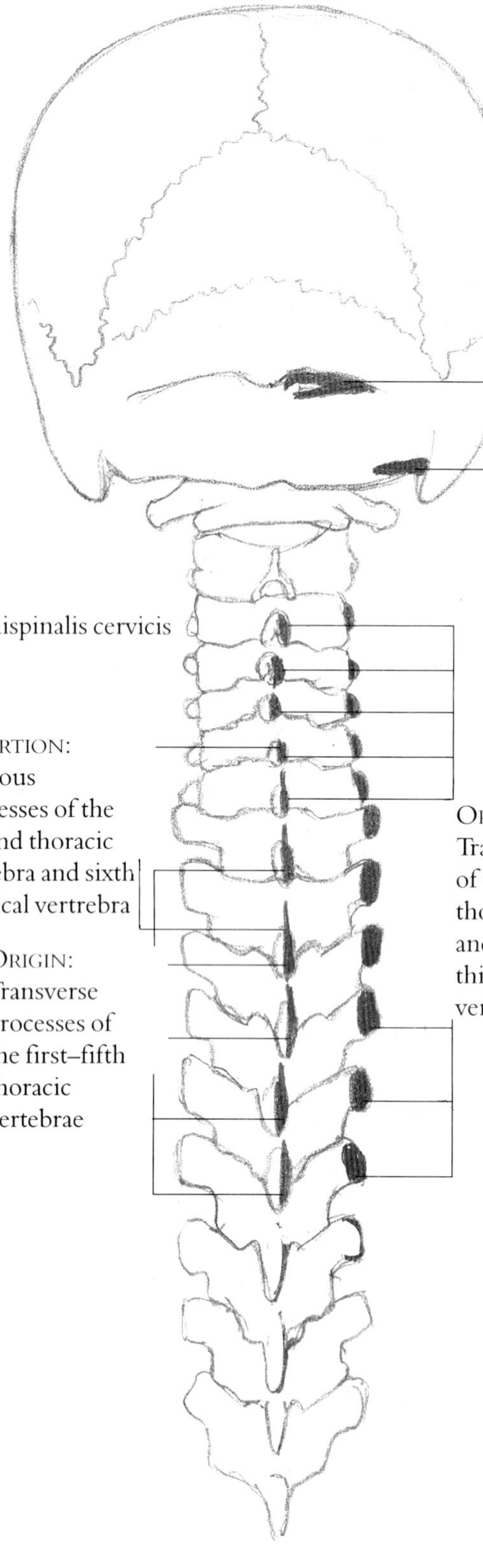

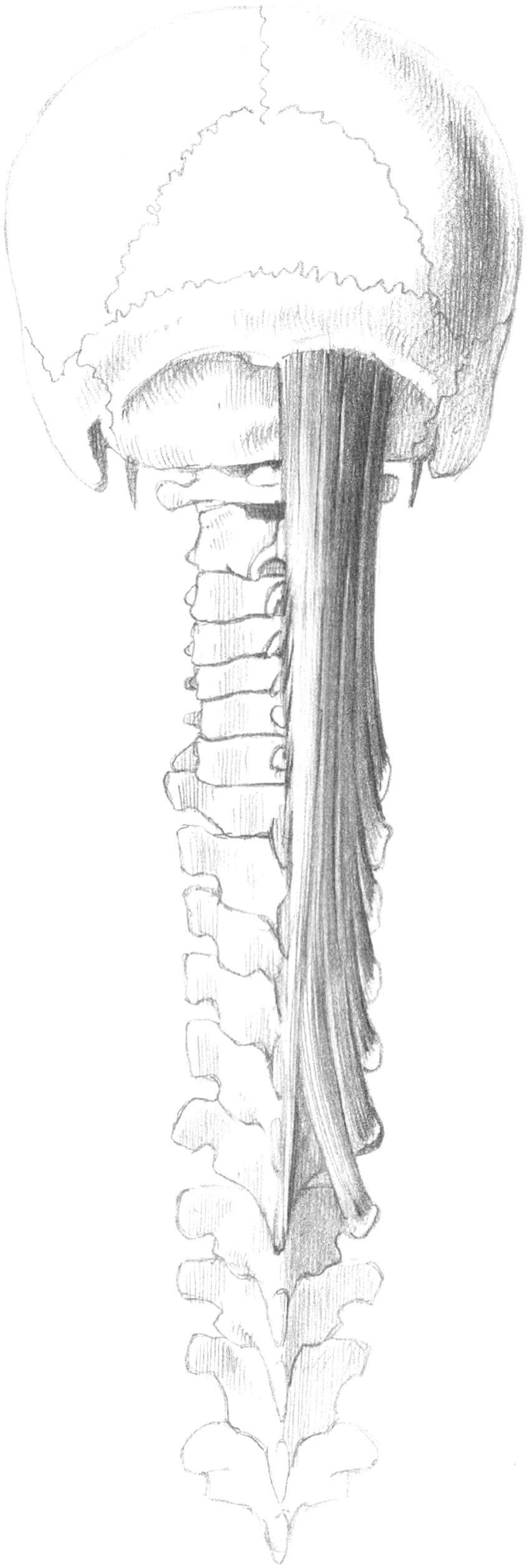

FUNCTION:
Rotates and flexes the head and neck

FUNCTION:
Extends the head and rotates it laterally

dorsal aspect

Fig. 206
The Iliocostal Muscle of the Neck
(M. iliocostalis cervicis, *26/1*)

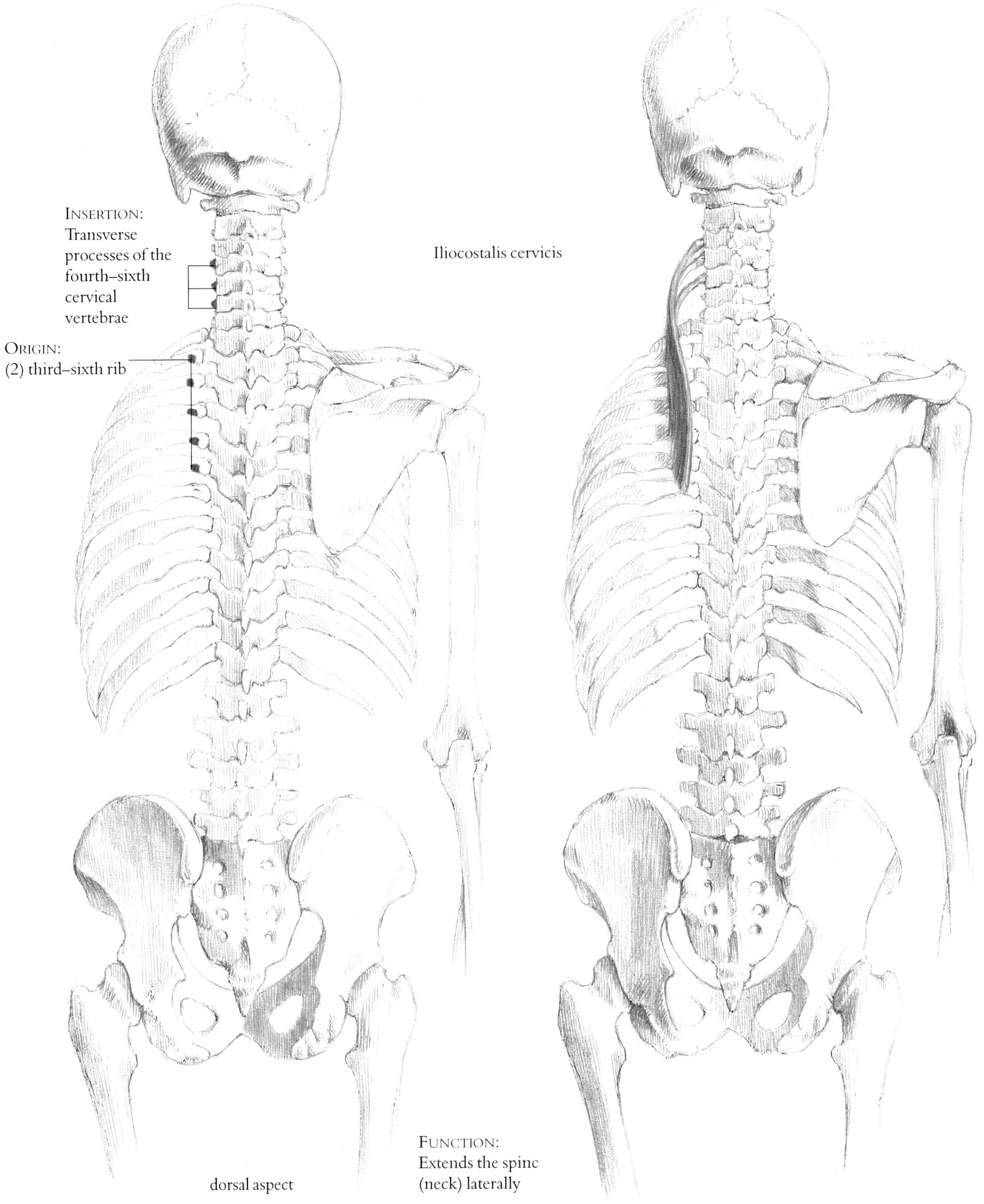

Fig. 207
The Serratus Posterior, *25*

posterior superior
(superior part)

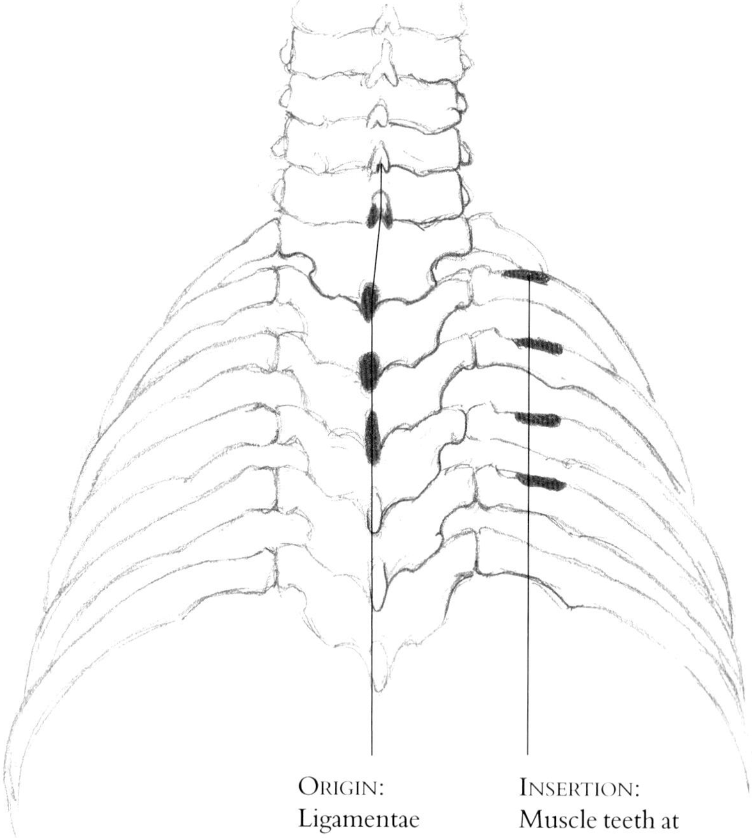

ORIGIN:
Ligamentae supraspinale, spinous processes of the last two cervical and first two thoracic vertebrae

INSERTION:
Muscle teeth at the corners of the second–fifth ribs

FUNCTION:
Lifts the ribs

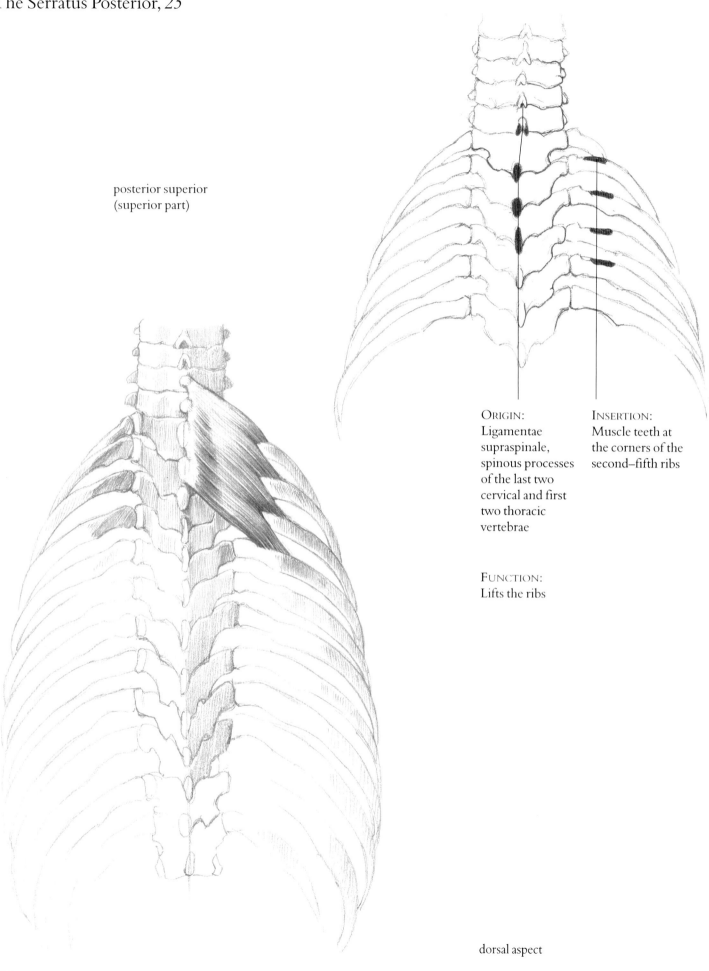

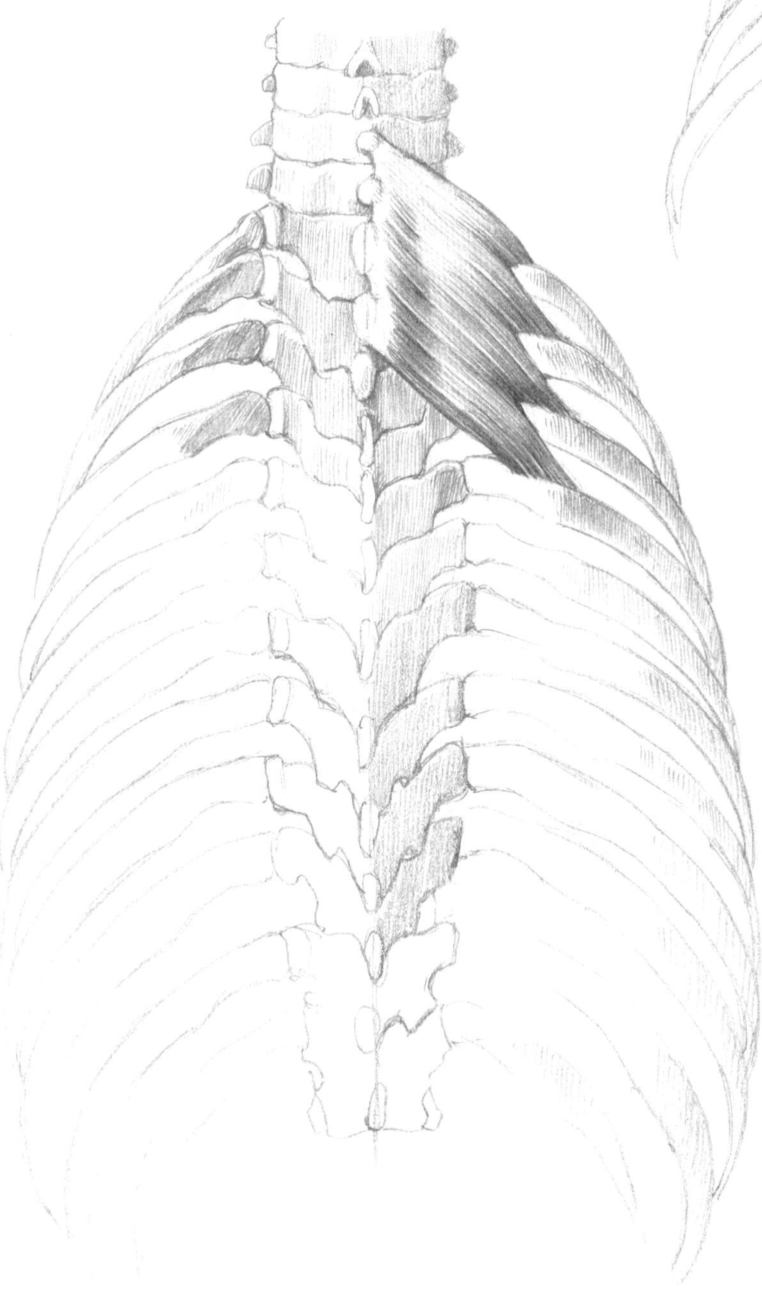

dorsal aspect

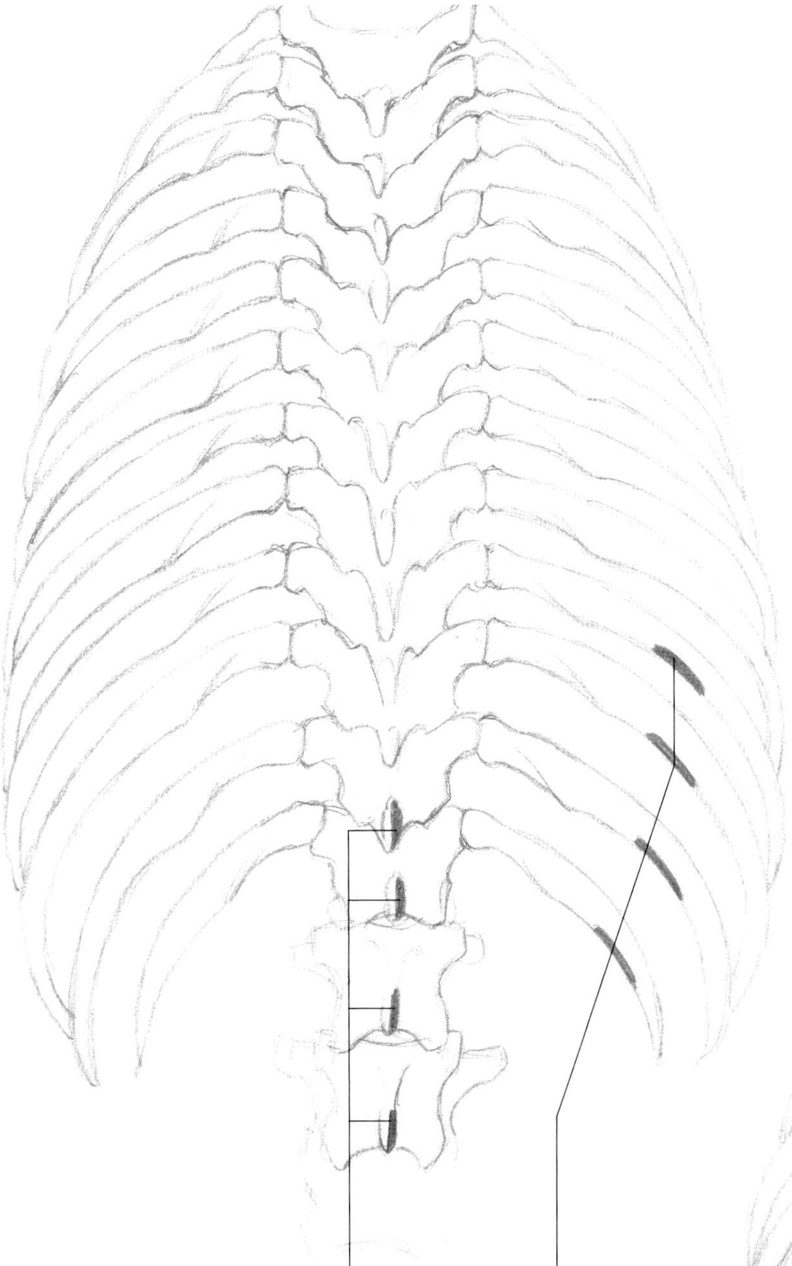

posterior inferior
(inferior part)

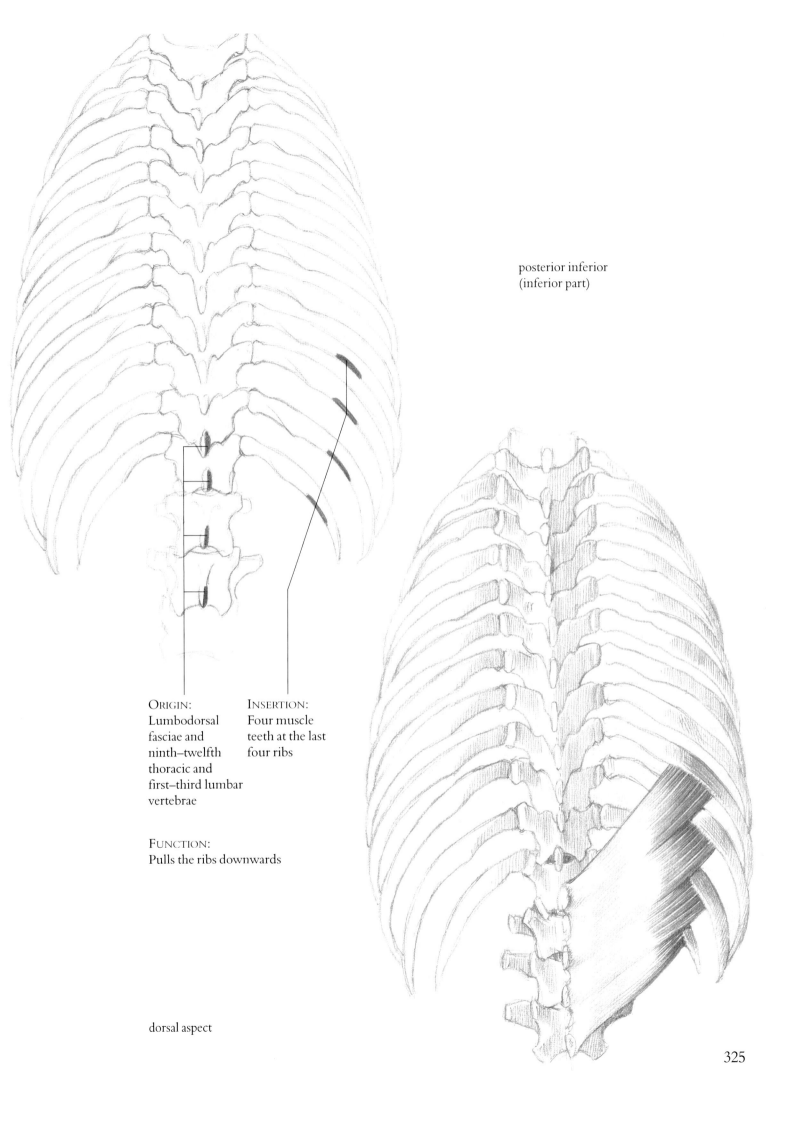

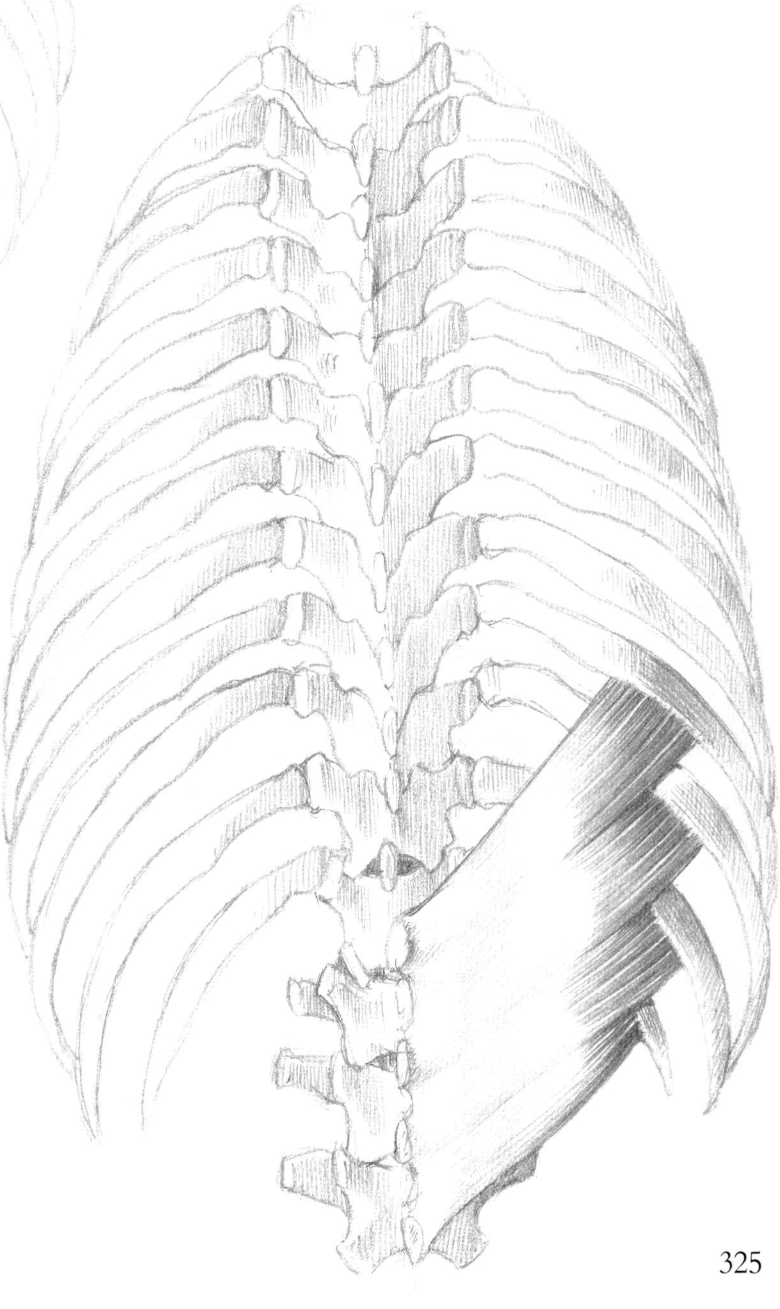

ORIGIN:
Lumbodorsal fasciae and ninth–twelfth thoracic and first–third lumbar vertebrae

INSERTION:
Four muscle teeth at the last four ribs

FUNCTION:
Pulls the ribs downwards

dorsal aspect

Fig. 208
The Spinal Muscle of the Back
(M. spinalis thoracis, *26/3*)

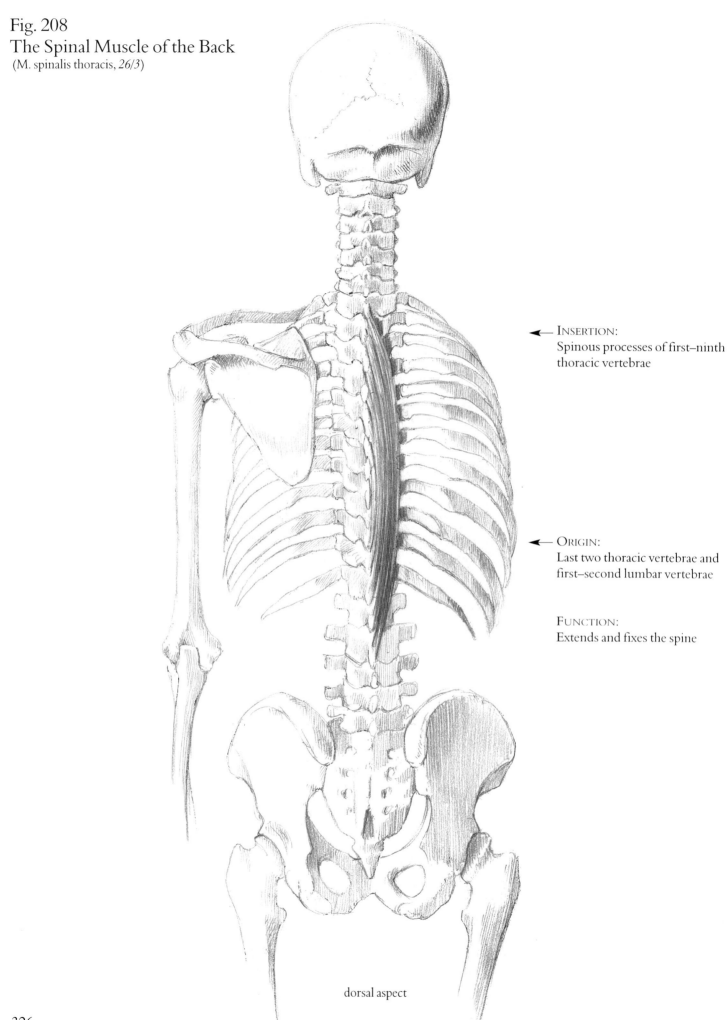

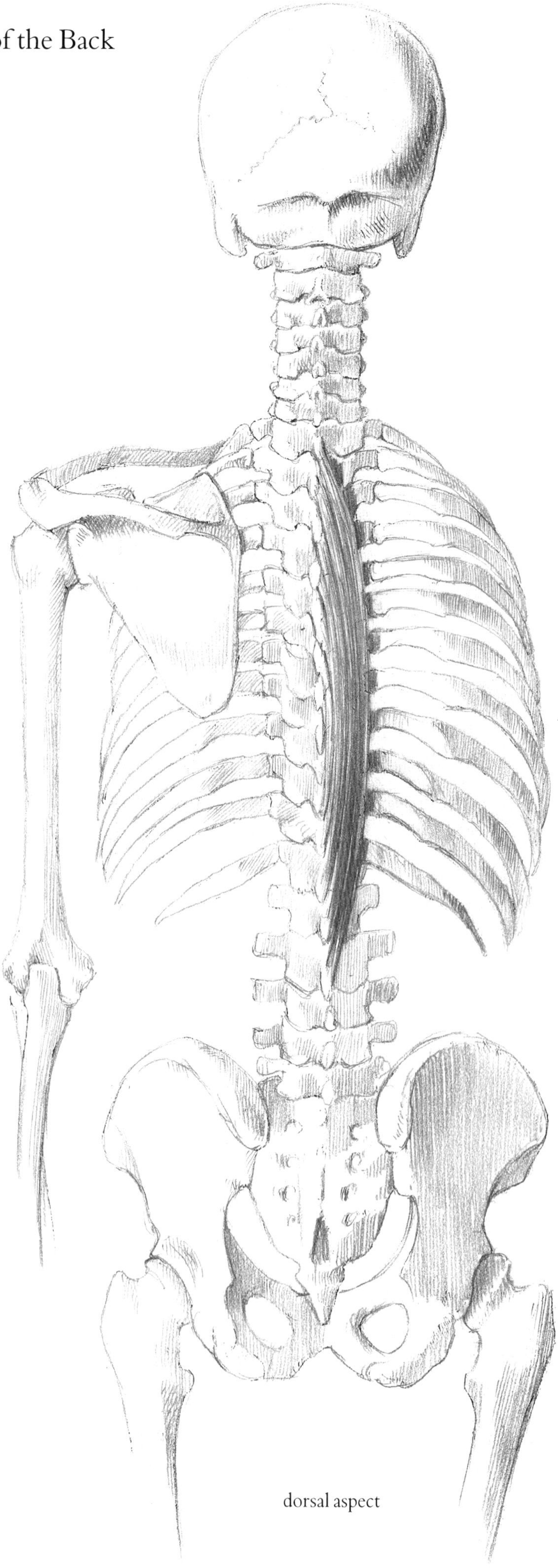

INSERTION:
Spinous processes of first–ninth thoracic vertebrae

ORIGIN:
Last two thoracic vertebrae and first–second lumbar vertebrae

FUNCTION:
Extends and fixes the spine

dorsal aspect

Fig. 209
The Iliocostalis, *26/1*

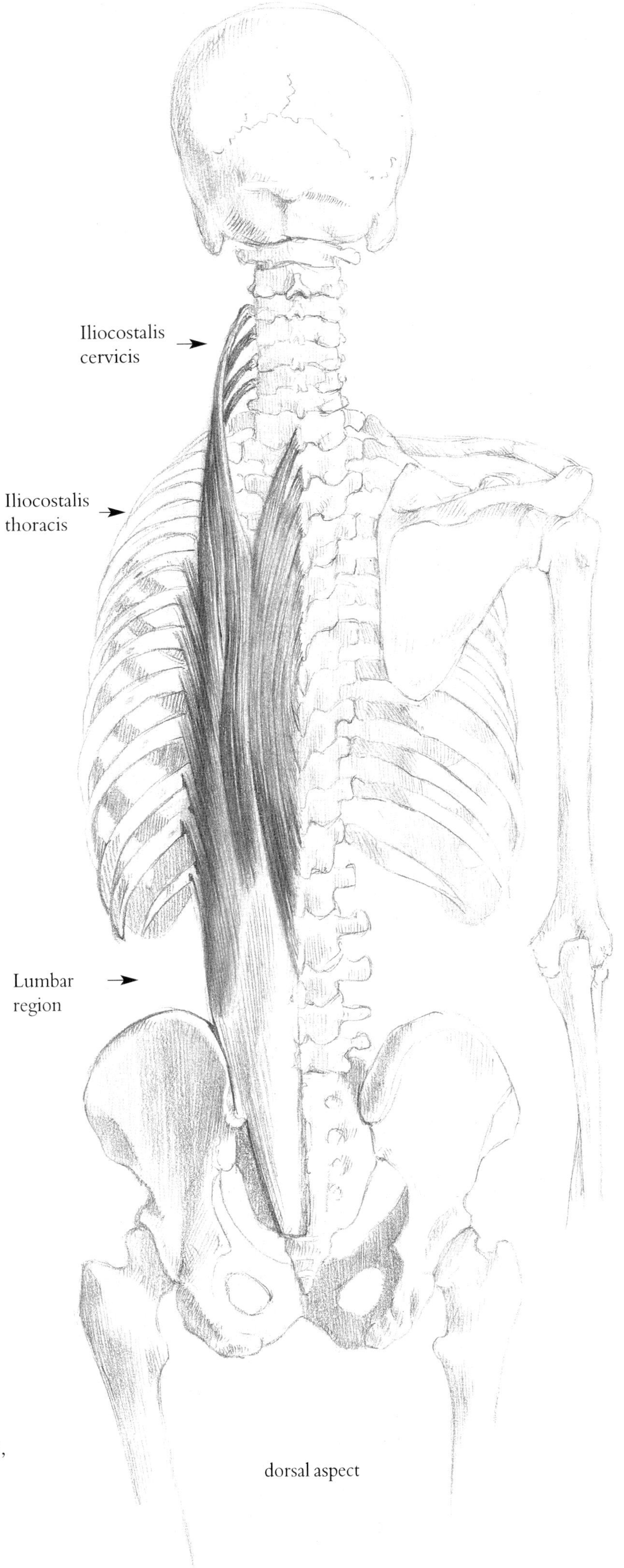

INSERTION:
Corners of fifth–twelfth ribs (lumbar region); transverse processes of fourth–seventh cervical vertebrae (thoracic part)

ORIGIN:
Dorsal surface of the sacrum (lumbar region); iliac ala with muscle at the seventh–twelfth ribs (thoracic part)

FUNCTION:
Extends the spine (on one side), flexes it laterally

dorsal aspect

Fig. 210
Muscle Studies

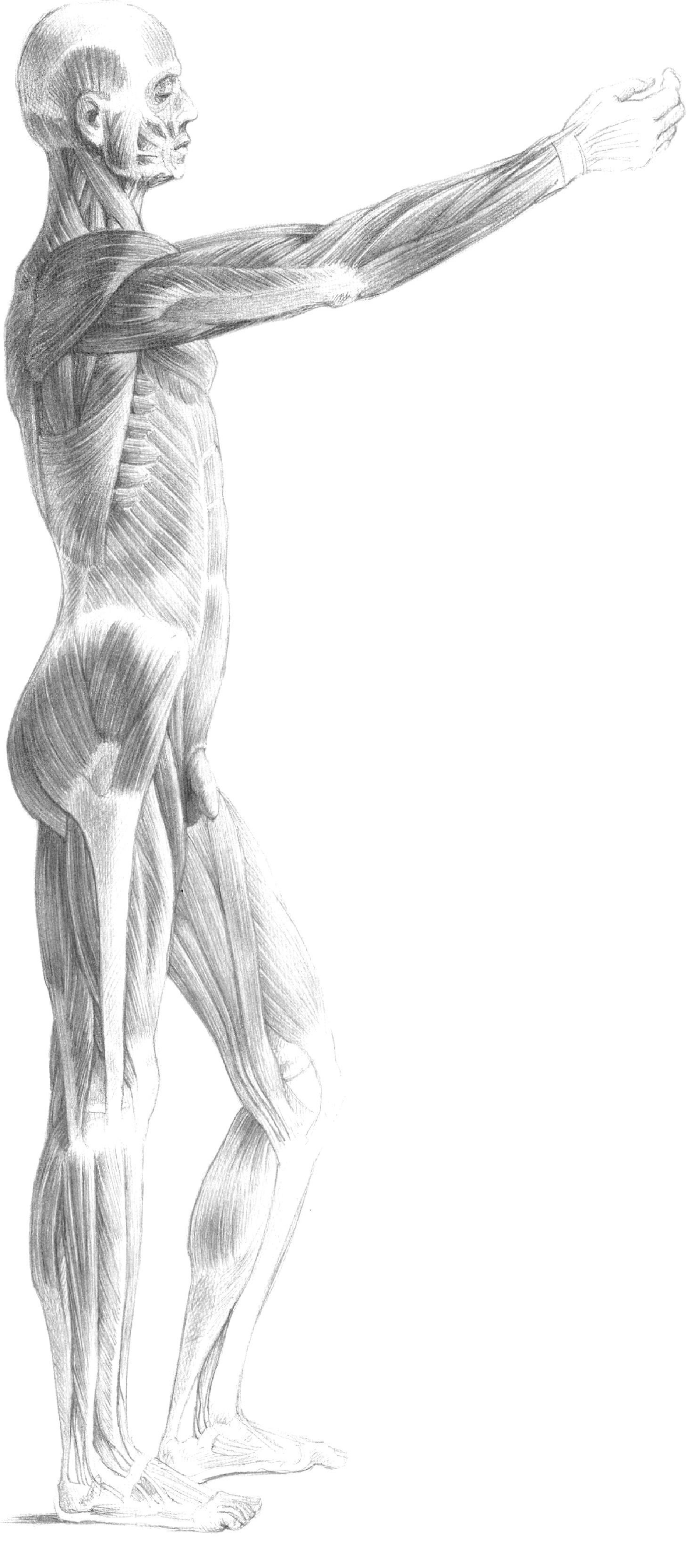

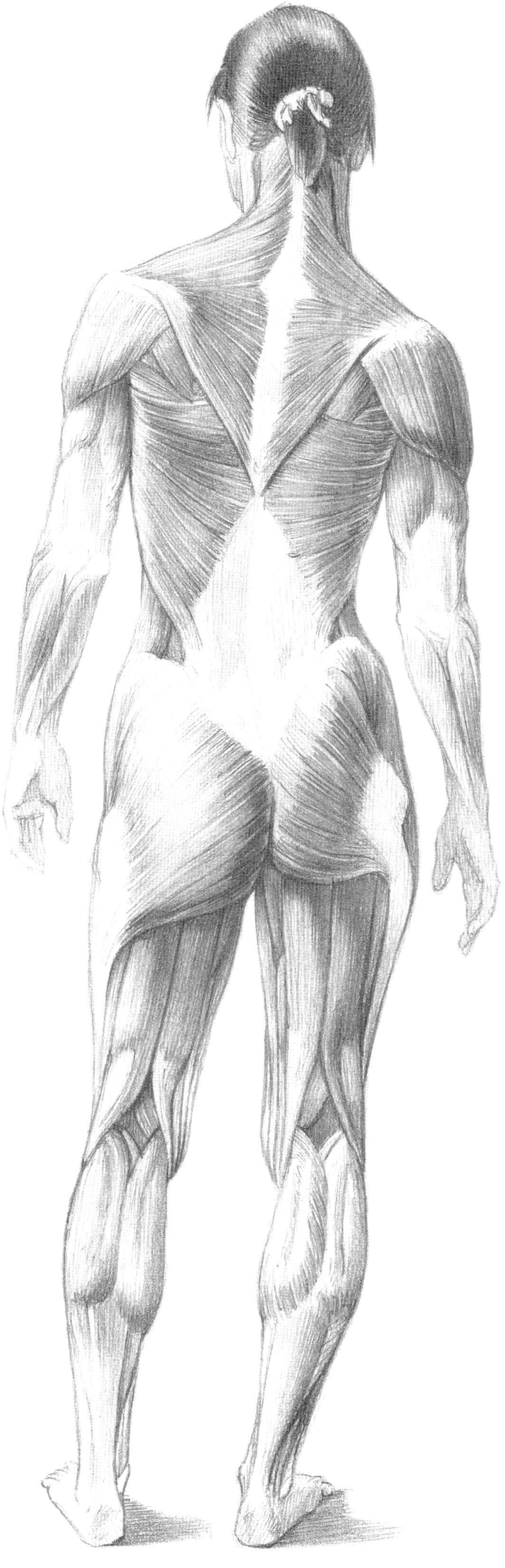

Fig. 211
The Bones and Muscles of the Thoracic and Shoulder Girdle

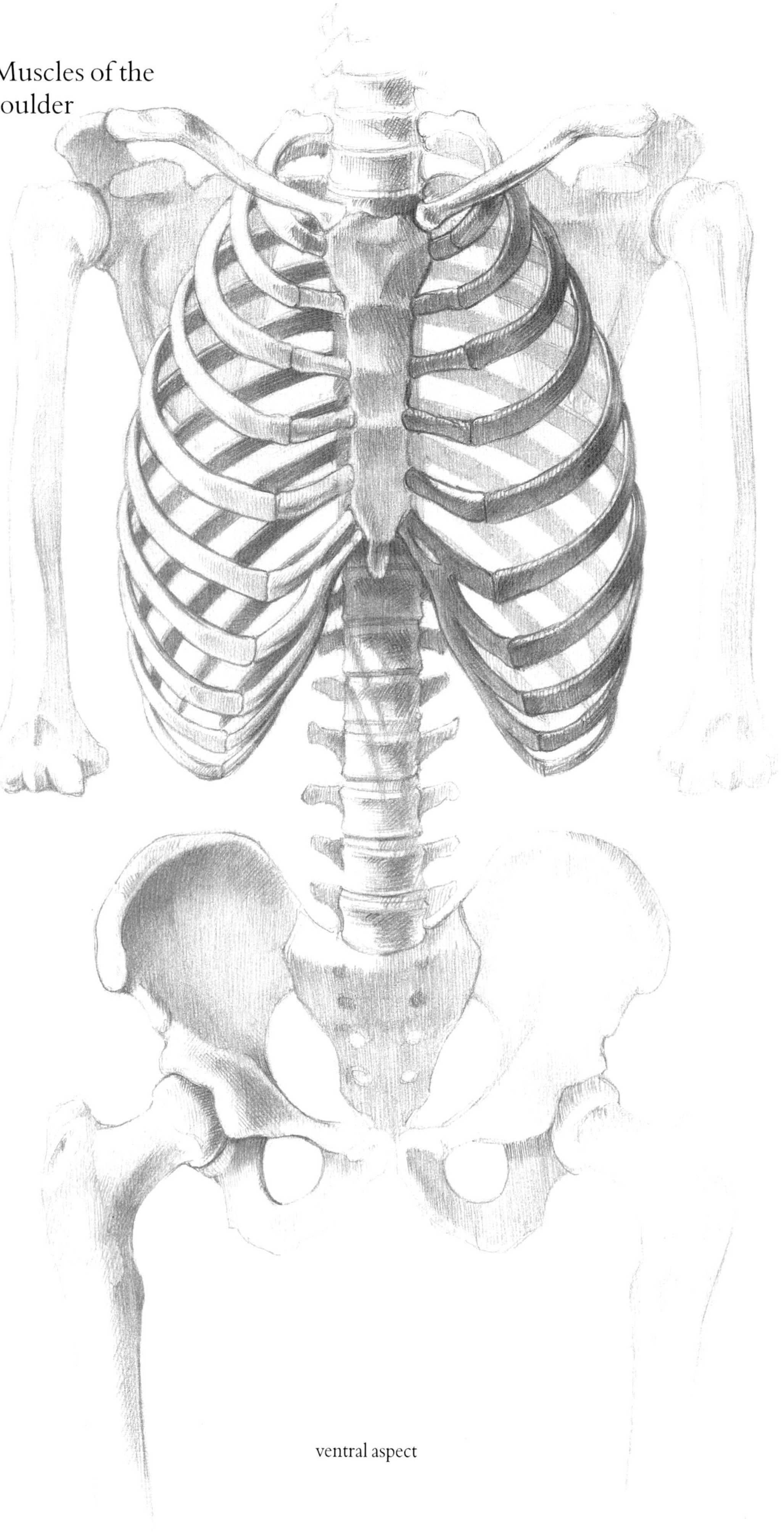

ventral aspect

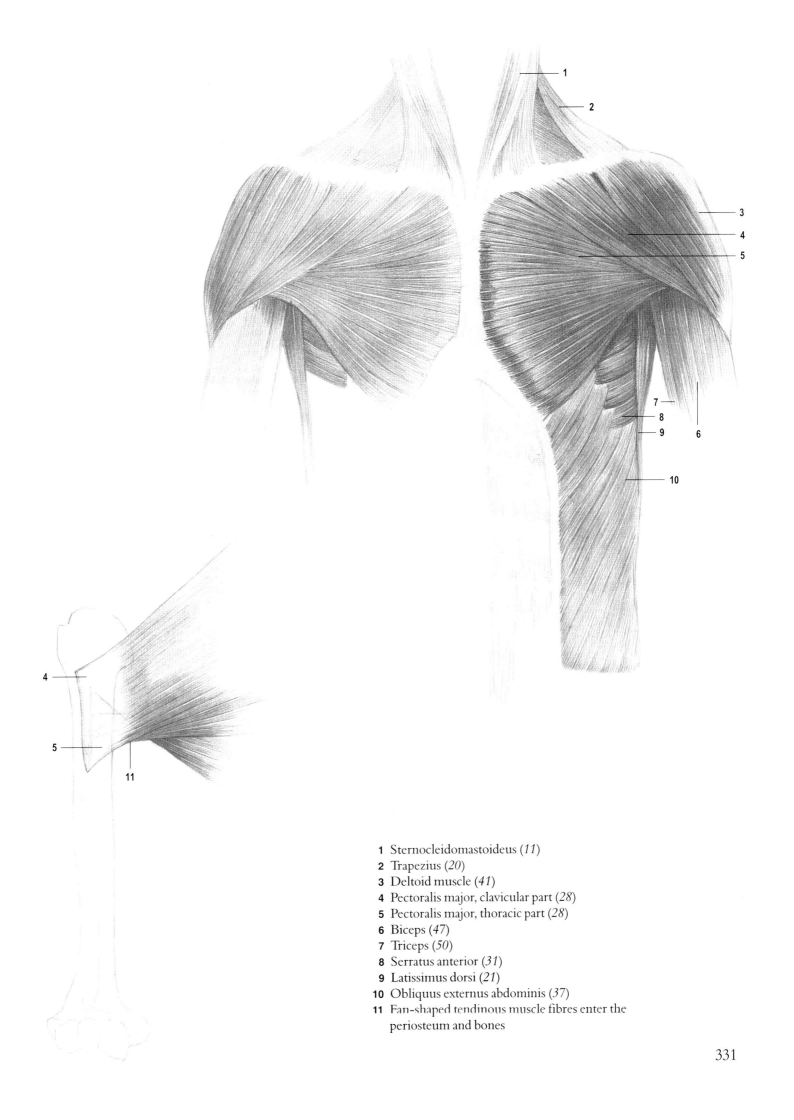

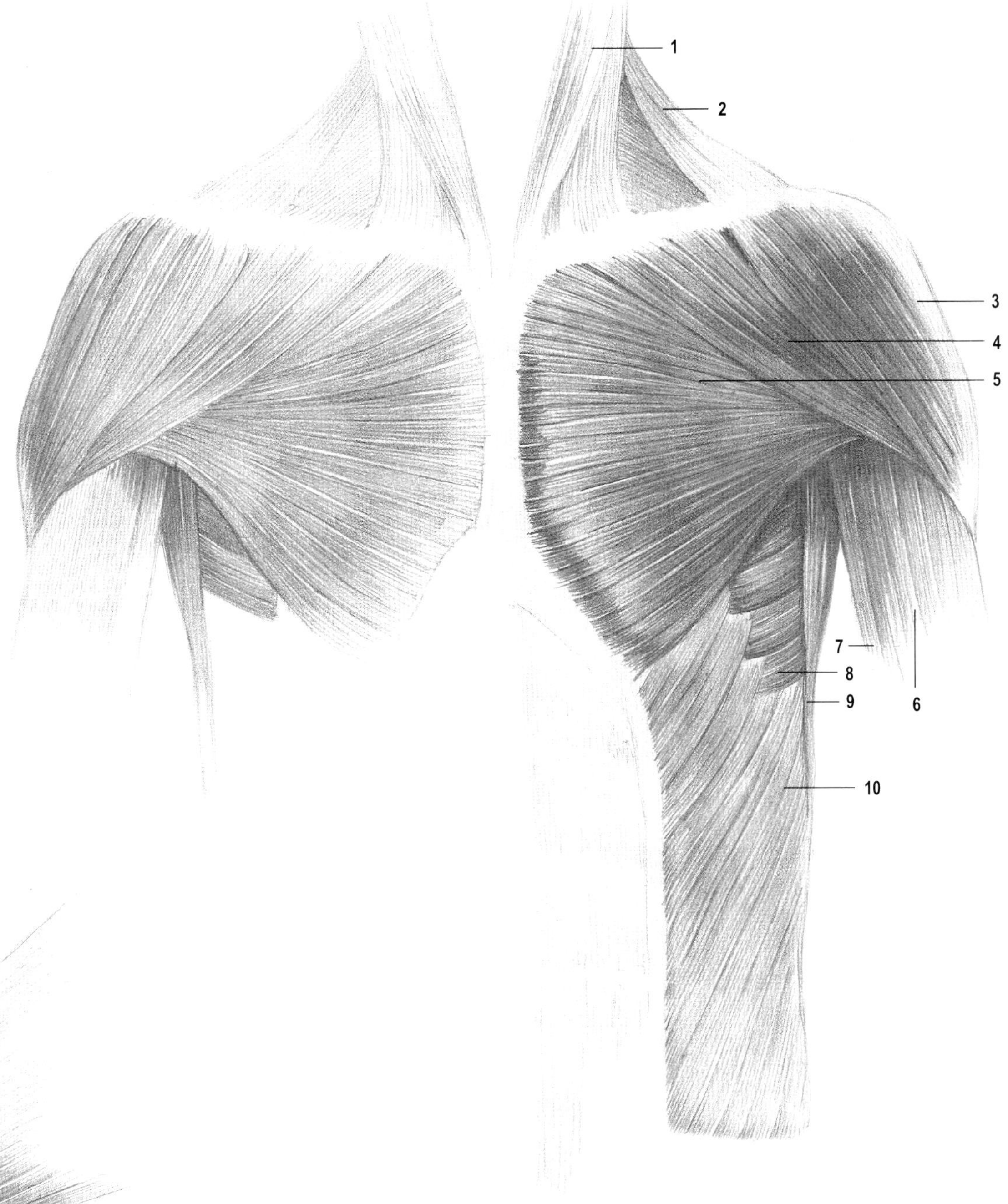

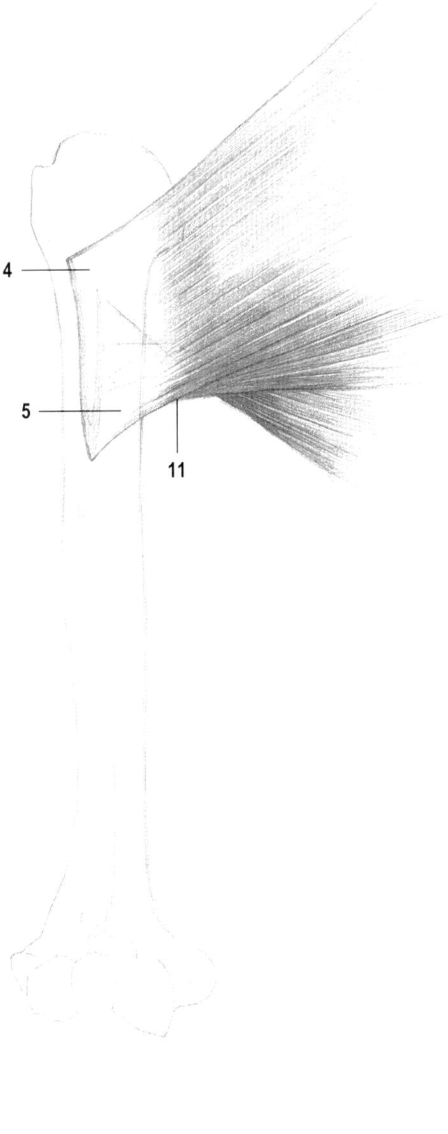

1 Sternocleidomastoideus (*11*)
2 Trapezius (*20*)
3 Deltoid muscle (*41*)
4 Pectoralis major, clavicular part (*28*)
5 Pectoralis major, thoracic part (*28*)
6 Biceps (*47*)
7 Triceps (*50*)
8 Serratus anterior (*31*)
9 Latissimus dorsi (*21*)
10 Obliquus externus abdominis (*37*)
11 Fan-shaped tendinous muscle fibres enter the periosteum and bones

Fig. 212

The Greater Pectoral Muscle

(M. pectoralis major, *28*)

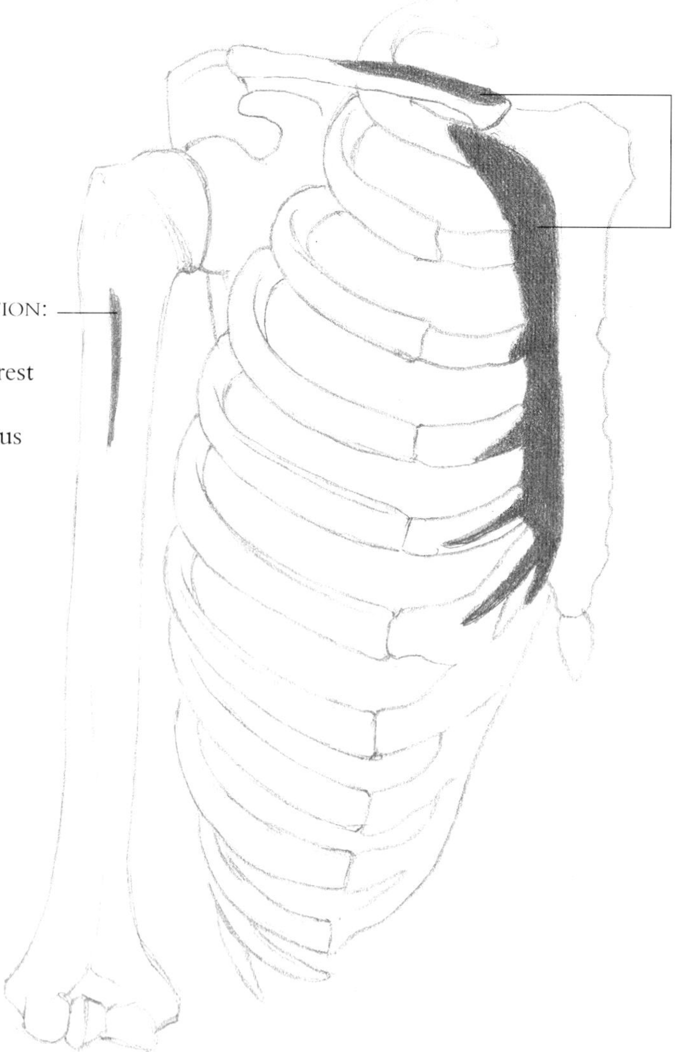

ORIGIN:
Sternal region of the clavicle (clavicular part); sternum-rib part; first–sixth rib cartilage, fascia recta (abdominal part)

INSERTION:
Lateral bone crest of the humerus

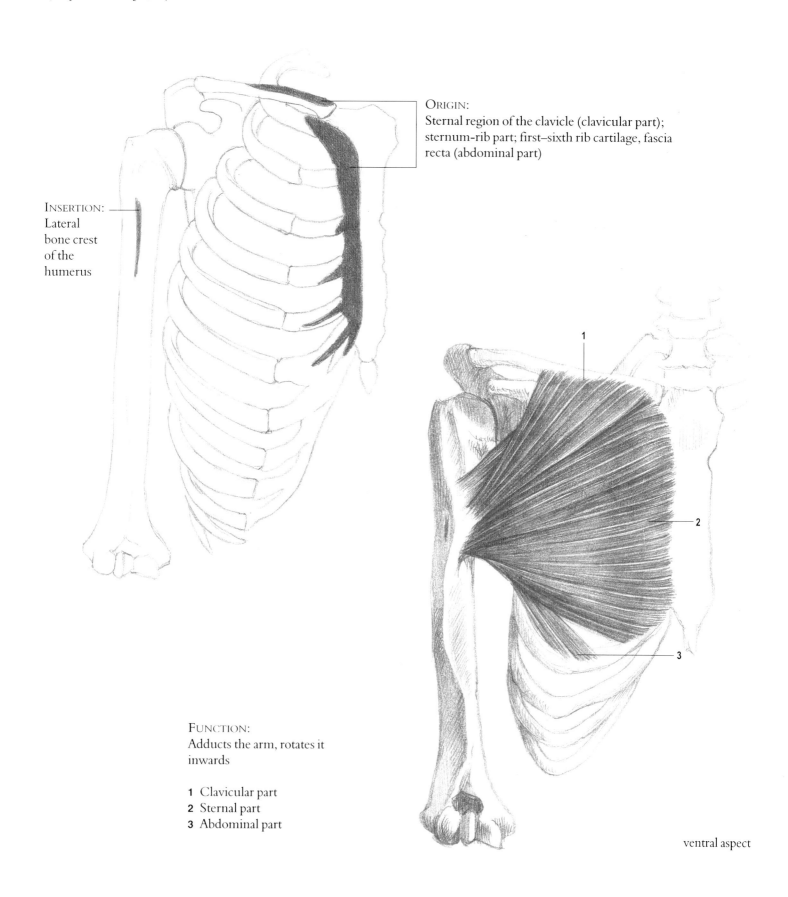

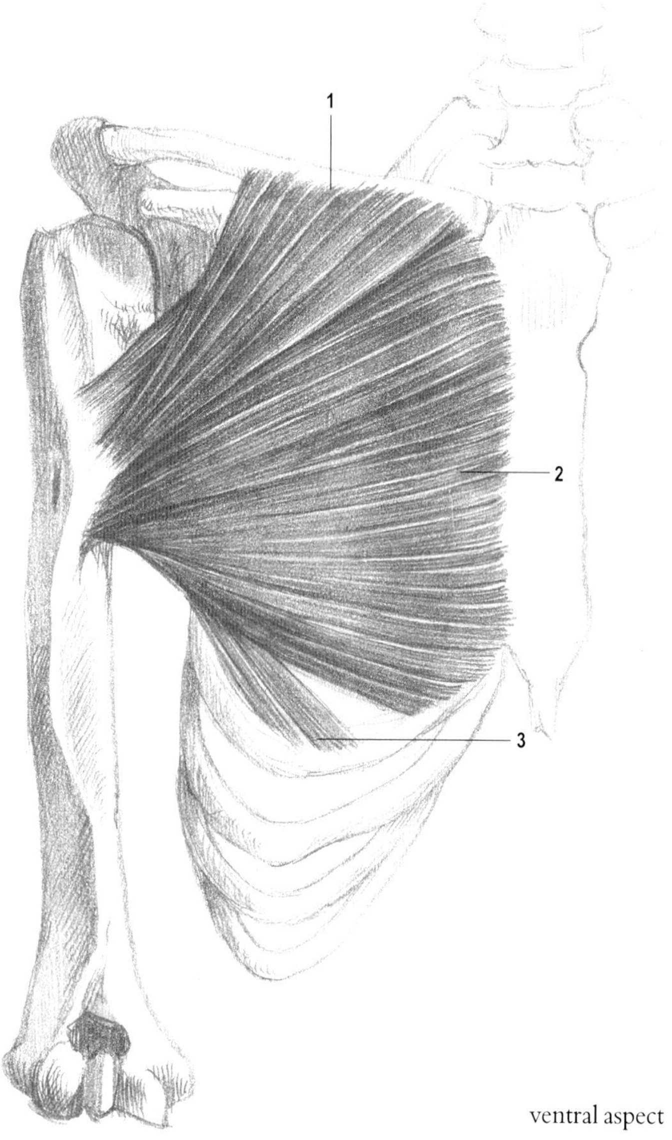

FUNCTION:
Adducts the arm, rotates it inwards

1 Clavicular part
2 Sternal part
3 Abdominal part

ventral aspect

Fig. 213
The Smaller Pectoral Muscle
(M. pectoralis minor, *29*)

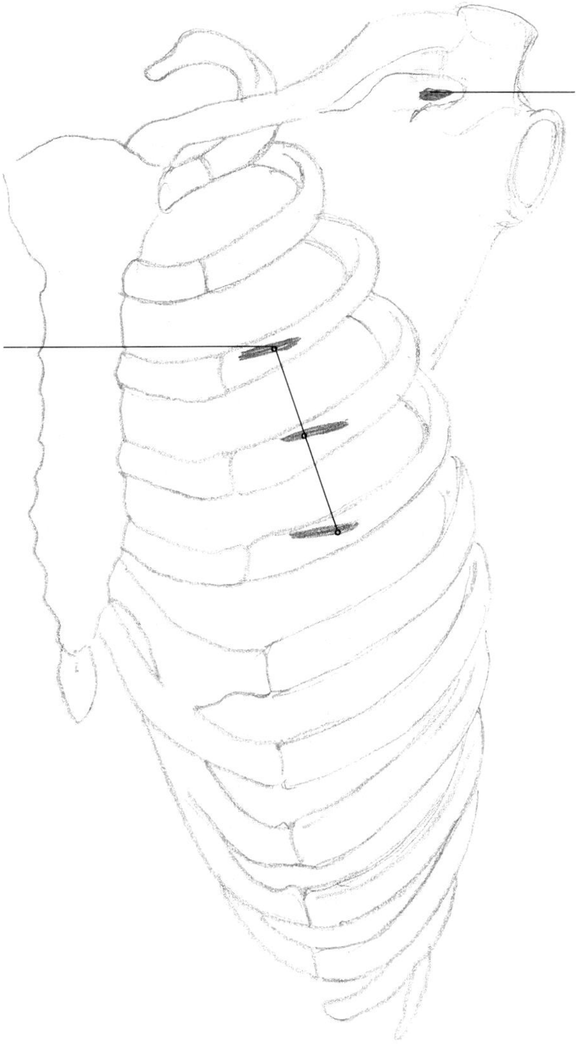

INSERTION:
Coracoid process of the shoulder blade

ORIGIN:
External surface of the (*2*) third-fifth ribs

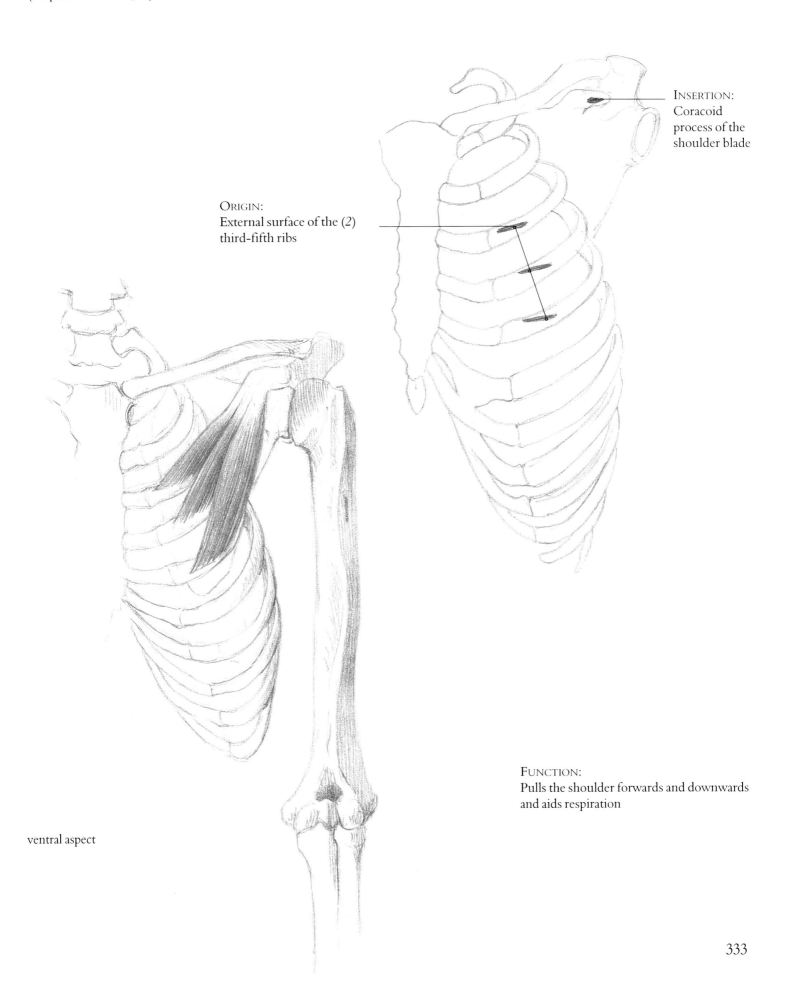

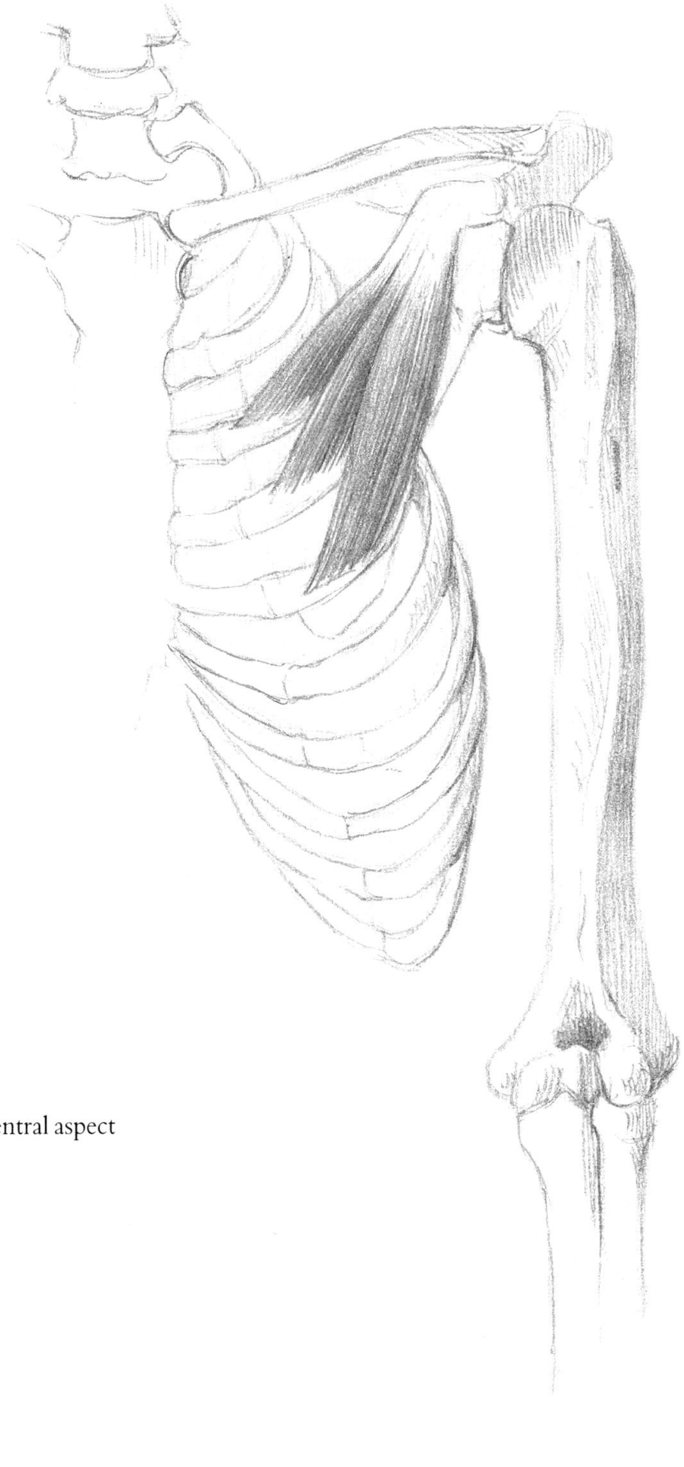

FUNCTION:
Pulls the shoulder forwards and downwards and aids respiration

ventral aspect

Fig. 214
The Muscles of the Shoulder Girdle and Humerus

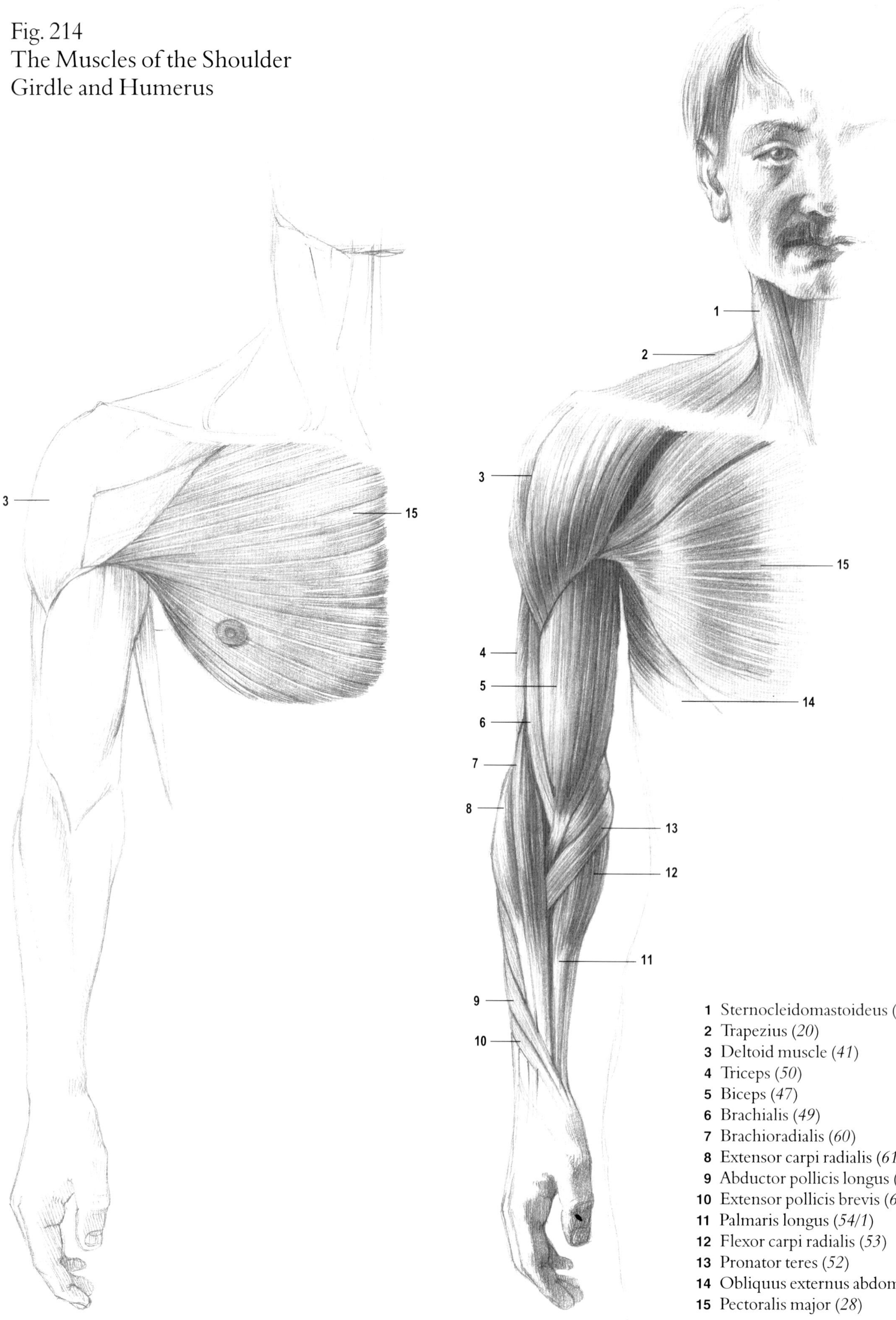

1 Sternocleidomastoideus (*11*)
2 Trapezius (*20*)
3 Deltoid muscle (*41*)
4 Triceps (*50*)
5 Biceps (*47*)
6 Brachialis (*49*)
7 Brachioradialis (*60*)
8 Extensor carpi radialis (*61*)
9 Abductor pollicis longus (*66*)
10 Extensor pollicis brevis (*67*)
11 Palmaris longus (*54/1*)
12 Flexor carpi radialis (*53*)
13 Pronator teres (*52*)
14 Obliquus externus abdominis (*37*)
15 Pectoralis major (*28*)

Fig. 215
The Serratus Anterior, *31*

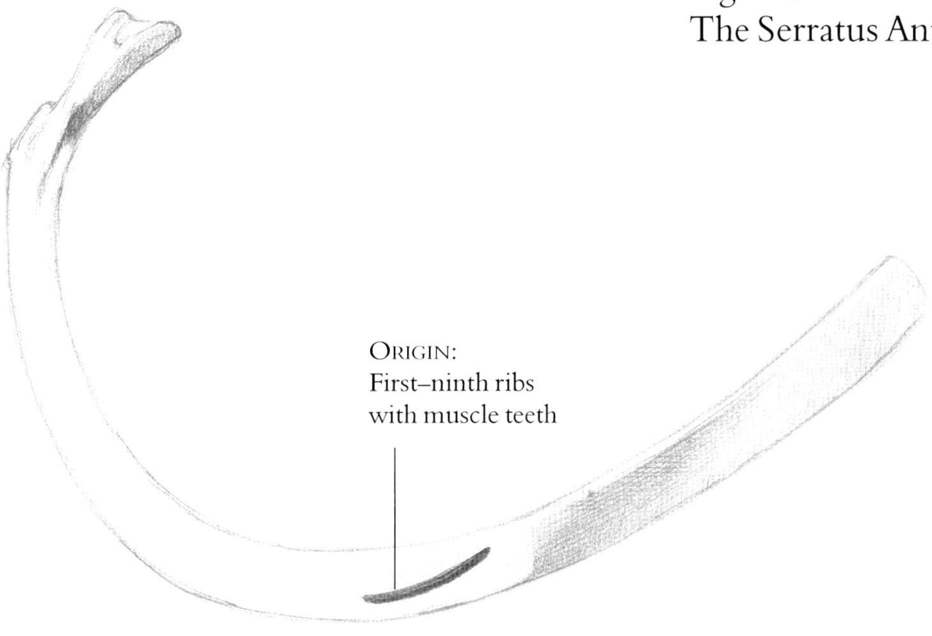

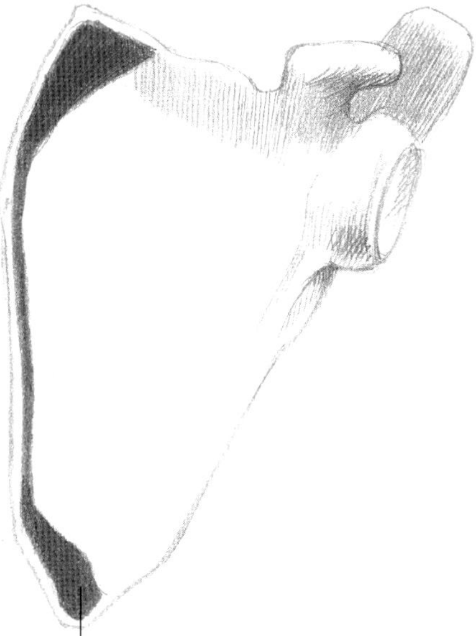

INSERTION:
On the medial edge of the
shoulder blade and
its superior and
inferior corners

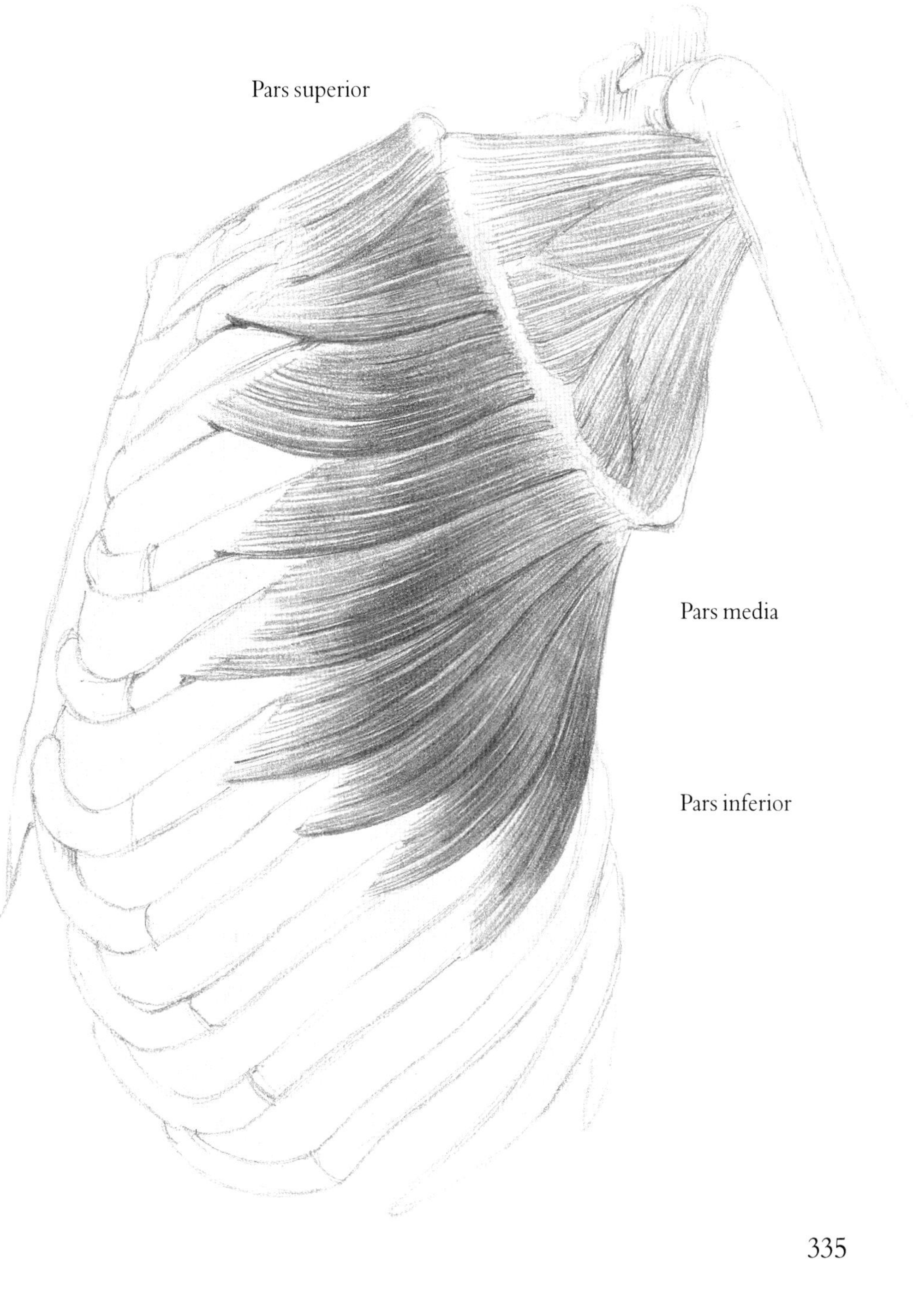

lateral aspect
(with shoulder blade and
shoulder joints pulled back)

The Serratus Anterior

(cont.)

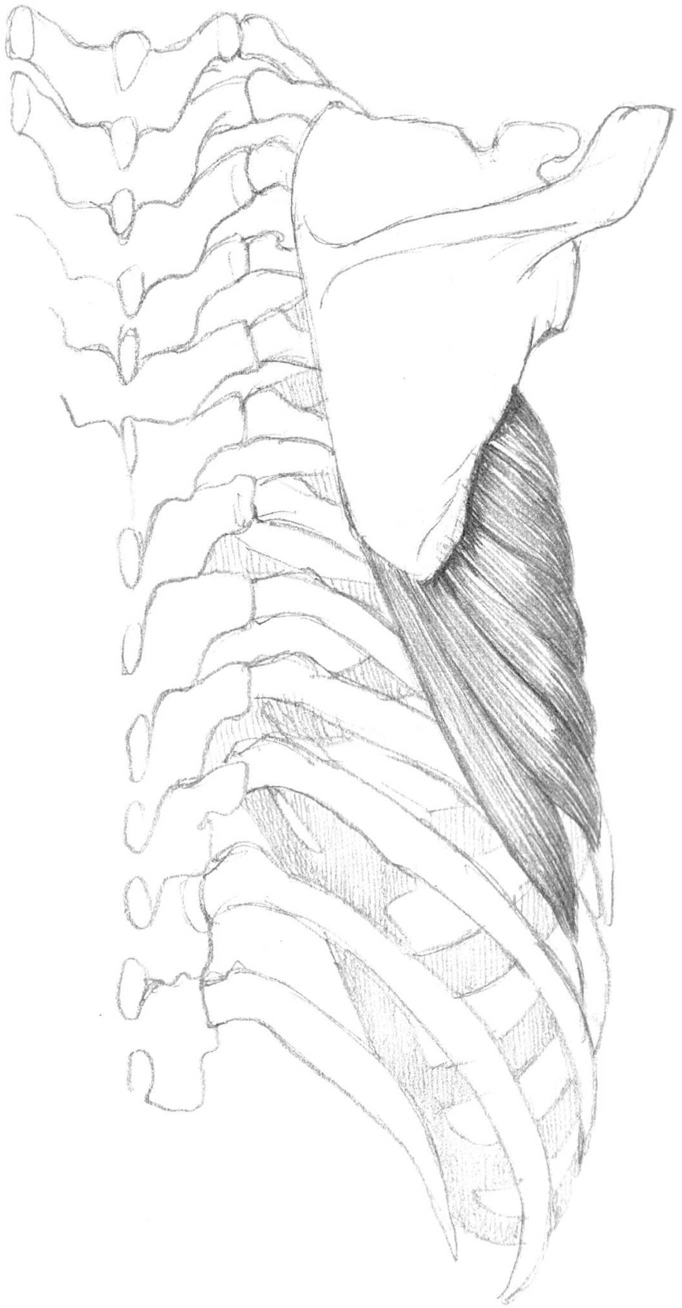

dorsal aspect

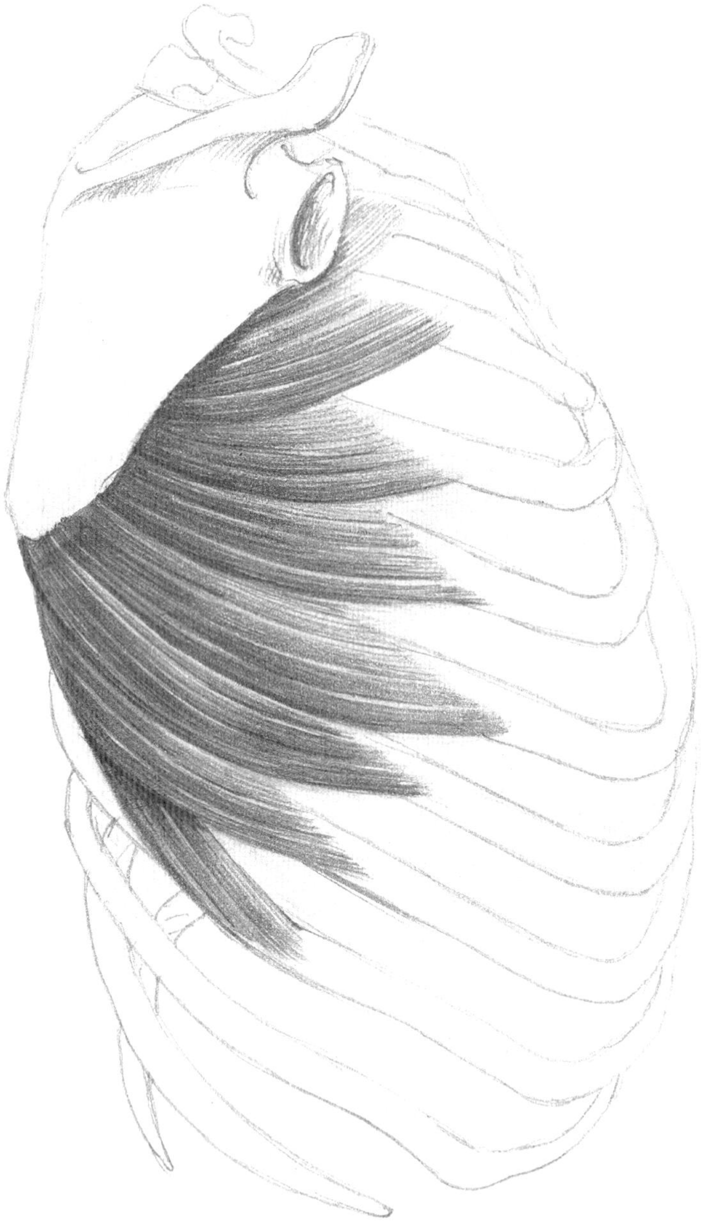

lateral aspect

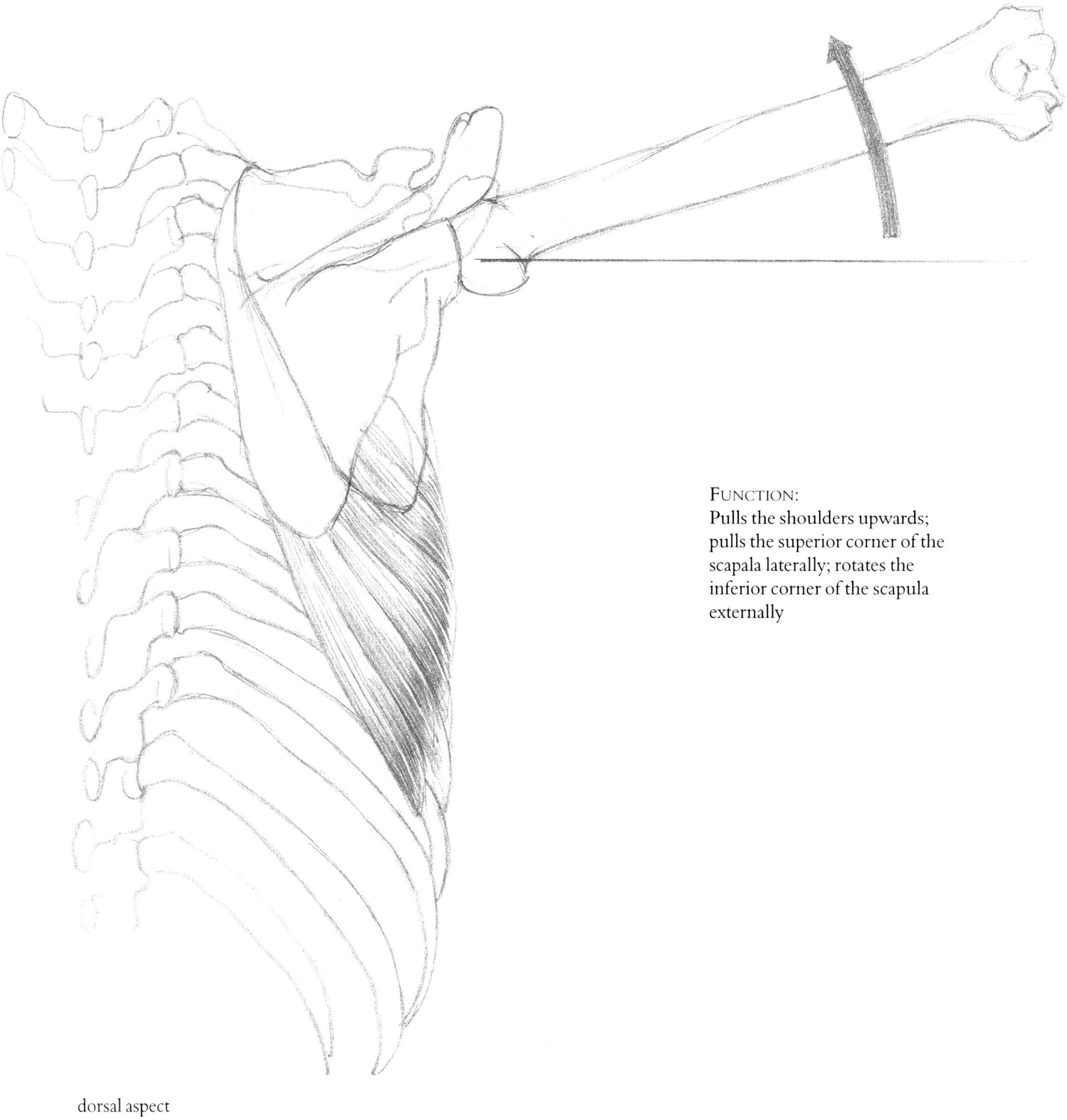

FUNCTION:
Pulls the shoulders upwards; pulls the superior corner of the scapala laterally; rotates the inferior corner of the scapula externally

dorsal aspect

Fig. 216
The Bones and Muscles of the Trunk

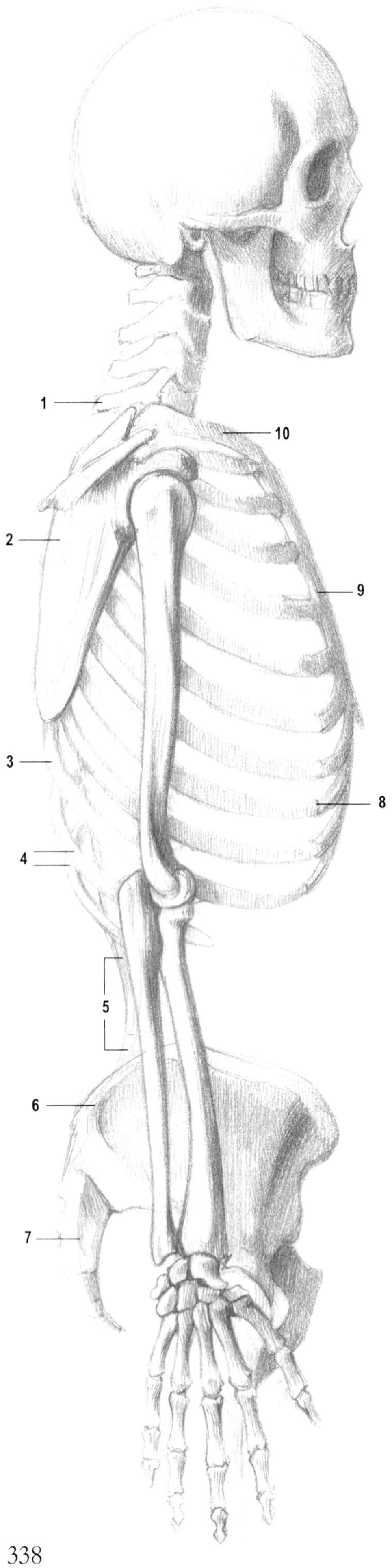

Bones

1 Sixth cervical vertebra
2 Shoulder blade
3 Tenth rib
4 Free ribs
5 Lumbar verterbrae
6 Os ilium
7 Sacrum
8 Costal arch
9 Sternum
10 Clavicle

Muscle

1 Sternocleidomastoideus (*11*)
2 Trapezius (*20*)
3 Deltoid muscle (*41*)
4 Latissimus dorsi (*21*)
5 Gluteus maximus (*81*)
6 Tensor fasciae latae (*84*)
7 Obliquus externus abdominis (*37*)
8 Serratus anterior (*31*)
9 Pectoralis major (*28*)

a Clavicle
b Mammary gland
c Nipple
d Umbilicus
e Anterior iliac spine
f Posterior iliac spine

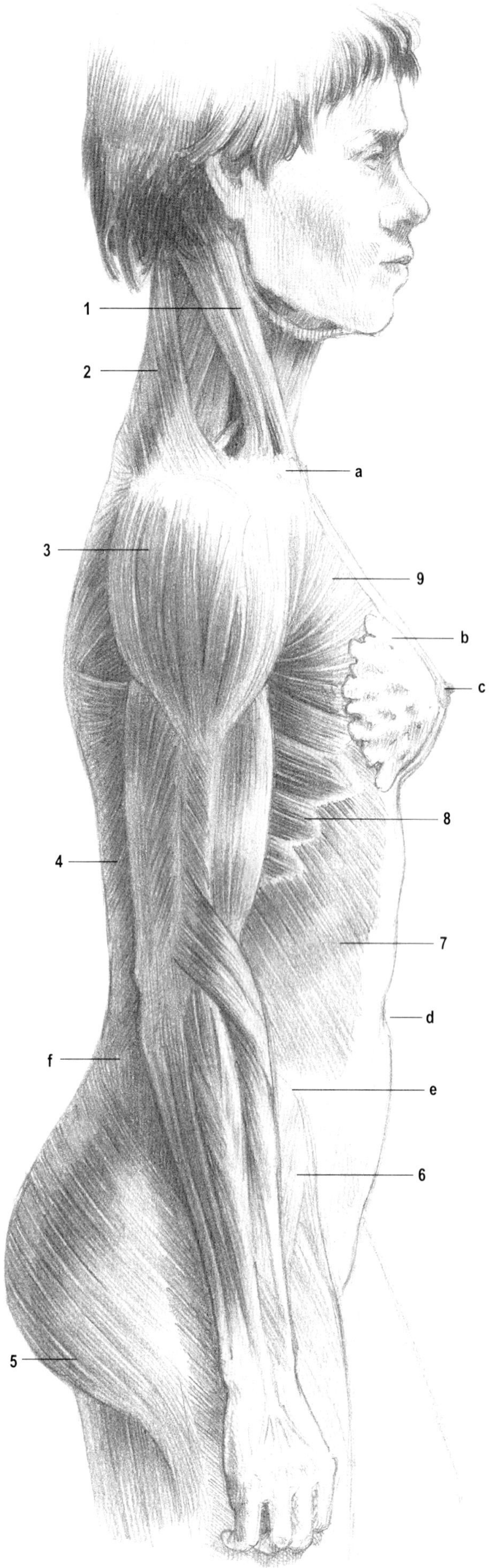

Fig. 217
The Superficial Muscles of the Trunk

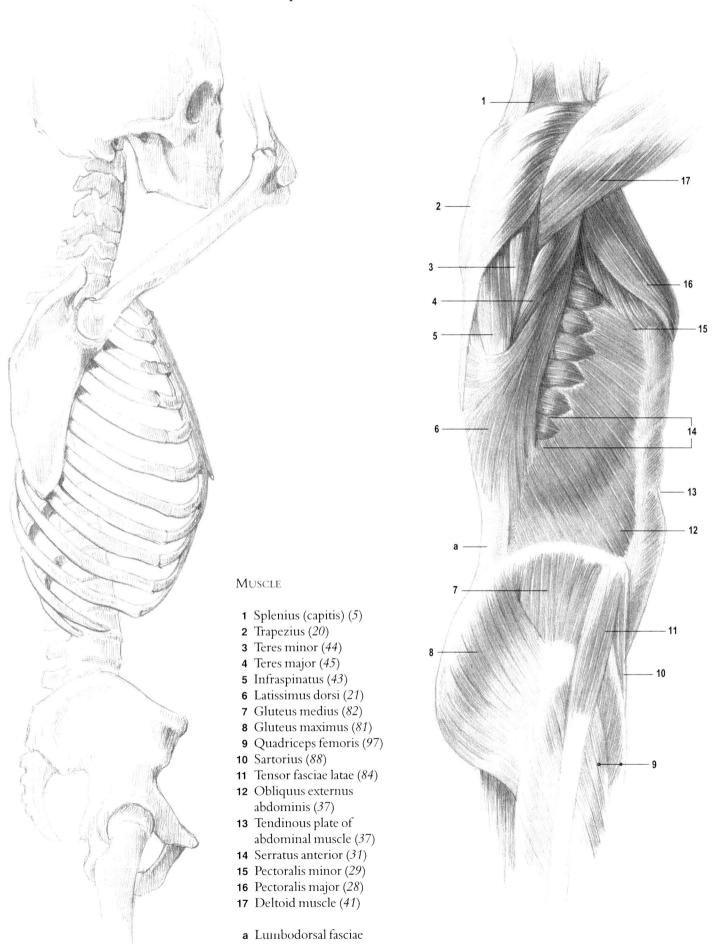

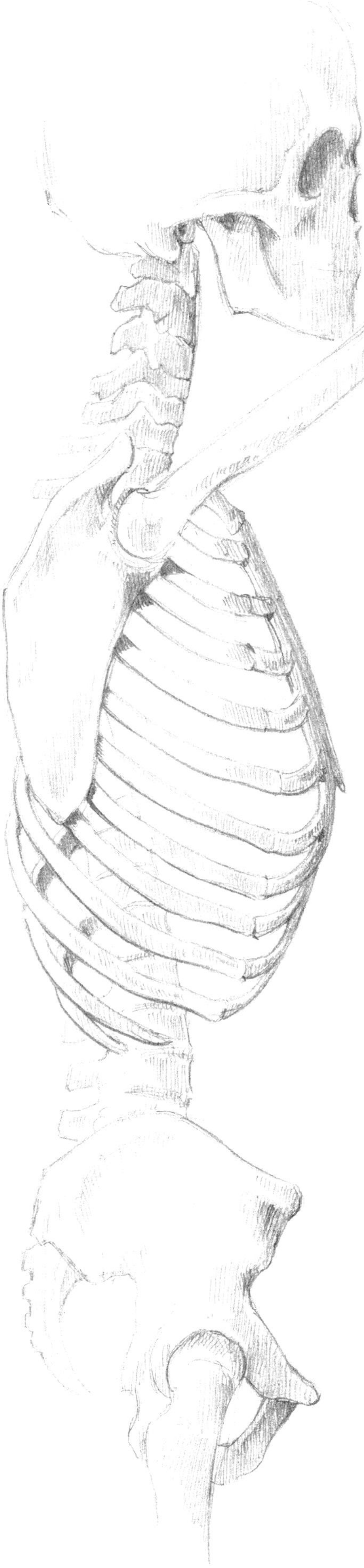

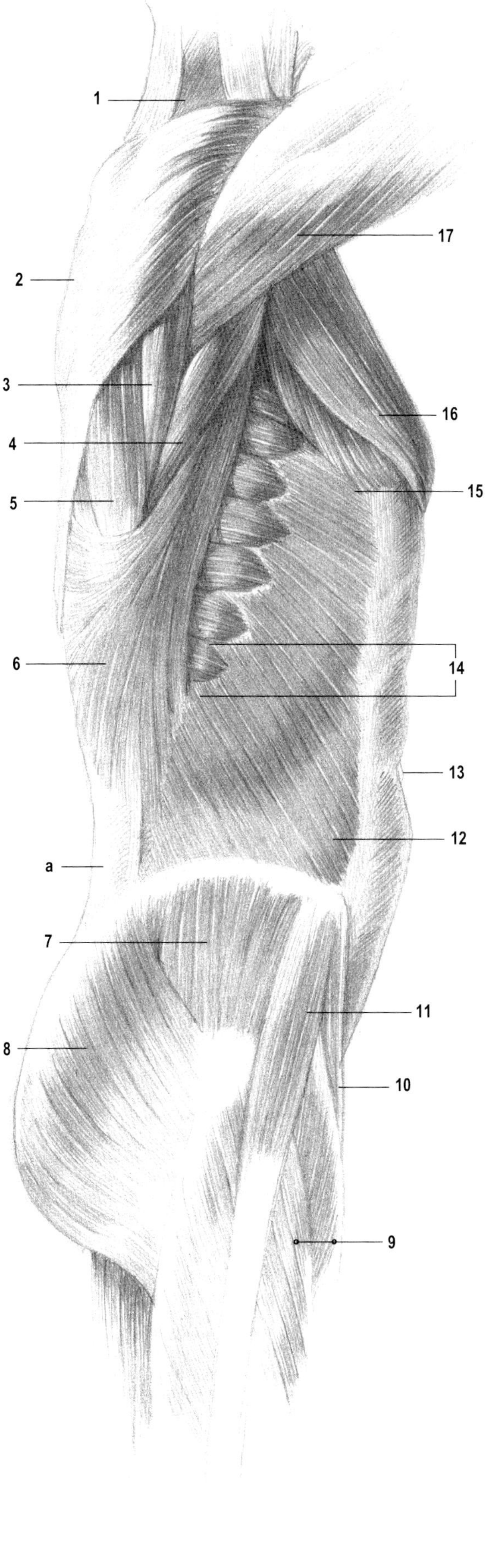

Muscle

1 Splenius (capitis) (*5*)
2 Trapezius (*20*)
3 Teres minor (*44*)
4 Teres major (*45*)
5 Infraspinatus (*43*)
6 Latissimus dorsi (*21*)
7 Gluteus medius (*82*)
8 Gluteus maximus (*81*)
9 Quadriceps femoris (*97*)
10 Sartorius (*88*)
11 Tensor fasciae latae (*84*)
12 Obliquus externus abdominis (*37*)
13 Tendinous plate of abdominal muscle (*37*)
14 Serratus anterior (*31*)
15 Pectoralis minor (*29*)
16 Pectoralis major (*28*)
17 Deltoid muscle (*41*)

a Lumbodorsal fasciae

Fig. 218
Studies in the Muscles
of the Thorax

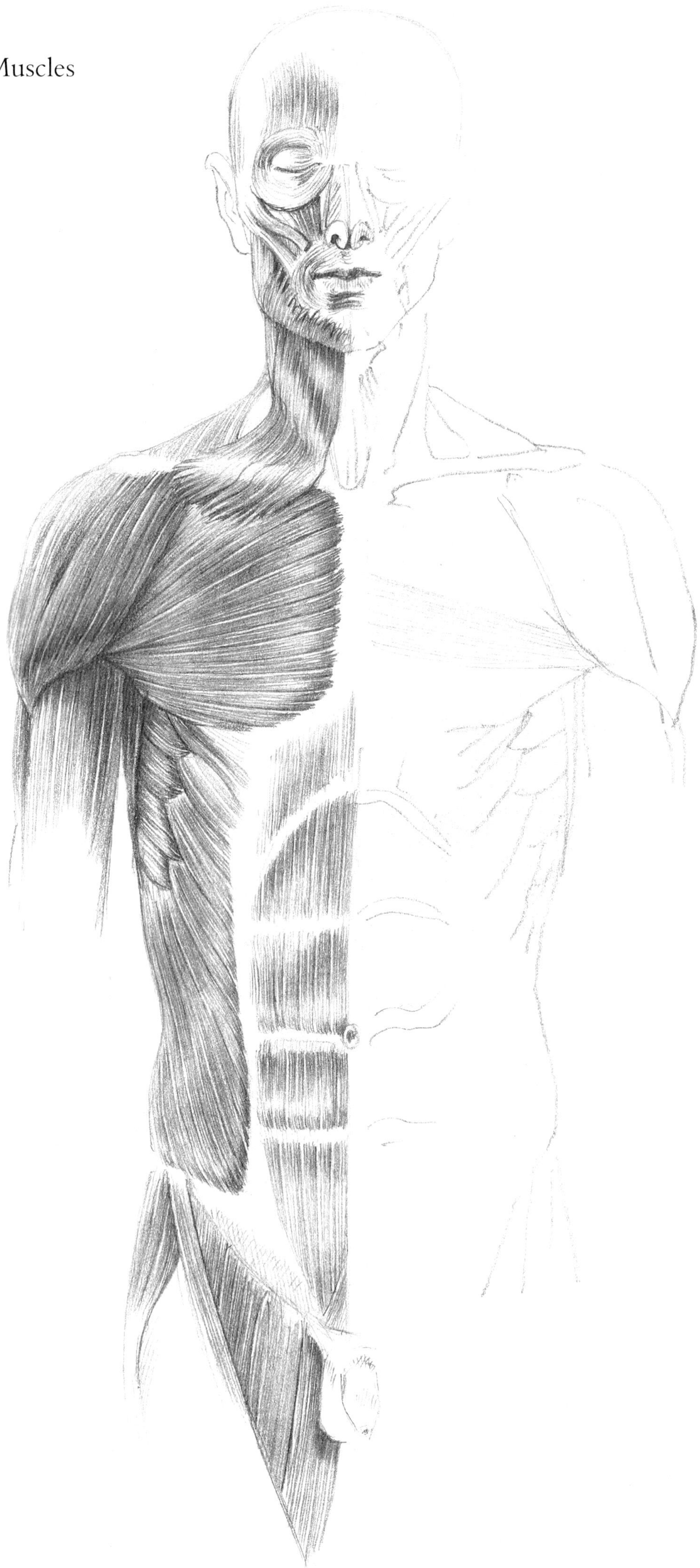

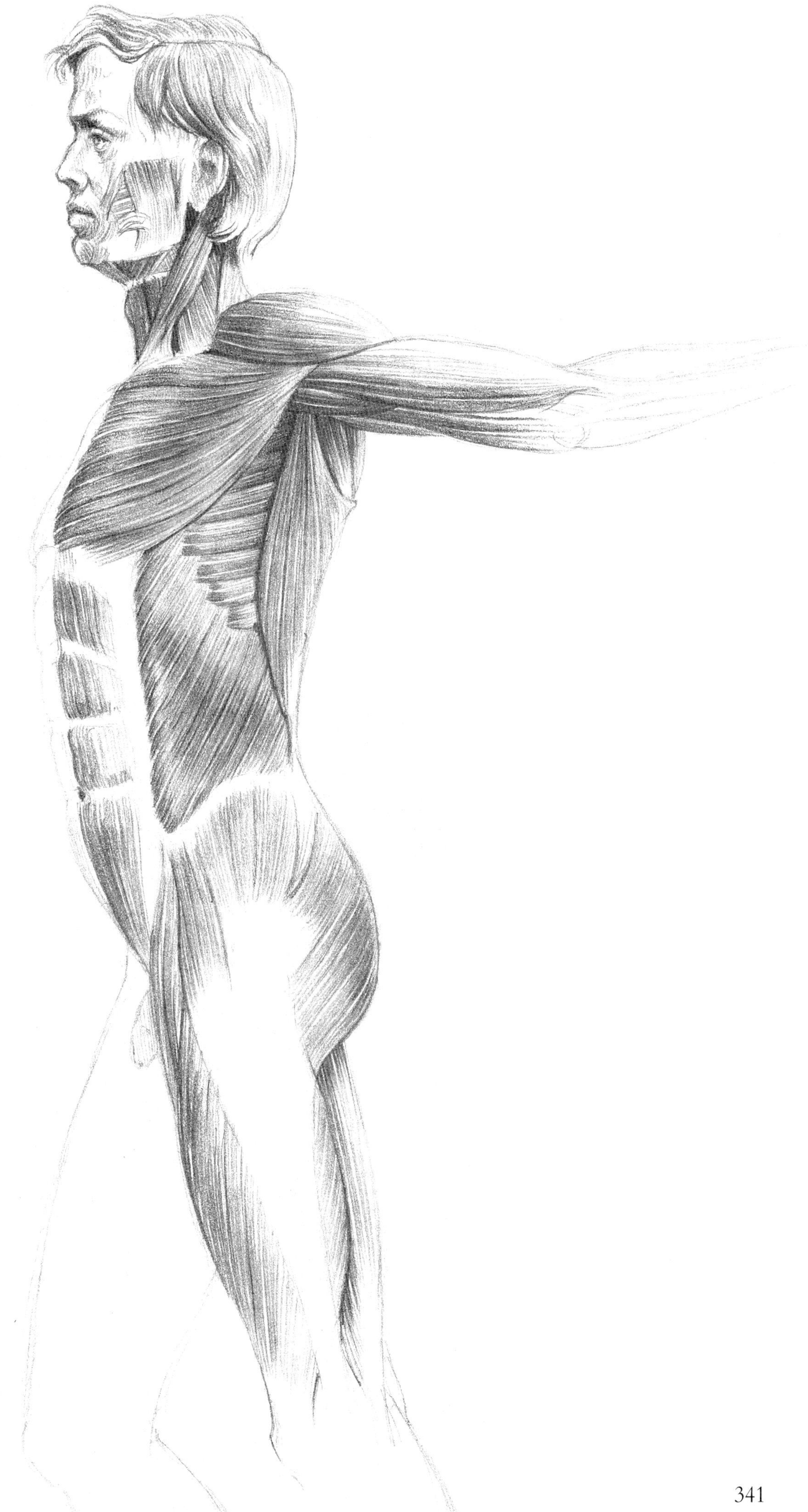

Fig. 219
The Muscles of the Thoracic Cavity and Abdomen

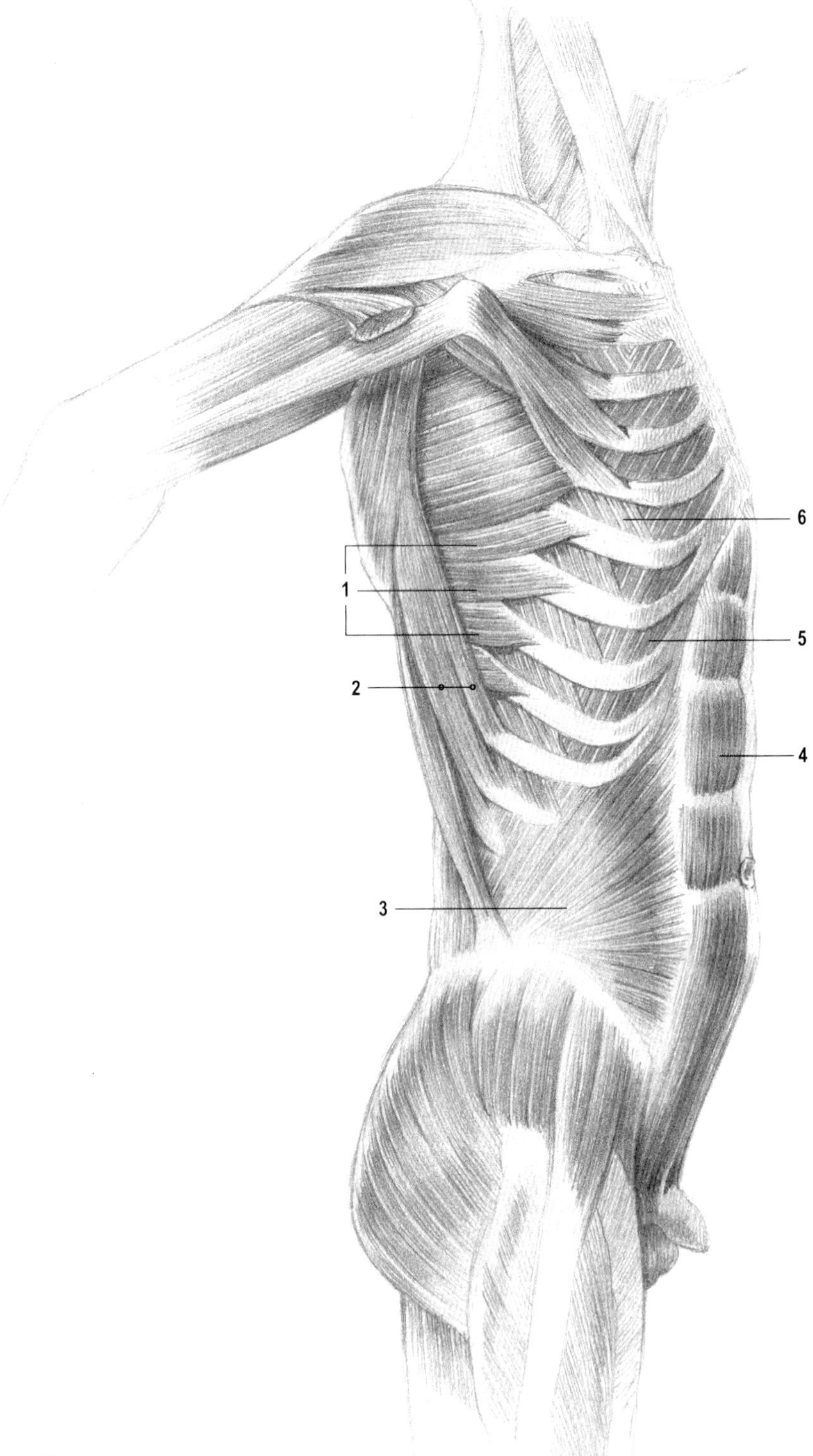

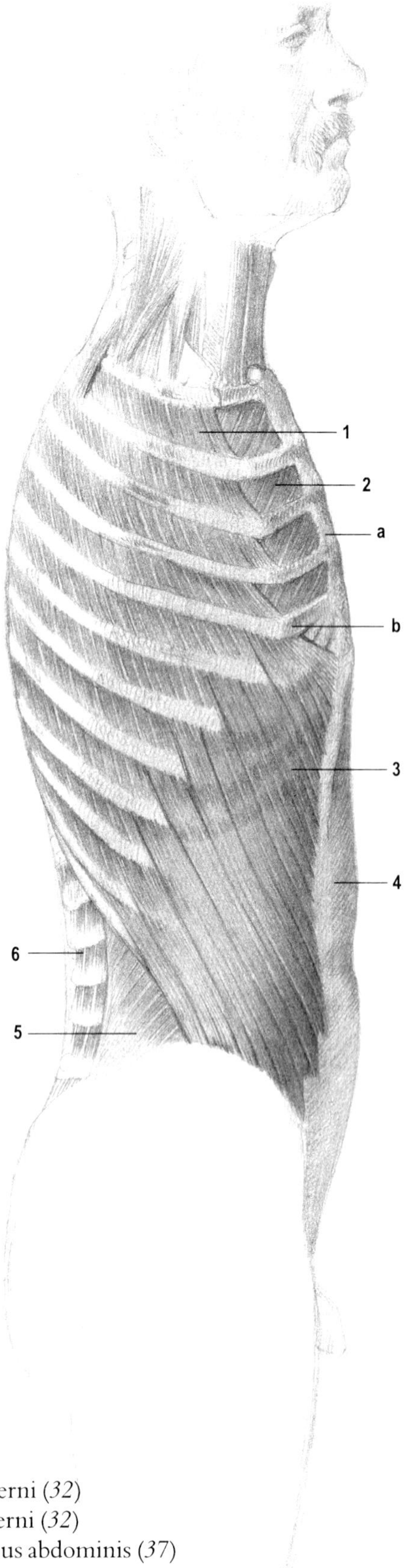

SUPERFICIAL LAYER

1 Serratus anterior (*31*)
2 Latissimus dorsi (*21*)
3 Obliquuus internus abdominis (*38*)
4 Rectus abdominis (*35*)
5 Intercostales interni (*32*)
6 Intercostales externi (*32*)

DEEP LAYER

1 Intercostales externi (*32*)
2 Intercostales interni (*32*)
3 Obliquus externus abdominis (*37*)
4 Facsia recta
5 Obliquus internus abdominis (*38*)
6 Intertransversal muscles (*26/7*)

a sternum
b cartilage of fifth rib

Fig. 220
The Abdominal Muscles

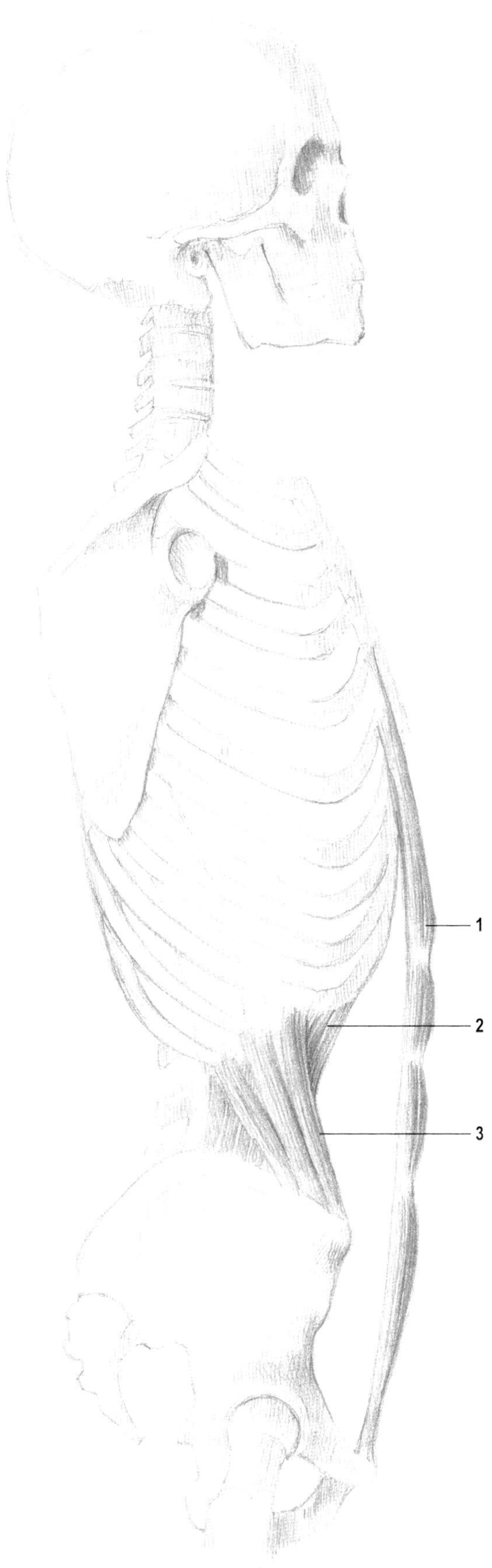

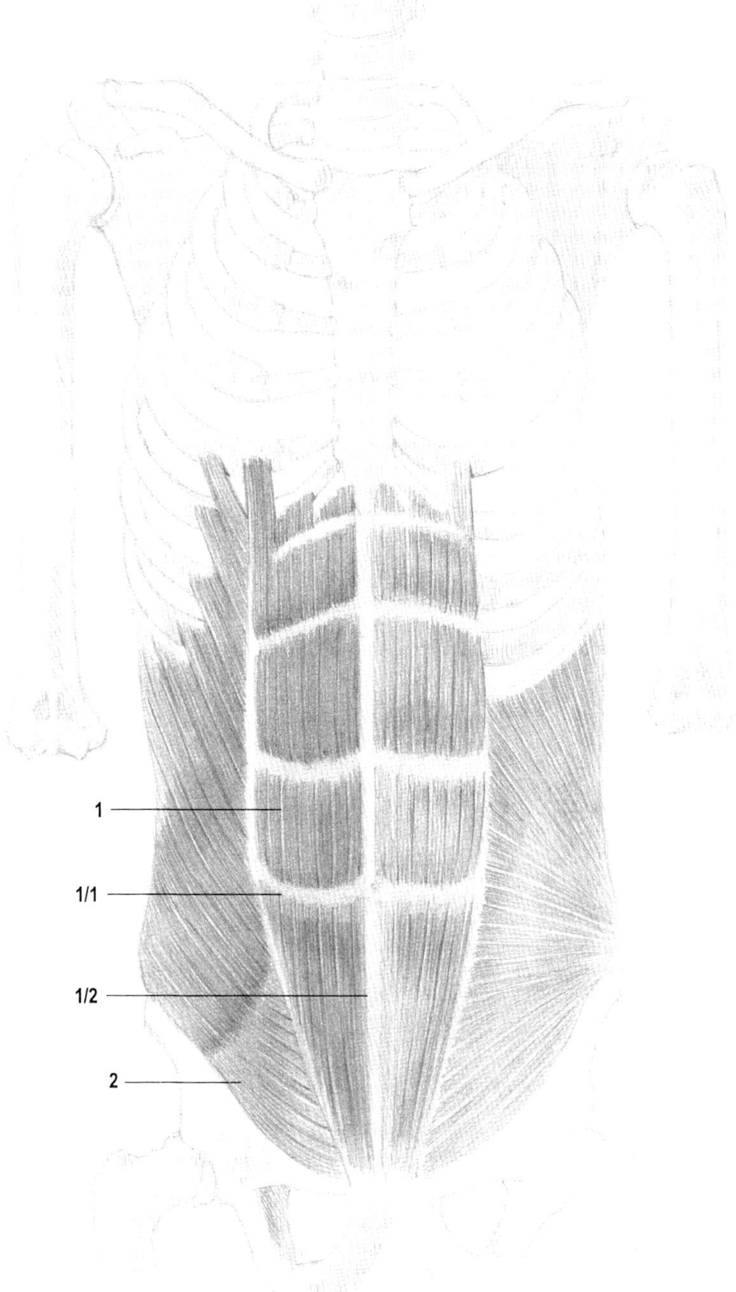

1 Rectus abdominis (*35*)
  1/1 Intermediate tendons
  1/2 Linea alba
2 Obliquus internus abdominis (*38*)
3 Quadratus lumborum (*80*)

These muscles form the lateral and anterior abdominal wall. In fit young people the edges and intermediate tendons of the rectus abdominis and the linea alba (median line of the abdomen) are clearly visible under the skin. The costal arches can be found at the edges of these muscles and inferiorly the iliac crest. The four layers of abdominal muscles lie on top of each other and their fibres traverse each other.

Fig. 221
The External Oblique Muscle of the Abdomen
(M. obliquus externus abdominis, *37*)

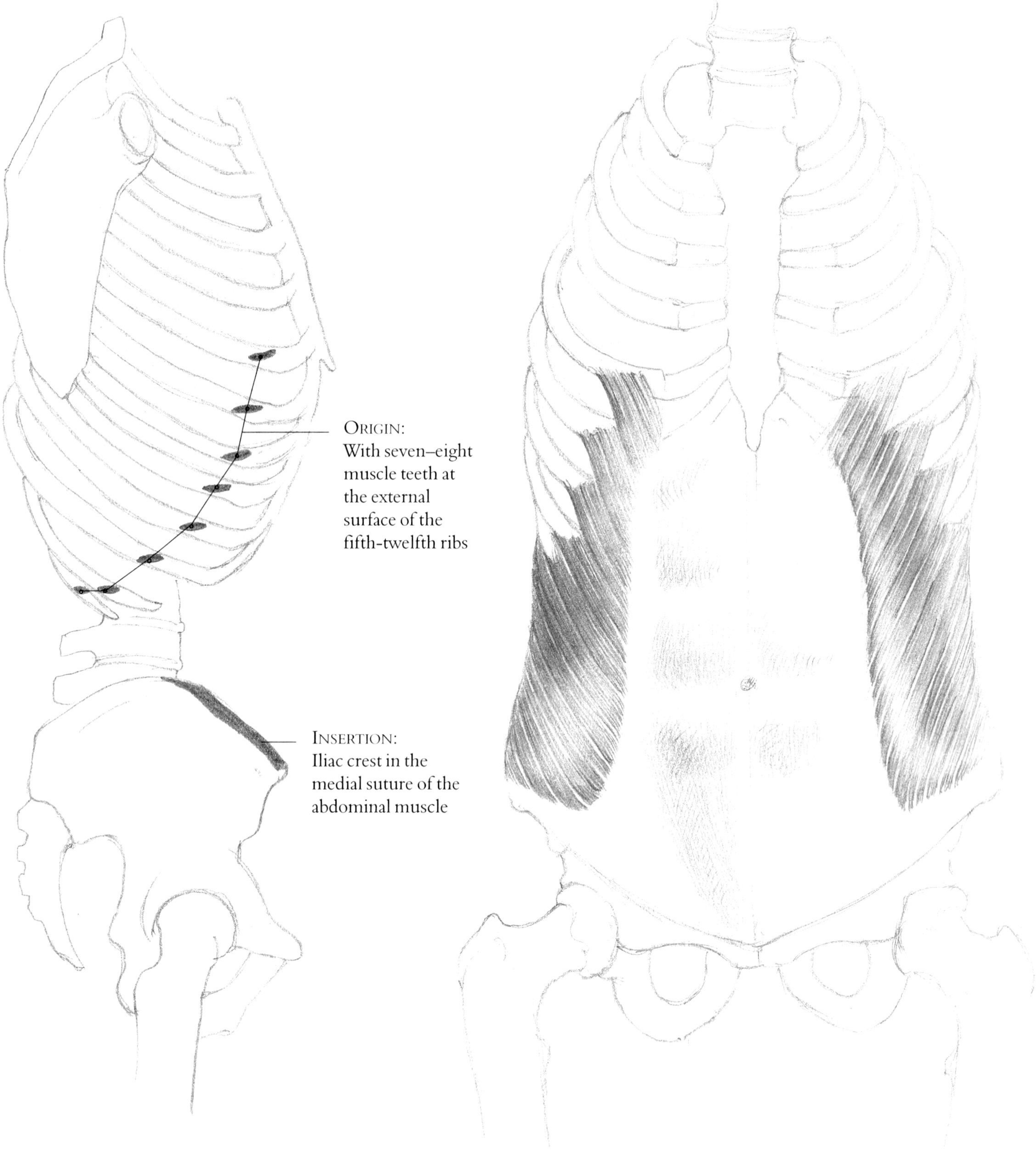

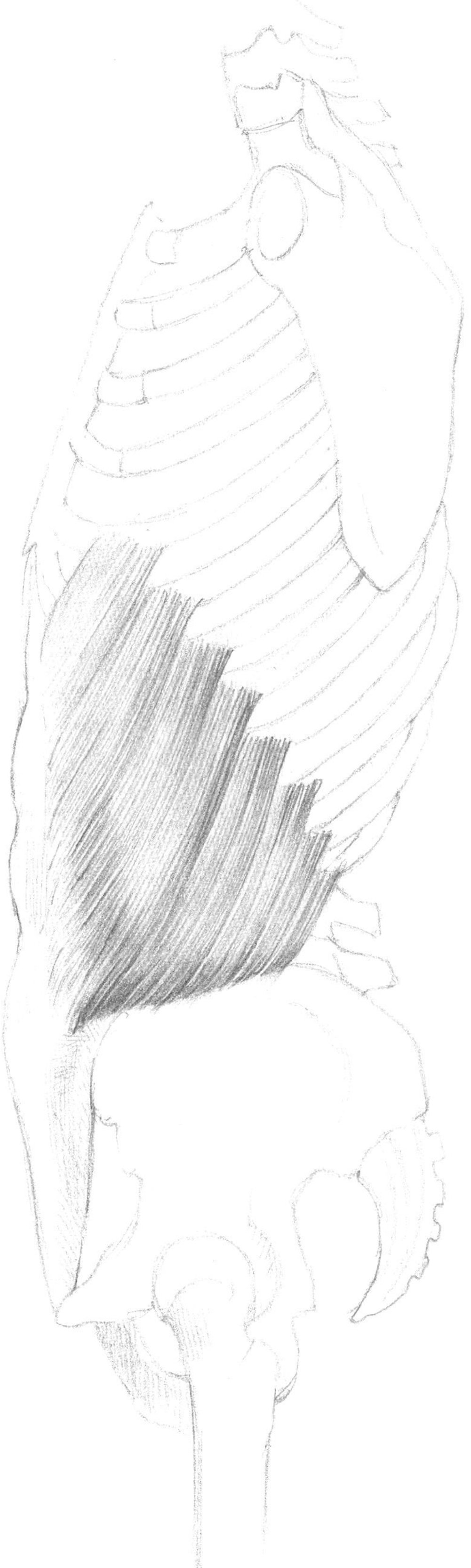

FUNCTION:
Supports the organs of the lower abdomen, expels stools and pushes during birth; flexes and rotates (one side of) the trunk

Fig. 222
The External Intercostal Muscles
(Mm. intercostales externi, *32*)

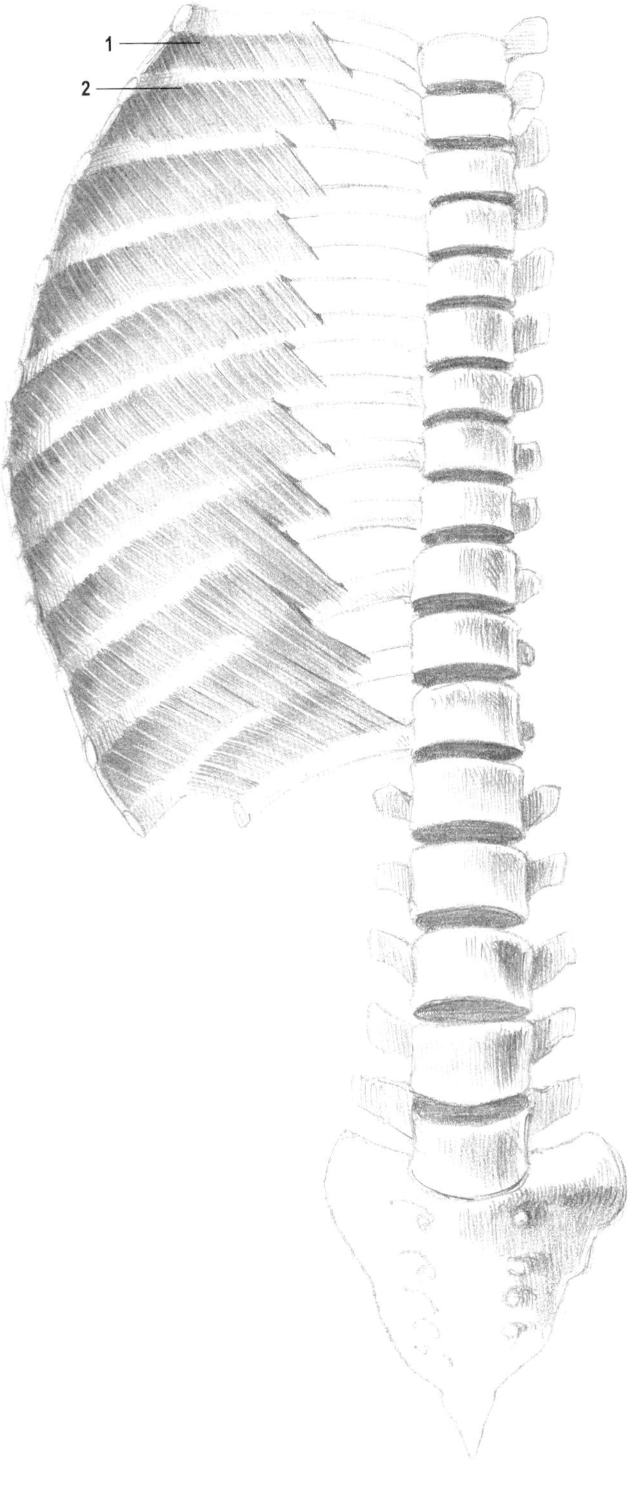

ORIGIN:
Cranial edge of the ribs (**2**)

INSERTION:
Caudal edge of the following ribs (**1**)

FUNCTION:
Lower the ribs (breathing out)

Fig. 223
The Internal Oblique Muscle of the Abdomen
(M. obliquus internus abdominis, *38*)

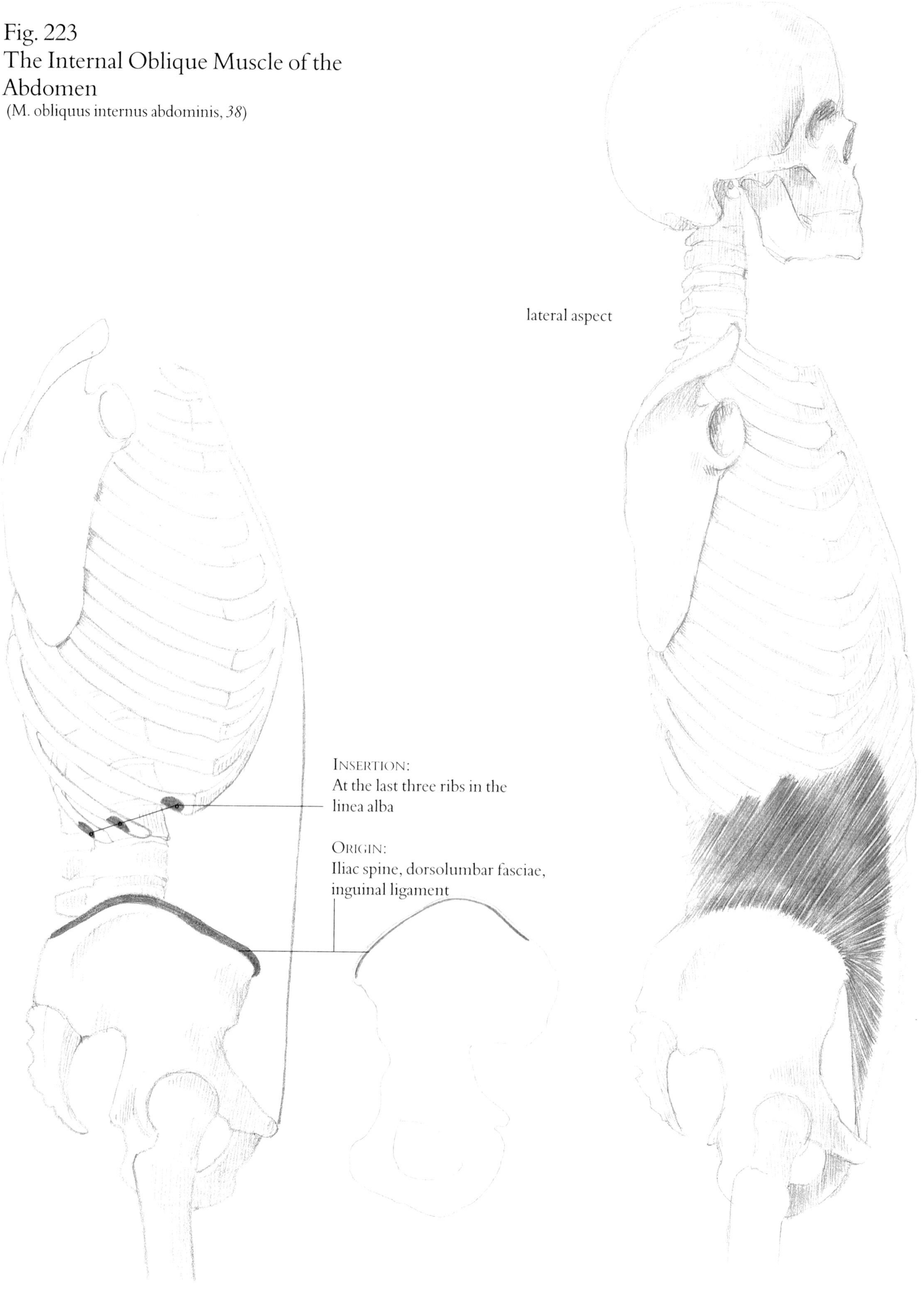

ventral aspect

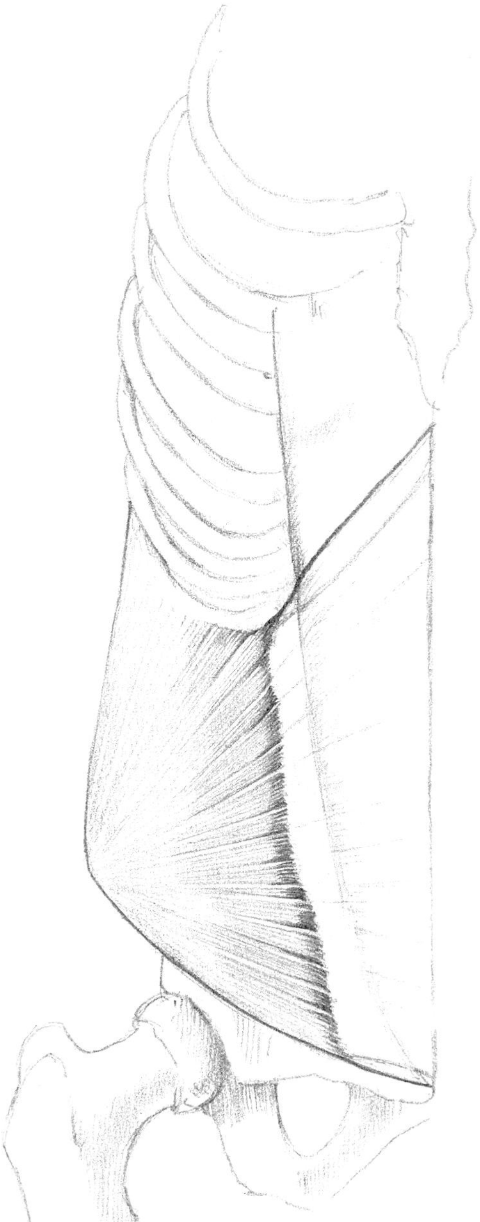

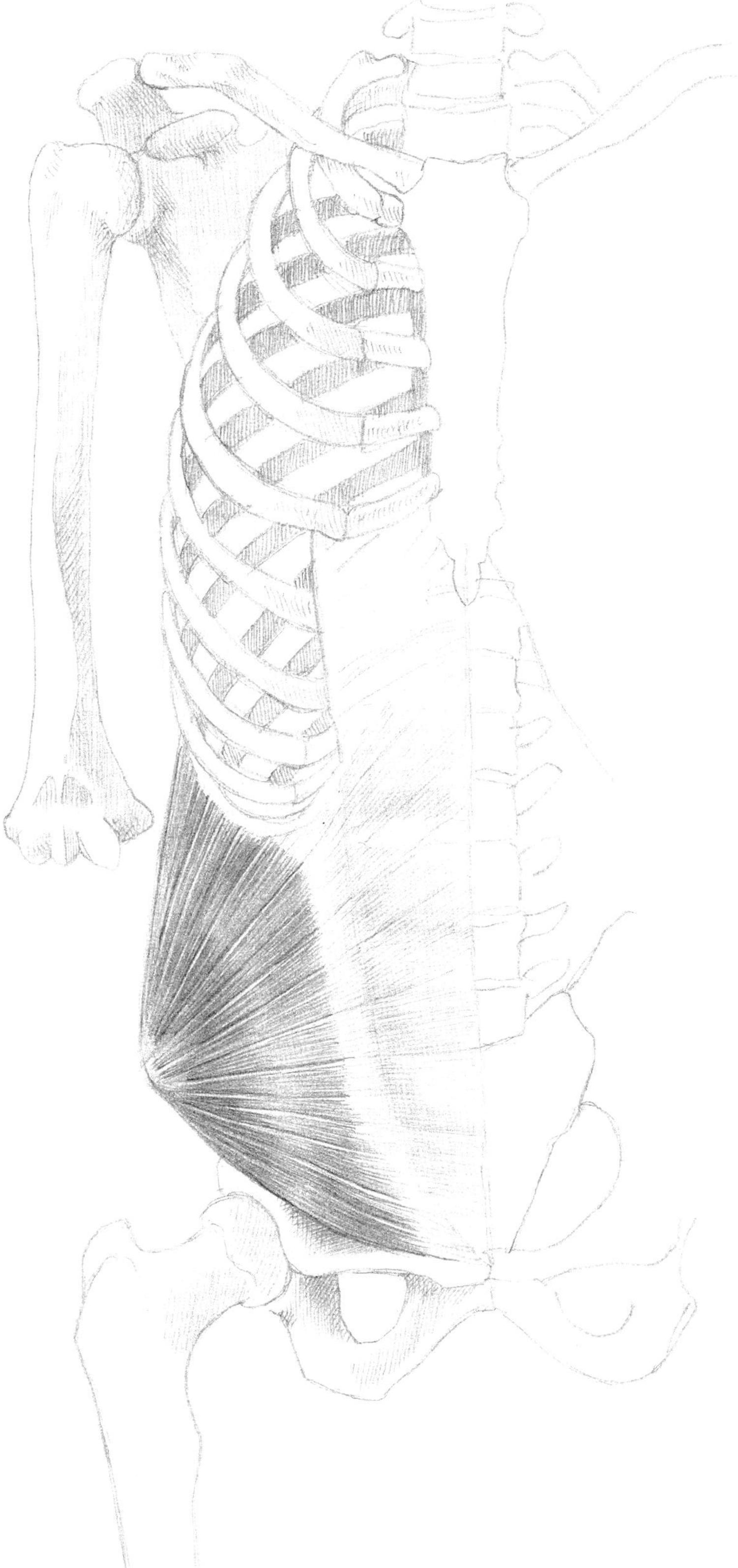

FUNCTION:
Supports the organs of the lower abdomen, flexes the trunk laterally

Fig. 224
The Straight Abdominal Muscle
(M. rectus abdominis, *35*)

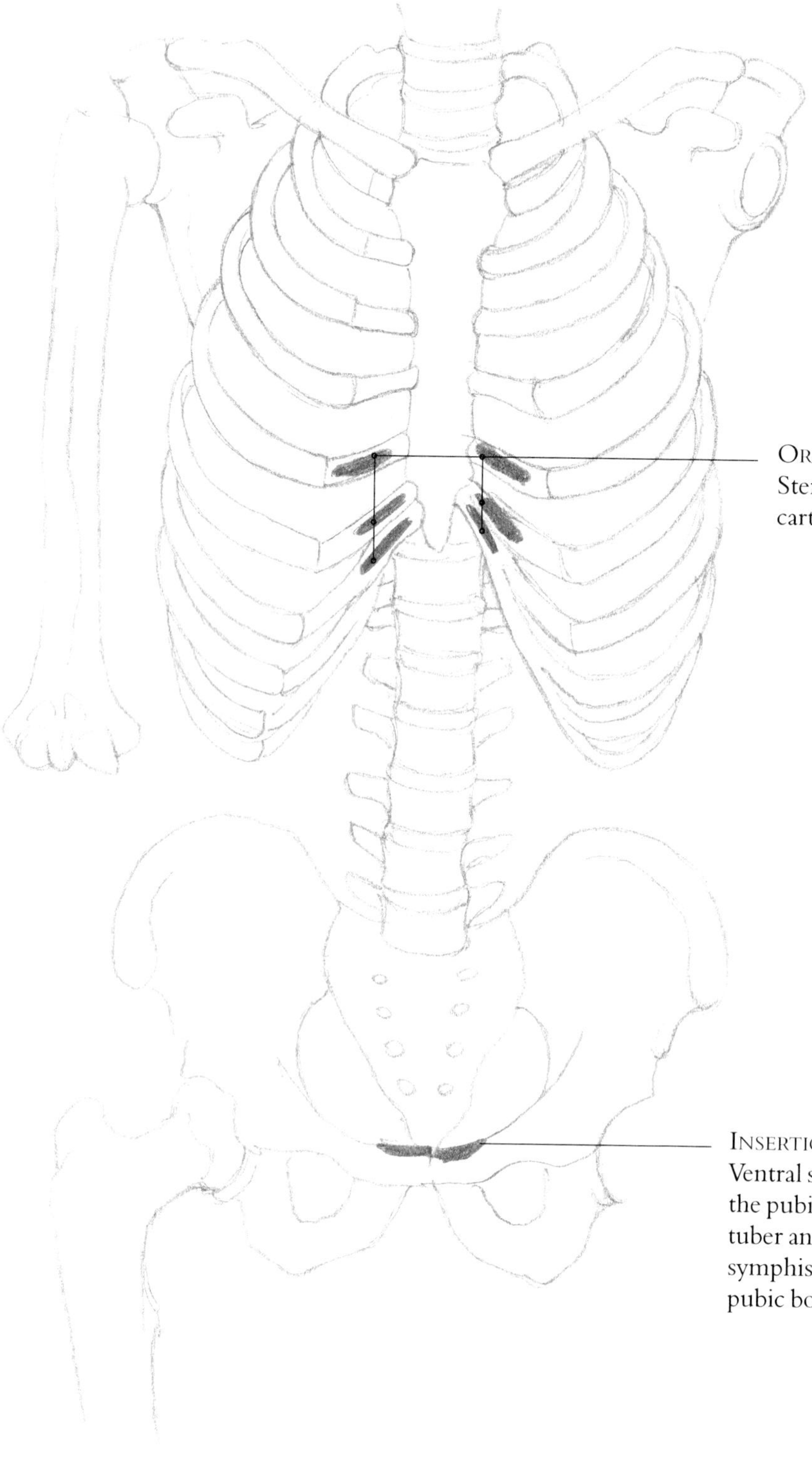

ORIGIN:
Sternal cartilage, fifth–seventh rib cartilage, costochondral ligament.

INSERTION:
Ventral surface of the pubic bone, tuber and symphisis of the pubic bone

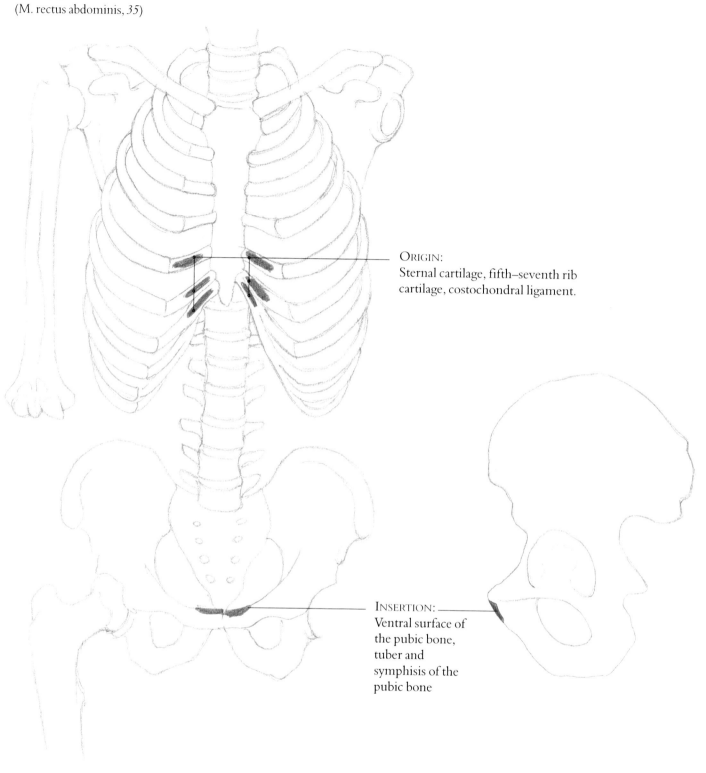

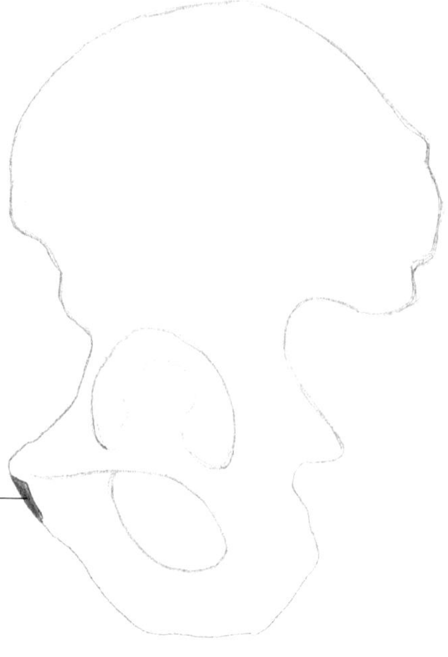

ventral aspect

lateral aspect

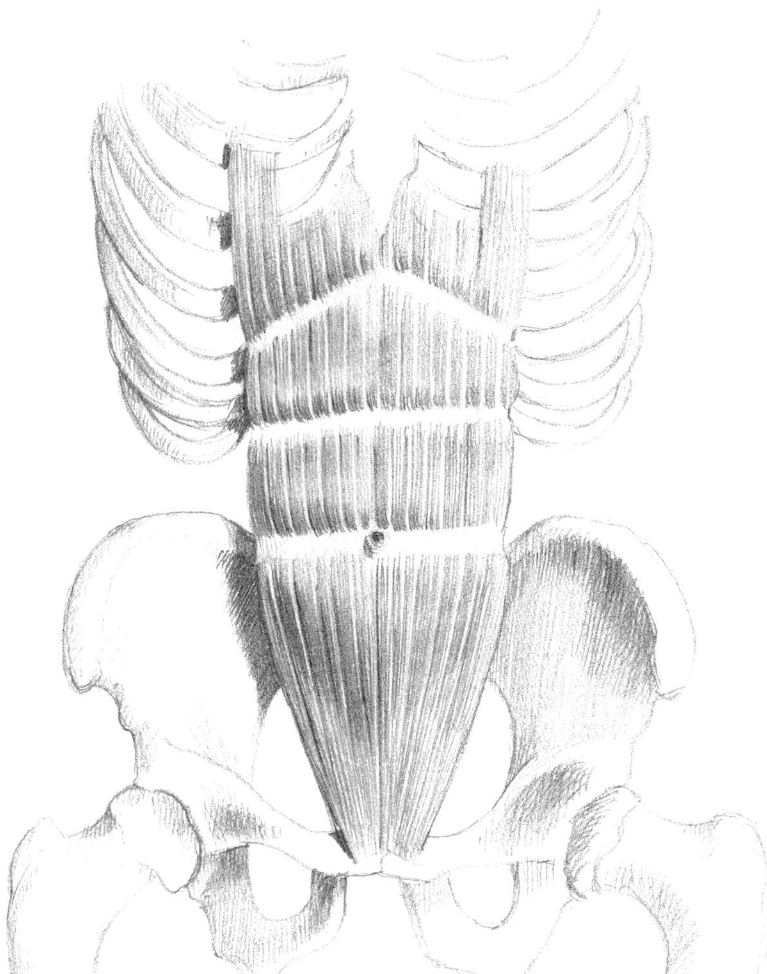

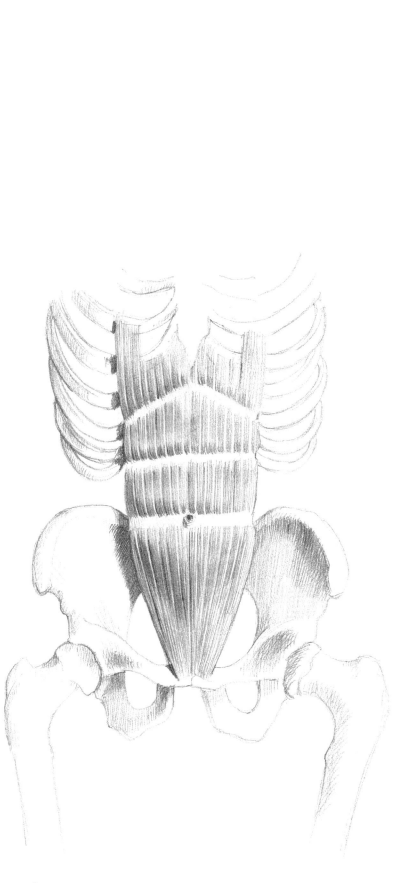

ventral aspect

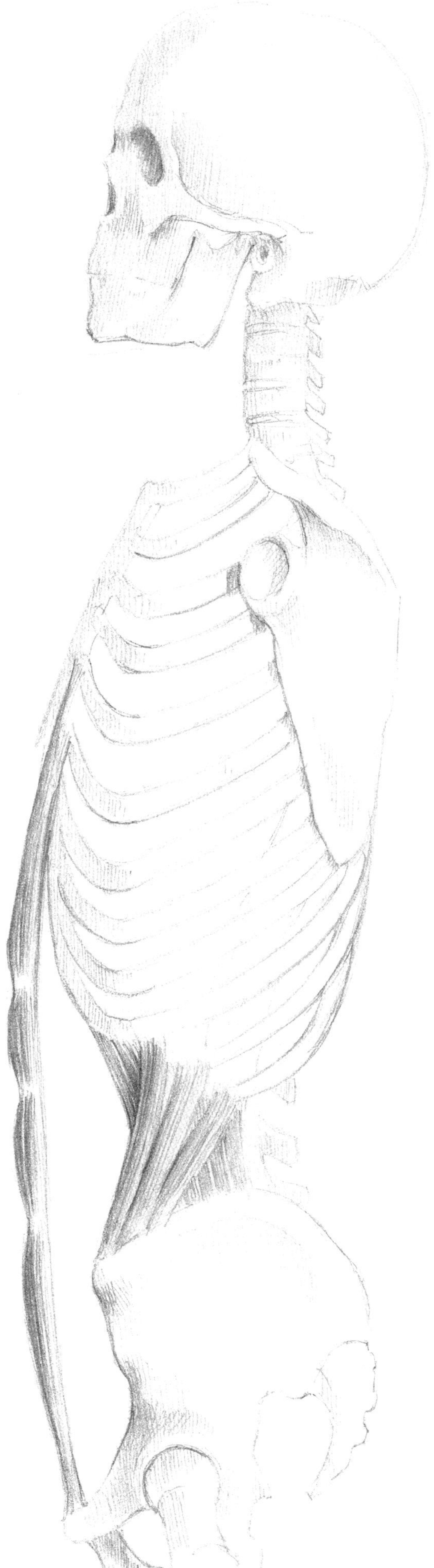

lateral aspect

FUNCTION:
Supports the organs of the lower abdomen, expels stools and pushes during birth, flexes the trunk

Fig. 225
The Pyramidal Muscle
(M. pyramidalis, *36*)

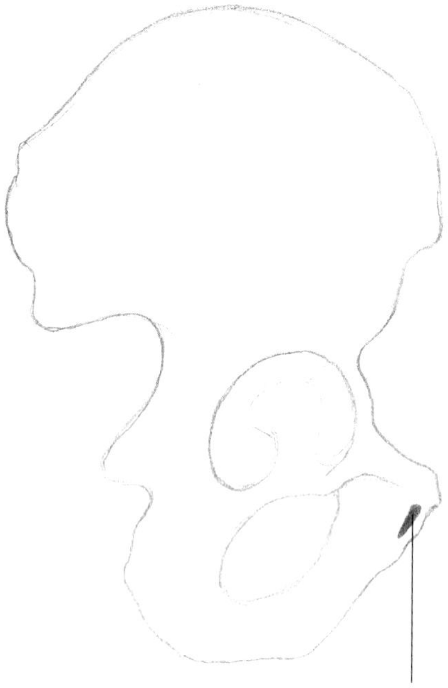

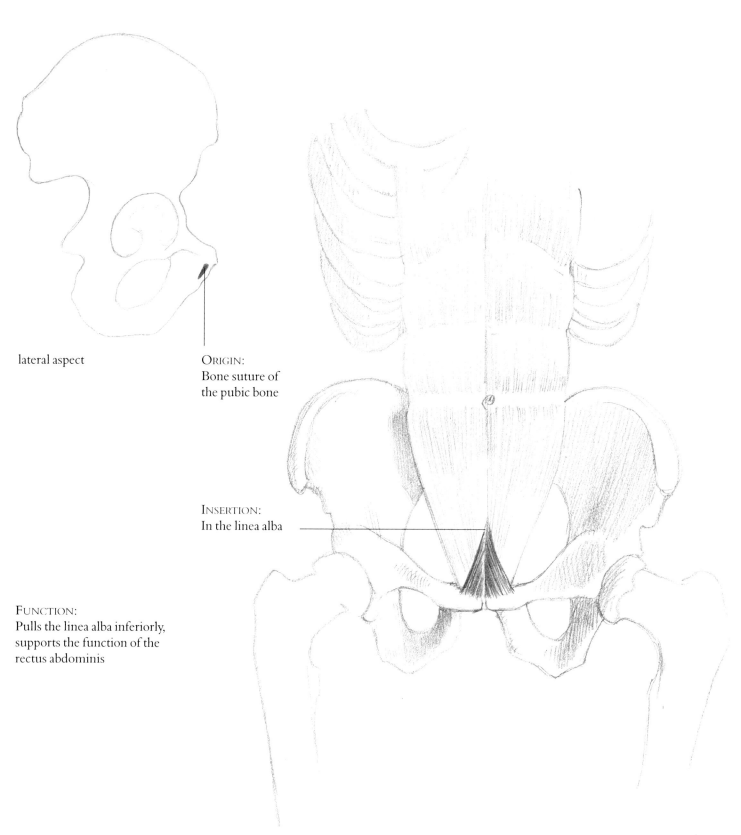

lateral aspect

ORIGIN:
Bone suture of the pubic bone

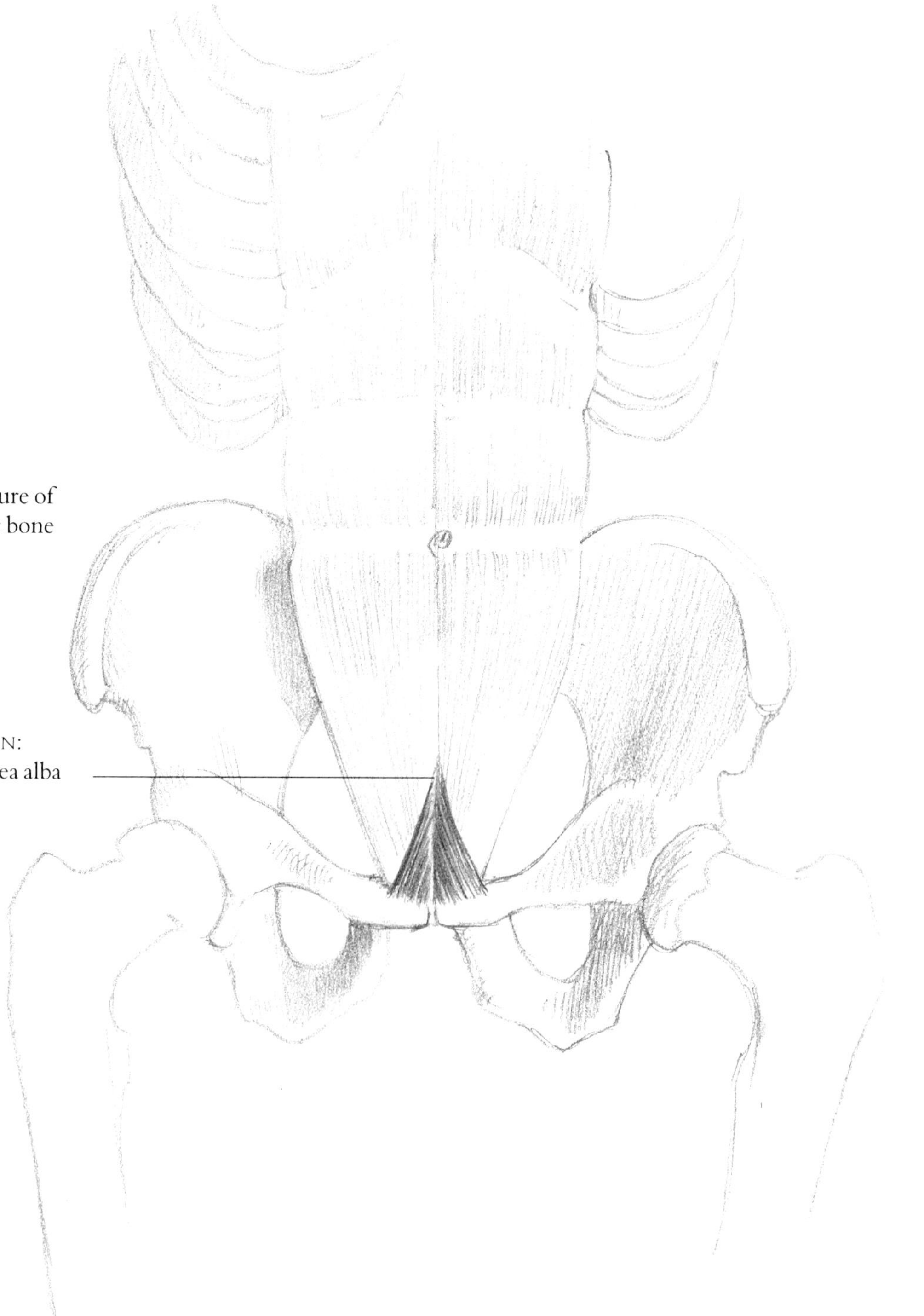

INSERTION:
In the linea alba

FUNCTION:
Pulls the linea alba inferiorly, supports the function of the rectus abdominis

ventral aspect

Fig. 226
The Transverse Muscle of the Abdomen
(M. transversus abdominis, *39*)

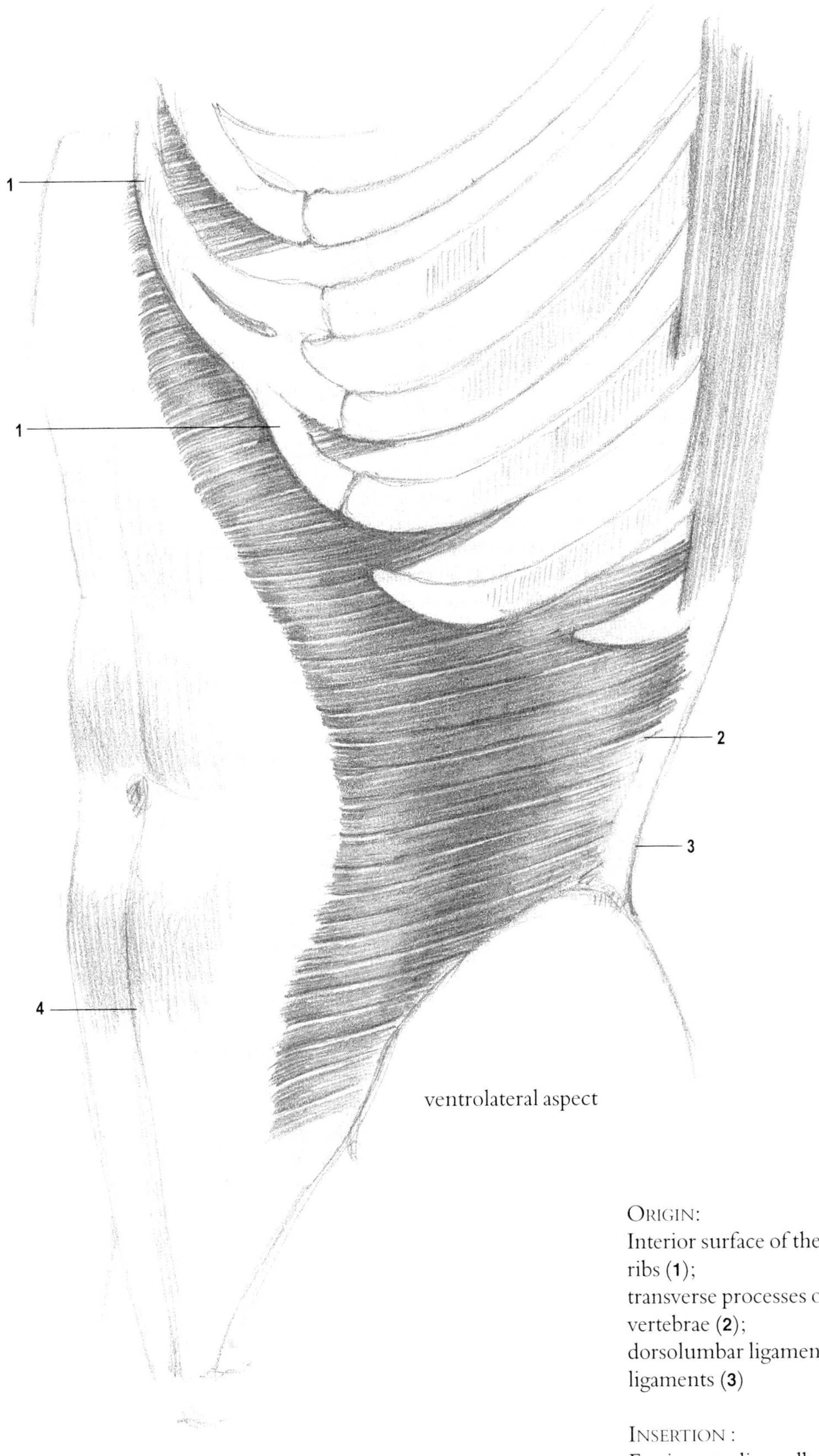

ventrolateral aspect

ORIGIN:
Interior surface of the cartilage of the last six ribs (**1**);
transverse processes of the lumbar vertebrae (**2**);
dorsolumbar ligaments, lateral inguinal ligaments (**3**)

INSERTION:
Fascia recta, linea alba (**4**)

FUNCTION:
Expels stools and pushes during birth

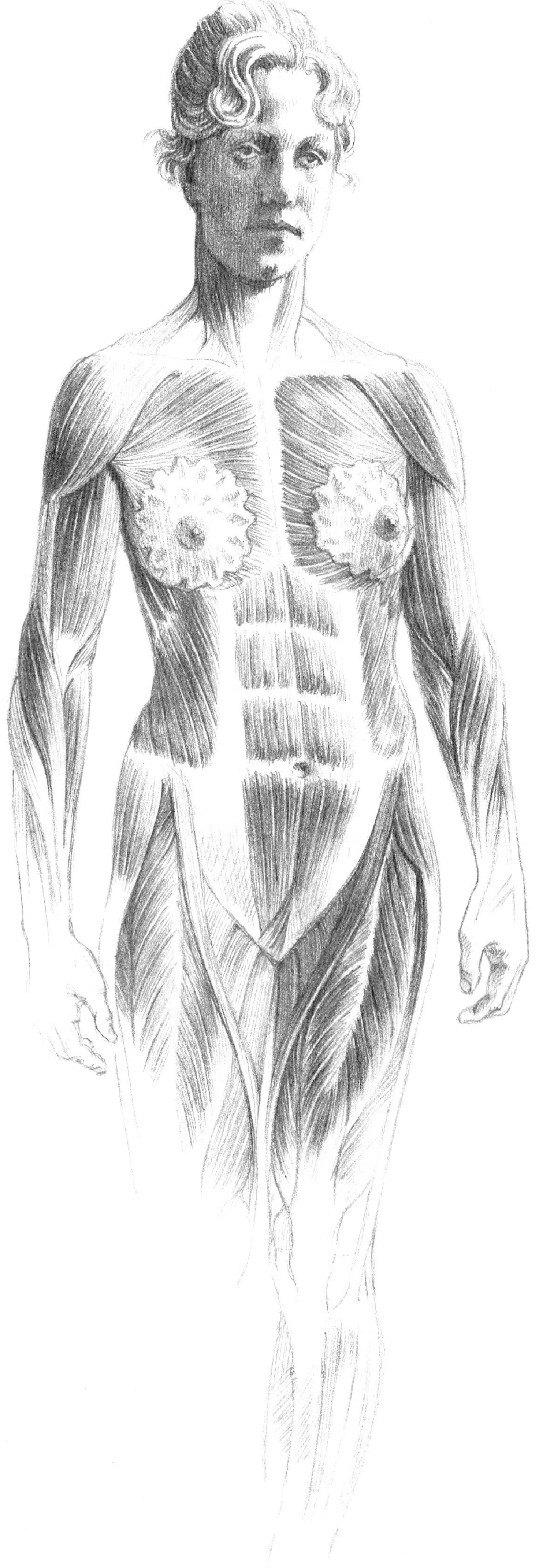

Fig. 227
The Body Axis

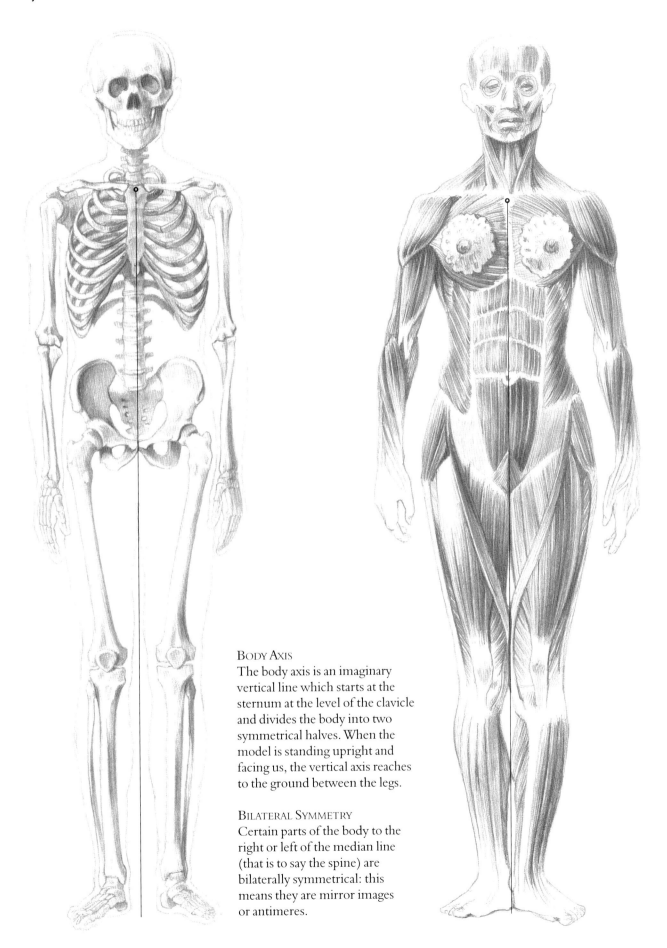

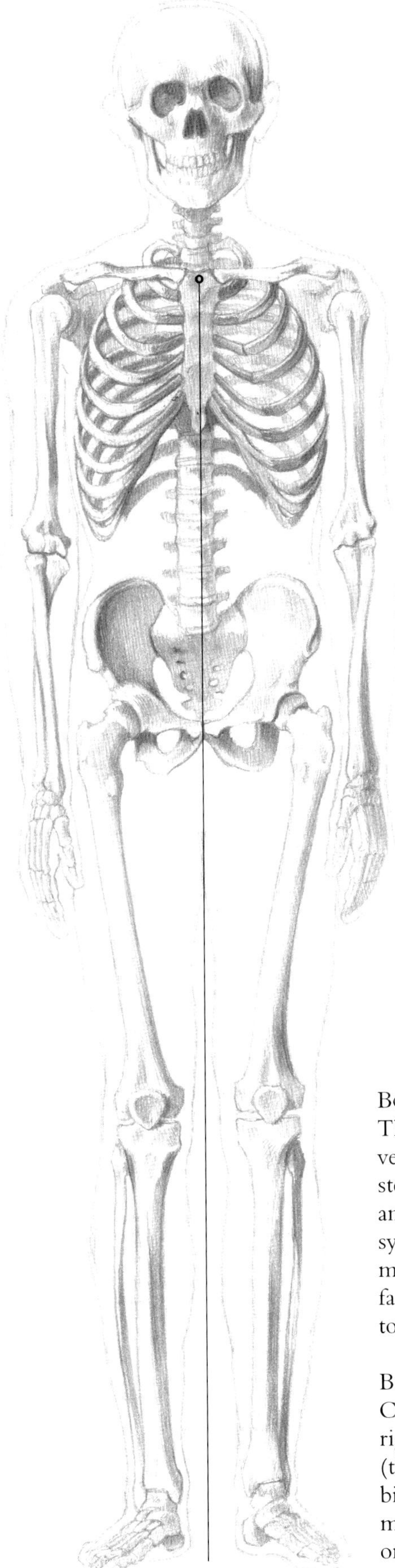

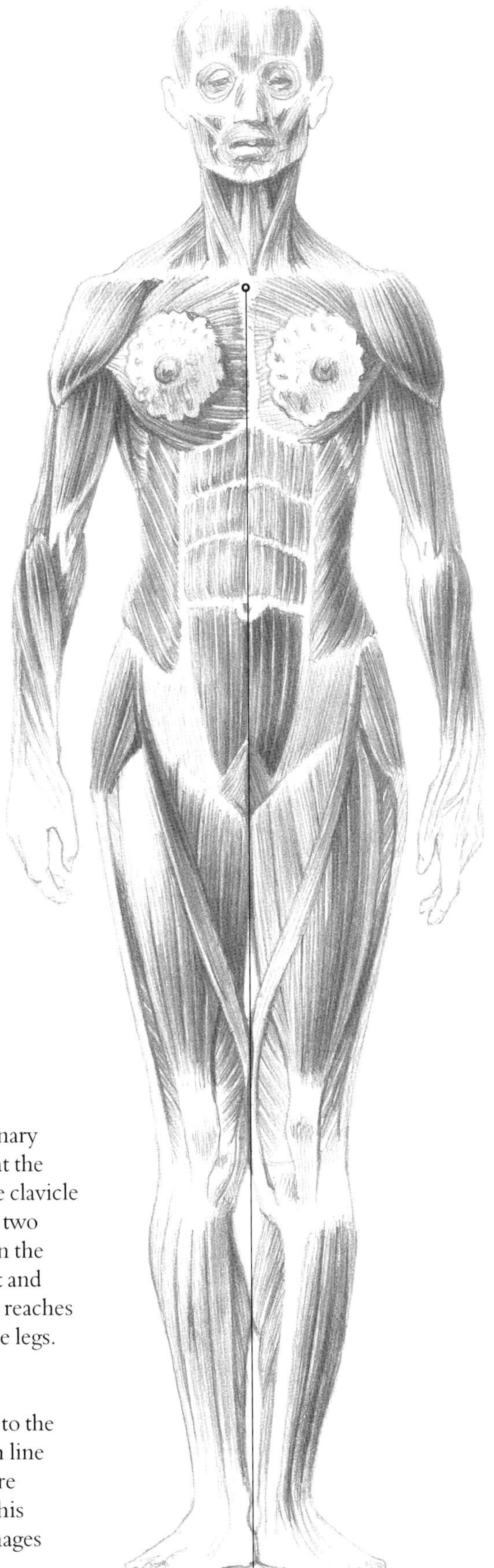

BODY AXIS
The body axis is an imaginary vertical line which starts at the sternum at the level of the clavicle and divides the body into two symmetrical halves. When the model is standing upright and facing us, the vertical axis reaches to the ground between the legs.

BILATERAL SYMMETRY
Certain parts of the body to the right or left of the median line (that is to say the spine) are bilaterally symmetrical: this means they are mirror images or antimeres.

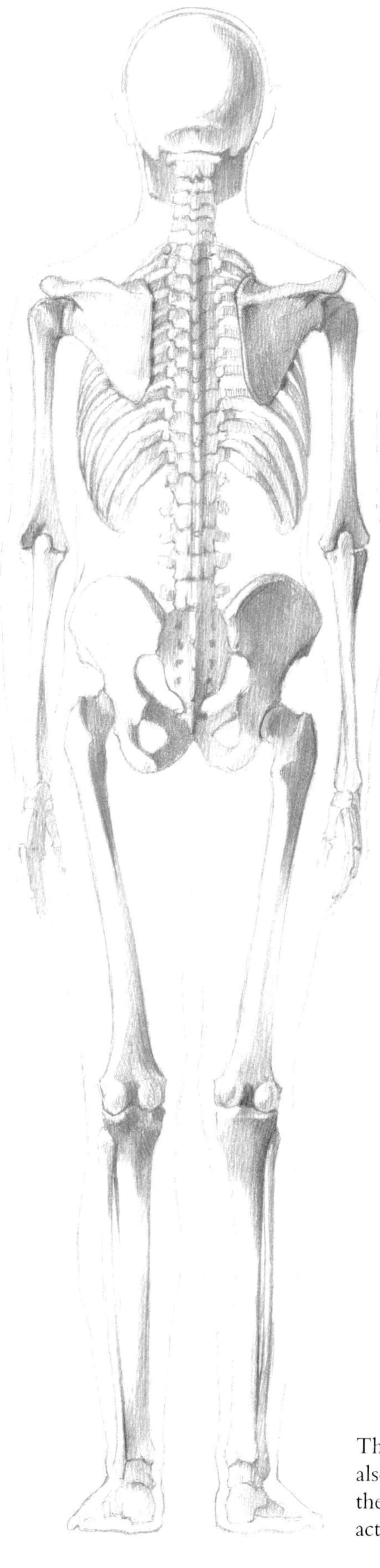

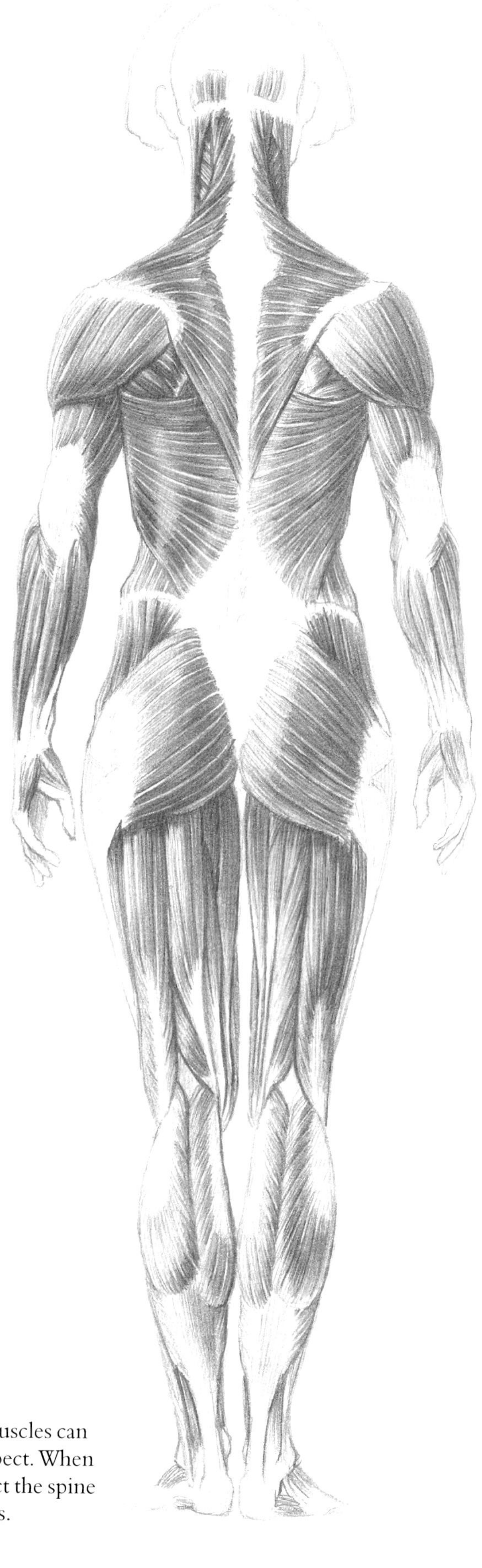

The symmetry of bones and muscles can also be seen in the posterior aspect. When the figure is drawn in this aspect the spine acts as an imaginary vertical axis.

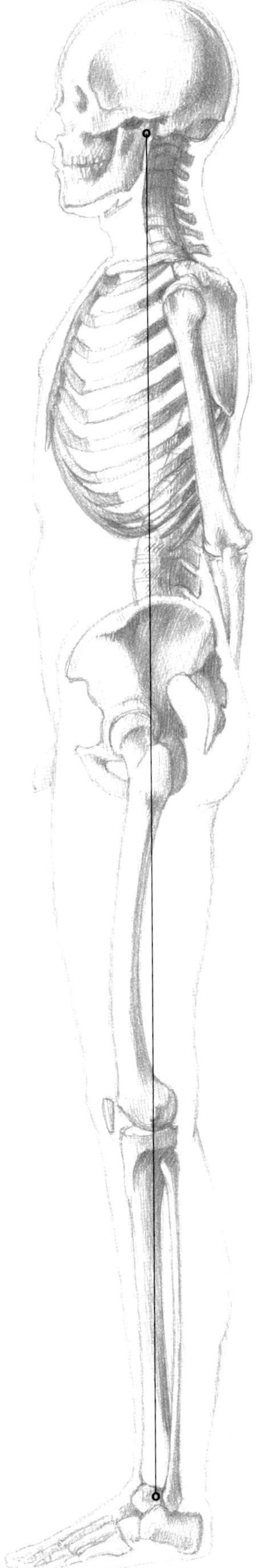

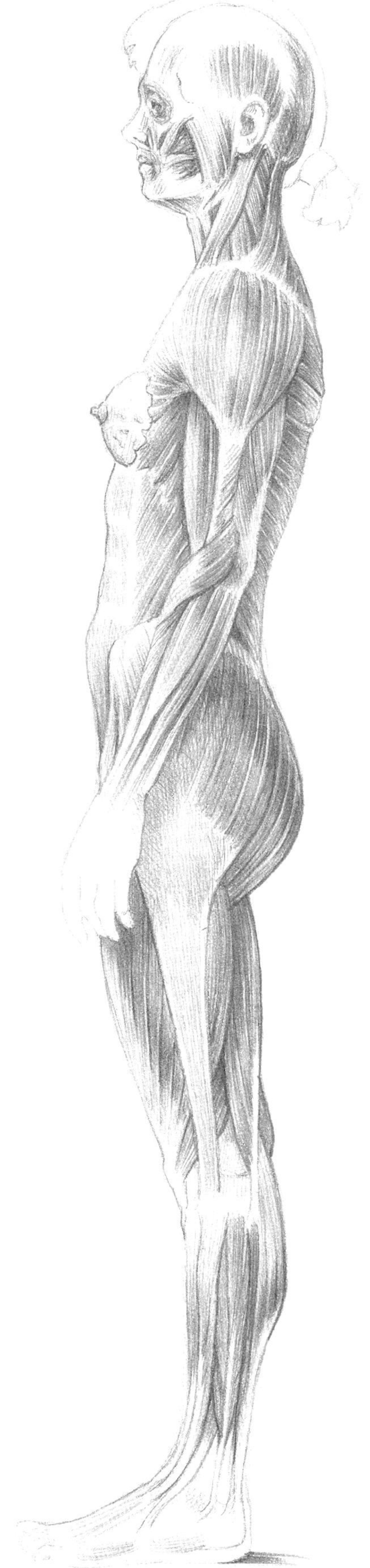

When the model is viewed in its lateral aspect, it is possible to draw an imaginary vertical line from the ear to the ankle. This imaginary vertical line makes it easier to make the figure stand straight, and prevents it from falling forwards or backwards.

Fig. 228
Ontogenesis

The genesis and development of an individual is known as ontogenesis. It begins with conception and ends in death. Variations in the development rates of the various organs mean that the proportions of certain body parts undergo change. For example, since the nervous system develops quickly and early, a new-born child will have a big head, whereas the trunk and limbs are short. After birth the bones of the limbs develop: firstly, they grow in length, and then in width. The number of muscle

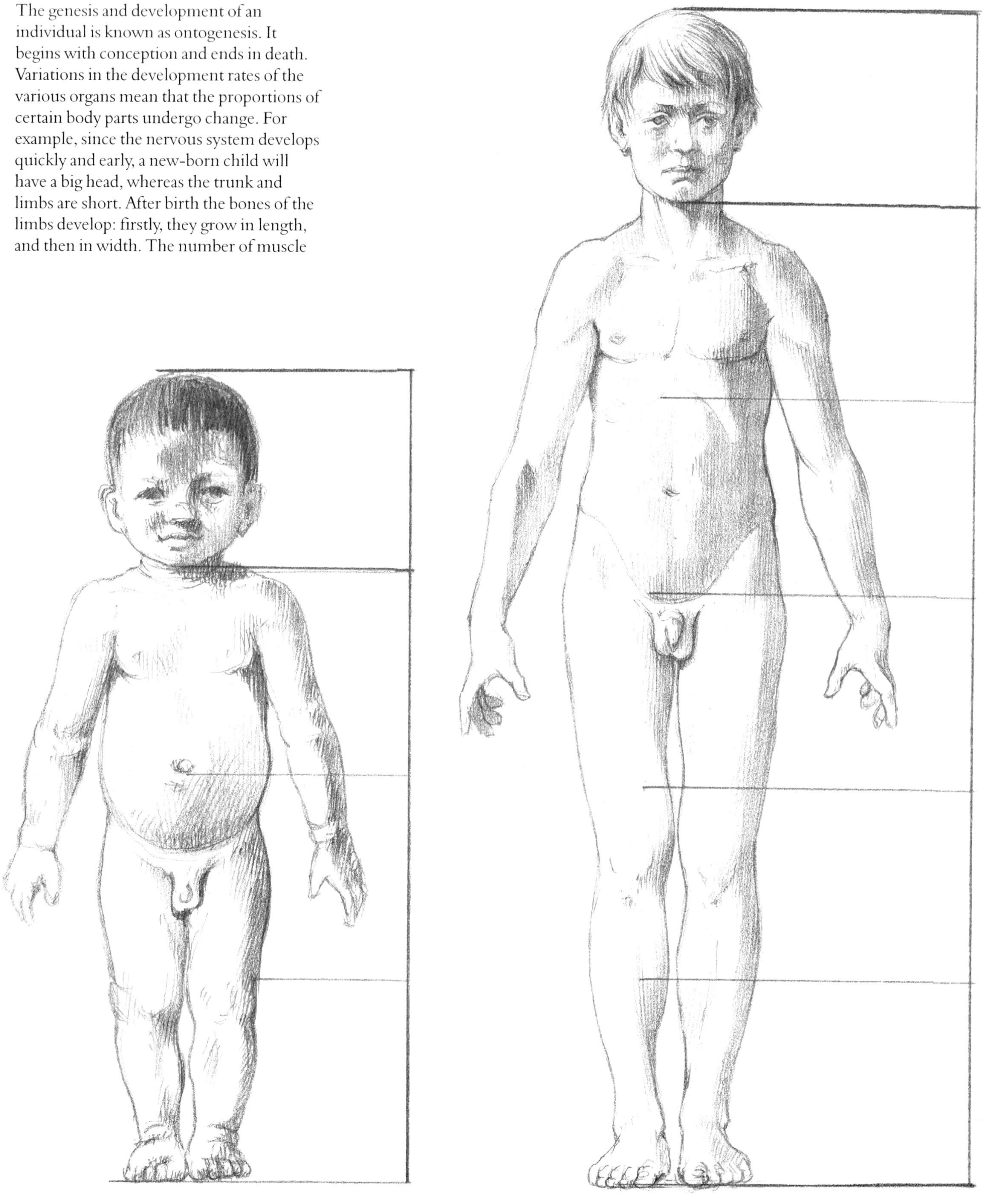

fibres is fixed; however, their length and width and their ultimate form depend on the development of the bones. Finally, the sexual organs assume their ultimate form and size after puberty – towards the end of the growth phase. A grown organism assumes its basic position when it stands straight and still. When we change the proportions of the body parts we start from this position.

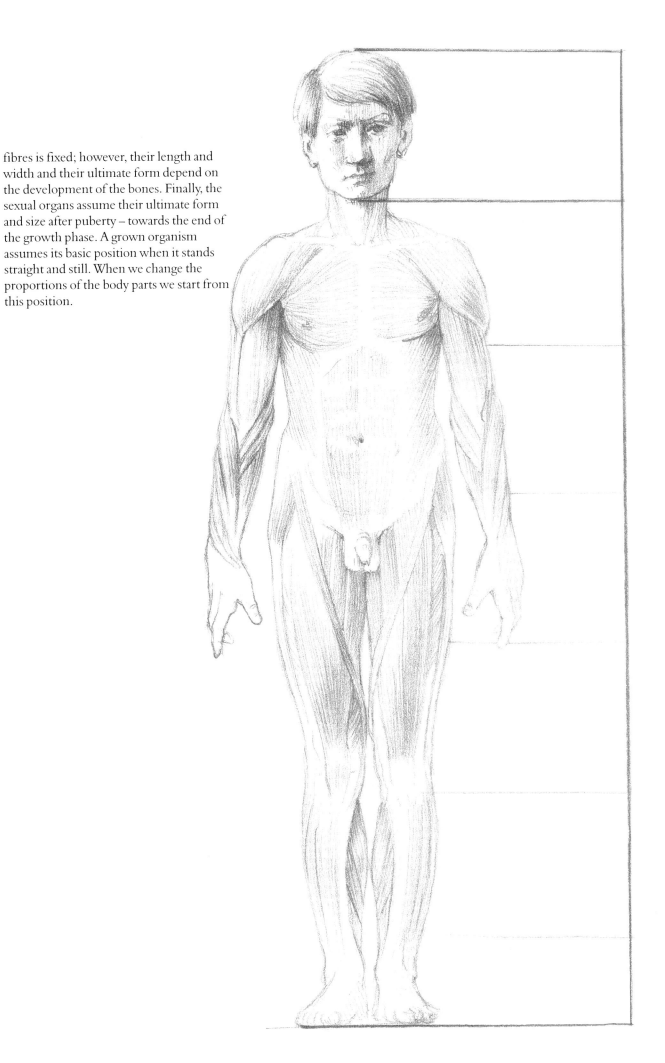

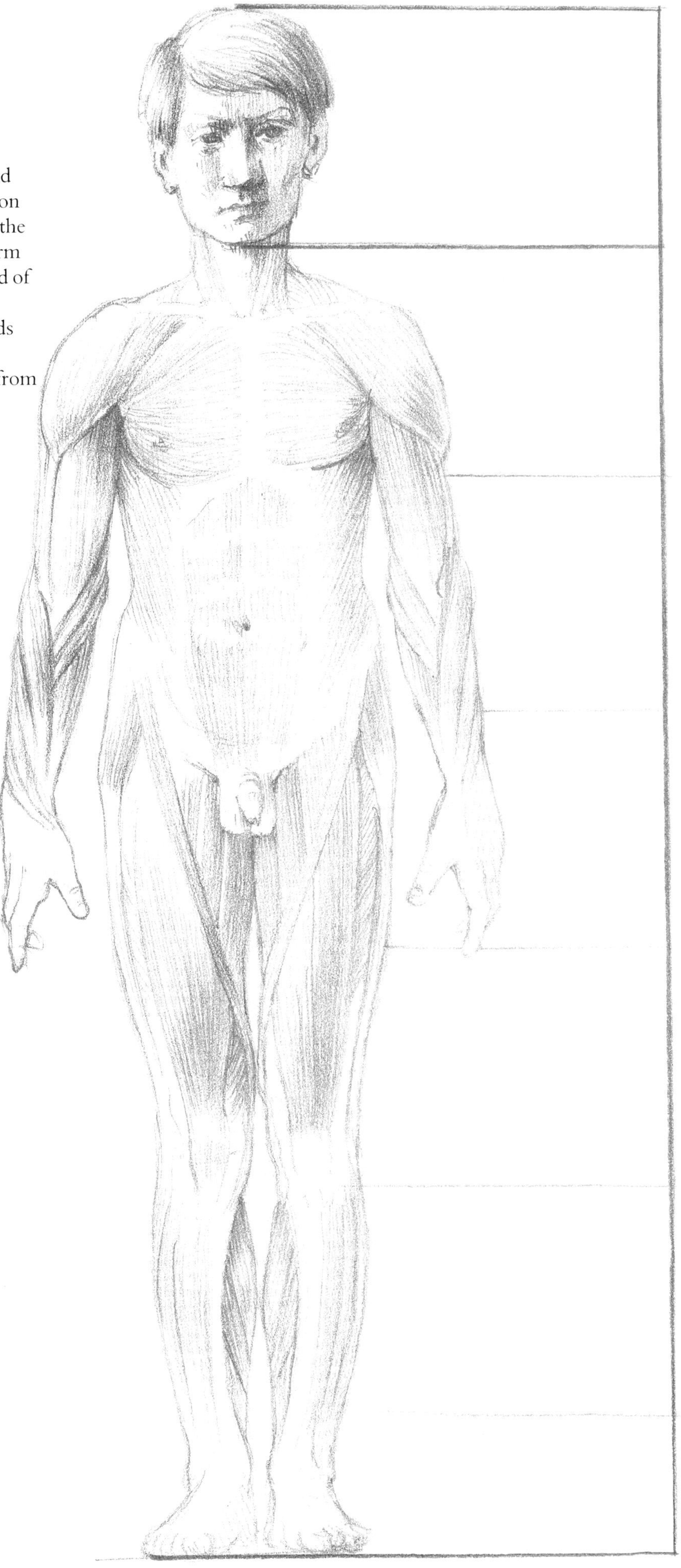

Fig. 229
The Principal Horizontal Axes

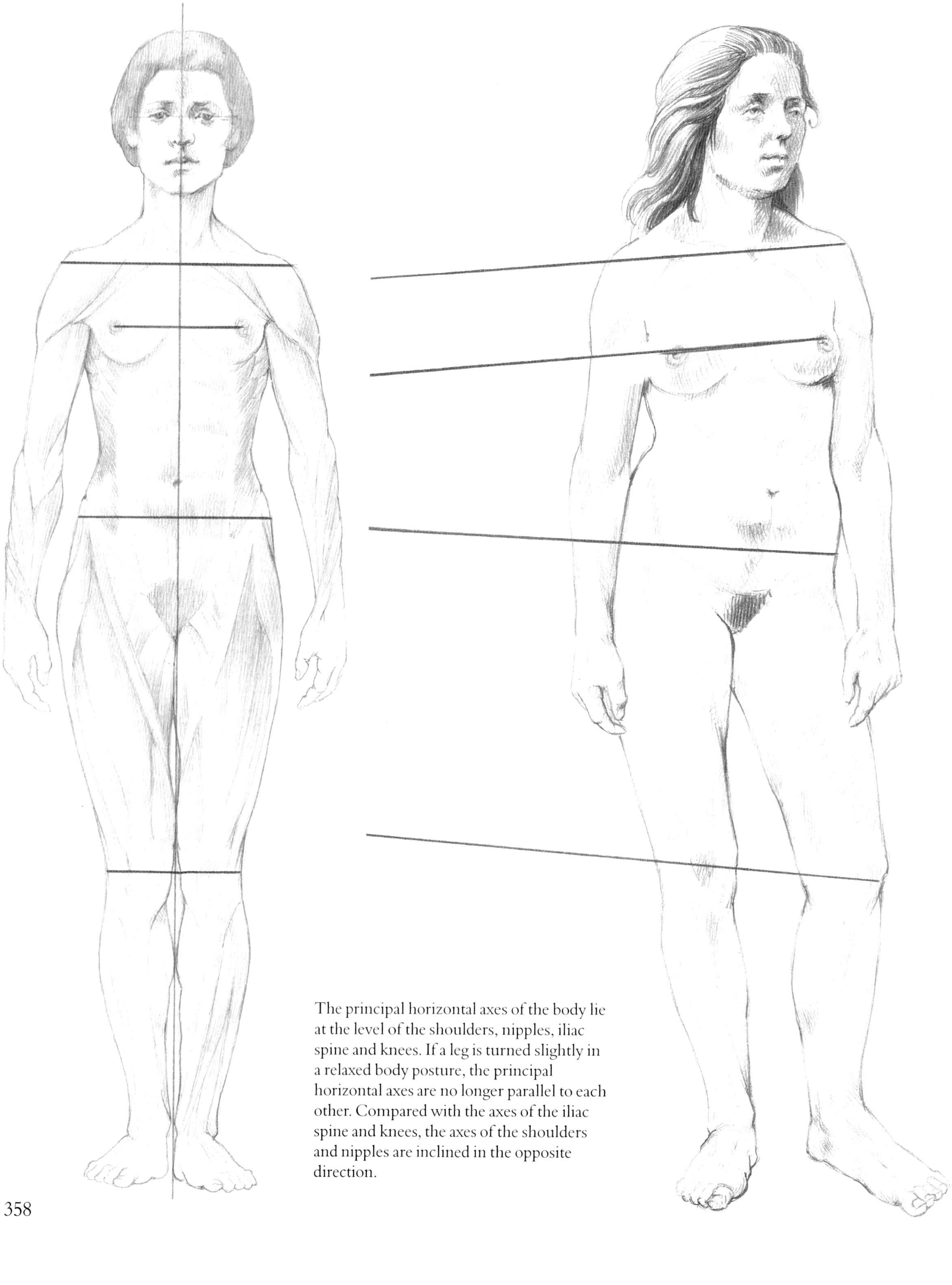

The principal horizontal axes of the body lie at the level of the shoulders, nipples, iliac spine and knees. If a leg is turned slightly in a relaxed body posture, the principal horizontal axes are no longer parallel to each other. Compared with the axes of the iliac spine and knees, the axes of the shoulders and nipples are inclined in the opposite direction.

Fig. 230
Metameres

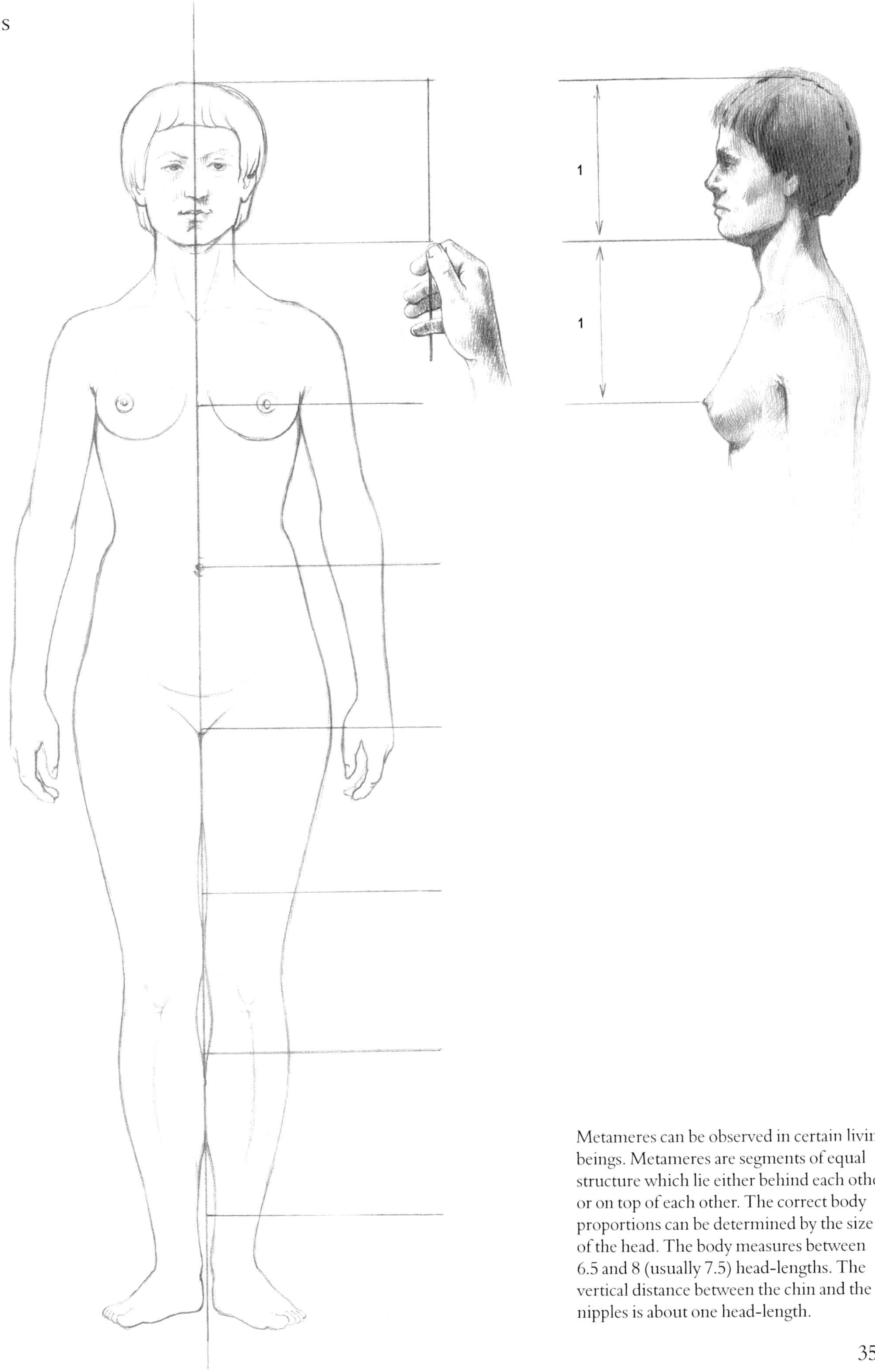

Metameres can be observed in certain living beings. Metameres are segments of equal structure which lie either behind each other or on top of each other. The correct body proportions can be determined by the size of the head. The body measures between 6.5 and 8 (usually 7.5) head-lengths. The vertical distance between the chin and the nipples is about one head-length.

Fig. 231
Gender

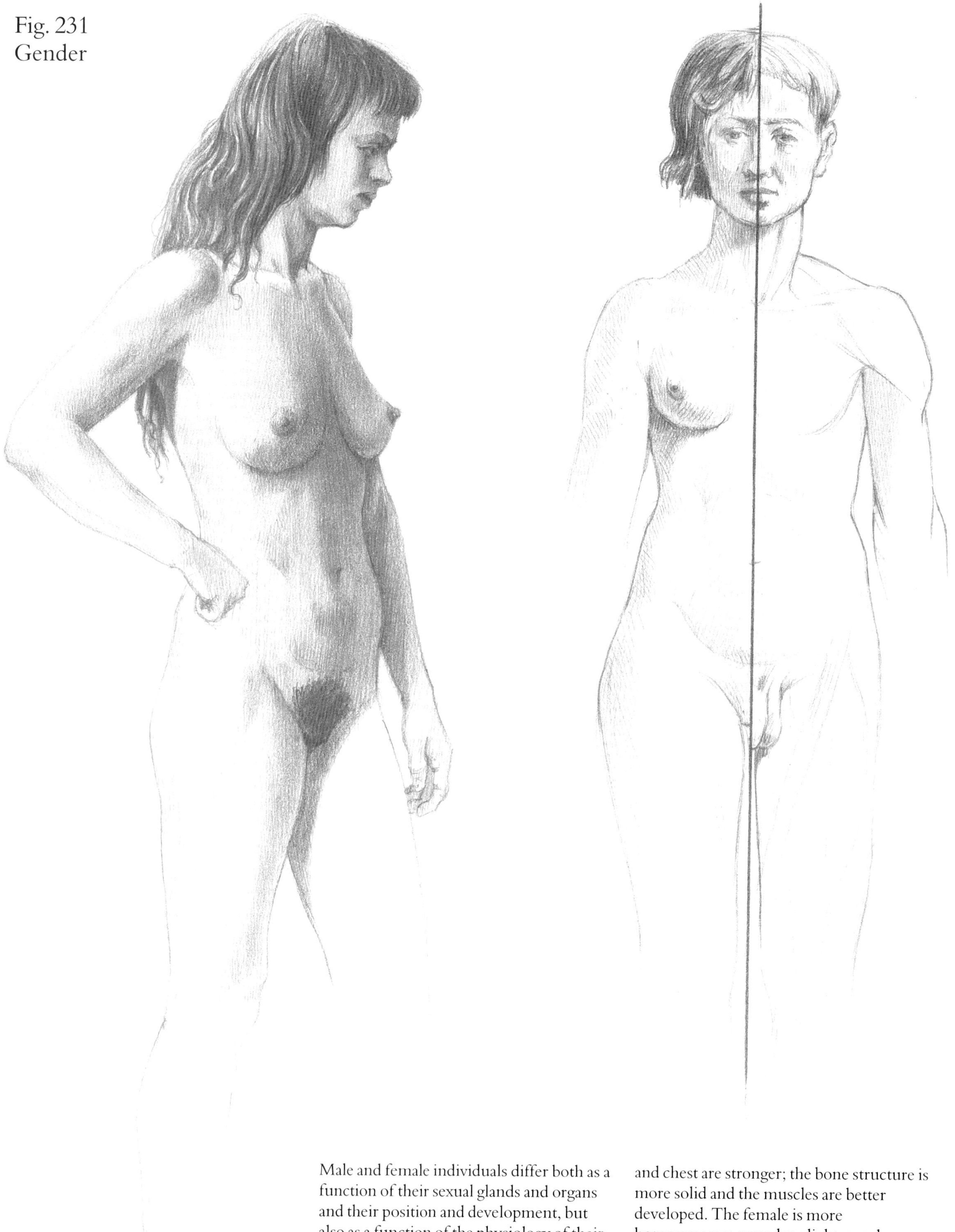

Male and female individuals differ both as a function of their sexual glands and organs and their position and development, but also as a function of the physiology of their behaviour. The male possesses a heavier structure: the body itself is heavier; his head and chest are stronger; the bone structure is more solid and the muscles are better developed. The female is more homogeneous, rounder, slighter and slimmer. The pelvis is broader and lower.

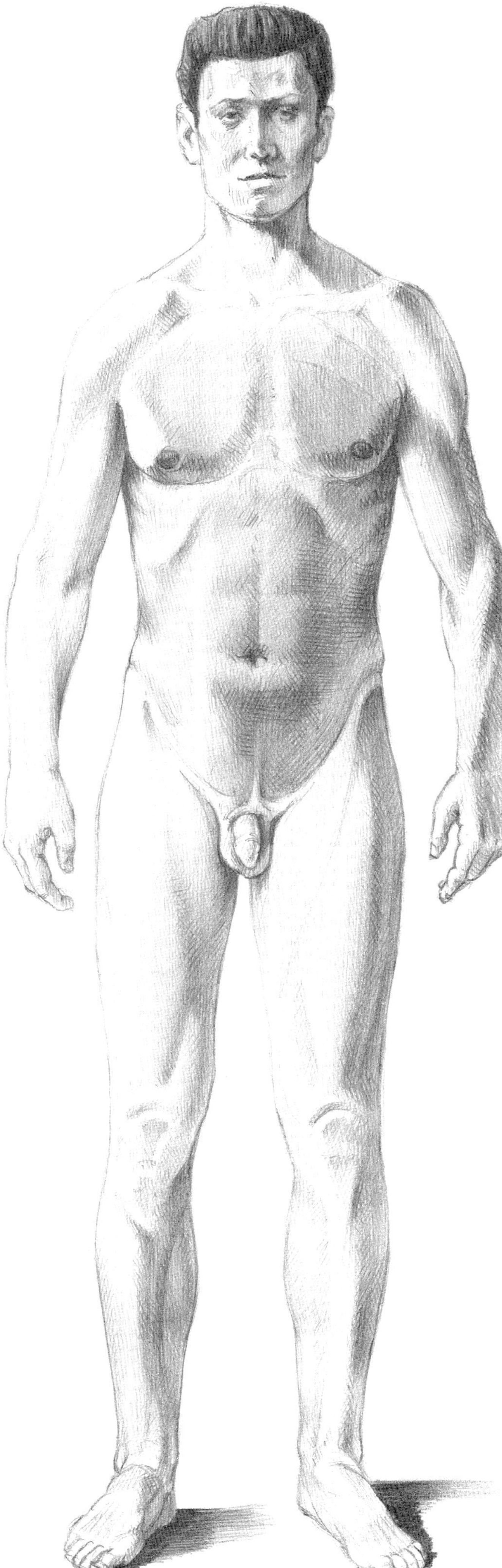

Fig. 232
Constitution

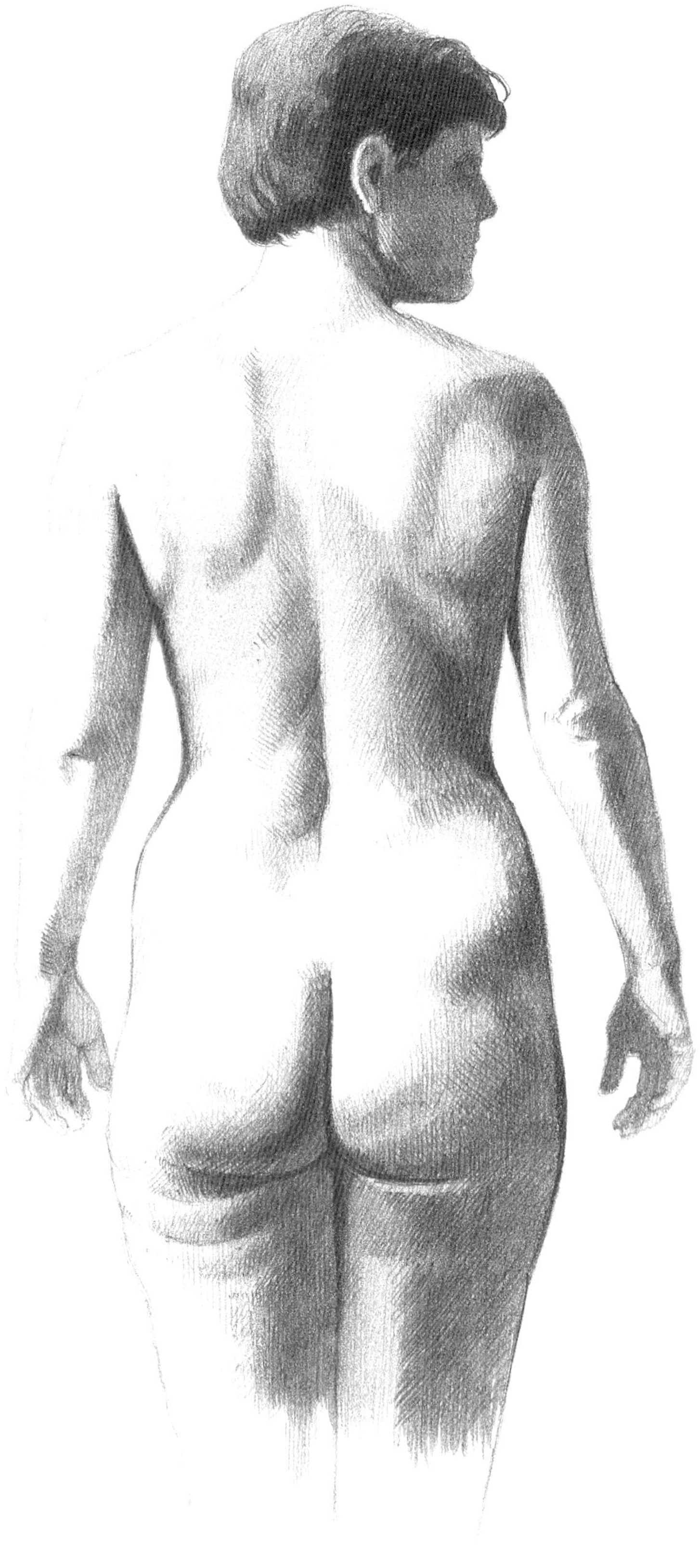

The constitution is the sum of all the morphological and functional characteristics of an individual and thus embraces the body structure and qualities such as temperament, resistance and adaptability. There are three different types of constitution: leptosomic (asthenic), athletic (muscular) and pyknic (stocky).

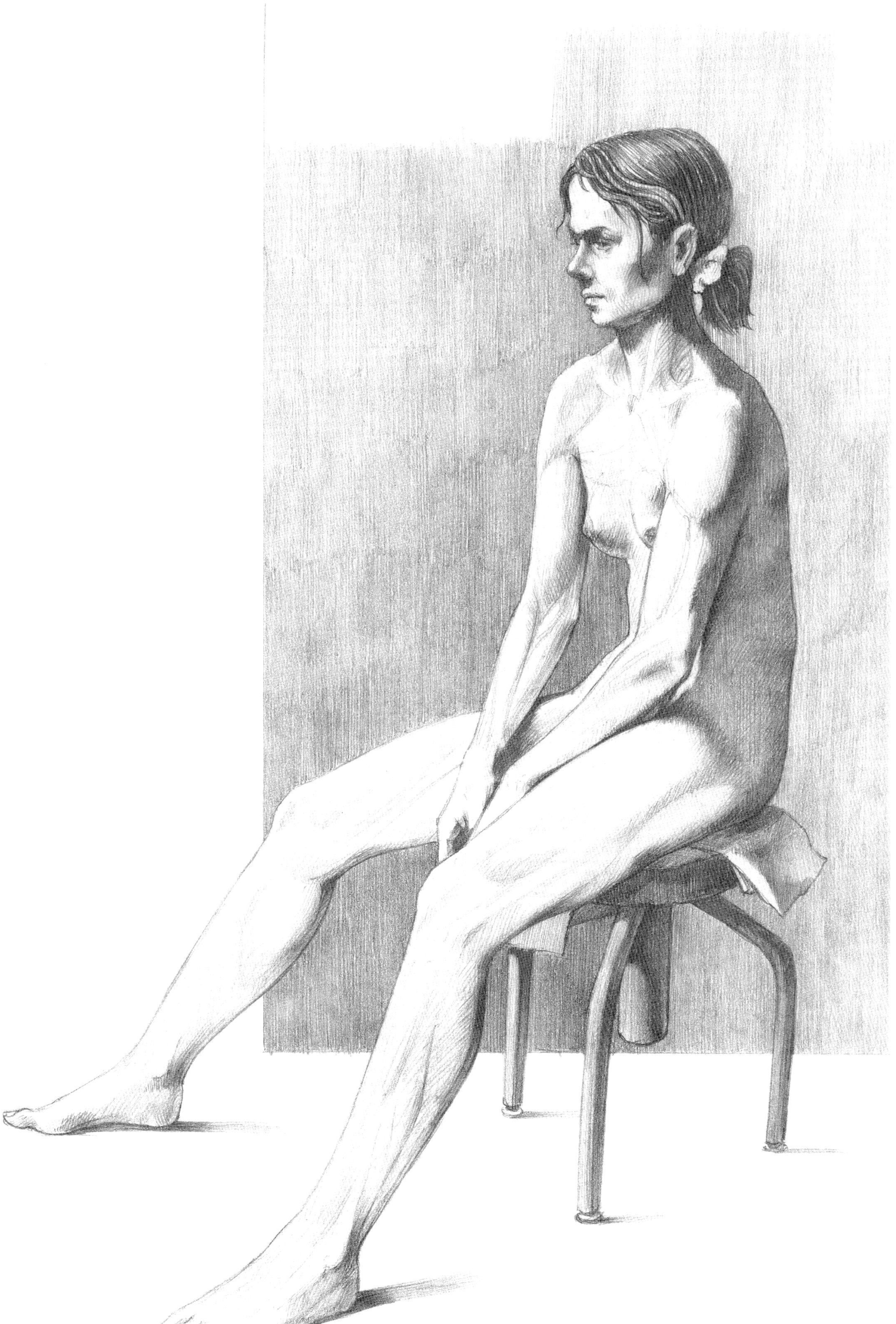

Fig. 233
Body Proportions

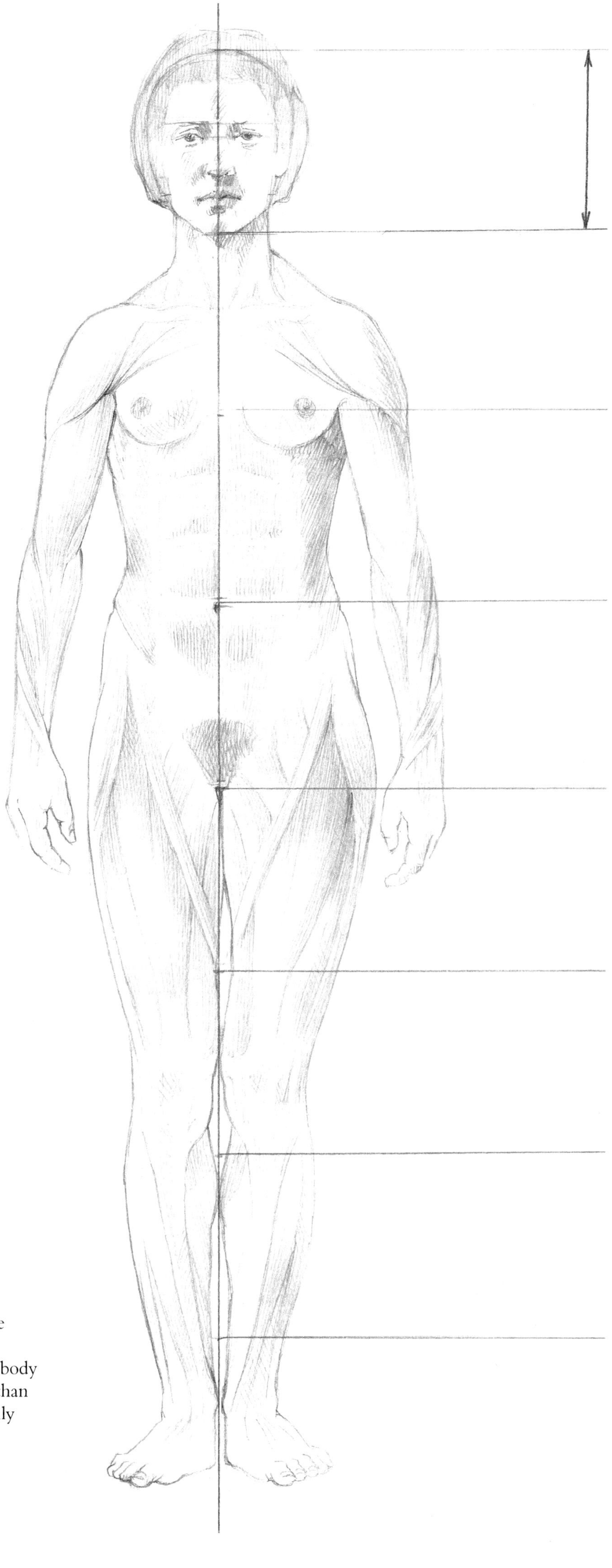

Correct body proportions can be determined by the size of the head. The body measures between 6.5 and 8 (usually 7) head-lengths. In relation to body size, women tend to have larger heads than men. Their body size measures generally between 6 and 7.5 head-lengths.

Fig. 234
Condition

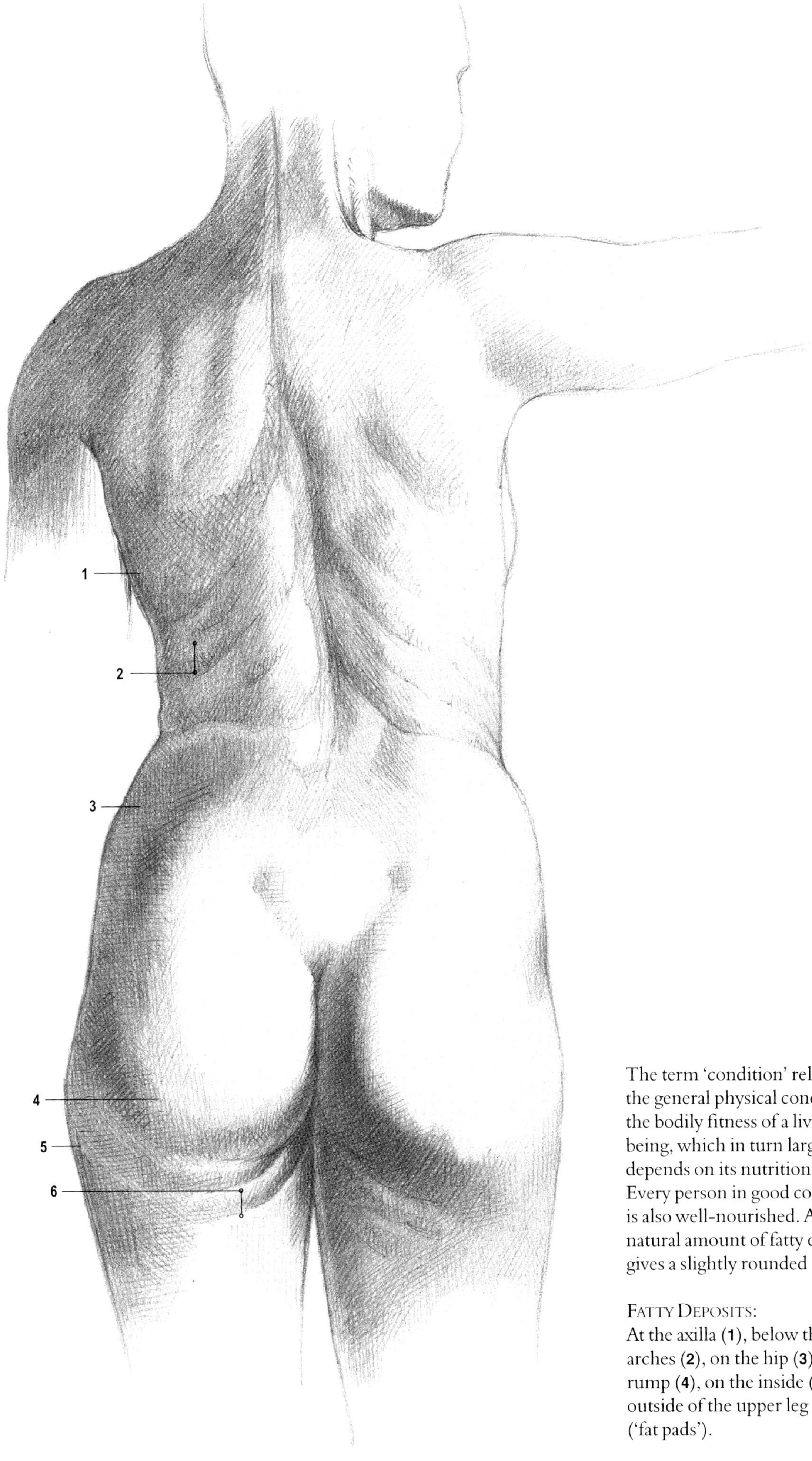

The term 'condition' relates to the general physical condition, the bodily fitness of a living being, which in turn largely depends on its nutritional state. Every person in good condition is also well-nourished. A natural amount of fatty deposits gives a slightly rounded form.

FATTY DEPOSITS:
At the axilla (**1**), below the costal arches (**2**), on the hip (**3**), on the rump (**4**), on the inside (**5**) and outside of the upper leg (**6**) ('fat pads').

Fig. 235
The Eminentiae, Sulci and Fossae of the Thorax

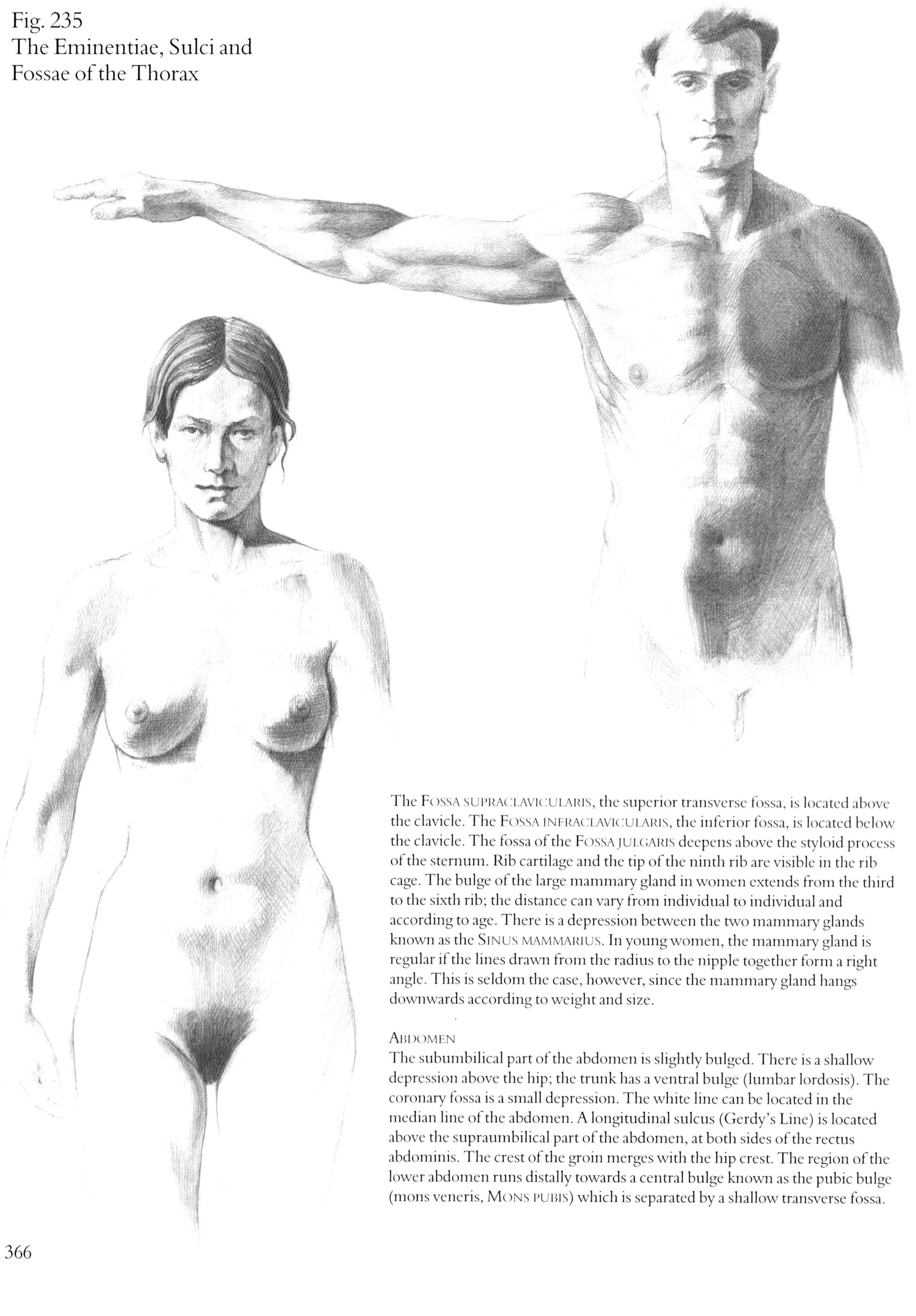

The FOSSA SUPRACLAVICULARIS, the superior transverse fossa, is located above the clavicle. The FOSSA INFRACLAVICULARIS, the inferior fossa, is located below the clavicle. The fossa of the FOSSA JULGARIS deepens above the styloid process of the sternum. Rib cartilage and the tip of the ninth rib are visible in the rib cage. The bulge of the large mammary gland in women extends from the third to the sixth rib; the distance can vary from individual to individual and according to age. There is a depression between the two mammary glands known as the SINUS MAMMARIUS. In young women, the mammary gland is regular if the lines drawn from the radius to the nipple together form a right angle. This is seldom the case, however, since the mammary gland hangs downwards according to weight and size.

ABDOMEN

The subumbilical part of the abdomen is slightly bulged. There is a shallow depression above the hip; the trunk has a ventral bulge (lumbar lordosis). The coronary fossa is a small depression. The white line can be located in the median line of the abdomen. A longitudinal sulcus (Gerdy's Line) is located above the supraumbilical part of the abdomen, at both sides of the rectus abdominis. The crest of the groin merges with the hip crest. The region of the lower abdomen runs distally towards a central bulge known as the pubic bulge (mons veneris, MONS PUBIS) which is separated by a shallow transverse fossa.

Fig. 236
The Eminentiae, Sulci, Fossae of the Dorsal Body Region

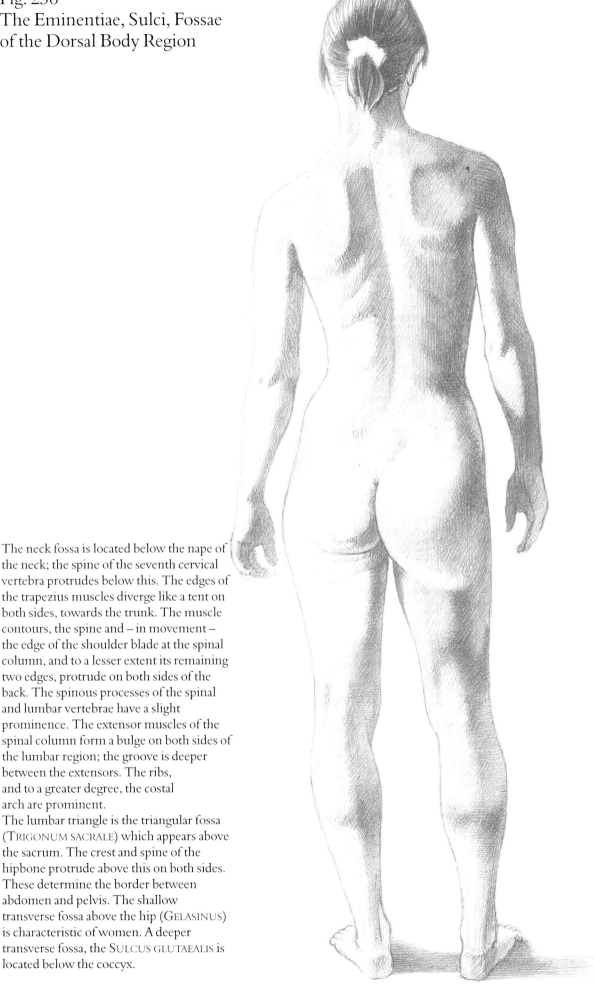

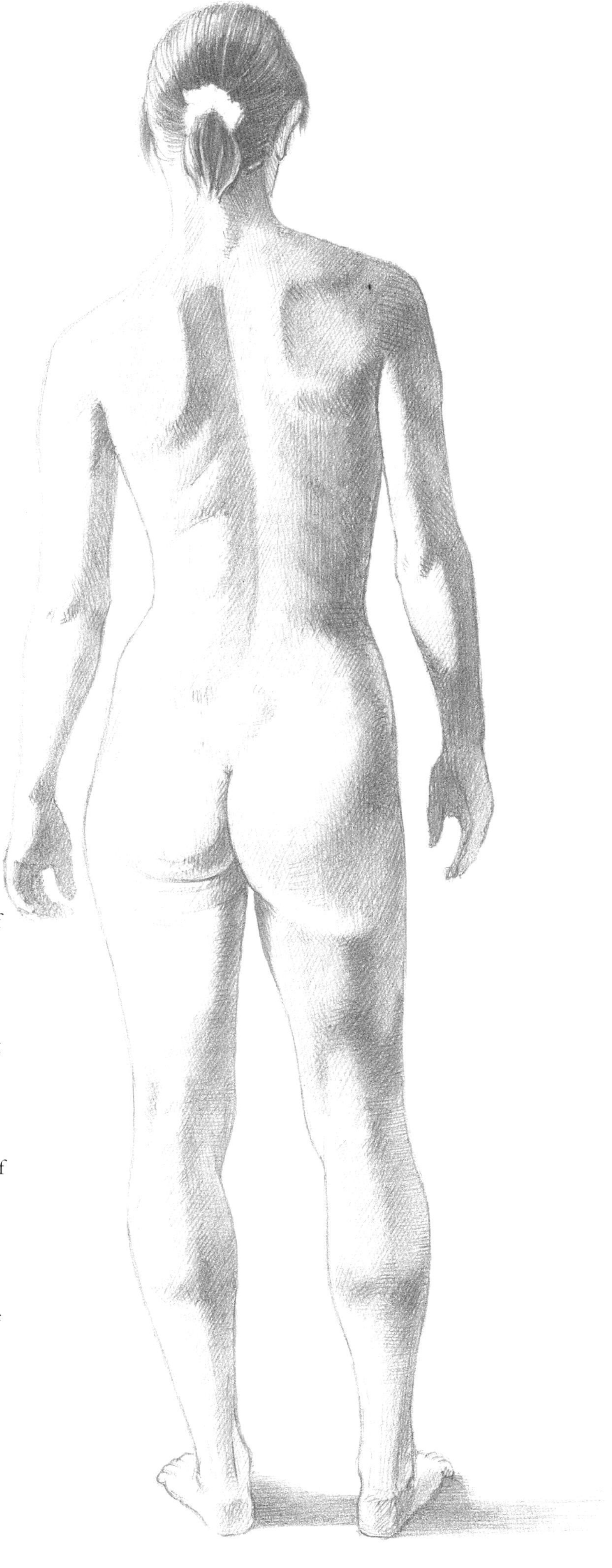

The neck fossa is located below the nape of the neck; the spine of the seventh cervical vertebra protrudes below this. The edges of the trapezius muscles diverge like a tent on both sides, towards the trunk. The muscle contours, the spine and – in movement – the edge of the shoulder blade at the spinal column, and to a lesser extent its remaining two edges, protrude on both sides of the back. The spinous processes of the spinal and lumbar vertebrae have a slight prominence. The extensor muscles of the spinal column form a bulge on both sides of the lumbar region; the groove is deeper between the extensors. The ribs, and to a greater degree, the costal arch are prominent.
The lumbar triangle is the triangular fossa (TRIGONUM SACRALE) which appears above the sacrum. The crest and spine of the hipbone protrude above this on both sides. These determine the border between abdomen and pelvis. The shallow transverse fossa above the hip (GELASINUS) is characteristic of women. A deeper transverse fossa, the SULCUS GLUTAEALIS is located below the coccyx.

THE NECK AND HEAD

Fig. 237
The Neck Muscles

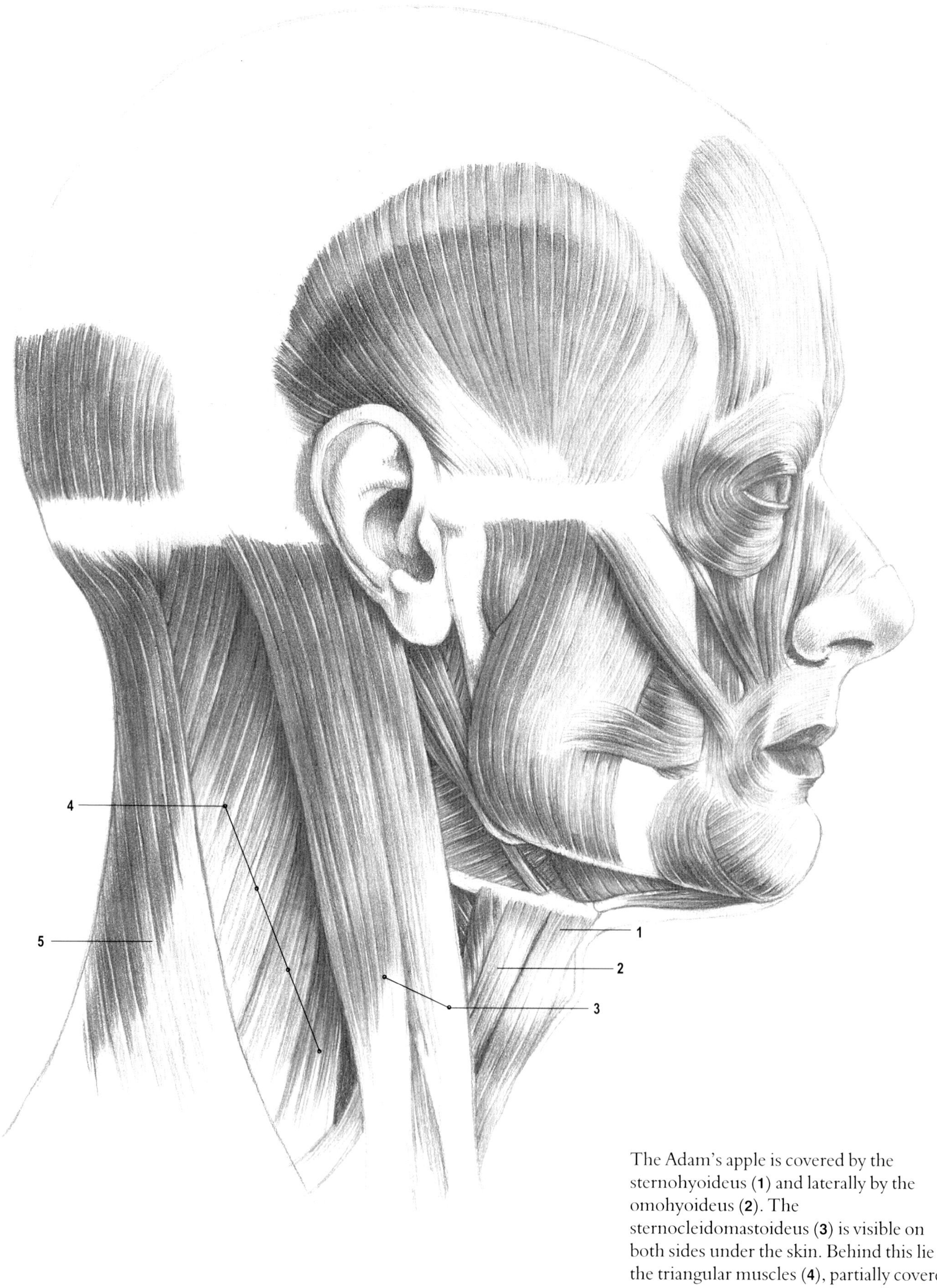

lateral aspect

The Adam's apple is covered by the sternohyoideus (**1**) and laterally by the omohyoideus (**2**). The sternocleidomastoideus (**3**) is visible on both sides under the skin. Behind this lie the triangular muscles (**4**), partially covered by the trapezius (**5**), which determines the posterior contours of the neck.

1 Sternocleidomastoideus (*11*)
2 Sternohyoideus (*12*)
3 Trapezius (*20*)
4 Levator scapulae (*24*)
5 Scalenus muscles (*8–10*)

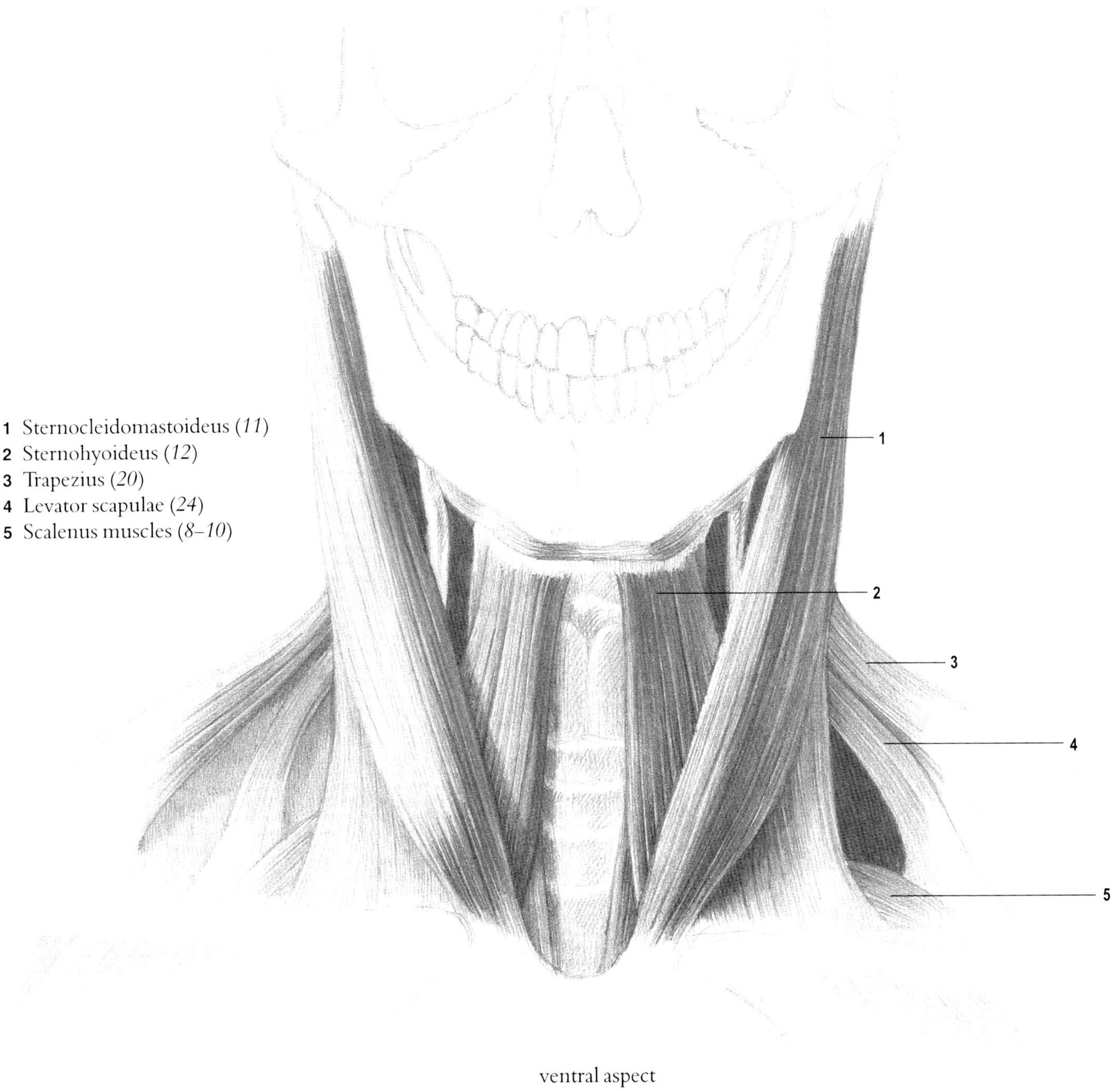

ventral aspect

Fig. 238
The Neck Muscles

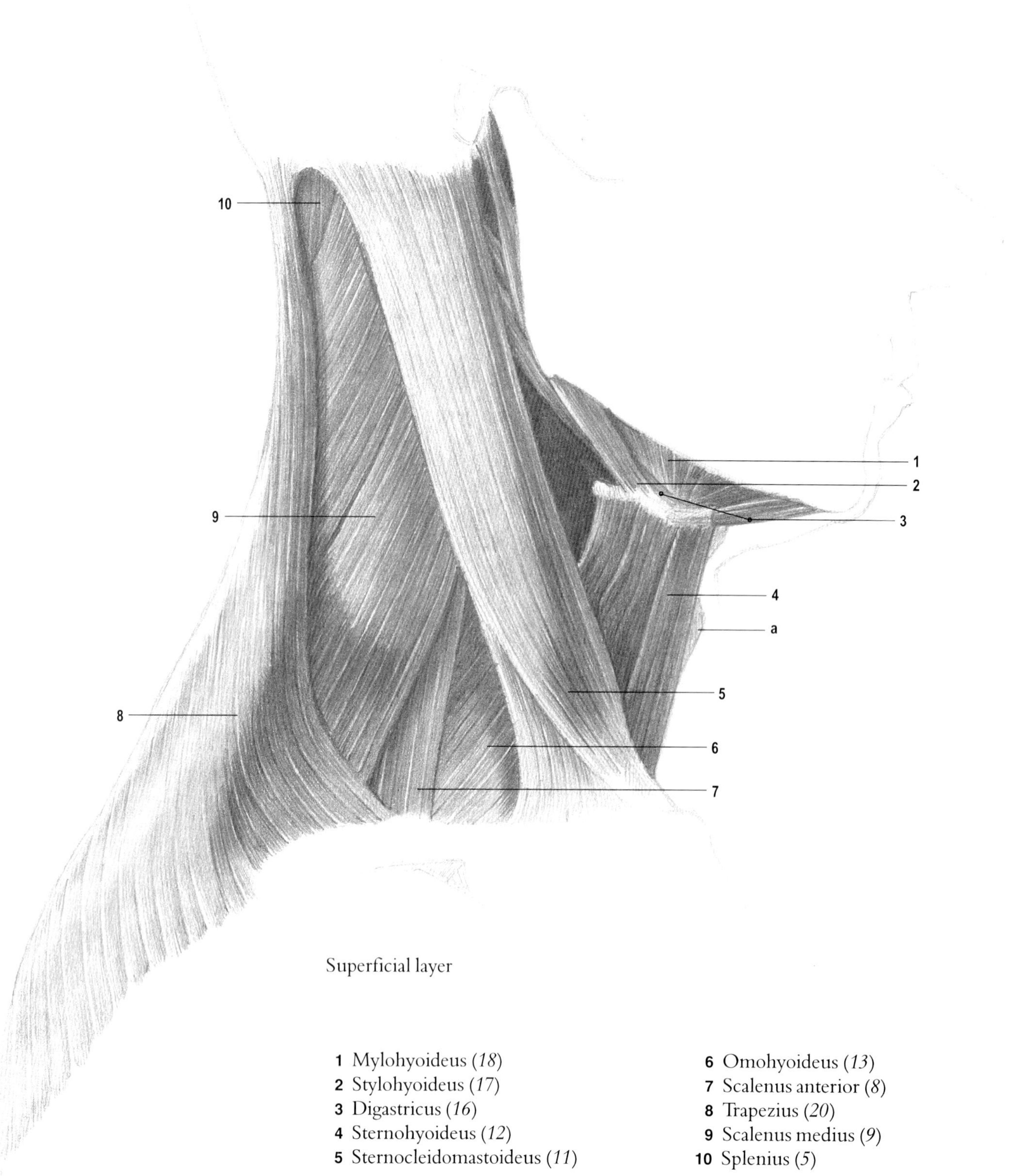

Superficial layer

1 Mylohyoideus (*18*)
2 Stylohyoideus (*17*)
3 Digastricus (*16*)
4 Sternohyoideus (*12*)
5 Sternocleidomastoideus (*11*)
6 Omohyoideus (*13*)
7 Scalenus anterior (*8*)
8 Trapezius (*20*)
9 Scalenus medius (*9*)
10 Splenius (*5*)

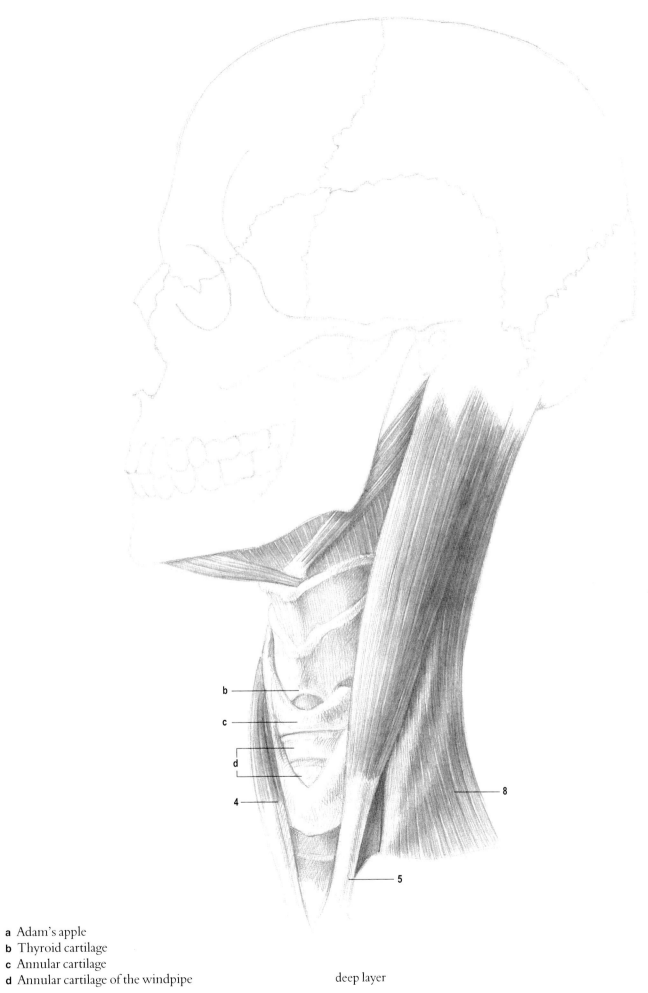

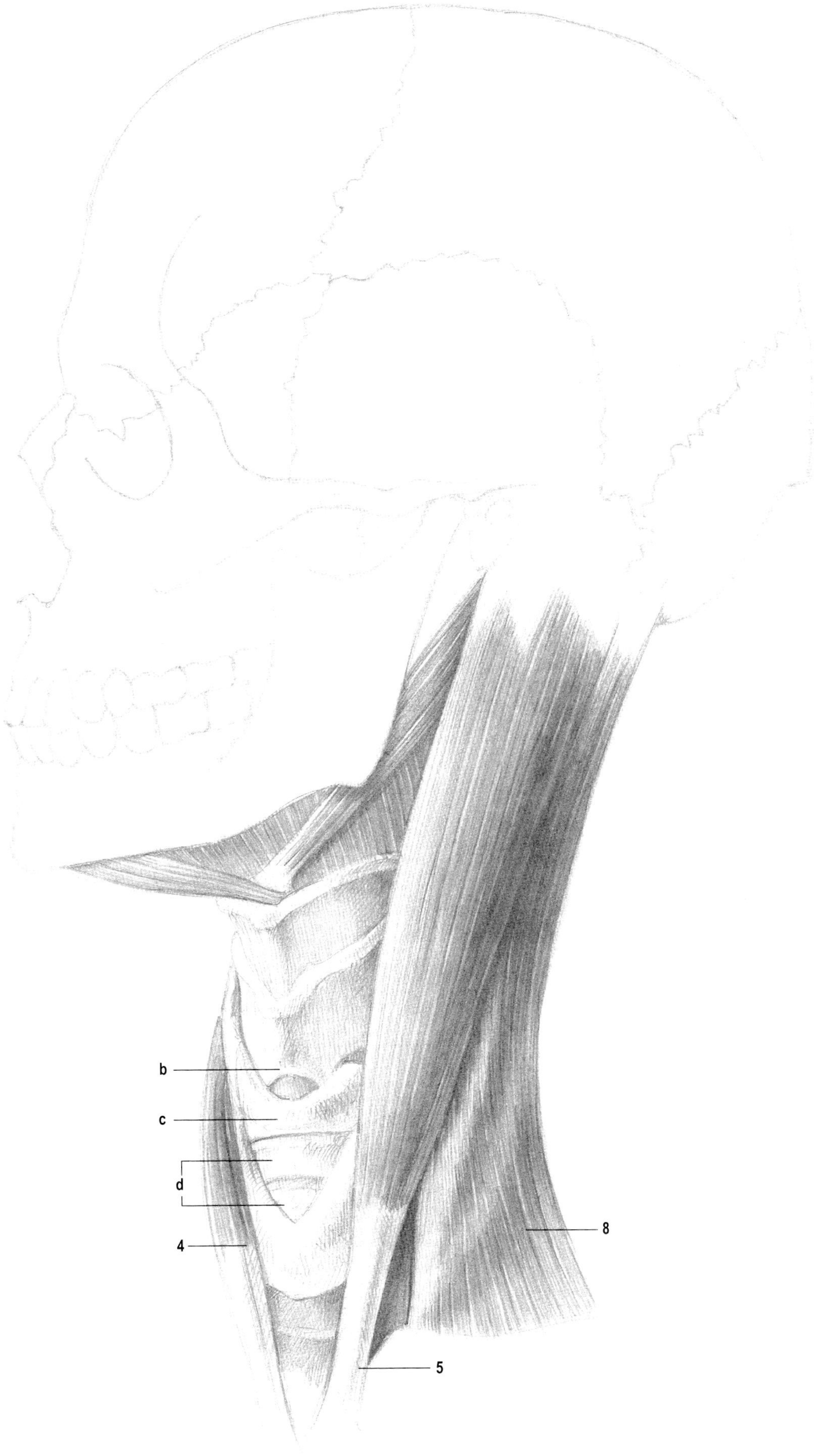

a Adam's apple
b Thyroid cartilage
c Annular cartilage
d Annular cartilage of the windpipe

deep layer

Fig. 239
Principal Muscles of the Neck and Head

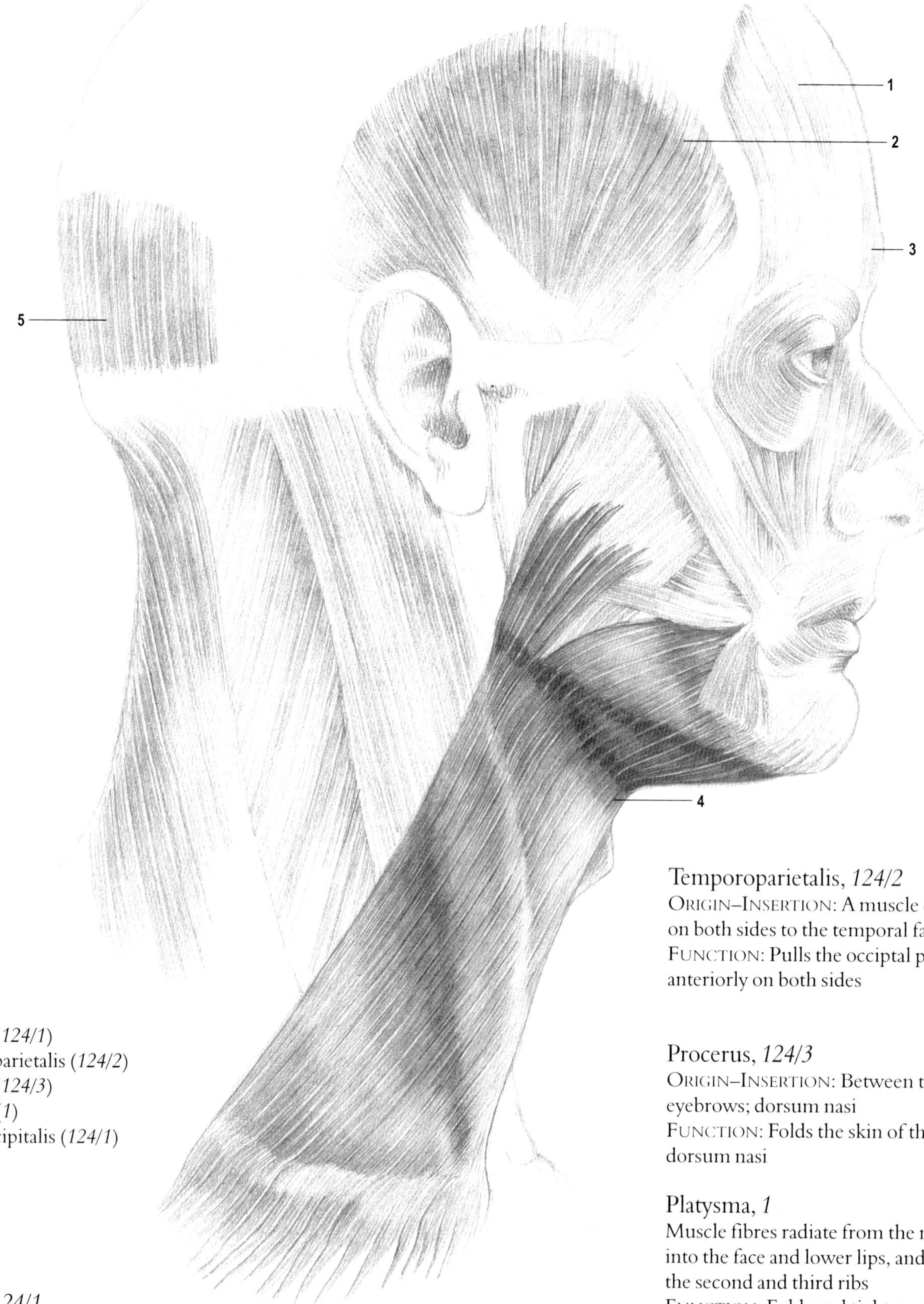

1 Frontalis (*124/1*)
2 Temporoparietalis (*124/2*)
3 Procerus (*124/3*)
4 Platysma (*1*)
5 Venter occipitalis (*124/1*)

Frontalis, *124/1*
ORIGIN–INSERTION: From the orbital arch running to the parietal bone
FUNCTION: Pulls the occiptal protuberance anteriorly and furrows the brow

Temporoparietalis, *124/2*
ORIGIN–INSERTION: A muscle cover runs on both sides to the temporal fasciae
FUNCTION: Pulls the occiptal protuberance anteriorly on both sides

Procerus, *124/3*
ORIGIN–INSERTION: Between the eyebrows; dorsum nasi
FUNCTION: Folds the skin of the dorsum nasi

Platysma, *1*
Muscle fibres radiate from the neck region into the face and lower lips, and extend to the second and third ribs
FUNCTION: Folds and tightens the skin of the neck

Fig. 240
The Neck Muscles

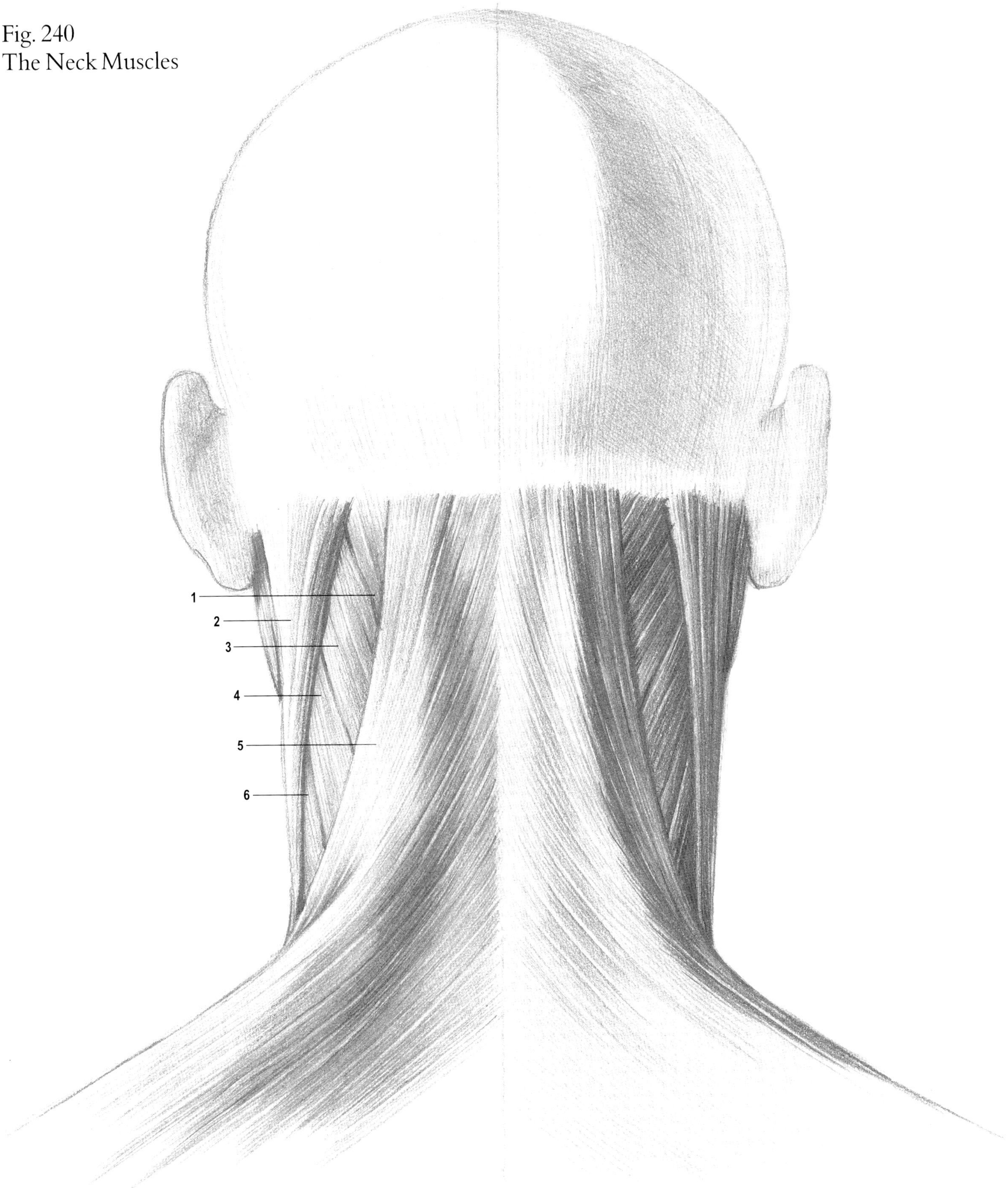

1 Splenius (*5*)
2 Sternocleidomastoideus (*11*)
3 Levator scapulae (*24*)
4 Scalenus medius (*9*)
5 Trapezius (*20*)
6 Scalenus anterior (8)

dorsal aspect

Fig. 241
The Muscles of the Head and Skull

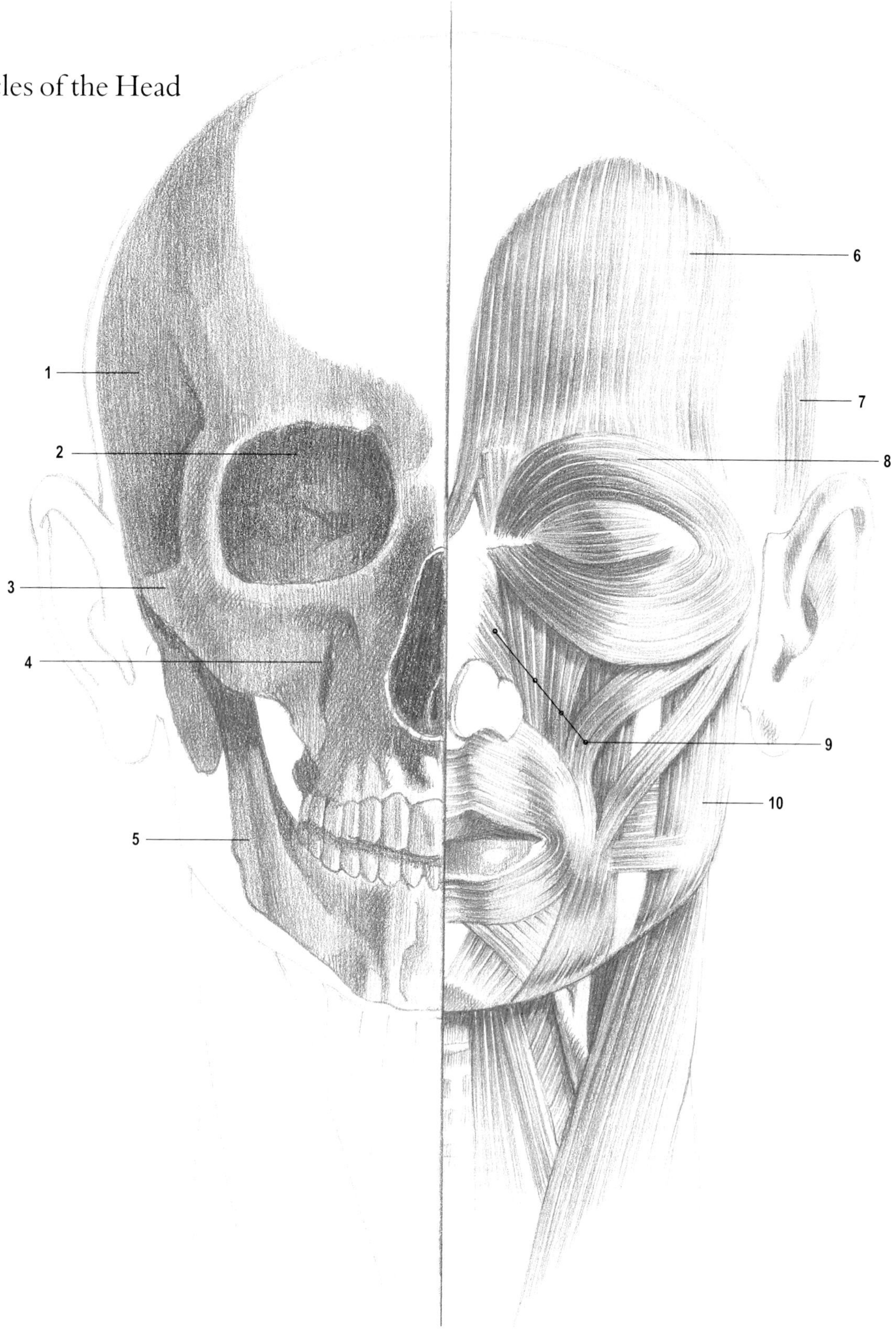

Viewed in the ventral aspect, the form of the head is determined by the forehead, temples and parietal bone (**1**), and by the form of the eye cavities (**2**), zygomatic bone (**3**), nose, upper jaw (**4**) and lower jaw (**5**). The os frontale is covered by the thin frontalis muscle (**6**) and the temporal bone by the temporalis muscle (**7**). The eye socket is surrounded by the orbicularis oculi (**8**). The muscles of the nose and lips (**9**) determine to a large extent facial expression. Only the body and branch of the lower jaw are covered by the masseter (**10**).

Fig. 242
The Occipitofrontalis, *124/1*
(M. occipito-frontalis, venter occipitalis, 124/1)

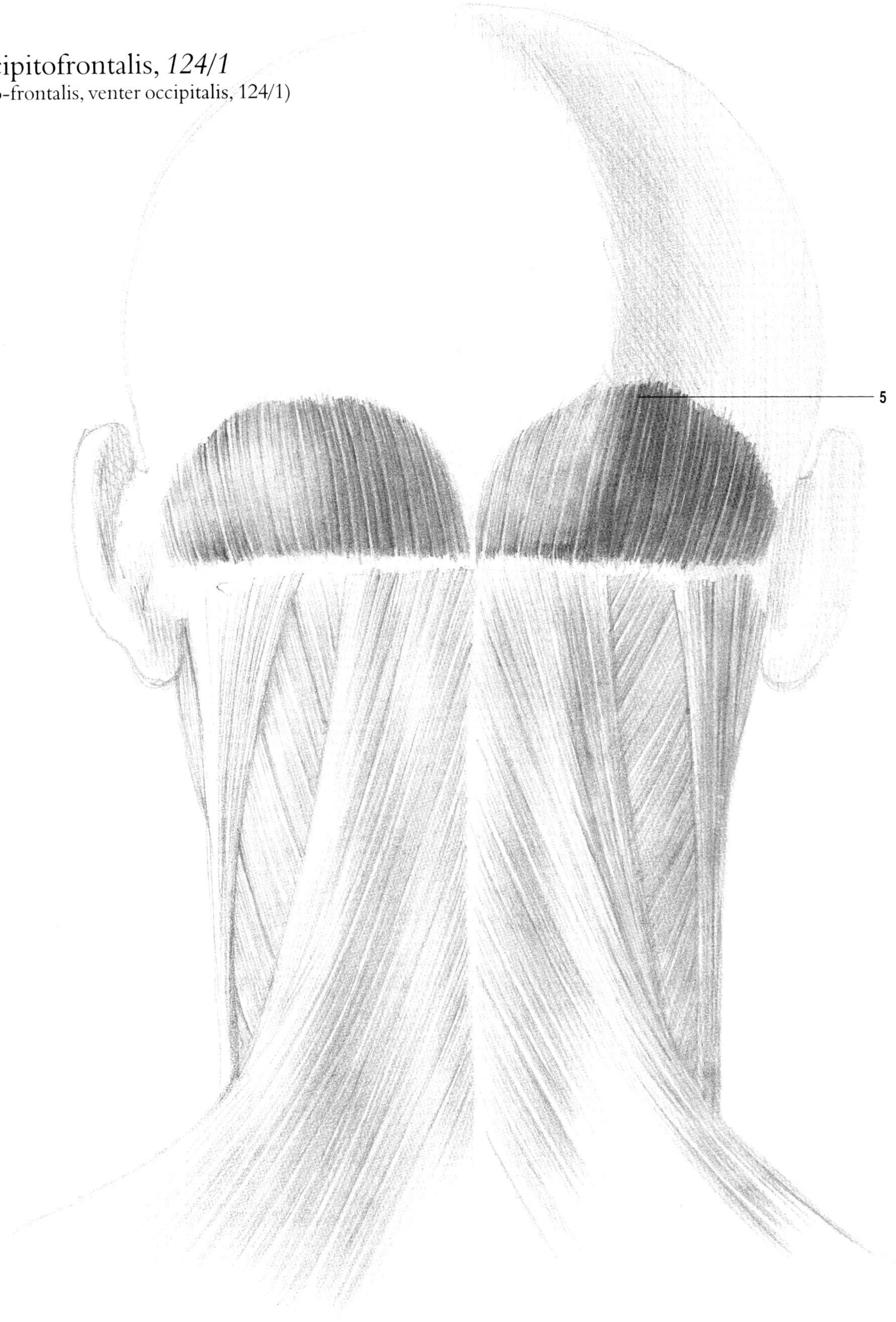

Origin:
Linear nuchae suprema

Insertion :
Occiptal protuberance

Function :
Pulls the occiptal protuberance posteriorly

Fig. 243
The Sternocleidomastoideus, *11*

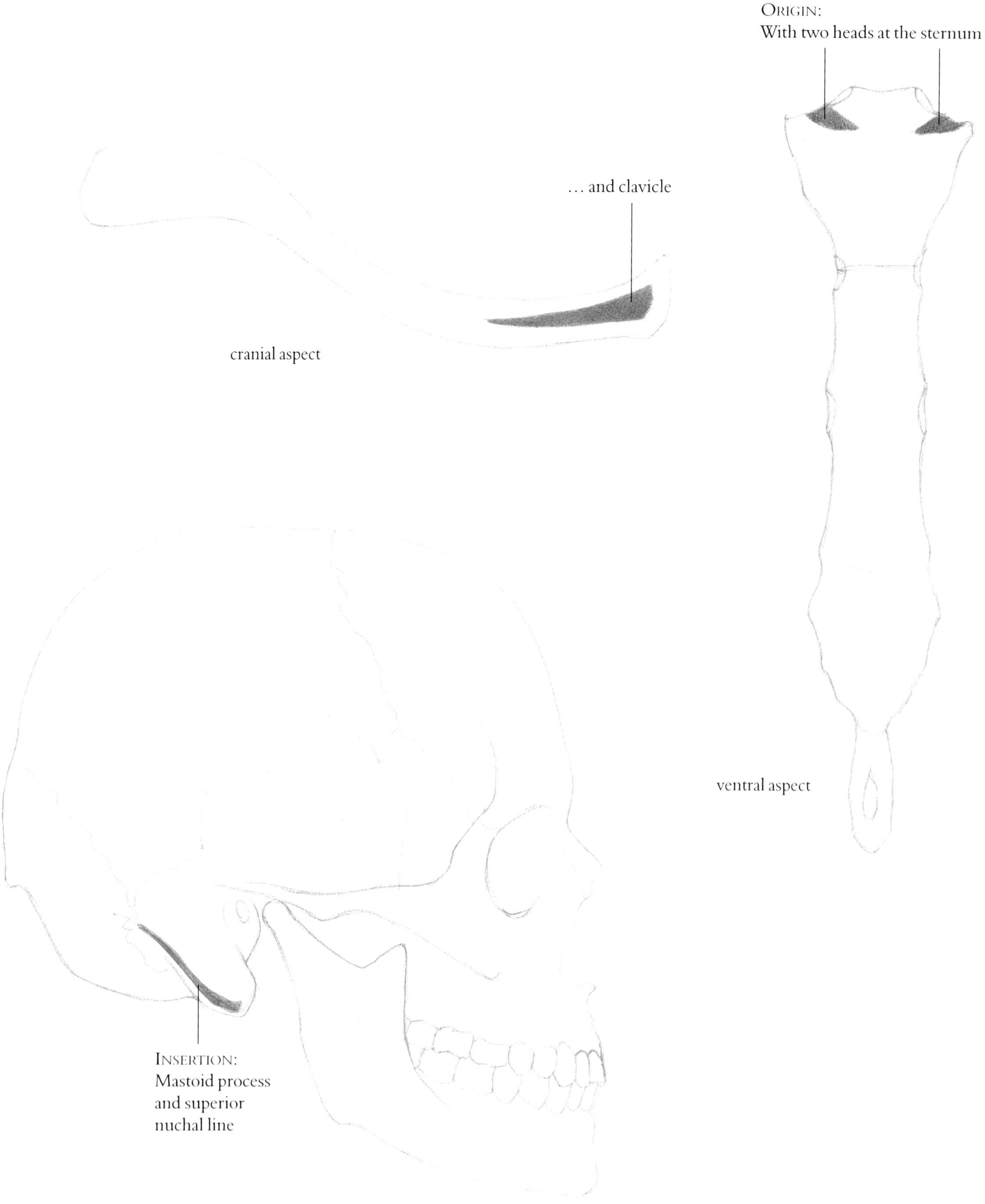

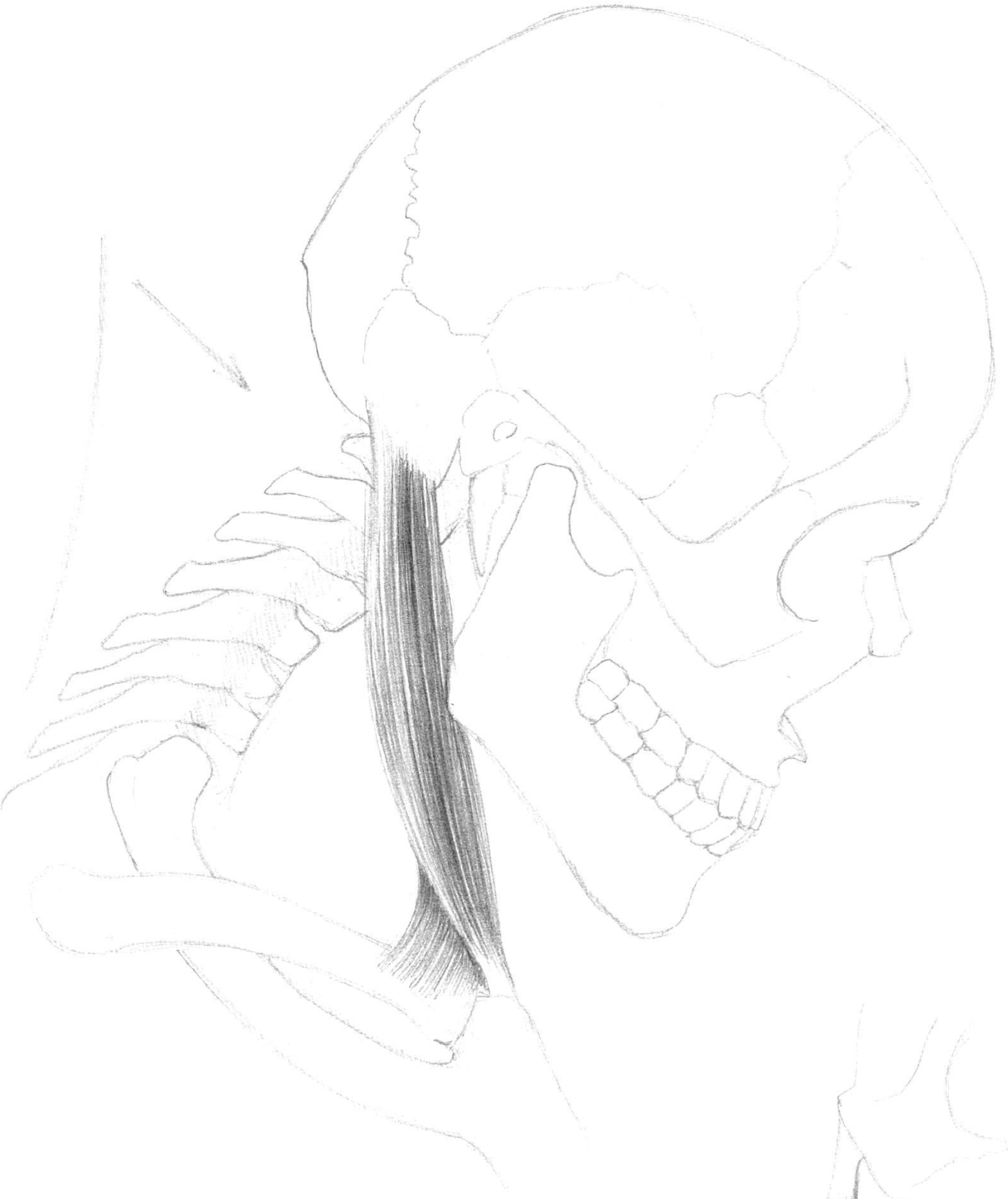

lateral aspect

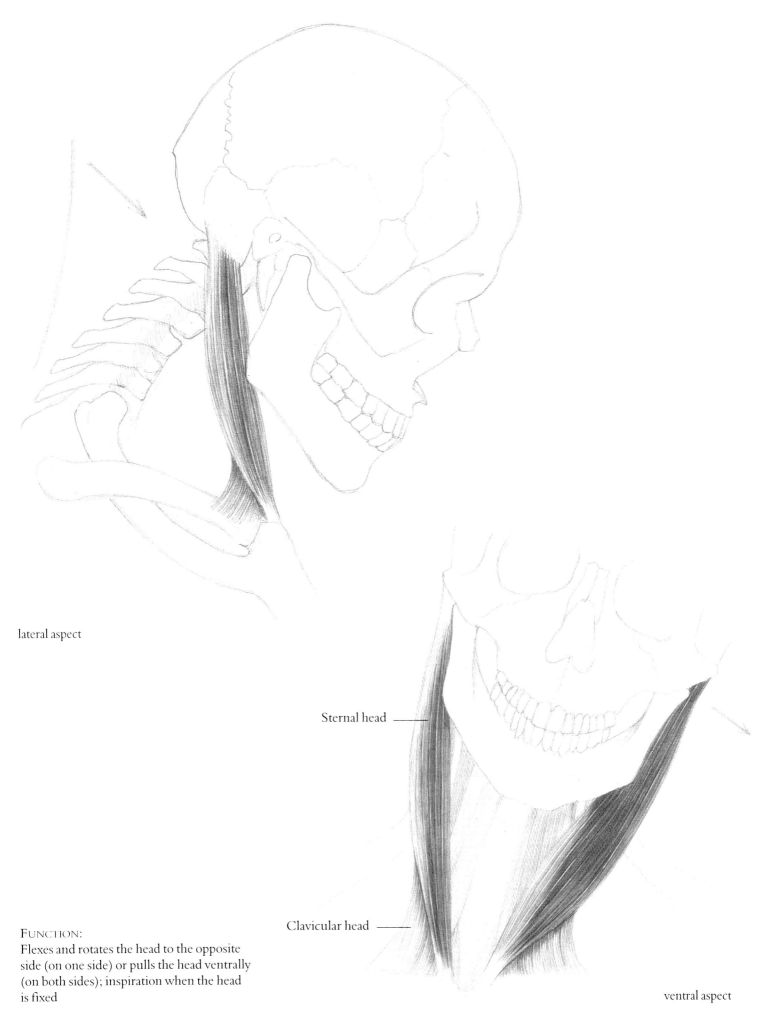

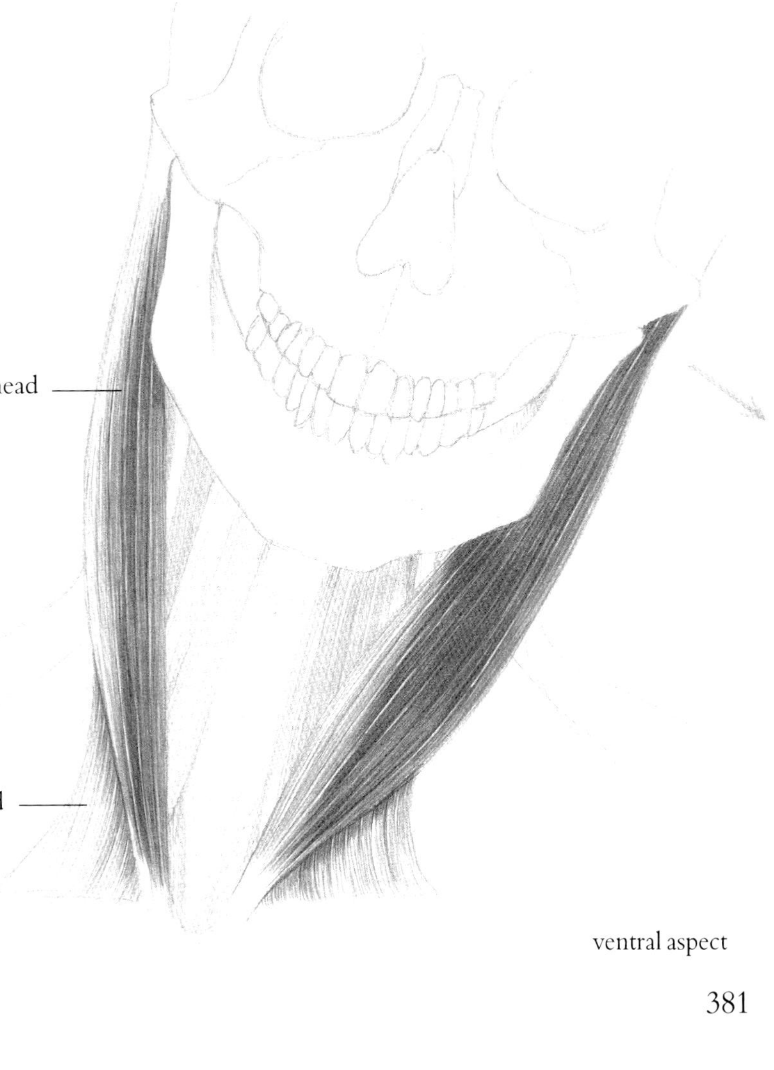

ventral aspect

FUNCTION:
Flexes and rotates the head to the opposite side (on one side) or pulls the head ventrally (on both sides); inspiration when the head is fixed

The Sternocleidomastoideus
(cont.)

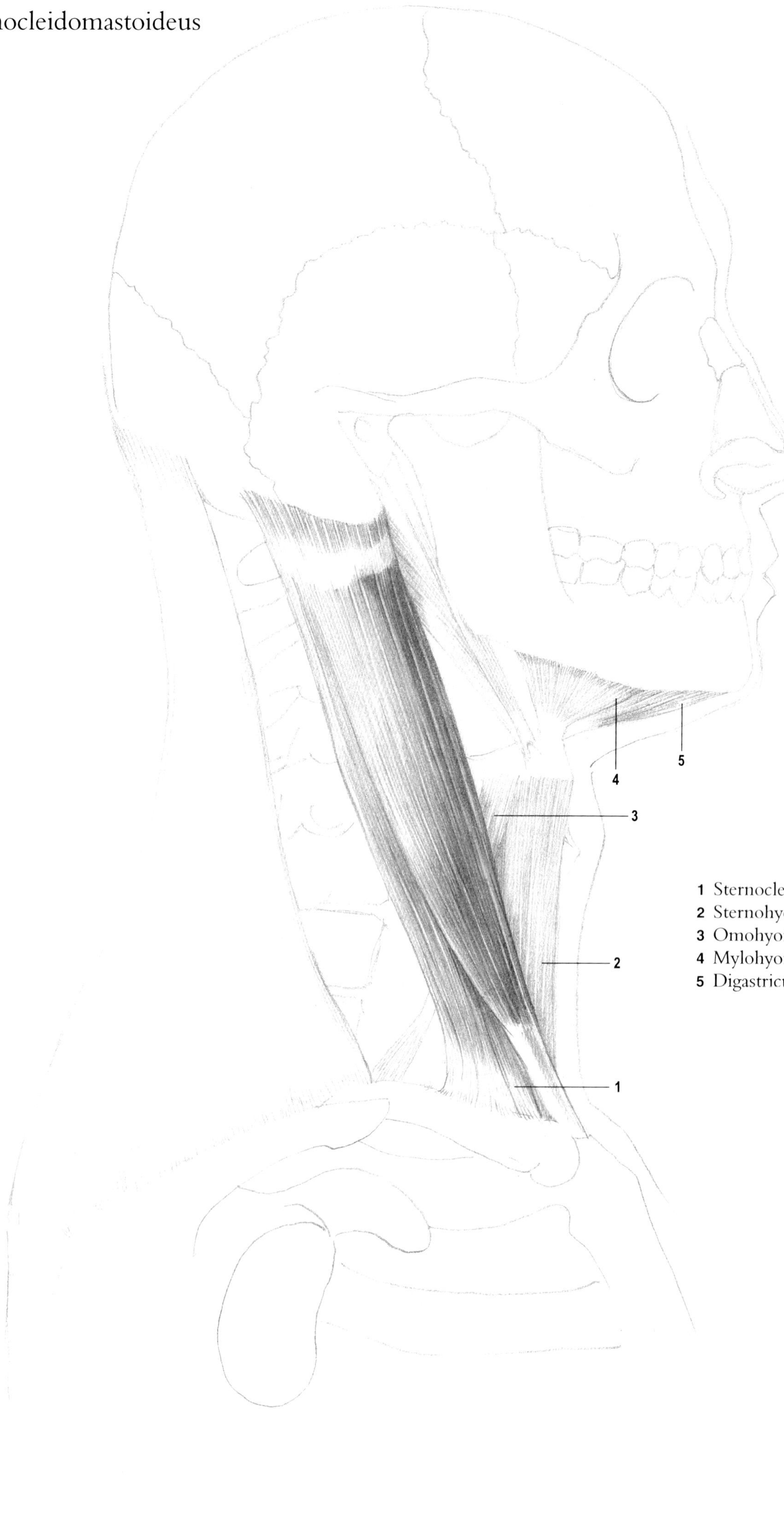

1 Sternocleidomastoideus (*11*)
2 Sternohyoideus (*12*)
3 Omohyoideus (*13*)
4 Mylohyoideus (*18*)
5 Digastricus (*16*)

Fig. 244
The Sternothyreoideus, *14*

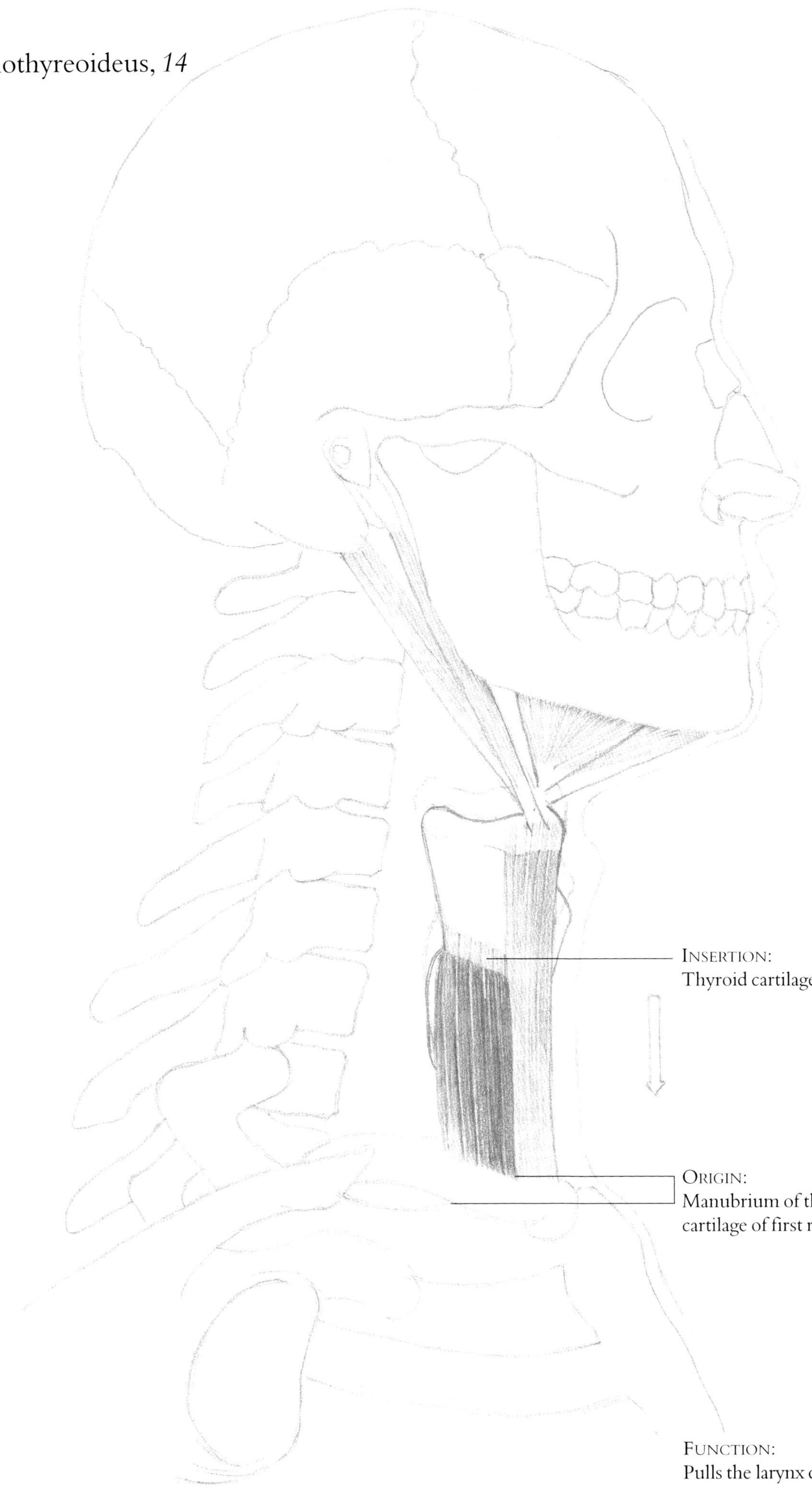

Fig. 245
The Thyreohyoideus, *15*

lateral aspect

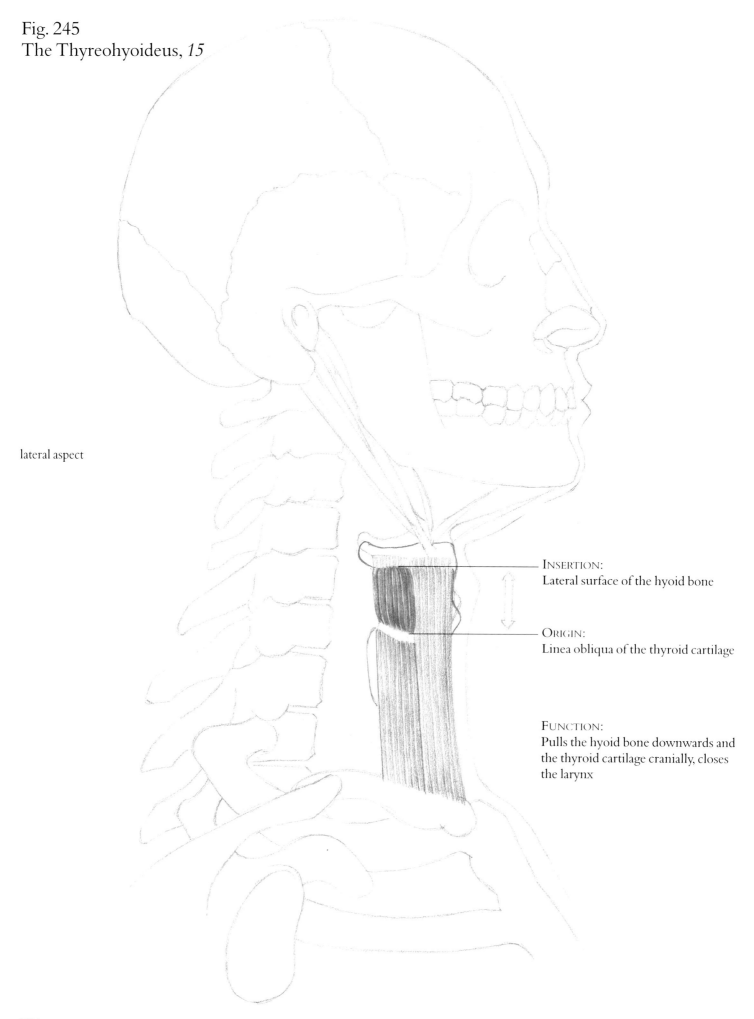

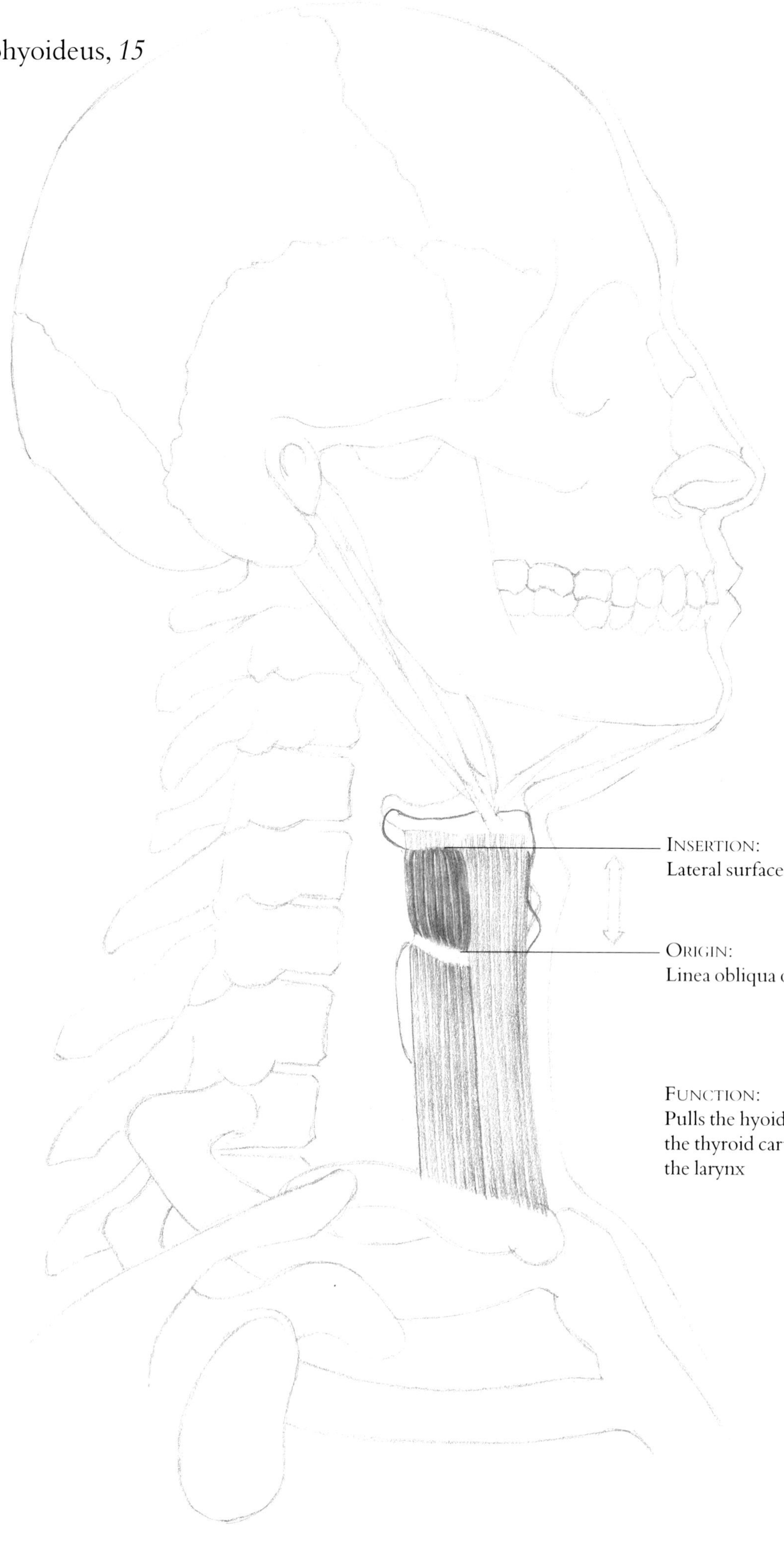

INSERTION:
Lateral surface of the hyoid bone
ORIGIN:
Linea obliqua of the thyroid cartilage
FUNCTION:
Pulls the hyoid bone downwards and the thyroid cartilage cranially, closes the larynx

Fig. 246
The Omohyoideus, *13*

lateral aspect

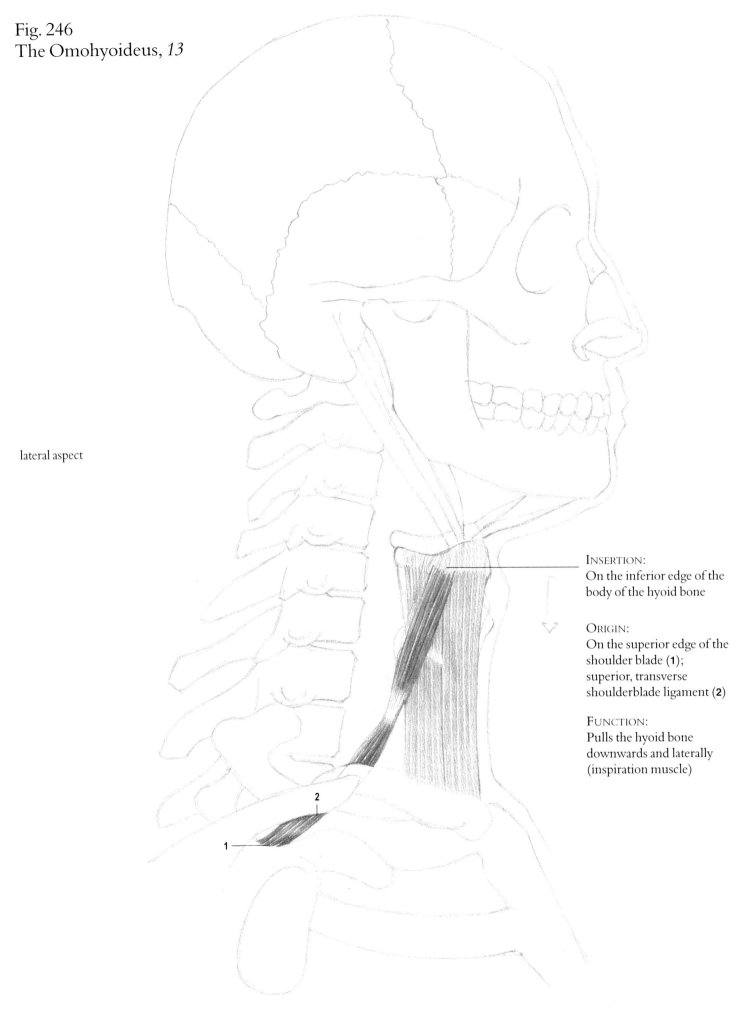

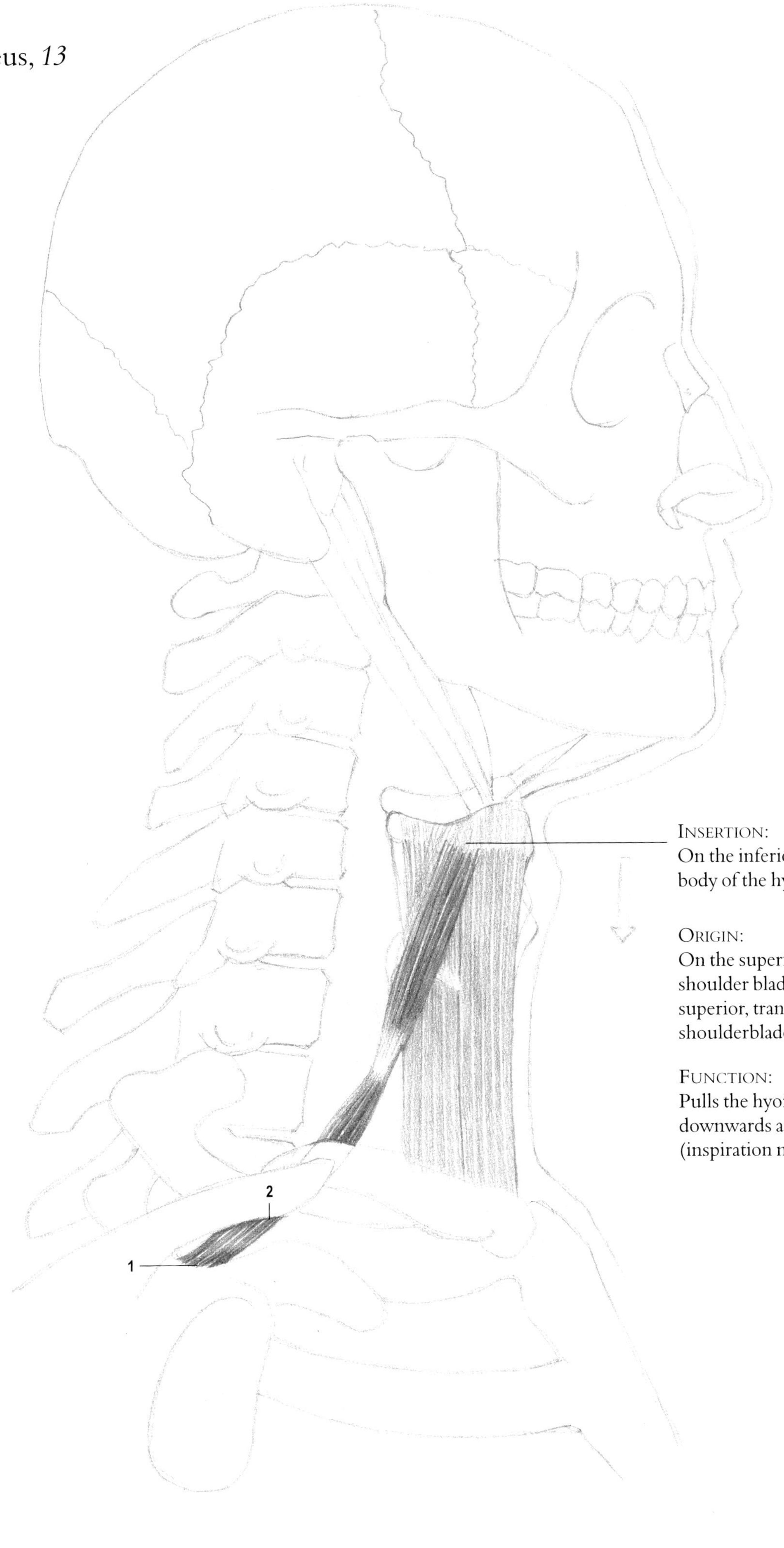

INSERTION:
On the inferior edge of the body of the hyoid bone

ORIGIN:
On the superior edge of the shoulder blade (**1**); superior, transverse shoulderblade ligament (**2**)

FUNCTION:
Pulls the hyoid bone downwards and laterally (inspiration muscle)

Fig. 247
The Sternohyoideus, *12*

lateral aspect

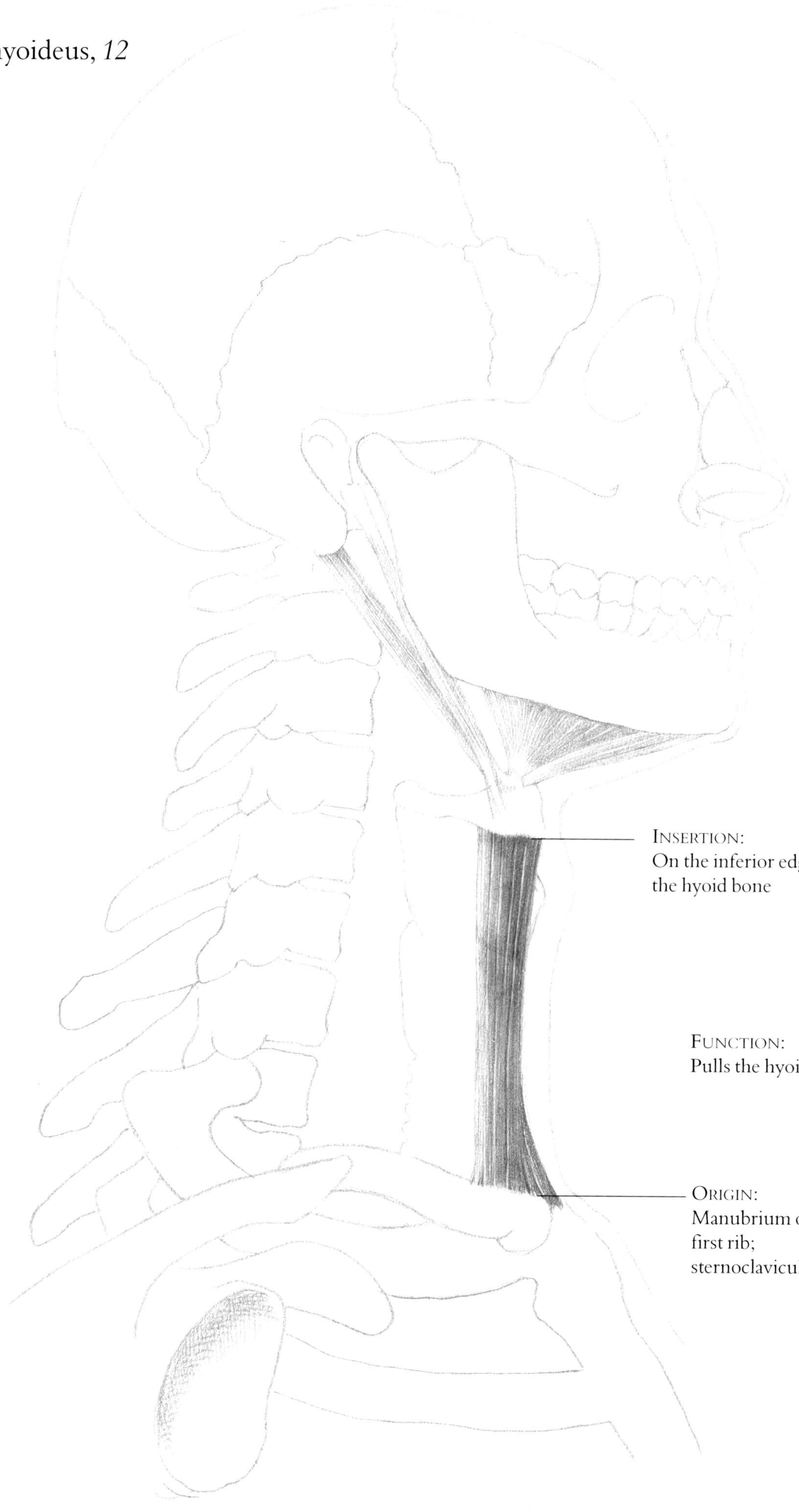

INSERTION:
On the inferior edge of the body of the hyoid bone

FUNCTION:
Pulls the hyoid bone downwards

ORIGIN:
Manubrium of the sternum;
first rib;
sternoclavicular joint

Fig. 248
The Cartilage of the Larynx

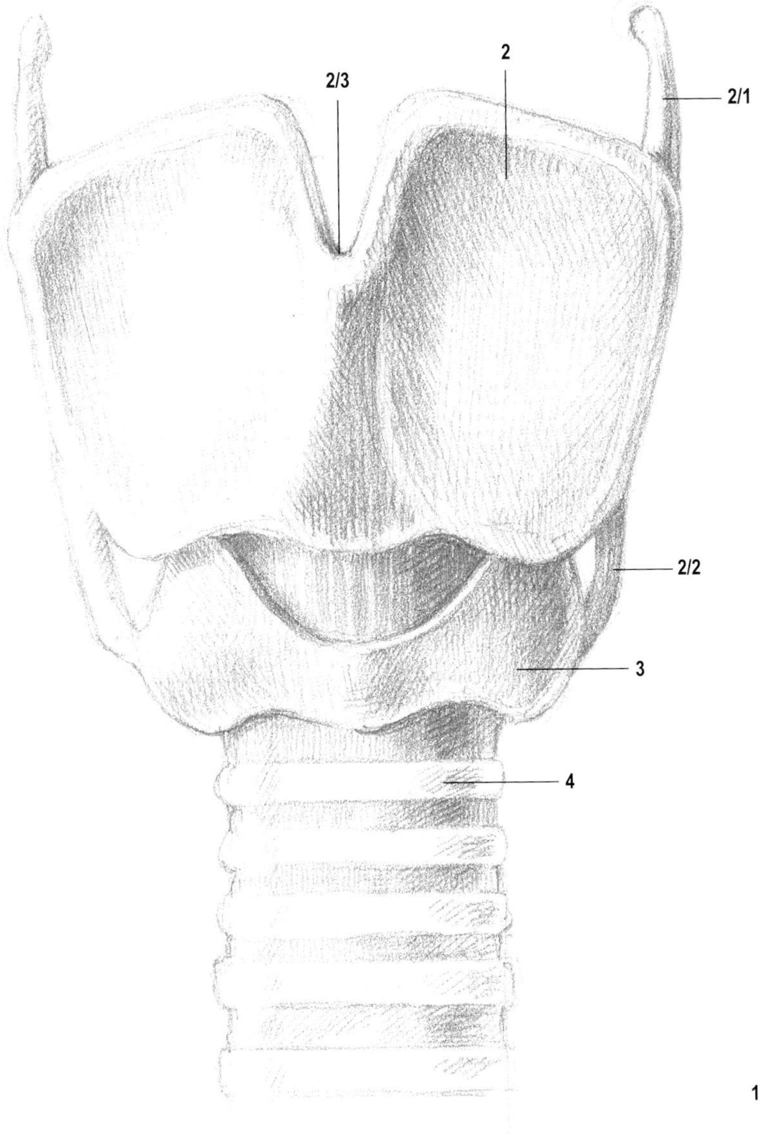

ventral aspect

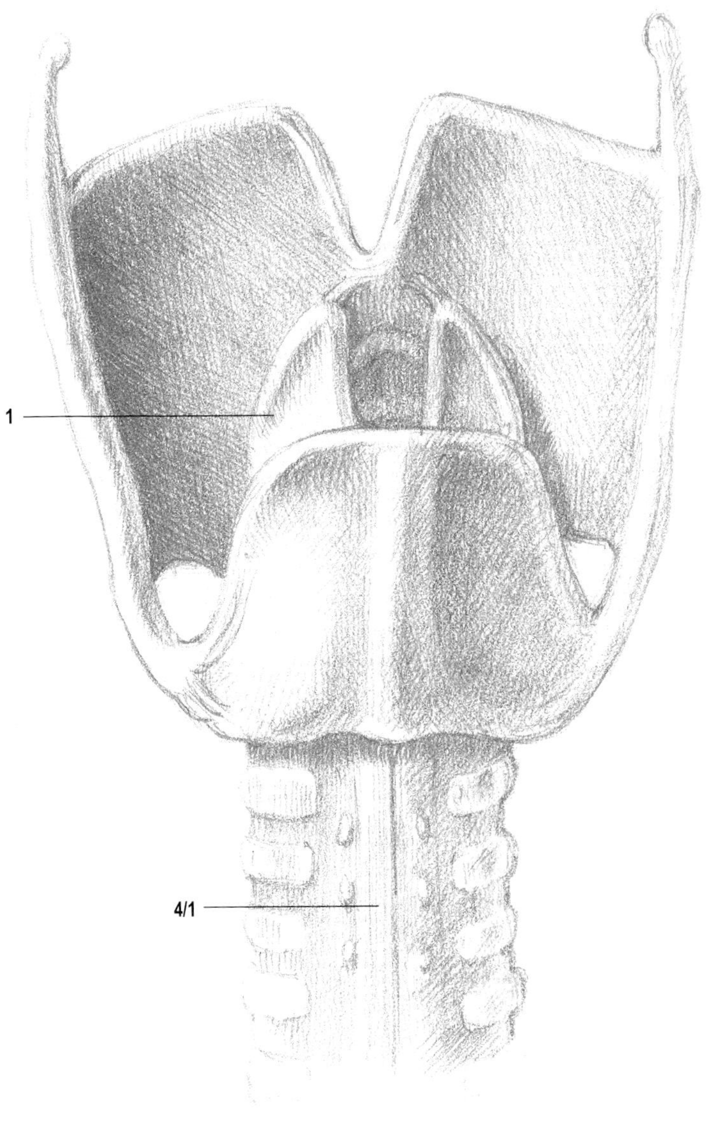

dorsal aspect

1 Arytenoid cartilage
2 Thyroid cartilage
  2/1 Thyroid cornu
  2/2 Cartilage cornu
  2/3 Incision in the superior edge of the thyroid cartilage
3 Annular cartilage of the larynx
4 Annular cartilage of the windpipe
  4/1 Membranous wall which connects the posterior ends of the thyroid cartilage

Fig. 249
The Superficial and Deep Muscles of the Larynx

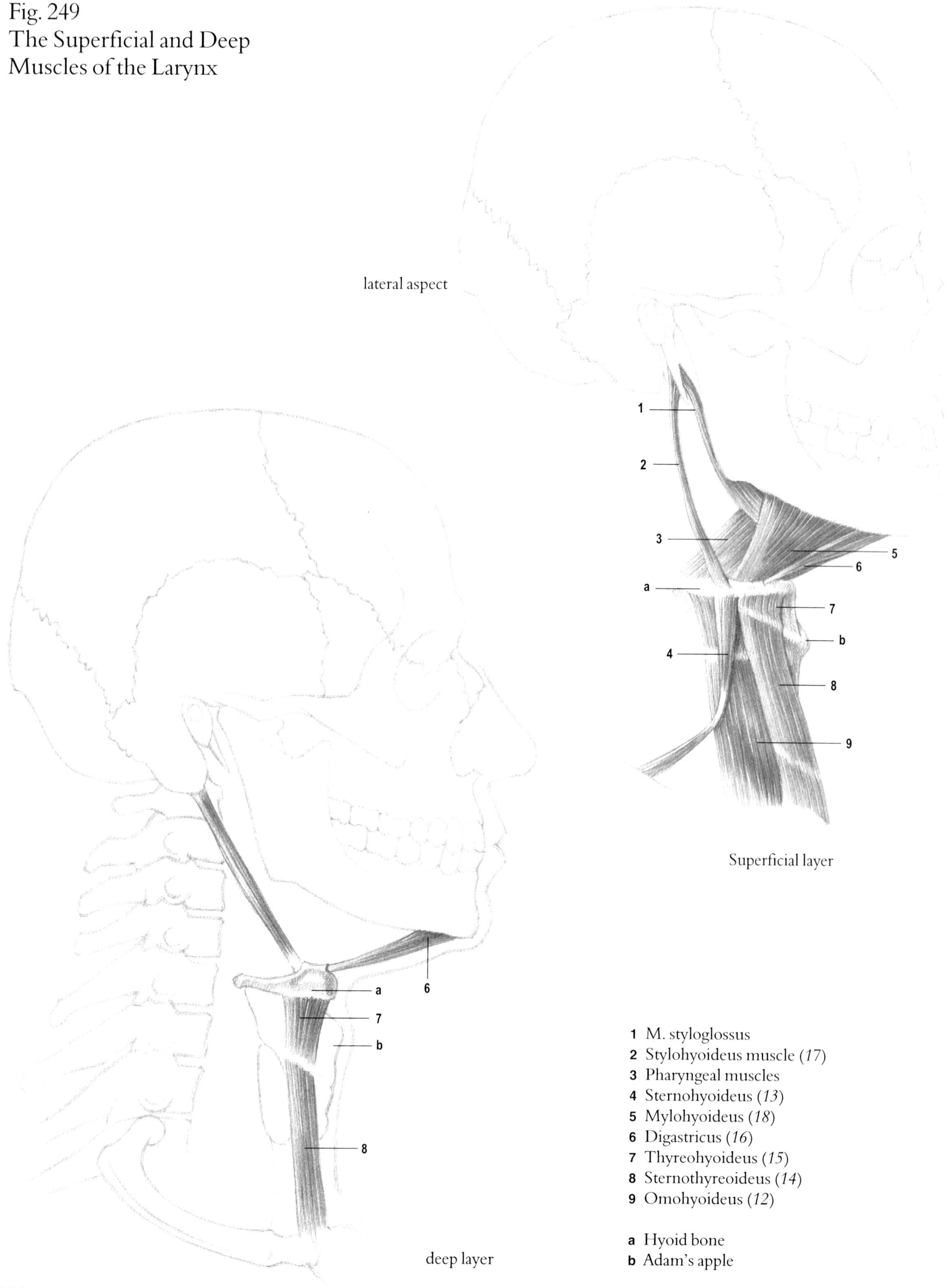

1 M. styloglossus
2 Stylohyoideus muscle (*17*)
3 Pharyngeal muscles
4 Sternohyoideus (*13*)
5 Mylohyoideus (*18*)
6 Digastricus (*16*)
7 Thyreohyoideus (*15*)
8 Sternothyreoideus (*14*)
9 Omohyoideus (*12*)

a Hyoid bone
b Adam's apple

Fig. 250
The Hyoid Bone

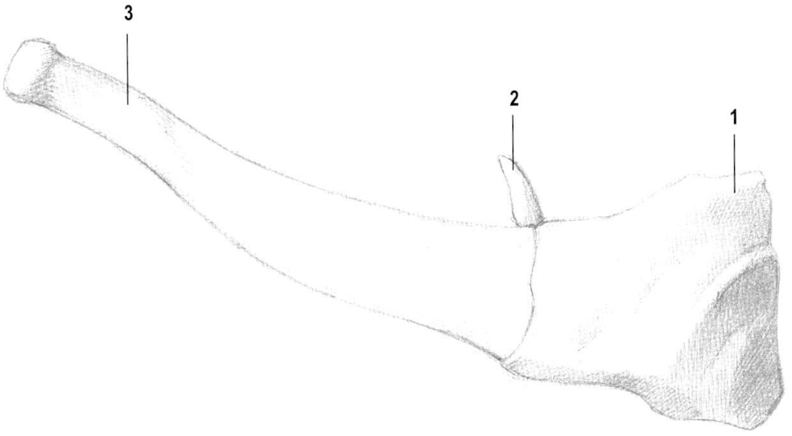

lateral surface

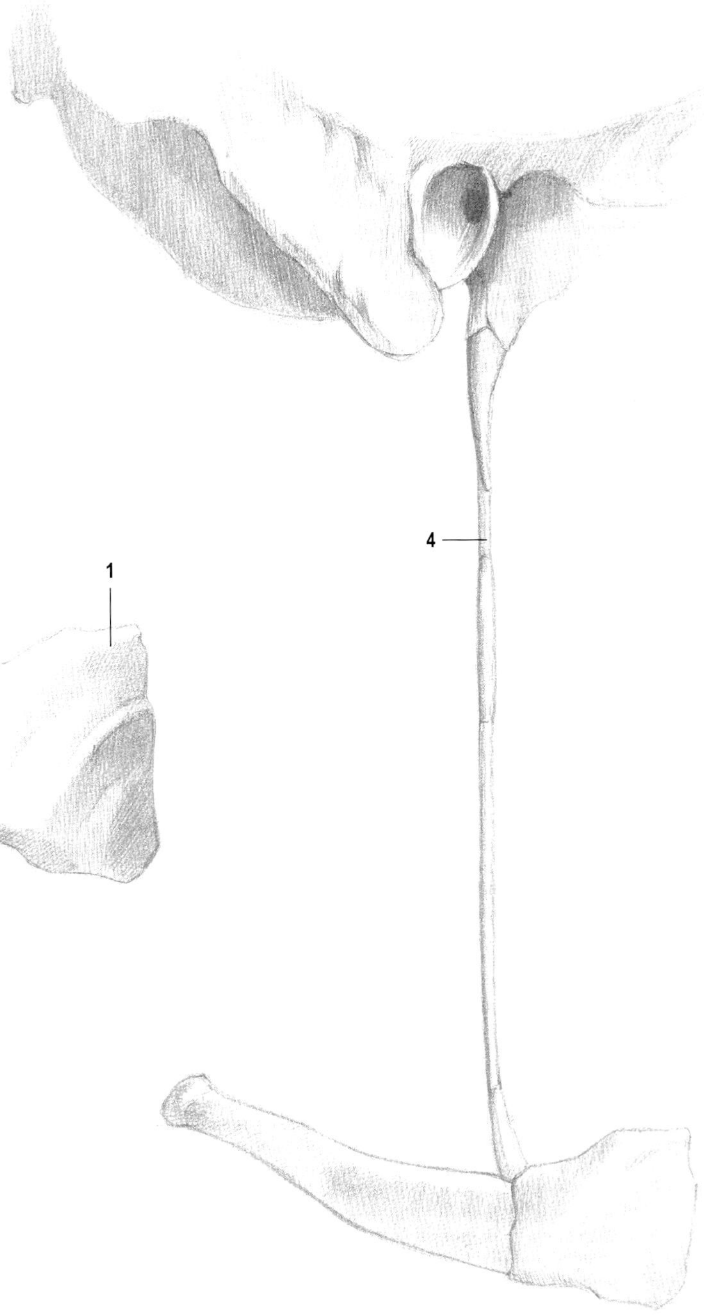

1 Body
2 Cornu minor
3 Cornu major
4 M. stylohyoideus (*17*)

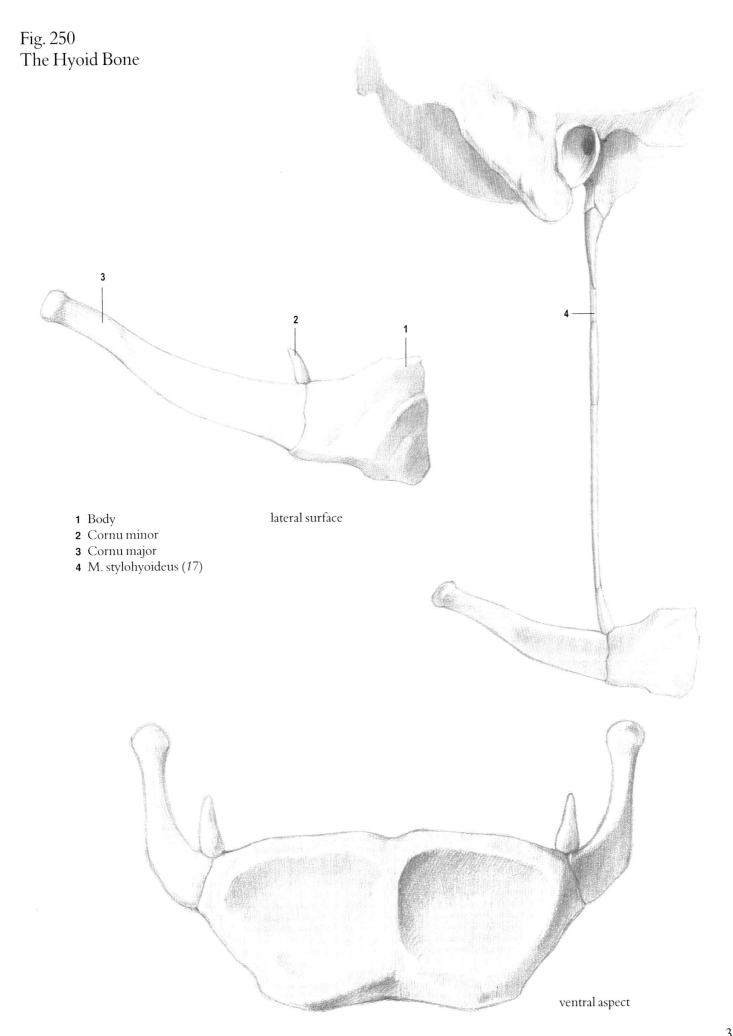

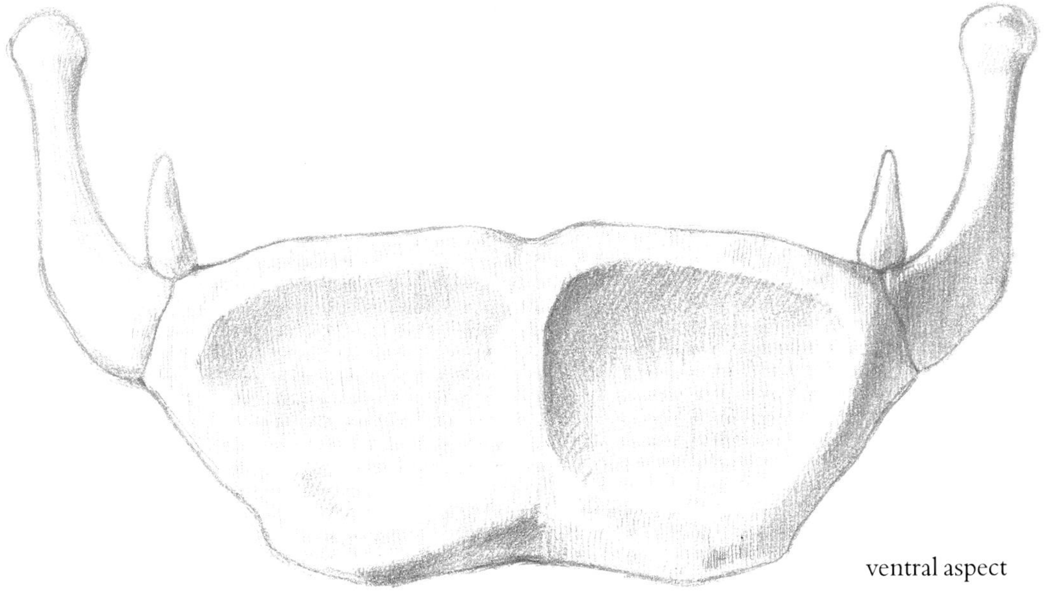

ventral aspect

Fig. 251
Movements of the Neck

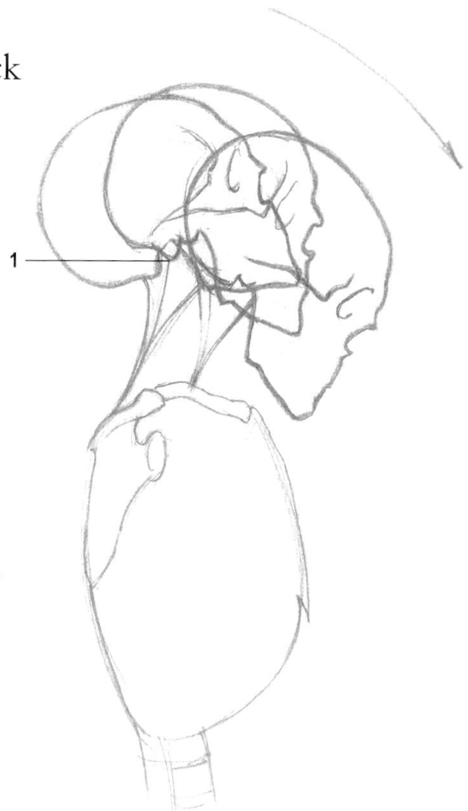

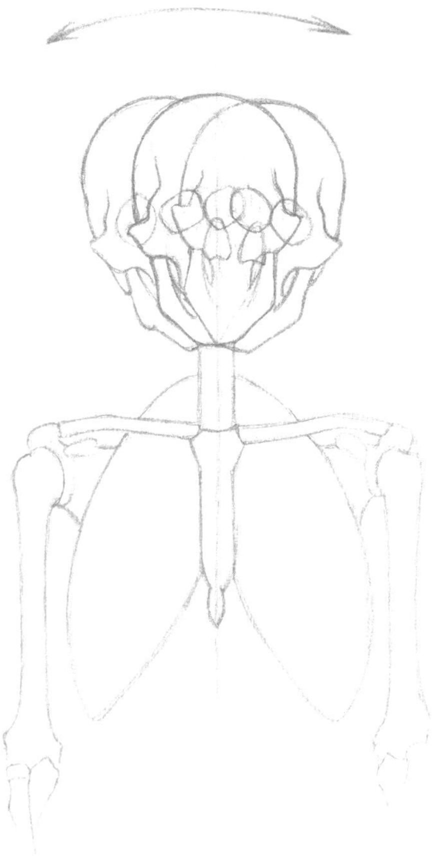

The neck can be flexed and extended in the superior (first) occipital joint (**1**). Rotational movements are made possible principally by the inferior (second) occipital joint (**2**). The head can be flexed laterally by synchronous movements of the joints between the cervical joints (**3**).

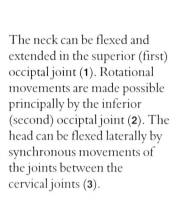

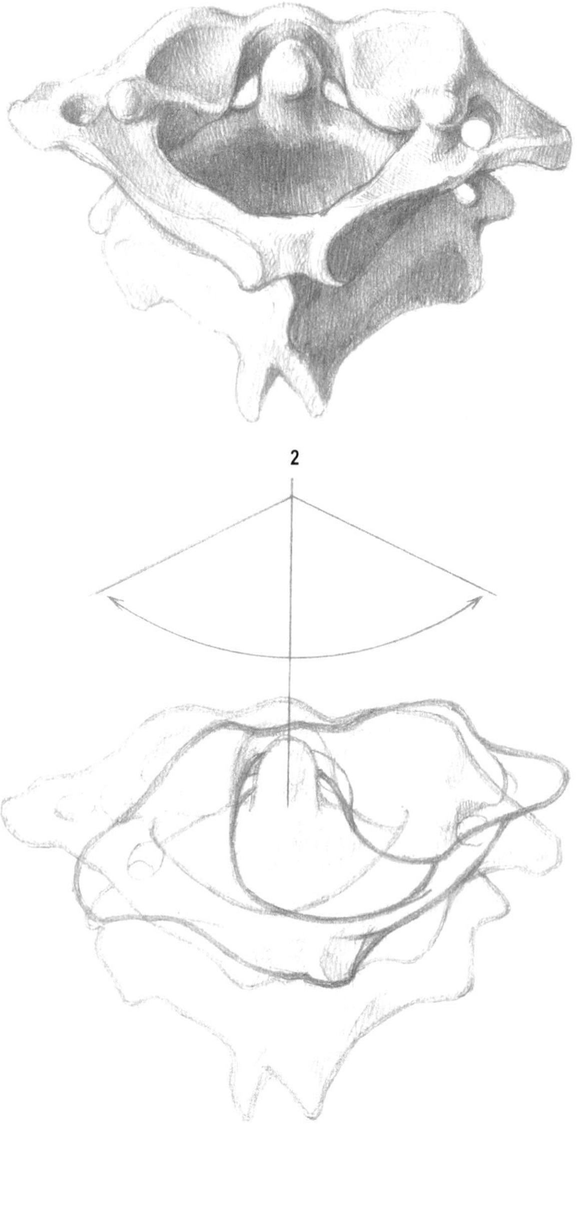

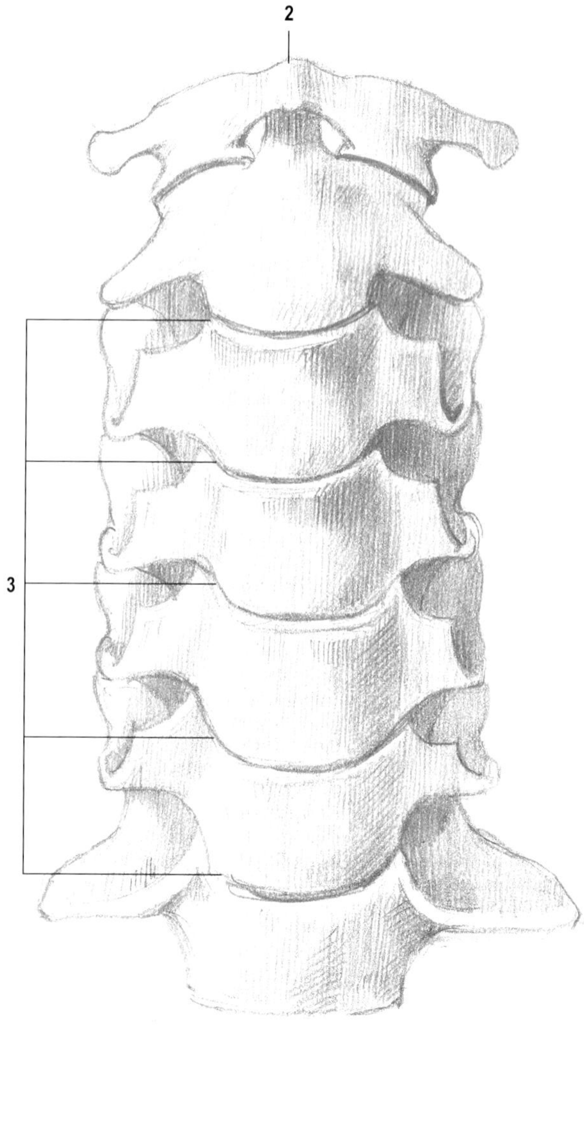

Fig. 252
Muscle Origins and Surfaces of Insertion

O= ORIGIN
I = INSERTION

POSTERIOR OCCIPITAL BONE

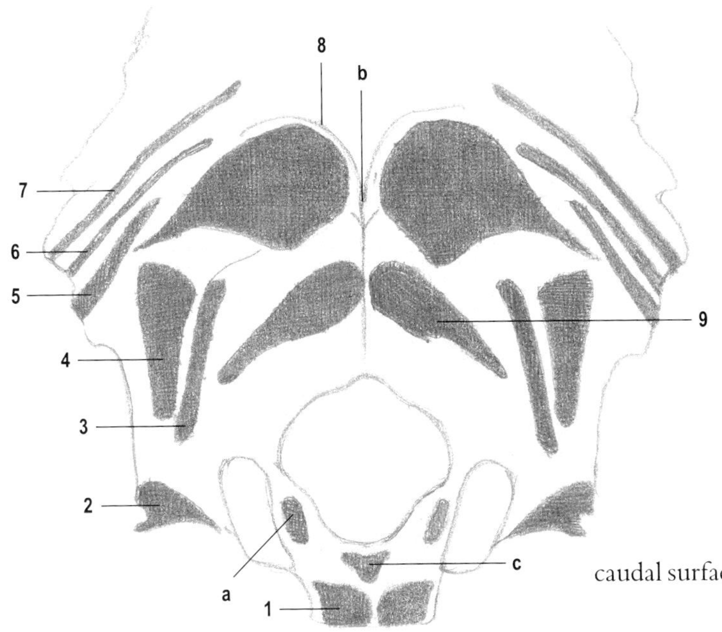

caudal surface

1 Rectus capitis anterior, I (*4*)
2 Rectus capitis posterior lateralis, I (*2*)
3 Rectus capitis posterior major, I (*2*)
4 Obliquus capitis superior, I (*3*)
5 Splenius (capitis), I (*5*)
6 Sternocleidomastoideus, I (*11*)
7 Epicranial muscles, O+I (*124/1*)
8 Trapezius (capitis), O (*20*)
9 Rectus capitis posterior minor, I (*2*)

a Alar ligament
b Nuchal ligament
c Tooth-shaped process ligament

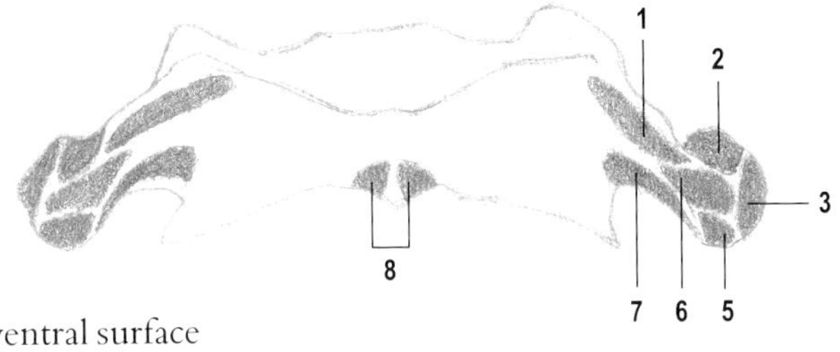

ventral surface

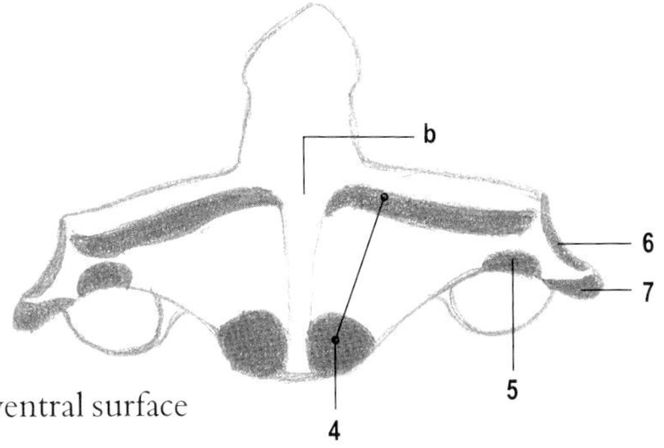

ventral surface

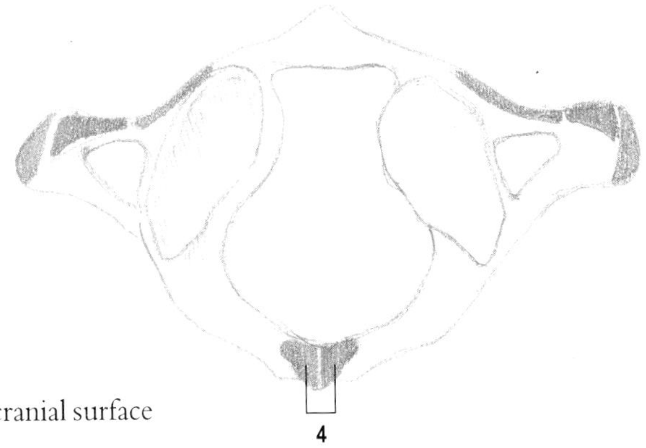

cranial surface

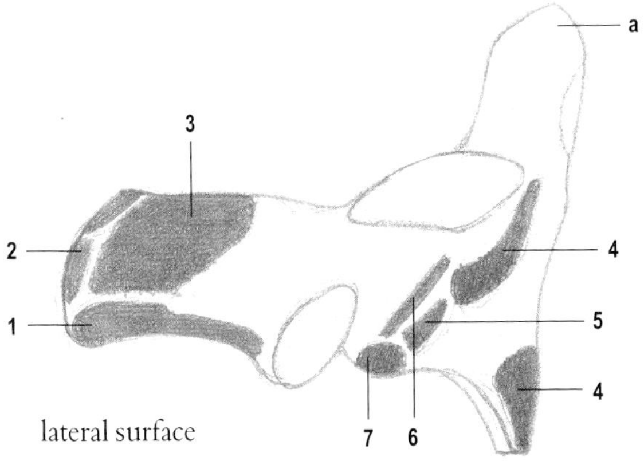

lateral surface

ATLAS
1 Rectus capitis anterior, O (*4*)
2 Rectus capitis posterior lateralis, O (*2*)
3 Obliquus capitis inferior, O (*3/1*)
4 Rectus capitis posterior minor, O (*2*)
5 Levator scapulae, O (*24*)
6 Splenius (capitis), I (*5*)
7 Intertransversal muscles (capitis anterior), I (*26/7*)
8 Longus colli, I (*6*)

AXIS
1 Semispinalis, a=O, b=I (*26/4*)
2 Rectus capitis posterior major, O (*2*)
3 Obliquus capitis inferior, O (*3/1*)
4 Longus colli, I (*6*)
5 Levator scapulae, O (*24*)
6 Intertransversal muscles (anterior *26/3*)
7 Splenius, I (*5*)

a Dens axis
b Vertebral axis

THE BONES AND MUSCLES OF THE HEAD

Fig. 253
The Skull

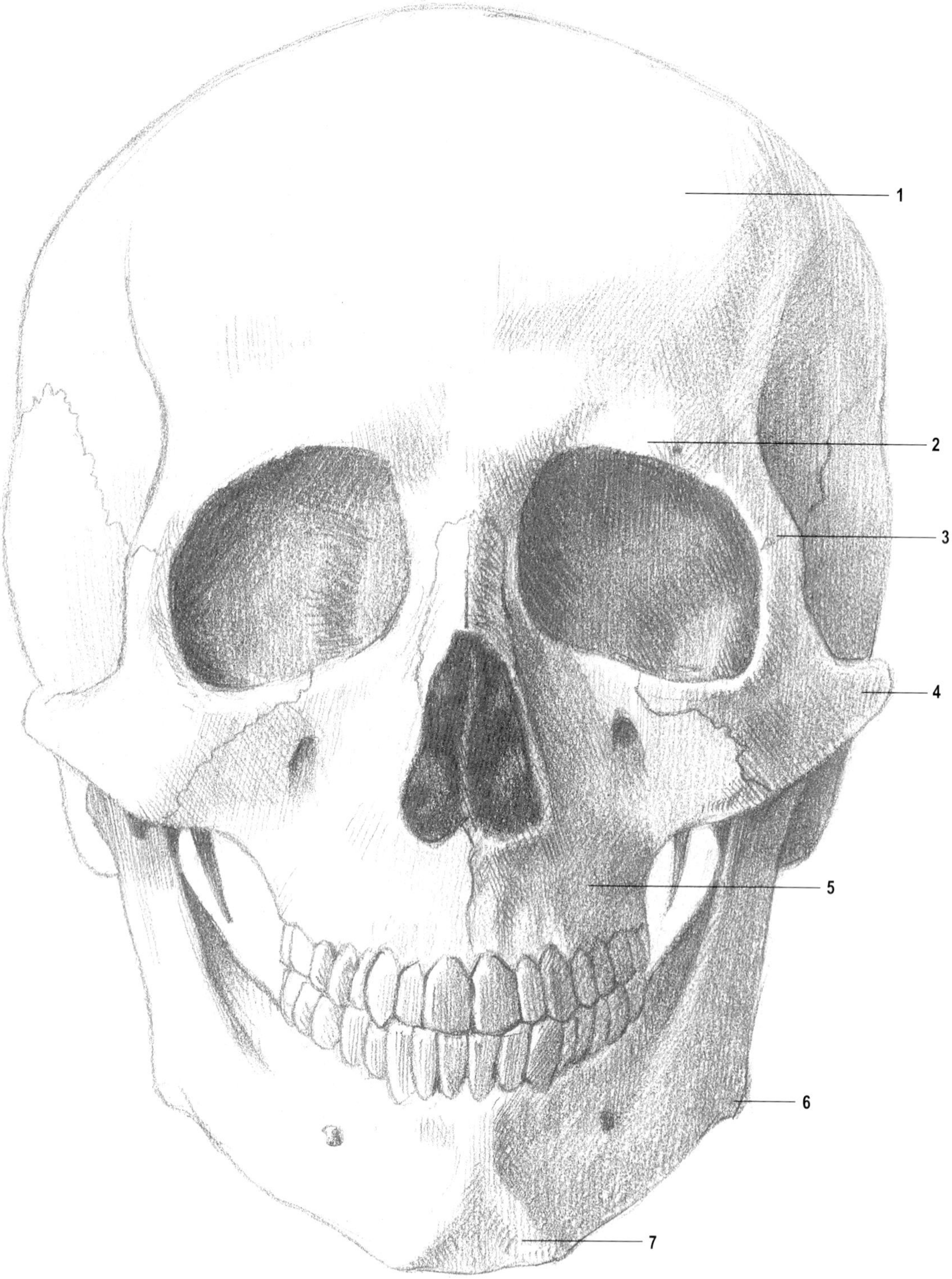

The bones of the forehead and face comprise the os frontale (**1**), supra-orbital margin (**2**), supra-orbital margin (**3**), zygomatic arch (**4**), upper jaw (**5**), corner of the lower jaw (**6**) and chin process (**7**). The form of the skull displays individual and ethnic differences.

Fig. 254
The Muscles of the Head

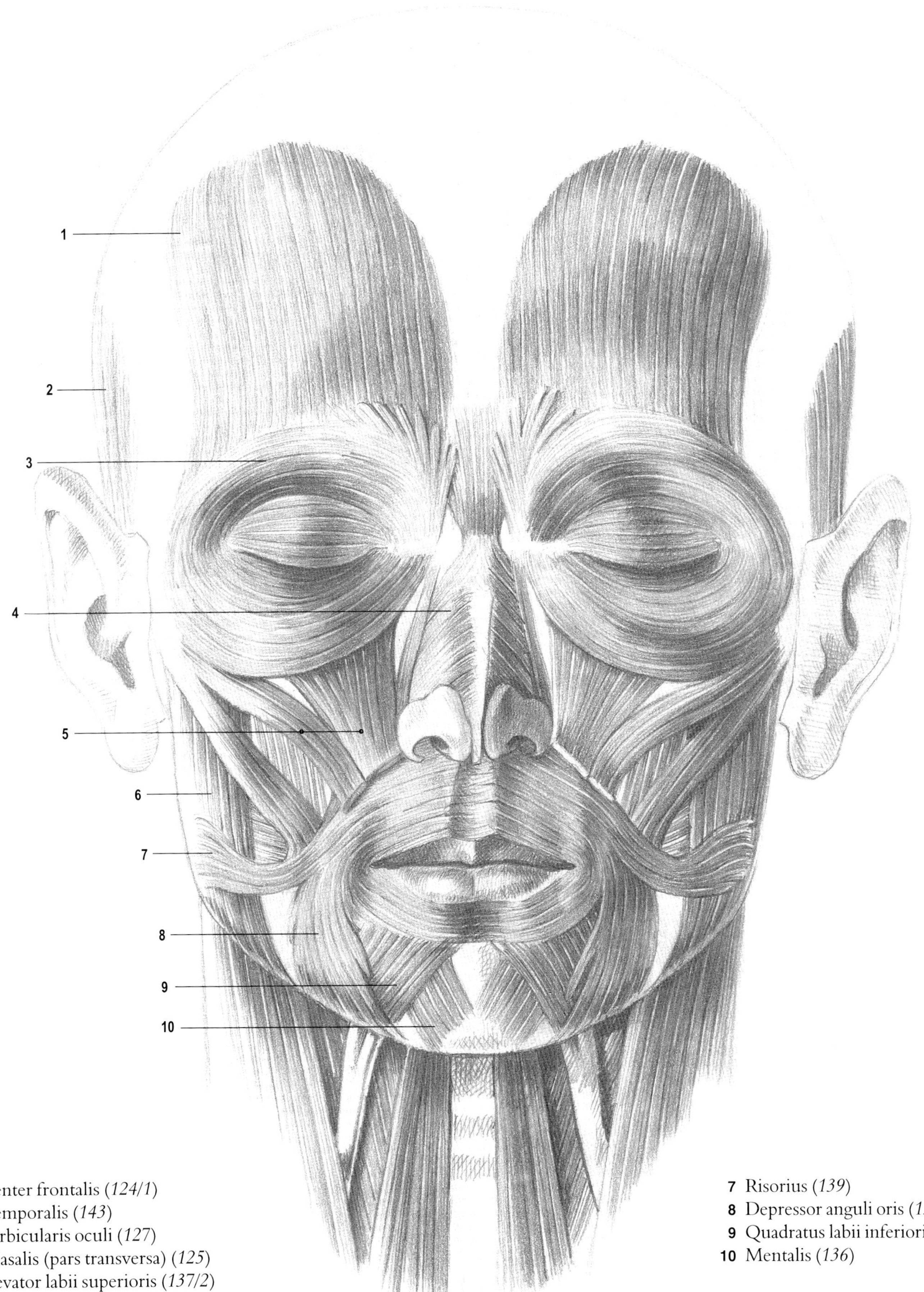

1 Venter frontalis (*124/1*)
2 Temporalis (*143*)
3 Orbicularis oculi (*127*)
4 Nasalis (pars transversa) (*125*)
5 Levator labii superioris (*137/2*)
6 Masseter (*142*)
7 Risorius (*139*)
8 Depressor anguli oris (*133*)
9 Quadratus labii inferioris (*135*)
10 Mentalis (*136*)

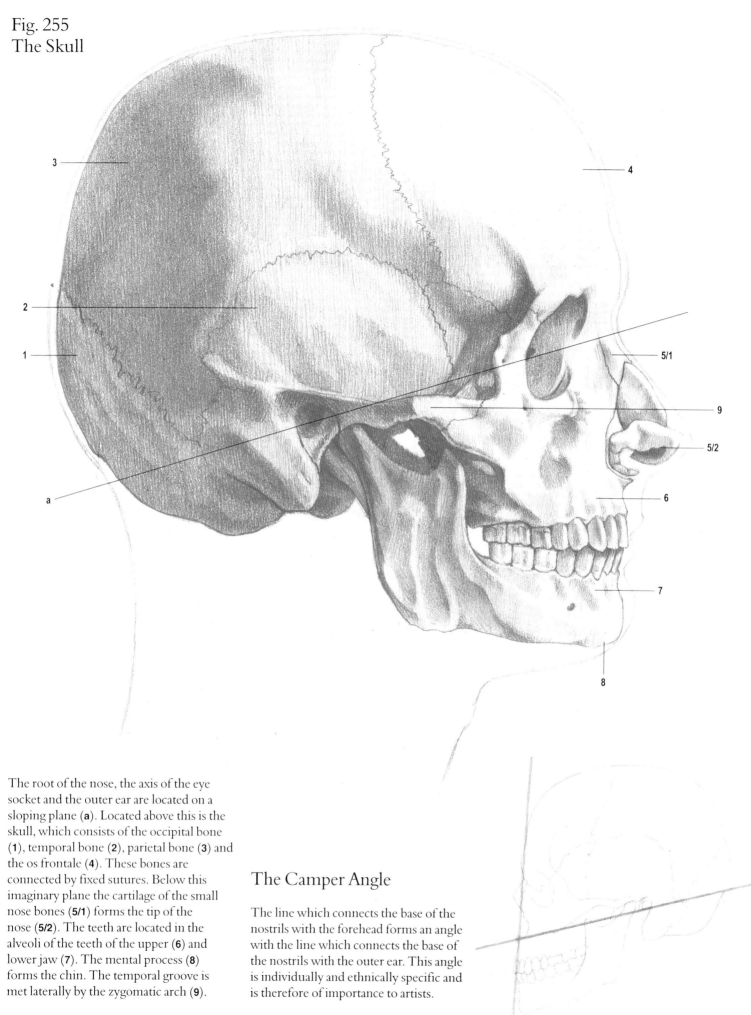

Fig. 255
The Skull

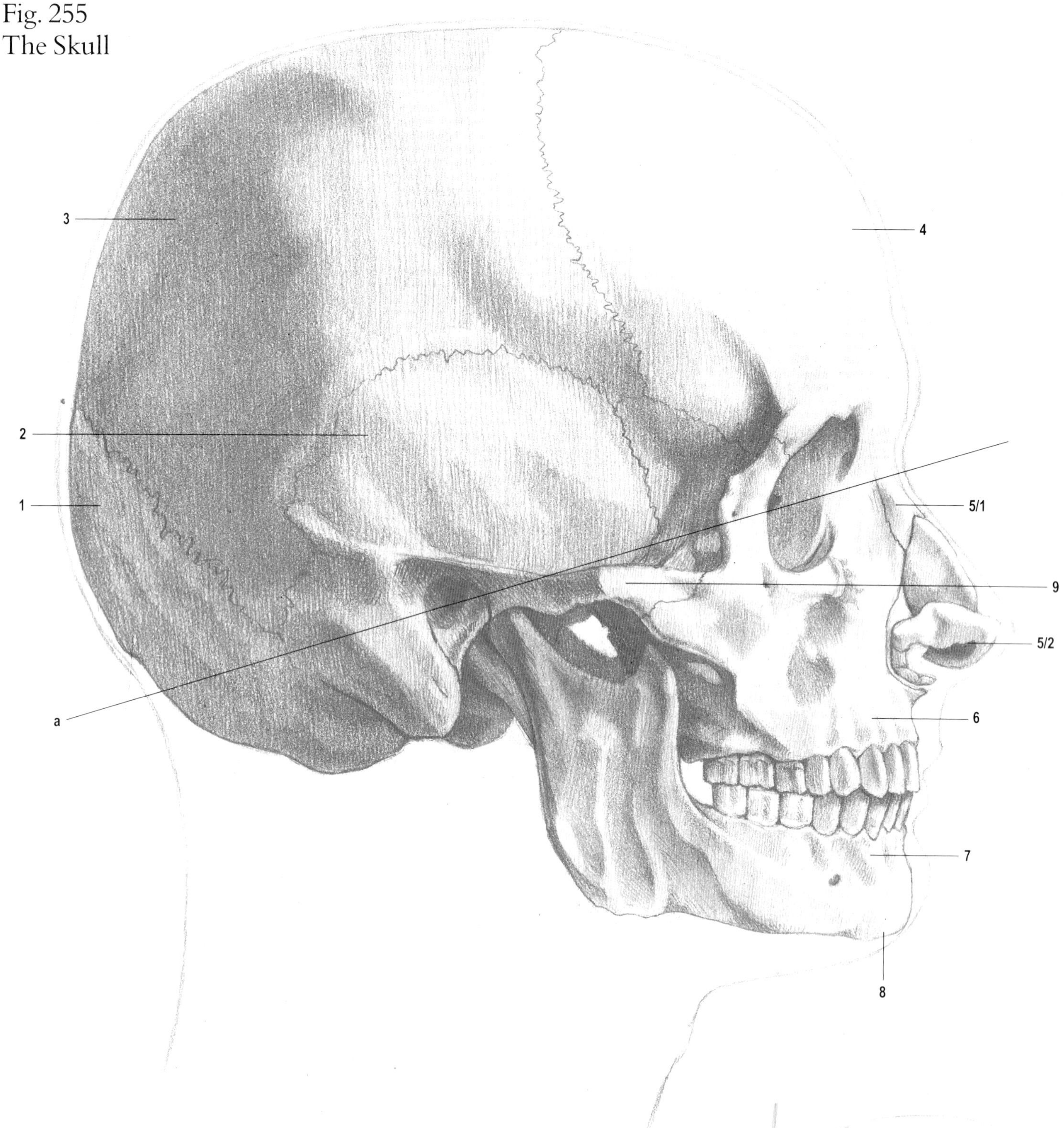

The root of the nose, the axis of the eye socket and the outer ear are located on a sloping plane (**a**). Located above this is the skull, which consists of the occipital bone (**1**), temporal bone (**2**), parietal bone (**3**) and the os frontale (**4**). These bones are connected by fixed sutures. Below this imaginary plane the cartilage of the small nose bones (**5/1**) forms the tip of the nose (**5/2**). The teeth are located in the alveoli of the teeth of the upper (**6**) and lower jaw (**7**). The mental process (**8**) forms the chin. The temporal groove is met laterally by the zygomatic arch (**9**).

The Camper Angle

The line which connects the base of the nostrils with the forehead forms an angle with the line which connects the base of the nostrils with the outer ear. This angle is individually and ethnically specific and is therefore of importance to artists.

Fig. 256
The Muscles of the Head

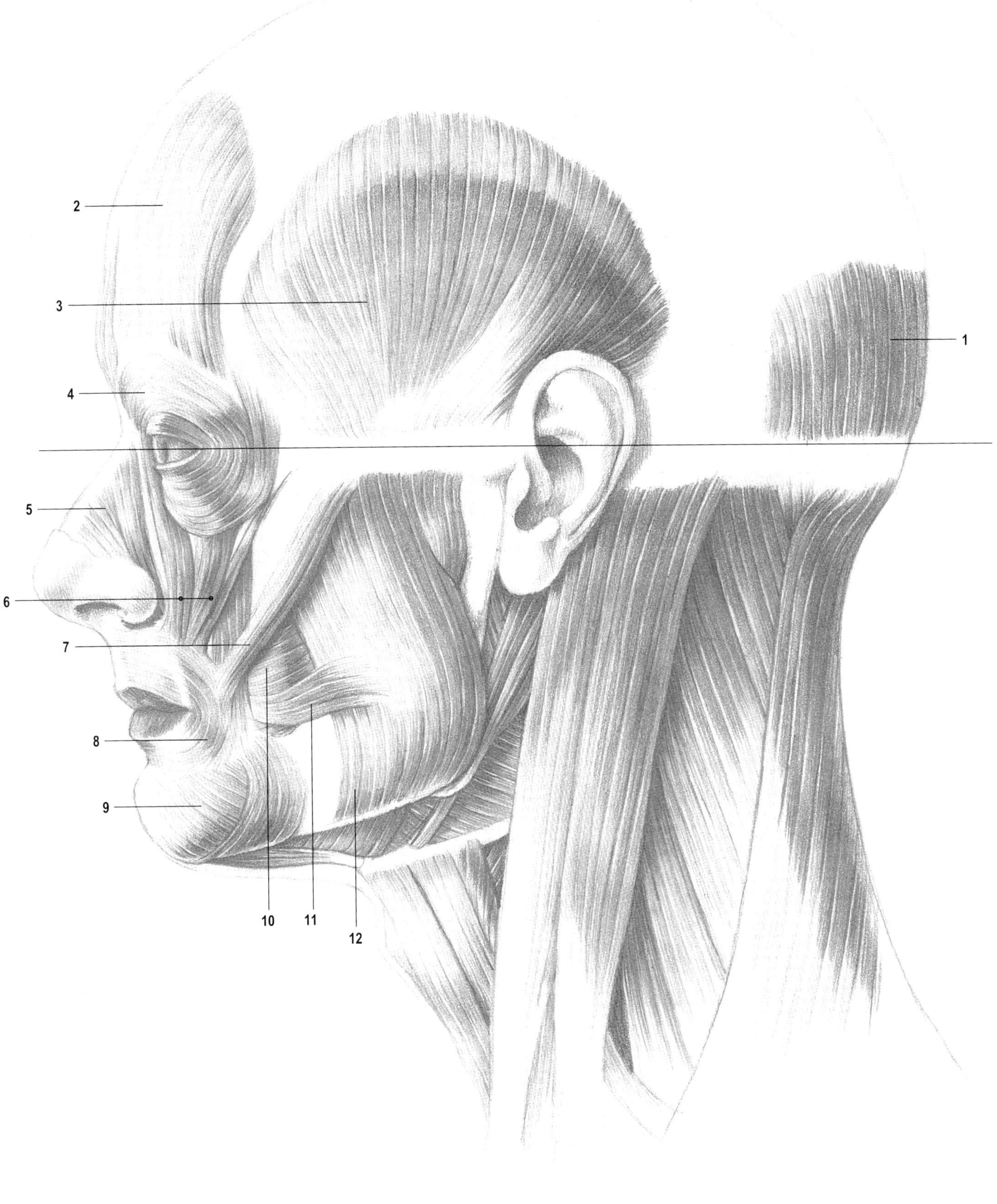

1 Occipitofrontalis (*124/1*)
2 Venter frontalis (*124/1*)
3 Temporalis (*143*)
4 Orbicularis oculi (*127*)
5 Nasalis (pars transversa) (*125*)
6 Levator labii superioris (*137/2*)
7 Zygomaticus major (*138*)
8 Orbicularis oris (*132*)
9 Mentalis (*136*)
10 Buccalis (*141*)
11 Risorius (*139*)
12 Masseter (*142*)

Fig. 257
The Bones of the Skull

The skull comprises 22 bones, of which eight exist in pairs. With the exception of the lower jaw these are firmly fixed to each other. These bones protect the brain, the mouth and the sensory organs. Some surround the air-filled collateral alveoli.

The Viscerocranium

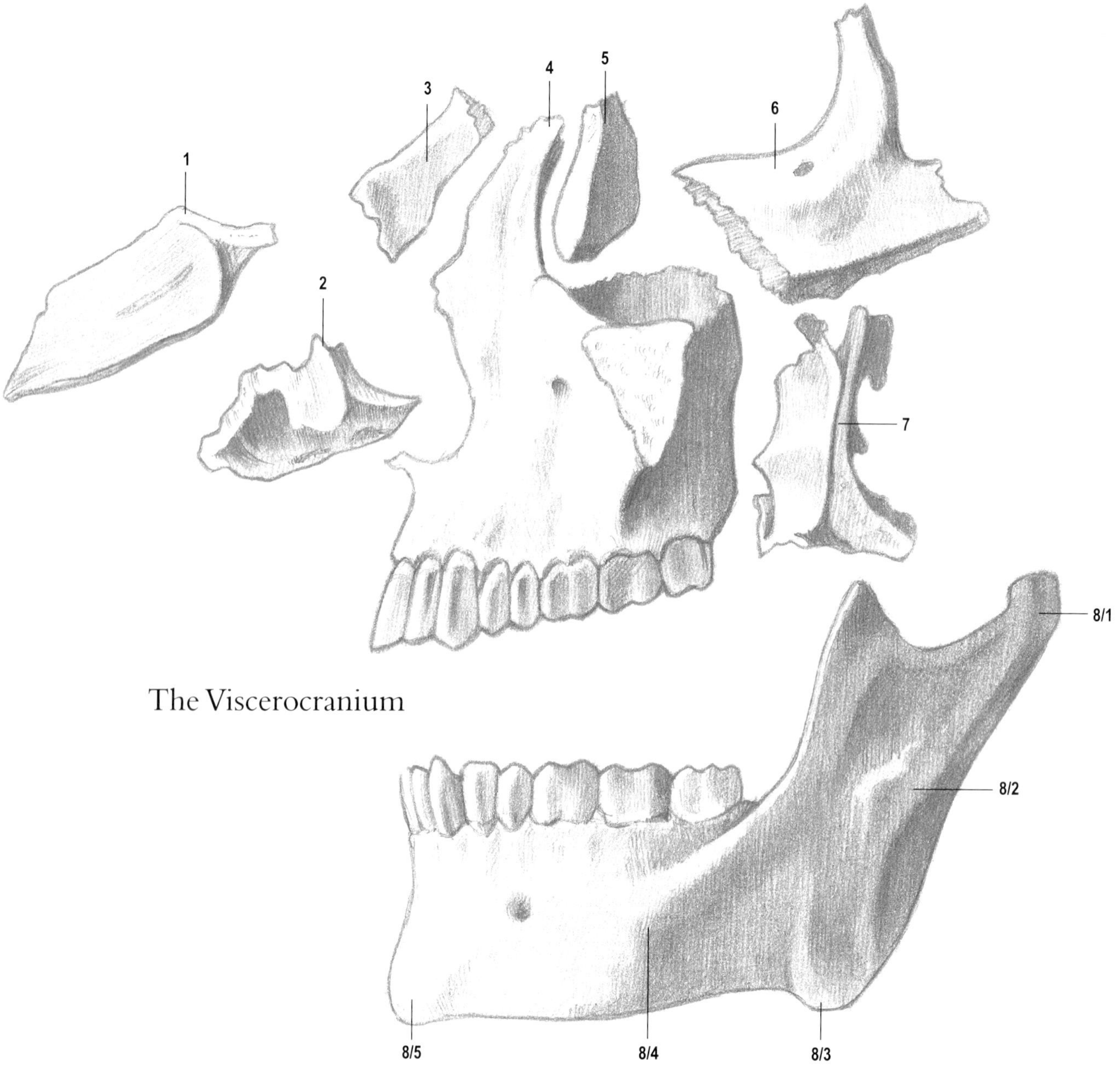

1 Nose bone
2 Lacrimal bone
3 Vomer
4 Upper jaw
5 Nasal septum
6 Palatal bone
7 Zygomatic bone
8 Lower jaw
8/1 Capitulum
8/2 Branch of the lower jaw
8/3 Angle of lower jaw
8/4 Body of lower jaw
8/5 Mental process

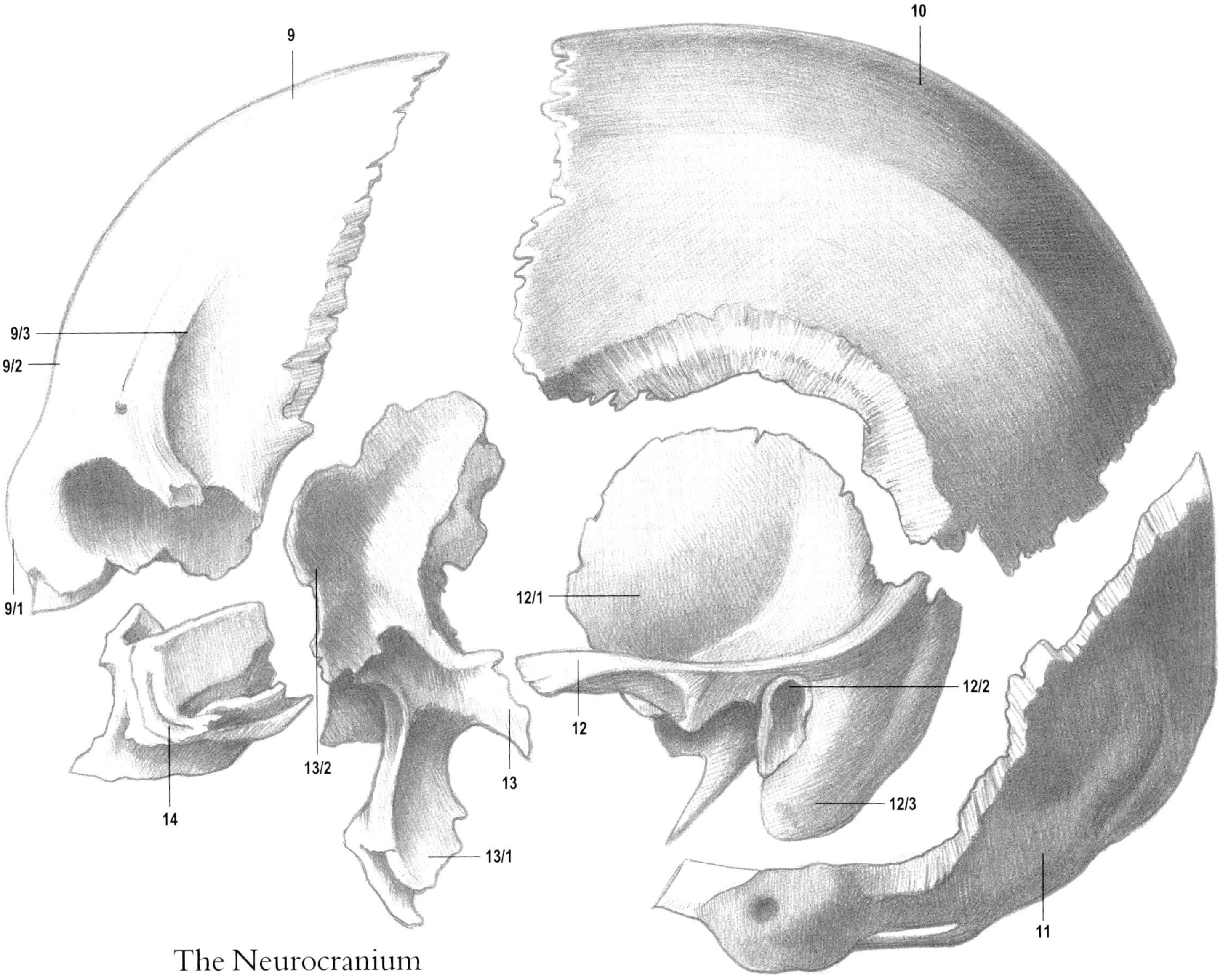

The Neurocranium

9 Os frontale
9/1 Supraciliar crest
9/2 Frontal protuberance
9/3 Temporal part
10 Parietal bone
11 Occipital bone
12 Temporal bone
12/1 Squama
12/2 Outer ear
12/3 Mastoid process
13 Body of the sphenoidal bone
13/1 Lesser ala of the sphenoidal bone
13/2 Greater ala of the sphenoidal bone
14 Cribriform bone

Fig. 258
The Skull

Os Frontale

1 Protuberance of os frontale
2 Temporal groove
3 Zygomatic arch
4 Supraciliar arch
5 Supraciliar crest
6 Supraciliar incisura
7 Glabella
8 Nose bone

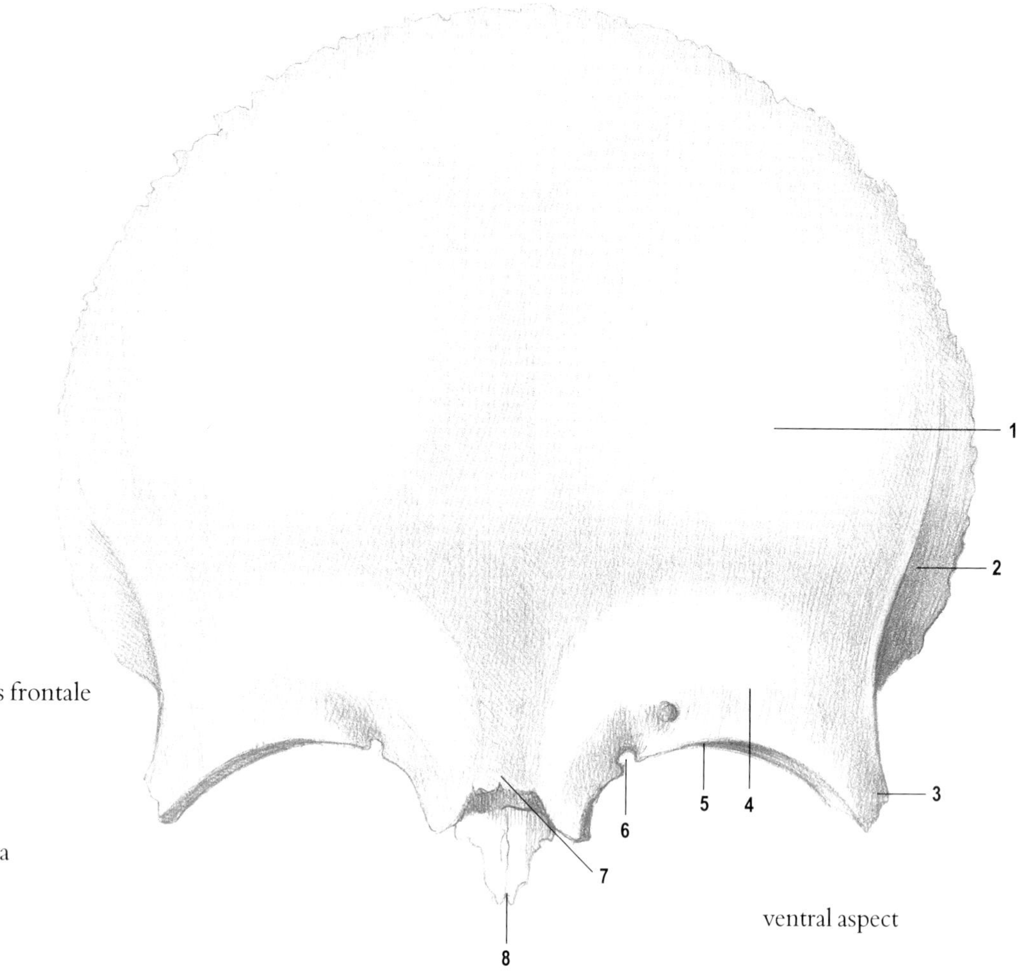

ventral aspect

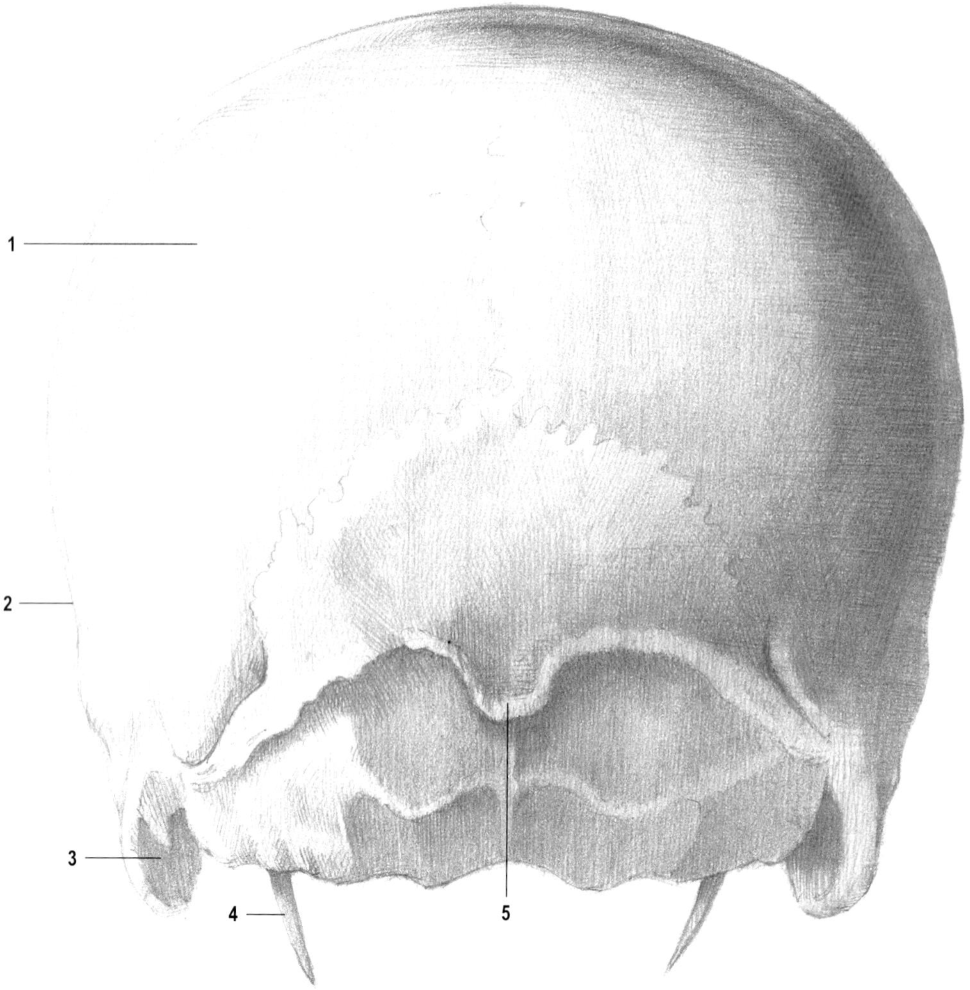

dorsal aspect

Occipital Bone

1 Parietal bone
2 Temporal bone
3 Mastoid process
4 Styloid process
5 Occipital protuberance and cranial linea nuchae

Fig. 259
Skull Sutures

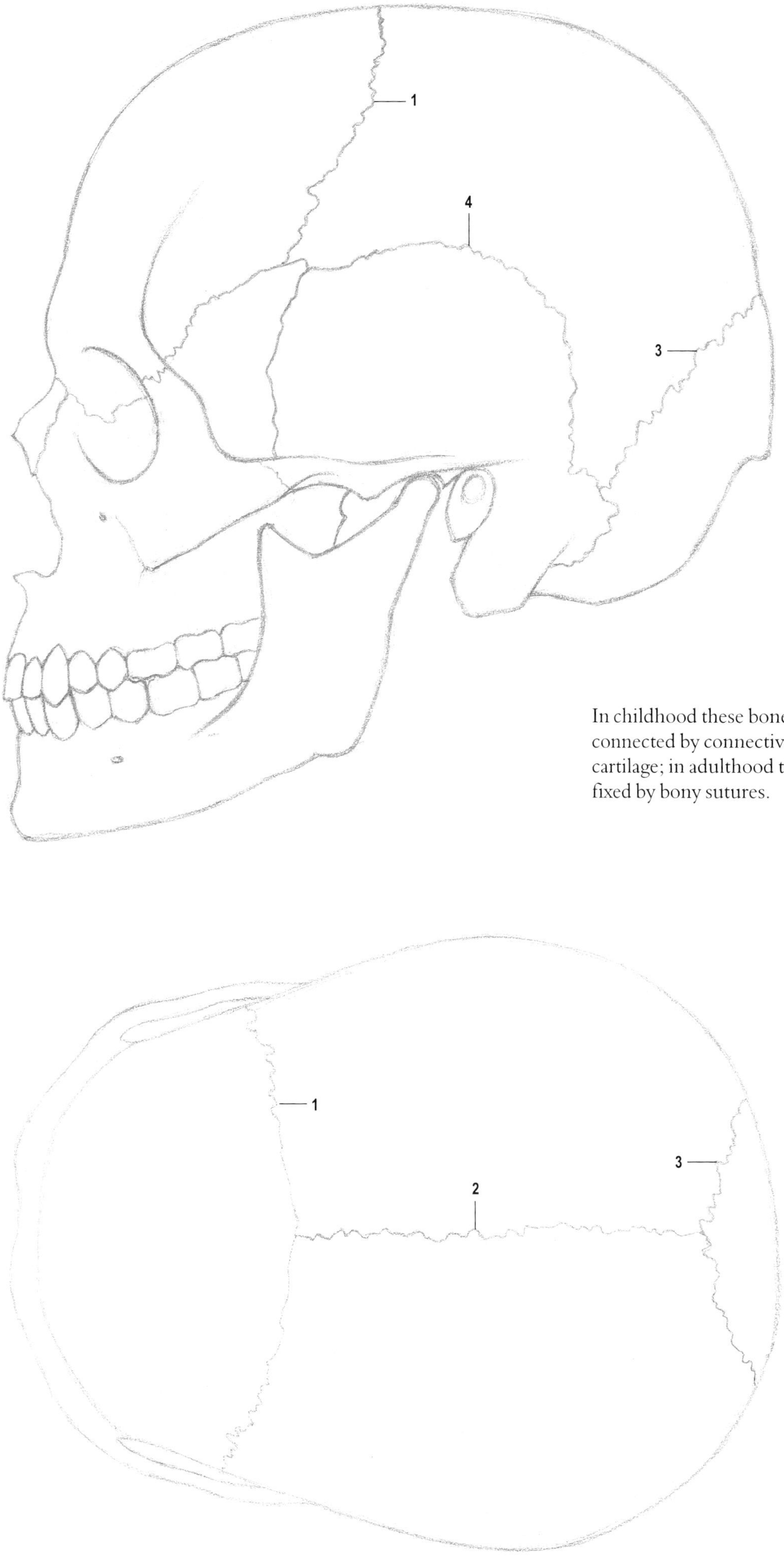

lateral aspect

In childhood these bones are connected by connective tissue or cartilage; in adulthood they are fixed by bony sutures.

cranial aspect

1 Coronal suture
2 Sagittal suture
3 Lambdoid suture
4 Squamous skin

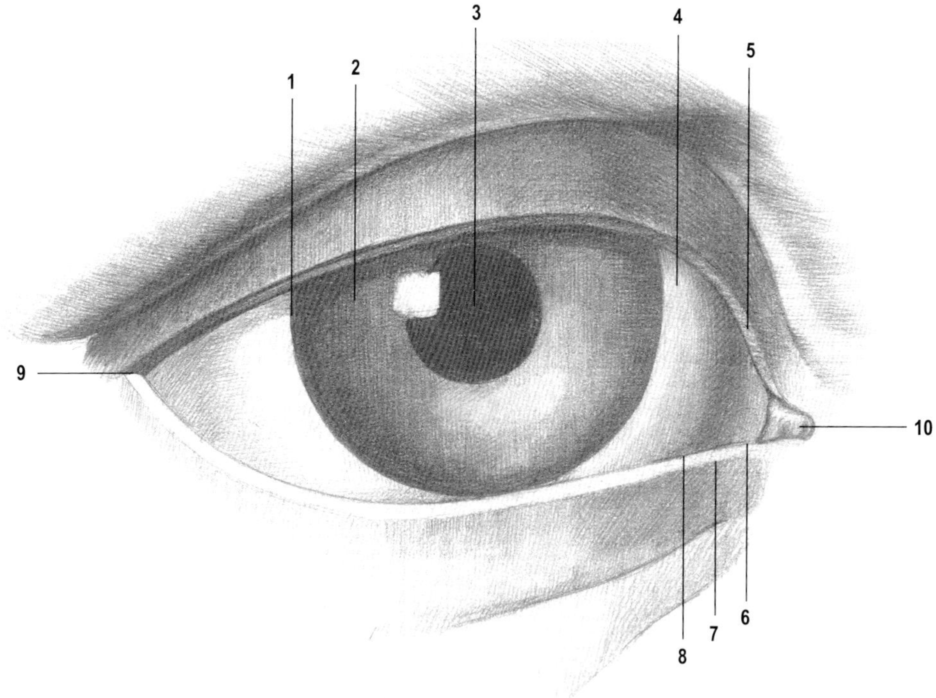

The iris (**2**), which determines the colour of the eye, can be seen through the cornea (**1**), with its black pupil in the center (**3**). When the lid is open the dermis appears white (**4**). The larger cranial lids (**5**) and the lower caudal lids (**6**) are folded. The lid glands (**8**) are located on the inside edge. On the outside edge (**7**) the eyelashes can be seen. The outer angle of the eye is pointed (**9**), whereas the inner angle is rounded (**10**).

Fig. 260
The Eye Muscles

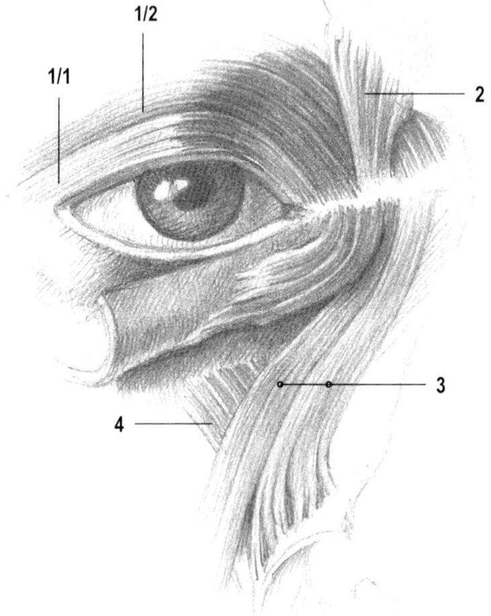

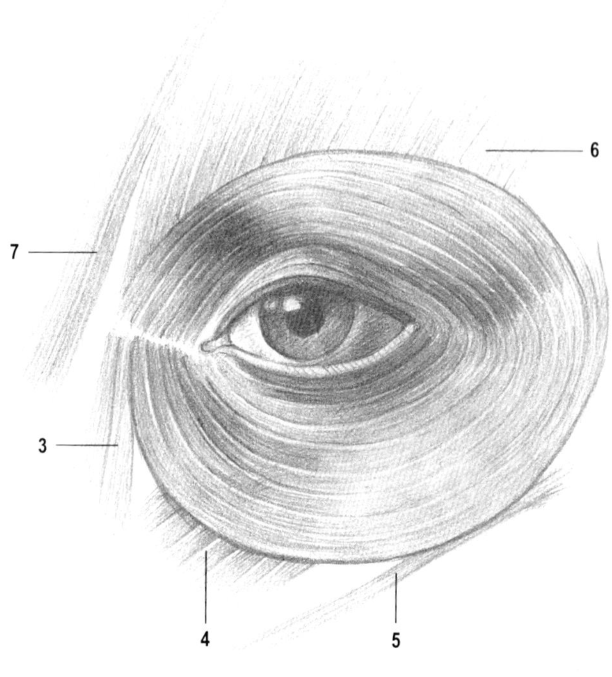

1 Orbicularis oculi (*127*)
1/1 Lid part
1/2 Eye socket part
2 Corrugator supercilii (*128*)
3 Levator labii superioris (medial part) (*137/1*)
4 Levator labii superioris (lateral part) (*137/2*)
5 Zygomaticus minor (*137/3*)
6 Venter frontalis (*124/1*)
7 Procerus (*124/3*)

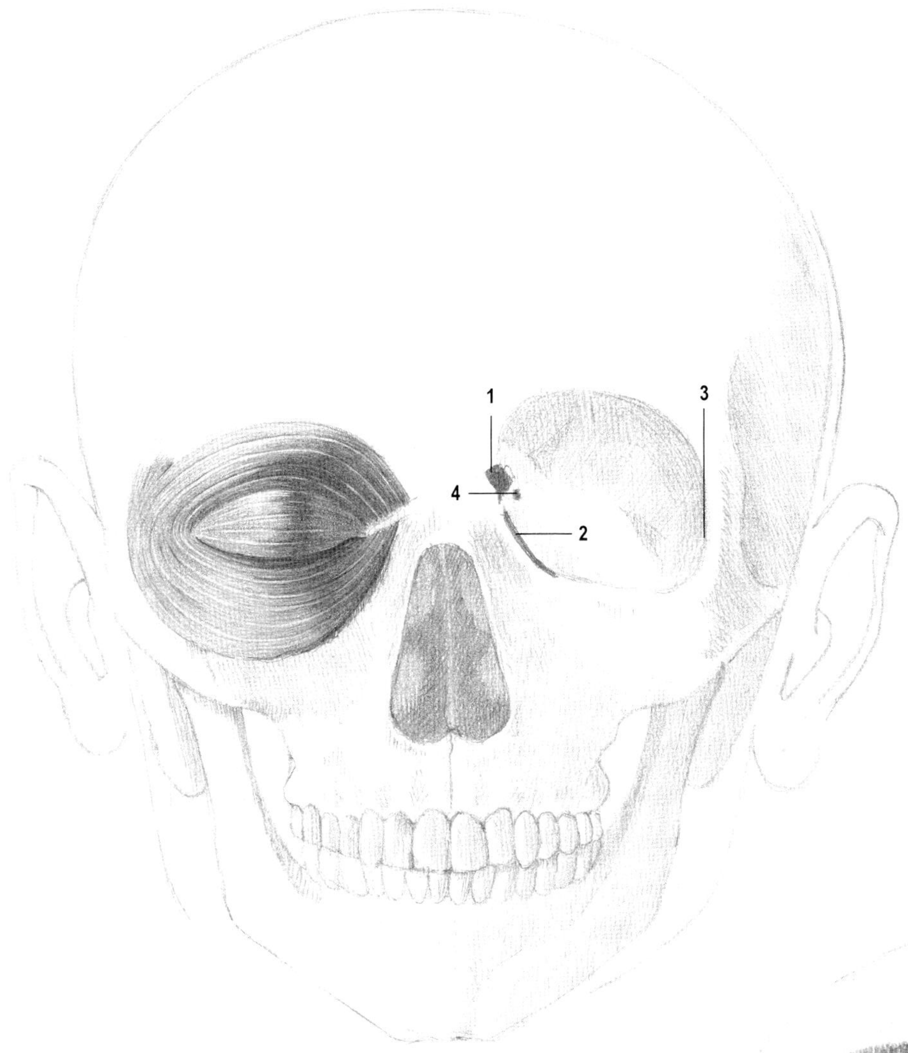

Fig. 261
The Orbicular Muscle of the Eye

(M. orbicularis oculi, *127*)

ORIGIN:
Medial corner of the eye (**1**), nasal part of the os frontale (**2**); maxillary process (**3**), cranial lid ligament (pars palpebralis); tear duct groove (**4**)

INSERTION:
Outer ligament of the corner of the eye

FUNCTION:
Folds the eyebrows, closes the eyelids; closes the eyes (pars orbicularis); furrows the brow (in sadness or pain); expands the tear duct

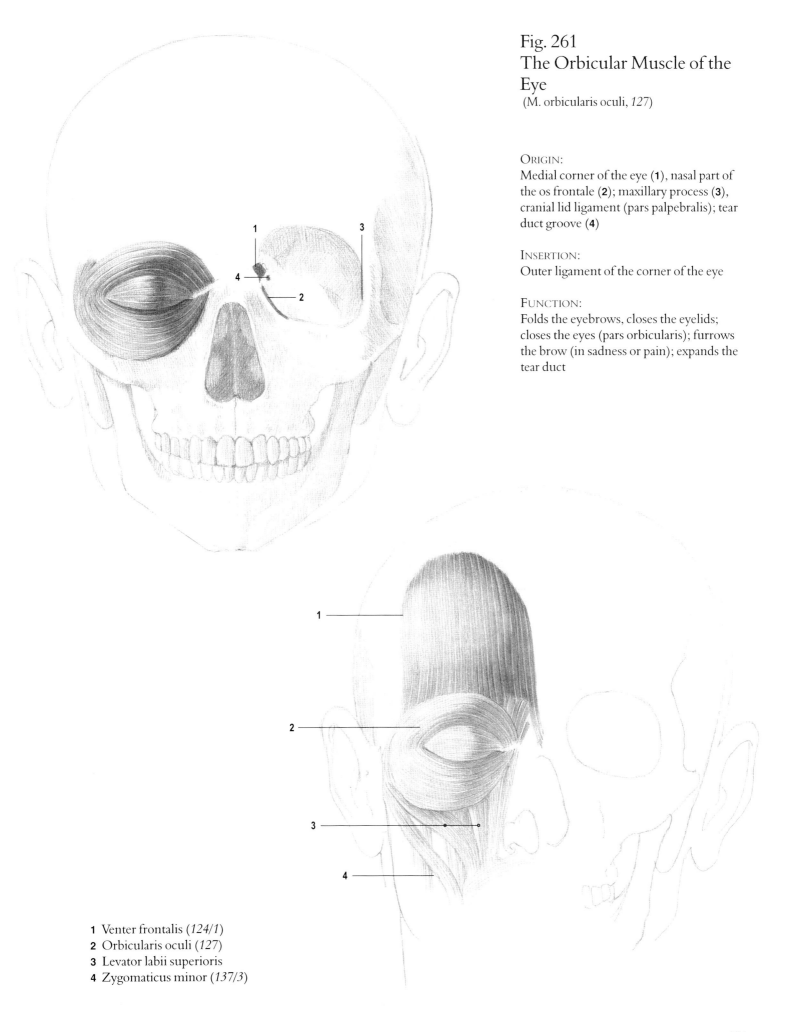

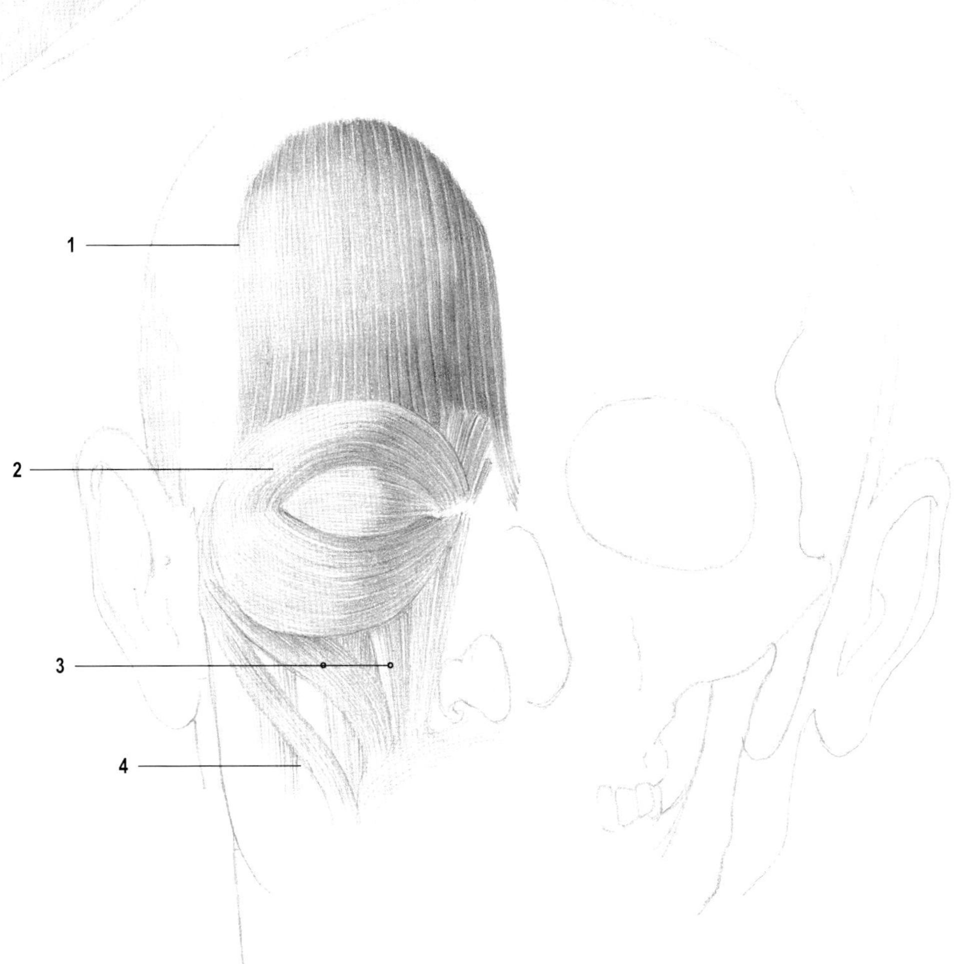

1 Venter frontalis (*124/1*)
2 Orbicularis oculi (*127*)
3 Levator labii superioris
4 Zygomaticus minor (*137/3*)

Fig. 262
The Superciliary Corrugator Muscle
(M. corrugator supercilii, *128*)

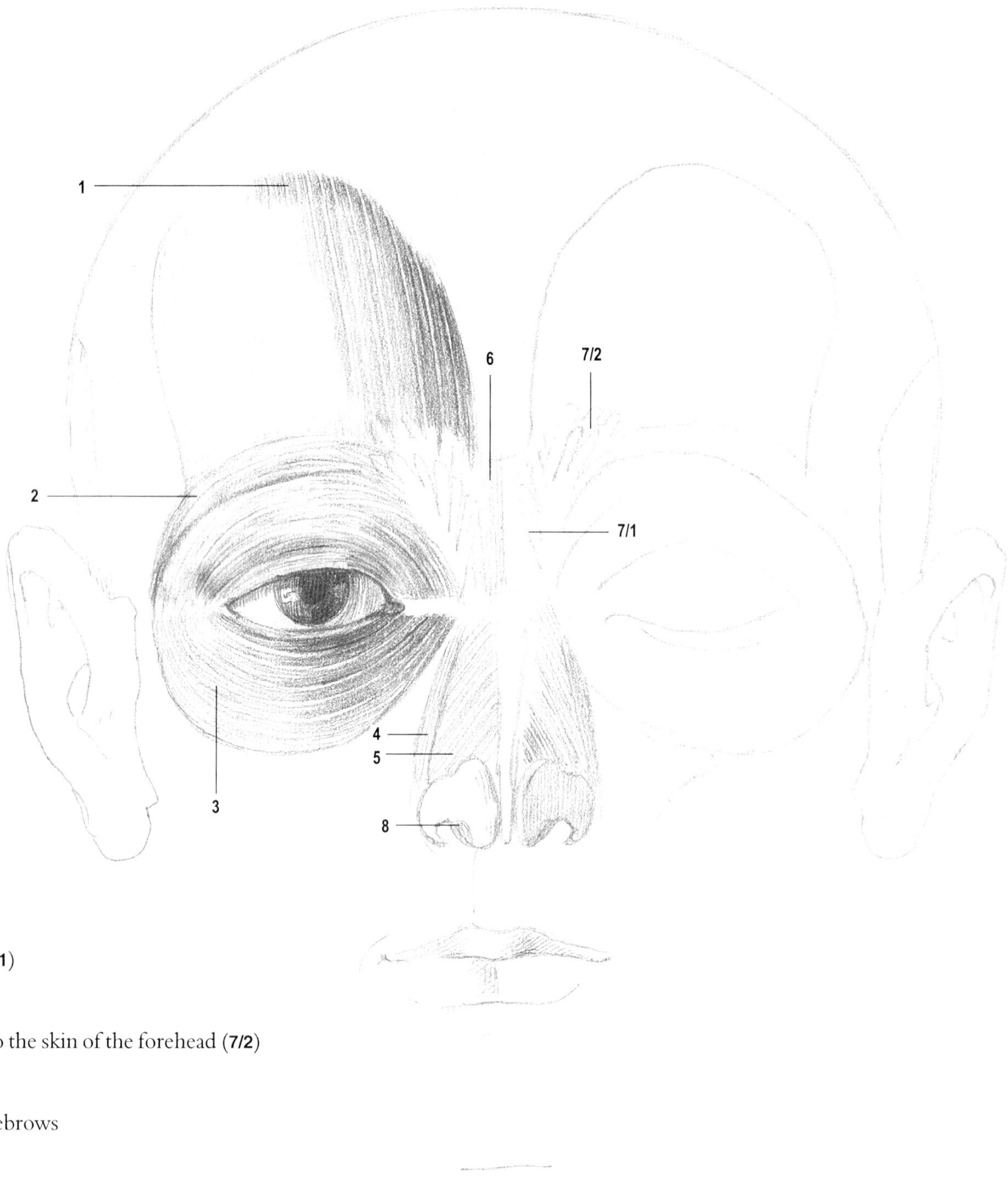

ORIGIN:
Forehead (**7/1**)

INSERTION:
Radiates into the skin of the forehead (**7/2**)

FUNCTION:
Folds the eyebrows

1 Frontalis (*124/1*)
2 Orbicualris oculi (pars orbitalis) (*127*)
3 Orbicualris oculi (pars palpebralis) (*127*)
4 Levator labii superioris (nasal part) (*137/1*)
5 Nasalis (transverse part) (*125*)
6 Procerus (*124/3*)
7 Corrugator supercilii (*128*)
8 Nasalis (pars alaris) (*125*)

Fig. 263
Eye Studies

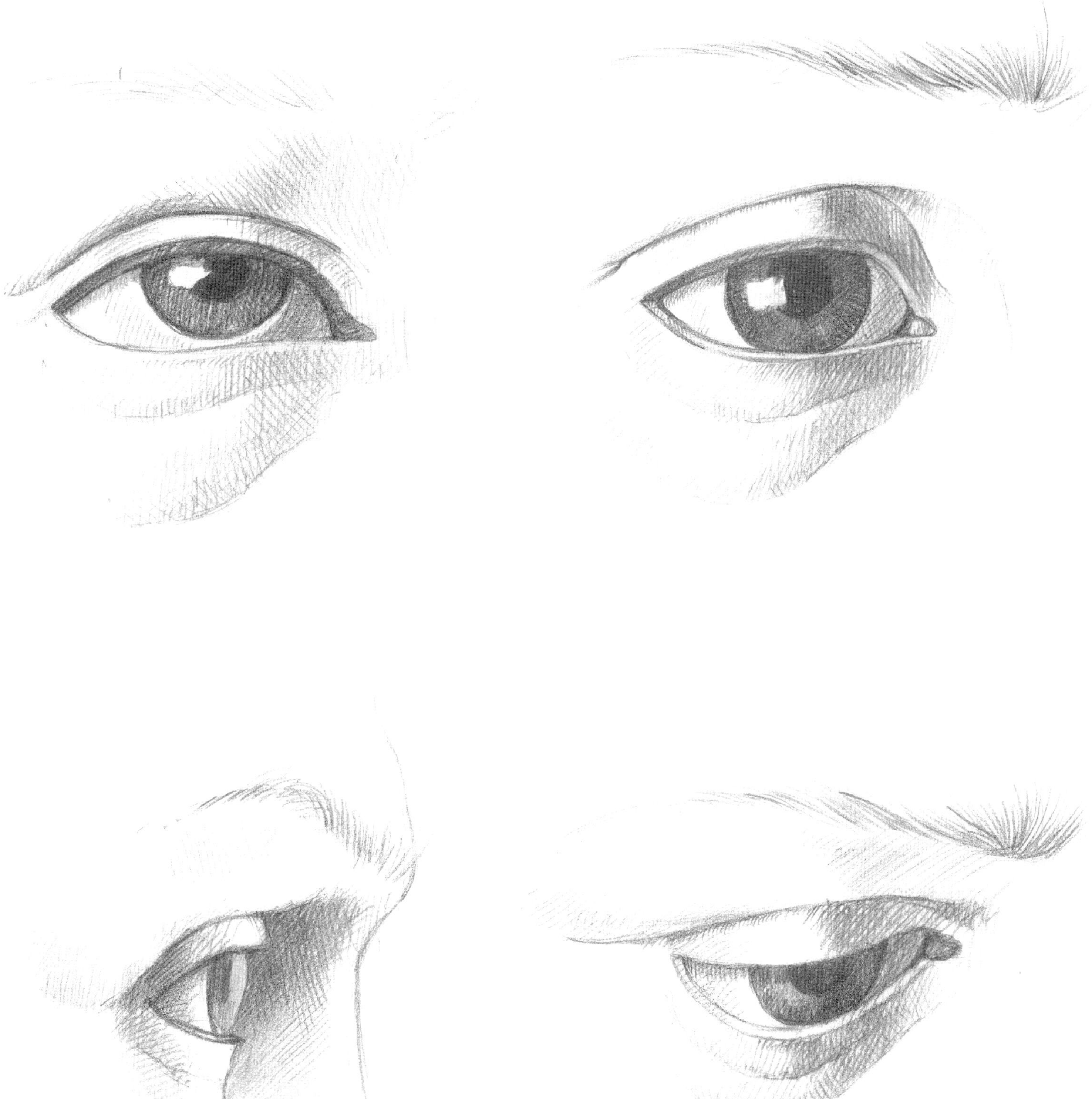

THE NOSE

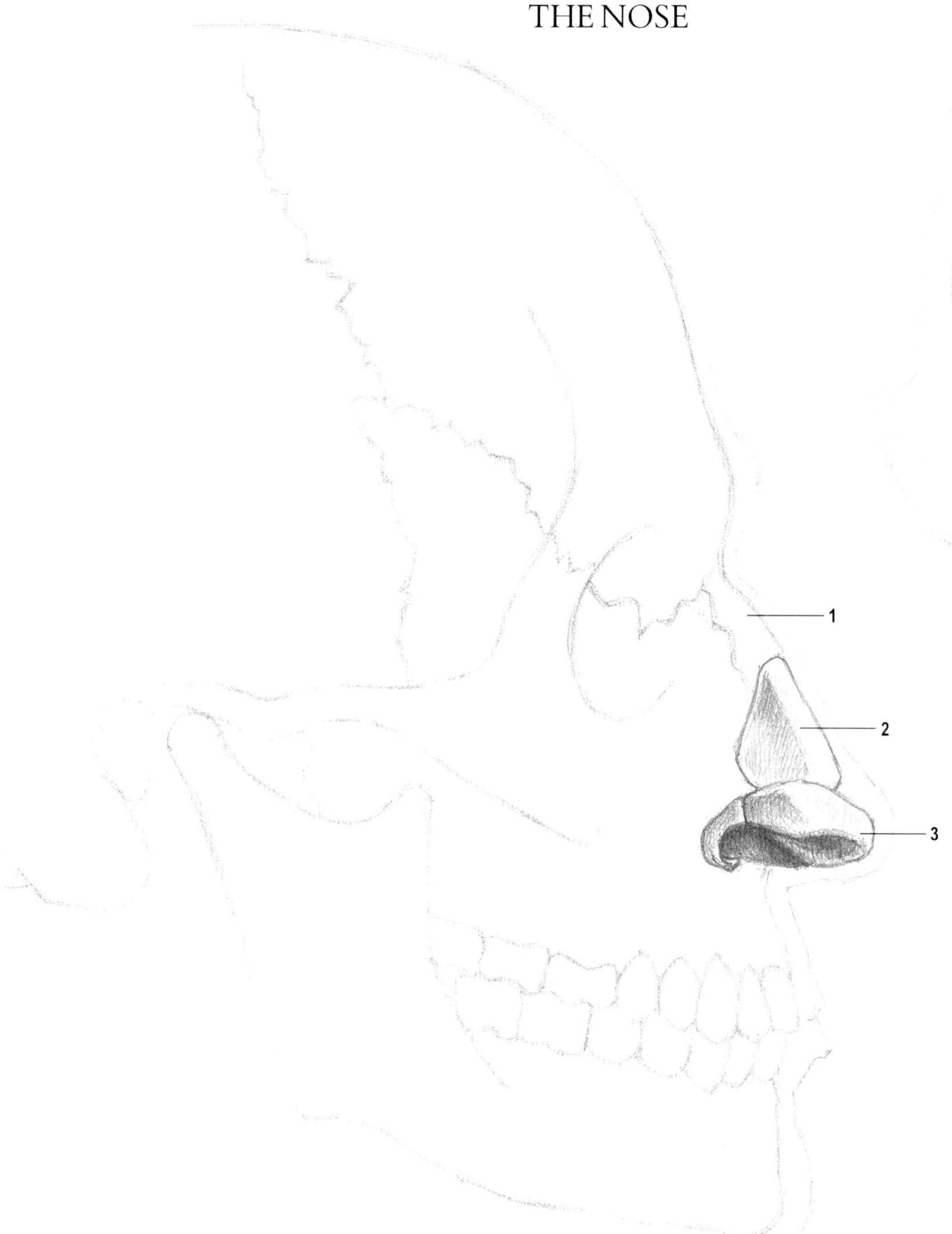

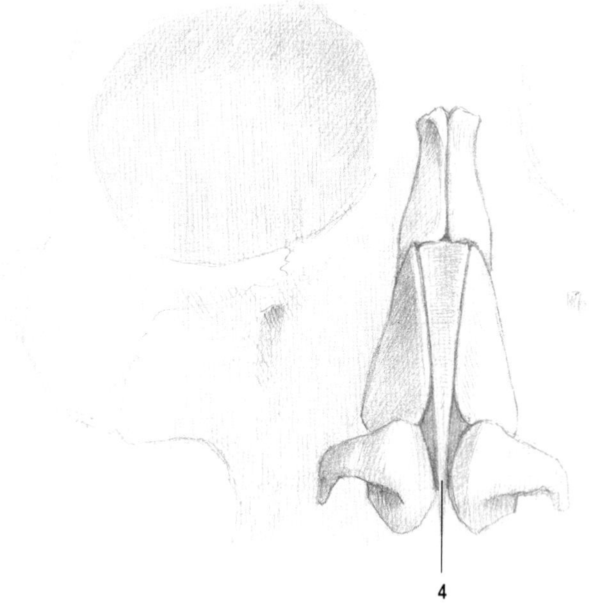

The bone structure of the nose consists of the nose bone (**1**), which is connected to the lateral (**2**) and alar cartilage (**3**). The cartilagenous septum (**4**) separates the right nostril from the left nostril.

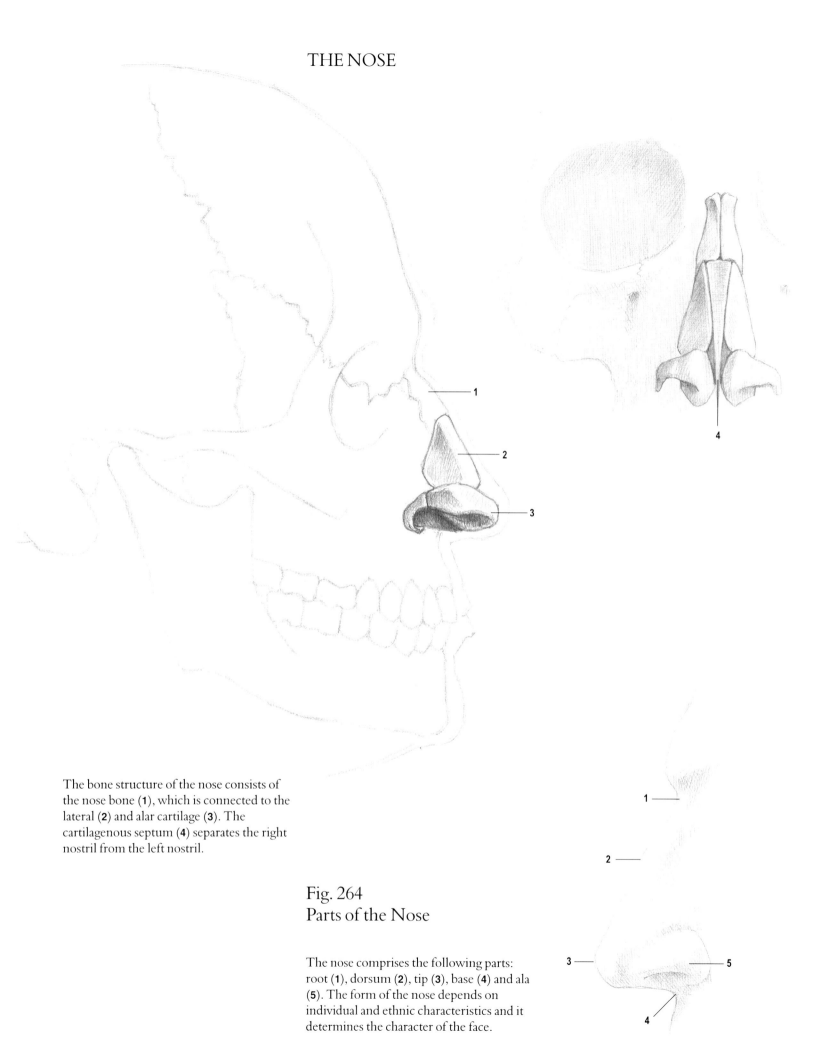

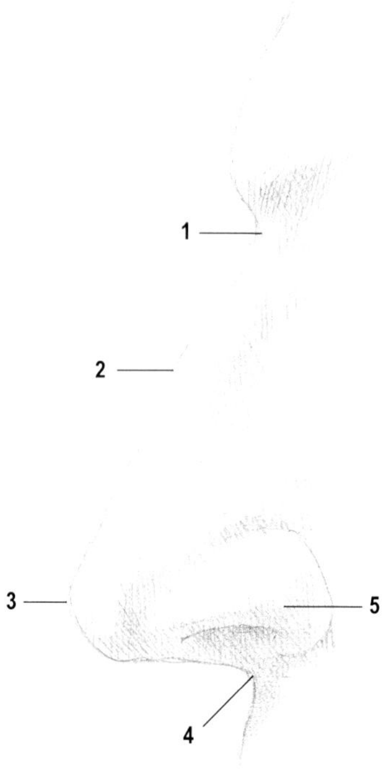

Fig. 264
Parts of the Nose

The nose comprises the following parts: root (**1**), dorsum (**2**), tip (**3**), base (**4**) and ala (**5**). The form of the nose depends on individual and ethnic characteristics and it determines the character of the face.

Fig. 265
The Muscles of the Nose

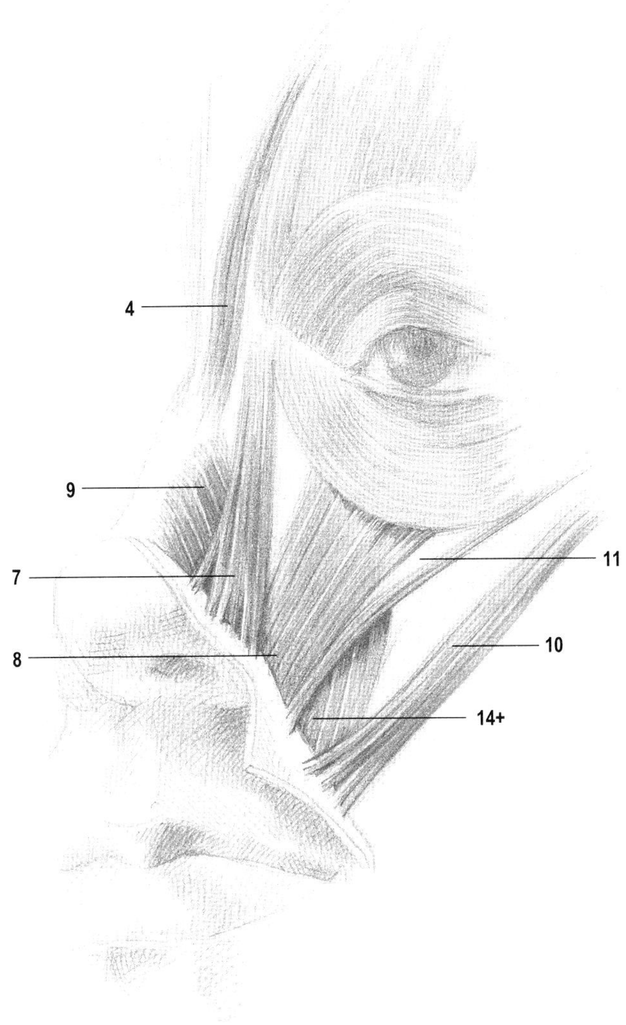

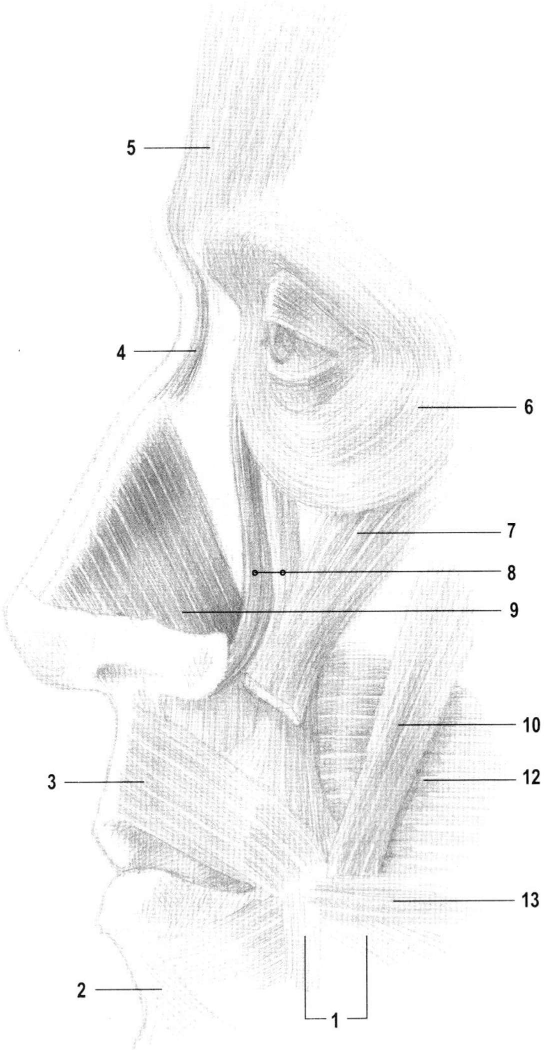

1 Triangularis (*133*)
2 Mentalis (*136*)
3 Orbicularis oris (*132*)
4 Procerus (*124/3*)
5 Temporoparietalis (*124/2*)
6 Orbicularis oculi (*127*)
7 Levator labii superioris (lateral part) (*137/2*)
8 Levator labii superioris (medial part) (*137/2*)
9 Nasalis (transverse part) (*125*)
10 Zygomaticus major (*138*)
11 Zygomaticus minor (*137/3*)
12 Buccalis (*141*)
13 Risorius (*139*)
14 Levator anguli oris (*140*)

Fig. 266
The Nose Muscle
(M. nasalis, *125*)

ORIGIN:
Infundibular process of the upper jaw (pars transversus, **1**); connects the plates of the parietal cartilage (ars alaris, **2**); anterior part of the nose, it merges with its contralateral part (pars septalis, **3**)

INSERTION:
Anterior part of the nose, it merges with its contralateral part (**4**)

FUNCTION:
Narrows the nasal cavity (transverse part); expands and lifts (**4**) the ala nasii (alar part), e.g. in breathing

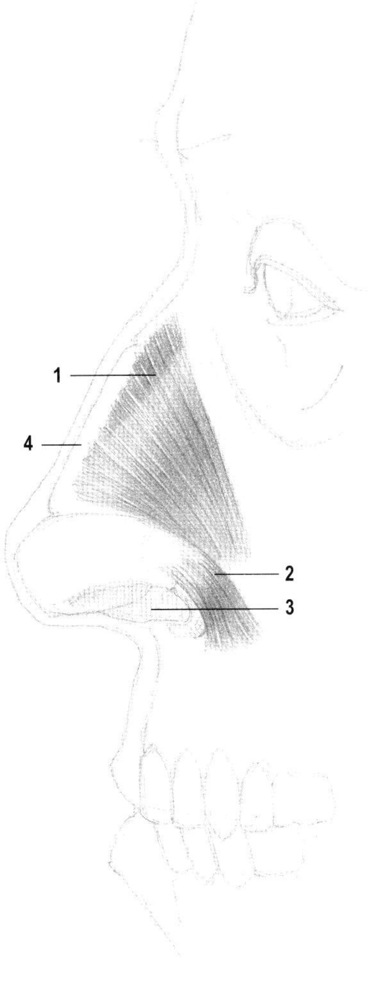

Fig. 267
The Muscles of the Nose and Lips

1 Orbicularis oculi (*127*)
2 Procerus (*124/3*)
3 Levator labii superioris (medial part) (*137/2*)
4 Nasalis (transverse part) (*125*)
5 Nasalis (septal part) (*125*)
6 Levator labii superioris (lateral part) (*137/2*)
7 Orbicularis oris (*132*)
8 Inferioris (*135*)
9 M. depressor anguli oris (*133*)
10 Risorius (*139*)
11 Zygomaticus major (*138*)
12 Buccalis (*141*)
13 Zygomaticus minor (*137/3*)
14 Venter frontalis (*124/1*)

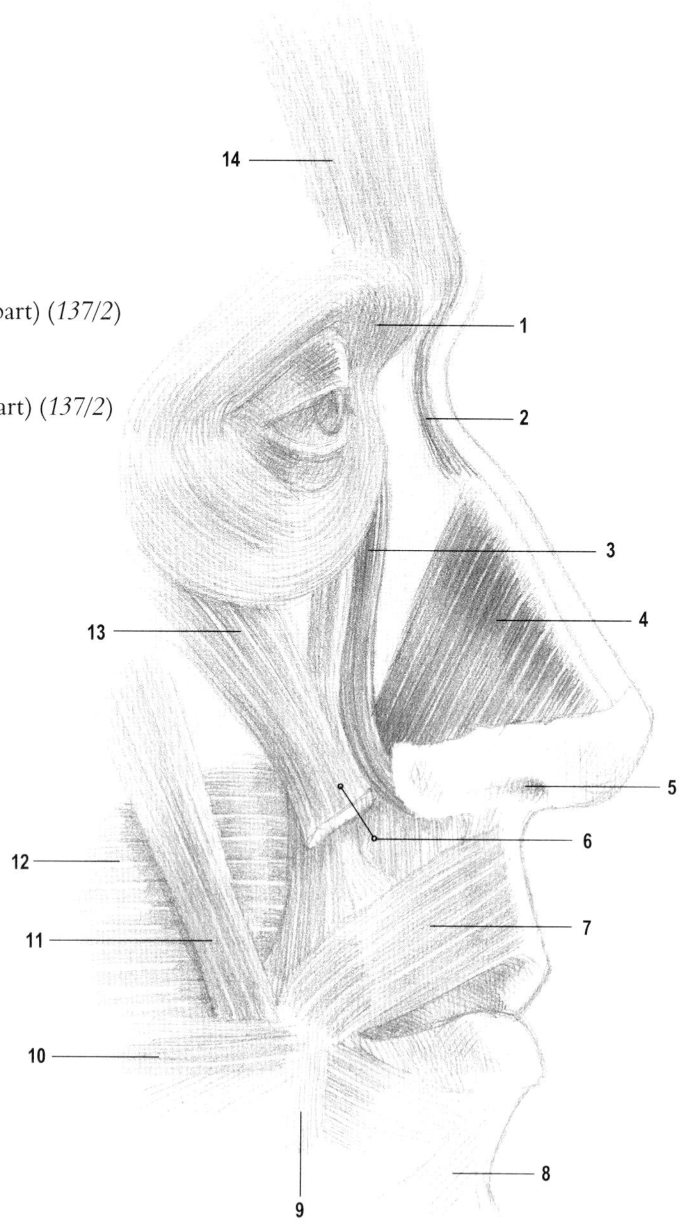

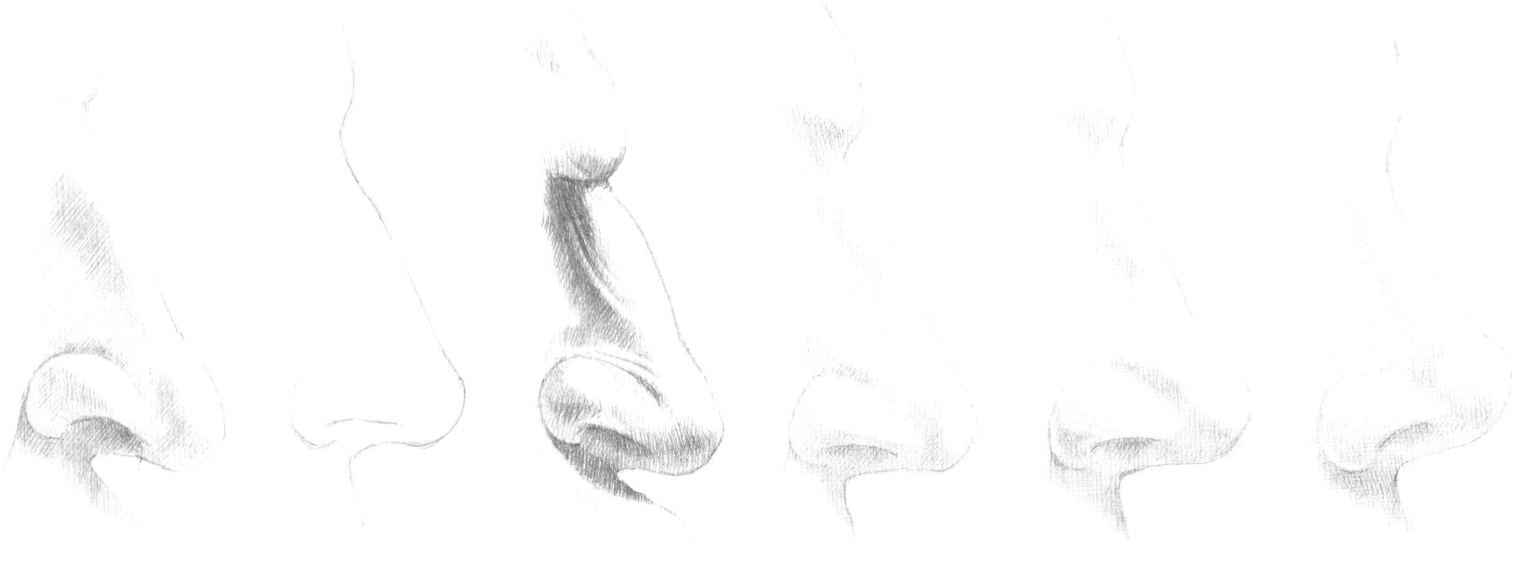

Fig. 268
The Buccinator Muscle, *141*

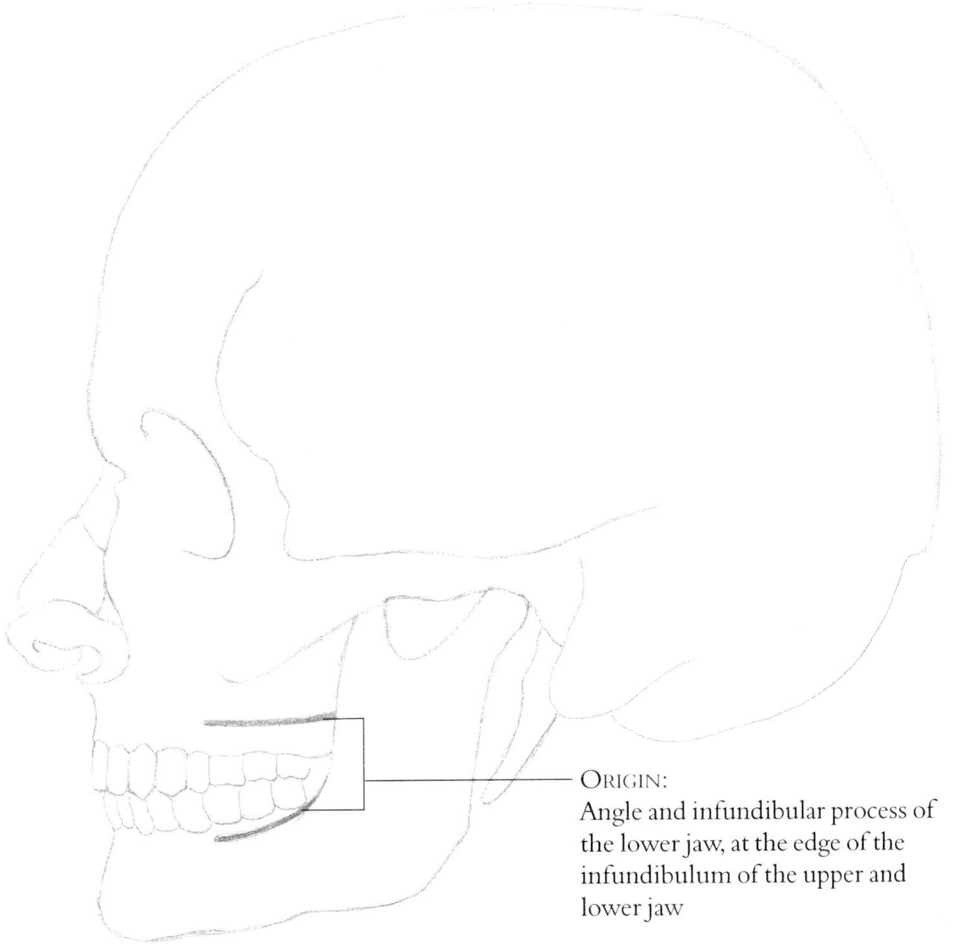

ORIGIN:
Angle and infundibular process of the lower jaw, at the edge of the infundibulum of the upper and lower jaw

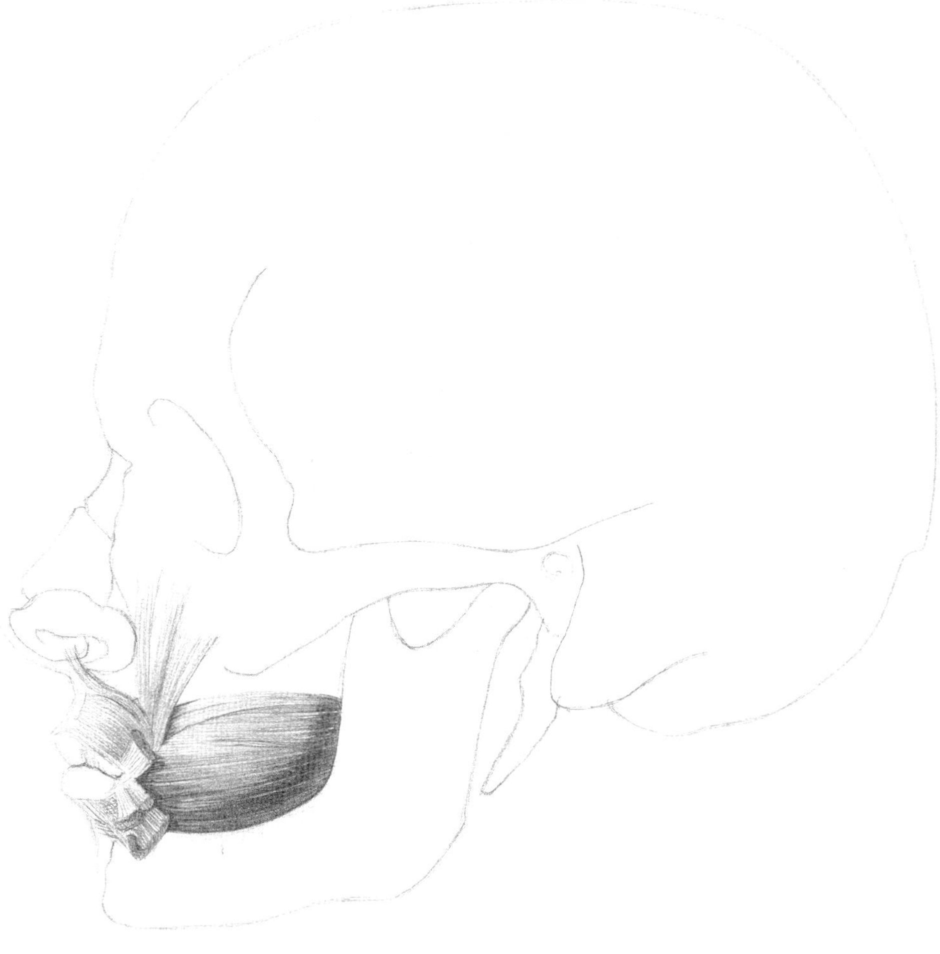

INSERTION:
Corners of the mouth

FUNCTION:
Pulls the corners of the mouth laterally and backwards

Fig. 269
The Chin Muscle
(M. mentalis, *136*)

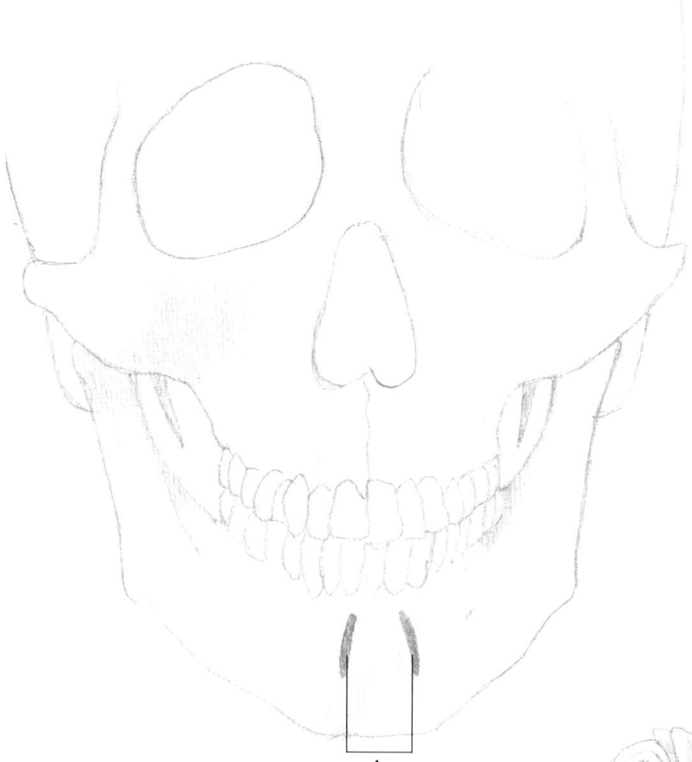

ORIGIN:
Incisura of the lower jaw, near the canine tooth (**1**)

INSERTION:
Skin in the chin region and Orbicularis oris (**2**)

FUNCTION:
Folds and extends the skin in the chin region

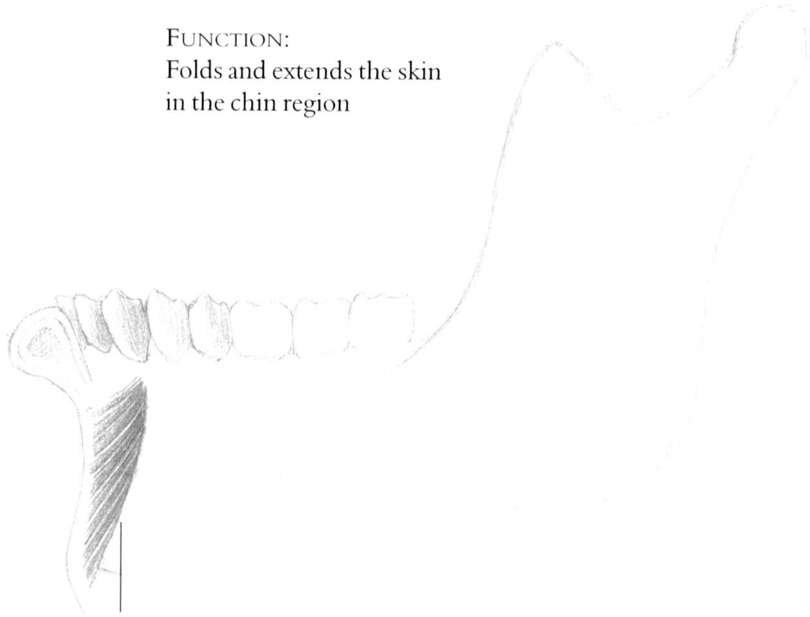

lateral side

1 Orbicularis oris (*134*)
2 Digastricus (*16*)
3 Mylohyoideus (*18*)
4 Transverse muscle of the chin
5 Mentalis (*138*)

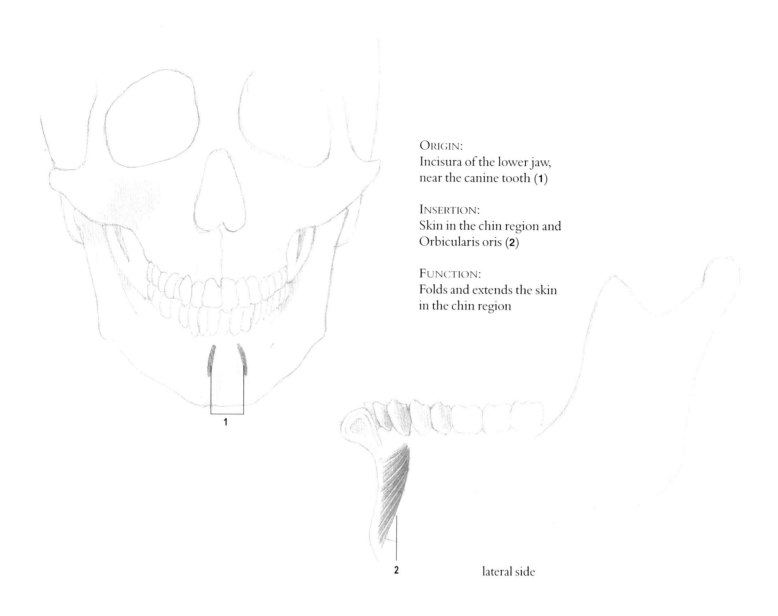

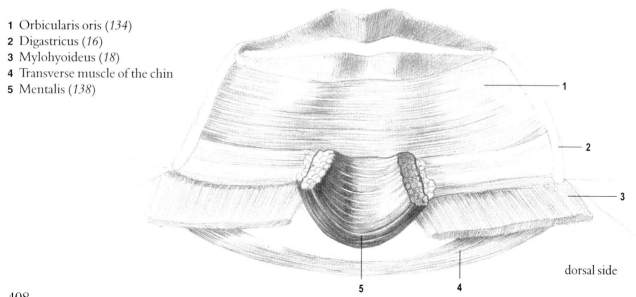

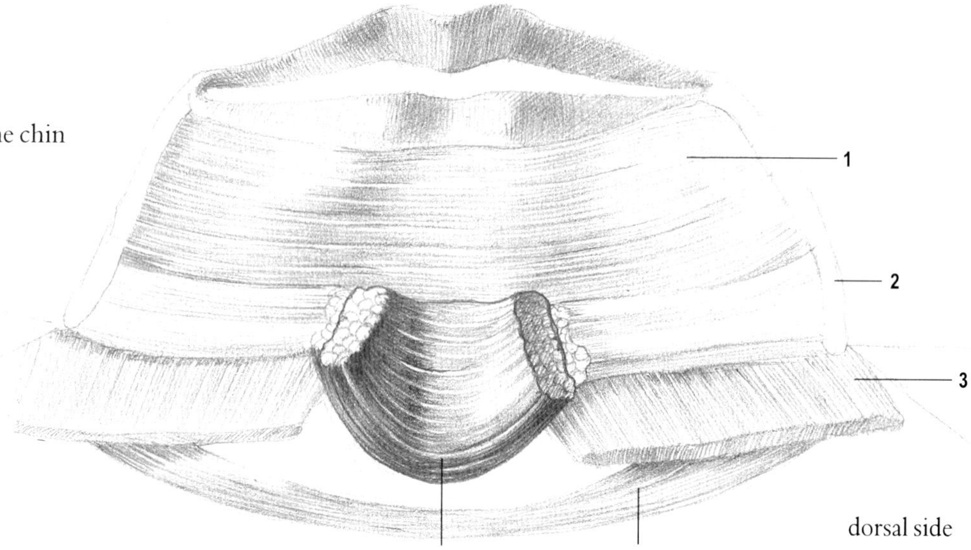

dorsal side

Fig. 270
The Triangular Muscle or Depressor Muscle of the Angle of the Mouth
(M. depressor anguli oris, *133*)

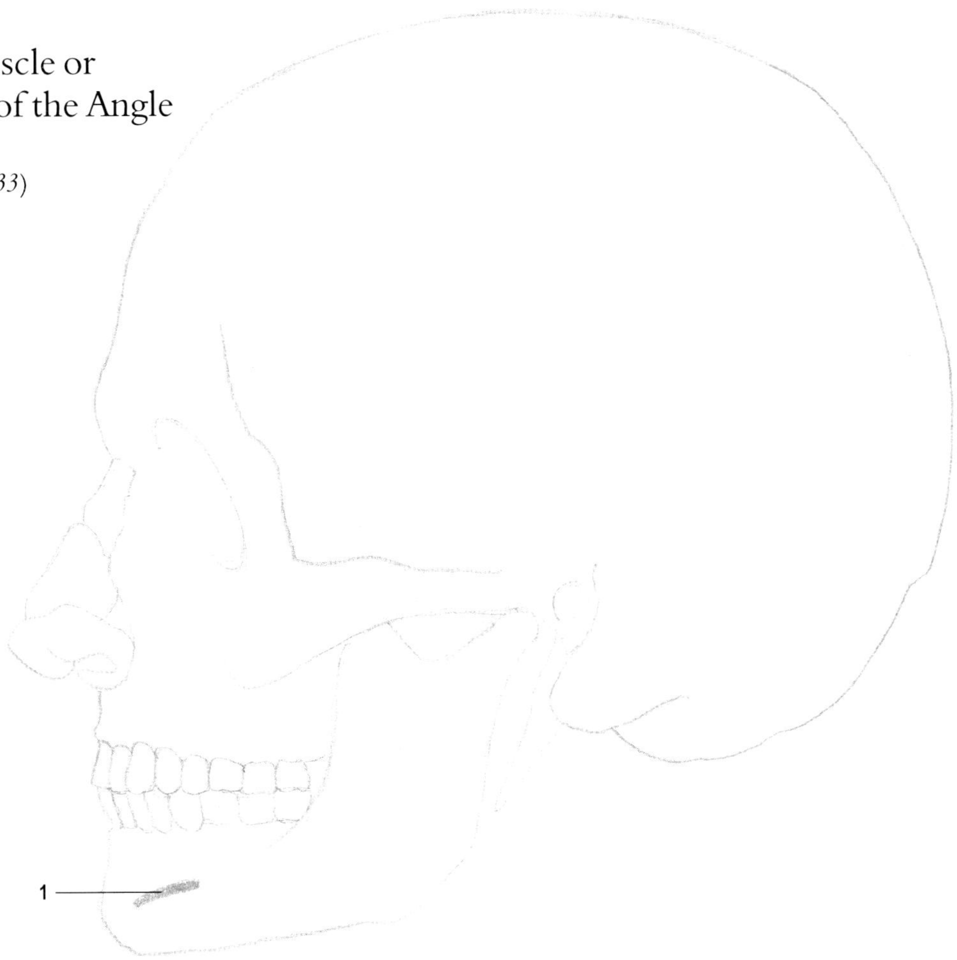

ORIGIN:
Mentum side of the lower jaw (**1**)

INSERTION:
Corners of the mouth (**2**)

FUNCTION:
Pulls the corners of the mouth downwards (sorrow, derision, hatred), folds the skin in the angle of the chin

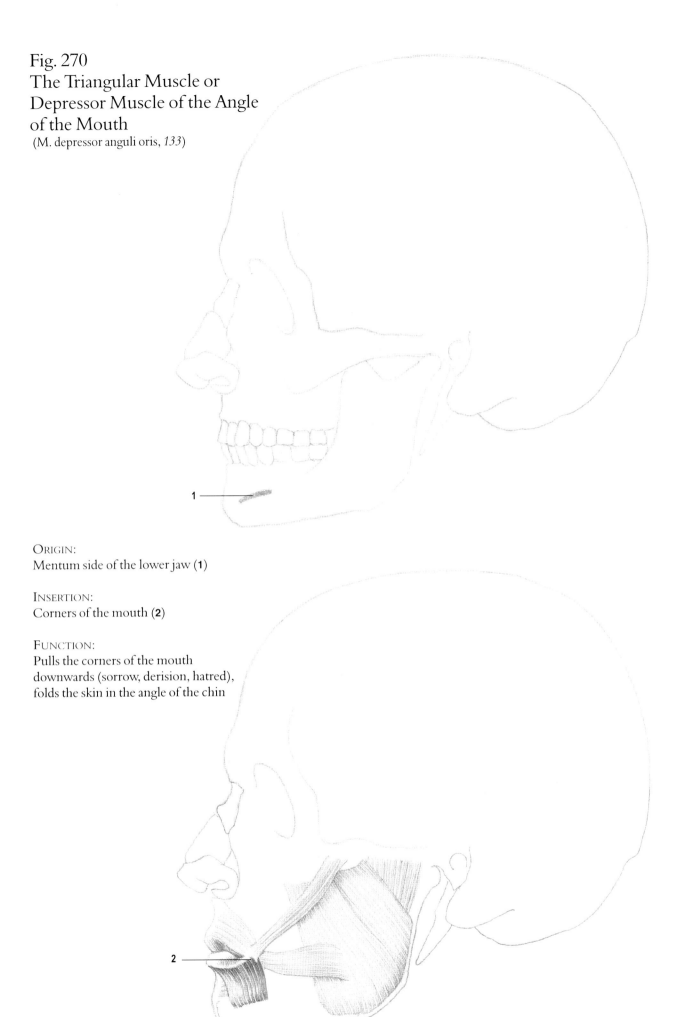

Fig. 271
The Levator Muscle of the Angle of the Mouth
(M. levator anguli oris, *140*)

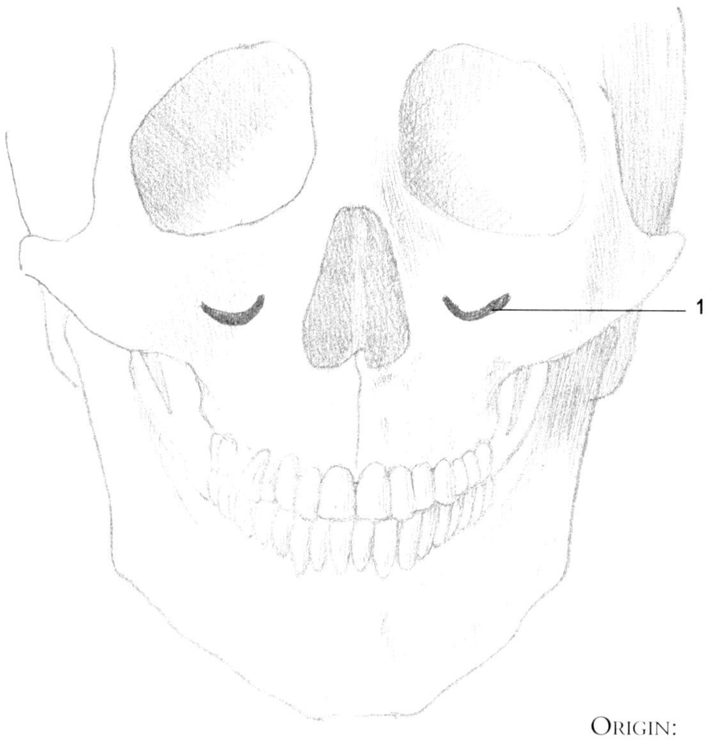

Origin:
Incisura of the upper jaw (**1**)

Insertion:
Corners of the mouth, orbicularis oris (**2**)

Function:
Folds the skin at the corners of the mouth (**3**)

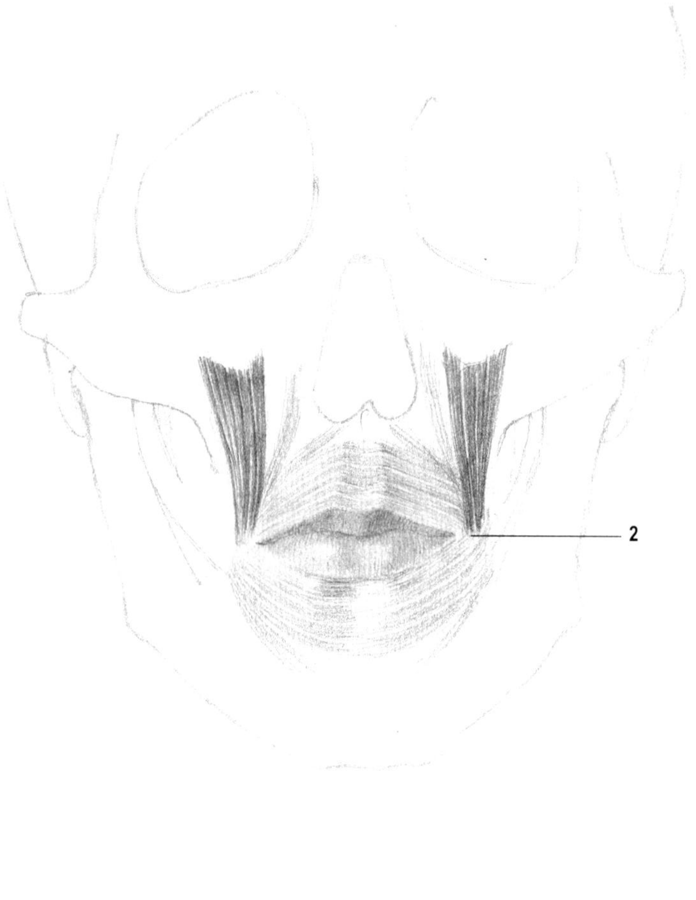

Fig. 272
The Levator Muscle of the Upper Lip and Nasal Wing *[1]*

(M. levator labii superior alaeque nasi, 137/1)

The Levator Muscle of the Upper Lip *[2]*

(M. levator labii superioris, 137/2)

The Lesser Zygomatic Muscle *[3]*

(M. zygomaticus minor, 137/3)

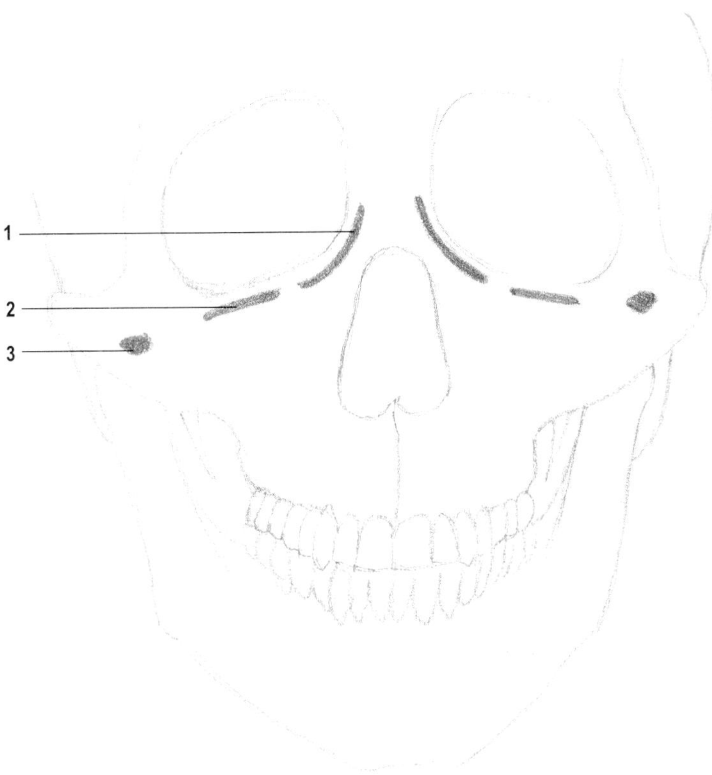

[1]
ORIGIN:
Frontal process of the upper jaw (**1**)

INSERTION:
Central part of the skin of the upper lip at the cartilage of the ala nasi

FUNCTION:
Lifts upper lip and ala of the nose

[2]
ORIGIN:
Edge of the eye socket (nose bone and lachrymal bone) (**2**)

INSERTION:
Skin of the upper lip and orbicularis oris

FUNCTION:
Lifts the upper lip laterally

[3]
ORIGIN:
Zygomatic bone (**3**)

INSERTION:
Lateral side of the upper lip

FUNCTION:
Lifts the upper lip

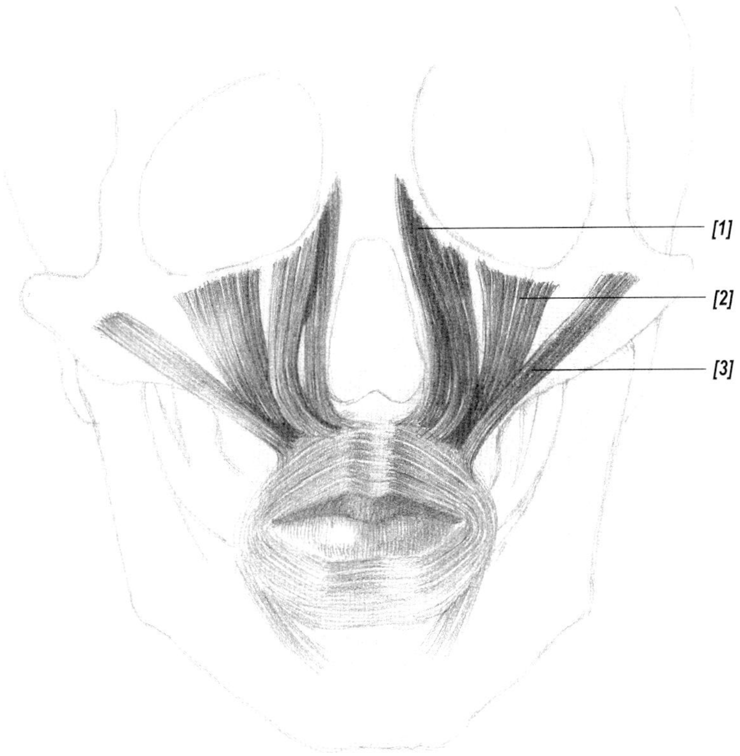

Fig. 273
The Depressor Muscle of the Lower Lip, *139*

(M. depressor labii inferioris, 135)

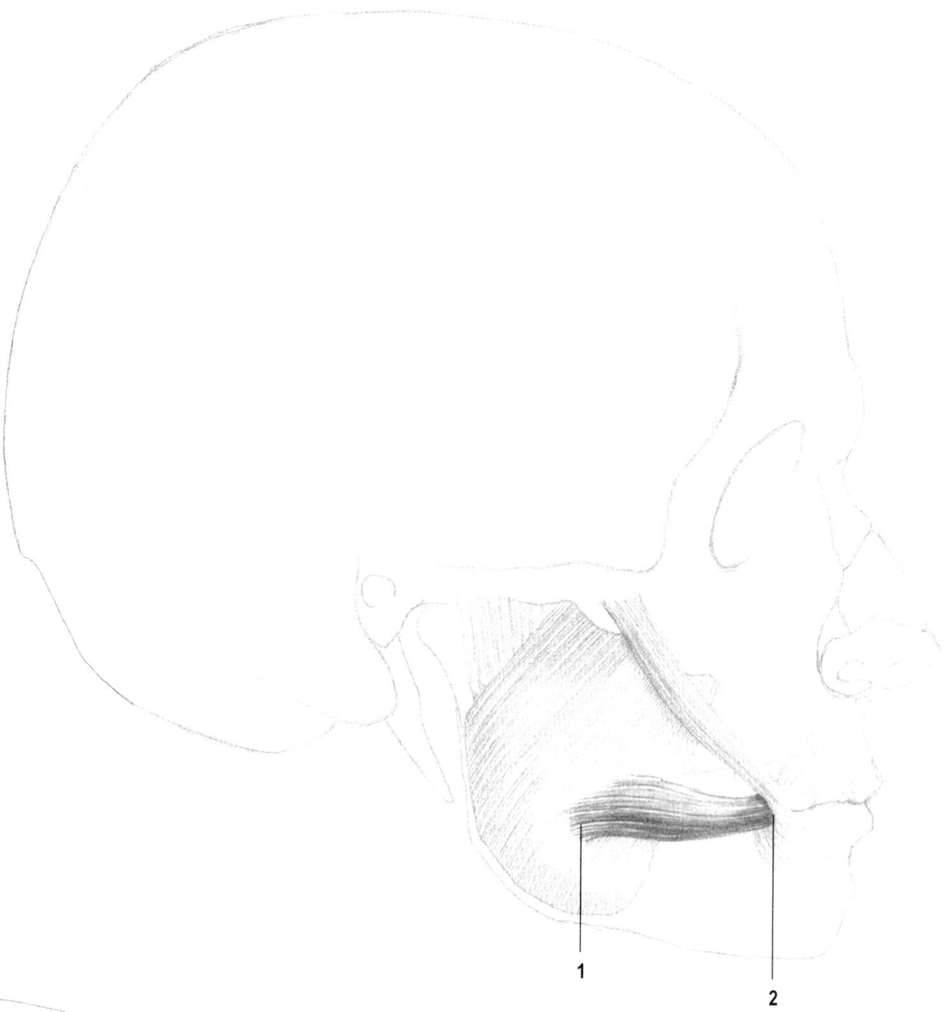

ORIGIN:
Fascia of the masseter (**1**)

INSERTION:
Skin of the corner of the mouth (**2**)

FUNCTION:
Pulls the corner of the mouth cranially

Fig. 274
The Quadratus/Depressor/Labii Inferioris, *135*

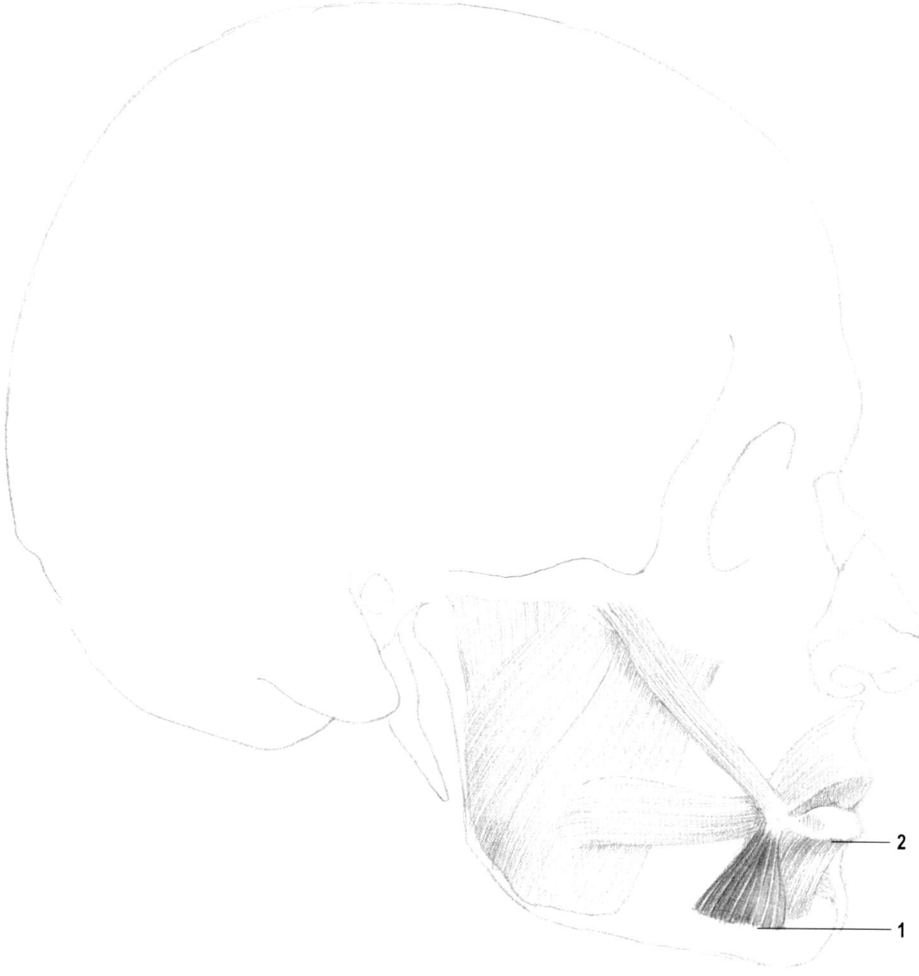

ORIGIN:
Inferior margin of the lower jaw (**1**)

INSERTION:
Lower lip part of the orbicularis oris (**2**)

FUNCTION:
Pulls the lower lip caudally

Fig. 275
The Zygomaticus major, *138*

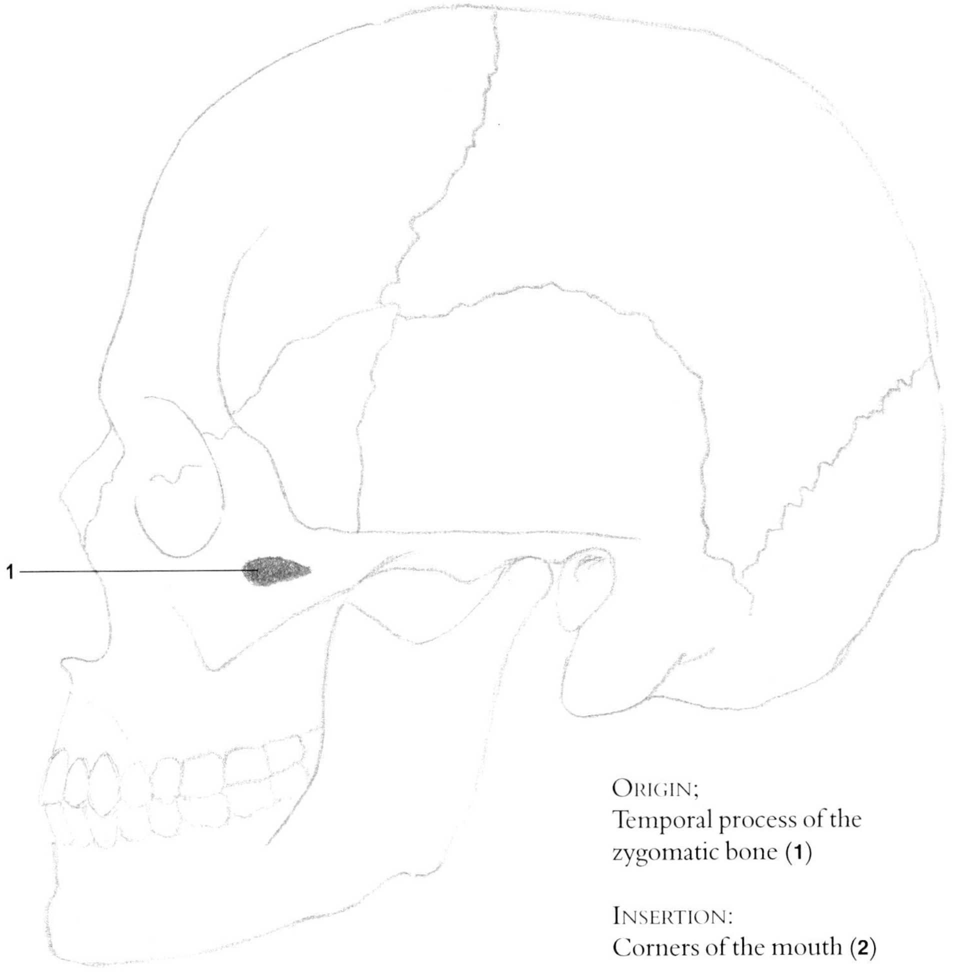

ORIGIN;
Temporal process of the zygomatic bone (**1**)

INSERTION:
Corners of the mouth (**2**)

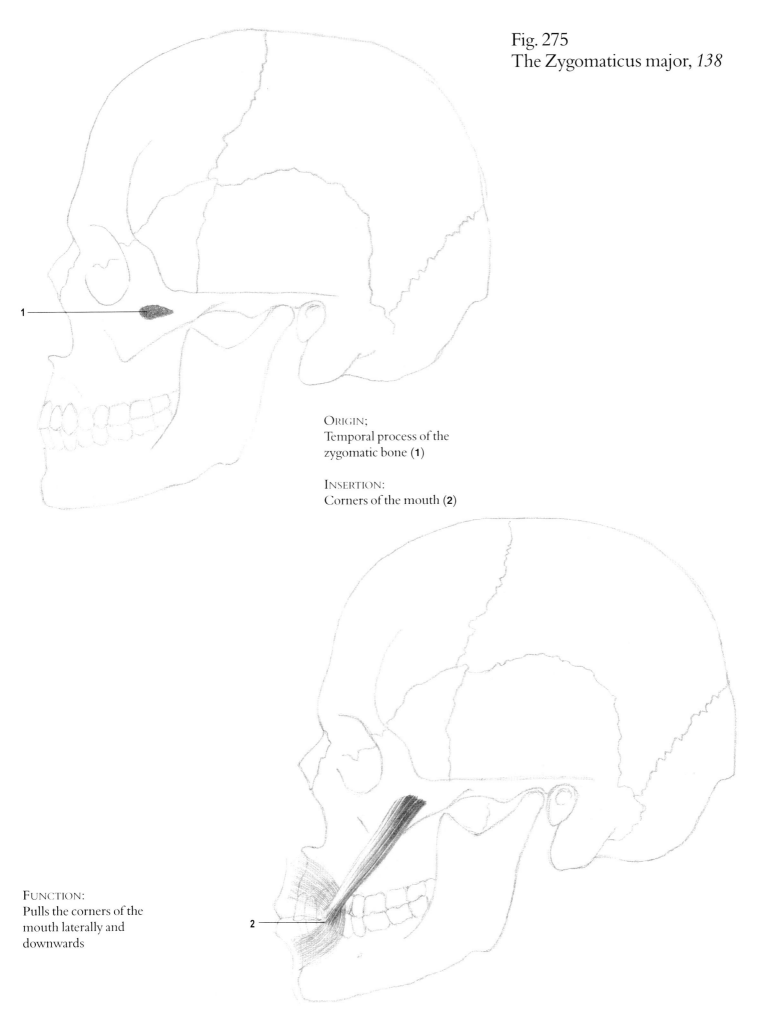

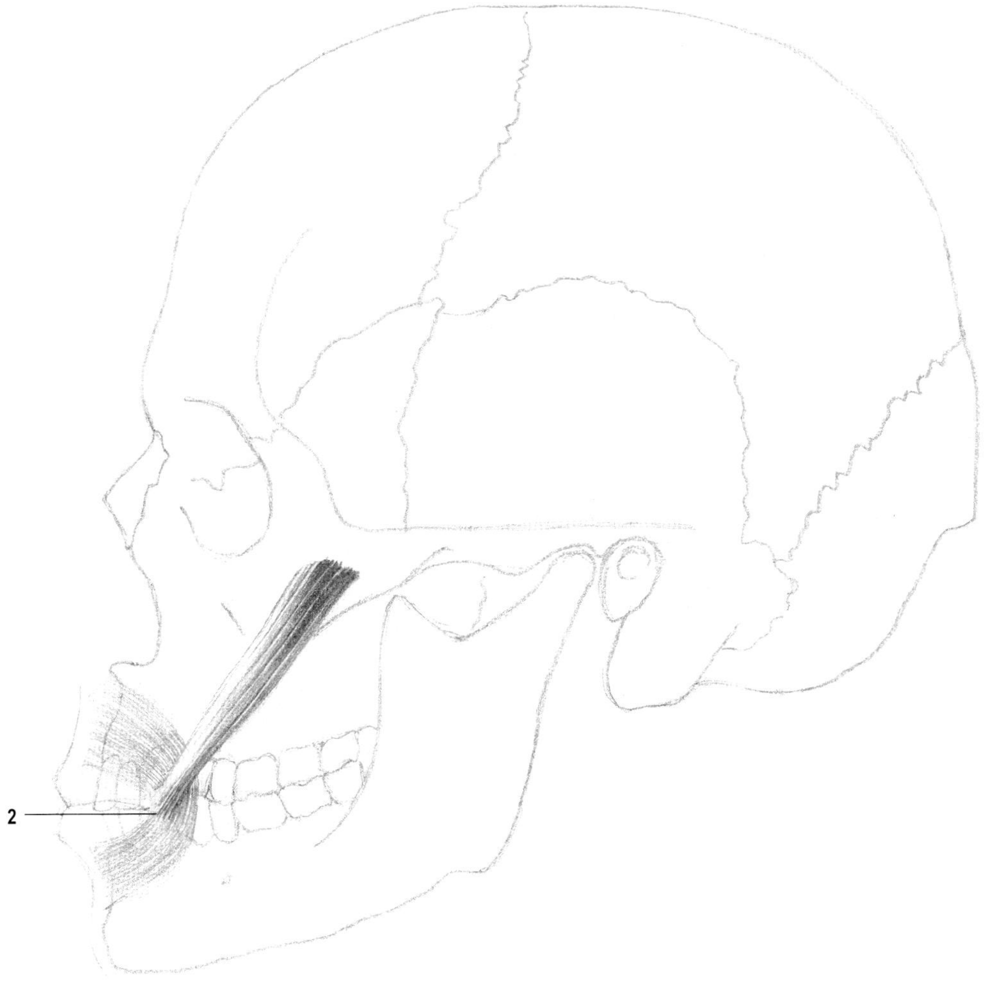

FUNCTION:
Pulls the corners of the mouth laterally and downwards

Fig. 276
The Lip Muscles

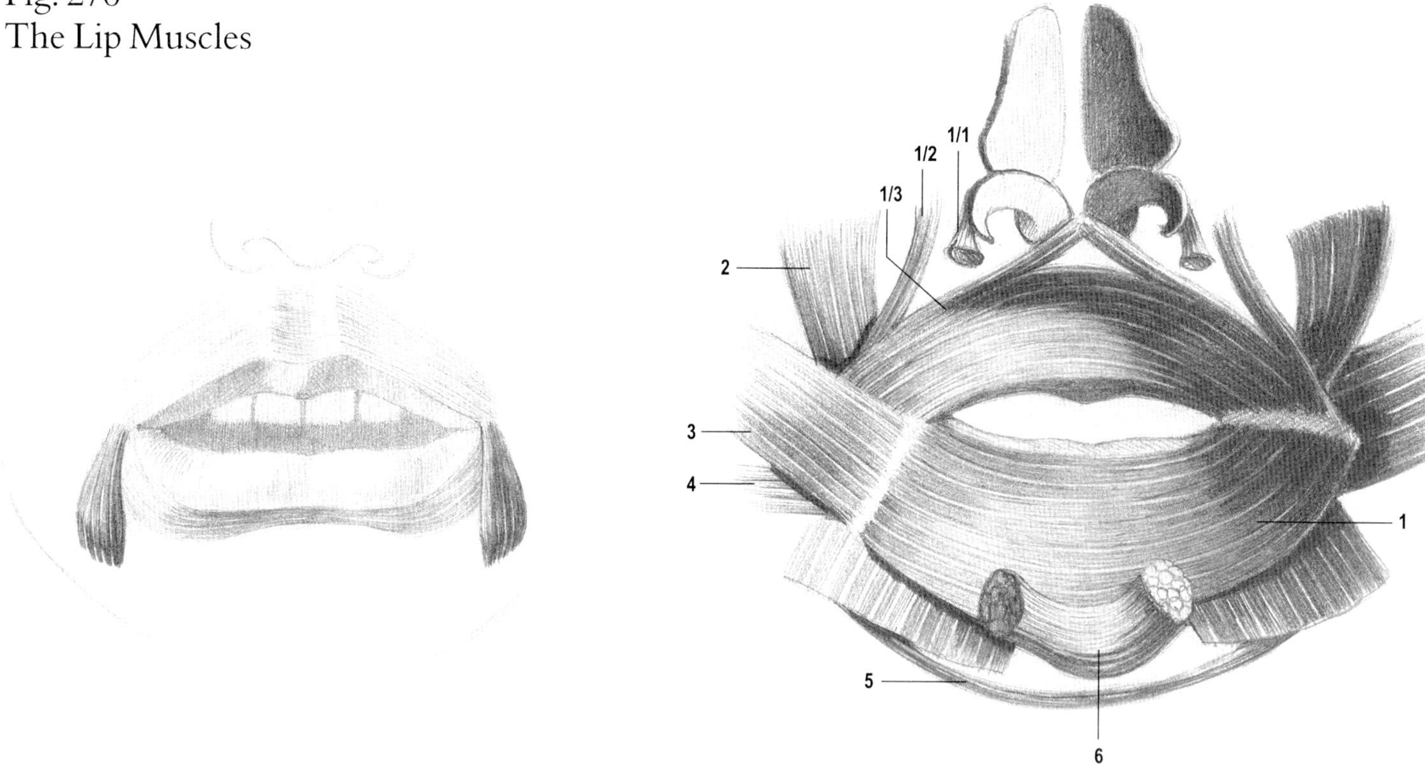

ventral side

dorsal side

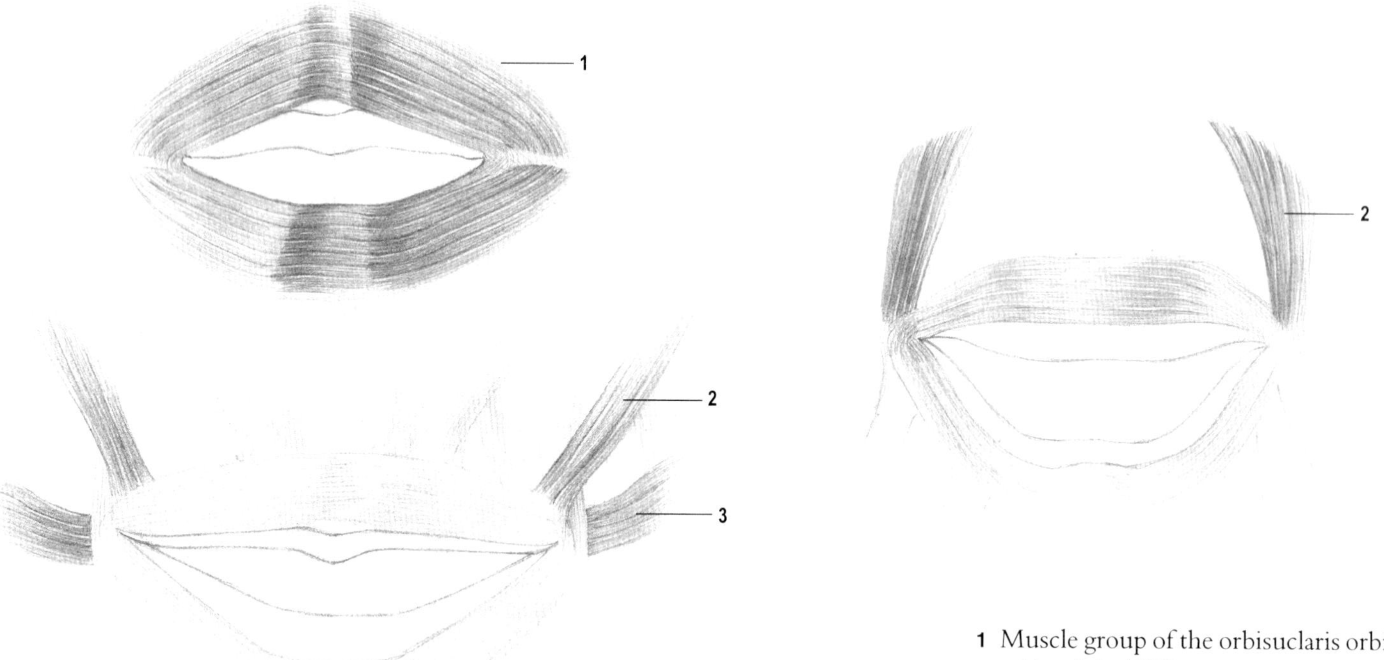

1 Muscle group of the orbisuclaris orbis with origin (*132*)
  1/1 at the cartilage of Triangularis
  1/2 at the cartilagenous septum
  1/3 at the upper jaw
2 Levator anguli oris (*140*)
3 Risorius (*139*)
4 M. depressor anguli oris (*133*)
5 Transversus menti (*134*)
6 Mentalis (*136*)

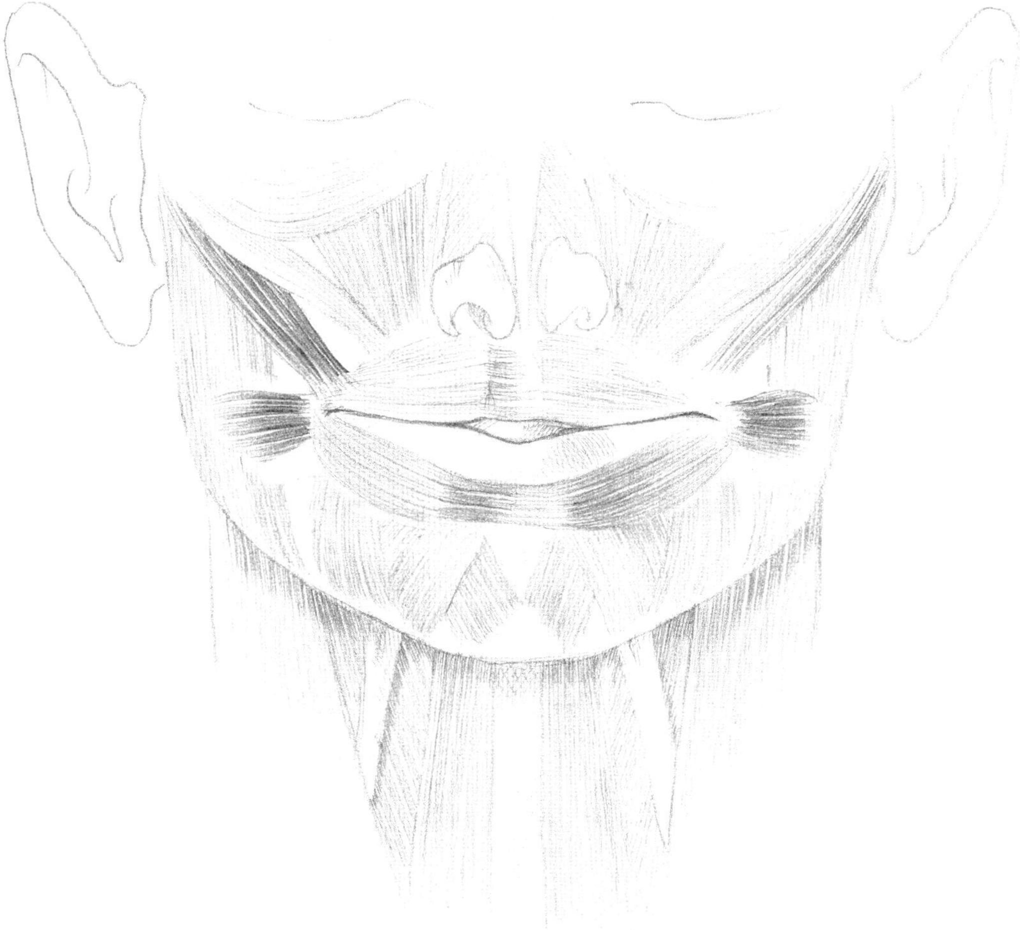

Fig. 277
The Orbicular Muscle of the Mouth
(M. orbicularis oris, *132*)

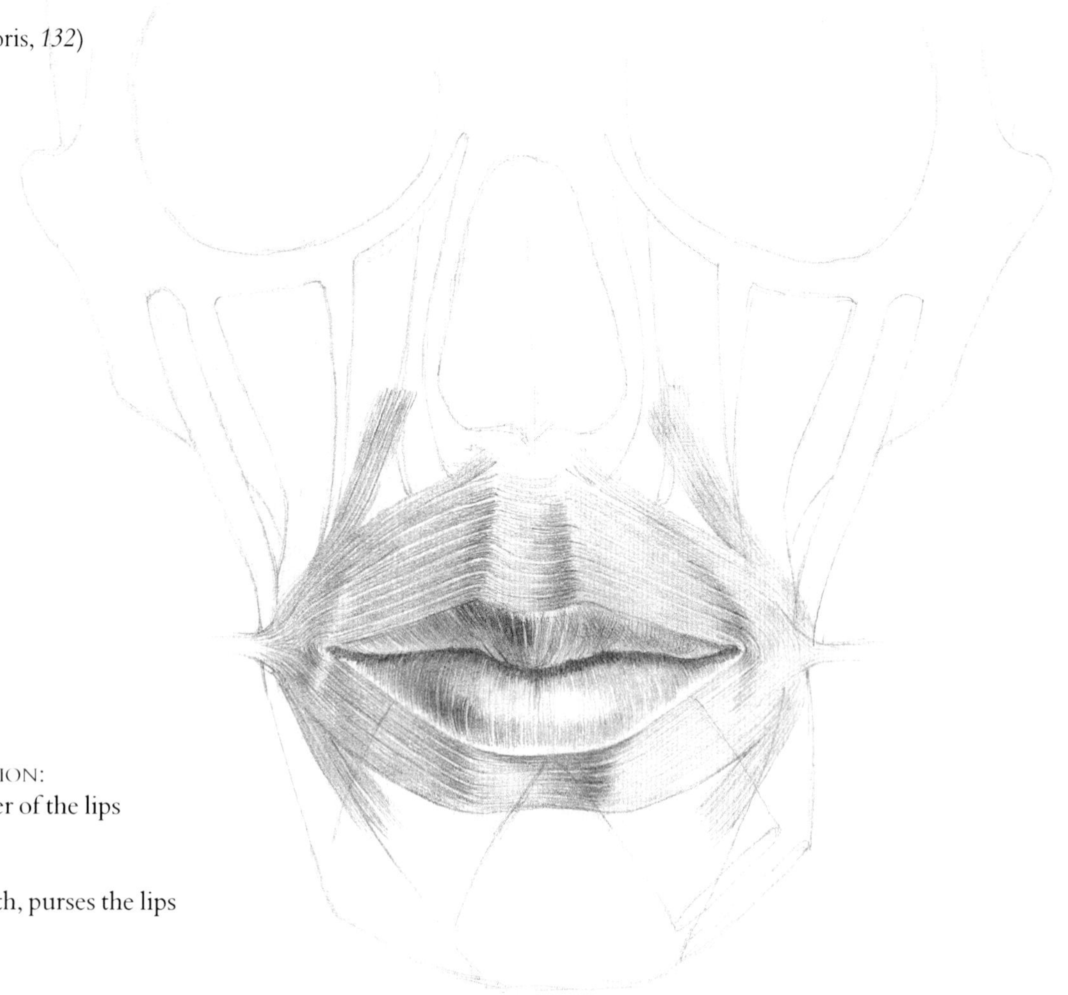

ORIGIN–INSERTION:
Forms the border of the lips

FUNCTION:
Closes the mouth, purses the lips

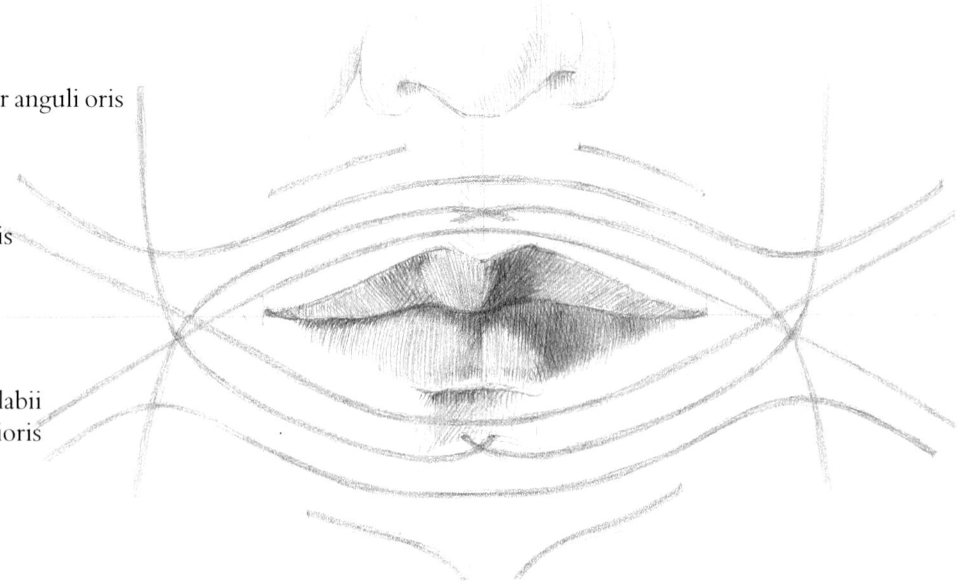

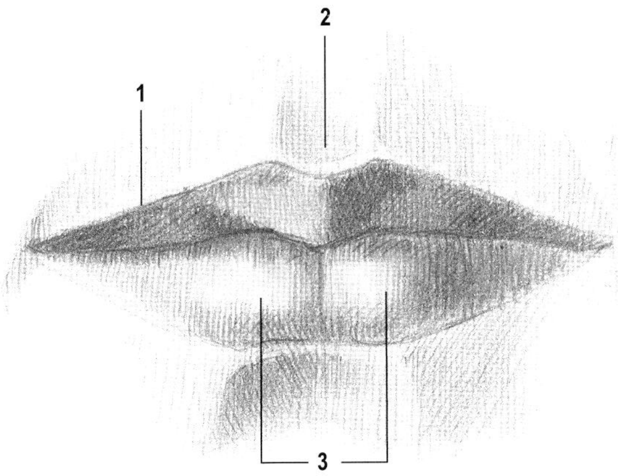

Fig. 278
The Mouth

At the edge of the lips (**1**) the skin merges with the mucous membrane of the buccal cavity. The lower edge of the philtrum (**2**) – the groove in the middle of the upper lip – passes between the two bulges (**3**) of the lower lip.

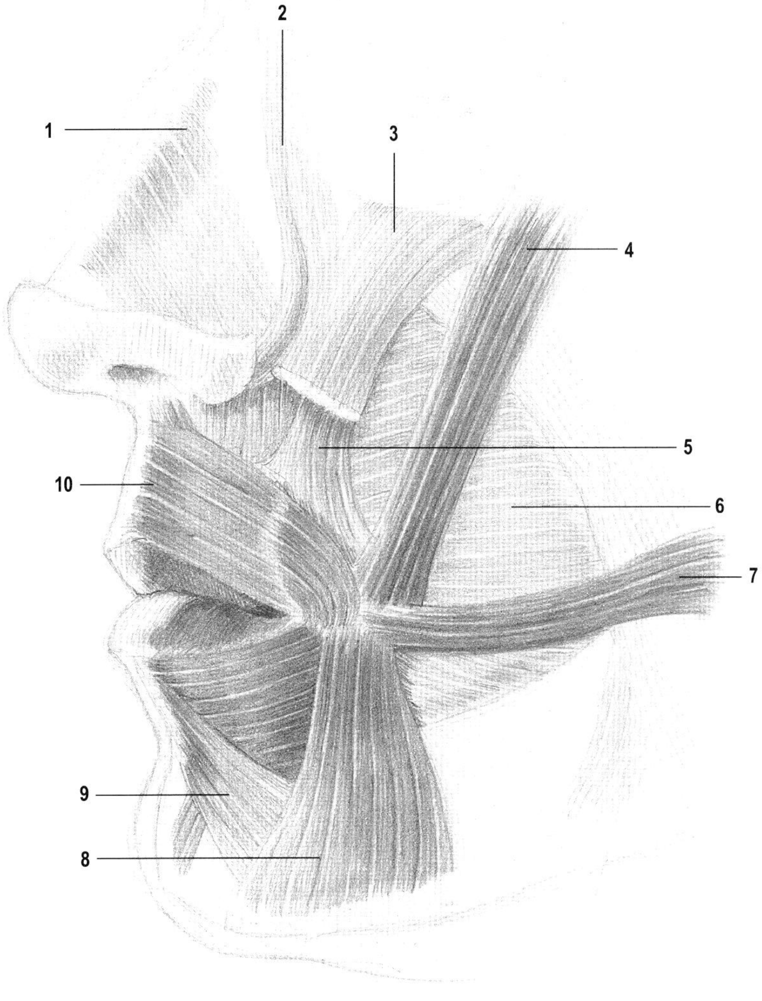

Fig. 279
The Muscles of the Nose and Lips

1 Nasalis (transverse part) (*125*)
2 Levator labii superioris (medial part) (*137/1*)
3 Levator labii superioris (lateral part) (*137/2*)
4 Zygomaticus major (*138*)
5 Levator anguli oris (*140*)
6 Buccalis (*141*)
7 Risorius (*139*)
8 Triangularis (*133*)
9 Quadratus labii inferioris (*135*)
10 Orbicularis oris (*132*)

Fig. 280
The Lips

The lips of people of African origin are an obvious example of ethnically determined lip forms (**1**). The lips and the chin have always been a key individual characteristic, for instance the prominent fat lip of the Habsburgs, known as the Leopoldinum. The prominent chin angle which accompanies these lips possesses one central and at times two lateral protuberances (mentum geminum). (**2**); at times, a horizontal sulcus (**3**) divides the chin into an upper and lower part (the chin of the Bourbons).

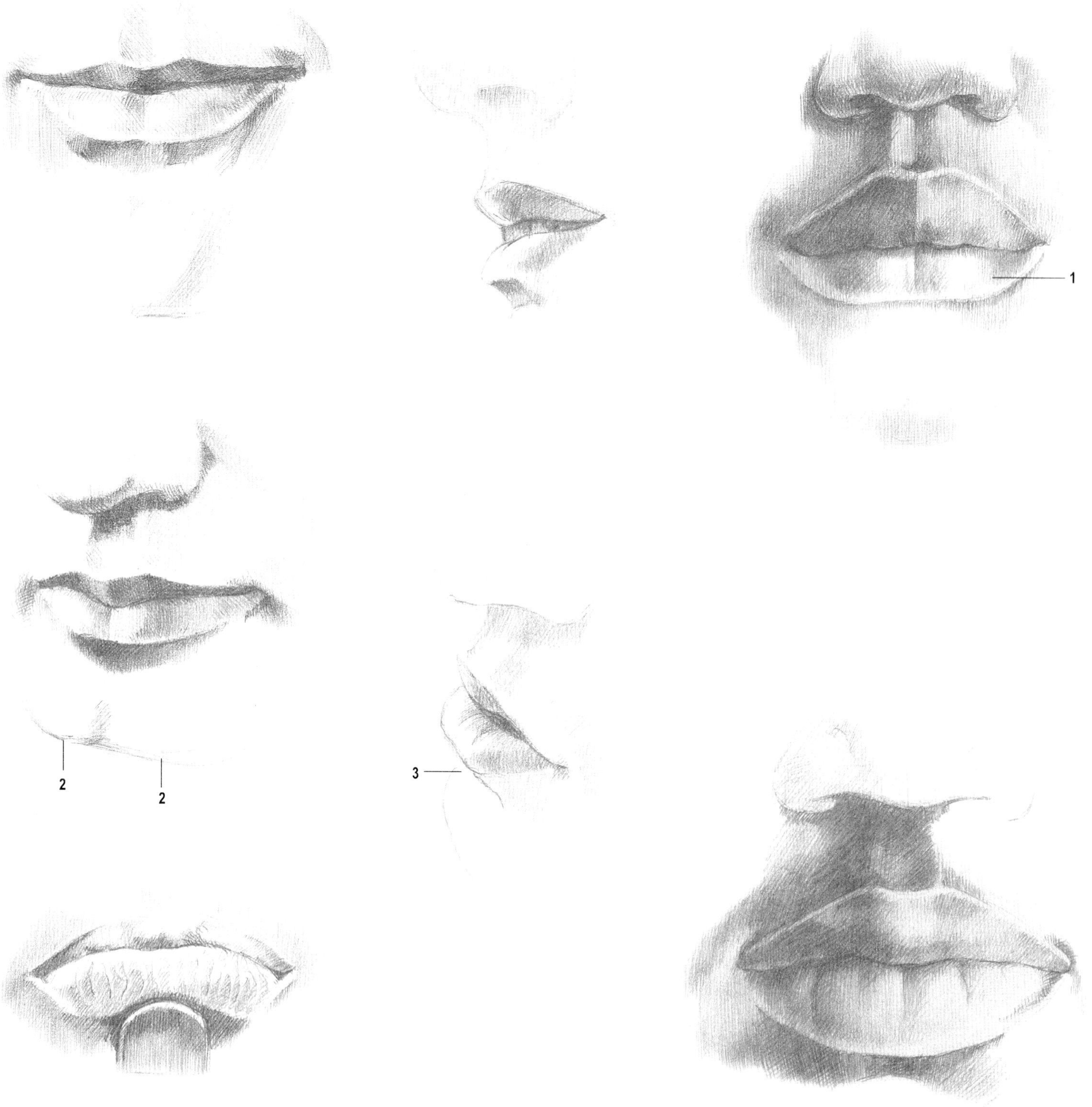

The ear muscle channels sound waves into the outer ear. Its elliptical form is the result of cartilage.

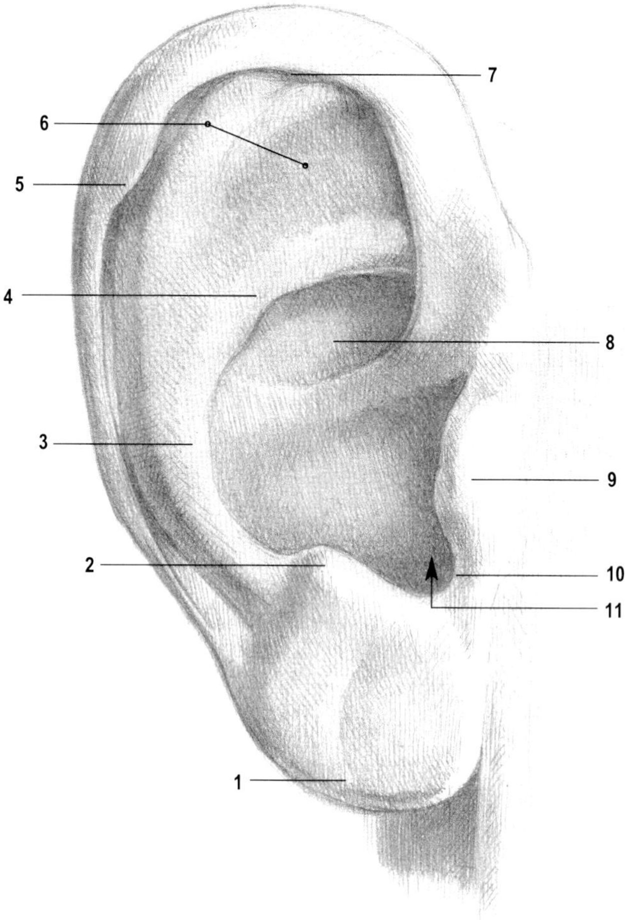

Fig. 281
The Ear Muscles

1 Ear lobes
2 Antitragus
3 Anthelix
4 Crus of the anthelix
5 Darwin's Ear
6 Triangular groove
7 Helix (snail)
8 Groove of the scutum
9 Tragus
10 Tragus incisura
11 Auditory canal

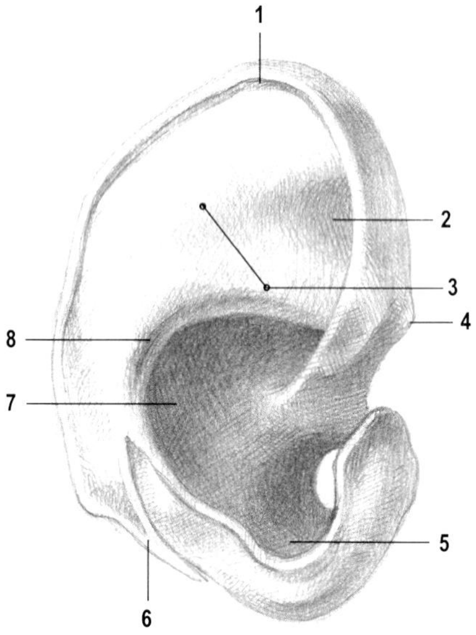

Fig. 282
The Cartilage of the Ear

1 Helix (snail)
2 Triangular groove
3 Two branches of the anthelix
4 Spine of the helix
5 Cartilage of the auditory canal
6 Tail of the helix
7 Scutum
8 Anthelix

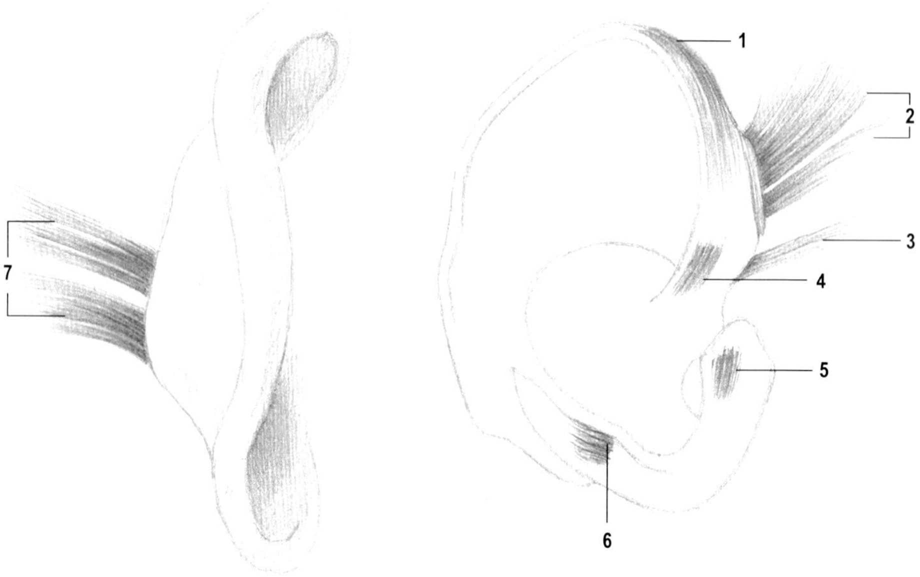

Fig. 283
The Ear Muscles

The muscles of the ear pull the root of the outer ear to the centre (**1**), to the front (**2**, **3**), and backwards (**7**). Muscles **4–6** pull the outer ear together and flex it.

1 Helicis major
2 Auricularis superior
3 Auricularis anterior
4 Helicis minor
5 Tragicus
6 Antitragicus
7 Auricularis posterior

Fig. 284
Ear Studies

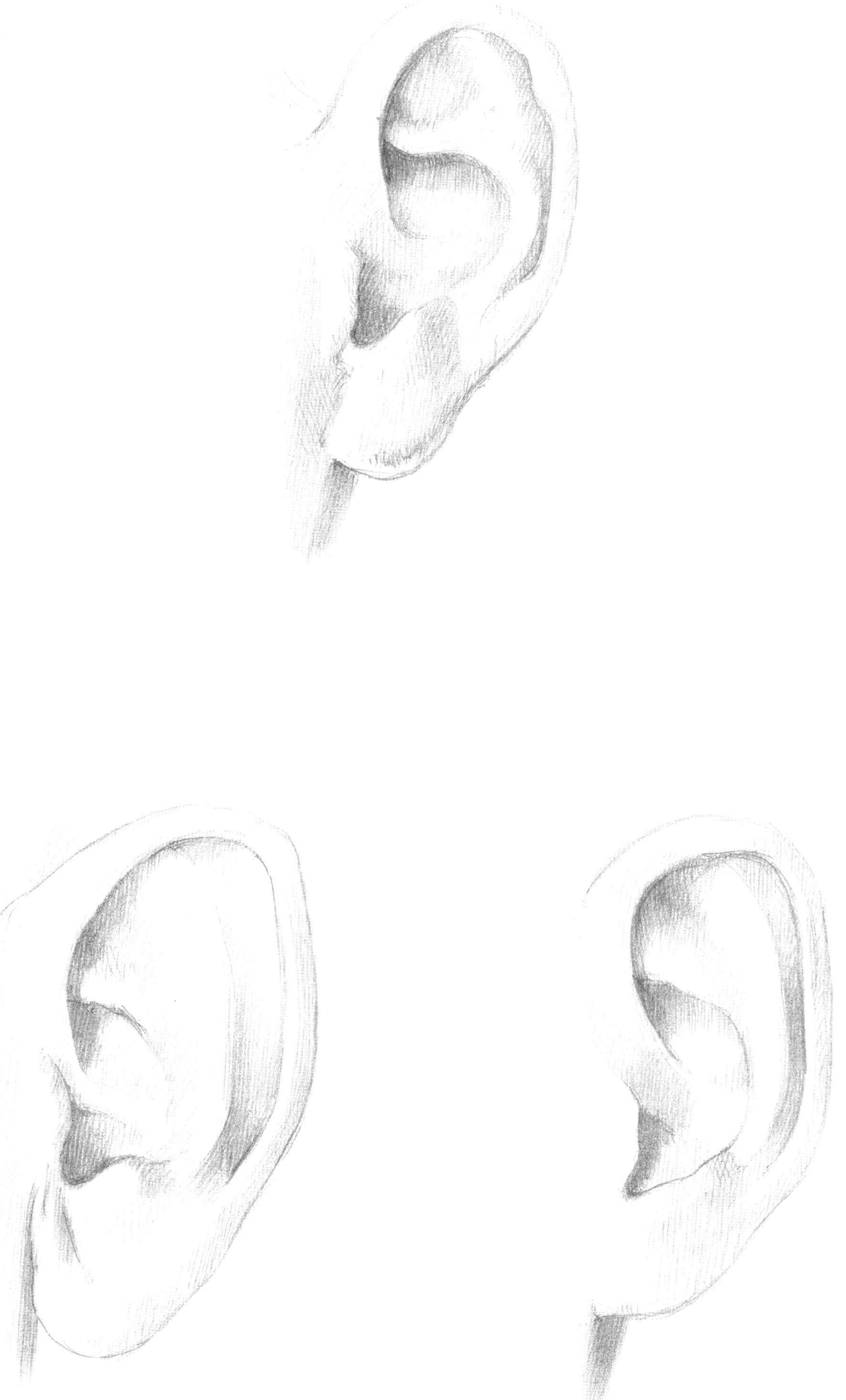

Fig. 285
Movements of the Jaw Joint

The protuberances of the lower jaw (**1**) form a joint with the articulatory groove (**2**) of the zygomatic process of the temporal bone. A cartilagenous plate corrects any irregularities between articulatory surfaces. The joint enables the opening, closure and chewing movements, and frontwards and backwards movements of the lower jaw. The movements of the lower jaw can be seen easily.

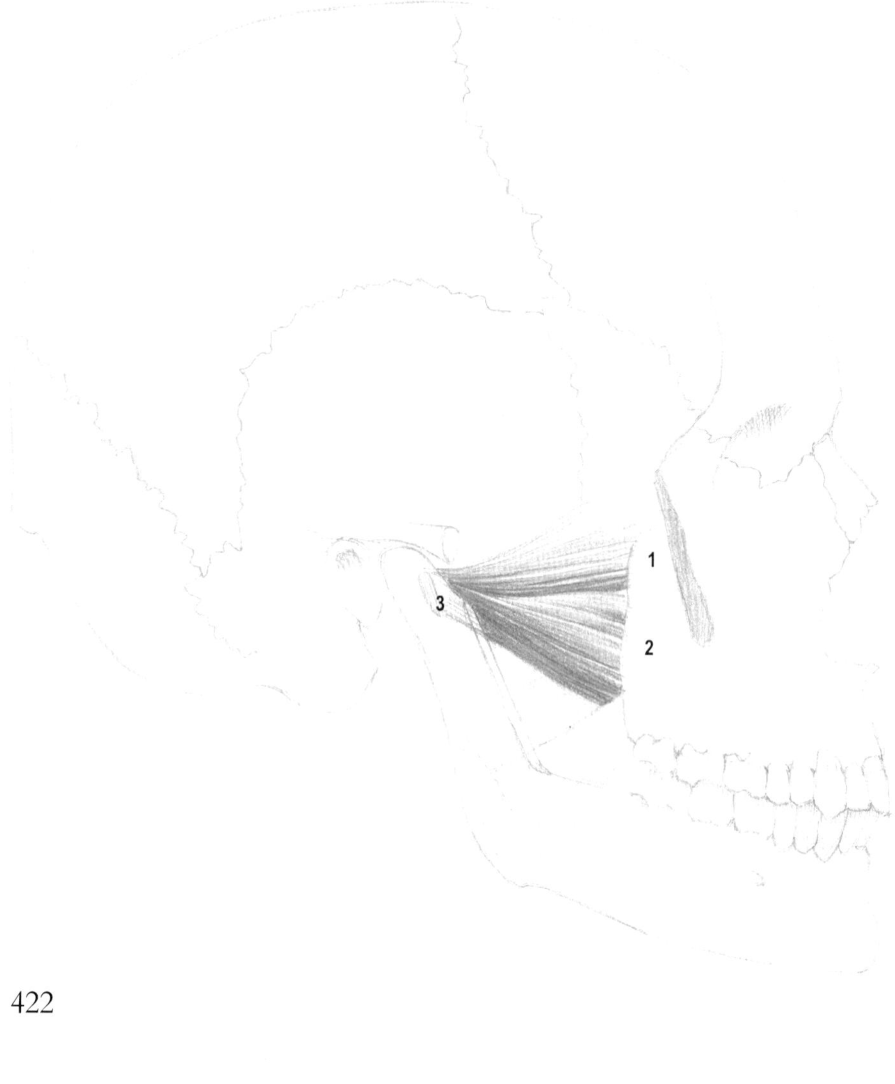

Fig. 286
The Pterygoideus Lateralis, *147*

ORIGIN:
Sphenoidal bone (cranial head, **1**); sphenoidal bone (caudal head, **2**)

INSERTION:
Surface of the lower jaw branch (**3**)

FUNCTION:
Pushes the lower jaw to the front and side; one-sided chewing movement

Fig. 287
The Pterygoideus Medialis, *144*

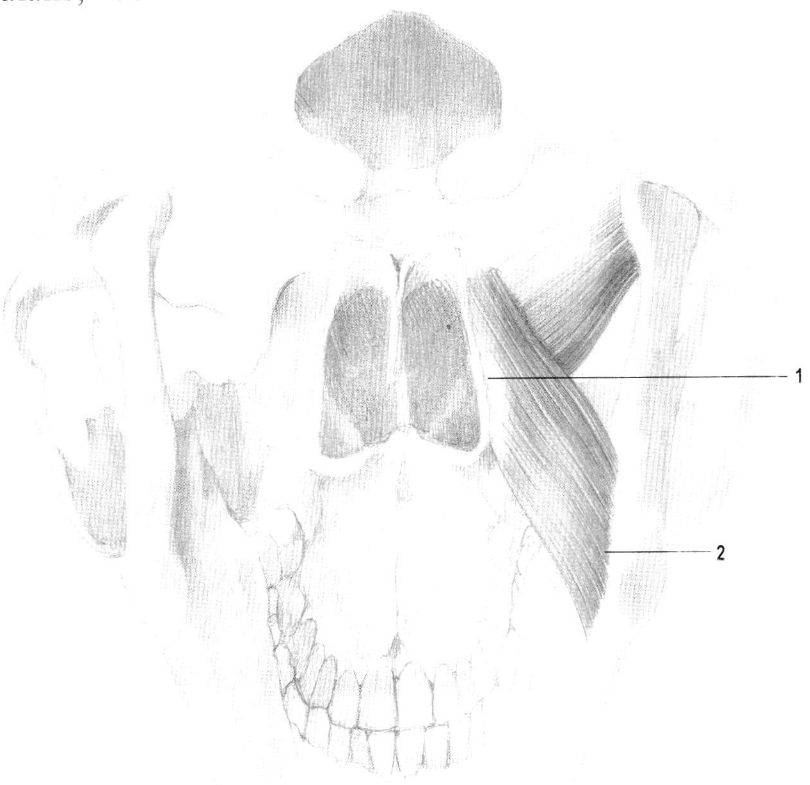

caudal aspect

ORIGIN:
Sphenoidal bone and palatal bone (**1**)

INSERTION:
Inner surface and corner of the lower jaw branch (**2**)

FUNCTION:
Lifts the lower jaw and pulls it to the front and side

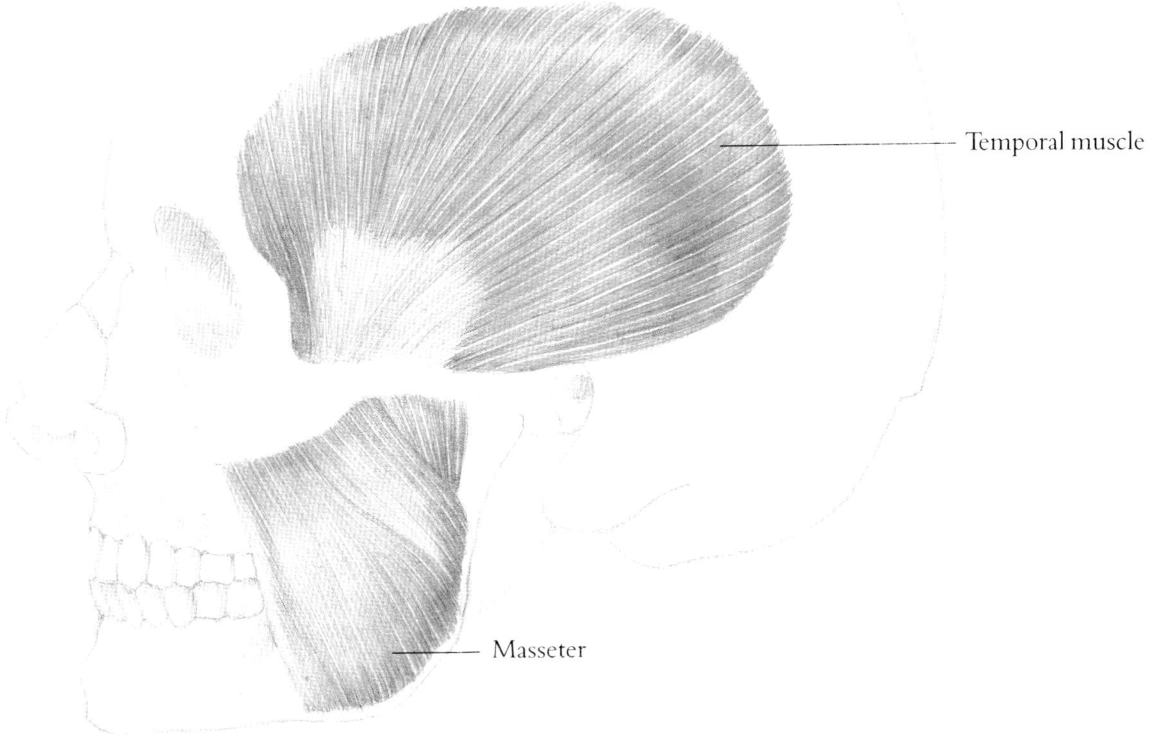

lateral aspect

Fig. 288
The Masseter, *142*

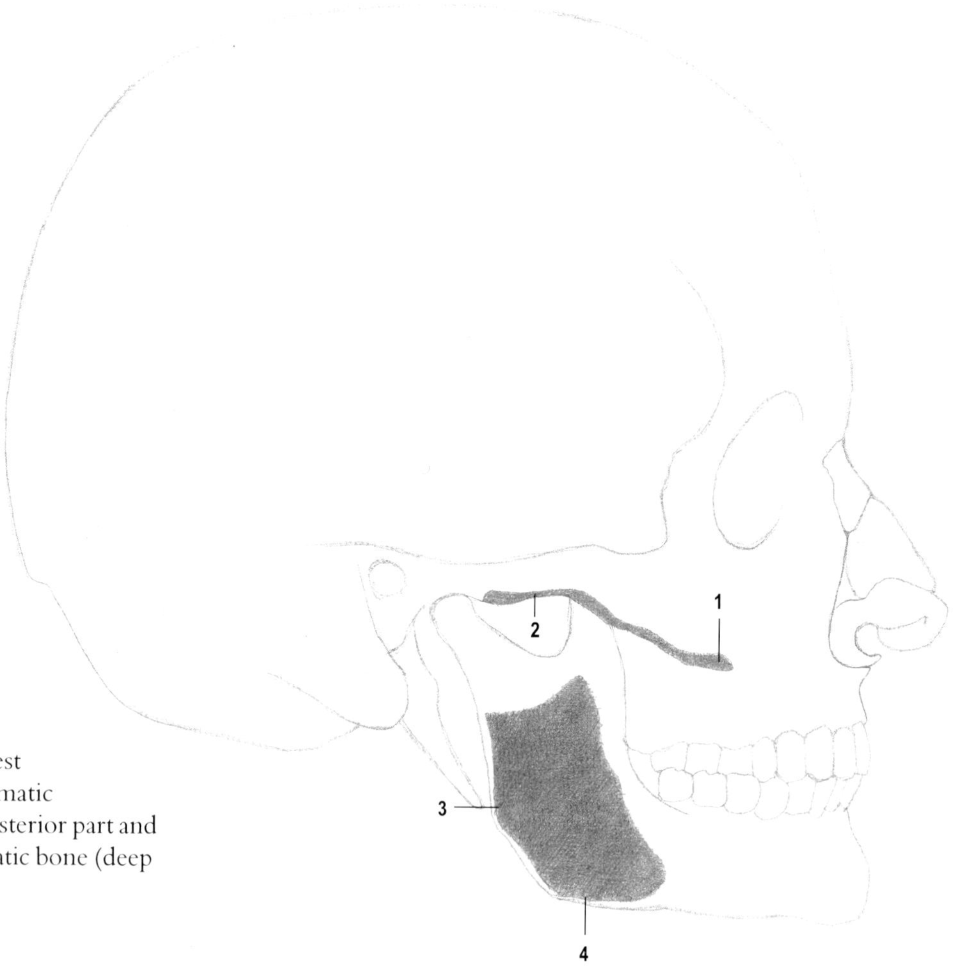

ORIGIN:
Anterior part of the facial crest and caudal ridge of the zygomatic bone (superficial part, **1**); posterior part and caudal surface of the zygomatic bone (deep masseter, **2**)

INSERTION:
Corner of the lower jaw (superficial part, **3**); external surface of the lower jaw branch (**4**)

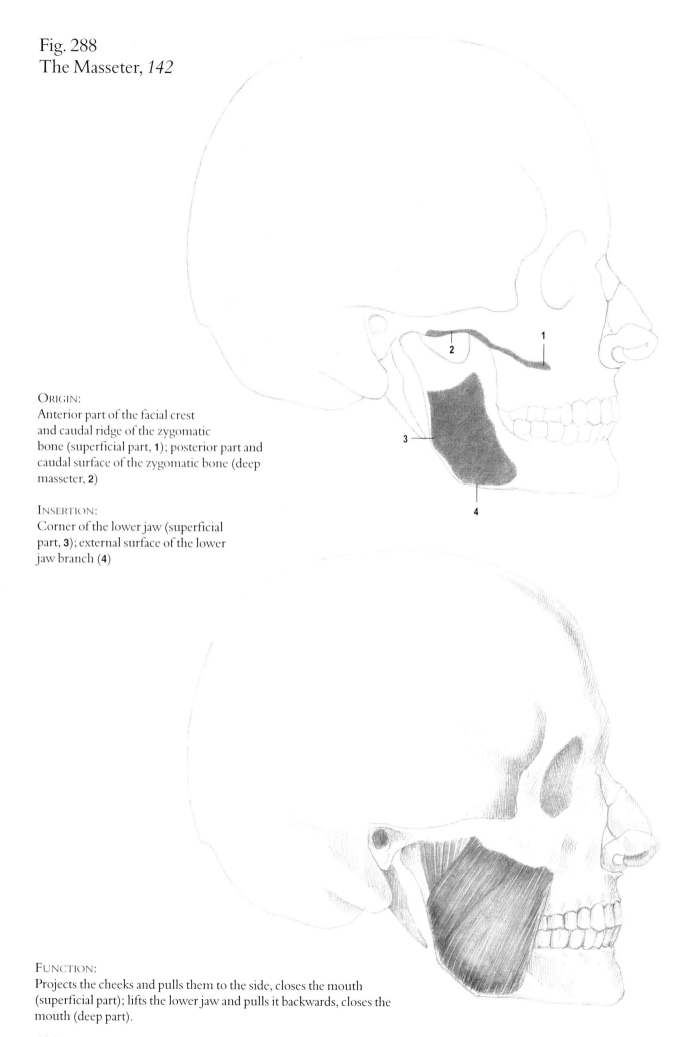

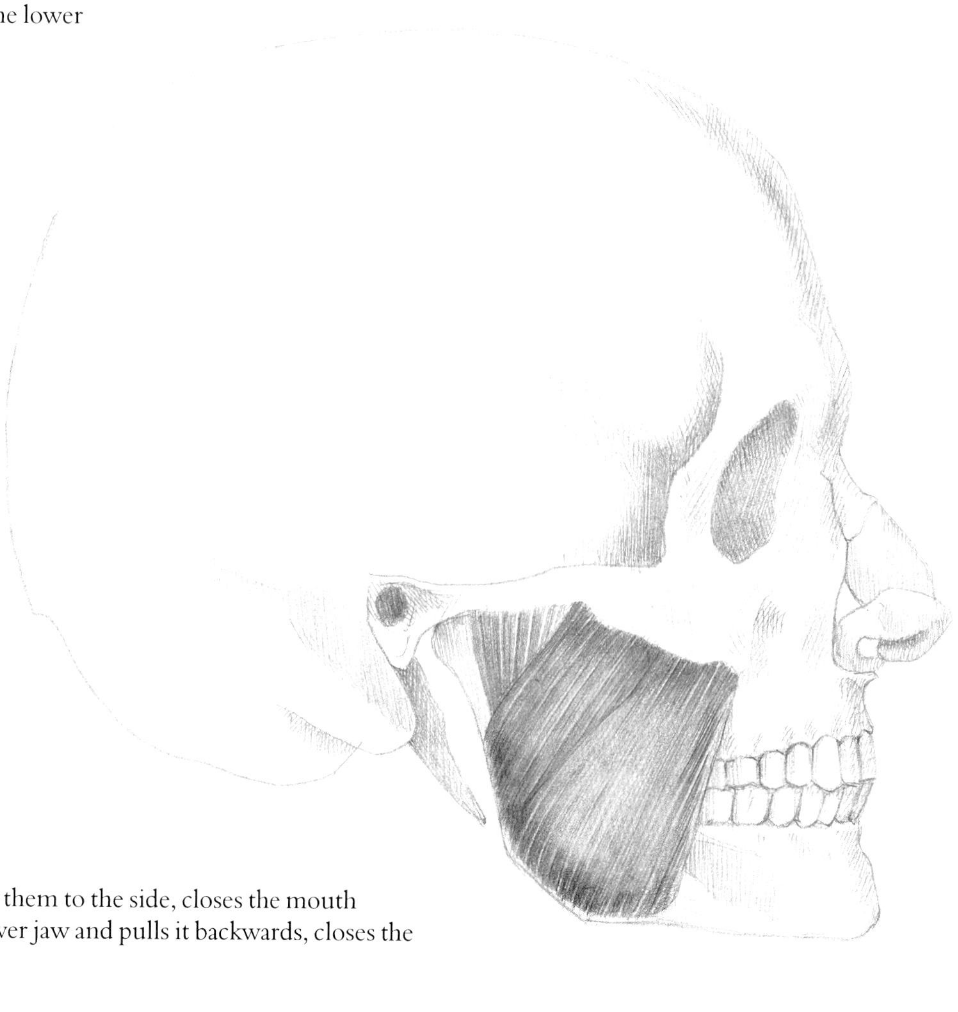

FUNCTION:
Projects the cheeks and pulls them to the side, closes the mouth (superficial part); lifts the lower jaw and pulls it backwards, closes the mouth (deep part).

Fig. 289
The Muscle Bundle Groups
of the Masseter

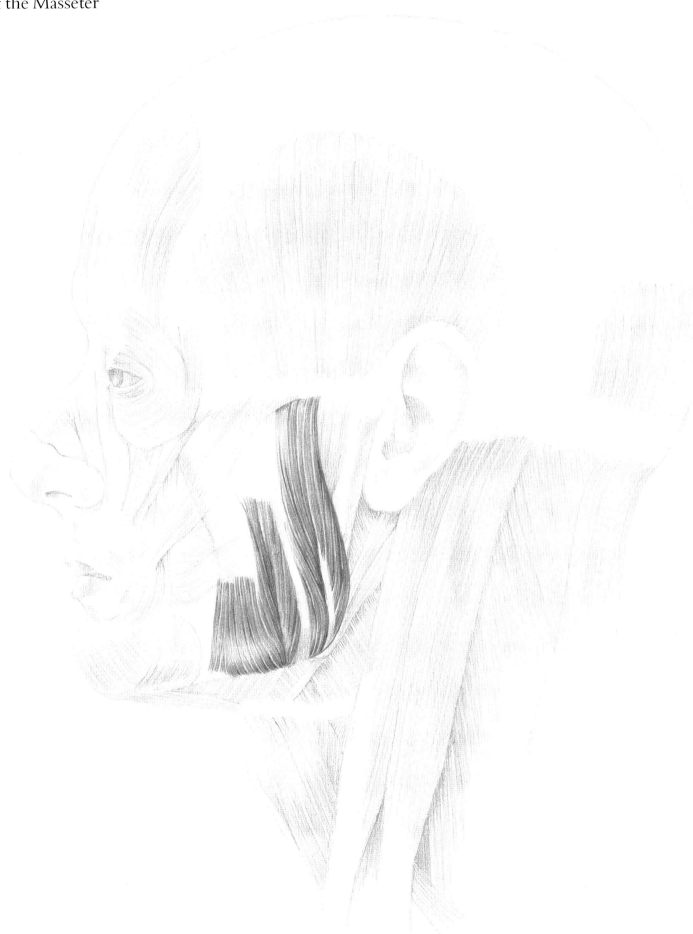

Fig. 290
The Temporalis, *143*

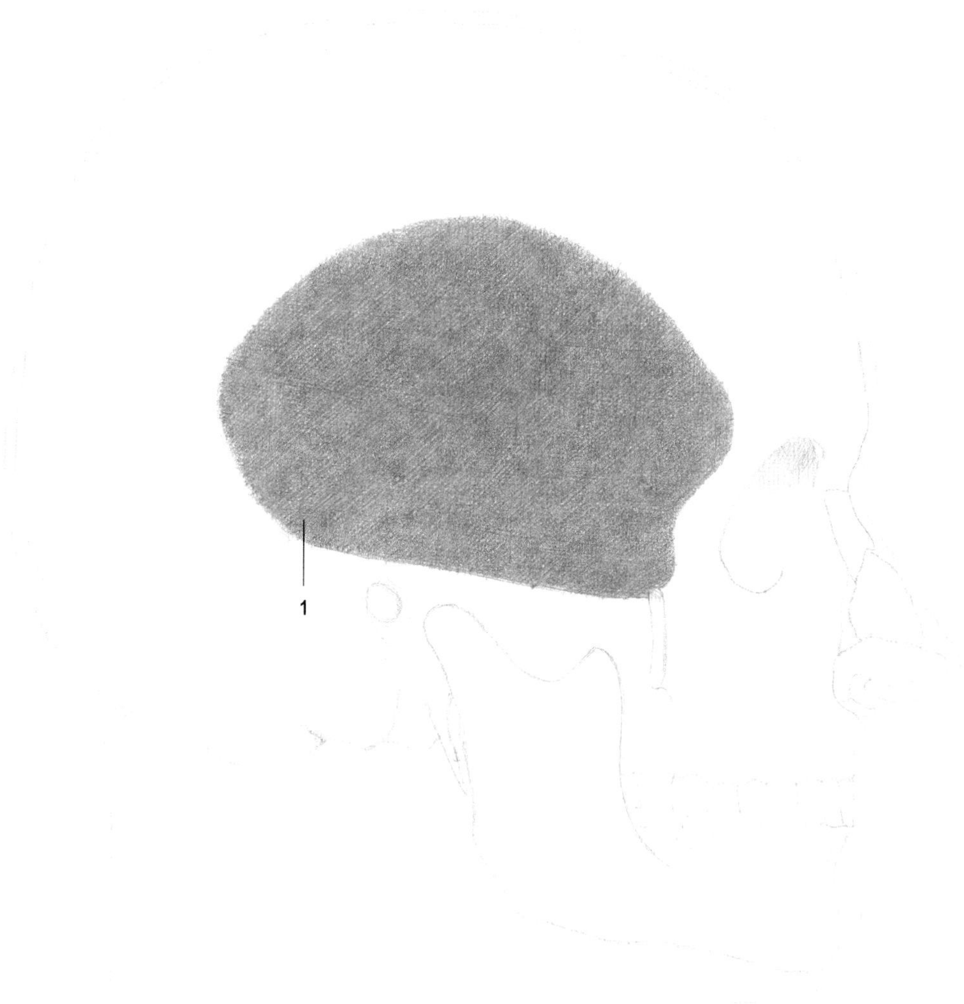

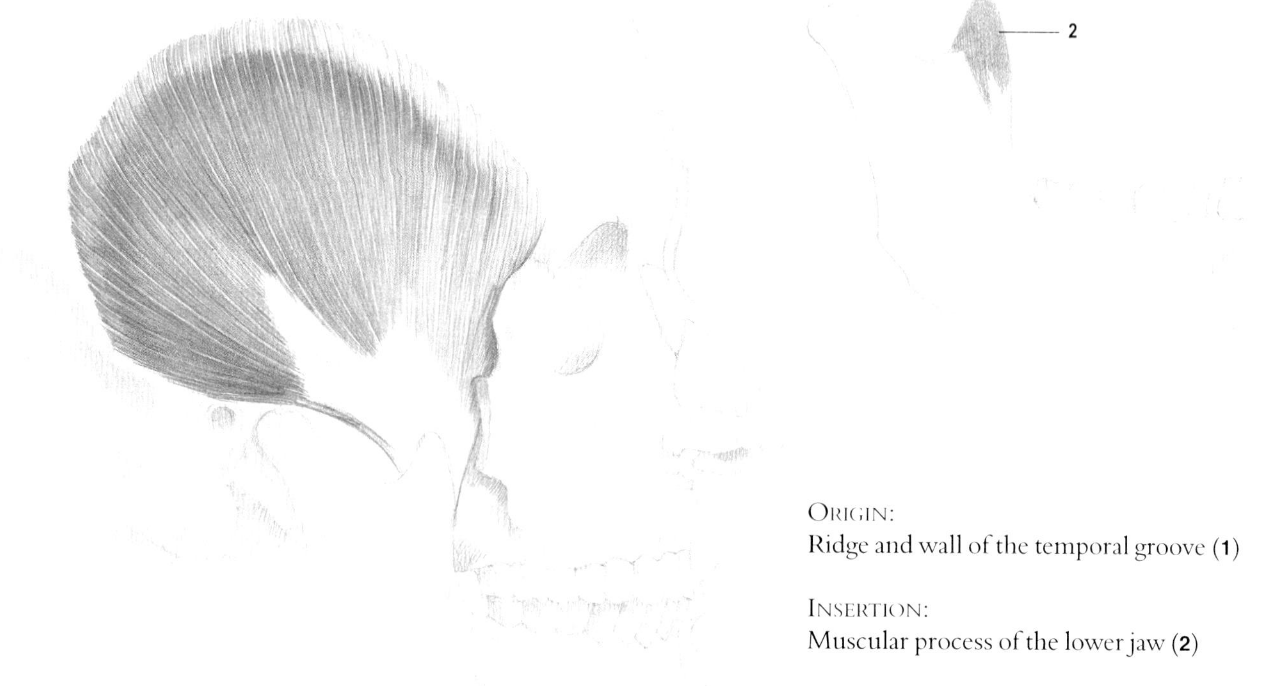

Origin:
Ridge and wall of the temporal groove (**1**)

Insertion:
Muscular process of the lower jaw (**2**)

Function:
Lifts the lower jaw and pulls it back, closes the mouth

Fig. 291
The Digastricus, *16*

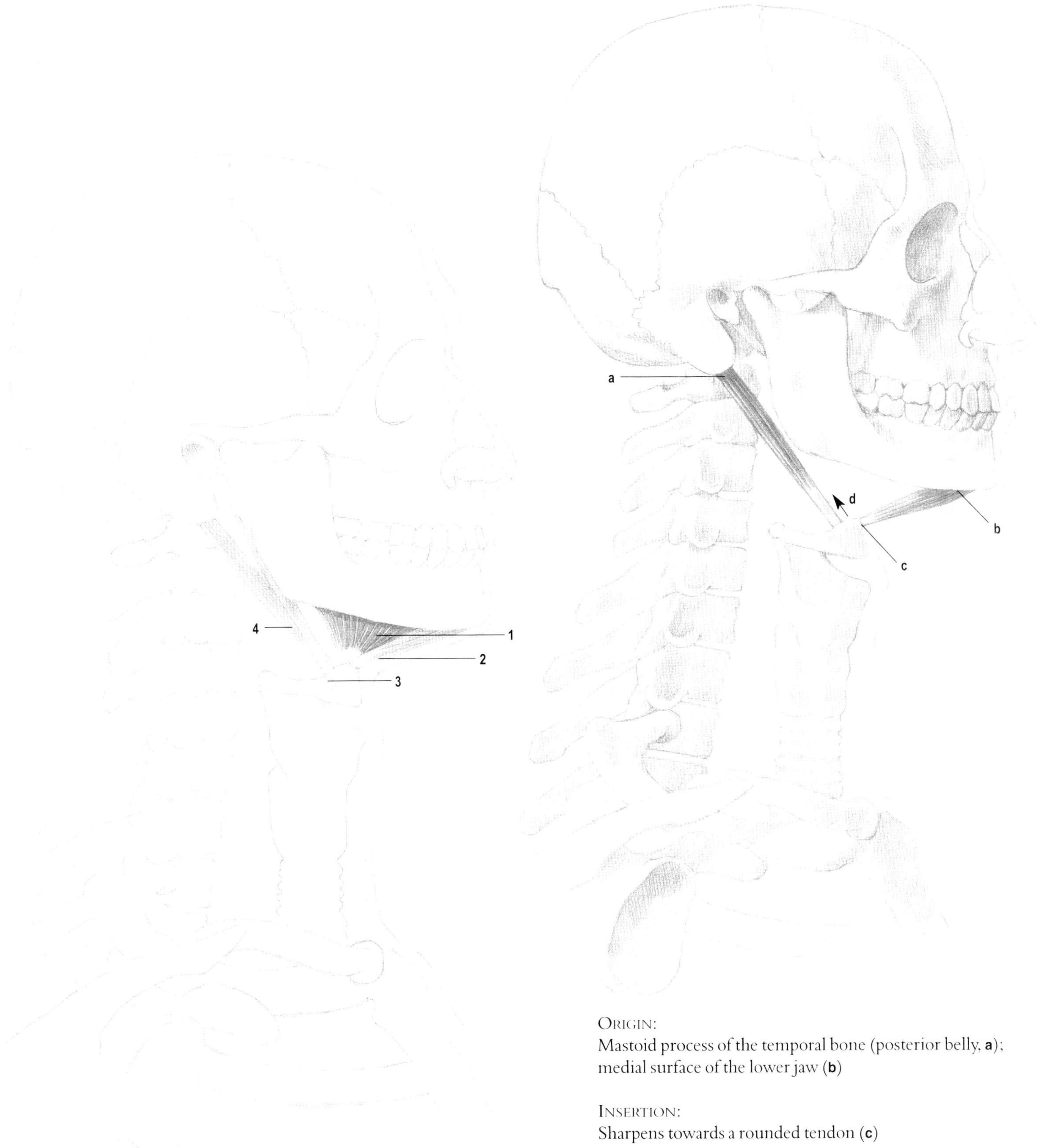

ORIGIN:
Mastoid process of the temporal bone (posterior belly, **a**); medial surface of the lower jaw (**b**)

INSERTION:
Sharpens towards a rounded tendon (**c**)

FUNCTION:
Pulls the lower jaw backwards, opens the mouth aperture, pushes the hyoid bone to the front, back and upwards (**d**)

1 Mylohyoideus, (*18*)
2 Geniohyoideus, (*19*)
3 Stylohyoideus, (*17*)
4 Digastricus, (*16*)

Fig. 292
The Mylohyoideus, *18*

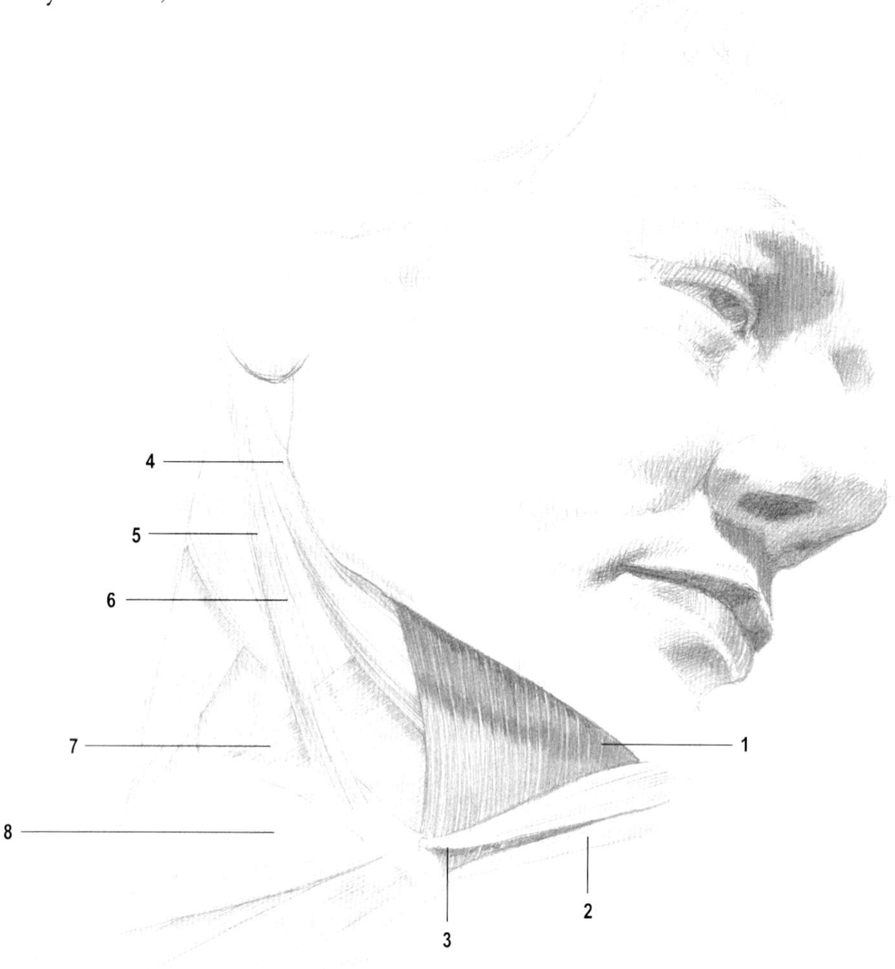

ORIGIN:
Crest of the lower jaw (**a**)

INSERTION:
Side of the body of the hyoid bone and central tendon (connects the medial sides of the branches of the lower jaw) (**b**)

FUNCTION:
Lifts the hyoid bone, pushes it forward and presses it against the hard palate

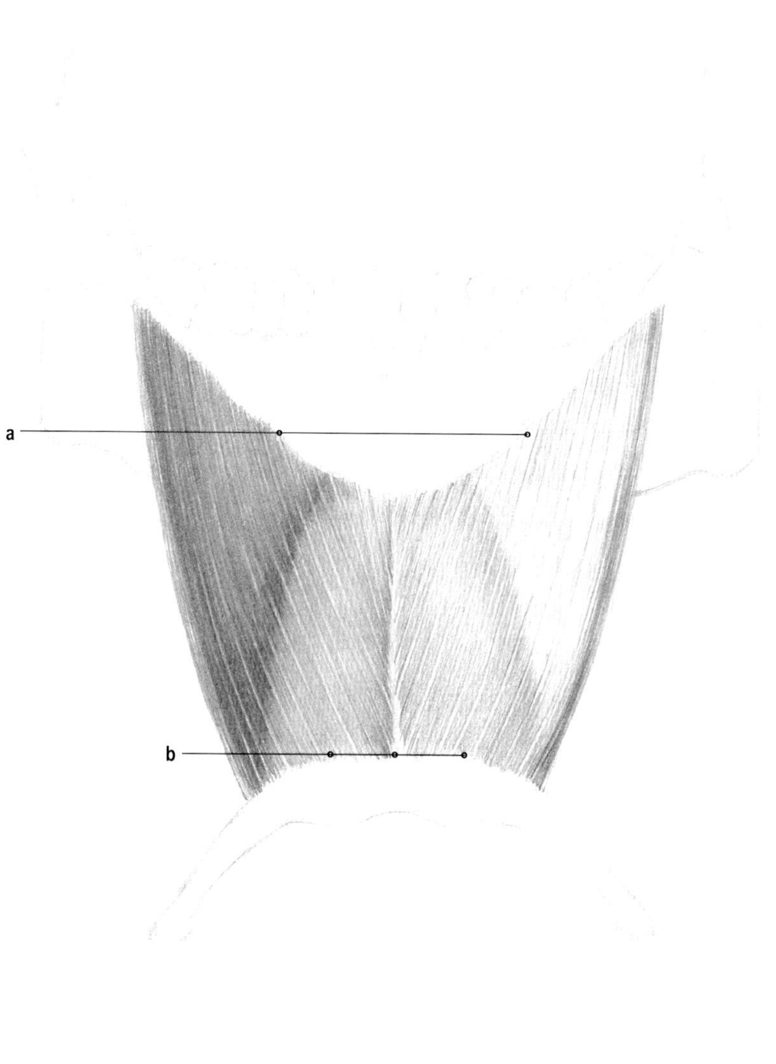

1 Mylohyoideus (*18*)
2 Digastricus (*16*)
3 Geniohyoideus (*19*)
4 Stylohyoideus (*17*)
5 Digastricus, posterior belly (*16*)
6 Stylohyoideus (*17*)
7 Extensor pharyngeum medialis
8 Thyreohyoideus (*15*)

Fig. 293
The Stylohyoideus, *17*

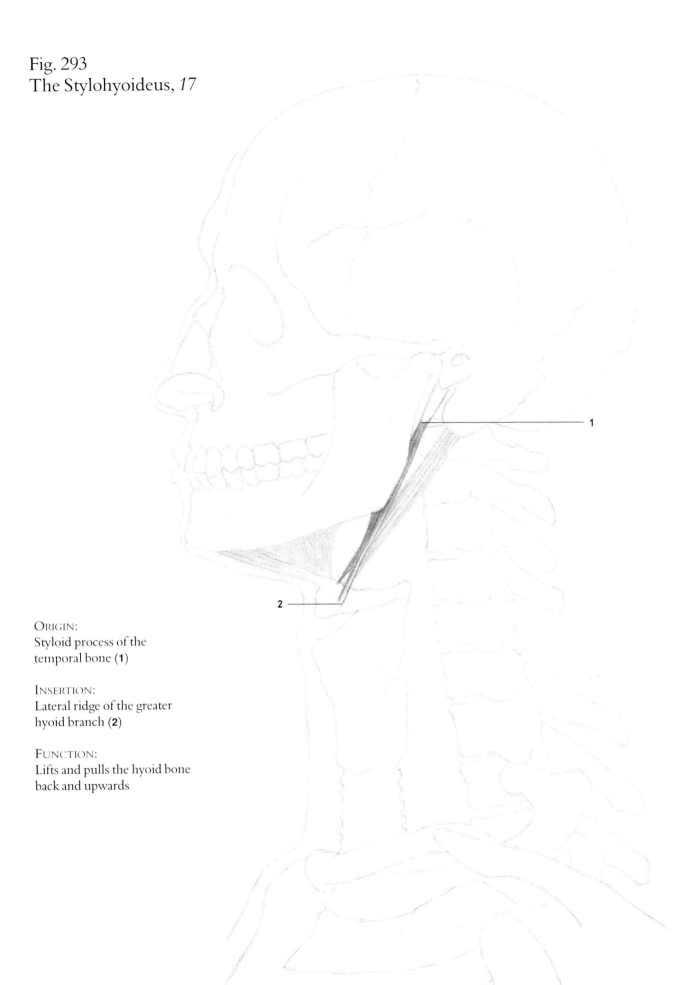

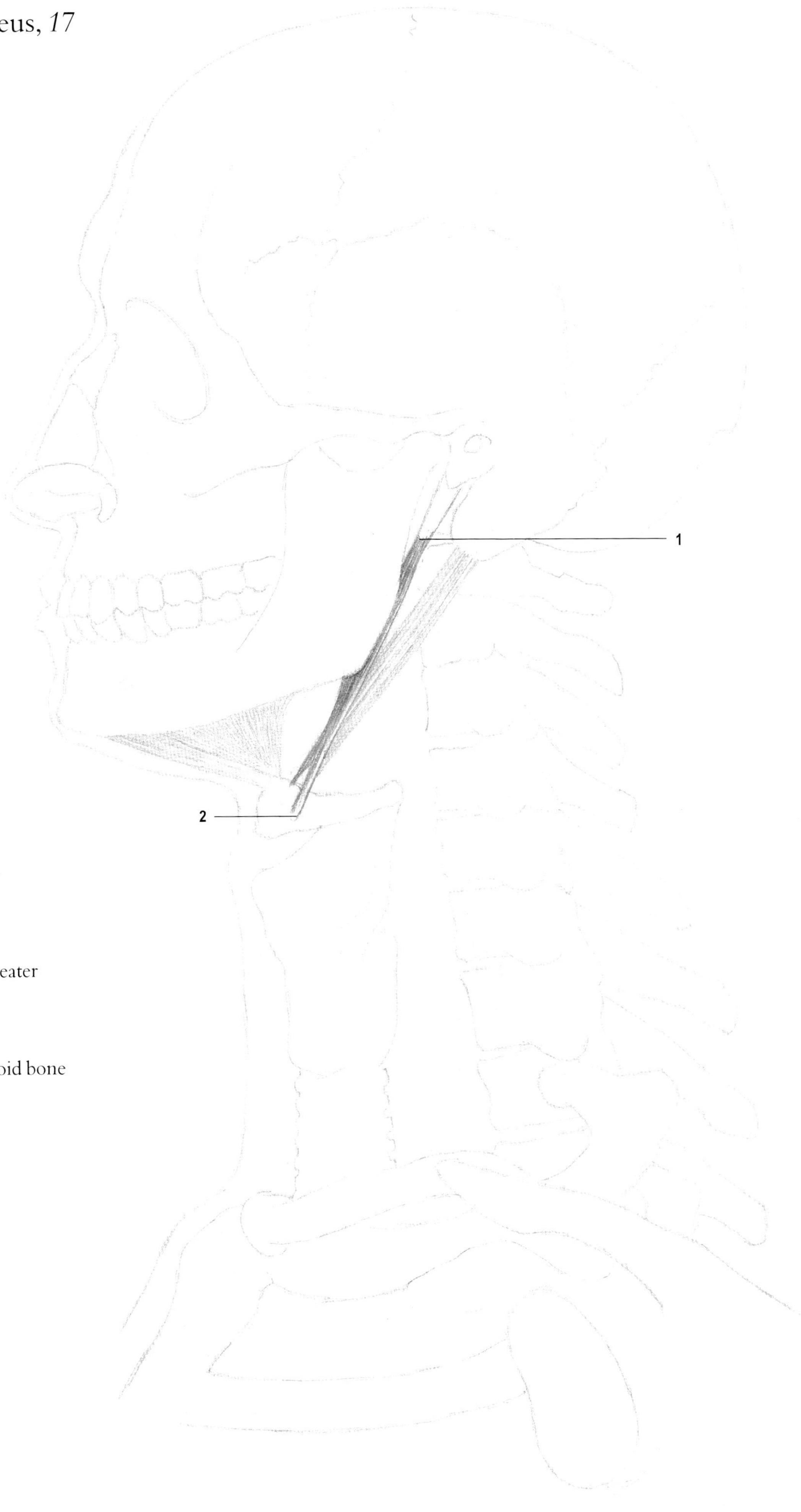

ORIGIN:
Styloid process of the temporal bone (**1**)

INSERTION:
Lateral ridge of the greater hyoid branch (**2**)

FUNCTION:
Lifts and pulls the hyoid bone back and upwards

Fig. 294
The Geniohyoideus, *19*

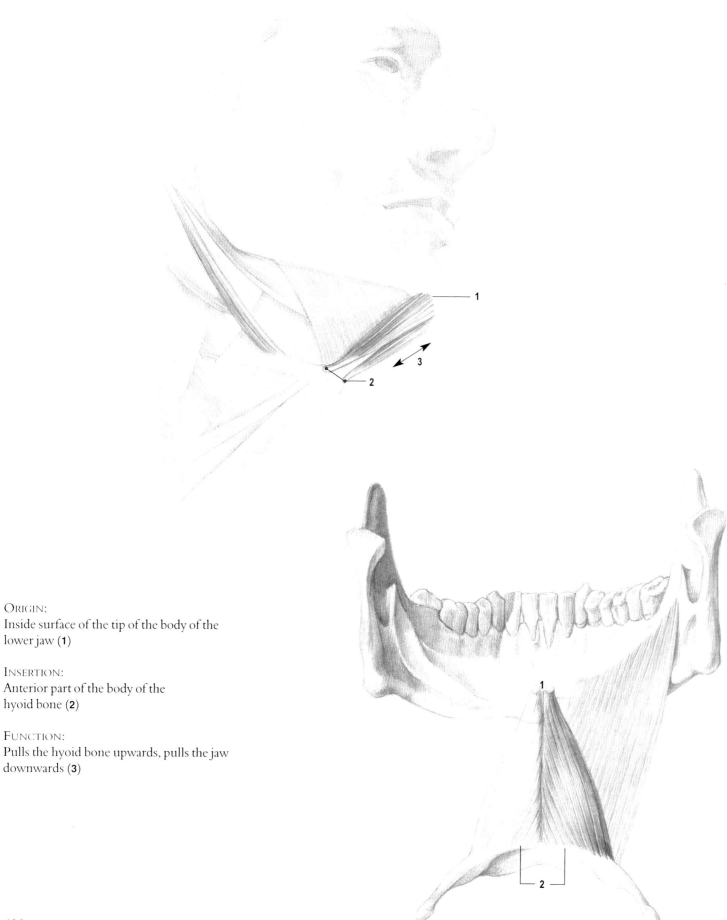

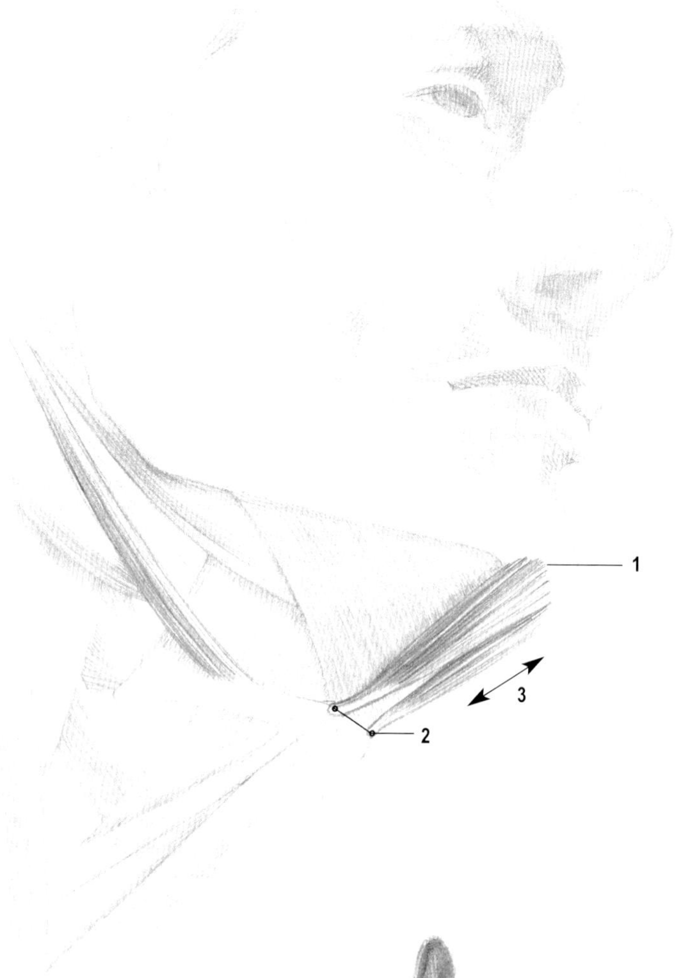

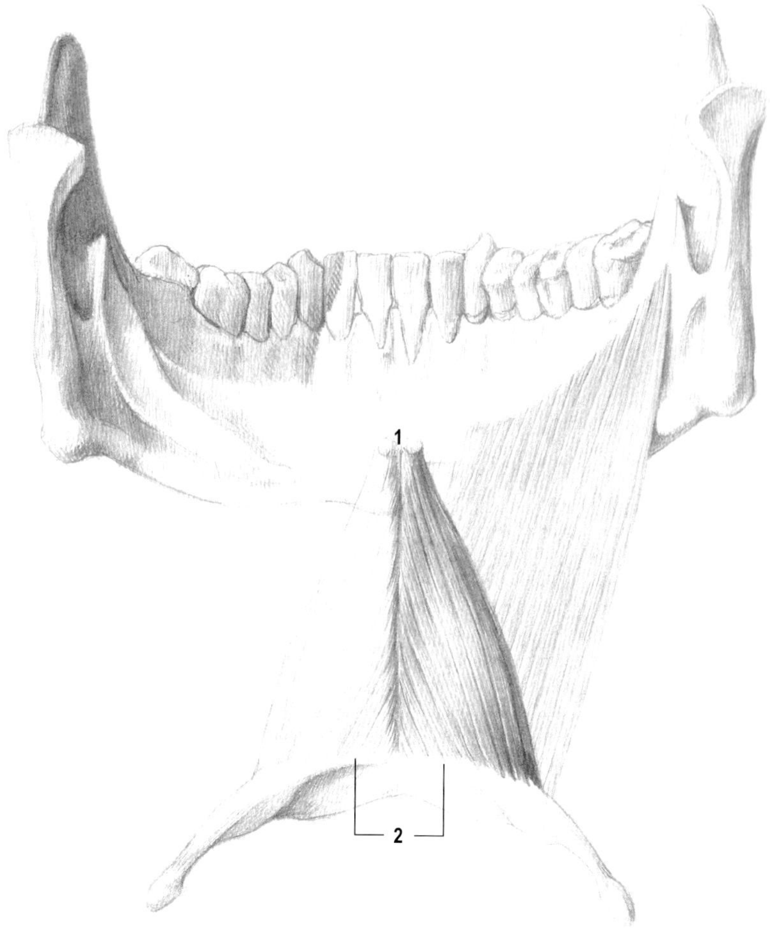

ORIGIN:
Inside surface of the tip of the body of the lower jaw (**1**)

INSERTION:
Anterior part of the body of the hyoid bone (**2**)

FUNCTION:
Pulls the hyoid bone upwards, pulls the jaw downwards (**3**)

Fig. 295
The Jaw Joints

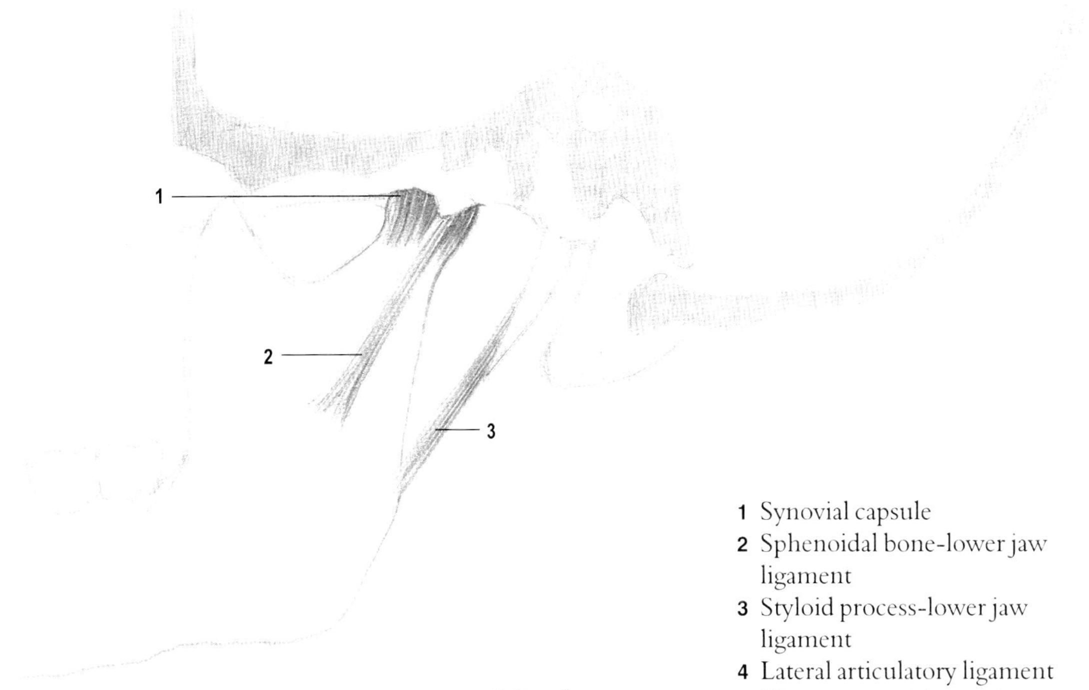

medial surface

1 Synovial capsule
2 Sphenoidal bone-lower jaw ligament
3 Styloid process-lower jaw ligament
4 Lateral articulatory ligament
5 Fibrous synovial capsule

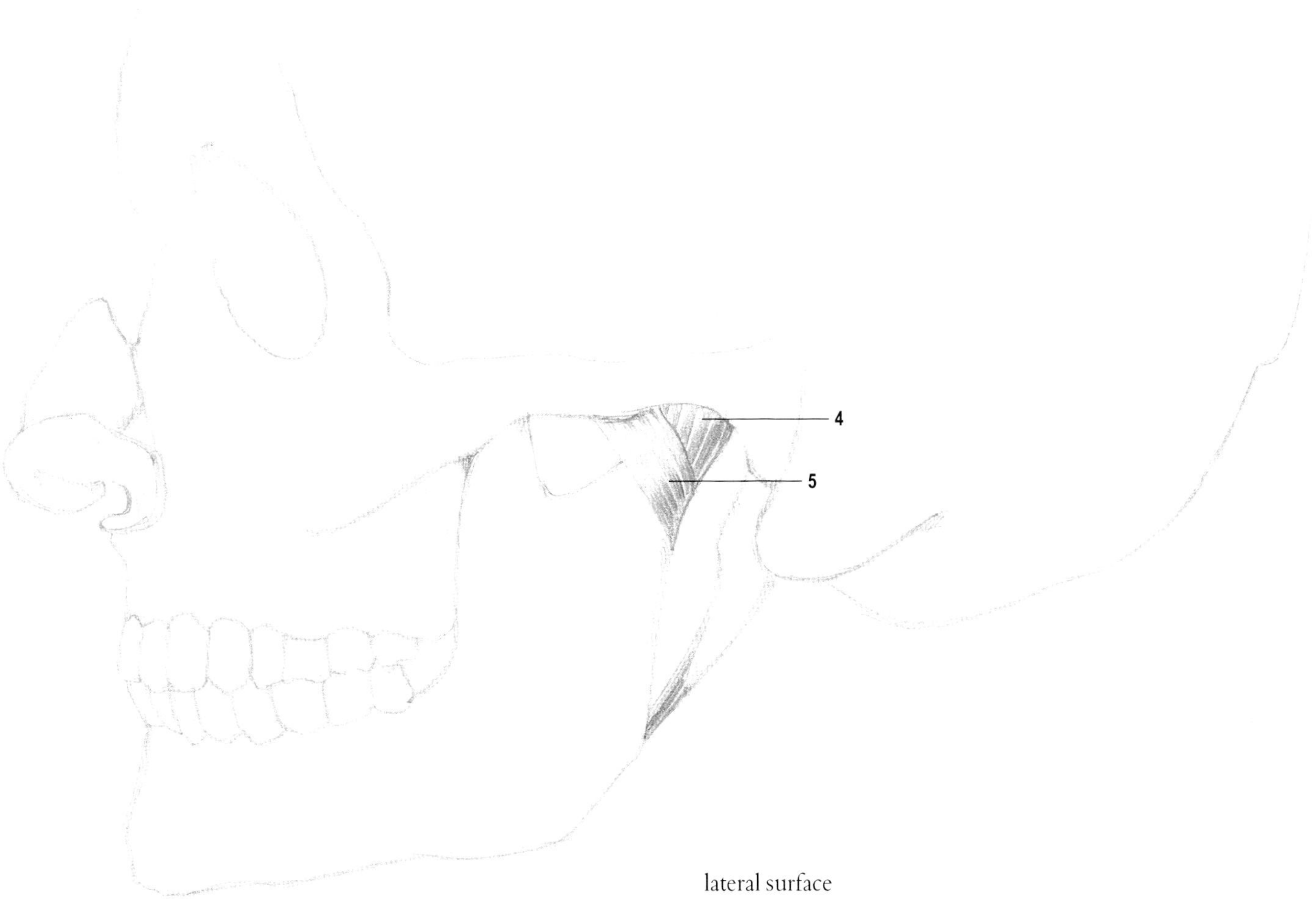

lateral surface

Fig. 296
Origins and Insertions of the Skull Muscles

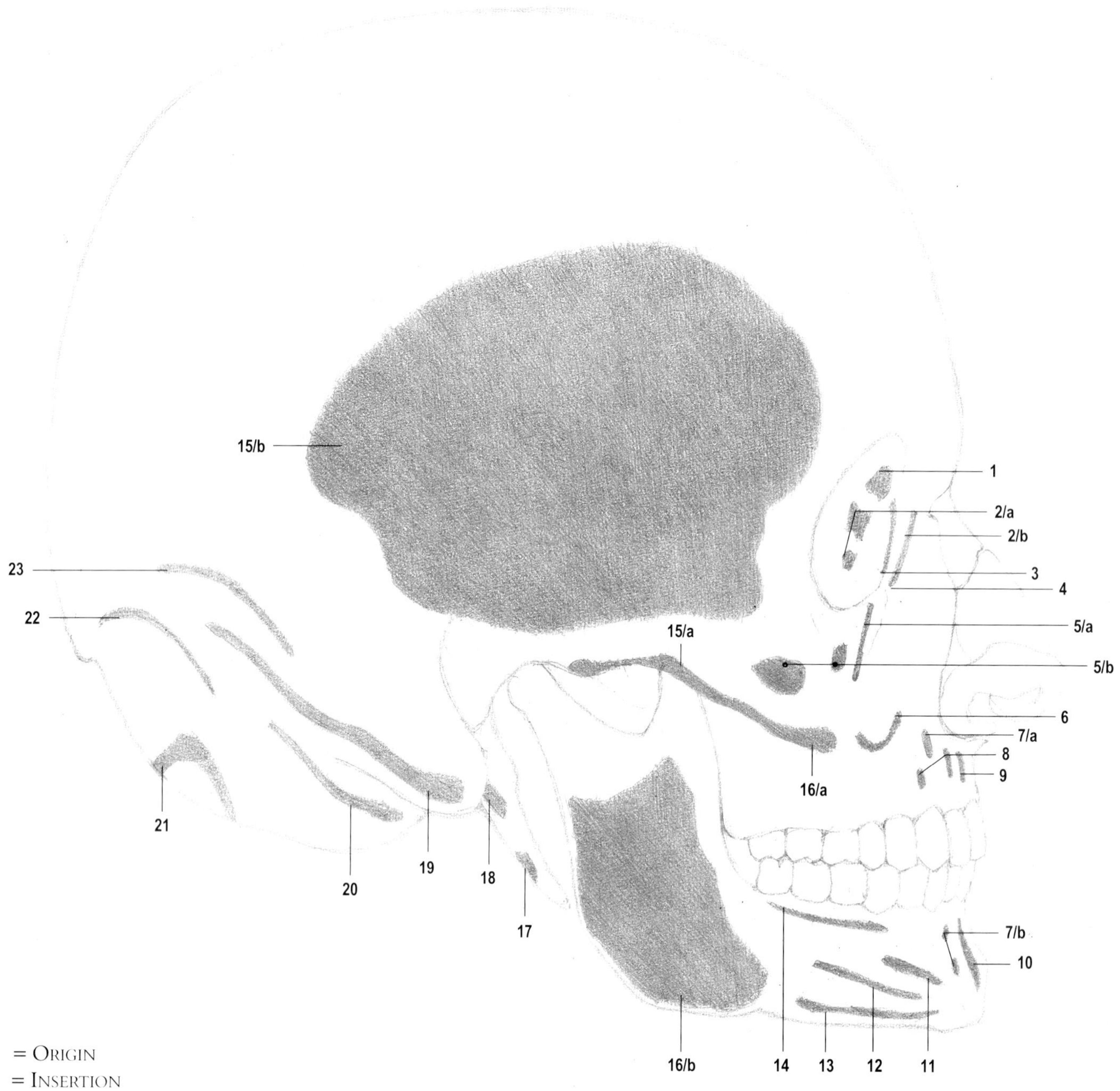

O = Origin
I = Insertion

Lateral Surface

1 Corrugator spercilii, O (*128*)
2 Orbicularis oculi, pars orbicularis (**a**), pars palpebralis (**b**), pars lacrimalis, O (*127*)
3 Levator labii superioris alaeque nasi, O (*137/1*)
4 Levator labii superioris, O (*137/2*)
5 Zygomaticus major (**a**) and minor (**b**), O (*138, 137/3*)
6 Levator anguli oris, O (*140*)
7 Orbicularis oris superioris (**a**) and inferioris (**b**) (pars labialis, *132*), O
8 Nasalis, O (*125*)
9 Depressor septi, O (*126*)
10 Mentalis, O (*136*)
11 Quadratus labii inferioris, O (*135*)
12 Triangularis, O (*133*)
13 Platysma, O+I (*1*)
14 Buccalis, **a** =O, **b** =I (*141*)
15 Temporalis, **a** =O, **b** =I (*143*)
16 Masseter, **a** =O, **b** =I (*142*)
17 Styloglossus, O
18 Stylopharyngeus, O
19 Sternocleidomastoideus, I (*11*)
20 Splenius, I (*5*)
21 Semispinalis (capitis), I (*26/4*)
22 Occipitofrontalis, O+I (*124/1*)
23 Trapezius (capitis), I (*20*)

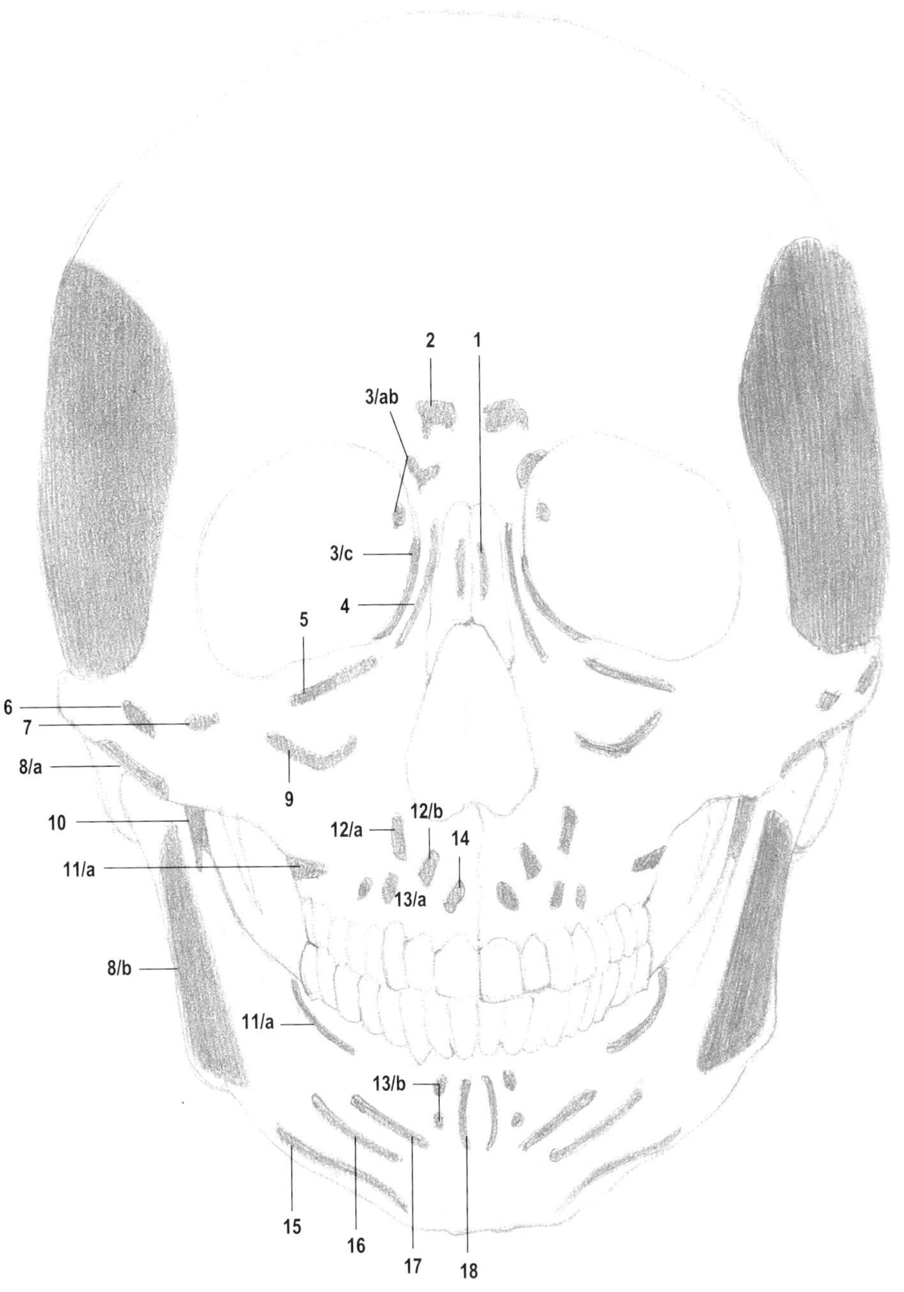

VENTRAL SURFACE

1 Procerus, O (*124/3*)
2 Corrugator supercilii, O (*128*)
3 Orbicularis oculi, pars orbicularis (**a**), pars palpebralis (b), pars lacrimalis, O (*127*)
4 Levator labii superioris alaeque nasi, O (*137/1*)
5 Levator labii superioris, O (*137/2*)
6 Zygomaticus major, O (*138*)
7 Zygomaticus minor, O (*137/3*)
8 Masseter, **a** = O, **b** = I (*142*)
9 Levator anguli oris, O (*140*)
10 Temporalis, I (*143*)
11 Buccalis, **a** = O, **b** = I (*141*)
12 Nasalis (pars alaris) (**a**), pars transversa (**b**), O (*125*)
13 Orbicularis oris superioris (**a**) and inferioris (**b**), O (*132*)
14 Depressor septi, O (*126*)
15 Platysma, O + I (*1*)
16 Triangularis, O (*133*)
17 Quadratus labii inferioris, O (*135*)
18 Mentalis, O (*136*)

Fig. 297
Origins and Insertions at the Base of the Skull

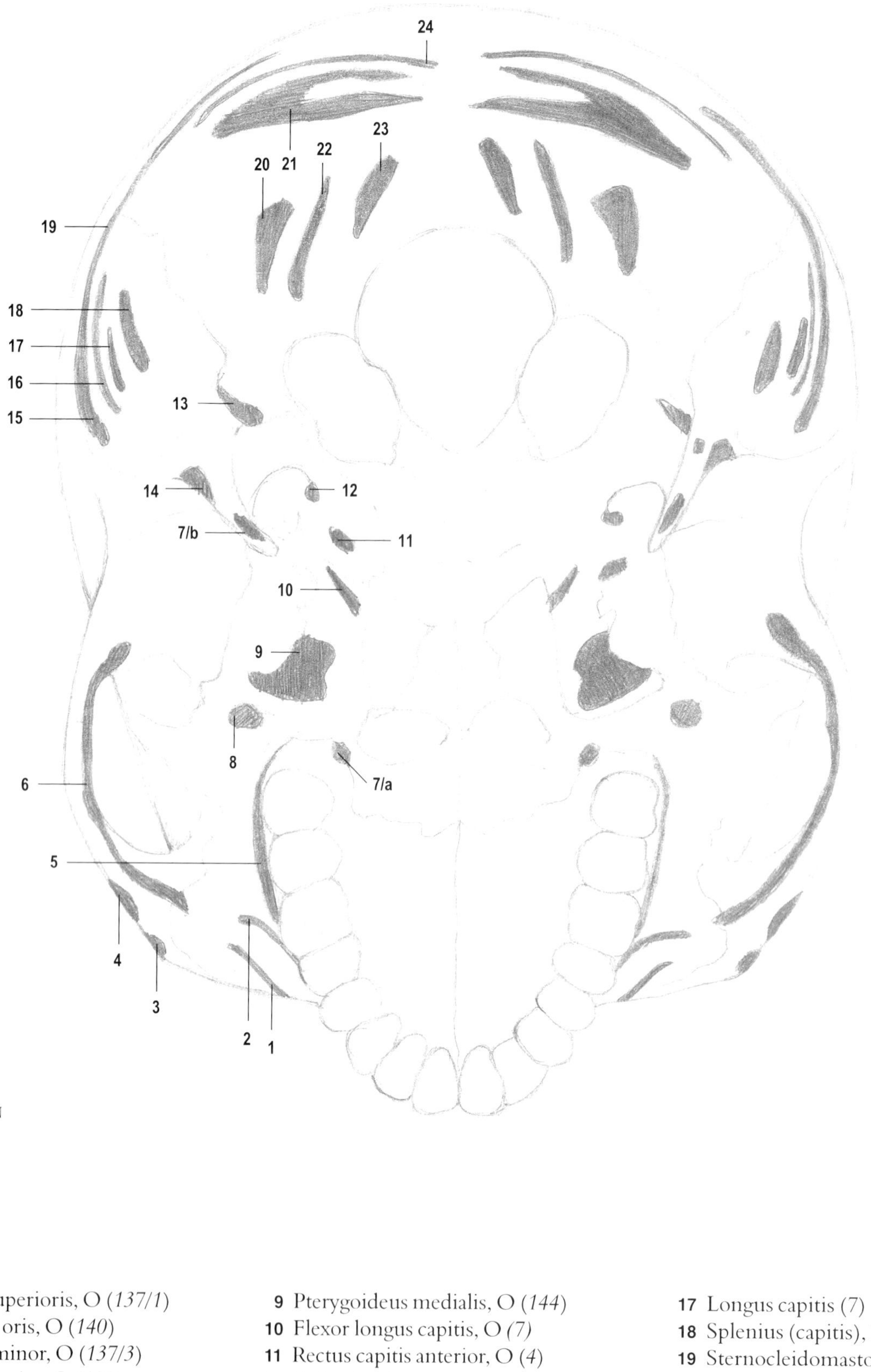

O = Origin
I = Insertion

Base of Skull

1 Levator labii superioris, O (*137/1*)
2 Levator anguli oris, O (*140*)
3 Zygomaticus minor, O (*137/3*)
4 Zygomaticus major, O (*138*)
5 Buccalis, O (*141*)
6 Masseter, O (*142*)
7 Soft palate tensor, O
8 Pterygoideus lateralis, O (*145*)
9 Pterygoideus medialis, O (*144*)
10 Flexor longus capitis, O (*7*)
11 Rectus capitis anterior, O (*4*)
12 Soft palate levator, O
13 Stylohyoideus, O (*17*)
14 Stylopalatal muscle, O
15 Rectus capitis posterior lateralis, O (*2*)
16 Digastricus, O (*16*)
17 Longus capitis (*7*)
18 Splenius (capitis), I (*5*)
19 Sternocleidomastoideus, I (*11*)
20 Obliquus capitis superior, I (*3*)
21 Semispinalis (capitis), I (*26/4*)
22 Rectus capitis anterior major, I (*2*)
23 Rectus capitis anterior minor, I (*2*)
24 Trapezius, I (*20*)

Fig. 298
Origins and Insertions of the Lower Jaw and Hyoid Muscles

Lower Jaw

1 Buccalis, O (*141*)
2 Temporalis, I (*143*)
3 Pterygoideus lateralis, I (*145*)
4 Pterygoideus medialis, I (*144*)
5 Mylohyoideus, I (*18*)
6 Digastricus, I (*16*)
7 Geniohyoideus, I (*19*)

Hyoid Bone

8 Genioglossus, O
9 Sternohyoideus, I (*12*)
10 Omohyoideus, I (*13*)
11 Stylohyoideus, I (*17*)
12 Hyopharyngeus, O
13 Thyreohyoideus, O (*15*)

Lower jaw

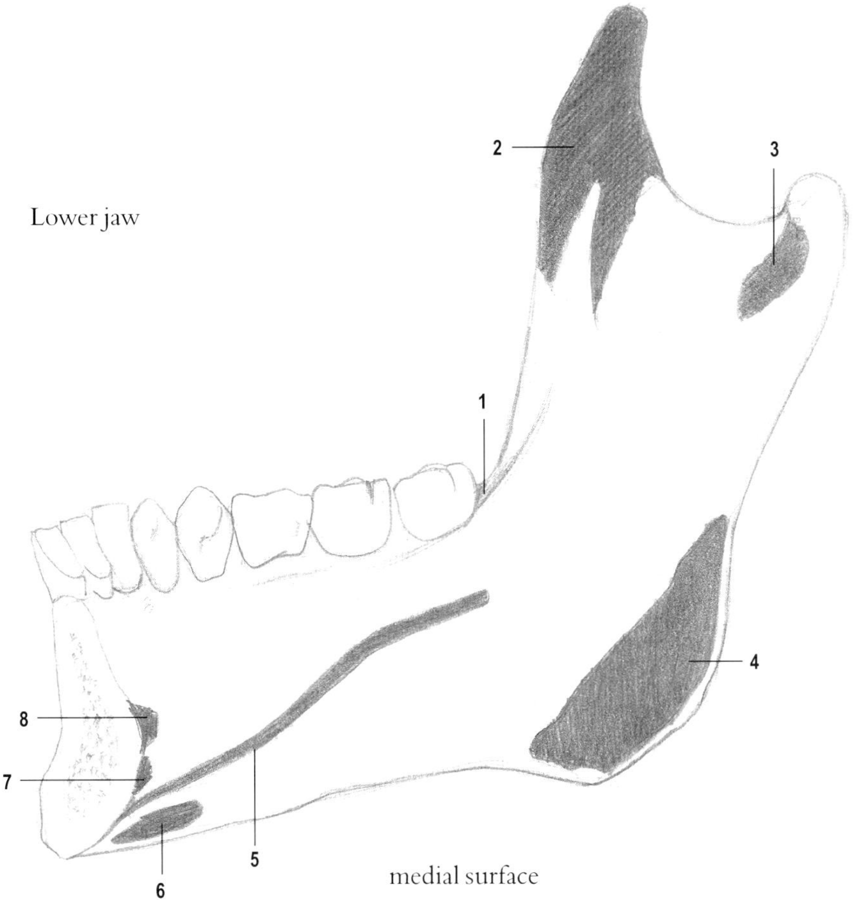

medial surface

Hyoid bone

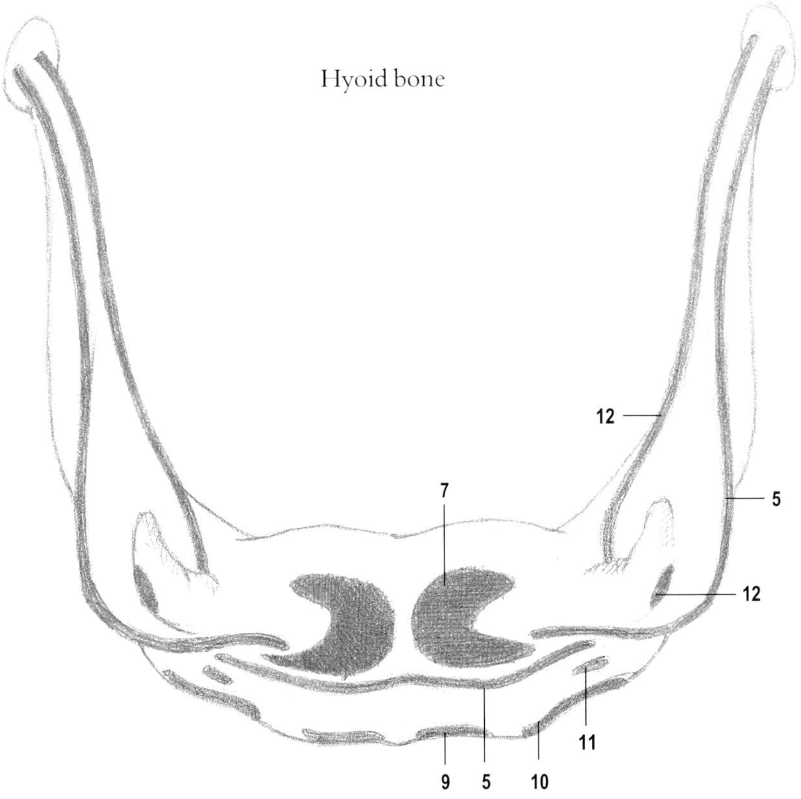

cranial aspect

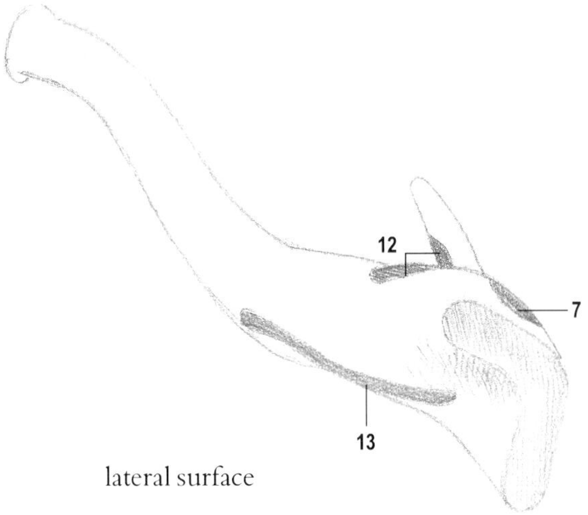

lateral surface

Fig. 299
The Head

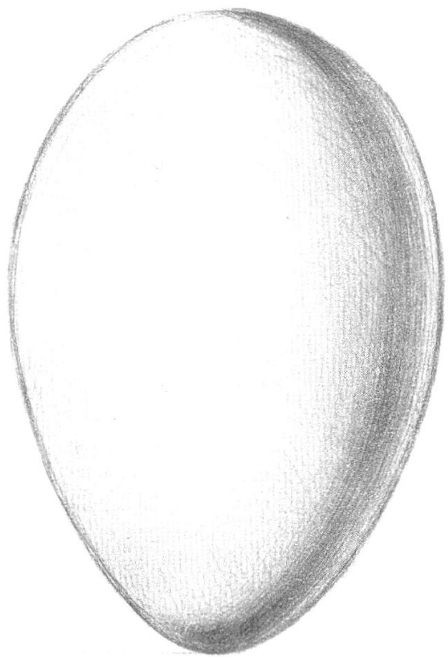

The human head is ovoid.

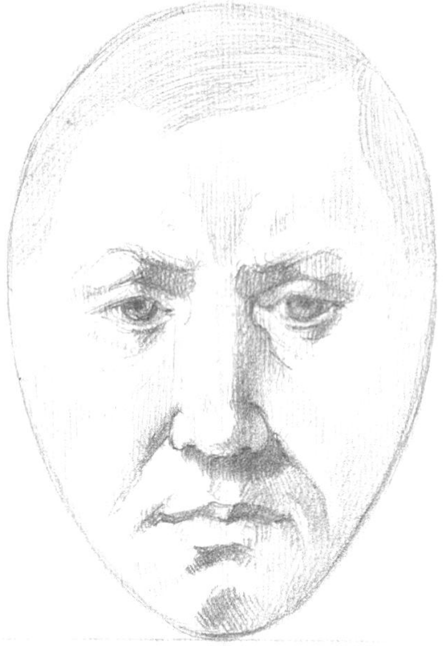

When the head is bowed, the three facial regions appear in different sizes. This is due to the laws of perspective.

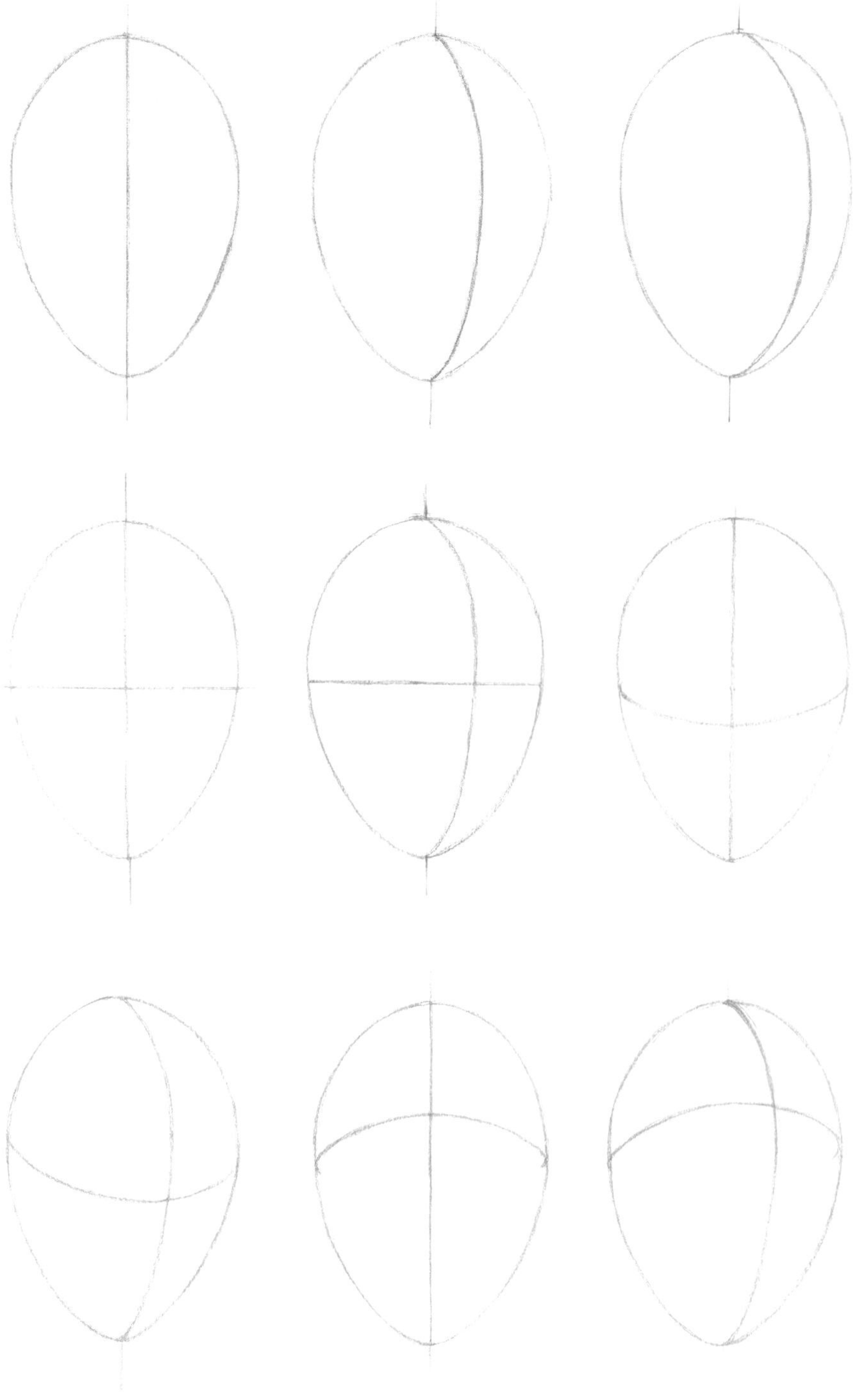

When the head is rotated in various directions the vertical and horizontal axes become ellipses.

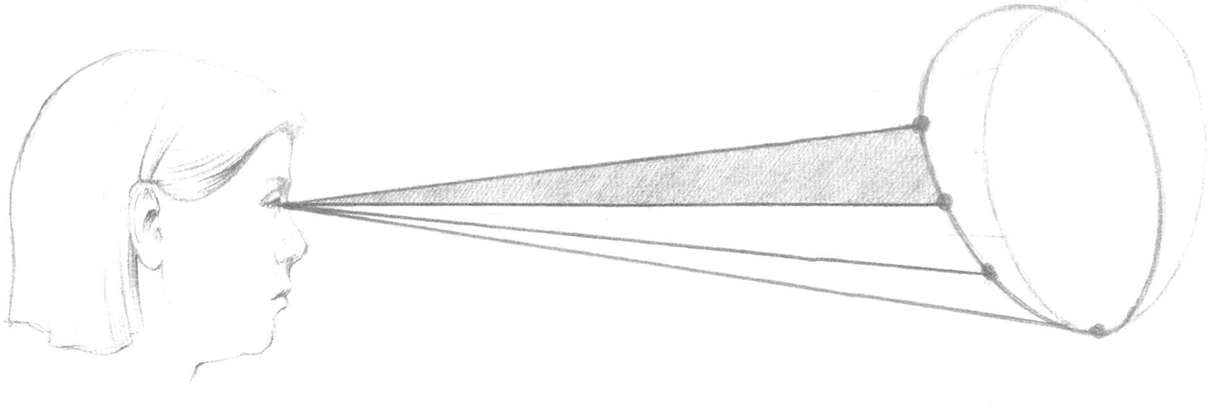

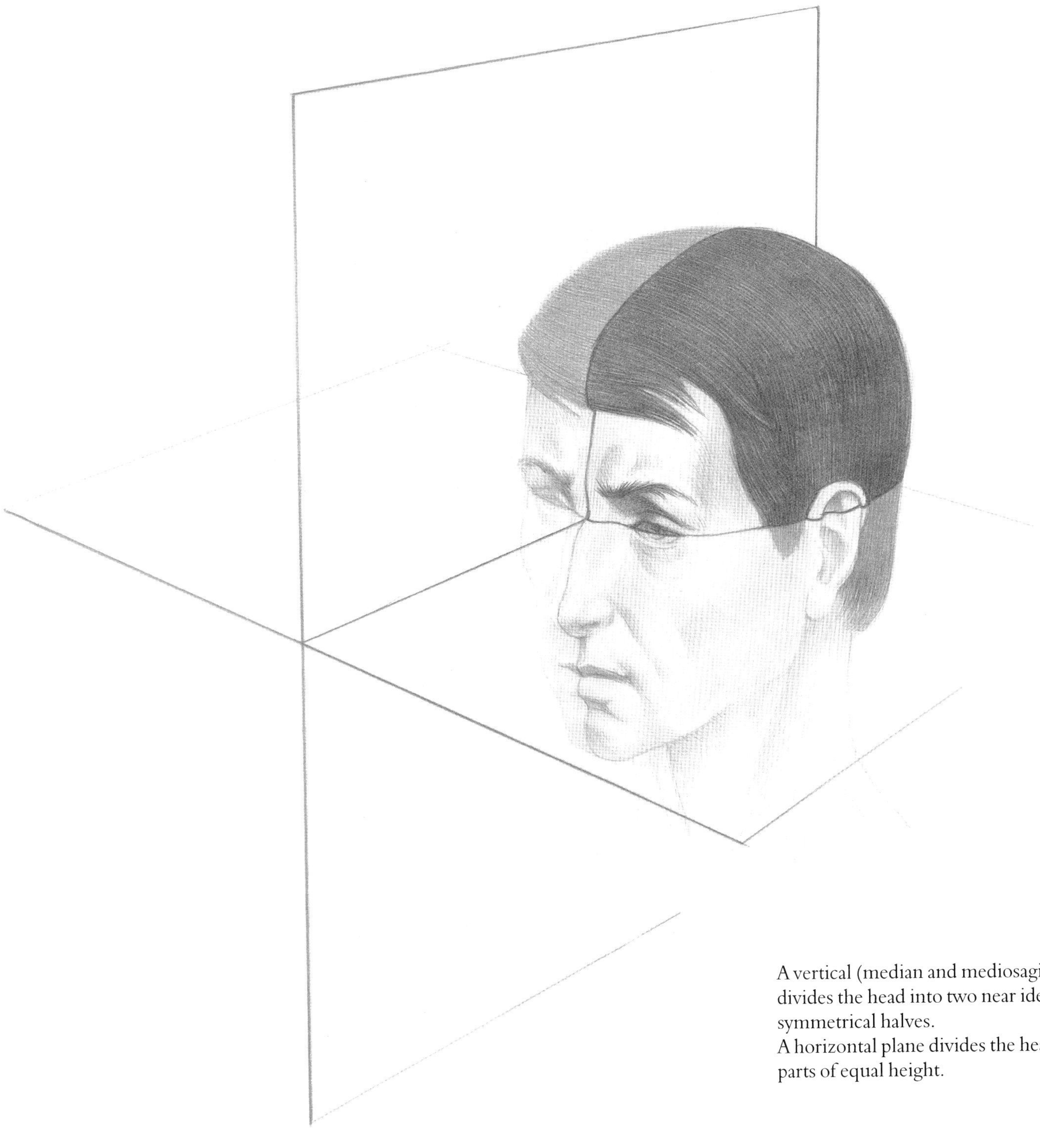

A vertical (median and mediosagittal) plane divides the head into two near identical and symmetrical halves.
A horizontal plane divides the head into two parts of equal height.

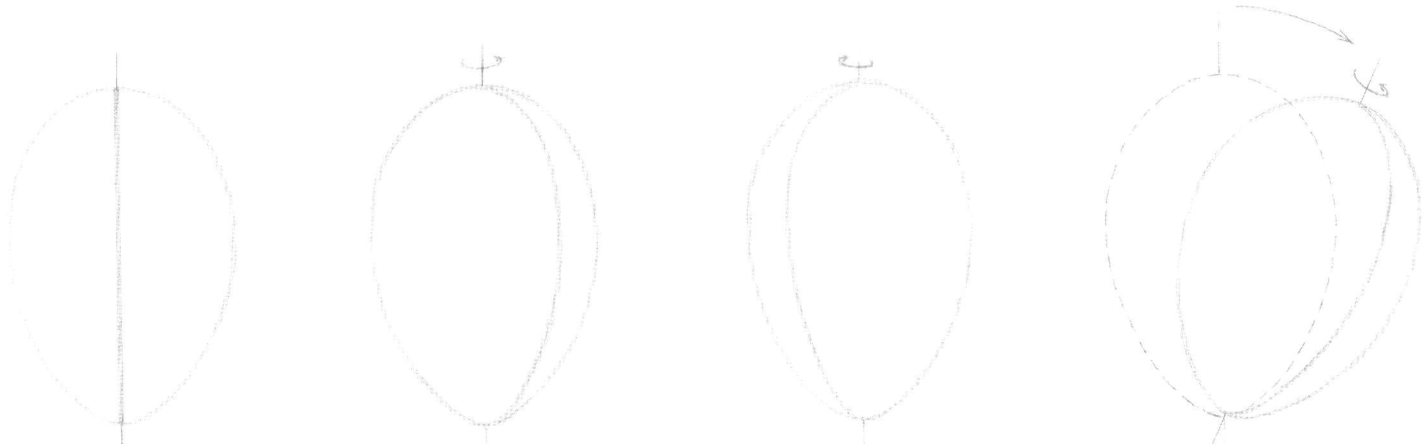

Fig. 300
Proportions of the Head to Individual Parts of the Body

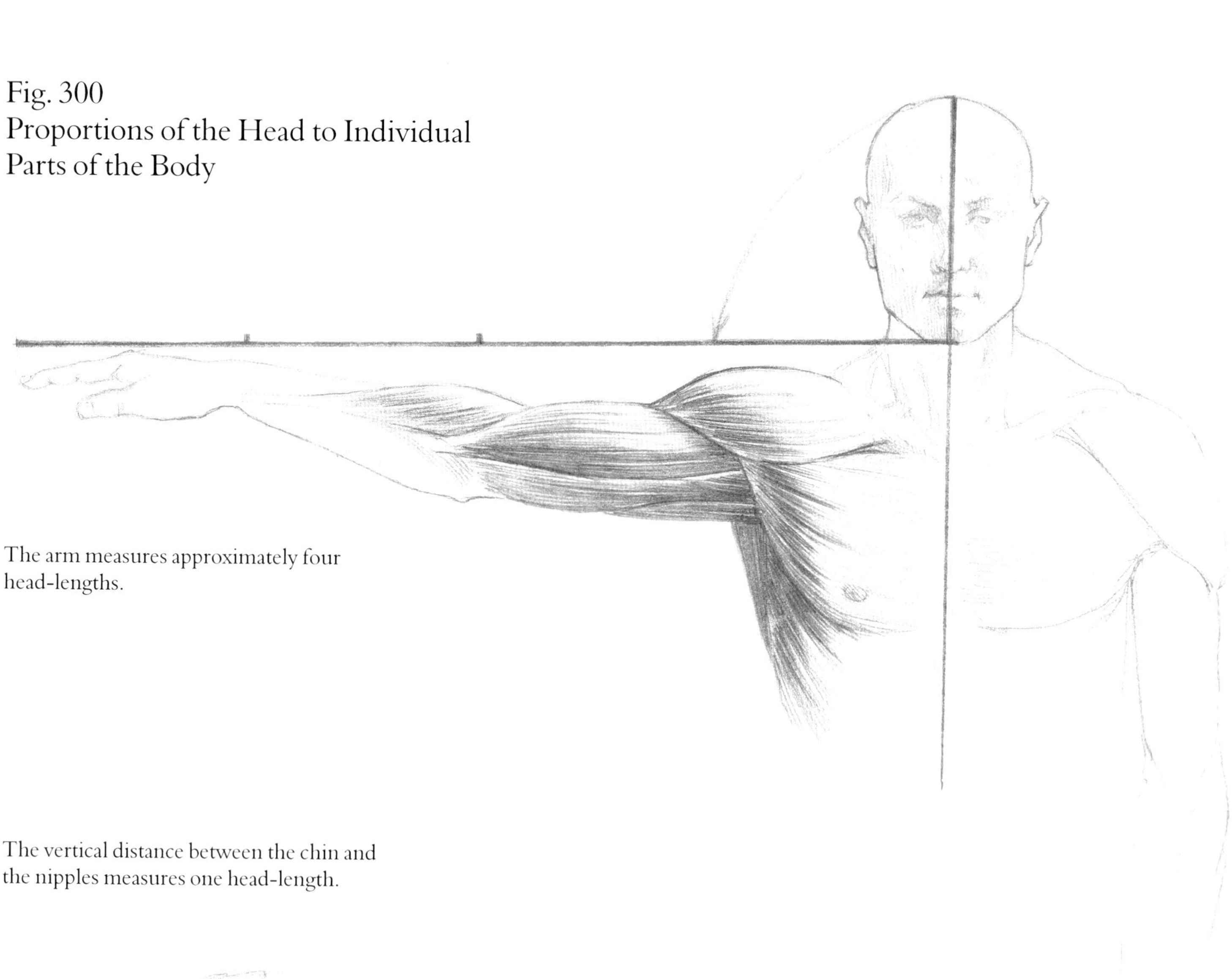

The arm measures approximately four head-lengths.

The vertical distance between the chin and the nipples measures one head-length.

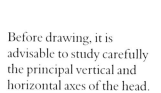

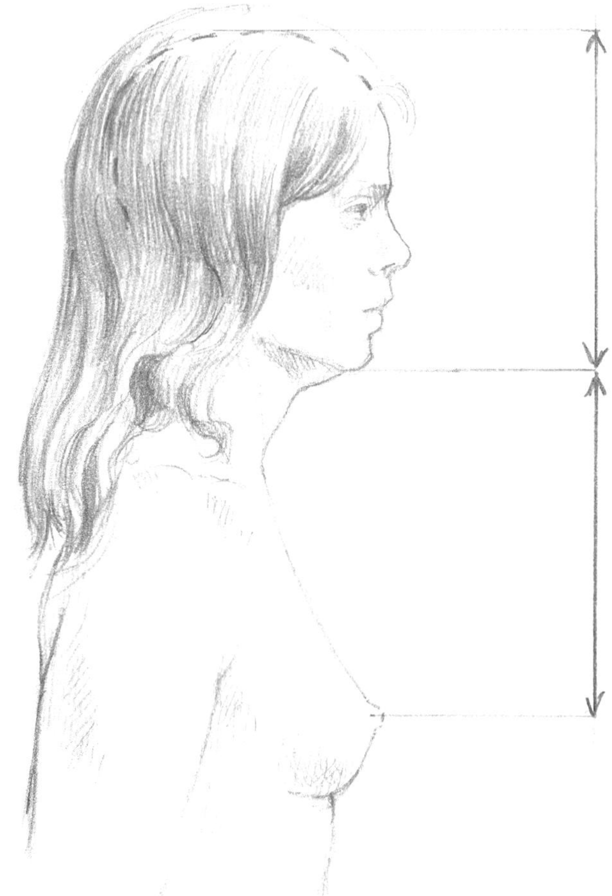

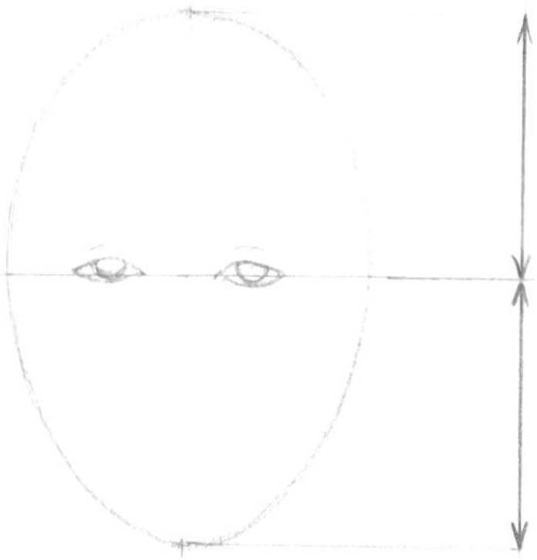

Before drawing, it is advisable to study carefully the principal vertical and horizontal axes of the head.

Fig. 301
Proportions of the Face

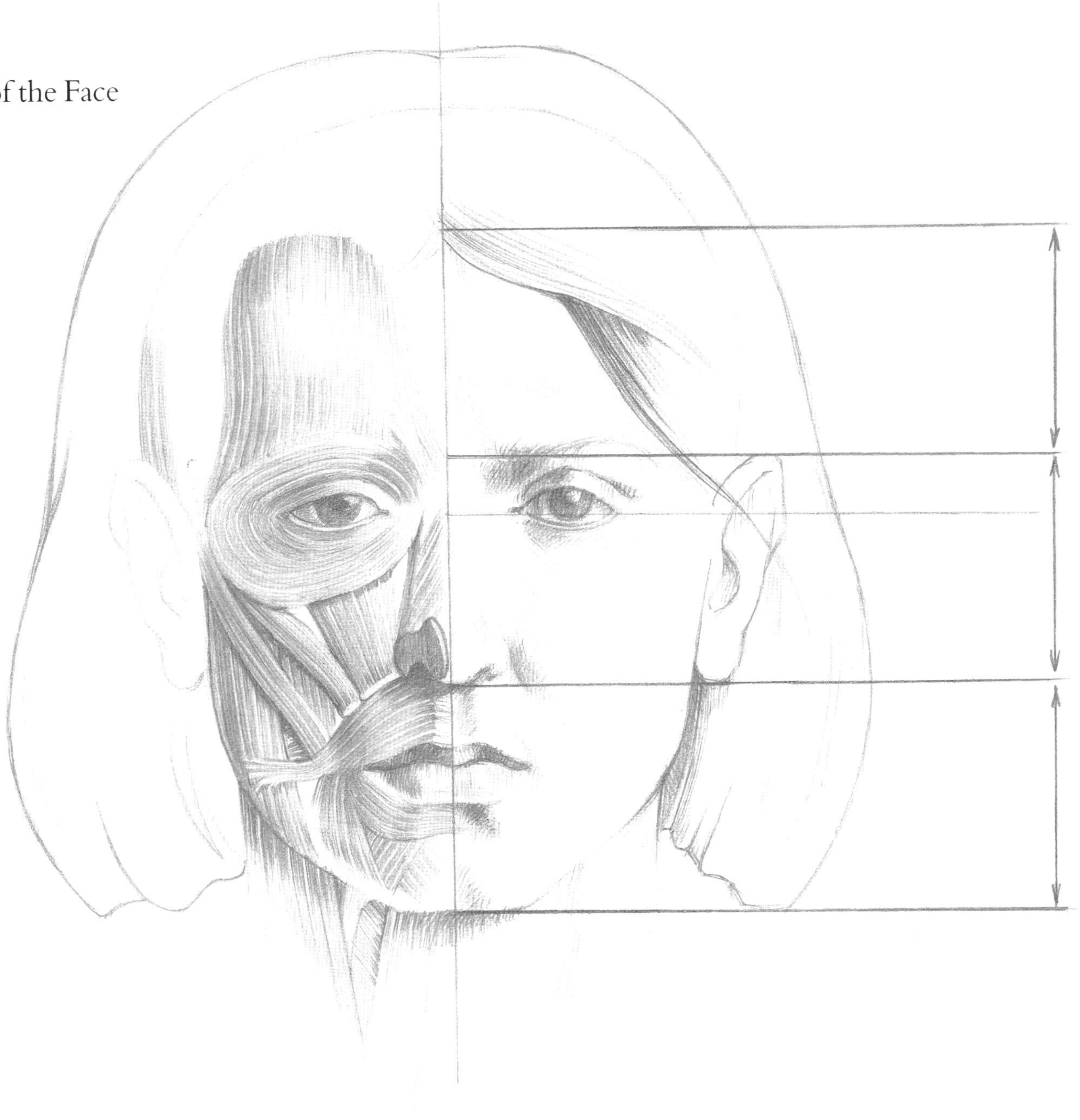

The height of a man's head, taken together with the neck, measures proportionally $^1/_8$ of the total body height. In women, it accounts for $^1/_7$–$^1/_6$ (20–22cm) of total body size. The weight of the head measures approximately $^1/_{17}$ of body weight (3.6–4kg). The bigger the body the smaller the head becomes proportionally. The head of a child is proportionally larger than that of an adult.

Facial Proportions

The forehead, nose and lips, and angle of the chin possess equal heights. The distance between the forehead protuberances is as large as the distance between the root of the nose and the tip of the chin. In the illustrations of the ancient Greek sculptors the distance between the inner corners of the eyes, the ala of the nose and the corners of the mouth are equal. However, in most people the mouth is wider than the distance between the ala of the nose and also than the distance between the corners of the eyes. The bilateral lines which connect the corners of the eyes to the edge of the ala of the nose and the corners of the lips, deviate laterally. The upper part of the dordum nasi is deeper in men; the corners of the lips do not extend beyond the canines.

The Forehead

A low forehead – depending in part on the extent of hair growth – gives the face a younger appearance (consider the paintings of Greek goddesses). A headband and virginal wreath in portraits of women in art and photography convey the same impression. A steep, broad forehead is the sign of deep thinking, a sloping forehead (such as in the silhouette of Robespierre) is known as the form of an animal forehead. When the cranial eyebrow arches and chin are pronounced, the face appears coarse.

Proportions of the Face
(cont.)

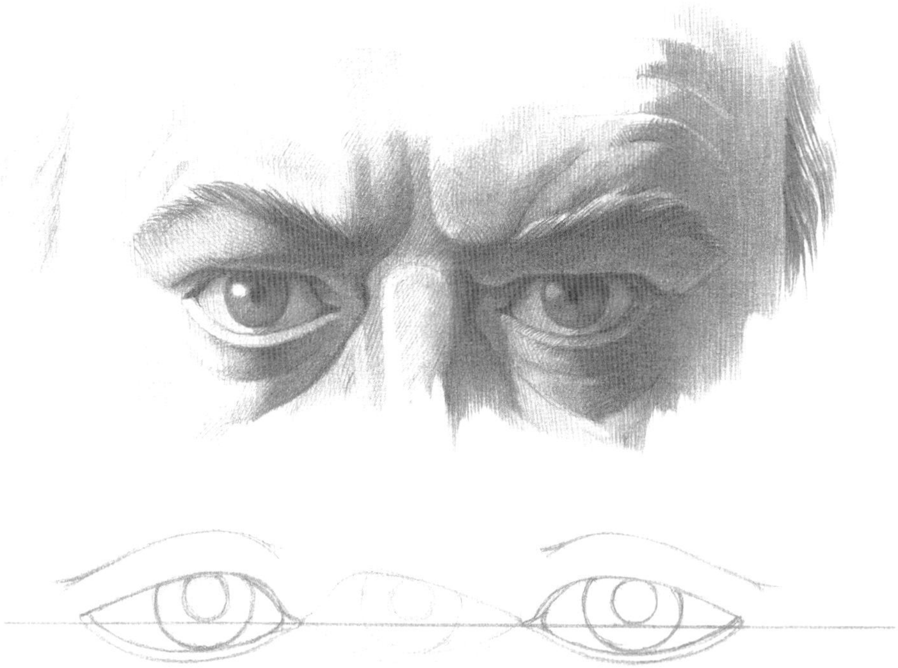

The distance between the eyes equals the width of one eye.

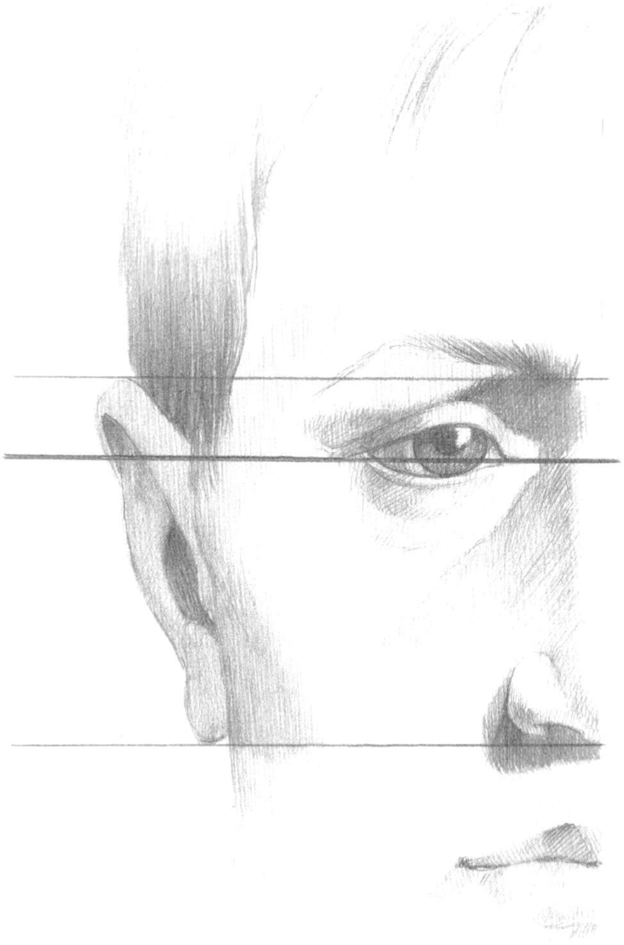

The ear is located in the region between the eyebrows and the tip of the nose.

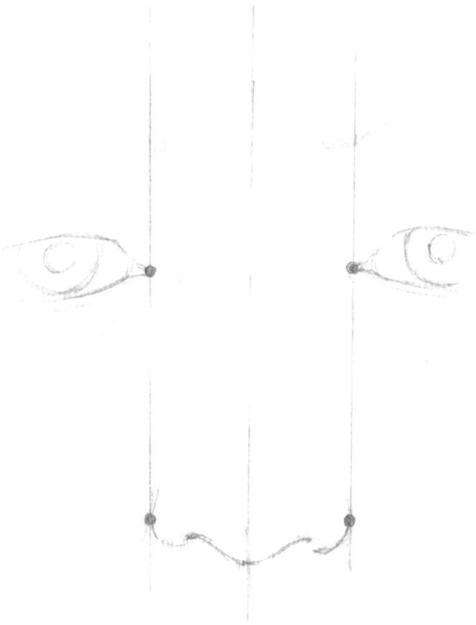

The width of the nose equals the distance between the inner corners of the eyes.

When the head is rotated, the parallel horizontal lines become parallel ellipses.

When the head is only slightly rotated, the size of the eyes and the distance between them seem to undergo change. The closer eye appears bigger, the more distant eye appears smaller.

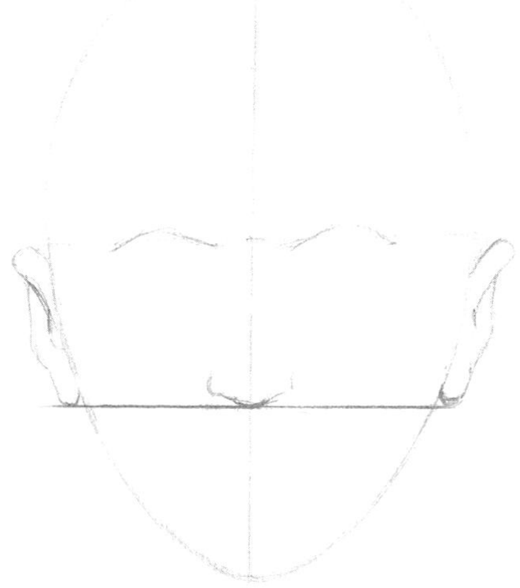

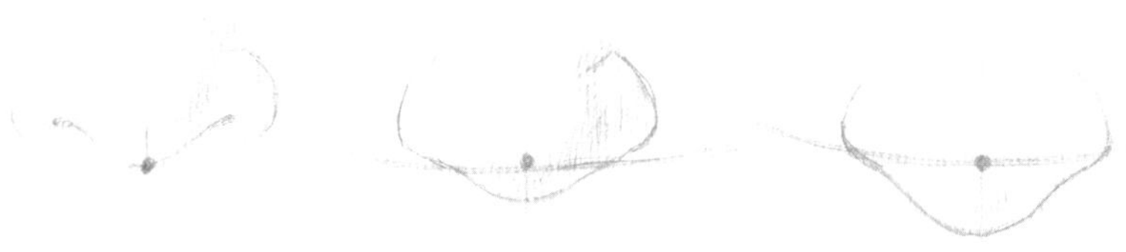

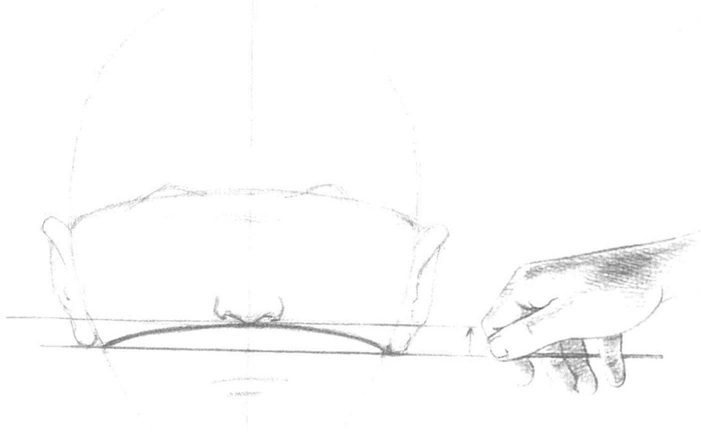

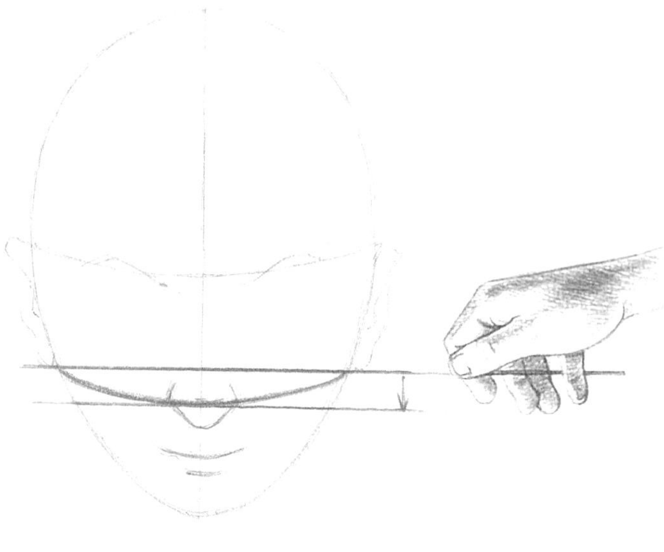

When the head is bowed, the region between the tip of the nose and the lower edge of the earlobe appears smaller. It is also partially obscured by the nose.

Fig. 302
The Grooves of the Face

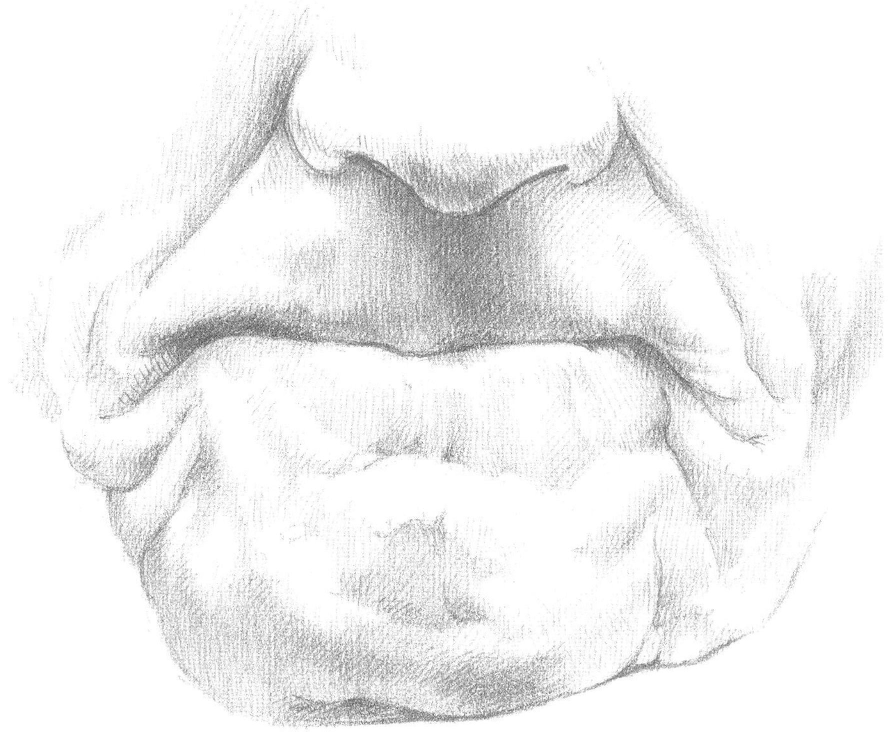

The grooves of the face are either inherited or acquired (SENILE FOLDS). Inherited folds include the fold below the cranial and caudal eye bulge (SULCUS ORBITOPALPEBRALIS SUPERIOR ET INFERIOR); the fold at the edge of the cranial and caudal edge of the eyelid; the fold at the nose and lips (SULCUS NASOLABIALIS), which runs from the ala of the nose to the cheeks; the caudal lip groove (SULCUS LABIOBUCCALIS), which extends in an arch from the corners of the mouth to the edge of the jaw. The jaw possesses either an even form (PROTUBERANTIA MENTALIS) or is divided by a vertical groove (MENTUM GEMINUM). The chin sulcus (SULCUS MENTOLABIALIS) encircles the lower lip in the shape of an arch. The groove beneath the nose (PHILTRUM) extends from the nasal cavities to the edge of the upper lip. Its prominent lower bulges meet the bulge of the lower lip. The "smile lines" become deeper at the side of the corners of the mouth when the risorius contracts. The chin groove (also considered to be an expression of female beauty) becomes deeper at the edge of the angle of the chin. The skin muscle of the forehead possesses transverse, arch-shaped folds. When people become older these folds – which radiate from the outer corners of the mouth – are called crow's feet. When the skin loses elasticity, radial folds appear at the lips.

Fig. 303
Male Head Hair

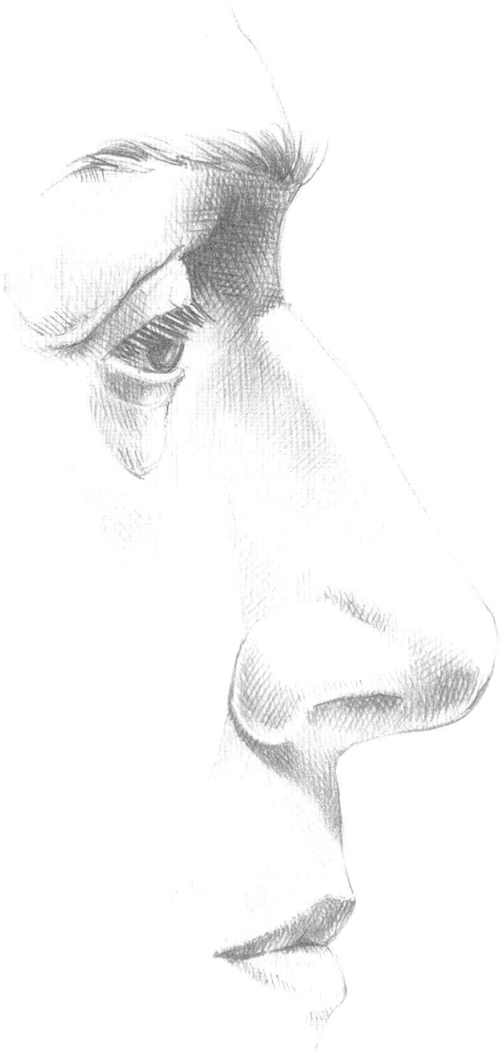

Indigenous peoples and certain peoples in South America do not have moustaches or beards. In eastern peoples the beard is the hallmark of great esteem. Its form and length or size vary greatly depending on fashion.

Hair occurs on the head in the form of head hair, eyebrows, the beard and moustache. Hairs extend from the forehead to the nape of the neck. The anterior edge of the hair forms a pentagon in the middle of the forehead, at the zygomatic process of the os frontale, and on both sides above the region behind the ears.

The eyebrow is an arch-shaped strip of hair which extends from the root of the nose to the zygomatic process. In southern peoples, eyebrows can meet in the middle. Long eyebrows make the face appear sombre.

MOVEMENT

This section is a summary of the principal concepts and characteristics of body positions (standing, sitting, lying), forms of movement (walking, running, jumping), or forms of action (throwing, hitting, kicking) from the standpoint of artistic illustration.

Fig. 304
The Body's Center of Gravity

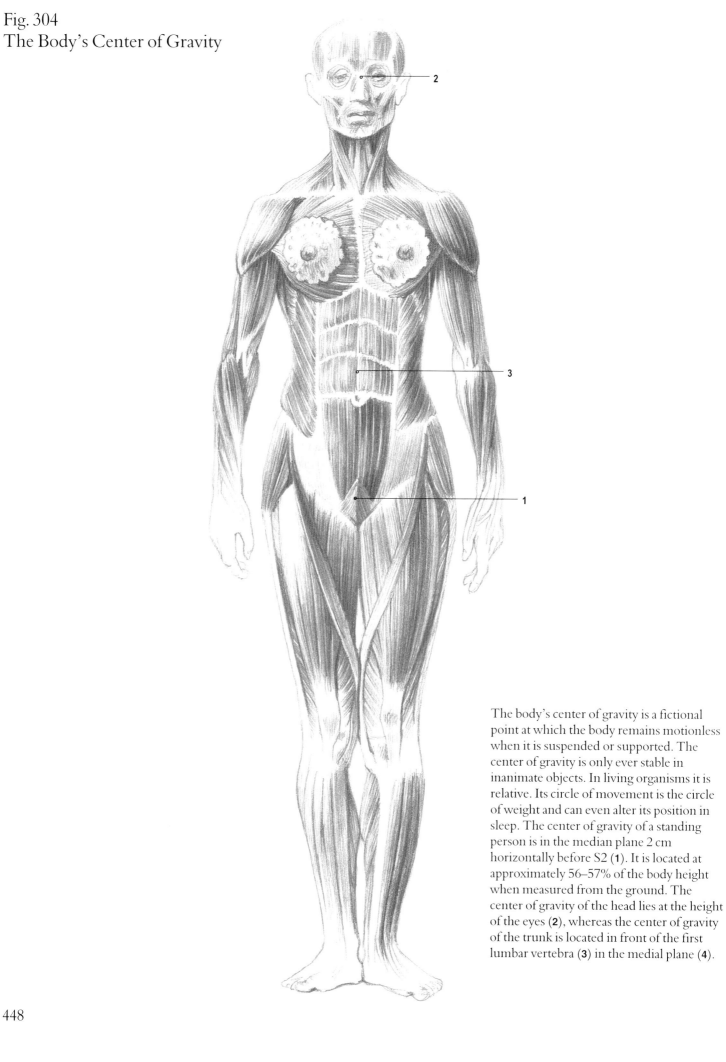

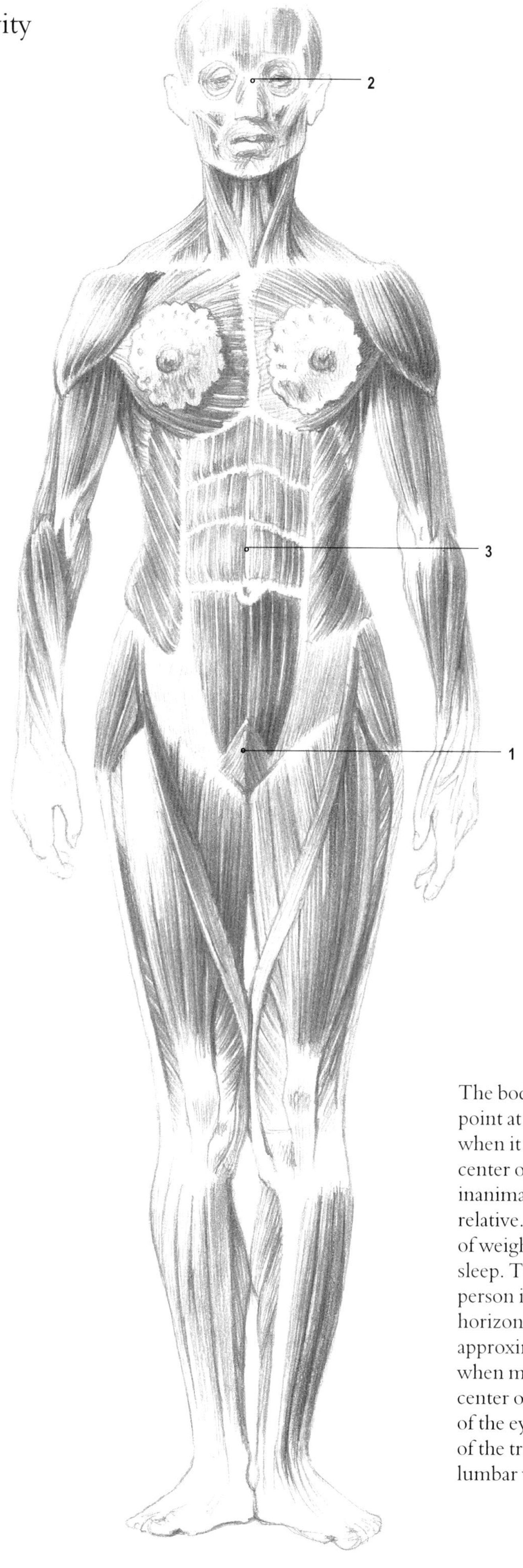

The body's center of gravity is a fictional point at which the body remains motionless when it is suspended or supported. The center of gravity is only ever stable in inanimate objects. In living organisms it is relative. Its circle of movement is the circle of weight and can even alter its position in sleep. The center of gravity of a standing person is in the median plane 2 cm horizontally before S2 (**1**). It is located at approximately 56–57% of the body height when measured from the ground. The center of gravity of the head lies at the height of the eyes (**2**), whereas the center of gravity of the trunk is located in front of the first lumbar vertebra (**3**) in the medial plane (**4**).

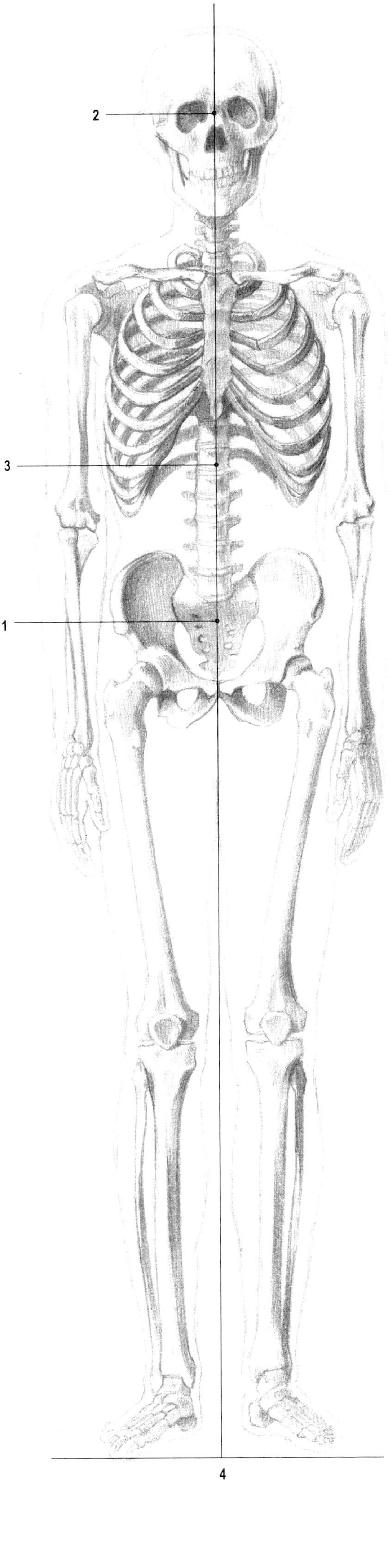
2
3
1
4

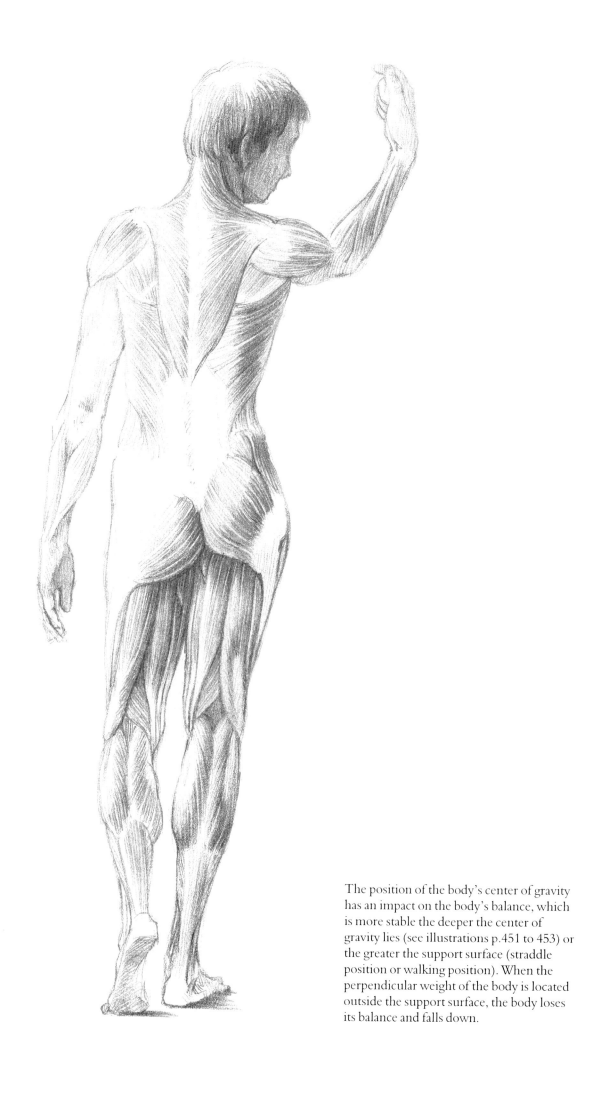

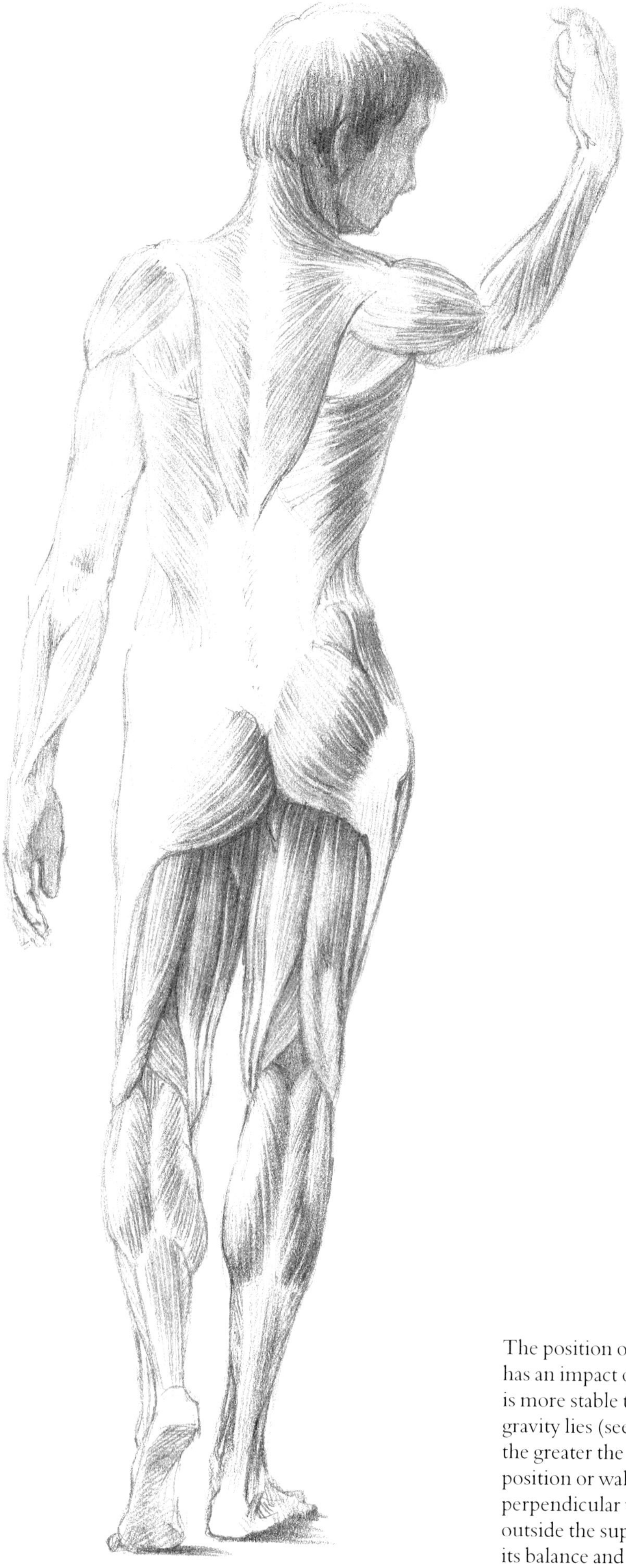

The position of the body's center of gravity has an impact on the body's balance, which is more stable the deeper the center of gravity lies (see illustrations p.451 to 453) or the greater the support surface (straddle position or walking position). When the perpendicular weight of the body is located outside the support surface, the body loses its balance and falls down.

Fig. 305
Surface Support

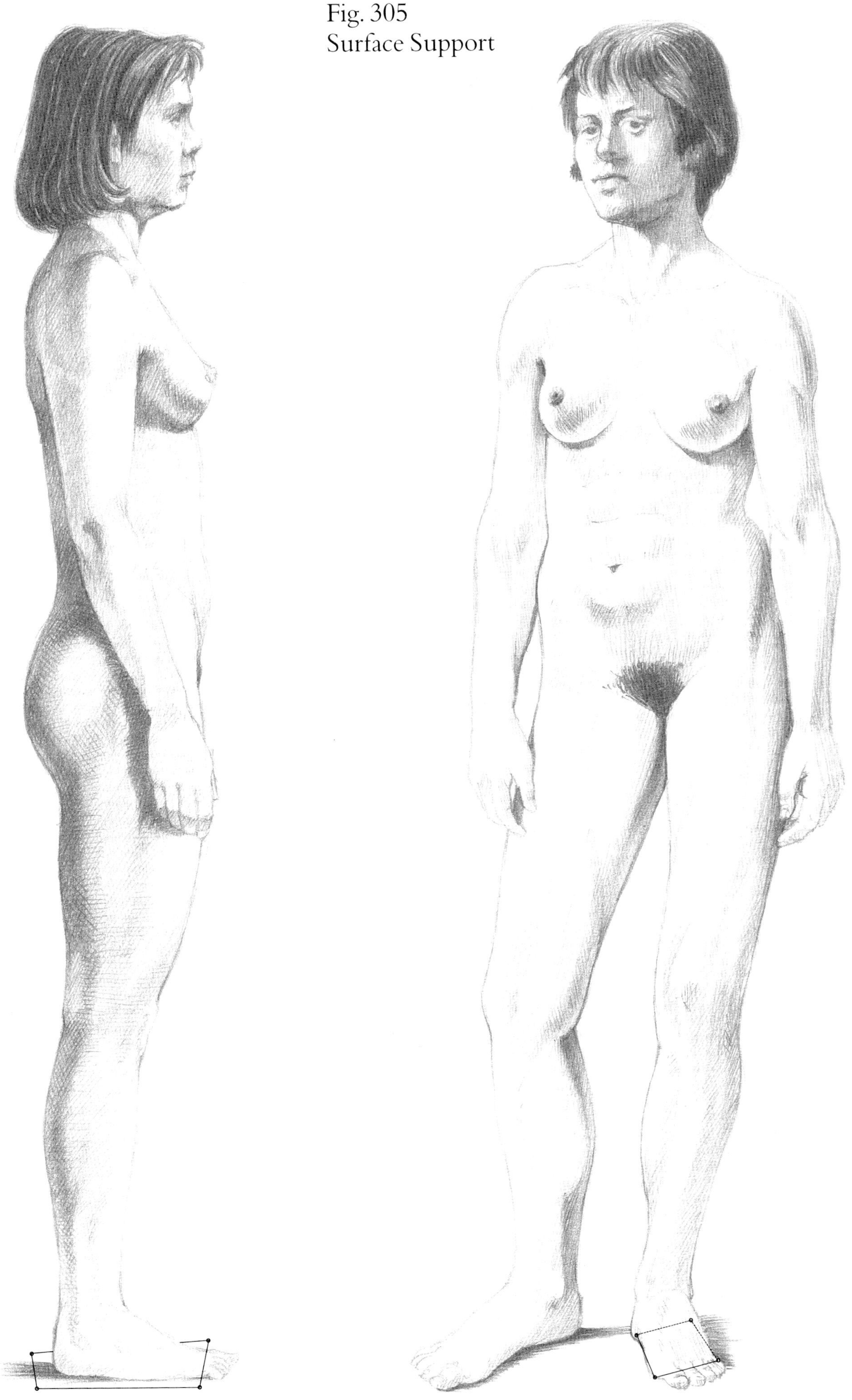

Fig. 306
Standing

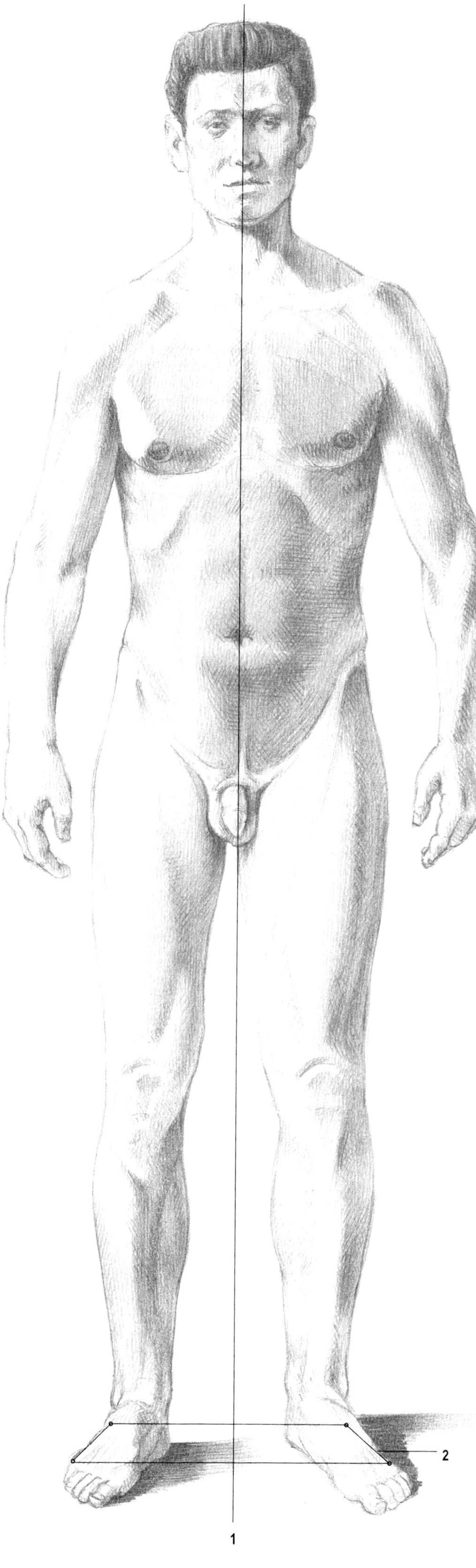

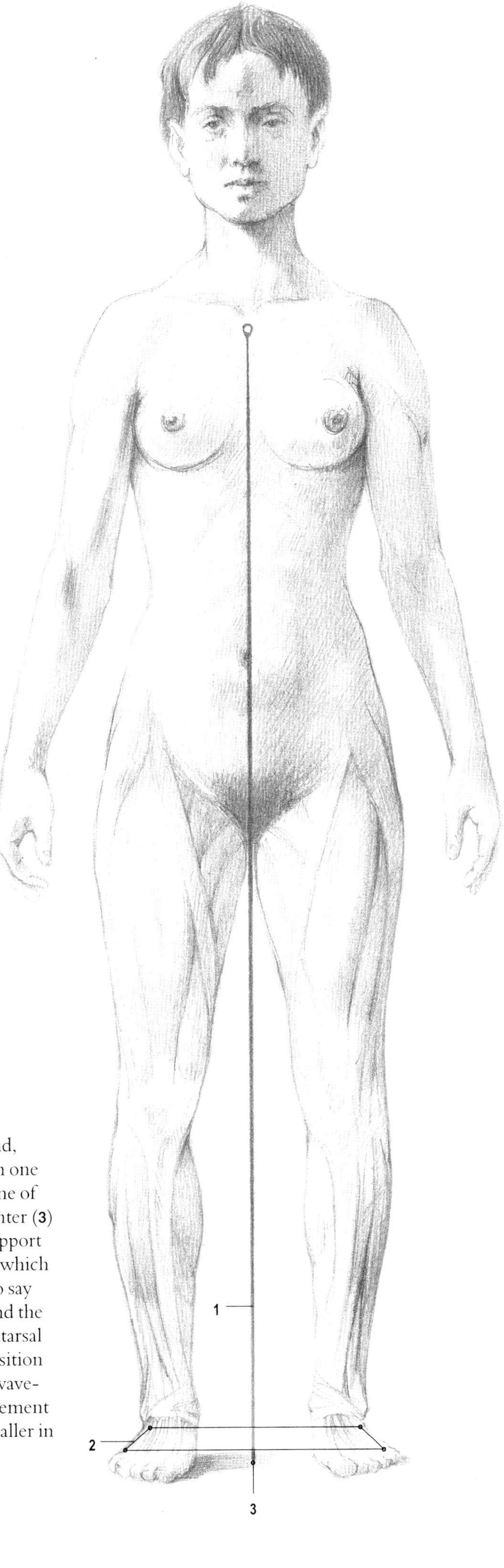

In a standing position, the body (head, neck, trunk and legs) is positioned in one plane – the line of gravity (**1**). The line of gravity reaches the ground in the center (**3**) of the surface of support (**2**). The support surface is the area between the lines which connect the support points, that is to say the two condyles of the heel bone and the small head of the first and fifth metatarsal bones. In a comfortable standing position the body can also effect a compact, wave-like swinging movement. This movement is larger in the median plane and smaller in the lateral plane.

The standing position is achieved by the antigravitational muscles and their antagonists (adductors and abductors), by means of their tone and also through their active contraction.

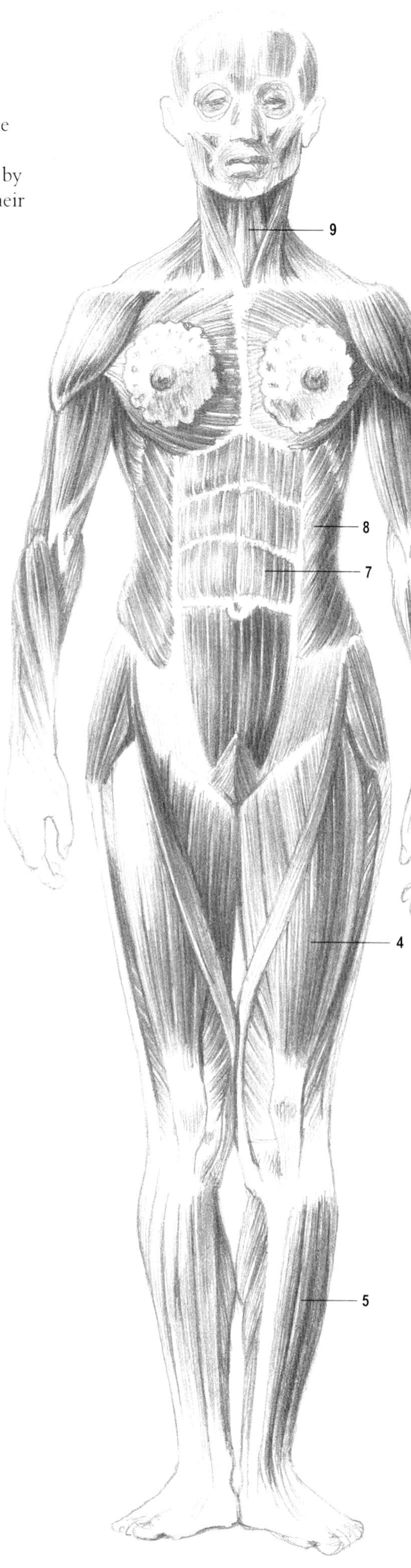

ANTIGRAVIATIONAL MUSCLES

1 Head and neck extensors
2 Trunk extensor
3 Extensor of the hip joint
4 Extensor of the knee joint
5 Extensor of the ankle joint

ANTAGONISTS

6 Iliac and lumbar muscles
7 Abdominal muscles
8 Serratus anterior
9 Flexors of the head and neck
10 Flexors of the knee joint
11 Flexors of the foot

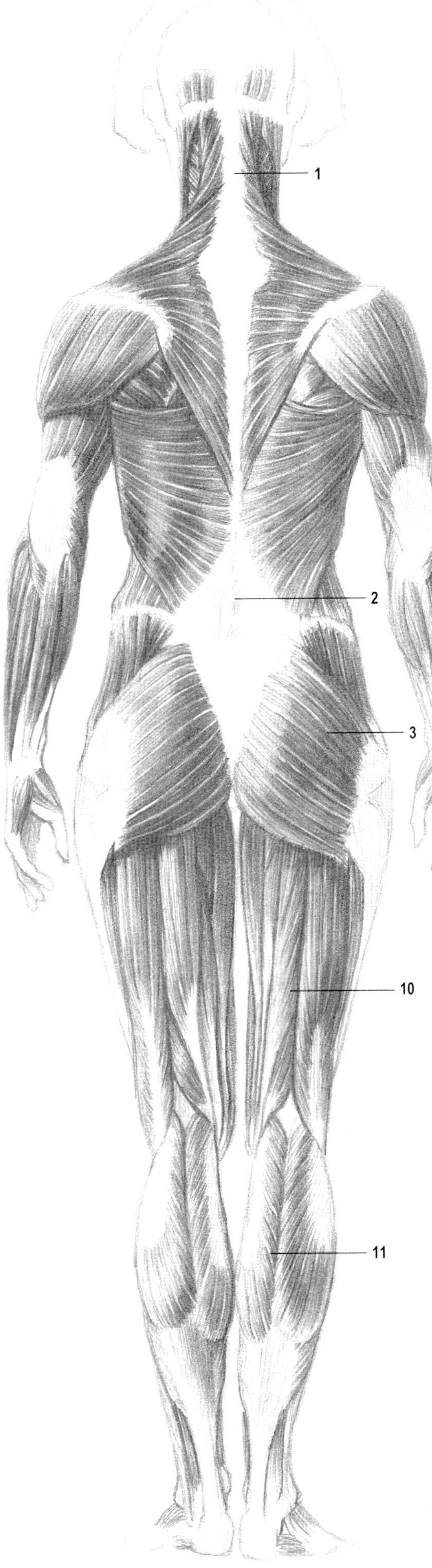
1
2
3
10
11

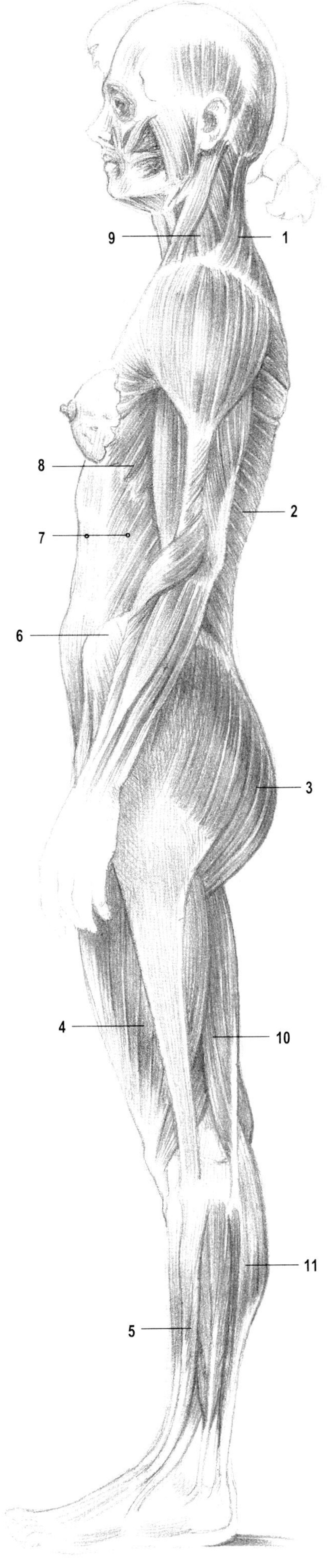
9
1
8
2
7
6
3
4
10
11
5

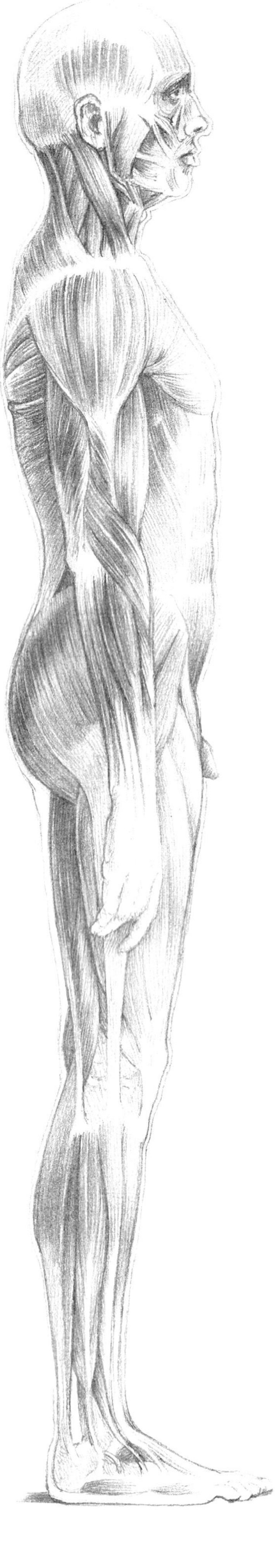

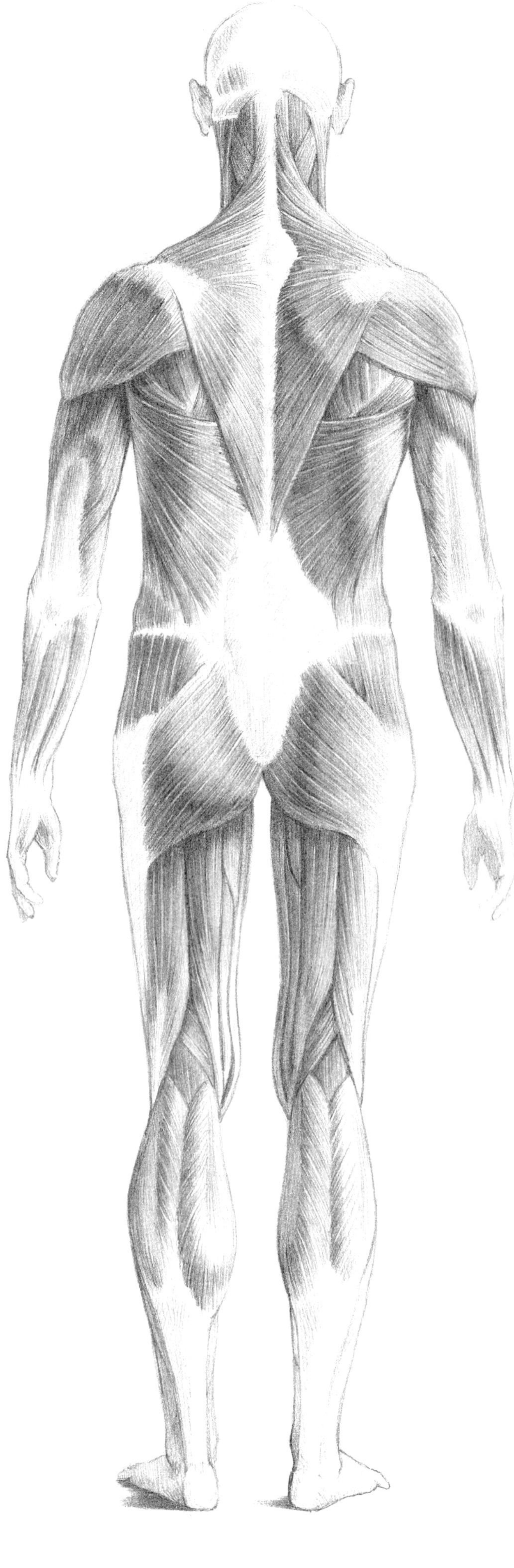

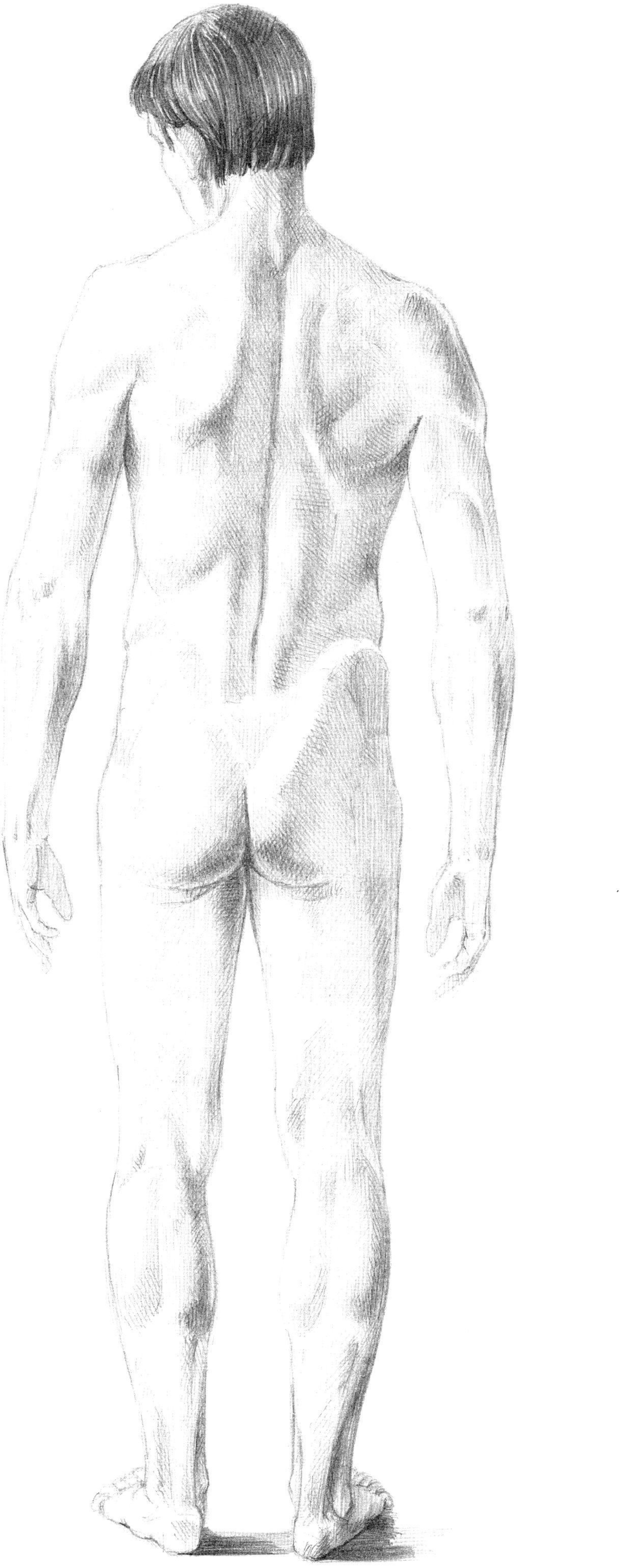

Fig. 307
Changes in Body Position

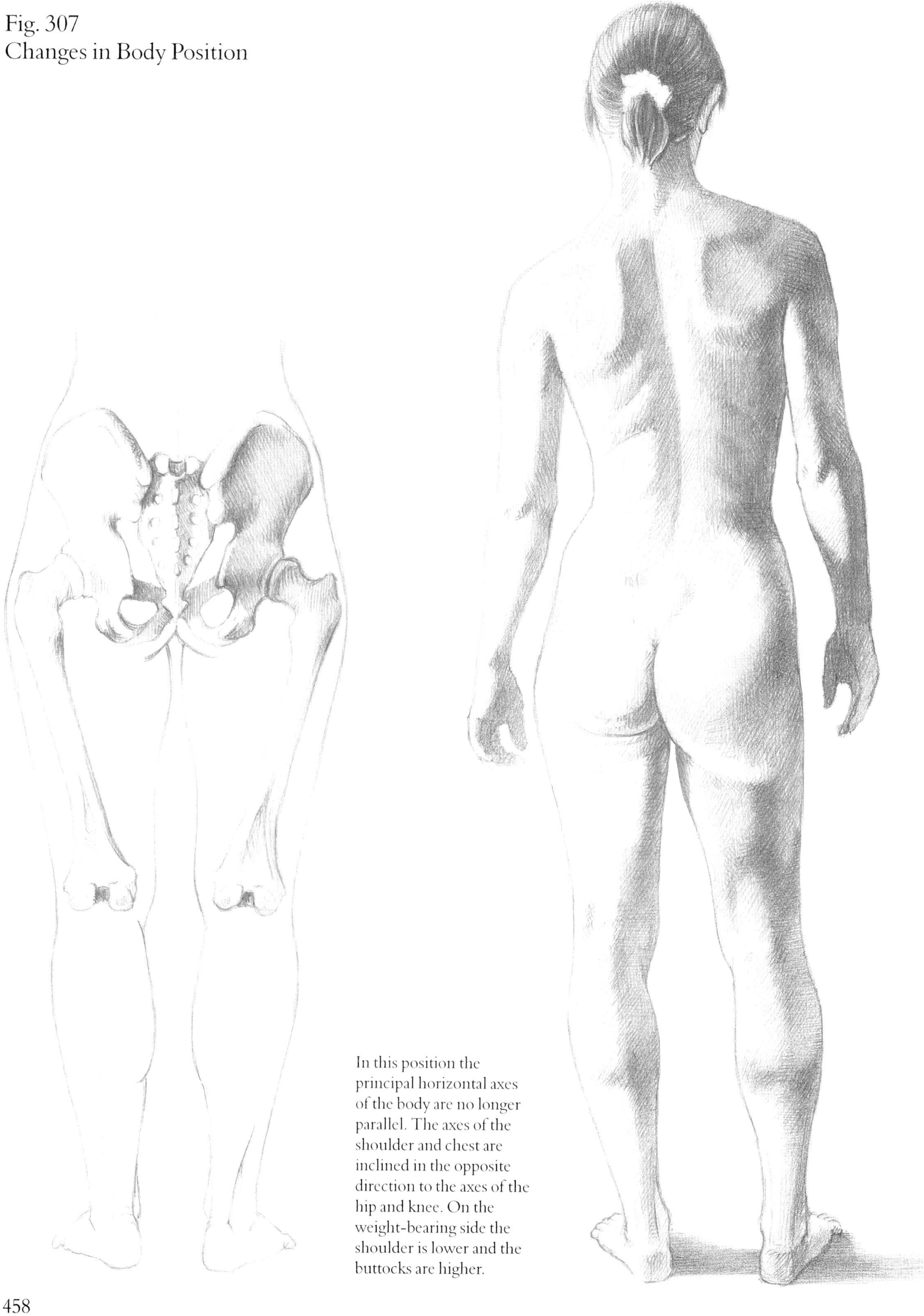

In this position the principal horizontal axes of the body are no longer parallel. The axes of the shoulder and chest are inclined in the opposite direction to the axes of the hip and knee. On the weight-bearing side the shoulder is lower and the buttocks are higher.

Walking is a rhythmic, alternating movement of the lower limbs, in which the body always maintains contact with the ground by means of the feet. The movements which accompany walking are the wave-like swaying of the trunk, the swaying of the arms and the movement of the head. These movements assist in walking. In a diagonal stride the shoulders and hips move in opposite directions. Here, the shoulder and pelvic movements are contrary.

Fig. 308
Walking

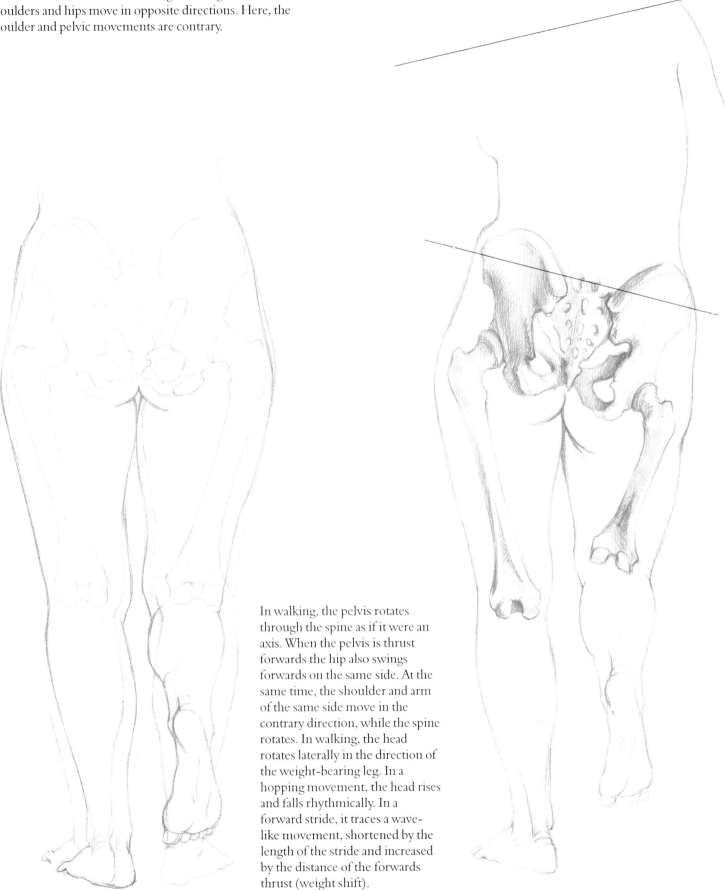

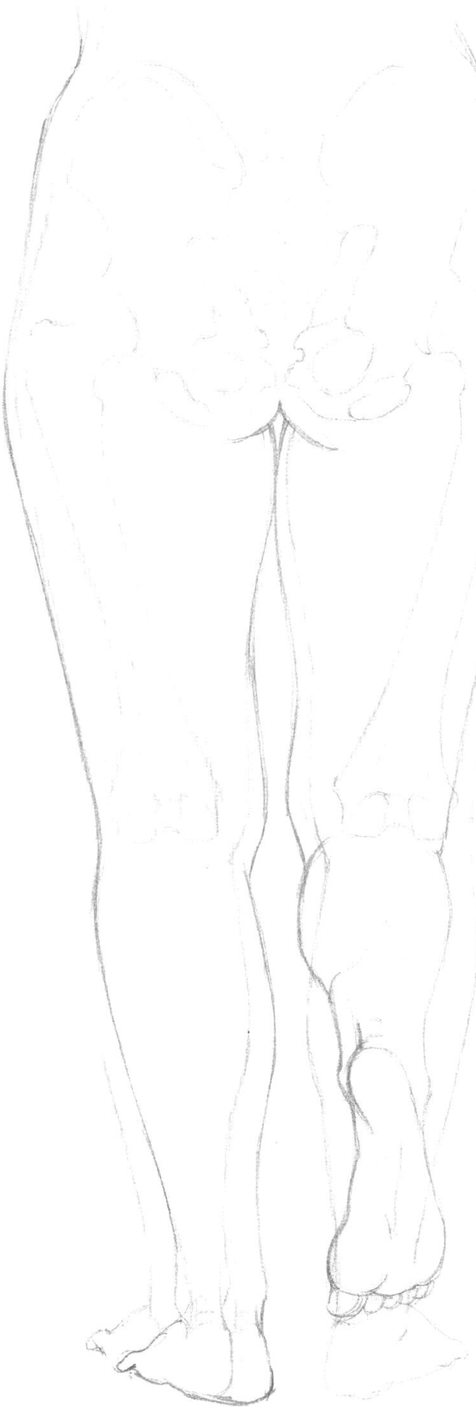

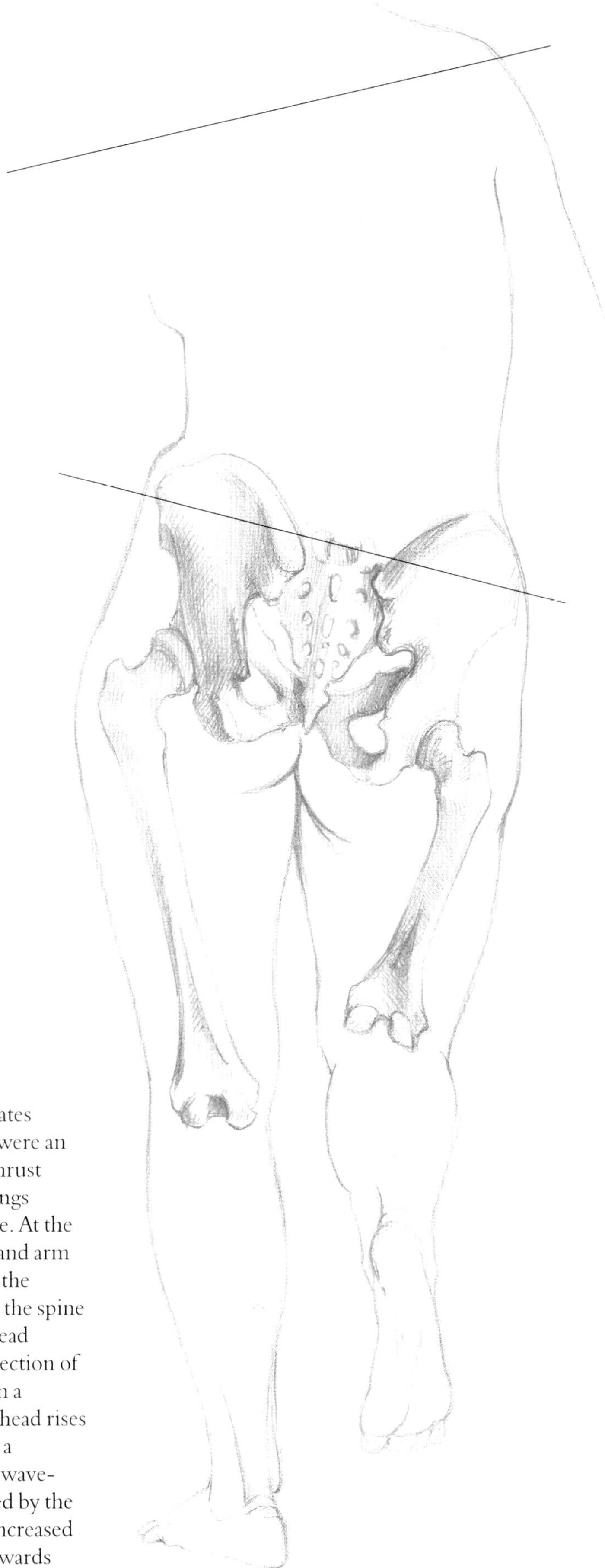

In walking, the pelvis rotates through the spine as if it were an axis. When the pelvis is thrust forwards the hip also swings forwards on the same side. At the same time, the shoulder and arm of the same side move in the contrary direction, while the spine rotates. In walking, the head rotates laterally in the direction of the weight-bearing leg. In a hopping movement, the head rises and falls rhythmically. In a forward stride, it traces a wave-like movement, shortened by the length of the stride and increased by the distance of the forwards thrust (weight shift).

Fig. 309
The Basic Element of Walking

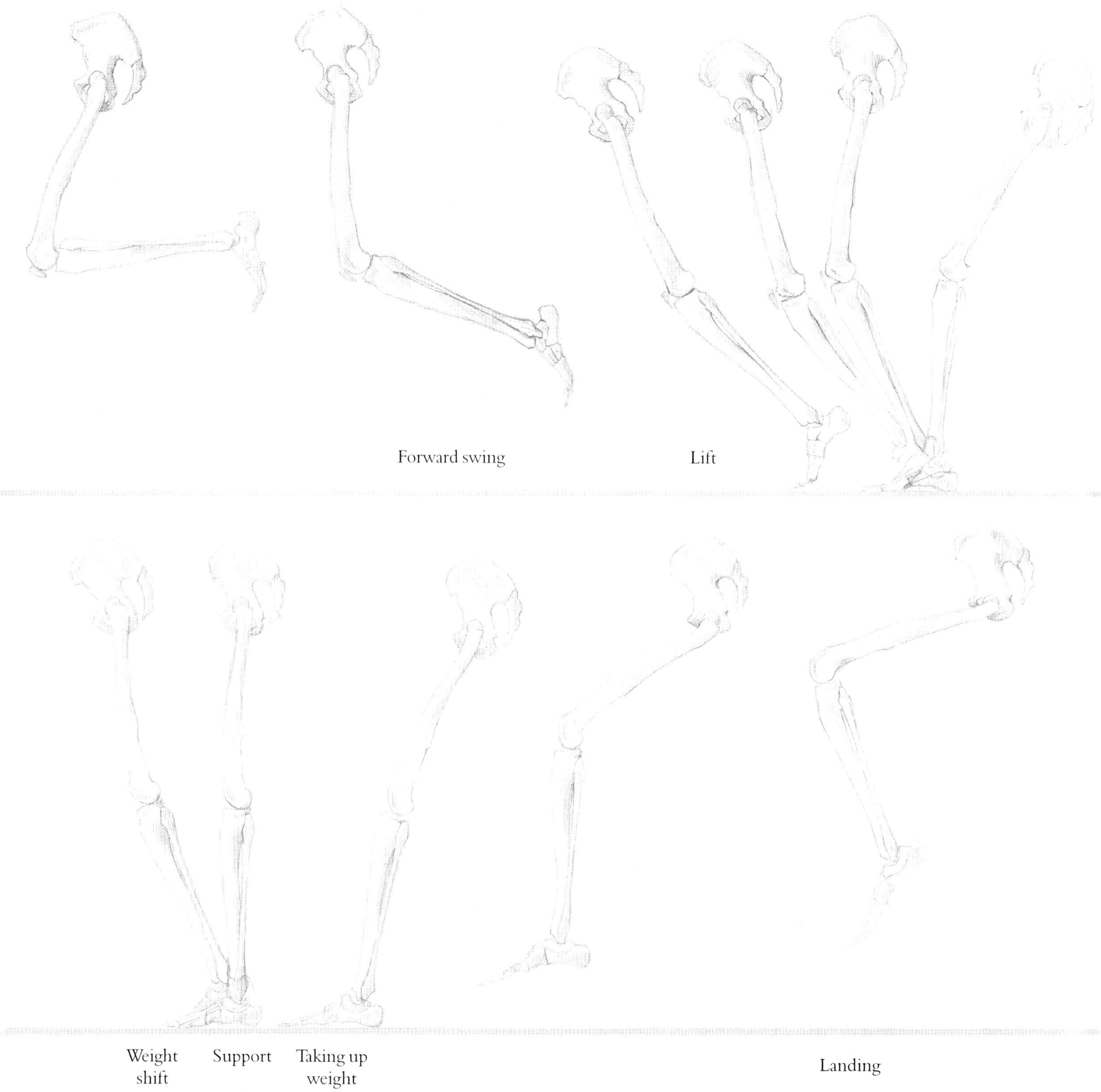

In walking, the trunk is alternately carried by one leg at a time (the standing leg), while the other leg swings past it (swinging leg). At the moment when the rear leg lifts from the ground, the proceeding leg makes contact with the ground firstly, with the heel, thus becoming the standing leg. In this phase of double support both feet touch the ground. Starting with the heel the entire sole of the foot is now placed on the ground and the leg assumes a vertical position. In order to bring the trunk forward the standing leg leans forward, the foot rolls through the toes off the ground. The standing leg becomes the swinging leg and is drawn forward with a flexed knee.

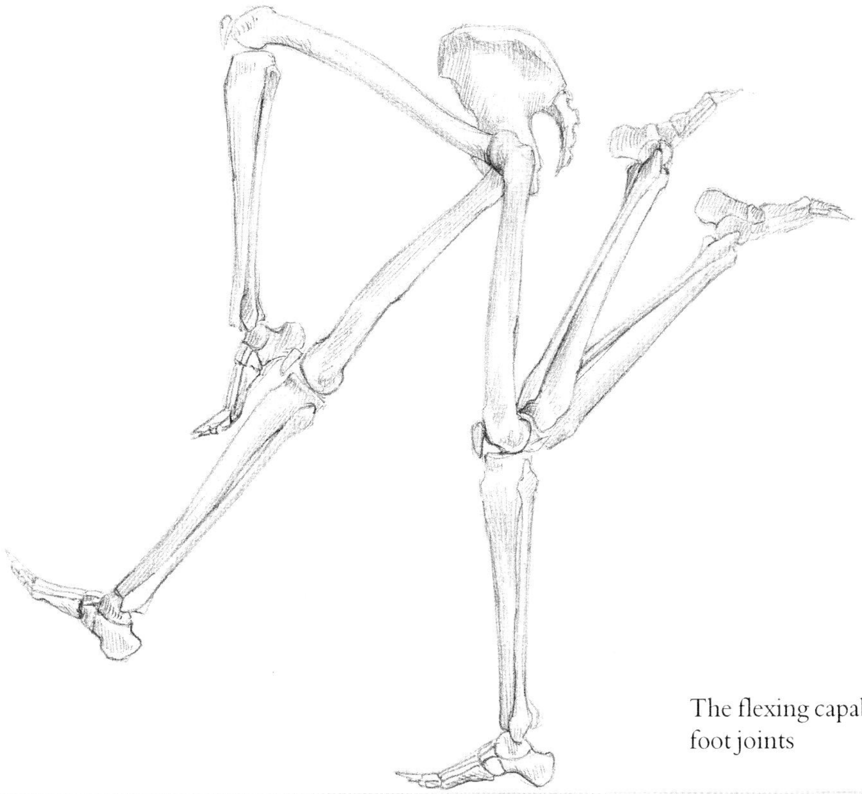

The flexing capabilities of the hip, knee and foot joints

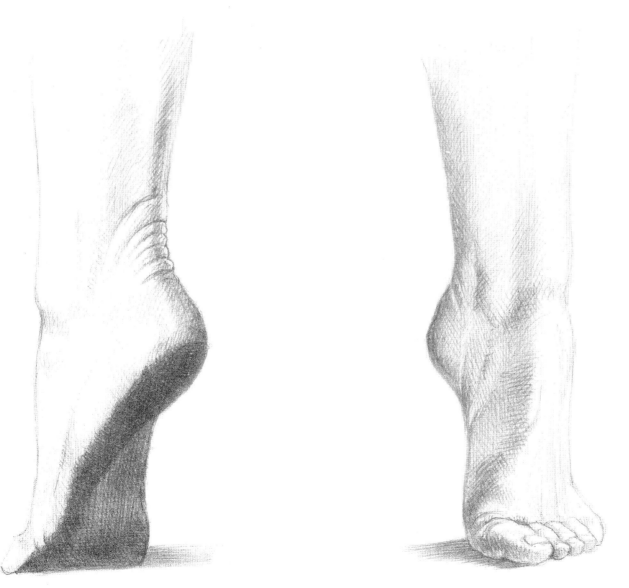

Fig. 310
Walking and Running

Walking consists of two continuously repeated tread cycles or, in other words, a twin tread of the right and left foot. This means that when weight transfer takes place the body gains double support since both feet are on the ground at this moment. One foot will be in the final phase of weight transfer.

Running consists of a sequence of jumps. It requires a characteristic use of muscle strength. The impulsion of the body rises with the number of steps per unit of time, so that with increasing steps the point is reached at which both legs leave the ground. The accompanying movements (of the hips, shoulders and arms) and the flexion of the joints become greater in range. The trunk leans forward by 55 to 60 degrees. The head is held stiff or rotates only slightly.

Fig. 311
Hurdling

Fig. 312
Long Jump

Fig. 313
Triple Jump

Fig. 314
Studies in Movement

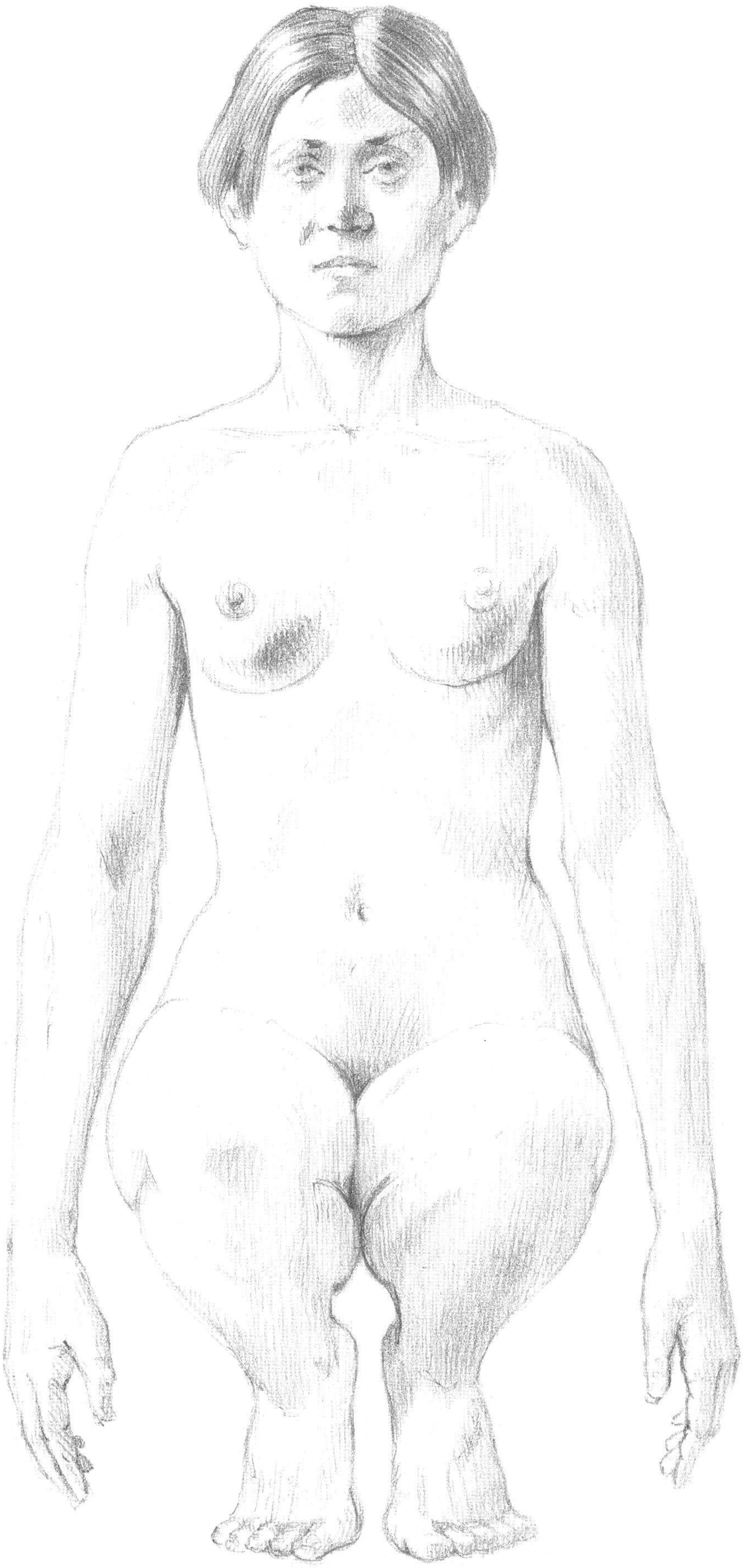

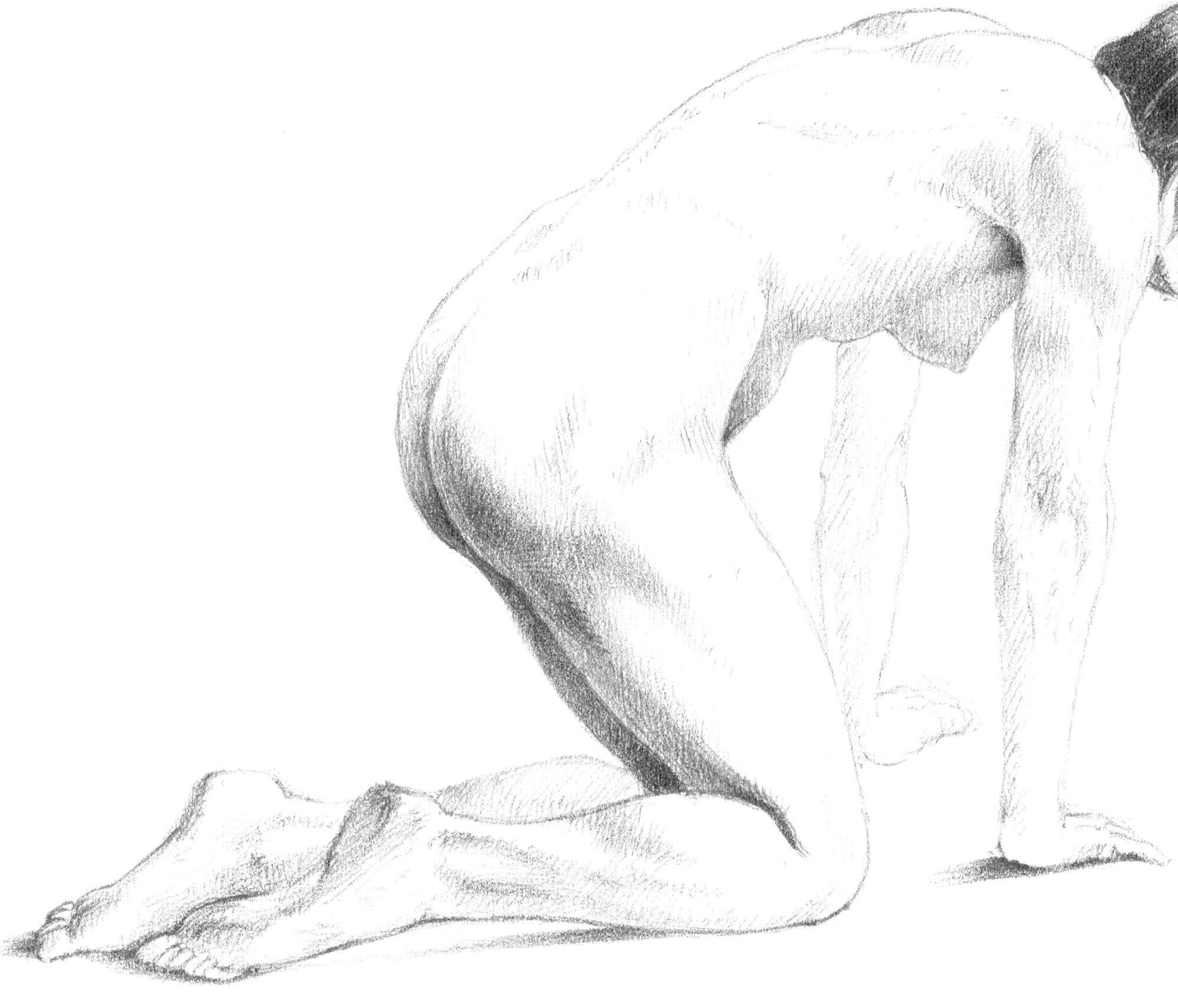

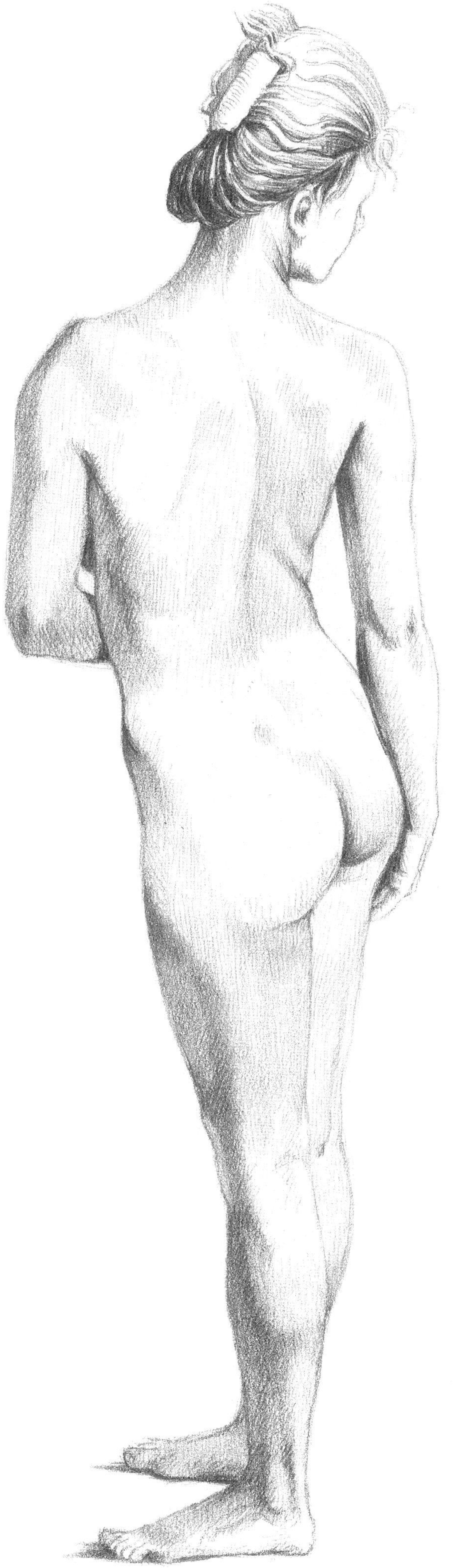

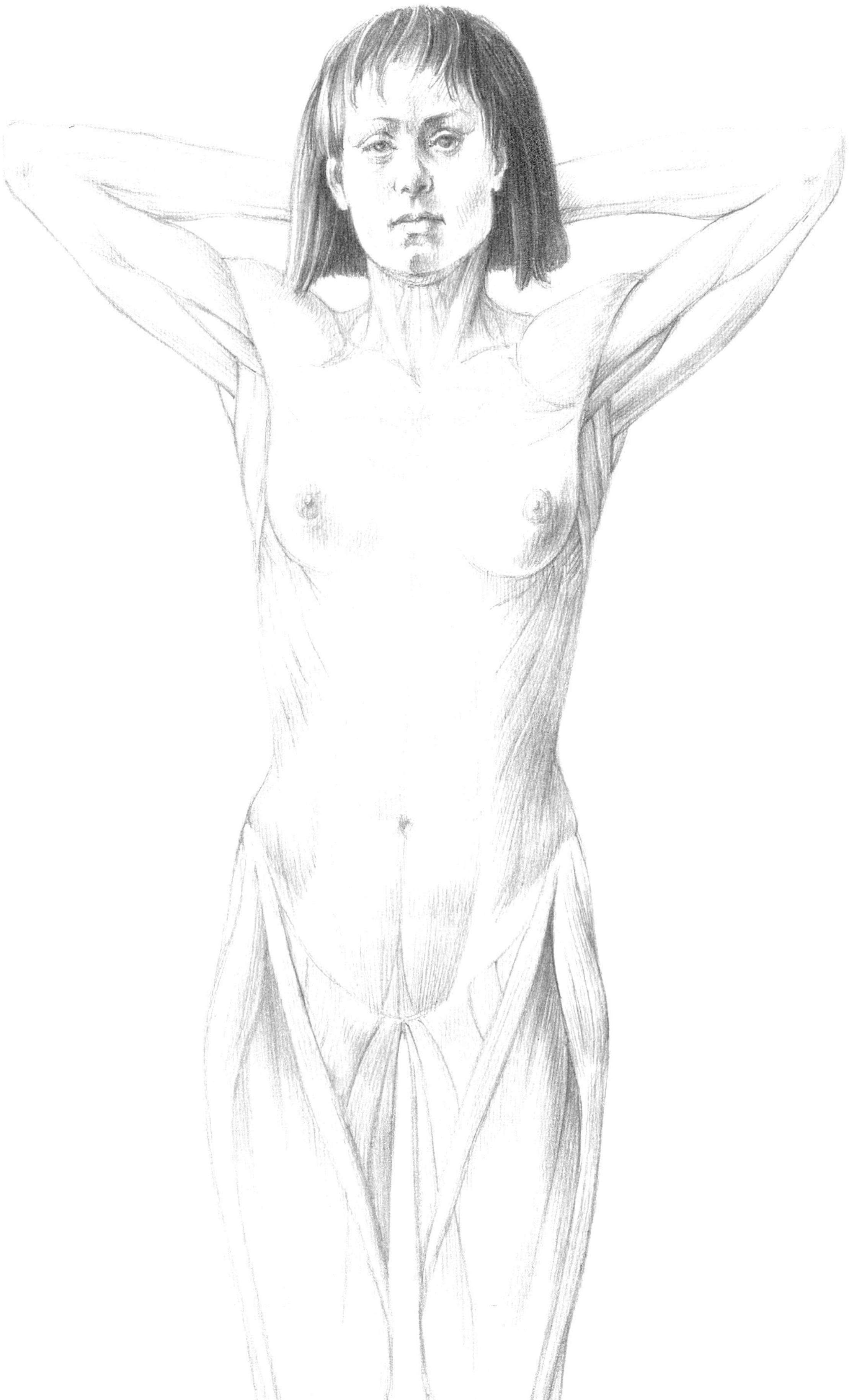

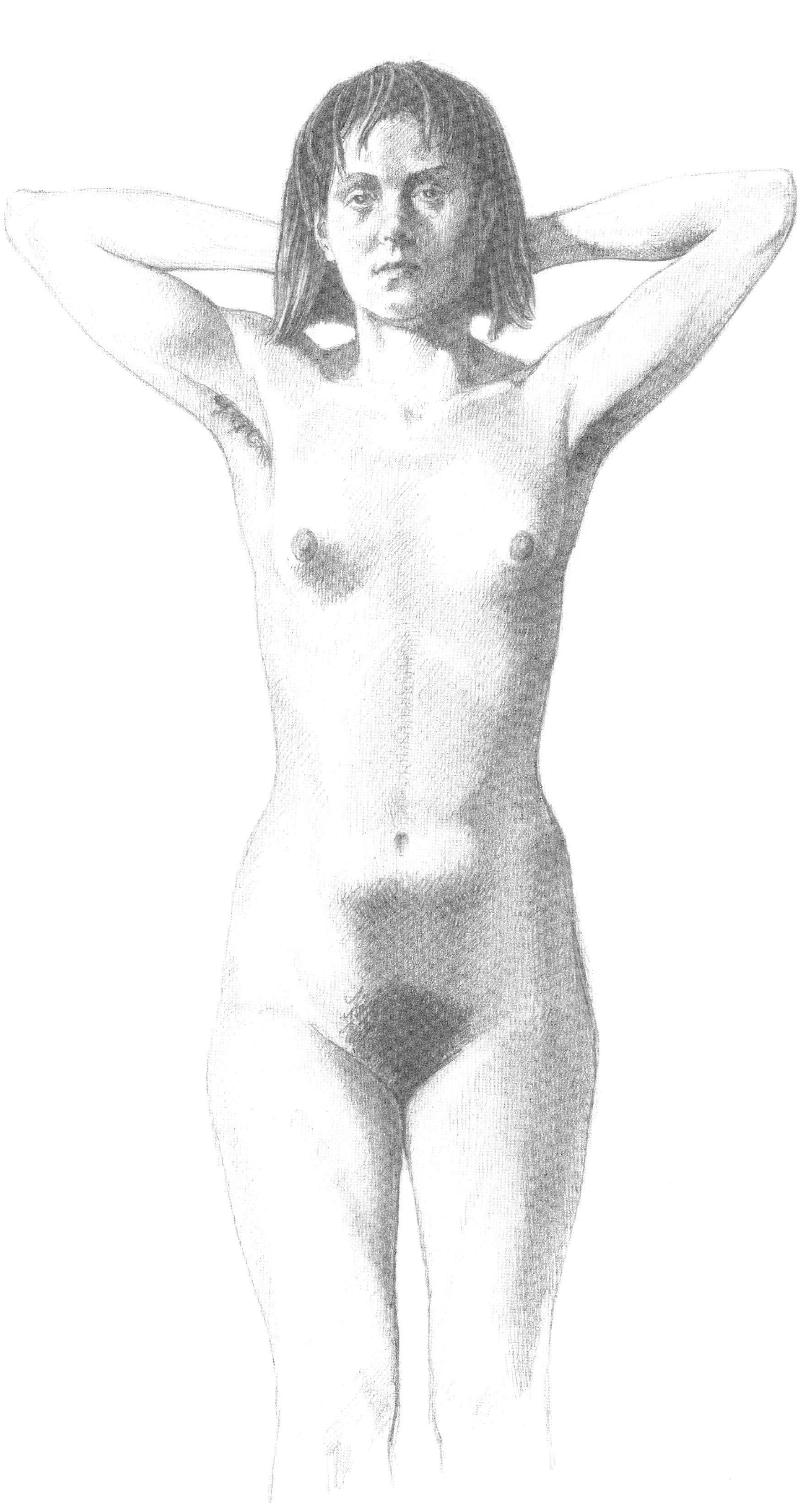

Table of Muscles

The Names, Origins, Insertions and Functions of Muscles

The physical form of the human being is largely determined by the size and length of the muscles. Most muscles are attached to the skeleton at two points. Human beings can move when, as a result of contraction, these points are brought closer together. This point, which is usually fixed, is called the ORIGIN, whereas the point which moves the bone like a lever is known as the INSERTION.

The italicised figures in brackets in the illustrated section of the atlas relate to the following table. The names of the muscles are given in Latin and, where possible, an English equivalent.

The Muscles of the Neck

	Name	Origin	Insertion	Function
1	**Platysma**	Facial skin in the region of the corners of the mouth and the cheek	Skin above the second and third rib	Folds and tightens the skin of the neck
2	**Rectus capitis posterior**			
	• **major** (straight posterior muscle of the head)	With short tendon on the spinous process of the axis	Linea nuchalis inferior	Extends the head and pulls it caudally
	• **minor**	Tuberculum of the atlas	Linea nuchalis inferior	Inclines the head backwards
	• **lateralis**	Transverse process of the atlas	Jugular process of the linea nuchalis inferior	
3	**Obliquus capitis superior**	Transverse process of the atlas and spinous process of the atlas	Inferior crest of the nape and arcus dorsalis atlantis	Fixes, rotates or pulls the head to the side
3/1	**Obliquus capitis anterior**	Spinous process and arcus of the axis	Transverse process of the atlas	Rotational movement of atlas and head
4	**Rectus capitis inferior**	Massa lateralis atlantis	Above the condyle of the occipital bone	Flexes the head
5	**Splenius** • **capitis**	Neck ligament and spinous process of the third–seventh cervical and first–third thoracic vertebrae	Mastoid process of the temporal bone and lateral crest of the nape	Rotates the head and pulls the neck to the side
	• **cervicis**	Supraspinal ligament and spinous process of the third–fifth thoracic vertebrae	Lateral bulges of the first and second cervical vertebrae	Extends the head and pulls the neck to the side
6	**Longus colli** (long neck muscle)	Anterior and lateral surface and lateral process of the bodies of the fourth–first thoracic and seventh–fifth cervical vertebrae	Bodies of the third–fourth cervical vertebrae and anterior protuberance of the atlas	Flexes the cervical column to the side and rotates the neck
7	**Longus capitis** (long head muscle)	Transverse processes of the third–sixth cervical vertebrae	Lateral surface of the tuberculum pharyngeum	Flexes the head, rotates it to the side
8	**Scalenus anterior**	Transverse processes of the third–sixth cervical vertebrae	Anterior surface and protuberance of the first rib	Flexes and rotates the cervical column to the side, lifts the ribs when head and neck are in a fixed position
9	**Scalenus medius**	Transverse processes of the third–sixth cervical vertebrae	Outer surface and posterior ridge of the first rib	Flexes the spinal column
10	**Scalenus posterior**	Transverse processes of the fifth–seventh cervical vertebrae	Anterior ridge and outer surface of the second rib	Lifts the second rib; rotates head and neck to the side when the second rib is in a fixed position
11	**Sternocleidomastoideus**	Ensiform process of the sternum; clavicle, sternocalvicular joint	At the inferior ridge of the body of the hyoid bone	Pulls the hyoid bone caudally

The Muscles of the Larynx and Tongue

12	**Sternohyoideus**	Manubrium of the sternum; clavicle, sternocalvicular joint	At the inferior ridge of the body of the hyoid bone	Pulls the hyoid bone caudally

	Name	Origin	Insertion	Function
13	**Omohyoideus**	On the superior ridge of the shoulder blade	On the inferior ridge of the body of the hyoid bone	Pulls the hyoid bone caudally and laterally (inspiration muscle)
	• venter inferior	On the superior ridge of the shoulder blade	Tendon of the lower abdomen	
	• venter superior	Tendon of the lower abdomen	Muscle crest of the body of the hyoid bone	
14	**Sternothreyoideus**	Manubrium of the sternum and first rib cartilage	Cartilage of the scutum	Pulls the larynx caudally
15	**Thyreohyoideus**	Linea obliqua of the cartilage of the scutum	Lateral surface of the hyoid bone	Pulls the hyoid bone caudally and the cartilage of the scutum cranially; closes the larynx
16	**Digastricus** (lower jaw muscle)			
	• venter posterior	Mastoid process of the temporal bone	Narrows to connect with rounded tendon	Pulls the lower jaw dorsally, opens the mouth
	• venter anterior	Medial surface of lower jaw	Narrows to connect with rounded tendon	Pulls the hyoid bone forwards and upwards
17	**Stylohyoideus**	Hyoid process of the temporal bone	Lateral edge of the greater branch of the hyoid bone	Lifts and pulls the hyoid bone backwards and upwards
18	**Mylohyoideus**	Crest of the lower jaw	Side of the body of the hyoid bone and middle tendon	Lifts the hyoid bone; pulls the lower jaw downwards and forwards, presses it against the hard palate
19	**Geniohyoideus**	Inner surface of the tip of the lower jaw	Anterior part of the body of the hyoid bone	Pulls the hyoid bone forwards and the lower jaw downwards

The Muscles of the Trunk

The Muscles of the Shoulder Girdle

	Name	Origin	Insertion	Function
20	**Trapezius**			
	• capitis	On the line of the nape of the neck	Shoulder blade part of the clavicle	Rotates the head and pulls the shoulder blade backwards or towards the spine; lifts the head;
	• colli	Neck ligament and spinous processes of the cervical vertebrae	Lateral part of the shoulder blades and acromion	pulls the shoulder blade towards the spine, pulls the shoulder
	• dorsi	Supraspinal ligament	Inner part of the shoulder ridge	blade downwards
21	**Latissimus dorsi**	Spinous processes of the fifth–seventh thoracic vertebrae, dorsolumbar fascia; with muscle teeth on the last three ribs	Tuberculum minus and intertubercular channel of the humerus	Pulls the lifted arm downwards and the hanging arm dorsally and medially; pulls the shoulder blade caudally and dorsally
22	**Rhomboideus major**	Spinous processes of the second–fifth thoracic vertebrae	Spinal ridge and inferior angle of the shoulder blade	Together with muscle 31 pulls the shoulder blade towards the spinal column and cranially, fixes the shoulder blade

	Name	Origin	Insertion	Function
23	**Rhomboideus minor**	Neck ligament and spinous processes of the first thoracic and last two cervical vertebrae	Spinal ridge (medial and caudal ridge) of the shoulder blade	Fixes the shoulder blade
24	**Levator scapulae**	With four muscle teeth on the transverse processes of the first–fourth cervical vertebrae	Cranial angle of the shoulder blade	Lifts the shoulder blade
25	**Serratus**			
	• posterior superior	Supraspinal ligament, spinous processes of the last two cervical and first two thoracic vertebrae;	Muscle teeth on the angles of the second–fifth ribs	Lifts the ribs
	• posterior inferior	Lumbodorsal fascia, tenth–twelfth thoracic and first lumbar vertebrae	Four muscle teeth on the last four ribs	Pulls the ribs caudally

The Muscles of the Spinal Column

	Name	Origin	Insertion	Function
26	**Erector spinae**			
26/1	**Iliocostalis**			Extends (one side) the spinal column, flexes it to the side
	• lumborum	Dorsal surface of the sacrum; ala of the os ilium, dorsolumbar fascia	Angles of the fifth–twelfth rib	
	• thoracis	With muscle teeth on the third–twelfth rib	Transverse processes of the fourth–seventh cervical vertebra	
	• cervicis	Third–sixth rib	Transverse processes of the fourth–sixth cervical vertebrae	
26/2	**Longissimus**			Flexes the spinal column sideways and backwards and extends it, rotates the spine to the side (on one side); pulls the head dorsally (on both sides) and rotates it
	• thoracis	Dorsal surface of the sacrum; spinous processes of the lumbar vertebrae and transverse processes of the last thoracic vertebrae	Lumbar vertebrae; transverse processes of the thoracic vertebrae; angles and bulges of the ribs	
	• cervicis	Transverse processes of the first and second thoracic vertebrae	Transverse processes of the first–fifth cervical vertebrae	
	• capitis	Capitula of the third–seventh cervical vertebrae	Mastoid process of the temporal bone	
26/3	**Spinalis**			
	• thoracis	Last two thoracic vertebrae	Spinous processes of the third–ninth thoracic vertebrae	Extends and fixes the spinal column
	• cervicis	Spinous processes of the first–second thoracic and sixth–seventh cervical vertebrae	Spinous processees of the second–fourth cervical vertebrae	
	• capitis	Spinous processes of the seventh cervical vertebra and first thoracic vertebra	Between superior and inferior linea nuchae of the occipital bone	

	Name	Origin	Insertion	Function
6/4	**Semispinalis**			
	• thoracis	Transverse processes of the sixth–tenth thoracic vertebrae	Spinous processes of the sixth–seventh cervical and first–fifth thoracic vertebrae	Fixes and extends the spinal column and rotates it to the side
	• cervicis	Spinous processes of the first–fifth thoracic vertebrae	Spinous processes of the second thoracic vertebra and sixth cervical vertebra	Flexes head and neck
	• capitis	Transverse processes of the fourth–sixth thoracic vertebrae and capitula of the third–seventh cervical vertebrae	Lineae nuchae superior et inferior	Extends the head and rotates it to the side
6/5	**Multifidus**	Ventral surface of the sacrum; transverse processes of the lumbar, thoracic and the fourth–seventh cervical vertebrae	Sides of the spinous processes of the cervical, thoracic and lumbar vertebrae	Fixes the spinal column and rotates it to the side
6/6	**Rotatores**			
	• cervicis	Transverse processes of the cervical vertebrae	Spinous processess (short) or sides of the spinous processes of the vertebrae	Rotates the head;
	• thoracis	Transverse processes of the thoracic vertebrae	Spinous processess (short) or sides of the spinous processes of the vertebrae	Rotates and fixes the spinal column
	• lumborum	Mastoid processes of the lumbar vertebrae	Spinous processess (short) or sides of the spinous processes of the vertebrae	
6/7	**Intertransversarii**			
	• lateral lumborum	Transverse processes of the lumbar vertebrae	Neighbouring transverse processes of the lumbar vertebrae	Fixes, flexes (one side), extends (two sides) the spinal column; supports the lateral movements of the spinal column
	• medial lumborum	Mastoid process of the lumbar vertebrae	Neighbouring mastoid and accessory processes of the lumbar vertebrae	
	• thoracis	Transverse processes of the thoracic vertebrae	Neighbouring transverse processes of the thoracic vertebrae	
	• cervicis posteriores	Connect neighbouring dorsal heads of the transverse process	Neighbouring transverse processes of the thoracic vertebrae	
	• cervicis anteriores	Connect neighbouring ventral heads of the transverse process	Neighbouring transverse processes of the thoracic vertebrae	
7	**Levatores costarum**			
	• breves	Transverse processes of the seventh cervical and first–eleventh thoracic vertebrae; sides of the spinous processes of the cervical, thoracic and lumbar vertebrae	Inferior ribs	Lift the ribs; flex, rotate and extend the spinal column, maintain the sopinal column in an upright position
	• longi	Transverse processes of the thoracic vertebrae	Last two ribs	

	Name	Origin	Insertion	Function

The Muscles of the Throat

28	**Pectoralis major**			
	• **pars clavicularis**	Sternal part of the clavicle	Lateral bony crest of the humerus	Adducts the arm, rotates it inwards; pulls the trunk upwards and the arm downwards
	• **pars sternocostalis**	Body and ensiform process of the sternum		
	• **pars abdominalis**	First–sixth rib cartilage, fascia recta		
29	**Pectoralis minor**	Dorsal surface and inside of the second–fifth rib	Coracoid process of the shoulder blade	Pulls the shoulder to the front and downward; rotates the shoulder blade inwards ("pull up" auxiliary breathing muscle)
30	**Subclavius**	Cartilage of the first rib	Clavicle	Pulls the clavicle downwards and inwards and fixes it
31	**Serratus anterior**			
	• **pars superior**	First–second rib with muscle teeth	On the medial ridge of the shoulder blade and its superior angle	Pulls the shoulder blade towards the spinal column
	• **pars media**	Second–fourth rib	On the medial ridge of the shoulder blade and its superior angle	Pulls the superior angle of the shoulder blade downwards
	• **pars inferior**	Fifth–ninth rib	Inferior angle of the shoulder blade	Rotates the inferior angle of the shoulder blade outwards
32	**Intercostal muscles**			
	• **externi**	Inferior ridge of the ribs	Superior ridge of the following ribs	Lifts the ribs and pulls them forwards
	• **interni**	Superior ridge of the ribs	Inferior ridge of the ribs	Lowers the ribs
33	**Subcostal muscles**	Inner surface of the angle of the last rib	Superior two/three ribs	Support for muscle 32
34	**Transversus thoracis**	Inside surface of the sternal cartilage, body of the sternum	With muscle teeth on the cartilage of the second–sixth rib	Pulls the ribs downwards, narrows the thoracic cavity

Abdominal Muslces

35	**Rectus abdominis**	Cartilage of the sternum, cartilage of the fifth–seventh rib; costochondral ligament	Ventral surface of the pubic bone, tuber and symphysis of the pubic bone	Supports the organs of the lower abdomen, helps evacuate stools and in birth; flexes the trunk
36	**Pyramidalis**	Bone suture of the pubic bone	Linea alba	Pulls the linea alba downwards, supports the function of the rectus abdominis

	Name	Origin	Insertion	Function
37	**Obliquus externus abdominis**	With seven–eight muscle teeth on the ventral surface of the fifth–twelfth ribs	Iliac crest in the median suture of the abdominis	Supports the organs of the lower abdomen, aids evacuation and childbirth; flexes, rotates (one side) the trunk
38	**Obliquus internus abdominis**	Iliac spine, dorsolumbar fascia, inguinal ligament	On the last two ribs, in the linea alba	Supports the organs of the lower abdomen, aids in evacuation and childbirth; flexes, rotates (one side) the trunk like obliquus externus, flexes the trunk sideways
39	**Transversus abdominis**	Inside surface of the cartilage of the last six ribs, transverse processes of the lumbar vertebrae, dorsolumbar ligament, lateral inguinal ligament	Costal arch, fascia recta, linea alba	Aids evacuation and childbirth
40	**Cremaster** (Levator testiculae)	A muscle ligament between muscles 38 and 39	Wall of the scrotum	Lifts the testicles

The Muscles of the Upper Extremity

The Muscles of the Shoulder Girdle

	Name	Origin	Insertion	Function
41	**Deltoid muscle**			
	• **pars clavicularis**	Clavicle	Deltoid protuberance of the humerus	Abducts (90 degrees) the upper arm and pulls it caudally, rotates it outwards and fixes the shoulder joint
	• **pars scapularis**	Shoulder bone	Deltoid protuberance of the humerus	
42	**Supraspinatus**	In the groove above the spina scapulae	Below the acromion in the intertubercular sulcus of the humerus	Abducts (90 degrees) the upper arm
43	**Infraspinatus**	In the groove below the spina scapulae	Lateral protuberance of the humerus	Rotates the arm backwards and outwards, pulls it caudally and draws it forward
44	**Teres minor**	Lateral ridge of the shoulder blade	Great humeral protuberance	Rotates the humerus outwards, pulls it downwards and draws it forward
45	**Teres major**	Inferior ridge and angle of the shoulder blade	Medial surface of the humerus	Adducts the upper arm, rotates the arm outwards
46	**Subscapularis**	Inner groove of the shoulder blade	Medial protuberance and bony crest of the humerus	Rolls the arm inwards, splays and adducts it

	Name	Origin	Insertion	Function

The Muscles of the Elbow Joint

	Name	Origin	Insertion	Function
47	**Biceps**		Protuberance on the proximal end of the radius, tendinoid bundle merges with the radial hand extensor	Flexes the arm at the elbow joint, extends the shoulder joint, fixes both together with the radis of the hand, rotates the arm outwards
	• caput longum	Tuberositas of the shoulder blade		
	• caput brevis	Coracoid process of the shoulder blade		
48	**Coracobrachialis**	Coracoid process	Medial side of the humeral shaft	Extends the shoulder joint, lifts the arm and rotates the arm inwards
49	**Brachialis**	Ventral-middle surface of the humerus	Muscle protuberance of the ulna and coracoid protuberance of the radius	Flexes the lower arm at the elbow joint
50	**Triceps**			Extends the lower arm at the elbow; pulls the medial part of the shoulder blade upwards; when the shoulder blade is in a fixed position pulls the upper arm backwards
	• caput longum	Infraglenoid protuberance of the shoulder blade	Process of the ulnar protuberance	
	• caput mediale	Dorsal-medial surface of the humeral shaft	Process of the ulnar protuberance	
	• caput laterale	Superior third of the humerus	Process of the ulnar protuberance	
51	**Anconeus**	Dorsal surface and lateral condylar extensor of the humerus	Lateral surface of the hamate process and dorsal surface of the shaft of the radius	Extends and fixes the elbow joint

The Muscles of the Carpal and Metacarpal Joints and the Finger Joints

	Name	Origin	Insertion	Function
52	**Pronator teres**			
	• caput humerale	Flexor tuberosity of the humerus, coronoid protuberance of the ulna	Ventral surface of the radial shaft	Rotates the lower arm inwards
	• caput ulnare	Flexor tuberosity of the humerus, coronoid protuberance of the ulna	Ventral surface of the radial shaft	Pronates the radius, flexes the lower arm at the elbow joint
53	**Flexor carpi radialis**	Flexor tuberosity of the humerus, on the radius	Second–third metacarpal bones (palmar side)	Flexes the hand towards the radial side, rotates it inwards
54	**Palmaris**			
54.1	**Palmaris longus**	Extensor tuberosity of the humerus	Palmar transverse ligament (palmaraponeurosis), vagina tendinis, flexoris digitorum	Extends the palmar transverse ligament, flexes the hand in the carpal joint
54.2	**Palmaris brevis**	In the center of the palmar transverse ligament	On the superior side of the palmar transverse ligament and on the lateral edge of the hand	Extends the palmar transverse ligament

	Name	Origin	Insertion	Function
55	**Flexor carpi ulnaris**			
	• caput humerale	Flexor tuberosity of the humeral head	Pisiform bone and pisiform-hamate bone ligament, fifth metacarpal bone	Rotates the carpal joint laterally; flexes the hand towards the ulnar side; flexes the elbow
	• caput ulnare	Medial surface of the superior third of the ulna		
56	**Flexor digitorum superficialis**			
	• caput humeroulnare	Medial ligament of the elbow joint	Two bundles on the middle phalanx of the second and fifth finger	Flexes the fingers in the middle and root joint; flexes the elbow joint
	• caput radiale	Medial ridge of the hand in the tendon below the coronoid process of the radius		
57	**Flexor digitorum profundus**	Ventral-medial surface of the superior third of the ulna	Traverses the tendon of the superficial flexor digitorum and inserts on the distal phalanx of the second–fifth finger	Flexes the fingers in the final joints
58	**Flexor pollicis longus**	Ventral surface of the superior third of the radius, interosseous ligament	Distal phalanx of the first finger	Flexes the carpus and the thumb at the base and end joint
59	**Pronator quadratus**	Medial surface of the ulna	Dorsal and lateral surface of the radius	Rotates the lower arm and hand inwards
60	**Brachioradialis**	Inferior third of the humerus and its extensor tuberosities	Lateral surface of the inferior third of the radius and styloid process of the radius	Flexes the elbow joint; pronates or supinates the bones of the lower arm
61	**Extensor carpi radialis**			
	• longus	Bony crest of the extensor tuberosity of the humerus	Posterior surface of the second metacarpal bones	Extends and adducts the carpus
	• brevis	Extensor tuberosity of the humerus	Dorsal surface of the second metacarpal bone	
62	**Extensor digitorum**	Extensor tuberosity of the humerus	Distal phalanx of the second–fifth finger	Extends the second–fifth fingers
63	**Extensor digiti minimi**	Extensor tuberosity of the humerus and transverse ligament of the elbow joint	First and third phalanx of the fifth finger	Extends the fifth finger
64	**Extensor carpi ulnaris**	Extensor tuberosity of the humerus and dorsal surface of the ulna	Dorsal surface of the fifth metacarpal bone	Rotates the carpus outwards (with muscle 55), flexes the carpus (with muscle 61)
65	**Supinator**	Extensor tuberosity of the humerus and collateral ligament of the elbow joint	Lateral surface of the superior third of the radius	Rotates the lower arm outwards (supination)
66	**Abductor pollicis longus**	Dorsal surface of the radius, interosseous ligament	Dorsal surface of the first metacarpal bone	Extends and splays the thumb

	Name	Origin	Insertion	Function
67	**Extensor pollicis brevis**	Radius and interosseous ligament on the central part of the lower arm	Proximal phalanx of the first finger	Extends the proximal phalanx of the thumb
68	**Extensor pollicis longus**	Dorsal surface of the middle third of the ulna and interosseous ligament	Proximal phalanx of the first finger	Adducts and extends the thumb
69	**Extensor indicis**	Distal end of the ulna and interosseous ligament	Tendon of the Extensor digitorum, distal phalanx of the second finger	Extends the second finger and pulls it inwards towards the carpus

The Short Muscles of the Hand

	Name	Origin	Insertion	Function
70	**Abductor pollicis brevis**	On the navicular bone and retinaculum flexorum	Proximal phalanx of the first finger	Flexes the first finger, abducts it, rotates the palm of the hand outwards
71	**Flexor pollicis brevis**	Greater and lesser multangular bone, transverse ligament of the carpus	Proximal phalanx of the first finger	Flexes and adducts the proximal phalanx of the first finger
72	**Opponens pollicis**	Greater multangular bone, transverse ligament of the carpus	Lateral-dorsal surface of the first metacarpal bone	Places the thumb in opposition to the other fingers and the palm
73	**Adductor pollicis**	Capitate bone (loxic head), third metacarpal bone (diagonal head)	Proximal phalanx of the first finger	Flexes the first finger and places it in opposition to the other fingers
74	**Lumbricals**	Metacarpal bones, sides of the tendon at the triangular ligament	On both tendons of the deep flexor digitorum	Flex the base phalanx and extend the middle and end phalanx of the fingers
75	**Interosseous muscles** dorsal (4) palmar (3)	Middle part of the metacarpal bones lateral (second finger) or medial (fourth–fifth fingers)	On the semsamoid bones of the second–fifth fingers on the triangular ligament	Splay the fingers (dorsal), draw the fingers together and extend them (palmar)
76	**Abductor digiti minimi**	Lateral surface and ligaments of the coracoid bone	Proximal phalanx of the fifth finger	Abducts the fifth finger
77	**Flexor digiti minimi**	Coracoid process on the transverse ligament	Proximal phalanx of the fifth finger	Flexes the fifth finger
78	**Opponens digiti minimi**	Coracoid process on the transverse ligament	On the bodies of the five metacarpal bones	Places the fifth finger in opposition to the thumb

The Muscles of the Pelvis and Lower Extremity

No.	Name	Origin	Insertion	Function
79	**Iliopsoas**			
79/1	**Iliacus**	Ventral surface of the sacrum and os ilium	Medial surface of the shaft of the femur on the lesser trochanter	Flexes and rotates the hip joint outwards and rotates the spinal column
79/2	**Psoas major**	Last thoracic and bodies of the first–fourth lumbar vertebrae	Medial surface of the bone shaft on the lesser trochanter	Rotates the knee outwards and stabilizes the hips
79/3	**Psoas minor**	Body of last thoracic and first lumbar vertebra	On the iliac crest, eminentia iliopubica and femur	Rotates the knee outwards, flexes and stabilizes the hips
80	**Quadratus lumborum**	Crista iliaca, on the ligament between the os ilium and lumbar vertebrae	Inside surface of the twelfth rib and transverse processes of the first–fourth lumbar vertebrae	Pulls the last rib caudally, flexes the lumbar vertebrae to the side (expiration muscle)

The Gluteus Muscles

No.	Name	Origin	Insertion	Function
81	**Gluteus maximus**	Dorsal surface of the iliac crest, deep trunk fascia, sacrum	Greater trochanter of the femur on the broad femural fascia	Extends, splays and rotates the leg outwards in the hip joint; fixes the hip joint
82	**Gluteus medius**	Superior and inferior line of the os ilium	Greater trochanter of the femur	Abducts the upper leg, rotates it inwards and outwards
83	**Gluteus minimus**	Superior and inferior line of the os ilium	Greater trochanter of the femur	Rotates the leg at the hip joint

The Rear Upper Thigh Muscles

No.	Name	Origin	Insertion	Function
84	**Tensor fasciae latae**	Iliac crest and anterior superior iliac spine	Iliaco-pubic ligament and broad femural fascia	Tenses the femural fascia; flexes the hip joint, extends the leg in the knee joint
85	**Biceps femoris**			
	• caput longum	Superior and inferior groove of the protuberance of the os sedentarium	Head of the fibula	Flexes the leg in the knee joint, extends the hip joint;
	• caput breve	Linea aspera of the femur		Rotates the femur outwards, flexes the knee joint
86	**Semitendinosus**	Protuberance of the os sedentarium, fused with muscle 85	Medial surface of the protuberance of the shin	Flexes and rotates inwards the lower leg in the knee joint, extends the hip joint
87	**Semimembranosus**	Protuberance of the os sedentarium	On the medial protuberance of the os sedentarium	Rotates inwards the lower leg and flexes it in the knee joint

	Name	Origin	Insertion	Function

The Inner Muscles of the Upper Leg

	Name	Origin	Insertion	Function
88	**Sartorius**	Anterior iliac spine	Medial surface of the crest of the shin	Flexes and rotates the upper leg outwards in the hip joint; rotates the lower leg inwards during flexing of the knee
89	**Gracilis**	Inferior branch of the pubic bone	On the medial surface and medial protuberance of the shin	Adducts the upper leg, flexes the knee, rotates the lower leg towards the centre
90	**Pectineus**	Body and crest of the pubic bone	Bony crest of the femur, synovial capsule of the hip	Adducts the upper leg, flexes and rotates it outwards
91	**Piriformis**	Inner surface of the sacrum, on the os ilium and broad pelvic ligament	Greater trochanter of the femur	Pulls the upper leg backwards, abducts and rotates it outwards
92	**Adductor**			
	• longus	Fusion of the superior and inferior branches of the body of the pubic bone	Inner lip and shaft of the femur	Adducts and rotates the upper leg inwards in its flexion and extension in the hip joint
	• brevis	Body and inferior branch of the pubic bone	Superior lip of the shaft of the upper leg;	Adducts, rotates the upper leg outwards
	• magnus	Protuberance of the os sedentarium	Superior third of the femur	Adducts extends and rotates the upper leg outwards
	• minimus	Inferior branch of the protuberance of the pubic bone	Inner lip and superior third of the femur	Adducts the upper leg, flexes and rotates it outwards in the hip joint

The Deep Muscles of the Hip Joint

	Name	Origin	Insertion	Function
93	**Obturatorius internus**	Medial ridge and ligament of the hip cavity	With a long tendon on the greater trochanter of the femur	Rotates the upper leg outwards
94	**Obturatorius externus**	Lateral ridge and ligament of the hip cavity	On the collateral groove of the trochanter of the upper leg	Rotates the upper leg outwards
95	**Gemelli muscles** **• superior** **• inferior**	Crest and protuberance of the os sedentarium	Below the groove next to the trochanter of the upper leg	Rotates the upper leg outwards
96	**Quadratus femoris**	Superior external surface of the protuberance of the os sedentarium	Bony crest between the trochanters of the upper leg	Rotates the upper leg outwards

The Muscles of the Knee Joint

	Name	Origin	Insertion	Function
97	**Quadriceps femoris**			
97/1	**Rectus femoris**			
	• caput rectum	Below the protuberance of the hip on the anterior inferior iliac spine, anterior surface of the azetabulum	Patella, on the crest of the shin	Extends the knee joint, flexes the upper leg in the hip joint

	Name	Origin	Insertion	Function
97/2	**Vastus medialis**	Medial lip of the shaft of the upper leg	Patella, on the crest of the shin	Extends the knee joint
97/3	**Vastus lateralis**	inferior lip of the shaft of the upper leg; on the ventral inferior ridge of its greater trochanter	With a straight patellar ligament on the shin crest	Extends the knee joint
97/4	**Vastus intermedius**	Lateral surface on the inferior third of the upper leg	Shin crest	Extends the lower leg in the knee joint
98	**Articularis genus**	Front of femural shaft	Capsule of the knee joint	Tightens the synovial capsule

The Extensor Group of the Ankle Joint

	Name	Origin	Insertion	Function
99	**Tibialis anterior**	Lateral surface and condyles of the shin, interosseous ligament	First metatarsal bone or proximal phalanx of the first toe, medial sphenoidal bone	Extends the foot in the foot joint and rotates it outwards
100	**Extensor hallucis longus**	Medial surface of the fibula and interosseous ligament	Distal phalanx of the first toe	Extends and adducts the first toe
101	**Extensor digitorum longus**	On the proximal articulatory heads of the shin, anterior surface of the fibula and interosseous ligament	Distal phalanx of the second–fifth toes	Extends toes second–fifth, dorsal flexion of the foot
102	**Fibularis tertius**	On the inferior-lateral third of muscle 101	Dorsal surface of the fifth metatarsal bone	Pronation and dorsal flexion of the foot
103	**Fibularis longus**	Head of the fibula and superior two thirds of the shaft of the fibula	With a long tendon on the first and second metatarsal bone and medial sphenoidal bone	Flexes (plantar flexion) the tarsus pedis joints
104	**Fibularis brevis**	Lateral surface and ventral ridge of the fibula	Protuberance of the fifth metatarsal bone	Flexes the foot in the joint, lifts the fibular and descends the tibial edge of the foot

The Flexor Muscles of the Ankle Joint

	Name	Origin	Insertion	Function
105	**Triceps surae**			
105/1	**Gastrocnemius**			
	• caput mediale	On the medial condyle of the femur	Protuberance of the heel bone and Achilles tendon	Plantar flexion of the foot, lifts the heel; flexes flexes the knee joint
	• caput laterale	On the lateral condyle of the femur		
105/2	**Soleus**	Head of the fibula, dorsal side and bony crest of the shin, interosseous ligament	Protuberance of the heel bone and Achilles tendon	Plantar flexion of the foot, lifts the heel

Name	Origin	Insertion	Function
106 **Plantaris**	Lateral protuberance of the ligament of the femural condyle and loxic popliteal ligament	Deep lower leg fascia and Achilles heel	Plantar flexion of the foot; pronates and flexes the knee joint
107 **Popliteus**	Lateral ligament protuberance of the femoral condyle and synovial capsules	Inferior-medial surface of the proximal head of the shin	Flexes the knee, rotates the lower leg inwards during flexing of the knee

The Muscles of the Toes

Name	Origin	Insertion	Function
108 **Tibialis posterior**	Dorsal surface of the shin, medial surface of the fibula and interosseous ligament Tuberosity of the navicular	bone, medial and middle sphenoidal bone; second–fifth metatarsal bones	Plantar flexion of the foot, lifts the tibial edge of the foot
109 **Flexor digitorum longus**	Dorsal surface of the shin	Distal phalanx of the second–fifth toe	Flees the end phalanges of the toes, plantar flexion
110 **Flexor hallucis longus**	Dorsal surface of the inferior third of the shin, interosseous ligament and posterior intermuscular septum	Distal phalanx of the first toe	Flexes the first toe, supinates the foot, lifts the heel
111 **Extensor digitorum brevis**	Ventral and lateral surface of the heel	Lateral surface of the tendon of the second–fourth foot flexors	Extends the toe joints
112 **Extensor hallucis brevis**	Ventral surface of the heel	Proximal phalanx of the first toe	Extends the joints of the first toe
113 **Lumbricals**	Ventral side of the long tendon of the toe flexor	Toe flexors and proximal phalanx of the second–fifth toe	Flex the toe joints
114 **Dorsal interosseous muscles**	Lateral surface of the metatarsal bones	Extensor tendon and proximal phalanx of the second–fifth toe	Splay the toes
115 **Plantar interosseous muscles**	Tibial edge of the second–fifth metatarsal bones	Proximal phalanx of the second–fifth toes	Plantar flexion of the toes
116 **Abductor hallucis**			
• **caput mediale**	Small medial protuberance of the heel protuberance	Proximal phalanx of the first toe	Flexes and abducts the first toe
• **caput laterale**	Deep plantar fascia		
117 **Flexor hallucis brevis**	Plantar surface of the sphenoidal bone, long plantar fascia	With two heads on the sesamoid bone and proximal phalanx of the first toe	Pulls the big toe plantarly
118 **Adductor hallucis** • **caput transversum** • **caput obliquum**	Tarsus pedis-metatarsal ligament and long plantar ligament, proximal end of the first–fourth metatarsal bones	Lateral sesamoid bone and proximal phalanx of the first toe	Adducts and opposes the big toe, fixes the arch of the foot
119 **Abductor digiti minimi**			
• **caput superficiale**	Lateral process of the protuberance of the heel	Lateral ridge of the proximal phalanx of the fifth toe, proximal surface of the fifth metatarsal bone	Lifts and fixes the arch of the foot
• **caput profundum**	Medial process of the protuberance of the heel		

Name	Origin	Insertion	Function
120 **Flexor digiti minimi brevis**	Long plantar ligament, fifth metatarsal bone	Proximal phalanx of the fifth toe	Flexes the fifth toe
121 **Opponens digiti minimi**	Covers the lateral edge of the fifth toe	Covers the lateral edge of the fifth toe	Pulls the small toe towards the tibia and plantarly
122 **Flexor digitorum brevis**	Medial process of the protuberance of the heel; plantar surface ligament of the foot	Four tendons on the middle phalanx of the second–fifth toes	Flexes the middle phalanx of the second–fifth toes
123 **Quadratus plantae**	Two heads on the dorsal surface of the protuberance of the heel, plantar surface ligament	Tendons of the long toe flexor	Supports muscle 109 in the plantar flexion of the end phalanges

The Muscles of the Head

124 **Epicranii**			
124/1 **Occipitofrontalis**			
• **venter frontalis**	Orbital arch	Parietal bone	Pulls the scalp forwards, furrows the brow
• **venter occipitalis**	On the superior chin line	Root of the mastoid process	Pulls the scalp backwards and smoothes the forehead
124/2 **Temporoparietalis**	A muscle cover on both sides of the temporal fascia		Tightens the scalp and skin
124/3 **Procerus**	Between the eyebrows and dorsum nasi		Folds the skin of the dorsum nasi

The Muscles of the Nose

125 **Nasalis**			
• **pars transversa**	Infundibular process of the upper jaw		
• **pars alaris**	Connects the plates of the parietal cartilage		
• **pars septalis**	Anterior part of the nose (fuses with its contralateral part)		
126 **Depressor septi**	Anterior part of the nose (fuses with its contralateral part)		

The Muscles of the Eyelid

	Name	Origin	Insertion	Function
127	**Orbicularis oculi**			
	• pars orbicularis	Medial canthus, nasal part of the os frontale	Lateral ligament of the canthus	Folds the eyebrow, pulls the eyelids together, closes them
	• pars palpebralis	Process of the upper jaw, eyelid ligament	Lateral ligament of the canthus	Furrows the brow (in sorrow or pain)
	• pars lacrimalis	Sulcus of the tear lachrymal sack	Lateral ligament of the canthus	Expands the lachrymal sack
128	**Corrugator supercilii**	Os frontale	Radiates into the skin of the os frontale	Furrows the brow

The Muscles of the Ear

	Name	Origin	Insertion	Function
129	**Auricularis anterior**	Fascia of the temporal bone	Ventral ear crest	Pulls the outer ear forwards
130	**Auricularis superior**	Fascia of the temporal bone and scalp	Dorsal-superior part of the outer ear	Pulls the outer ear downwards
131	**Auricularis posterior**	Mastoid process of the temporal bone	Inner surface of the outer ear	Pulls the outer ear backwards

The Muscles of the Lips

	Name	Origin	Insertion	Function
132	**Orbicularis oris** **• pars marginalis** **• pars labialis**			Forms the border of the lips, closes the mouth, tightens the lips
133	**M. depressor anguli oris**	Mental side of the lower jaw	Corner of the mouth	Pulls the corners of the mouth downwards (sadness, derision or hatred), folds the skin of the chin
134	**Transversus menti**	Holds the triangular muscles of both sides together		
135	**M. depressor labii inferioris**	Inferior edge of the lower jaw	Lower lip part of the orbicularis oris	Pulls the lower lip downwards
136	**Mentalis**	Incisura of the lower jaw near the canine	Skin of the chin region and orbicularis oris	Folds and extends the skin in the chin region
137	**Quadratus labii superioris**			
137/1	**Levator labii superioris alaeque nasi pars medialis-pars lateralis**	Nasal process of the os frontale	Middle part of the skin of the upper lip, ala nasi	Lifts the upper lips and ala nasi
137/2	**Levator labii superioris**	Nose and lachrymal bone	Skin of the upper lip and orbicularis oris	Lifts the lateral part of the upper lip

	Name	Origin	Insertion	Function
7/3	**Zygomaticus minor**	Zygomatic bone	Outer side of the upper lip	Lifts the upper lip
38	**Zygomaticus major**	Temporal process of the sygomatic bone	Corner of the mouth	Pulls the corners of the mouth upwards and laterally
39	**Risorius**	Fascia of the masseter	Skin of the corner of the mouth	Pulls the corner of the mouth laterally
40	**Levator anguli oris**	Incisura of the upper jaw	Corner of the mouth, orbicularis oris	Folds the skin of the corner of the mouth
41	**Buccinator**	Angle and infundibular process of the lower jaw, on the edge of the infundibulum of the upper and lower jaw	Corner of the mouth	Pulls the corner of the mouth sideways and backwards; presses food between the teeth

The Mastication Muscles

	Name	Origin	Insertion	Function
142	**Masseter**			
	• **pars superficialis**	Ventral part of the face crest and inferior edge of the zygomatic arch	Angle of the lower jaw and posterior edge of the branch of the lower jaw	Pulls the lower jaw upwards and forwards, closes the mouth;
	• **pars profunda**	Dorsal part and inferior surface of the zygomatic arch		Lifts the lower jaw and pulls it backwards, closes the mouth (sign of anger)
143	**Temporalis**	Edge and wall of the temporal sulcus	Muscular process of the lower jaw	Lifts the lower jaw and pulls it backwards, closes the mouth (sign of anger)
144	**Pterygoideus medialis**	Upper jaw, pyramidal process of the palatal bone	Medial surface and corner of the branch of the lower jaw	Lifts and pulls the lower jaw forwards and sideways
145	**Pterygoideus lateralis**			
	• **caput dorsale**	Bony crest and sulcus below the temporal sulcus and of the sphenoidal bone	Joint and capitulum of the lower jaw	Pulls the lower jaw forwards, the one-sided movement is a grinding movement
	• **caput ventrale**	Inferior temporal sulcus, palatal bone		